2019 IEEE 31st International Conference on Microelectronics (MIEL 2019)

Nis, Serbia
16 – 18 September 2019

IEEE Catalog Number: CFP19432-POD
ISBN: 978-1-7281-3420-8

**Copyright © 2019 by the Institute of Electrical and Electronics Engineers, Inc.
All Rights Reserved**

Copyright and Reprint Permissions: Abstracting is permitted with credit to the source. Libraries are permitted to photocopy beyond the limit of U.S. copyright law for private use of patrons those articles in this volume that carry a code at the bottom of the first page, provided the per-copy fee indicated in the code is paid through Copyright Clearance Center, 222 Rosewood Drive, Danvers, MA 01923.

For other copying, reprint or republication permission, write to IEEE Copyrights Manager, IEEE Service Center, 445 Hoes Lane, Piscataway, NJ 08854. All rights reserved.

*** *This is a print representation of what appears in the IEEE Digital Library. Some format issues inherent in the e-media version may also appear in this print version.*

IEEE Catalog Number:	CFP19432-POD
ISBN (Print-On-Demand):	978-1-7281-3420-8
ISBN (Online):	978-1-7281-3419-2
ISSN:	2159-1660

Additional Copies of This Publication Are Available From:

Curran Associates, Inc
57 Morehouse Lane
Red Hook, NY 12571 USA
Phone: (845) 758-0400
Fax: (845) 758-2633
E-mail: curran@proceedings.com
Web: www.proceedings.com

2019 IEEE 31ˢᵗ INTERNATIONAL CONFERENCE ON MICROELECTRONICS

PROCEEDINGS

Niš, Serbia
September 16ᵗʰ-18ᵗʰ, 2019

organized by

 **IEEE Serbia and Montenegro Section - ED/SSC Chapter**

in cooperation with

 Serbian Academy of Sciences and Arts - Branch in Niš
Faculty of Electronic Engineering, University of Niš

under the co-sponsorship of

IEEE Electron Devices Society

under the auspices of

Serbian Ministry of Education, Science and Technological Development

Society for ETRAN

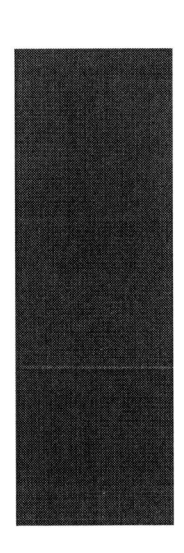

2019 IEEE 31st INTERNATIONAL CONFERENCE ON MICROELECTRONICS

PROCEEDINGS

Niš, Serbia
September 16th-18th, 2019

organized by
IEEE Serbia and Montenegro Section - ED/SSC Chapter

in cooperation with

Serbian Academy of Sciences and Arts - Branch in Niš
Faculty of Electronic Engineering, University of Niš

under the co-sponsorship of

IEEE Electron Devices Society

under the auspices of

Serbian Ministry of Education, Science and Technological Development

Society for ETRAN

CONTENTS

Mini-Colloquium on Nano- and Flexible- Electronics

On the ESD Protection and Non-Fatal ESD Strike on Nano CMOS Devices
H. Wong, S. Dong and Z. Chen 3

CMOS-Compatible Gas Sensors
L. Filipovic and S. Selberherr 9

Tutorial on Power Devices and Modules

High-Fidelity Predictive Simulation of High Power Devices and Modules at the Rim of the Safe-Operating Area and Beyond
G. Wachutka 19

Plenary Session

The Fundamental Current Mechanisms in SiC Schottky Barrier Diodes: Physical Model, Experimental Verification and Implications
S. Dimitrijev, J. Nicholls, P. Tanner, and J. Han 31

Design Techniques for Wireless Sensor Network Nodes Powered by Ambient Energy Harvesting
Z. Prijić, A. Prijić and Lj. Vračar 37

Hardware/Software Co-Design of Wireless LAN Transceiver: A Case Study
Z. Stamenković, K. Tittelbach-Helmrich, M. Krstić, M. Stojčev, and B. Dimitrijević 45

Session
Device Physics, Technology and Characterization

On the Ballistic Transport in Si Nano-Devices
G. Golan, M. Azoulay and J. Bernstein 55

Impact of γ Radiation on Charge Trapping Properties of Nanolaminated HfO_2/Al_2O_3 ALD Stacks
D. Spassov, A. Paskaleva, V. Davidović, S. Djorić-Veljković, S. Stanković, N. Stojadinović, Tz. Ivanov, and T. Stanchev 59

Determination Method for Interface State Densities Adapted to Ultrathin Dielectrics
N. Novkovski and A. Skeparovski 63

Feasibility of Applying an Electrically Programmable Floating-Gate MOS Transistor in Radiation Dosimetry
S. Ilić, A. Jevtić, S. Stanković, and V. Davidović 67

Heavy-Ion-Induced Single Event Burnout in SiC Schottky Diodes: Safe Operating Area
P. S. Gromova, G. G. Davydov, A. S. Tararaksin, A. S. Kolosova, D. V. Boychenko, V. N. Vyuginov, and V. V. Luchinin 71

Fabrication and Characterization of High-k Dielectrics Based Gate Stacks/MOS Capacitors for Advanced CMOS Devices
R. Vaid and R. Gupta 75

In-depth Structure and Electrical Characteristics Study of HfOx-based Resistive Random Access Memories (ReRAMs)
B. Attarimashalkoubeh and Y. Leblebici 79

Strong Electric Monopulses in Nonuniformly Doped Nitride Films under Negative Differential Conductivity
V. Grimalsky, S. Koshevaya, J. Escobedo-A., and J. Sanchez-S. 83

Arrays of Bowtie Plasmonic Nanoantennas for Field Enhancement in MOEMS
M. Obradov, Z. Jakšić, I. Mladenović, D. Tanasković, and O. Jakšić 87

Reviewing MXenes for Plasmonic Applications: Beyond Graphene
Z. Jakšić, M. Obradov, O. Jakšić, D. Tanasković, and D. Vasiljević Radović 91

Nanocrystalline Porous Nickel Ferrite Ceramics for Humidity Sensing Applications
D. L. Sekulic, Z. Z. Lazarevic and N. Z. Romcevic 95

D-mode pHEMT 0.5 um Process Characterization to Wide-Band LNA Design
D. I. Sotskov, N. A. Usachev, V. V. Elesin, A. G. Kuznetsov, K. M. Amburkin, G. V. Chukov, M. I. Titova, and N. M. Zidkov 99

Practical Evaluation of Optocouplers TID-Hardness Research Method Using an X-ray Unit
S. V. Novikov, M. E. Cherniak, R. K. Mozhaev, and D. V. Boychenko 103

Proton Accelerator's Direct Ionization Single Event Upset Test Procedure
A. O. Akhmetov, G. S. Sorokoumov, A. A. Smolin, D. V. Bobrovsky, D. V. Boychenko, A. Y. Nikiforov, and A. E. Shemyakov 107

Poster Session
Device Physics, Technology and Characterization

Reliability of Various Type of Gas-filled Surge Arresters Under DC Discharge
E. Živanović, S. Veljković, M. Živković, and M. Pejović 113

Effect of Rare-Earth Ions on Electrical Properties of BaTiO₃ Ceramics
V. Paunović, M. Đorđević, V. Mitić, and Z. Prijić 117

**Generation of Harmonics of Terahertz Radiation in Paraelectrics
in a Wide Temperature Range**
V. Grimalsky, S. Koshevaya, J. Escobedo-A., and Y. Gomez-B...................................121

**Study of Nanoporous Anodic Aluminum Oxide as a Template Filled with
Piezoelectric Materials**
T. Tsanev, M. Aleksandrova and V. Videkov 125

**An AlSiN Nanocomposite Film with Improved Mechanical Parameters
for Multifunctional Applications**
L. Kolaklieva, R. Kakanakov, T. Cholakova, H. Bahchedzhiev, and V. Chitanov 129

**Artificial Neural Network for Composite Hardness Modeling of Cu/Si Systems
Fabricated Using Various Electrodeposition Parameters**
I. Mladenović, J. Lamovec, V. Jović, M. Obradov, K. Radulović,
D. Vasiljević Radović, and V. Radojević 133

Modeling Errors of the MISFET-Based Sensors' Characteristics
B. Podlepetsky and N. Samotaev..137

Flexible Oxide-Polymeric Composites for Piezoelectric Energy Harvesting
M. Aleksandrova, G. Kolev, Y. Vucheva, I. Pandiev, and K. Denishev..........................141

Pt Resistive Film Sensors
Z. Stanimirović and I. Stanimirović...145

High-Voltage Pulse Trimming of Thick-Film Resistors – Some Modelling Aspects
I. Stanimirović and Z. Stanimirović..149

SnO₂-Pd as a Gate Material for the Capacitor Type Gas Sensor
N. Samotaev, K. Oblov, A. Litvinov, and M. Etrekova..153

Rapid Prototyping of MOX Gas Sensors in Form-Factor of SMD Packages
N. Samotaev, K. Oblov, A. Ivanova, A. Gorshkova, and B. Podlepetsky.........................157

**Analysis of Intrinsic Stochastic Fluctuations of the Time Response of Adsorption-Based
Microfluidic Bio/Chemical Sensors: the Case of Bianalyte Mixtures**
I. Jokić, Z. Djurić, K. Radulović, M. Frantlović, and P. M. Krstajić 161

Modeling Noise and Stability of Affinity-Based MEMS, NEMS and NOEMS Sensors of Ternary Gas Mixtures
O. Jakšić, I. Jokić, Z. Jakšić, M. Obradov, D. Tanasković,
D. Randjelović, and D. Vasiljević Radović..165

Effects of Endspoiling on Microchannel Plate Performance
A. Stanković, I. Zlatković, R. Nikolov, B. Brindić, and D. Pantić 169

Electrical Characteristics of $Ag_{10}(As_{40}S_{30}Se_{30})_{90}$ as Resistive Switching Material for Potential Application in Memory Devices
K. O. Čajko, D. L. Sekulić, D. M. Petrović, T. B. Ivetić, and S. R. Lukić–Petrović........173

Modelling of ΔV_T in NBT Stressed P-Channel Power VDMOSFETs
N. Mitrović, D. Danković, Z. Prijić, and N. Stojadinović 177

Comparison of Radiation Characteristics of HfO_2 and SiO_2 Incorporated in MOS Capacitor in Field of Gamma and X Radiation
S. Stanković, D. Nikolić, N. Kržanović, L. Nadjdjerdj, and V. Davidović......................181

Reduced Low Dose Rate Sensitivity (RLDRS) in Bipolar Devices
V. Pershenkov, A. Bakerenkov, A. Rodin, V. Felitsyn, V. Telets, and V. Belyakov 185

Blocking of Impacts of Single Ionizing Particles by CMOS C-Element in Two-Phase Systems
V. Ya. Stenin, Yu. V. Katunin and K. A. Petrov...189

Nonparametric Statistical Analysis of Radiation Hardness Threshold Variation in CMOS IC Wafer Lots Series with the Aim of Process Monitoring
Yu. I. Bogdanov, N. A. Bogdanova, D. V. Fastovets, Y. M. Moskovskaya,
A. V. Sogojan, and A.Y. Nikiforov...193

Radiation Hardness Evaluation of LEDs Based on InGaN, GaN and AlInGaP Heterostructures
D. S. Ukolov, N. A. Chirkov, R. K. Mozhaev, and A. A. Pechenkin 197

Simulation of Errors Impulses from Single Ionizing Particles in CMOS Triple Majority Gates
Yu. V. Katunin, V. Ya. Stenin and A. G. Prozorova 201

Influence of Temperature over Impedance of Different Inkjet Printed Patterns and Substrates
B. Nikolova, G. Nikolov, E. Gieva, I. Ruskova, and M. Mladenov 205

Virtual System for Measurement of Inkjet Printed Resistive and Capacitive Structures
G. Nikolov, E. Gieva, B. Nikolova, and I. Ruskova 209

Investigation on Body Potential in Cylindrical Gate-All- Around MOSFET
M. Kessi, A. Benfdila, A. Lakhelef, L. Belhimer, and M. Djouder 213

Session
Circuit and System Design and Testing

SAR ADC Architecture with Fully Passive Noise Shaping
D. Osipov, A. Gusev, V. Shumikhin, and St. Paul..219

Improving Magnitude Response of Comb Two-Stage Structure Using Simple Multiplierless Filters
G. Jovanovic Dolecek 223

Hardware Implementation of Selected Statistical Quantities for Applications in Automotive V2I Communication System
M. Banach, R. Długosz and T. Talaśka 227

Absorptive Filters in the Realization of RCIED Activation Jamming
A. Lebl, M. Mileusnić, B. Pavić, and J. Radivojević...231

Multi-Rate Signal Processing with the Use of Filter Banks Composed of Parallel FIR Filters
M. Banach and R. Długosz 235

Algorithm for Restructuring of Structurally Synthesized BDDs
L. Jürimägi and R. Ubar...239

Design of Non-Metastable SRAM Cells in 28 nm CMOS Technology
F. Crescioli, L. Frontini, V. Liberali, and A. Stabile 243

A Highly Parametrizable Chisel HCL Generator of Single-Path Delay Feedback FFT Processors
V. M. Milovanović and M. L. Petrović 247

Experimental Estimation of Input Offset Voltage Radiation Degradation Rate in Bipolar Operational Amplifiers
A. Bakerenkov, V. Pershenkov, V. Felitsyn, A. Rodin, V. Telets, V. Belyakov,
A. Zhukov, and N. Gluhov 251

Linear Slot Array Centrally Fed by CPW T-junction for 5G Applications
M. Milijic and B. Jokanovic 255

Advanced Electro-Optical Analysis of Photoplethysmogram Signal
L. Evdochim, D. Dobrescu and L. Dobrescu...259

Poster Session
Circuit and System Design and Testing

VHDL-AMS Model Development for Digitally Programmable Monolithic Instrumentation Amplifiers
I. Pandiev 265

Verification of VHDL-AMS Simulation Model for Digitally Programmable Monolithic Instrumentation Amplifiers
I. Pandiev 269

Multi-Phase Ring-Coupled Oscillator for TDC Using a Differential Inverter with an Oscillation Frequency Booster Circuit
T. Shima, S. Kozuki, T. Otsuka, and N. Retdian..273

A Programmable Current-Mode Digital-to-Analog Converter with Correction of Nonlinearity of Input-Output Characteristics
J. Dalecki, R. Długosz, T. Talaśka, and G. Fischer 277

A Parallel Adaptive LMS FIR Filter Realized in CMOS Technology
R. Długosz, T. Talaśka, T. Nikolić, and G. Nikolić...281

Performance Evaluation of Block-Based Adaptive Algorithms
T. Nikolić, T. Talaśka, G. Nikolić, and R. Długosz..285

Comparative Analysis of Layout-Aware Fault Injection on TMR-based DMA Controllers
P. Chernyakov, A. Skorobogatov, A. Zvyagin, E. Emin, I. Danilov, A. Balbekov,
A. Shnaider Khazanova, and M. Gorbunov 289

Approximate Adder with Reduced Error
P. Balasubramanian, D. L. Maskell and K. Prasad 293

Simulation of Ternary CMOS Schemes for Many-Valued Logic Systems
A. A. Krasnyuk and A. G. Prozorova...297

Optimization of Hsiao Decoders by Circuit-Level Minimization
K. Petrov, I. Danilov, A. Shnaider Khazanova, and M. Gorbunov 301

Differential Input Area Efficient Current Comparator
A. R. Serazetdinov and E. V. Atkin...305

Estimation of Errors of Integrated Hydrogen Sensors based on MISFET with structure Pd–Ta_2O_5 –SiO_2 –Si
B. Podlepetsky and A. Kovalenko 309

Analysis and Design of Power Processing Circuits for Thin Film Piezoelectric Energy Harvesters on Flexible Polyethylene Terephthalate Substrates
I. Pandiev, M. Aleksandrova and G. Kolev..313

Towards Portable Thermal Vacuum Sensor - Consideration of Electrical Building Blocks and Compact Housing
D. V. Randjelović, P. Poljak, M. Sarajlić, M. Vorkapić, M. Frantlović,
D. Tanasković, and B. Popović 317

Electrical Characterization of Microbial Fuel Cells – Method and Preliminary Results
D. V. Randjelović, O. M. Jakšić, B. Popović, K. Joksimović, S. Miletić,
P. Poljak, and V. Beškoski...321

Simulation Results of 2.45 GHz Coaxial Antenna with a Ring Slot for Microwave Ablation of a Cancer
K. Cocic, A. Davidovic and D. L. Sekulic...325

Dual Mode Ion Mobility Spectrometer High Voltage Formation Circuit
E. Gromov, M. Matusko, Y. Shaltaeva, V. Pershenkov, V. Belyakov, A. Golovin, E. Malkin,
I. Ivanov, and V. Vasilyev..329

Fast Switching of the Polarity of Dual Mode Ion Mobility Spectrometer
V. Pershenkov, V. Belyakov, Y. Shaltaeva, E. Malkin, A. Golovin, I. Ivanov, V. Vasilyev,
M. Matusko, and E. Gromov 333

Microcontroller's Sensitivity to Voltage Pulse Series in Comparison with a Single Voltage Pulse
A. N. Shemonaev, K. A. Epifantsev, P. K. Skorobogatov, and A. Y. Nikiforov....................337

Correlation between Temperature and Dose Rate Dependences of Input Bias Current Degradation in Bipolar Operational Amplifiers
A. Bakerenkov, V. Pershenkov, V. Felitsyn, A. Rodin, V. Telets, V. Belyakov, A. Zhukov,
and N. Gluhov..341

Software and Hardware System for Charge Coupled Devices with Interline Transfer of Charge Parameters Monitoring During Radiation Tests
V. P. Lukashin, M. E. Cherniak, A. O. Akhmetov, A. Y. Nikiforov, and A. V. Ulanova 345

Comparative Assessment of Digital and UHF Optoelectronic Transceivers Radiation Hardness
R.K. Mozhaev, M.E. Cherniak, A.A. Pechenkin, A.V. Ulanova, and A.Y. Nikiforov 349

Analysis of a Bridgeless Single Stage PFC based on LLC Resonant Converter for Regulating Output Voltage
S. Esmailirad, R. Beiranvand, S. Salehirad, and S. Esmailirad 353

Author Index 357

MINI-COLLOQUIUM ON NANO- AND FLEXIBLE- ELECTRONICS

978-1-7281-3420-8/19 $31.00 © 2019 IEEE

On the ESD Protection and Non-Fatal ESD Strike on Nano CMOS Devices

H. Wong, S. Dong and Z. Chen

Abstract – Electrostatic discharge (ESD) has been one of the major causes for the failure of electronic equipment and components and have attracted quite significant research efforts in minimizing the losses induced. Much tougher challenge comes up in the nano CMOS era. For the device technology itself, the aggressive scaling on gate length, high-k replacement of gate oxide, and the reduction of supply voltage have made the design window of ESD protection device be ever narrower. New ESD protection devices are yet to be developed for the 10 nm technology and beyond. For system and application level, the mobile devices we used right now are much vulnerable as they are more frequent to be exposed to various sources of ESD and power surges. Fatal ESD strike protection is always the primary design specification and should have been mostly fulfilled. The effects of non-fatal ESD strike has not attracted much attention yet. Recent experiment showed that the non-fatal ESD strikes at gate and drain can cause significant charge trapping and trap generation. It resulted in the device characteristic degradation and hence some reliability issues of CMOS circuits and the MOS-based ESD clamps. This review addresses all these issues in detail.

I. ELECTROSTATIC DISCHARGE AND DEVICE RELIABILITY

Electrostatic discharge phenomenon has long been recognized as the most severe hazard to electronic devices and particular precautions and measures have been taken during manufacturing, storage, transportation, and usage. Earlier active measure for ESD protection was mainly on the use board-level transient voltage suppressor (TVS) based on diodes. It soon found that it was not enough. A human body can accumulate static charges with voltage around 3500 V. The transient discharge of this voltage can damage almost every electronic devices in an integrated circuit (IC). Chip-level ESD protection has become the standard requirement now [1-3]. More stringent ESD protection scheme are yet to be developed not only due to the aggressively shrunked device structure but also due to ever frequent human-machine interaction of the widely used handheld smart terminals.

H. Wong is with Department of Electronic Engineering, City University of Hong Kong, Kowloon, Hong Kong, China, E-mail: H.Wong@cityu.edu.hk

S. Dong is with Institute of Photonics and Microelectronics, School of Information Sciences and Electronic Engineering, Zhejiang University, Hangzhou, China

Z. Chen is with Macronix Microelectronics (Suzhou), No.55 Su Hong Xi Road, Suzhou Industrial Park, Suzhou, China

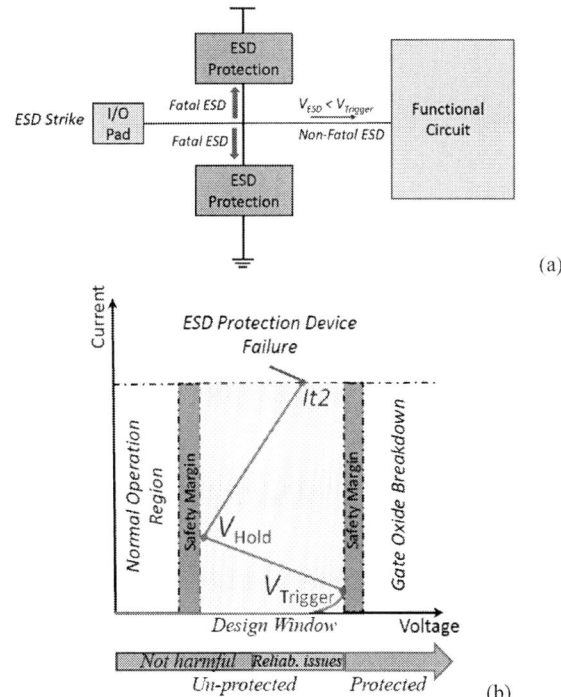

Fig. 1. (a) Schematic showing the ESD protection on an I/O pad which is designed to bypass the ESD current to ground or to supply; (b) illustration of the switching characteristics of ESD protection device and the key parameters constituting design window for an ESD clamp. Possible range and issues of unprotected non-fatal ESD strikes is shown at the bottom.

Considerable efforts have been devoted to preventing this kind of fatal damages [1-7]. The well-established ESD protection philosophy for CMOS circuits is illustrated in Fig 1. The ESD protection circuits are usually integrated with the I/O pads. Under normal operation, the ESD protection part is considered to be transparent though in reality it did introduce series/shunt resistance and parasitic capacitance. If an ESD discharge with voltage V_{Trigger} close to the breakdown voltage of the gate oxide, a special mechanism will be triggered to turn on the ESD clamp and it maintains functioning (provides that the voltage is not dropped below the holding voltage, V_{Hold}) so as to bypass the transient high current to the ground or to the V_{DD}. Gate oxide is the weakest part of integrated circuit, the typical breakdown field of silicon dioxide is around 10 MV/cm only and the breakdown voltages of high-k materials are

978-1-7281-3420-8/19 $31.00 © 2019 IEEE

less than half of that of the silicon oxide [8-9]. The ESD clamp may no long work if the conduction current exceeds *It2*. These three parameters $V_{Trigger}$, V_{Hold}, and *It2* together form the design window for specific ESD protection device [1, 4]. This ESD protection mechanism would allow transient voltages or ESD voltages that are less than the trigger voltage to strike on the functional circuit under protected. In principle, this non-fatal voltage should not cause any physical damage to the circuit but could still cause some reliability issues. Gate oxides (silicon dioxide and high-k oxides) can be significantly degraded due to charge trapping and trap generation after some constant-voltage or constant current stressing [10-13]. The energy of a non-fatal ESD strike could be well exceeds the one used in hot-carrier reliability study [10-11]. As will be discussed in Sec. II, experiment showed that a couple of non-fatal ESD strike can cause significant threshold voltage shift and that may lead to some soft failures of the integrated circuits. This issue has not taken seriously. With the aggressive downsizing of the gate length of MOS transistor to few nanometers range, the high-k gate oxide to subnanometer EOT range, and with the ever frequent man-machine interaction for handheld devices as well, device characteristic degradation due to non-fatal ESD will be an crucial reliability issue. Hence the design window needs to be re-defined especially for high robustness mobile devices with this concern. Section III highlights some recently proposed options for ESD protection in nano CMOS technology. The implications of non-fatal ESD strike induced reliability issues on these clamps will be discussed.

II. NON-FATAL ESD STRIKES ON NANO CMOS DEVICES

All kind of gate oxides including silicon dioxde, silicon oxynitirde and high-k metal oxides are found to have high amount of defects [1, 8-13]. The oxide/silicon interface also has high amount of interface state density. In addition, at the oxide/silicon interface there exists a lot of weak bonds or defect precursors which can be readily transformed into interface states with high-field or hot-carrier stressing [1,8-13]. The biases used for high-field or hot-carrier stressing are well below the level of ESD strike. Hence, it is expected that the ESD strike, in particular the multiple ESD strike, should be able to produce similar effects as in the high-field or hot-carrier stressing. There were few preliminary reports revealed that the ESD strike can cause the gate oxide to degrade [14-16].

Figure 2 compares the I-V characteristics of an NMOS with 1.0 V, 100 ns TLP pulse was applied to the drain terminal while the gate terminal was kept opened. The transistor gate length was 100 nm and it was fabricated using the 65 nm CMOS technology with silicon nitrided oxide [12] as the gate dielectric. This TLP voltage is well below the design window for the given technology. Significant reduction of threshold voltage and the increase of drain currents were observed. This effect can be

explained with the oxide charge trapping of high-energy channel electrons. The high-speed TLP pulse can induce the impact ionization in the high-field region of the reversely-biased drain junction [17-18]. The hot carriers generated in the impact ionization region can surmount the oxide/Si barrier and then localized in the defect sites of the oxide. As a result, a significant threshold voltage shift was observed due to the charge storage effect. This effect is similar the well-known hot-carrier stressing studies for MOS devices [10]. To trigger the impact ionization near the drain, the drain voltage must be large enough and it depends on the gate length also [17]. Figure 3 shows the effect of TLP voltage on the drain ESD strike for NMOS with gate length of 150 nm. Because of longer gate length, notable threshold voltage shift was only observed for V_{TLP} > 1.0 V but the change was quite significant once the impact ionization took place. For $V_{TLP} \sim 1.2$ V, about 17% of threshold voltage shift was registered for a single ESD strike. It is too large that a digital circuit will no longer function properly.

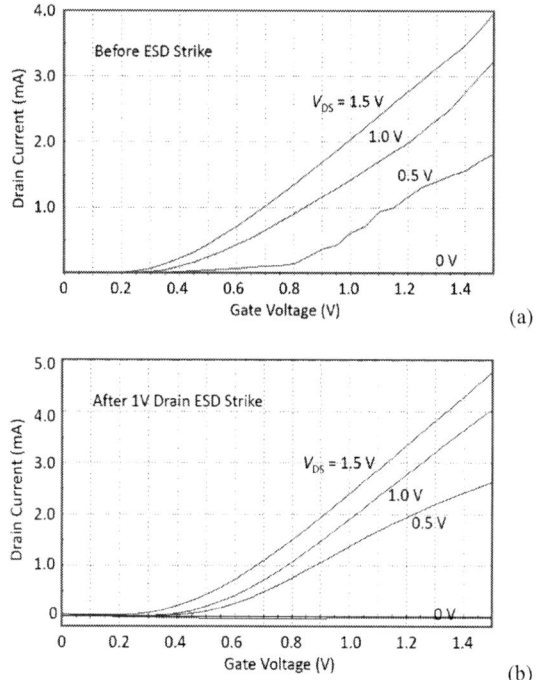

Fig. 2. Typical characteristic change of an NMOS transistor after a single drain ESD strike. After 1 V ESD pulse applied across drain-source. Reduction of threshold voltage and enhancement of currents were observed.

Yet it is unexpected that gate is more robust against non-fatal ESD strike. Provides that the TLP pulse is less that the breakdown voltage of gate oxide, the threshold voltage shift or the gate charge trapping is much smaller than the same level of drain ESD strike. In the same experiment, a notable shift could only be observed for gate V_{TLP} > 2.0 V. The experiment showed that the gate oxide still functioned well with 5V gate TLP strike [19]. Extensive study on the multiple non-fatal ESD strike on the gate was conducted [19]. Figure 4 shows the threshold voltage variation due to the multiple ESD strikes with 5 V TLP on the gate of a PMOS transistor. This voltage is sufficient to cause electron injection from the substrate. However, negative threshold voltage was registered which is inconsistent with the effect of substrate electron injection. It indicates that the TLP has resulted in the interface state generation at the oxide/Si interface. The threshold voltage shift increases as the gate TLP strike proceeds. Both charge trapping and interface state generation took place during the repeatedly gate strike and when the bulk oxide traps are fully filled after 20 times of strike, the interface state generation dominated the threshold voltage shift [19].

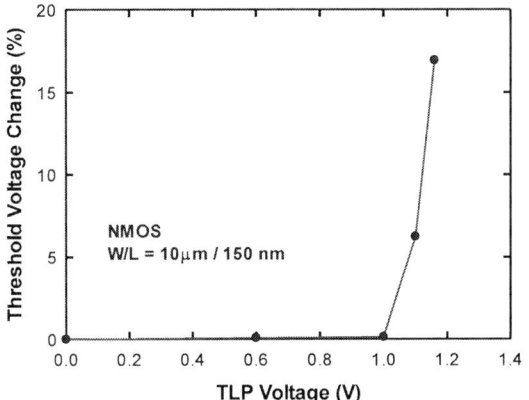

Fig. 3. Threshold voltage degradation of NMOS transistor at different levels of drain TLP strike. The gate length of the transistor is 150 nm.

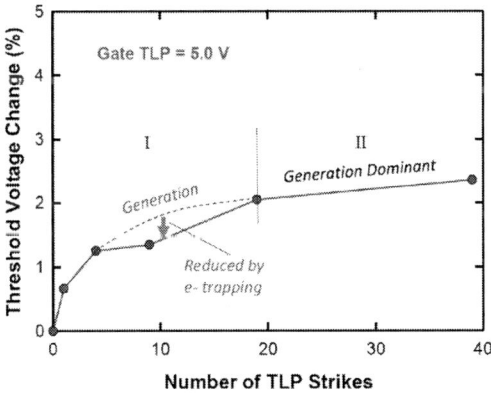

Fig. 4. Threshold voltage degradation due to the multiple ESD strikes at 5.0 V TLP on gate for a PMOS transistor with gate length of 65 nm.

In short, for non-fatal ESD strike, the drain strike will produce much significant threshold voltage shift due to the generation of high-energy electrons in the drain impact ionization region. For gate ESD strike, the change is much smaller even with much larger TLP pulse. However,

multiple strike on the gate can still produce significant charge trapping and interface state generation which would result in the reliability issues of CMOS circuits.

III. NANO CMOS ESD PROTECTION TECHNOLOGY

The design window of ESD protection has become much narrower in recent CMOS technology nodes. In a typical 28 nm CMOS technology we explored, the typical supply voltage is 1.1 V and the specified maximum gate voltage is 5.1 V. Considering 10 to 20% safety margin, the lower bound holding voltage should be in the range of 1.2 to 1.3 V and the trigger voltage should be in the range of 4 to 4.5 V. The key challenge should be the trigger voltage. Table I lists the characteristics of various ESD clamps realized with a typical 28 nm CMOS technology [20]. Grounded-Gate MOS transistor (GGMOS), and the Silicon-Control Rectifier (SCR) have been the common options for CMOS ESD clamps [20-22]. Although simple diode is no longer a feasible option but it may still be used for assisting the trigger voltage lowering of the SCR-based ESD clamps. The high robust SCR device usually has much higher trigger voltage. Several techniques have been developed to lower the trigger voltage of the SCR-based ESD clamps. By incorporating a two dimensional diode string which is also known as diode-triggered SCR or DTSCR, the trigger voltage can be lower to 1.86 V and the holding voltage of the same device can be lower to around 1.67 V which is in appropriate range for avoiding the possibility of latch-up of the 28 nm technology [20].

TABLE I
COMPARISON PARAMETERS OF DIFFERENT ESD PROTECTION DEVICE FABRICATED WITH 28 NM CMOS TECHNOLOGY.

Device Structure	Trigger Voltage (V)	Holding Voltage (V)	Robustness (mA/μm)	R_{ON} (Ω)
Diode	0.70	NA	36.7	3.8
GGNMOS	5.80	4.30	6.7	18.0
SCR	11.20	3.50	32.5	4.6
DTSCR	4.30	3.90	41.1	2.7
TD-DTSCR-TLP	1.86	1.67	53.7	4.1
TD-DTSCR VFTLP	2.18	NA	67.3	4.9

*All the devices are of 30 μm width. Reproduced from Ref. [20].

(a) *Simple Diode*: Diode was used in the early ESD protection, especially as TVS for board level protection [23]. Diode is simple but it in fact is not a good ESD protection device. First, diode does not have snapback so that the trigger voltage is the same as the holding voltage. The size must be large in order to conduct the large ESD current. Diode has notable leakage current and turn-on resistance also. The turn-on voltage or the trigger voltage of a diode is about 0.7 V. Higher voltage was obtained with series connection or by using a reversely biased Zener diode. However the leakage current in series-connected

978-1-7281-3420-8/19 $31.00 © 2019 IEEE

diode string will increase due to the Darlington effect. The trigger voltage from the multiple diode string could not fit the design window specification in recent CMOS technological nodes and the size of diode string is too large as compared with the CMOS functional blocks. Diode should be robust against non-fatal ESD strike. Though non-fatal ESD may result in defect generation and the reverse saturation current and the ideality factor will increase, it should have little effect on the ESD protection capability. Hence diode may still be used as an reliable auxiliary trigger circuit for SCR ESD clamp.

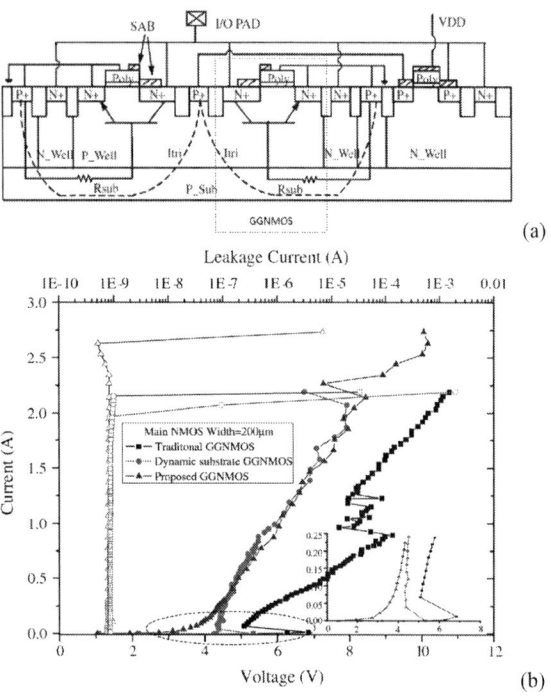

Fig. 5. (a) Schematic of a GGMOS with reduced trigger voltage for 65nm CMOS ESD protection; (b) Comparison of TLP characteristics for simple GGNMOS, dynamic substrate GGNMOS, and modified GGNMOS shown in (a) ESD clamps. Reproduced from Ref.[25].

(b) *Grounded-Gate MOS transistor (GGMOS)*: Being fully compatible with the CMOS process, GGMOS had been widely used in ESD protection for CMOS circuits. The GGMOS basically uses the breakdown characteristics resulting from the parasitic bipolar transistor constituting by drain, bulk, and source [24]. Though the device structure is more complicated, its size could be much smaller than the diode ones. The trigger and holding voltages of a GGMOS depend on the substrate doping concentration and channel length or the base width of the parasitic BJT. By modifying the effective substrate doping and the trigger mechanism, it is suggested that the GGMOS based ESD protection device can be used for 65 nm or even 45 nm process [25]. Figure 5(a) shows a modified GGMOS-based ESD protection circuit [25]. The key

component is the NMOS with gate connected to the ground as marked in red dashed box. The gate length of this transistor should be shorter enough so that the parasitic NPN could be turned on during the ESD strike. In the modified structure, the triggering of the parasitic NPN is via the PMOS with drain connected to the P+ diffusion between drains of the NMOS fingers. In this structure, the substrate resistance was enhanced with the additional N-well in between the N+ source and P+ bulk of the GGMOS. When the PMOS is turned on, it injects hole (substrate trigger current, I_{tri}) into the substrate to increase the substrate resistance and so as to reduce the trigger voltage. Figure 5(b) compare the characteristics of the two modified triggering schemes with the conventional GGMOS under an ESD strike with same GGMOS size [25]. For conventional GGMOS, the Vt1 value is about 6.84 V which is close to the gate oxide breakdown voltage of 65 nm technology. With dynamic substrate GGMOS [26], Vt1 reduced to 5.3 V. With the scheme given in Fig.5 (a), the substrate resistance is increased when the PMOS is turned on and Vt1 is reduced to 2.8 V which should be suitable for 65 nm technology. For 40 nm process and beyond, this trigger voltage is too large. The ESD design window is much narrower because of the low breakdown voltage for the introduction of high-k gate dielectric and the scaled supply voltage. Since the MOS transistor is the key component for the protection circuit and with the observations as reported in previous section, the ESD strike on drain can cause significant defect generation and charge trapping in the gate oxide. For long-term reliability issue, it will lower the turn-on voltage or known as trigger voltage walk-in for GGMOS. Trigger voltage walk-in was observed in LDMOS based ESD clamp [27]. It is believed that it should be due to the same mechanism at reported in Sec. II. For the extreme case, the gate oxide may be broken down and the GGMOS will be no longer functioning.

(c) *SCR-based Protection Devices*: Silicon-Controlled Rectifier (SCR) devices have been suggested to be one of the best candidates for ESD protection for its robustness and compactness. SCR has a much higher It2 than GGMOS and diode. It has smaller device size and lower parasitic capacitance. However, an intrinsic SCR has too large trigger voltage and was not considered as suitable candidate for ESD protection of CMOS circuits. For the narrowed design window of 28 nm process, the trigger voltage was too large. Recent efforts have demonstrated that the trigger voltage can be effectively reduced by incorporating an additional triggering path so called low-voltage triggered SCR (LVTSCR) or with a built-in parasitic MOS transistor. However, since the major conduction path of the LVTSCR is still through the SCR, the holding current in LVTSCR is too low and that latch-up may occur. To solve this issue, together to cope with the narrowed design window of 28 nm CMOS technology, two-dimensional diode-trigged SCR (TD-DTSCR) [20, 28-29] and double snapback SCR (DS-SCR) were proposed [30].

Figure 6(a) shows the cross-sectional view of the double snapback SCR (DS-SCR) reported recently [30]. This device constitutes with a GGMOS and a SCR. The embedded GGMOS is in the middle separating the anode and cathode of SCR. When an ESD strike on the anode, two current conduction paths as depicted in dash curves will be established. The current conduction over the GGMOS (red dashed curve) should be easier and this conduction will make the voltage drop in the P-well to increase. When the voltage drop of P-well exceeds the base-emitter voltage of the parasitic bipolar transistor, constituting by P-Well and N+ regions, the parasitic bipolar transistor induced drain breakdown will occur [24] and that results the first snapback. When the conduction current later on reaches certain level, the SCR will be turned on and second snapback will occur.

Figure 6(b) compares the TLP characteristics of conventional SCR (SSSCR), LVTSCR and DS-SCR [30]. The DS-SCR shows two snapbacks with lower trigger voltage and higher holding voltage. The first snapback was at trigger voltage of 6.1V and holding voltage of 4.1 V. The second snapback occured at a current of about 0.22 A and the holding voltage was reduced to 3.8 V. In the DS-SCR structure, the first device being exposed to the serve ESD strike was the GGMOS and the ESD strike was on the source/drain terminal. As discussed in Sec. II, even non-fatal ESD strike can still cause a transient high-field near the drain region which will result in hot-carrier generation. The charge trapping and trap generation in the gate oxide result and that make the GGMOS to be turned on earlier. This effect should also affect the second snapback produced by the SCR as the degradation of GGMOS will affect the SCR current level also. It is worth to have a detailed study on the trigger voltage walk-in behavior of DS-SCR so as to investigate the long-term stability of this kind of ESD clamp.

IV. CONCLUSION

Electrostatic discharge has become a challenging issue on nano CMOS technology because of the shrinking on device dimensions, the use of ultrathin high-k oxide, the reduction operation voltage, and the more frequent ESD strike due to mobile applications as well. The aggressive device downsizing results in much narrower design window for realizing the ESD protection devices. SCR based ESD clamps with some novel triggering regimes such as incorporating two-dimensional diode string or GGMOS could still be the good technological options as highlighted. The mobile applications lead to more frequent non-fatal ESD strike and that would cause, in particular under the drain/source ESD strike on nanoscale CMOS circuits, more significant device characteristic degradation. The non-fatal strike should also cause some reliability issues such as trigger voltage walk-in for ESD protection devices being based on GGMOS. The safety margin and the design methodologies for ESD protection device needs

to be reviewed in order to take the effects of non-fatal ESD strike on ESD device and functional circuit to be protected into account.

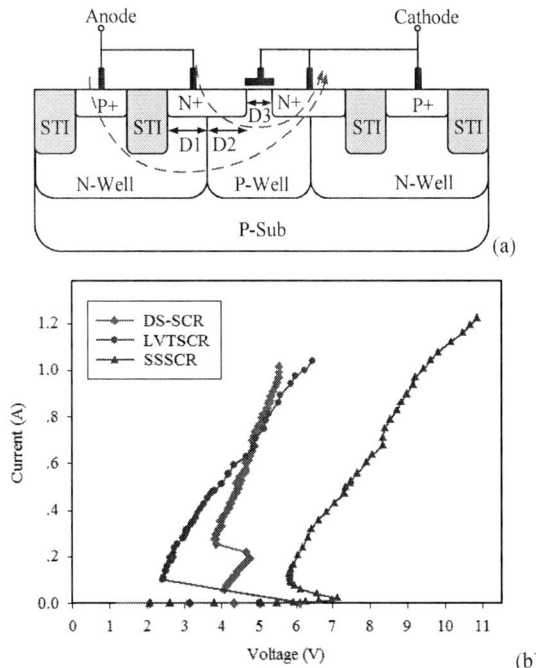

Fig. 6. (a) Cross-sectional view of DS-SCR structure; (b) comparison of TLP curves measured from DS-SCR, LVTSCR and SSSCR fabricated using the same 28 nm CMOS process. Reproduced from Ref. [30].

ACKNOWLEDGEMENT

The authors wish to thank the University Grant Council of Hong Kong which was supported under project no. CityU123613.

REFERENCES

[1] A. Amerasekera and C. Duvvury, *ESD in Silicon Integrated Circuits*, Wiley, 2002.

[2] J. E. Vinson and J. J. Liou, "Electrostatic discharge in semiconductor devices: an overview," *Proc. IEEE*, 1998, vol.86, pp. 399-420.

[3] M.-D. Ker, and K.-C. Hsu, "Overview of on-chip electrostatic discharge protection design with SCR-based devices in CMOS integrated circuits," *IEEE Trans. Device Mater. Reliab.*, 2005, vol.5, pp. 235-249.

[4] A. J. Walker, H. Puchner, and S.P. Dhanraj, "High-voltage CMOS ESD and the safe operating area," *IEEE Trans. Electron Devices*, 2009, vol.56, pp.1753-1760.

[5] C.-Y. Lin and Y.-L. Chiu. "Design of embedded SCR device to improve ESD robustness of stacked-device output driver in low-voltage CMOS technology." *Solid-State Electron.*, 2016, vol.124, pp.28-34.

[6] J. Zeng, S. Dong, H. Wong, T. Hu, X. Li, "Layout optimization of GGISCR structure for on-chip system level

ESD protection applications," *Solid-State Electron.*, 2016, vol.126, pp.152-157.

[7] F. Lu, R. Ma, Z. Dong, L. Wang, C. Zhang, C. Wang, Q. Chen, X. S. Wang, F. Zhang, C. Li, H. Tang, Y. Cheng and A. Wang, "A systematic study of ESD protection co-design with high-speed and high-frequency ICs in 28 nm CMOS." *IEEE Trans. Circuits Systems I: Regular Papers*, 2016, vol. 63, pp. 1746-1757.

[8] H. Wong, *Nano-CMOS gate dielectric engineering*, CRC Press, Boca Raton, 2012.

[9] H. Wong, Y.C. Cheng, "Instabilities of metal-oxide-semiconductor transistor with high-temperature annealing of its gate oxide in ammonia," *J. Appl. Phys.*, 1990, vol.67, pp. 7132-7138.

[10] H. Wong, Y.C. Cheng, "Generation of interface states at the silicon/oxide interface due to hot-electron injection," *J. Appl. Phys.*, 1993, vol.74, pp.7364-7368.

[11] B.L. Yang, H. Wong, Y.C. Cheng, "Study of process-dependent electron-trapping characteristics of thin nitrided oxides," *Solid-State Electron.*, 1994, vol.37, pp.481-486.

[12] H. Wong, V.M.C. Poon, C.W. Kok, P.J. Chan, V.A. Gritsenko, "Interface structure of ultrathin oxide prepared by N_2O oxidation," *IEEE Trans. Electron Devices*, 2003, vol.50, pp.1941-1945.

[13] B. Sen, H. Wong, V. Filip, H.Y. Choi, C.K. Sarkar, M. Chan, C.W. Kok, M.C. Poon, "Current transport and high-field reliability of aluminum/hafnium oxide/silicon structure," *Thin Solid Films*, 2006, vol.504, pp.312-316.

[14] J.-W. Lee and H. Tang, "Punchthrough effects on the electrostatic discharge robustness of ultrathin silicon films on insulator devices," *Appl. Phys. Lett.*, 2006, vol.89, art no. 103508.

[15] J. Wu and E. Rosenbaum, "Gate oxide reliability under ESD-like pulse stress," *IEEE Trans. Electron Devices*, 2004, vol.51, pp.1928-1531.

[16] C. Y. Lin, M. D. Ker, P. H. Chang and W. T. Wang, "Study on the ESD-induced gate-oxide breakdown and the protection solution in 28nm high-k metal-gate CMOS technology," in: *2015 IEEE Nanotechnology Materials and Devices Conference* (NMDC), Anchorage, 2015, pp. 1-4.

[17] H. Wong, "A physically-based MOS transistor avalanche breakdown model," *IEEE Trans. Electron Devices*, 1995, vol.42, pp.2197-2202.

[18] H. Wong, "Drain breakdown in submicron MOSFETs: a review," *Microelectron. Reliab.*, 2000, vol.40, pp.3-15.

[19] H. Wong, S. Dong, and Z. Chen, "Effects of non-fatal electrostatic discharge on nano CMOS devices," to be published.

[20] S. Parthasarathy, J. A. Salcedo, S. Herrera and J. J. Hajjar, "ESD protection clamp with active feedback and mis-trigger immunity in 28nm CMOS process," in *Proc IEEE Int. Reliab. Phys. Symp.*, Monterey, 2015, pp. EL.3.1-EL.3.5.

[21] X. Li, S. Dong, H. Jin, M. Miao, T. Hu, W. Guo, H. Wong, "28 nm CMOS process ESD protection based on diode-triggered silicon controlled rectifier," *Solid-State Electron.*, 2017, vol.137, pp.128-133.

[22] Y. Zhou, M. Miao, J. A. Salcedo, J. J. Hajjar and J. J. Liou, "Compact Thermal failure model for devices subject to electrostatic discharge stresses," *IEEE Trans. Electron Devices*, 2015, vol. 62, pp. 4128-4134.

[23] W.S. Tam, C.W. Kok, S.L. Siu, H. Wong, "Snapback breakdown ESD device based on zener diodes on silicon-on-insulator technology," *Microelectron. Reliab.*, 2014, vol.54, pp.1163-1168.

[24] H. Wong, "Modeling of the parasitic transistor-induced drain breakdown in MOSFETs," *IEEE Trans. Electron Devices*, 1996, vol.43, pp.2190-2196.

[25] F. Ma, Y. Han, B. Song, S. Dong, M. Miao, J. Zheng, J. Wu, K. Zhu, "Substrate-engineered GGNMOS for low trigger voltage ESD in 65 nm CMOS process," *Microelectron. Reliab.*, 2011, vol.51, pp.2124-2128.

[26] M.-D. Ker, T.-Y. Chen, "Substrate-triggered technique for on-chip ESD protection design in a 0.18-μm salicided CMOS process," *IEEE Trans. Electron Devices*, 2003, vol.50, pp.1050-1057.

[27] Z. Yu, H. Jin, S. Dong, H. Wong, J. Zeng, W. Wang, "Comparative study of reliability degradation behaviors of LDMOS and LDMOS-SCR ESD protection devices," *Microelectron. Reliab.*, 2016, vol.61, pp. 111–114.

[28] M. Mergens C. Russ K. Verhaege J. Armer P. Jozwiak R. Mohn, B. Keppens S. Trinh "Diode-triggered SCR (DTSCR) for RF-ESD protection of BiCMOS SiGe HBTs and CMOS ultrathin gate oxides" in: *IEEE Int. Electron Devices Meeting* Tech. Dig. pp. 21.3.1-21.3.4, 2003.

[29] W.Y. Chen, E. Rosenbaum and M.D. Ker, "Diode-triggered silicon-controlled rectifier with reduced voltage overshoot for CDM ESD protection," *IEEE Trans. Device Mater. Reliab.*, 2012, vol. 12, pp. 10-14.

[30] T. Hu, S. Dong, H. Jin, H. Wong, Z. Xu, X. Li, J.J. Liou, "A double snapback SCR ESD protection scheme for 28 nm CMOS process," *Microelectron. Reliab.*, 2018, vol.84, pp.20-25.

CMOS-Compatible Gas Sensors

L. Filipovic and S. Selberherr

Abstract— As transistor scaling along Moore's law approaches its physical limits, the semiconductor industry has been intensely working on functional integration of devices along the More-than-Moore approach. The integration of sensors, RF circuits, and other functionalities with electronics is enabled by innovations in packaging, three-dimensional integration, and most importantly through the fabrication of multiple components and features on silicon using established technology. With the application of semiconductor metal oxide (SMO) thin films, there is potential for the integration of gas sensors with processing electronics. This manuscript describes SMO gas sensors, their fabrication, and operating techniques which require high temperatures and therefore an integrated microheater. Microheaters require the fabrication of a membrane in order to isolate the high temperature component from other circuitry. Finally, the manuscript looks at recent achievements in engineering of SMO films and in understanding and modeling their sensing mechanism.

I. INTRODUCTION

Aggressive device scaling has been the driving force behind many advancements made in the microelectronics industry over the past several decades. As the miniaturization trend for complementary metal oxide semiconductor (CMOS) devices along the path of Moore's law [1], [2] continues, the typical transistor scaling is reaching its physical limits. This has caused the industry to look increasingly at different semiconductor materials and vertical transistor architectures, but also at the potential for the integration of multiple functionalities on a single chip or device, appropriately labeling this type of integration as Moore-than-Moore [3], [4]. The interest in the Internet of Things (IoT) and Internet of Everything (IoE) are clear indicators of this trend, where multi-featured electronics are packaged together. The first attempts at this multi-application packaging and integration was by electrically connecting different dies using bond wires. However, this approach is not ideal, since long bonding wires result in high resistance-capacitance (RC) delays, limiting the high frequency performance of the device and increasing the power dissipation. An increase in efficiency was recently achieved by applying three-dimensional (3D) integration using through silicon vias (TSVs).

With 3D integration, different dies are stacked on top of each other and their interconnections are managed using TSVs which provide an electrical contact between the front and back of a wafer. Using this approach,

L. Filipovic and S. Selberherr are with the Institute for Microelectronics, Technische Universität Wien, 1040 Vienna, Austria, E-mail: {filipovic|selberherr}@iue.tuwien.ac.at

the RC delay and package size can be significantly reduced. Ultimately, achieving a full integration of all necessary components on a single wafer as a System-on-Chip (SoC) would be the optimal solution. Using silicon as a substrate for functionalities beyond integrated circuits and transistors, including radio frequency (RF) components, sensors and actuators, or biochips, would allow for an efficient integration between CMOS and micro-electro-mechanical systems (MEMS) devices into a monolithic device. From the fabrication side, the integration of added functionalities must be performed after the CMOS front end of line (FEOL) devices are in place, meaning that the same rules as those existing for back end of line (BEOL) metalization apply. With this in mind, a part of the challenge is that the temperature of all additional processing steps used to fabricate the added functionalities must be kept below 450°C so as to not damage the FEOL devices.

In the following sections we introduce currently existing gas sensing mechanisms while concentrating on the semiconducting metal oxide (SMO) gas sensor due to the recent achievements made in its integration with CMOS technology. The rest of the manuscript is divided into three sections: First, the processing techniques used to fabricate this gas sensor are described, including the membrane release and sensing layer deposition. Subsequently, we discuss the features of the sensors critical to operation and reliability, the microheater, and the SMO sensing layer. Finally, we discuss the sensor operation in terms of the current understanding of the sensing mechanism of SMO films.

A. Mechanisms for gas sensing

While the human nose can detect many different odors in our environment, it is unable to detect many harmful gases and it fails outright, when there is a need to detect a specific gas concentration. Even though many harmful gases are regularly present in our environment, it is not until a certain critical concentration is reached that they become harmful to our health and well-being. The ability to detect specific gas concentrations electrically is a research field which has been studied extensively over many decades. Many applications and industries rely on - or could greatly benefit from - the development of efficient and affordable gas sensors including health and safety [5], [6], automotive and aviation [7], [8], environmental monitoring [9], [10], [11], [12], and chemical warfare detection [13], [14], [15], just to name a few. More recently, the push in sensor research has been towards device integration in portable electronics such as smart-watches, smart-rings, smart-phones, tablets and wearables [16], [17]. Alongside portable electronics, fabrication and process controls, laboratory analytics, and smart-homes can be made more affordable, if cheaper gas sensing devices and equipment was available. Here, we examine current mechanisms for gas sensing while paying extra attention to properties which affect a sensor's portability such as low power, low cost, and a small footprint.

Today, there is a large variety of gas sensing principles which have shown value to industry and research including semiconductor, optical, thermal, infrared (IR), quarz microbalance, catalytic, dielectric, electrochemical, electrolyte, and conductivity based ones [18], [19], [20], [21]. Of all the currently known gas sensing mechanisms it is the semiconductor - or more accurately the semiconducting metal oxide - sensor which provides the biggest advantage towards integration and portability. While the catalytic pellistor also enjoys low power consumption, cost, and a relatively small footprint, it's selectivity is very poor, while its sensitivity is poorer and response time is longer than that of an SMO sensor. The piezo-electric sensor, on the other hand, has an excellent sensitivity, accuracy, and response time, but its power consumption is a limitation to portability. The same can be said for the photo-ionization and IR adsorption sensors. The electro-chemical sensor requires a large footprint, while the sensitivity of a thermal pellistor is not up to par with an SMO sensor. All in all, the SMO sensor provides the most advantages over alternatives, especially when it comes to its sensitivity, response time, and potential for miniaturization and portability through very low power consumption, very low fabrication costs, and a very small footprint. The primary concerns about the operation of SMO gas sensors is their lack of selectivity and difficulty in removing the chemisorbed gas molecules after a sensing event.

B. Semiconducting metal oxide gas sensor

The sensing mechanism of an SMO film is based on a changing electrical resistance due to charge accumulation on its surface in the presence of a target gas. The gas molecule interacts with the film's surface and through chemisorption either donates or captures an electron to or from the sensing film, respectively. This makes it clear why selectivity is an issue, as there is no direct way to know precisely which molecule caused the difference in resistance, only that the resistance has changed. The selectivity can be artificially introduced through the use of a sensor array, where multiple sensors with different properties are activated simultaneously [22], [23], [24], [25], [26], [27]. The different properties can be created by having each sensor operate at a different temperature or by introducing a different dopant to each sensor, which is optimized for a particular gas molecule. The collected data can subsequently

be processed with a variety of methods such as neural networks [27] or machine learning [28] algorithms.

As alluded to in the previous paragraph, temperature plays an important role in the operation of SMO sensors. In fact, the sensing mechanism is activated only at elevated temperatures in the range between 250°C and 500°C. This provides enough energy to allow surface reactions to take place. The need for high temperature operation means that a microheater must be incorporated into the sensor structure, which in-turn must be isolated from the remaining circuitry. Not having proper thermal isolation would be a significant hindrance to the concept of integration. Therefore, the microheater is most commonly fabricated within a MEMS membrane, which hosts the microheater and provides a platform for the SMO film and sensing electrodes. A typical membrane stack is depicted in Fig. 1.

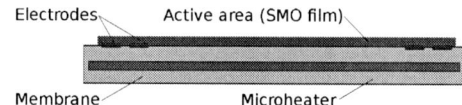

Fig. 1. SMO sensor membrane stack, depicting the "membrane" materials (SiO$_2$/Si$_3$N$_4$), the microheater, the electrodes, and the SMO sensing film, corresponding to the active area.

The formation of the aforementioned membrane is one of the most challenging steps in the fabrication of SMO sensors. There are two most commonly used MEMS membrane types: A closed or full membrane, where the active sensor area is isolated by removing the wafer underneath it and a suspended membrane, which is held in place by suspension beams above an air void. A perforated membrane, created using a sacrificial polyimide or using an additional photolithography step after a closed membrane is fabricated is an alternative option which has recently garnered some attention [29], [30]. In Fig. 1 we see that the membrane is composed of several layers, including the microheater, which could also have an additional heat spreading plate above it, the SMO film, electrodes, and the membrane stack. The membrane stack is composed of a combination of silicon dioxide (SiO$_2$) and silicon nitride (Si$_3$N$_3$) layers.

II. SENSOR FABRICATION

Many fabrication steps are required for the SMO sensor. However, the two which present the most challenges are the membrane release, meaning the creation of a void under the membrane and the deposition of the SMO film. The main concern with the membrane release is that the fabrication should be compatible with CMOS technology to ensure easy integration with CMOS electronics, cost efficient fabrication, and low power consumption [31]. When depositing the sensing film itself, the ability to integrate this step within the CMOS sequence, or as a post-processing step, while ensuring a film of high quality, is of utmost importance.

A. Membrane release

The fabrication of the suspended membrane is one of the most critical steps in the creation of SMO gas sensors. As discussed in the previous section the types of membranes which can be fabricated are the closed, suspended, and perforated types. In Fig. 2a, Fig. 2b, and Fig. 2c, several recently published gas sensor designs which apply the closed membrane from [32], perforated membrane from [33], and suspended membrane from [34], respectively, are shown.

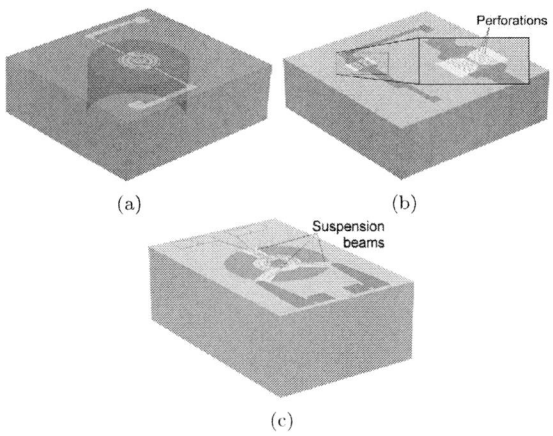

Fig. 2. Membrane types for gas sensors. (a) Closed membrane. (b) Perforated membrane. (c) Suspended membrane.

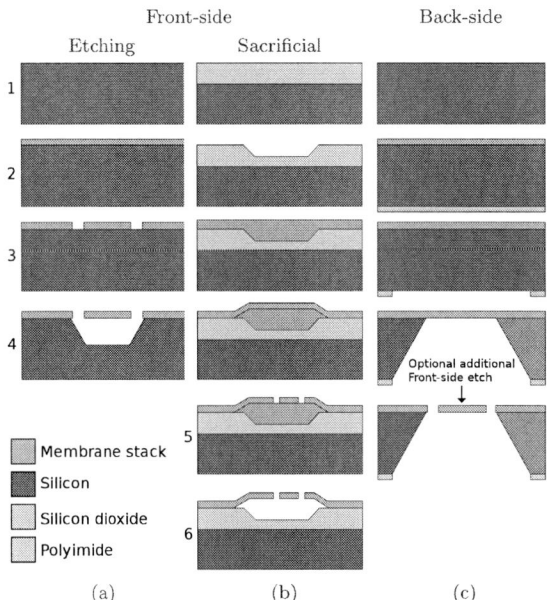

Fig. 3. Etching steps used to release the membrane for SMO gas sensors. (a) Front side etching for a suspended membrane. (b) Sacrificial Polyimide etching for a perforated membrane. (c) Back-side etching for a closed membrane.

The steps required to fabricate the three membrane types shown in Fig. 2 are depicted in Fig. 3. The suspended membrane can be formed by patterning the suspension beams and front-side etching using an etchant with potassium hydroxide (KOH), tetra-methyl ammonium hydroxide (TMAH), ethylenediamine pyrocatechol (EDP), or by using selective plasma etching [35]. A primary goal is that the membranes are made very thin, resulting in a reduced power consumption. Furthermore, the fabrication technique utilizing patterning and etching is suitable for CMOS fabrication, since it is all performed from the front-side and no high temperature is required. However, wet etching is not a clean process and care must be taken to ensure that no other sections of the wafer are harmed during this step.

The closed membrane can be formed by etching from the back-side of the wafer using wet chemical etchants KOH, TMAH, and EDP or by the application of deep reactive ion etching (DRIE), which is a cyclical etching technique whereby each cycle consists of isotropic polymer deposition in fluorocarbon (e.g. C_4F_8) followed by one or more anisotropic etch steps, most frequently performed in ion-enhanced SF_6/O_2 plasma [36], [37]. It should be noted that the closed membrane can be turned into a perforated one by adding an extra frontside step where holes are patterned trough the membrane, as shown in Step 5. of Fig. 3c. The main advantage of the closed membrane is its mechanical stability. Because it is not suspended by thin beams, this structure can sustain higher thermal stresses, and therefore also higher operating temperatures. High stress in the beams can result in cracking and delamination, leading to a reduced device lifetime [38]. However, closed membranes generally dissipate more power, since more material is available to carry the heat away laterally.

The perforated membrane combines the benefits of the suspended and closed membranes. The structure is formed by introducing a sacrificial polyimide layer, depositing a membrane on top of it, then etching away the sacrificial layer using pre-patterned holes through the membrane. In addition to providing access to the sacrificial layer during etching, the holes help to reduce lateral heat conduction and heat losses through the membrane, thereby reducing the total power dissipation [29], [33]. For this structure, sacrificial polyimide HD8820 is deposited in a pre-etched SiO_2 cavity, then cured for one hour to obtain the desired thickness, shown in Fig. 3b [33]. The perforated membrane can also be formed by etching holes in a closed membrane, shown in Step 5 of the back-side process in Fig. 3c.

The membrane stack is usually composed of thin Si_3N_4 and SiO_2 layers. The typical values of the postdeposition stresses of SiO_2 and Si_3N_4 are about 1GPa tensile and about 300MPa compressive, respectively. Placing a Si_3N_4 layer between two SiO_2 layers can help reduce the membrane stress to a value below 100MPa, which is essential to ensure its mechanical stability [39].

B. Sensing film deposition

Ever since the gas sensing capabilities of metal oxide films were discovered, different techniques have been successfully attempted to deposit thin and thick films in a way conducive to integration with CMOS foundry technology. The sensing films can be deposited using physical or chemical methods. The chemical methods can be separated according to whether the chemical interaction with the surface takes place while the depositant is in the gas phase or in the liquid phase. One physical process which has been applied to deposit quality SMO films is sputtering [40], [41].

The chemical processes which have been examined for thin SMO film deposition are given in Fig. 4, where chemical vapor deposition (CVD) and atomic layer epitaxy (ALE) are the gas processes mentioned. Spray pyrolysis is shown as a liquid phase deposition technique. While it is true that spray pyrolysis starts from a liquid source which is sprayed onto the surface where it is to be deposited, the actual deposition mechanism is more analogous to CVD. It was recently shown that the coverage around corners and edges of 3D structures are highly uniform using spray pyrolysis and that the spray direction does not influence the surface coverage or film thickness [42]. The first application of spray pyrolysis was carried out by Chamberlin and Skarman [43] in 1966 for the growth of CdS thin films for solar cells. Since then, it has been used for many SMO films, including SnO_x, In_2O_3, indium-tin-oxide (ITO), PbO, ZnO, ZrO_2, yttria-stabilized zirconia (YSZ), and many others [44], for solar cells, batteries, optoelectronic devices, and gas sensors [42].

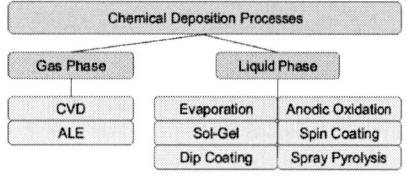

Fig. 4. Summary of chemical deposition techniques used for SMO film deposition.

The principal advantages of spray pyrolysis include:
• Integrable with CMOS technology since the deposition temperature is kept at about 400°C. While sputtering also allows this, it is a physical process which can result in non-uniform deposition and film thinning around corners and vertical walls, limiting the types of designs which can be used for the gas sensor.
• Cost effective. The necessary setup can be very straight-forwardly implemented as a post-CMOS processing step and does not require any cost-intensive equipment [45].
• Substrates with complex geometries can be coated and the resulting film has been shown to deposit very uniformly with a high quality for sensing [42], [45].

The main interest in spray pyrolysis stems from its cost effectiveness and possibility of integration in CMOS technology. The three steps which take place during spray pyrolysis deposition are summarized by:
1. Atomization of the precursor: The atomization is the act of applying pressure to push a liquid through a small nozzle, resulting in the formation of small droplets with an initial velocity, which compose the spray solution. The nozzle can be gas pressure based, ultrasonic, or electrostatic [46], while the resulting droplet size, rate of atomization, and droplet velocity are determined by the pressure applied and the nozzle diameter.
2. Aerosol transport of the droplet: The droplets travel through the air in liquid form towards their source due to several forces acting on them, finally reaching the surface. In the heated area, the droplets start to lose their volume, forming a precipitate, a vapor, or a powder, depending on the initial size of the droplet [42].
3. Decomposition of the precursor: This step initiates the film growth by interactions between the surface and the droplets which reach the surface in vapor form.

III. SENSOR ELEMENTS

The key elements of the SMO gas sensor are the sensing film itself and the microheater which, as mentioned previously, is necessary to heat the sensing film to an appropriately elevated temperature to initiate the sensing mechanism. The choices of materials used for the microheater and sensing film is critical to ensuring the proper selectivity towards a desired gas and high reliability. In this section we discuss the choice of materials and designs tested for the microheater and SMO sensing layer, recently attempted for gas sensor applications.

A. Microheater

Many materials have the potential to be used as a microheater, since the heating principle is based on Joule heating which is a phenomenon affecting many metals and semiconductors. Materials such as silicon carbide (SiC) [47], polysilicon [48], aluminum [49], copper [50], molybdenum [51], platinum [52], tungsten [32], nickel alloys [53], tantalum-aluminum (TaAl) [29], and many others have been used to heat the sensing film. There has been significant interest in SiC recently due to its high power density, potential for miniaturization, and low power consumption [54]. However, the use of SiC strays away from the typical silicon-based SiO_2/Si_3N_4 membrane stack, meaning its fabrication is more complex and not compatible with CMOS technology.

The initial materials used for microheater development were primarily those readily available in CMOS technology, such as polysilicon and aluminum [55]. However, these materials suffer from electromigration defects, when high current gradients are applied, and they have poor contact properties, a particular problem for high temperature operation. Today, platinum

978-1-7281-3420-8/19 $31.00 © 2019 IEEE

is a popular choice due to its chemical inertness at high temperatures and its ability to deal with high current densities [40]. However, platinum is also not ideal because of its positive temperature coefficient of resistance (TCR), which magnifies hotpot effects, and its high cost. Hotspots can lead to potential drift in the response, which negatively affects the long-term reliability of the device [56]. Tungsten was recently suggested as a microheater material and it appears to be the perfect choice due to its resistance to electromigration; unfortunately, it has a tendency to form an oxide at temperatures above 300°C, so care must be taken when sealing the microheater in the membrane. Research into nickel and nickel alloys is currently ongoing due to several positive aspects of these thin films, including a low coefficient of thermal expansion (CTE), resistance to humidity, and a high Young's modulus. Tantalum-aluminum is another recently studied option [29] and its advantage is the ability to maintain its mechanical strength at elevated temperatures and its negative TCR. In summary, a good microheater material is one which has a low thermal conductivity, high electrical resistivity, high melting point, low CTE, low Poisson's ratio, and high compatibility with CMOS technology [40].

In addition to testing different materials for the microheater, innovative geometries and designs have been attempted in order to ensure a uniform temperature distribution across the active area and low power consumption. Some designs involve the placement of an additional highly thermally conductive plate below or above the microheater, which is electrically inert, in order to help distribute the heat more evenly [36]. While this is effective, it introduces an additional photolithography step, increasing fabrication costs. Some typical designs and patterns used for microheaters, in order to ensure good temperature uniformity, are shown in Fig. 5. The layout pitch (line and separation widths) has a significant influence on the efficiency of the temperature distribution [21]. Minimizing the separation leads to an improved power uniformity and reduced power dissipation. Having a predictable and uniform temperature is of primary importance to give confidence in the sensor's response, as can be extruded from Fig. 6. However, care should be taken to make sure novel designs do not introduce intricacies which increase the sensor's fabrication complexity, cost, and power consumption.

Recently, the influence of the microheater quality on the sensing behavior has been reported in [33]. The authors show that by introducing slight openings along the length of the microheater, with the goal of improving temperature uniformity, not only does the mechanical stability improve, but the sensitivity of the sensor does as well. The vertical displacement reduced from about 175nm to about 70nm when operating at 500°C, and the sensitivity of the sensor towards CH_2O and ethanol improved by 15% and 5%, respectively, with-

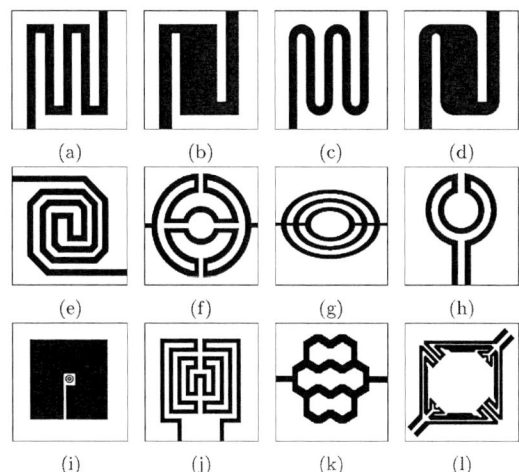

Fig. 5. Microheater geometries characterized and modeled over the last decades. These include the shapes: (a) Meander, (b) S-meander, (c) Curved, (d) S-curved, (e) Double spiral, (f) Drive wheel, (g) Elliptical, (h) Circular, (i) Plane plate, (j) Fin shape, (k) Honeycomb, and (l) Irregular.

out any increase in power dissipation which was about 10.5mW. Therefore, the microheater design must be treated as a core feature, influencing many aspects in the performance and reliability of a gas sensor.

B. Sensing film

The discovery of the sensing capabilities of SMO films has created a vision for the miniaturization, portability, and integration of gas sensors with CMOS electronics. In the 1950s Brattain and Bardeen [57] demonstrated that several semiconductor materials change their electrical resistivity when exposed to certain gas molecules, especially when heated to elevated temperature. The first device based on this effect was developed in the 1960s using a thin zinc oxide (ZnO) film [58], while by 1967 it was shown that adding small amounts of noble metal dopants, such as copper, platinum, rhodium, iridium, gold, or palladium can improve sensor performance [59].

Over the past decades, researchers have attempted to enhance gas sensor performance, including the sensitivity and selectivity of SMO films, by testing different SMO films and designing novel nanomaterials. Several SMO films which have been deposited using spray pyrolysis and used in various applications are given in Section II-B. For gas sensors in particular, ZnO, SnO_2, α-Fe_2O_3, CdO, $ZnSnO_4$, NiO, PbO, YSZ, WO_3, ITO, and In_2O_3, have been studied [60], [61]. From all the available materials tin oxide (SnO_2), an n-type wide band-gap semiconductor, has been investigated very frequently due to several excellent properties, including a high electron mobility of $160cm^2V^{-1}s^{-1}$ and high chemical and thermal stability [61]. Furthermore, SnO_2 has good sensitivity towards many gases and its deposi-

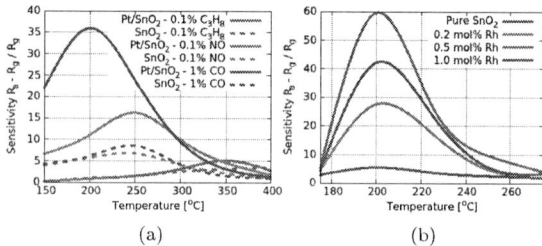

Fig. 6. Sensitivity of doped and undoped SnO_2 sensors toward different gases as a function of operating temperature. (a) Pt-doped SnO_2 response to CO, NO, and C_3H_8 and (b) Rh-doped SnO_2 response to 50ppm acetone.

tion can easily and cost-effectively be incorporated into CMOS technology. For these reasons, SnO_2 shows the most promise for an integrated gas sensor and has recently been commercialized by several vendors [62], [63].

In Fig. 6 the sensitivities of several doped and undoped SnO_2 based sensors are summarized as a function of temperature. Fig. 6a shows that the addition of platinum results in an improved sensor signal response towards CO, NO, and C_3H_8, cf. [64], but a shift in the optimal operating temperature is also observed. Meanwhile, Fig. 6b shows that the presence of rhodium (Rh) can improve the sensing performance towards acetone, cf. [61], but that there is a peak amount of Rh doping which will improve performance. Of note is that 0.5 mol% Rh gives an improved sensitivity to 1.0 mol% Rh. A similar phenomenon was observed by Mädler et al. [65] regarding Pt impurities in SnO_2. The authors showed that, while a 0.2wt% Pt concentration improved the sensitivity, increasing this to 2wt% Pt reduced the sensing response significantly, even below pure SnO_2 levels. One part of the reason for the low signal is likely due to a "localized" consumption of gas molecules by Pt without electron transfer, resulting in no changes in the SnO_2 film's resistivity.

IV. Sensing mechanism

As of yet, a comprehensive and complete understanding of the sensing mechanism of SMO films is not available. However, significant progress has recently been made in understanding the conductivity and surface adsorption effects in SnO_2. Sensing is based on the concept of reception - of an analyte gas on an SMO layer through a surface chemical reaction - and transduction - changes at the SMO layer surface, which influence the conductive properties of the film [66]. The conduction inside the SnO_2 layer can be described using drift-diffusion equations [67], which are commonly used to model charge transport in semiconductors [68]. The thicknesses of the SMO layers used in gas sensors are comparable to the mean free path of the charge carriers, meaning that the diffusion component can be ignored. Using electrons as majority carriers, as is the case for

the n-type semiconductor SnO_2, the conductivity σ is

$$\sigma = q \cdot n \cdot \mu_n, \qquad (1)$$

where q is the electron charge, n is the electron concentration, and μ_n is the electron mobility. Both n and μ_n can vary significantly with temperature, meaning that both the surface reaction and conductivity of the sensing film are influenced by temperature [68].

A. Surface adsorption

The sensitivity of any film is primarily driven by its surface-to-volume ratio, since sensing is the effective change in a volume behavior due to surface reactions taking place. While many different structures are currently under investigation, the typical SMO sensor which is in production today is based on a porous film. The film is comprised of many grains which expose their external surface to the target gas, thereby increasing the effective surface-to-volume ratio. In addition to the drift in the grains, the conductivity involves electron transport through grain-grain, grain-bulk, and grain-electrode interfaces [69]. Since the charge mobility does not change during molecular adsorption, the sensing mechanism depends purely on increasing or decreasing charge concentrations [70]. Therefore, the sensing takes place through charge manipulation by gas molecules adsorbed on the surface of the film or grains and the resulting formation a charge depletion region [21].

The sensing of a target gas (e.g. CO) on SnO_2 starts by oxygen being ionosorbed on the SnO_2 surface, taking one (O^-) or two (O^{-2}) electrons from the SnO_2 bulk and creating a depletion region around the grain, as depicted in Fig. 7a. Then, in the presence of CO gas, CO molecules react with the surface oxygen, releasing the electron back, reducing the thickness of the depletion region and creating CO_2 gas as a byproduct, as depicted in Fig. 7b [69], [70].

The discussion above assumes that sensing of reducing gases (e.g. CO) would only be enabled in the presence of oxygen. However, recent studies have shown that even when oxygen is depleted, the SMO film continues to show a sensing response, while no oxidation byproduct (e.g. CO_2) is found [71]. A proposed explanation is that CO gas molecules can directly interact with the surface atoms, thereby donating an electron to SnO_2 and forming an accumulation region, depicted in Fig. 7c. The most involved model for the sensing mechanism is presented in recent works by the group of Barsan and Weimar in [7], [70], [72], [73].

B. Influence of metal additives

As mentioned previously, noble metal additives have shown to increase the sensitivity and selectivity of SMO sensors. Using operando spectroscopy in [70], the authors show that nanosized Pt clusters are formed on the

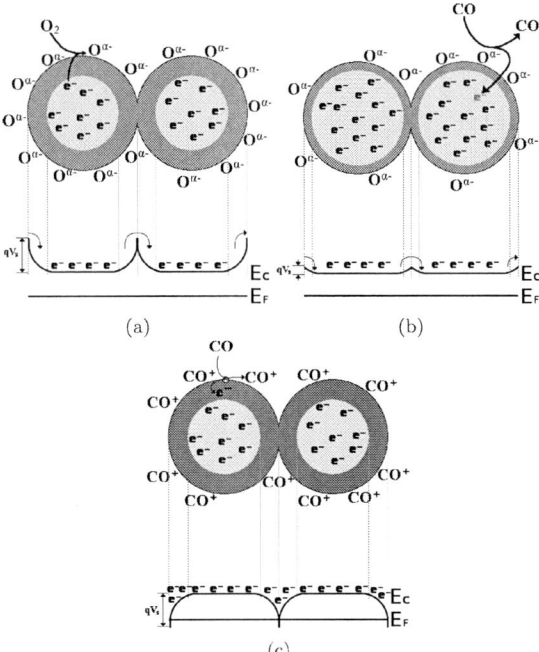

Fig. 7. Gas sensing and resulting band bending for a porous metal oxide, where the oxygen and reducing gas can penetrate to interact with grains. (a) Oxygen adsorbs on the surface, creating a depletion region, (b) CO reacts with oxygen, reducing the depletion region, and (c) after oxygen depletion, CO adsorbs on the surface, forming an accumulation region.

SnO_2 surface, acting as primary reaction sites for CO oxidation, thereby increasing the sensing effect. The amount of platinum determines whether CO oxidation or CO sensing will dominate, so simply adding more additive does not guarantee an improved response, which was discussed earlier and shown in Fig. 6b.

The ways in which a noble metal atom influences the sensing mechanism of an SMO film, including chemical and electrical sensitization, are described in some details in recent works, e.g. [7], [70], and a summary is given in Fig. 8. Chemical sensitization refers to the spill-over effect, which is attributed to metallic clusters which adsorb oxygen, and reducing gases, shown in Fig. 8a [74].

Electrical sensitization is based on the alignment of the Fermi levels of the SMO material and noble metal phase due to the different work function of the metal phase. The electrical contact between the two materials leads to a Fermi level alignment and therefore to surface band bending which is irrespective of the presence of any ionosorbed species, shown in Fig. 8b. Assuming no chemical or electrical interaction between the metal and the SMO film, the noble metal phase competes for adsorbed species with the SMO surface, decreasing the concentration of the target gas and hindering the sensing effect, as shown in the top of Fig. 8b [66]. However,

in the case of an oxidation reaction taking place at the interface of the metal and the SMO film, an increased sensing behavior is also possible [7].

The metal atoms or ions are also able to incorporate themselves into the SnO_2 lattice, thereby changing the electrical and chemical characteristics of the film. In Fig. 8c the locations where the metal atom can incorporate into the SMO film and their effects are shown [66]. If the valence state of the metal ion is different from the replaced cation, new acceptor or donor states have the potential of being introduced [7]. In particular, doping with Pt leads to an increase in SnO_2 conductivity and an increased amount of oxygen vacancies acting as adsorption sites, thereby improving sensitivity.

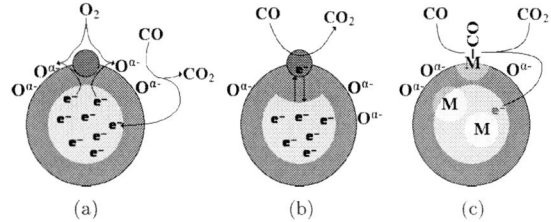

Fig. 8. Cross section of a grain within an SMO film, together with a noble metal doping. (a) The metal atom serves to spill over an oxygen molecule, slitting it into two atoms. (b) Fermi level control mechanism. (c) Possible donor/acceptor sites for the metal additive (M) are shown.

V. CONCLUSION

Significant progress has recently been achieved in the fabrication, design, and understanding of CMOS-compatible gas sensors based on semiconductor metal oxide films. These sensors are currently the most cost effective to fabricate, have excellent sensitivity towards many target gas species and environmental pollutants, operate at low power consumption, and their fabrication can be integrated in a mature CMOS foundry technology. The gas sensing mechanism involves the heating of a sensing film to several hundred degrees Celsius to activate the surface chemical reactions. The high temperature is provided using a microheater fabricated within a closed, suspended, or perforated membrane composed of several SiO_2 and Si_3N_4 layers.

In this manuscript we describe the fabrication process necessary to generate the three types of membranes as well as the deposition step used to introduce the SMO sensing film on top of this membrane stack. Furthermore, we summarize recent research to find the ideal microheater material and describe the geometrical designs which are meant to provide a uniform temperature distribution to the sensor. Finally, the sensing mechanism is discussed and recent achievements in understanding and modeling the surface reactions taking place during operation are described. The influence of noble metal additives on the sensing mechanism is also described.

REFERENCES

[1] G. E. Moore, *Electronics*, vol. 38, no. 8, 1965.

[2] M. M. Waldrop, *Nature News*, vol. 530, no. 7589, pp. 144–147, 2016.

[3] W. Arden *et al.*, "More-than-Moore white paper," 2010.

[4] G. Q. Zhang and A. van Roosmalen, Eds., *More than Moore: Creating High Value Micro/Nanoelectronics Systems*. Springer Science & Business Media, 2010.

[5] H. Li *et al.*, *IEEE Sensors Journal*, vol. 14, no. 10, pp. 3391–3399, 2014.

[6] R. A. Potyrailo, *Chemical Reviews*, vol. 116, no. 19, pp. 11 877–11 923, 2016.

[7] S. A. Müller *et al.*, *ChemCatChem*, vol. 10, no. 5, pp. 864–880, 2018.

[8] R. Moos *et al.*, *Sensors and Actuators B: Chemical*, vol. 83, no. 1-3, pp. 181–189, 2002.

[9] G. Martinelli *et al.*, *Sensors and Actuators B: Chemical*, vol. 55, no. 2-3, pp. 99–110, 1999.

[10] M. C. Carotta *et al.*, *Sensors and Actuators B: Chemical*, vol. 58, no. 1-3, pp. 310–317, 1999.

[11] G. F. Fine *et al.*, *Sensors*, vol. 10, no. 6, pp. 5469–5502, 2010.

[12] G. Wysocki *et al.*, *Applied Optics*, vol. 46, no. 33, pp. 8202–8210, 2007.

[13] A. A. Tomchenko, G. P. Harmer, and B. T. Marquis, *Sensors and Actuators B: Chemical*, vol. 108, no. 1-2, pp. 41–55, 2005.

[14] F. Wang, H. Gu, and T. M. Swager, *Journal of the American Chemical Society*, vol. 130, no. 16, pp. 5392–5393, 2008.

[15] R. Yoo *et al.*, *Sensors and Actuators B: Chemical*, vol. 221, pp. 217–223, 2015.

[16] G. Neri, *Chemosensors*, vol. 3, no. 1, pp. 1–20, 2015.

[17] E. Singh, M. Meyyappan, and H. S. Nalwa, *ACS Applied Materials & Interfaces*, vol. 9, no. 40, pp. 34 544–34 586, 2017.

[18] G. Korotcenkov, *Materials Science and Engineering: B*, vol. 139, no. 1, pp. 1–23, 2007.

[19] G. Eranna *et al.*, *Critical Reviews in Solid State and Materials Sciences*, vol. 29, no. 3-4, pp. 111–188, 2004.

[20] A. Dey, *Materials Science and Engineering: B*, vol. 229, pp. 206–217, 2018.

[21] L. Filipovic and A. Lahlalia, *Journal of The Electrochemical Society*, vol. 165, no. 16, pp. B862–B879, 2018.

[22] B. T. Marquis and J. F. Vetelino, *Sensors and Actuators B: Chemical*, vol. 77, no. 1-2, pp. 100–110, 2001.

[23] N. Barsan, D. Koziej, and U. Weimar, *Sensors and Actuators B: Chemical*, vol. 121, no. 1, pp. 18–35, 2007.

[24] K. T. Ng, F. Boussaid, and A. Bermak, *IEEE Transactions on Circuits and Systems I: Regular Papers*, vol. 58, no. 7, pp. 1569–1580, 2011.

[25] T. Konduru, G. C. Rains, and C. Li, *Sensors*, vol. 15, no. 1, pp. 1252–1273, 2015.

[26] L. A. Horsfall *et al.*, *Journal of Materials Chemistry A*, vol. 5, no. 5, pp. 2172–2179, 2017.

[27] D. Zhang *et al.*, *Sensors and Actuators B: Chemical*, vol. 240, pp. 55–65, 2017.

[28] Y.-H. Liao *et al.*, *Sensors*, vol. 19, no. 8, p. 1866(15), 2019.

[29] A. Lahlalia *et al.*, *Journal of Microelectromechanical Systems*, vol. 27, no. 3, pp. 529–537, 2018.

[30] O. Le Neel *et al.*, "Miniature gas analyzer," Patent US 20 180 017 513 A1, January, 2018.

[31] E. Lackner *et al.*, *Materials Today: Proceedings*, vol. 4, no. 7, pp. 7128–7131, 2017.

[32] S. Z. Ali *et al.*, *IEEE Sensors Journal*, vol. 15, no. 12, pp. 6775–6782, 2015.

[33] A. Lahlalia *et al.*, *Sensors*, vol. 19, no. 2, p. 374(14), 2019.

[34] A. Lahlalia, L. Filipovic, and S. Selberherr, *IEEE Sensors Journal*, vol. 18, no. 5, pp. 1960–1970, 2018.

[35] C. Dücsö *et al.*, *Sensors and Actuators A: Physical*, vol. 60, no. 1-3, pp. 235–239, 1997.

[36] R. Coppeta *et al.*, in *Sensor Systems Simulations*, W. D. van Driel, O. Pyper, and C. Schumann, Eds. Springer Nature Switzerland AG, 2020; Chapter 2, pp. 17–72.

[37] F. Lärmer and A. Urban, *Microelectronic Engineering*, vol. 67, pp. 349–355, 2003.

[38] A. I. Uddin, D.-T. Phan, and G.-S. Chung, *Sensors and Actuators B: Chemical*, vol. 207, pp. 362–369, 2015.

[39] L. Filipovic and S. Selberherr, *IEEE Transactions on Device and Materials Reliability*, vol. 16, no. 4, pp. 483–495, 2016.

[40] I. Simon *et al.*, *Sensors and Actuators B: Chemical*, vol. 73, no. 1, pp. 1–26, 2001.

[41] M. Frietsch *et al.*, *Sensors and Actuators B: Chemical*, vol. 65, no. 1-3, pp. 379–381, 2000.

[42] L. Filipovic *et al.*, *IEEE Transactions on Semiconductor Manufacturing*, vol. 27, no. 2, pp. 269–277, 2014.

[43] R. Chamberlin and J. Skarman, *Journal of the Electrochemical Society*, vol. 113, no. 1, pp. 86–89, 1966.

[44] J. B. Mooney and S. B. Radding, *Annual Review of Materials Science*, vol. 12, no. 1, pp. 81–101, 1982.

[45] E. Brunet *et al.*, *Sensors and Actuators B: Chemical*, vol. 165, no. 1, pp. 110–118, 2012.

[46] D. Perednis, "Thin film deposition by spray pyrolysis and the application insolid oxide fuel cells." Ph.D. dissertation, ETH Zürich, 2003.

[47] F. Solzbacher *et al.*, *Sensors and Actuators B: Chemical*, vol. 64, no. 1-3, pp. 95–101, 2000.

[48] S. Astié *et al.*, *Sensors and Actuators B: Chemical*, vol. 67, no. 1-2, pp. 84–88, 2000.

[49] N. Abedinov *et al.*, *Journal of Vacuum Science & Technology A: Vacuum, Surfaces, and Films*, vol. 19, no. 6, pp. 2884–2888, 2001.

[50] Y. S. Kim, *Sensors and Actuators B: Chemical*, vol. 114, no. 1, pp. 410–417, 2006.

[51] L. L. R. Rao *et al.*, *IEEE Sensors Journal*, vol. 17, no. 1, pp. 22–29, 2016.

[52] F. Mailly *et al.*, *Sensors and Actuators A: Physical*, vol. 94, no. 1-2, pp. 32–38, 2001.

[53] S. Roy, C. Sarkar, and P. Bhattacharyya, *Solid-State Electronics*, vol. 76, pp. 84–90, 2012.

[54] F. Solzbacher *et al.*, *Sensors and Actuators B: Chemical*, vol. 77, no. 1-2, pp. 111–115, 2001.

[55] K. Zhang, S. Chou, and S. Ang, *International Journal of Thermal Sciences*, vol. 46, no. 6, pp. 580–588, 2007.

[56] L. Xu *et al.*, *IEEE Sensors Journal*, vol. 11, no. 4, pp. 913–919, 2010.

[57] W. H. Brattain and J. Bardeen, *The Bell System Technical Journal*, vol. 32, no. 1, pp. 1–41, 1953.

[58] T. Seiyama *et al.*, *Analytical Chemistry*, vol. 34, no. 11, pp. 1502–1503, 1962.

[59] P. Shaver, *Applied Physics Letters*, vol. 11, no. 8, pp. 255–257, 1967.

[60] L. Filipovic and S. Selberherr, *Sensors*, vol. 15, no. 4, pp. 7206–7227, 2015.

[61] X. Kou *et al.*, *Sensors and Actuators B: Chemical*, vol. 256, pp. 861–869, 2018.

[62] Figaro USA, Inc., "TGS 8100."

[63] SGX SENSORTECH, "MiCS-6814 - MOS triple sensor."

[64] M. Saberi, Y. Mortazavi, and A. Khodadadi, *Sensors and Actuators B: Chemical*, vol. 206, pp. 617–623, 2015.

[65] L. Mädler *et al.*, *Journal of Nanoparticle Research*, vol. 8, no. 6, pp. 783–796, 2006.

[66] D. Degler, "Spectroscopic insights in the gas detection mechanism of tin dioxide based gas sensors," Ph.D. dissertation, Eberhard Karls Universität Tübingen, 2017.

[67] G. Tulzer *et al.*, *Nanotechnology*, vol. 24, no. 31, p. 315501(10), 2013.

[68] L. A. Caffarelli and A. Vasseur, *Annals of Mathematics*, vol. 171, pp. 1903–1930, 2010.

[69] N. Barsan and U. Weimar, *Journal of Electroceramics*, vol. 7, no. 3, pp. 143–167, 2001.

[70] D. Degler *et al.*, *Journal of Materials Chemistry A*, vol. 6, no. 5, pp. 2034–2046, 2018.

[71] N. Bârsan, M. Hübner, and U. Weimar, *Sensors and Actuators B: Chemical*, vol. 157, no. 2, pp. 510–517, 2011.

[72] N. Barsan, J. Rebholz, and U. Weimar, *Sensors and Actuators B: Chemical*, vol. 207, pp. 455–459, 2015.

[73] D. Degler *et al.*, *The Journal of Physical Chemistry C*, vol. 119, no. 21, pp. 11 792–11 799, 2015.

[74] G. N. Vayssilov *et al.*, *The Journal of Physical Chemistry C*, vol. 115, no. 47, pp. 23 435–23 454, 2011.

TUTORIAL ON POWER DEVICES AND MODULES

978-1-7281-3420-8/19 $31.00 © 2019 IEEE

978-1-7281-3420-8/19 $31.00 © 2019 IEEE   18

High-Fidelity Predictive Simulation of High Power Devices and Modules at the Rim of the Safe-Operating Area and Beyond

G. Wachutka

Abstract - The development of high-performance power devices is increasingly supported by predictive computer simulations on the basis of well-calibrated physical device models. Today´s challenge is to make virtual experiments and tests on the computer, which are qualitatively reliable and quantitatively accurate even for device structures that have never been built before, and under operational conditions that very rarely occur as long as the device is kept within the "safe operating area (SOA)". What we are interested in is to explore the rim of the SOA and even to go beyond it in order to study failure and, eventually, destruction mechanisms with a view to improving robustness and reliability of the devices with respect to a customer-defined "mission profile". In particular in the field of high power electronics, predictive high-fidelity computer simulations of "virtual desctruction" are of utmost importance. We will illustrate today´s state of the art with reference to selected real-life examples.

I. INTRODUCTION AND MOTIVATION

The continuous progress in power device technology is increasingly supported by power-specific modeling methodologies and dedicated simulation tools. These do not only enable the visualization of fabrication processes and operational principles, but also the detailed analysis of the device and system operation of competing design variants in a very early stage of the development process. Virtual fabrication, virtual experimentation and virtual test by computer simulations have become an integral part of the design methodology for electronic power devices and systems in order to realize cost-efficient and time-economizing development cycles.

A successful virtual design strategy requires modeling methodologies on different levels of abstraction and computational expense. Among the most important aspects to be focussed on, we identify the consistent treatment of electro-thermally coupled fields and coupled domains required for setting up physically-based models for high-fidelity computer simulations. Equally crucial is the reliable validation and accurate calibration of the models, which is the indispensable prerequisite for predictive simulation, in particular for "virtual experiments" at the rim of the safe-operating area close to the destruction of a device.

Gerhard Wachutka is the retired head of the Chair for Physics of Electrotechnology at the Technical University of Munich, Arcisstrasse 21, 80290 Munich, Germany. E-mail: wachutka@tep.ei.tum.de

The high relevance of virtual prototyping is reflected in today´s most crucial problem areas (cf. Fig.1):
- thermal design & management/cooling
- high-temperature operation
- packaging, housing, and interconnects
- vibrational design
- reliability/failure analysis
- robustness ("ruggedness") and stability (even against unlikely, but catastrophic events)
- EMC/EMI and signal integrity
- full system behavior w.r.t. mission profiles

Hence, nowadays the most challenging task of high power device modeling and simulation consists in exploring the safe-operating area, and this means the detailed analysis of irregular operating states and failure mechanisms.

II. SPECIFIC FEATURES OF POWER DEVICE AND MODULE MODELING

In today´s power electronics, we basically face the same problem as in micro-electronics, namely that the concept and design of a state-of-the-art power device consisting of some 10,000 microstructured cells, which work in parallel in a cell array for a specified action in a power circuit or module, is essentially governed by intricate trade-off considerations. Therefore a systematic improvement of the performance is hardly possible without being equipped with a profound expertise in the details of the operation of single power device cells and their collaborative interaction as part of a large scale power device with co-integrated additional functionalities such as state monitoring, overload detection and protection against overvoltage, overcurrent, or overheating including error detection and compensation by a smart control circuitry, as it is, for example, realized in smart power technologies (SPT) or in high-voltage integrated circuits (HVICs). Using predictive simulation, the required expertise can be gained faster and cheaper than by experimental investigations.

Evidently this requires the availability of efficient simulation platforms, which fit in with today's far advanced design environments used in the semiconductor industry and, in particular, are conform with the widely accepted bottom-up and top-down modeling hierarchies. Moreover, optimal prototyping of power modules requires the concurrent co-optimization of the active and passive electronic

978-1-7281-3420-8/19 $31.00 © 2019 IEEE

elements of a power component and their electrical and thermal interconnection by wire bonds, bus bars, base plate etc. as defined by the packaging and housing of the complete module. To this end, a comprehensive methodology for setting up physically-based (self-) consistent power device and full system models has been developed to enable the effort-economizing and yet accurate numerical co-simulation of individual power devices and full power modules built up of them. This modeling framework [1,2] includes the consistent treatment of electrothermally [3,4,5], thermomechanically [6], and electromagnetically [7] (inductively and capacitively) coupled fields and coupled energy and signal domains ("multi-physics" models, cf. Fig. 1) required for deriving electro-magneto-thermomechanical macromodels from the continuous field level, and it also provides a solid basis for the reliable validation and calibration of the models.

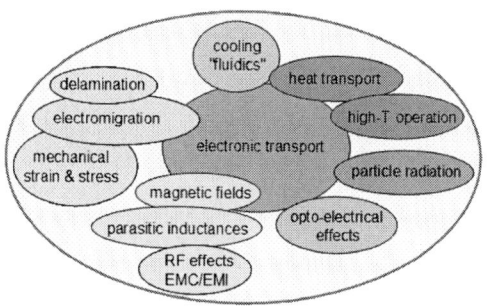

Fig. 1. Problem areas in the optimized design of power devices and their interrelation in coupled energy domain ("muliphysics") modeling.

The key to the predictive simulation of entire power modules is the use of model reduction techniques, as they allow for setting up physically-based system-level models, which are tailored in the sense that the descriptive complexity is largely reduced, but without losing the scalability and quantitative accuracy required for predictive simulation. The adequate formal representation of the full system description is provided in terms of a generalized ("coloured") Kirchhoffian network description in combination with an appropriate analog hardware description language such as SPICE, SABER, or VHDL-AMS. This makes it possible to code the models of all the individual system components in a generic and uniform way and to assemble the full power module model by linking the constituent parts on the same descriptive level.

As already stated, the complexity of power modules originates primarily from the complicated coupling between different energy and signal domains (Fig. 1) which, on the one hand, is the inherent property of any controllable electronic device, sensor element or other constituent part of a power module and, on the other hand, is an undesired detrimental property, when it occurs as parasitic cross-coupling between the power system components. Hence the accurate analysis of all relevant

kinds of physical coupling effects has a major impact on the optimization of power devices and modules and is, thus, one of the most crucial issues that has to be tackled in the computer-aided design of microdevices and microsystems. In this sense, the physically-based, but yet computationally tractable modeling of power devices and power modules is widely recognized as key technology and necessity, even though it may easily become quite involved. Therefore the computational effort as well as the time spent into model development, validation, calibration, and parameter extraction have to be carefully adjusted to the actual needs, as it has been proposed by the concept of "tailored modeling" [6,8]. Powerful user-friendly commercial power device simulation platforms are already available and enable an efficient and time-economizing computer-aided power device and system development.

III. VIRTUAL EXPERIMENTS AS MODEL CALIBRATION AID

The predictiveness of physical device models is decisively determined by the requirement of being "transparent", which means that all numerical model parameters allow for an intuitive interpretation as physical quantities such as geometrical dimensions, material properties, or other technologial data, and that the values of all these parameters have been extracted from test samples or prototypes by experimental characterization. Usually this process of "model calibration" is restricted to the analysis of the electrical terminal behavior, while the "innerelectronic" behavior is amenable to theoretical analysis or computer simulation only. Hence, the internal physical parameters can be determined by "inverse modeling" only and, therefore, their accuracy is often questionable. But with a view to optimizing the device properties a detailed and precise view into the interior of the real devices under real operation conditions is highly desirable.

This gives the impact to set up an experimental platform for the direct space- and time-resolved measurement of the basic state variables in the electrothermal semiconductor transport equations [1,5], namely the carrier distributions and the temperature profile. To this end, a laser-aided probing technique has been developed [9], which exploits the plasma-optical and thermo-optical effect, i.e. the sensitivity of the complex refractive index to the carrier concentrations and lattice temperature. The correct interpretation of the measured light absorption and deflection signals is quite involved and requires an accurate physical model of the measurement process itself [10,11]. Equipped with that, virtual numerical experiments have to be performed consisting of three parts (Fig. 2): electrothermal device simulation, calculation of the light propagation through the device under test (DUT), and simulation of the detector responses. Comparing the results of real and virtual experiment allows one to correctly evaluate the measured raw data and, at the same time, to calibrate the underlying physical device models.

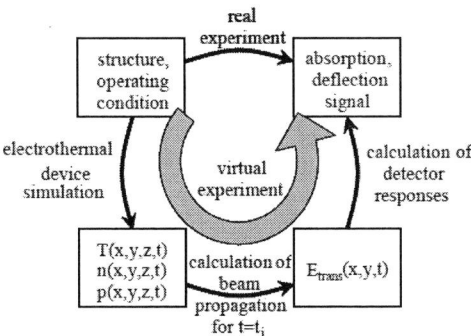

Fig. 2. Interrelation between real and virtual experiment, consisting of three parts.

The latter is of particular interest in the case of "unusual" semiconductor material like silicon carbide (SiC), which serves as industrial basis for high-temperature power devices. Therefore, devices made of SiC have been in the focus of the power device experimentalists [12] as well as in that of the theoretical analysts [5] for two decades of years; but nevertheless many material parameters like mobilities and carrier lifetimes are still not very reliable.

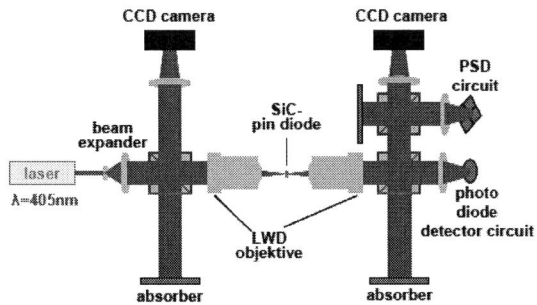

Fig. 3. Schematic view of the experimental setup: optical path, electrical lighting, and imaging cameras.

The real experimental setup (Fig. 3) comprises a laser source, an optical set-up with the device under test, and two detector units for light absorption (photo-diode) and beam deflection ("PSD"). The device under test (SiC pin-diode) is biased by strong current pulses with variable pulse height. This drives the DUT in high injection conditions, under which the electron and hole distributions form a quasi-neutral plasma ($n(x) \approx p(x)$) in the low-doped n^--region. According to the Drude model, the local light absorption coefficient is proportional to a weighted sum of the electron and hole densities. In addition, the spatial gradients in the carrier and temperature profiles lead to a deflection of the laser beam (Fig. 4), which complicates the correlation between measured absorption signals and real carrier profiles in the intrinsic layer. For retrieving the carrier profiles from the measured raw data, the virtual experiment is essential (Fig. 2).

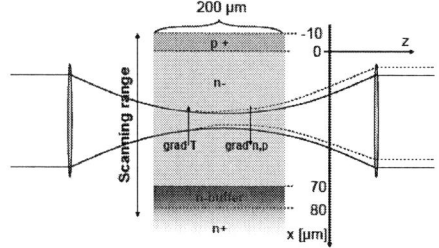

Fig. 4. Plasma-optical and thermo-optical effect causing the attenuation and deflection of a coherent light beam, which traverses a semiconductor region under high injection conditions perpendicularly to the direction of current flow.

Typical exemplary real and virtual detector responses in the absorption sensor are displayed depicted in Fig. 5; they are in good overall agreement with each other. This clearly demonstrates that, after a careful readjustment of uncertain device parameters, the real experiment is properly represented by the virtual one. The latter, in turn, makes it possible to reveal the origin of the qualitative differences between the light absorption profile and the charge carrier profiles, in particular in the intrinsic region close to the pn-junction and close to the n-buffer. These can be attributed to an enhanced deflection of the laser beam caused by the gradients of carrier density and temperature.

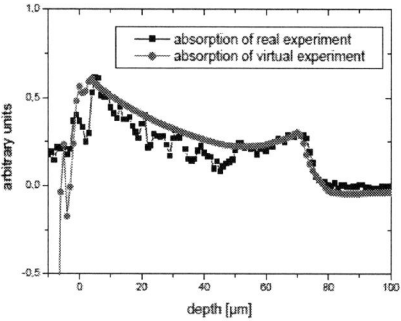

Fig. 5. Absorption detector response in the real and in the virtual experiment. The pn-junction is located at 0 µm.

IV. VIRTUAL TESTING ALONG THE RIM OF THE SAFE OPERATING AREA

Achieving the largest possible ruggedness of high power devices against harsh operating conditions is one of the optimization targets during design. Its quantitative description is based on the concept of the (static or dynamic) safe-operating area (SOA), which is defined by the maximum set of (voltage, current) pairs for which the device operates without failure. Determining the boundaries of the SOA is a major concern in industrial applications, as there is an increasing demand for power devices which can be controlled even at the rim of the SOA and where the transient occurrence of dangerous states such as, e.g., avalanche multiplication is tolerated, as long as it does not become destructive. A common approach are experimental stress tests, but these have the obvious disadvantage that they end with the destruction of the device under test in most cases. Therefore one attempts to replace the real stress tests by "virtual tests" based on predictive high-fidelity computer simulations. However, this poses a big challenge to the simulation methodology, because the operating states to be modelled are typically unstable and amenable to a full field-theoretical electrothermally coupled treatment only.

As an illustrative example, we consider cell arrays comprising many thousands of parallel IGBT cells, which are encompassed by an edge termination structure to attain the maximum breakdown voltage also along the boundary of the cell array, since IGBT cell arrays tend to break down at their periphery. Therefore a detailed functional understanding and an optimum design of the edge termination are crucial to achieve the largest possible safe-operating area.

We studied the behavior of two alternative edge termination structures, "junction termination extension" (JTE) [13] and "variation of lateral doping" (VLD) [14,15], in the avalanche regime by numerical simulations and measurements with a view to assessing their ruggedness against such harsh operating conditions. The analyzed parts of a cell array consist of a pn+n-p-layered structure and a subsequent JTE or VLD region (Fig. 6). JTE or VLD, respectively, is a p-doped region, with uniform or with gradually decreasing doping towards the right border.

We investigated the time-dependent internal device behavior in the vicinity of the VLD and the JTE edge termination, respectively, under avalanche breakdown conditions. The device region considered (Fig. 6) is a cylindrical ring, where the radius of curvature is chosen in accordance with the real structure. Actually, it represents one of the four corners of the IGBT cell array, which constitute the weakest parts of the cell array as confirmed by experimental findings. The virtual stress test consists in ramping the blocking voltage up to a level slightly above the breakdown voltage and, then, keeping it constant for approximately 100 µs. The response of the two device variants is completely different.

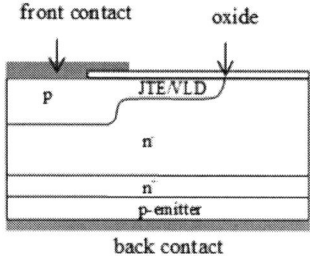

Fig. 6. Simplified schematic structure of the edge terminations of an IGBT cell array. JTE or VLD, respectively, is a p-doped region, with uniform or with gradually decreasing doping towards the right border.

In the JTE structure we find an uninterrupted current flow through the device (Fig. 7). Impact ionization occurs localized in the JTE region near the surface and, consequently, the current rises due to avalanche multiplication.

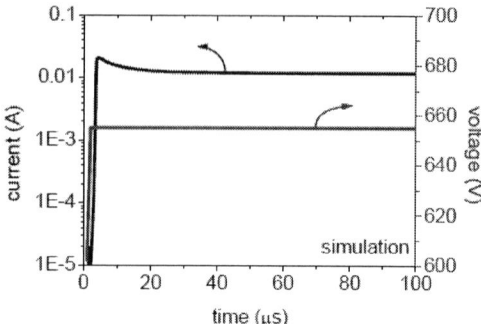

Fig. 7. Current transient as obtained from the JTE structure after ramping the blocking voltage up to a level slighty above the avalanche breakdown voltage.

A vertical current filament develops underneath the location of avalanche muliplication. However, by the special design of the JTE structure, the profile of the electric field is, throughout the device, only slightly distorted by the avalanche-generated electrons and holes. The power dissipation in the filament leads to an increase of the local temperature which, in turn, reduces the impact ionization rate; as a consequence, the current slightly decreases with time (Fig. 8) and stabilizes at a constant stationary value. Hence, the current filament eventually arrives at a stable stationary state, staying at the same location and carrying a constant current.

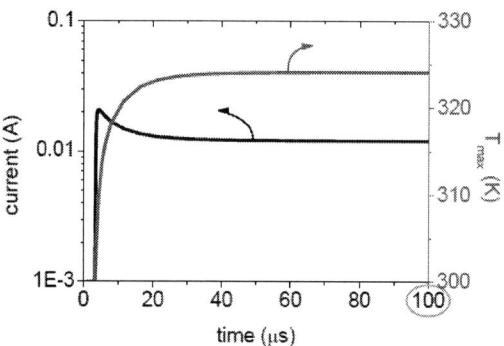

Fig. 8. Correlation between temperature transient and current transient as obtained from the JTE structure under avalanche breakdown conditions.

A completely different behavior is observed in the VLD structure: we find time-periodic peaks in the avalanche-generated current flowing through the VLD structure (Fig. 9), the period of which decreases with increasing applied voltage.

978-1-7281-3420-8/19 $31.00 © 2019 IEEE

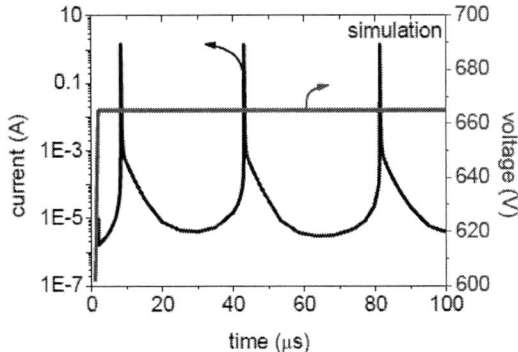

Fig. 9. Current waveform as obtained from the VLD structure, exhibiting sharp periodic current peaks after it is driven into the avalanche regime (the bias voltage is kept constant in the simulation).

These current peaks are caused by periodically evolving and self-extinguishing current filaments: In the VLD structure, avalanche multiplication occurs at the bending of the p-body. The avalanche-generated electron-hole pairs and additional holes injected from the p-emitter at the rear side distort the static profile of the electric field drastically. In particular, we observe a local steepening of the peak of the electric field (not shown here) which, in turn, causes a strong localized increase of the avalanche-driven current.

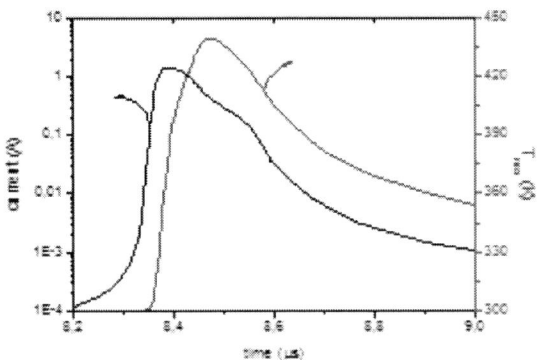

Fig. 10. Detailed view of the initial current peak in Fig. 9 (VLD structure), together with the evolution of the maximum temperature in the device.

The result is a current filament again, for which we find the following time-periodic mechanism: Because of the nearly abrupt onset and steep rise of the avalanche-driven current, the current attains its maximum without a significant change of temperature in the filament (see Fig. 10). But since the maximum current is by a factor of 50 higher than in the JTE structure and stronger localized, the dissipated power density increases by the same amount, which leads to a subsequent very rapid rise of the filament temperature up to a level (440 K) that is much higher than that reached in the JTE structure (320 K). As a consequence,

the impact ionization rate falls very quickly below the multiplication threshold and the filament is destabilized and decays. After a certain period during which the hot spot around the filament cools down, the process of filament formation will restart again. Thus, in contrast to the JTE structure, the VLD structure does not run into an asymptotic stationary state, but orbits in a periodic limiting cycle (Fig.11). By this periodic sequence of short heating and long cooling the dissipated heat is quite effectively distributed in the bulk of the device and, thus, it is protected from running into destruction.

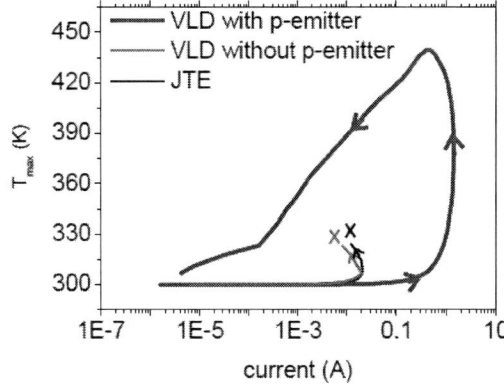

Fig. 11. Current-temperature trajectories of the time-variant operating points for the VLD and JTE structure.

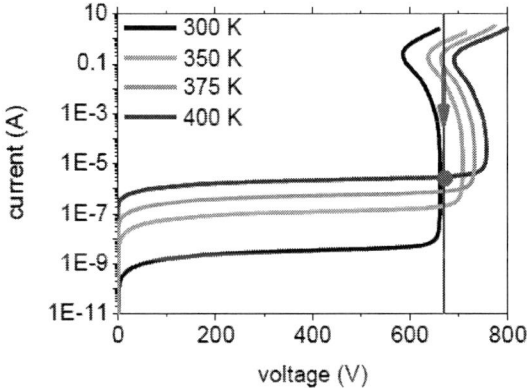

Fig. 12. Temperature-dependent blocking characteristics for the VLD structure, showing pronounced negative differential resistance (NDR) behavior.

Based on our computer simulations we could prove that the completely different transient behavior of the two structures under avalanche breakdown conditions is the result of electrothermal interaction effects. In [15] it is discussed in detail that it is sufficient to consider the temperature dependence of the static I-V-characteristics (cf. Fig. 12) in order to explain the qualitatively different transient behavior and, thus, may be used as a-priori criterion which transient behavior can be expected.

V. VIRTUAL TESTING IN THE *TERRA INCOGNITA* BEYOND THE RIM OF THE SOA

A. Demonstrator: High Power Devices Attacked from Space by Cosmic Rays

Silicon high power devices are endangered by a certain very small, but finite likelyhood to run into thermal destruction after the impact of a cosmic particle [16]. The electric field profile present in the blocking state of such devices with vertical current flow (such as IGBTs, GTOs, PiN-diodes etc.) has a trapezoidal shape. The slope of the field profile inside the base zone (space charge zone) is determined by the doping concentration of the n--base material (see Fig. 13a). For a specified maximum blocking voltage well below the rim of the safe-operating area (SOA) the electric field near the blocking pn-junction attains a peak value typically in the order of 100 kV/cm.

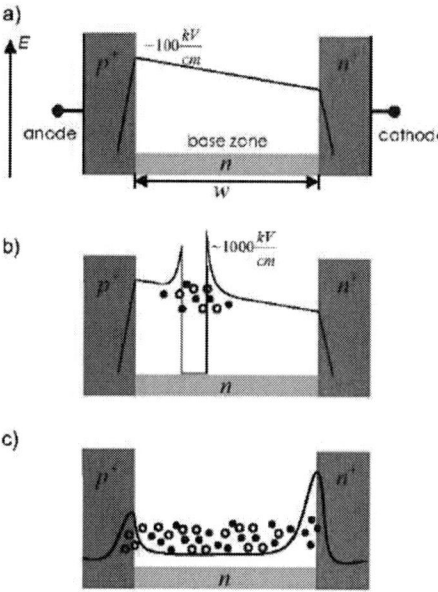

Fig. 13. Schematic device structure, electric field profile, and carrier densities a) before, b) about 40 ps after, and c) about 1 ns after the impact of a cosmic particle.

When a cosmic particle, e.g. a neutron, penetrates the high field region of the device structure, there is a certain finite probability that it undergoes a nuclear reaction, in which light-weight ions with kinetic energies between 10 and 100 MeV are generated. These spallation products produce electron-hole pairs along their path through the device. As a consequence, a strongly localized, highly concentrated filamentary electron-hole plasma is built up causing a local breakdown of the electric field. But at the rim of the plasma filament, the different diffusivities of electrons and holes lead to two spots with very high space charge density at the ends of the plasma filament

which, in turn, generates extremely steep electric field peaks about one order of magnitude higher than the static field strength (Fig. 13b). Triggered by these field peaks, impact ionization sets on, by which additional carriers are generated. This carrier multiplication leads to a very fast extension of the plasma filament through the device and short-circuits it. The described mechanism is well-known in plasma physics as so-called "streamer". Decisive for the ignition of the streamer is the peak value of the static electric field, while the average static electric field determines the total charge generated during the run through the device [17]. Once the plasma filament is fully developed from anode to cathode, carriers start diffusing in lateral directions and move to the contacts, so that eventually the plasma filament may extinguish or not, depending on the total charge deposited within the streamer. This situation is shown schematically in Fig. 13c and illustrated by the numerical simulation displayed in Fig. 14. The neutral plasma filament makes the electric field in the base zone locally break down, while electron-hole recombination in the highly doped contact regions leads to strongly localized space charge layers and, consequently, to very steep electric field peaks. This happens in a timescale of 10 ns, where strongly localized self-heating can lead to the thermal destruction of the device.

We performed three-dimensional electro-thermal device simulations to study these phenomena in detail. It is notable that even for such extreme operating conditions our physical models are predictive, as they have been calibrated with reference to ion irradiaton experiments [18].

The simulations refer to proton irradiation experiments, which demonstrate the influence of certain design parameters (related to the base zone) on the robustness against cosmic radiation. Coupling electric transport to heat dissipation allows us to calculate the temperature distribution with high spatial resolution, thus delivering a deep physical insight and understanding of the failure and destruction mechanisms.

B. Experimental Findings

Non-destructive ion irradiation experiments were performed for exploring the robustness of industrial power diodes against cosmic radiation. This approach has the advantage that no nuclear reaction statistics needs to be taken into account, but that instead well-defined test conditions are preserved (energy of ions → energy loss function → number of electron-hole pairs generated) [19]. A PiN-diode with a volume breakdown voltage higher than 1.2 kV was investigated. The maximum charge generated by one single oxygen ion with a kinetic energy of 30.6 MeV was measured as a function of the blocking voltage. Similarly to [17], we see strong charge multiplication above a critical blocking voltage, which is about 650 V in the device under test. The diode fails when the blocking voltage exceeds 900V.

978-1-7281-3420-8/19 $31.00 © 2019 IEEE

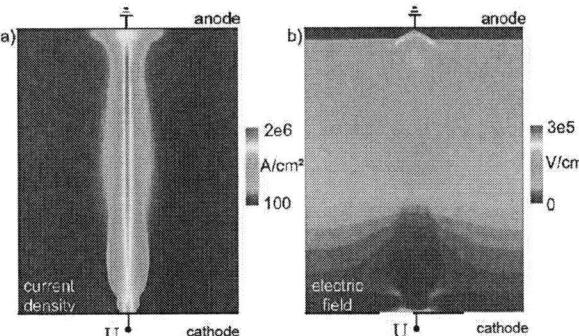

Fig. 14. Simulated spatial distribution of the electrical current density (a) and corresponding spatial distribution of the absolute electric field (b) 1 ns after the impact of a cosmic particle.

Fig. 15 shows the total charge Q generated by an incident ion as a function of the blocking voltage U. The measured data (crosses) are compared with the results of simulation (squares). We see a good agreement w.r.t. the critical voltage for "streamer" formation as well as w.r.t. the amount of generated charge near the failure voltage. It is possible to simulate even the behavior at high voltages were the device fails in the experiment.

As already mentioned above, the situation after nano-seconds is of decisive importance for the failure of the device. Figure 14a shows the spatial distribution of the charge carriers 1 ns after the impact of an ion. The plasma channel which short-circuits the diode is visible. The carrier density is in the order of $10^{18} cm^{-3}$. About 1 ns after the formation of the plasma channel, the lateral diffusion of carriers starts diminishing the carrier density, and the plasma channel vanishes after about 10 ns. In this period the electric field breaks down inside the channel and the voltage mainly drops near the contacts, which can be recognized as field steepening in these regions (cf. Fig. 14b).

These findings will be explained and discussed in the following.

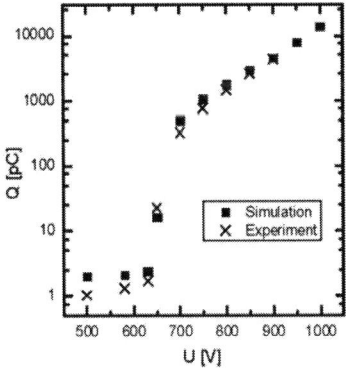

Fig. 15. Maximum measured total charge Q as a function of the blocking voltage U generated after an impact of an ^{26}O ion with a kinetic energy of 30.6 MeV (crosses) compared to the simulated total charge (squares).

C. Simulation Model

Our transient simulations are based on the drift- diffusion transport model with well-established models of the carrier mobilities, impact ionization, and trap-recombination. The calibration of the employed models is reported in [18]. A physically correct accurate description of local self-heating during the formation of the streamer requires an electro-thermally self-consistent calculation scheme [3]. However, the attempt to calculate the self-heating in the high-field regions near the contacts up to the melting point of silicon (about 1700 K) fails, because the impact ionization coefficients inside the hotspot are not known yet for temperatures higher than 800 K, and a straight-forward extrapolation of the common models beyond this temperature limit would lead to non-physical results.

An alternative way to get a reasonable approximation of the temperature peak in the region near the cathode contact consists in an a-posteriori estimation methode: The heat diffusion equation is solved after each time step of a transient isothermal electrical simulation. This decoupling of the electrical and the thermal energy domain offers a fast and effective way to estimate self-heating in strongly localized high-field regions [20].

D. Numerical Analysis of the Destruction Process

The above-described electrothermal simulation scheme allows us to calculate the spatial temperature distribution at any time during the transient simulation. Fig. 16 shows the temperature distribution at the time when the temperature attains its maximum value (about 20 ns after the impact of a cosmic particle). We recognize a hotspot caused by the simultaneous occurrence of a high electric field and a high current density near the cathode contact.

Fig. 17 displays the transient behavior of the total collected current and the peak temperature, after the impact of a cosmic at the time t = 0. The peak temperature lags behind the total current with a delay of about 15 ns and attains more than 1500 K for a short while. The variance of the hardness against cosmic radiation resulting from different diode designs can be extracted by plotting the peak temperature in the hotspot versus the blocking voltage and using this relation as criterion for the quality of different design variants. The influence of the doping concentration (i.e. ohmic resistivity) in the base zone is illustrated in Fig. 18.

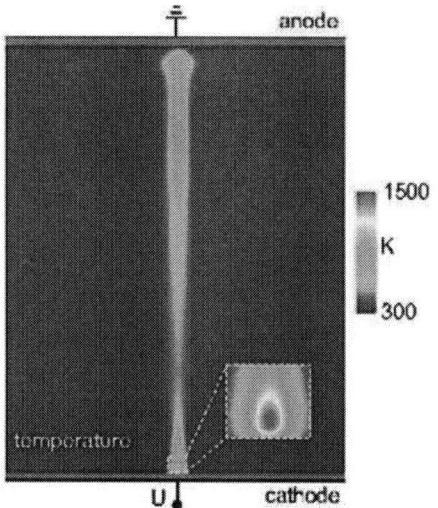

Fig. 16. Simulated spatial distribution of the a-posteriori calculated temperature 20 ns after the impact of a cosmic particle.

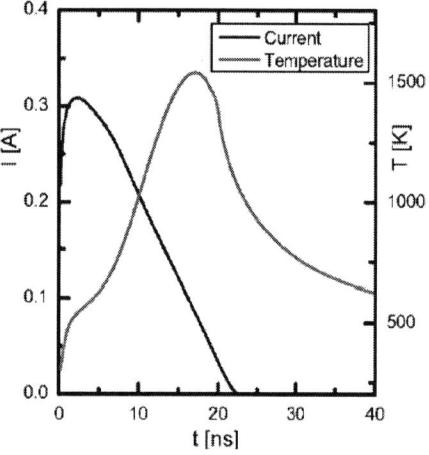

Fig. 17. Total collected current and peak temperature as determined by the a-posteriori estimation method as a function of time after the impact of a cosmic particle.

Fig. 18a shows the static electric field for four different resistivities when a fixed blocking voltage is applied. As the slope of the field profile increases with decreasing resistivity, the peak value of the static electric field must increase to sustain the given total voltage. Hence, the formation of a streamer sets on at a higher blocking voltage, when the base doping concentration is reduced. This explains the largely varying peak temperatures

at low reverse voltages in Fig. 18b. But at higher reverse voltages, when the peak temperature exceeds the melting point of silicon, the voltage is higher than the threshold voltage for streamer ignition at all tested devices. The resulting peak temperature shows only a slight dependence on the doping concentration in the base region, because the static electric field has a similar average value. The shift of the temperature vs. voltage relation in the range of the silicon melting point for different doping concentrations explains the measured improvement of about 50 V, when the base resistivity is varied from 90 Ωcm to 160 Ωcm.

In Fig. 19 three PiN-diodes with different widths of the base zone are compared. The electric field profiles in Fig. 19a demonstrate that the peak value of the static electric field as well as the decrease of the average value, when the base zone is elongated. So the relation of peak temperature vs. blocking voltage (Fig. 19b) is shifted towards higher voltage for a larger width of the base zone. The shift of about 100 V caused by a variation of the base zone width from 75 µm to 95 µm conforms very well with the data obtained from proton irradiation measurements.

VI. CONCLUSIONS

Coupled energy-domain, continuous-field device modeling and coupled-domain system simulation have proven to be indispensable for the development of highly optimized power devices, modules, and systems with competitive performance. Computer-aided design is reducing the number of costly trial-and-error steps and decreases the turn-around times of development cycles. Easy-to-use predictive "TCAD tool boxes" for power devices and systems have today become an integral part of design methodology. Including the functionality for automated parameter identification by closed-loop simulation and, thus, the capability of easy model calibration as well as providing the capability of high-fidelity simulation of irregular operating states at the rim of the safe-operating area is largely desired, already attempted, and in part realized. To this end, we developed a comprehensive and consistent modeling strategy for setting up physics-based "transparent" models, model calibration, device and system analysis, design optimization, reliability and robustness analysis etc. The practicability and efficiency of such a comprehensive, but modular and hierarchical approach has already been demonstrated in the power electronics community by numerous examples.

Now continuing efforts must be made to transform these results in robust, easy-to-use software packages, which are ready for use in existing professional TCAD environments. This implies that, on the device level, software tools must be developed which allow for the efficient interfacing of different single-domain simulators in such a way that - with a view to the specific problem areas of high power technology - new advanced, "tailored" coupling schemes can be realized by a flexible control of the solution process.

978-1-7281-3420-8/19 $31.00 © 2019 IEEE

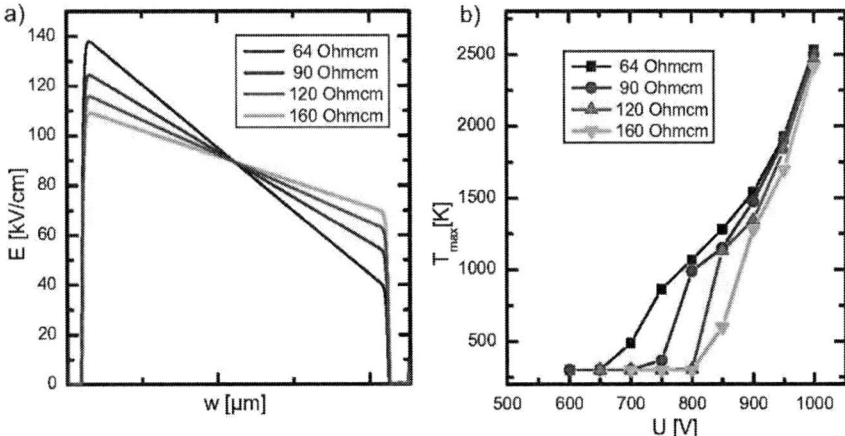

Fig. 18. a) Static electric field profile of reverse-biased PiN-diodes with four different levels of the base zone doping. b) Peak temperature as determined by the a-posteriori estimation method as a function of the blocking voltage for the PiN-diodes in a)

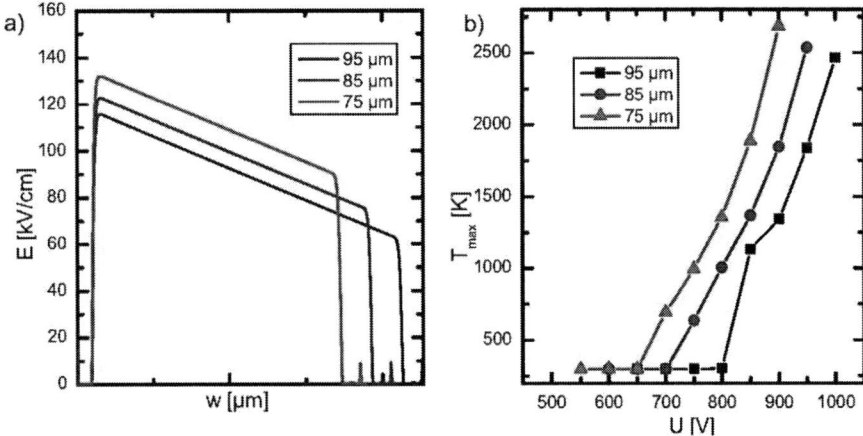

Fig. 19. a) Static electric field profile of reverse-biased PiN-diodes with three different widths of the base zone. b) Peak temperature as determined by the a-posteriori estimation method as a function of the blocking voltage for the PiN-diodes in a).

ACKNOWLEDGEMENT

The author would like to acknowledge the scientific and technical assistance from his research group and staff and from our industrial cooperation partners, in particular the contributions from Dipl.-Ing. M. Aigner, Dipl.-Phys. F. Hille, Dr. U. Knipper, Dipl.-Ing. A. Korzenietz, Prof. Dr. G. Schrag, Dr. G. Sölkner, Dr. D. Werber, and Dr. F. Wittmann.

REFERENCES

[1] G. Wachutka: "Coupled Field Modeling of Microdevices and Microsystems", in *Simulation of Semiconductor Processes and Devices*, Kobe, Japan, Sept. 4-6, 2002, pp. 9-14.

[2] G. Wachutka: "Unified Framework for Thermal, Electrical, Magnetic, and Optical Semiconductor Device Modeling", *COMPEL* vol. 10, No. 4, 1991, pp. 311-321.

[3] G. Wachutka: "Rigorous Thermodynamic Treatment of Heat Generation and Conduction in Semiconductor Device Modelling", *IEEE Trans. on CAD of ICAS*, CAD-9, 1990, pp. 1141-1149.

[4] G. Wachutka: "Consistent Treatment of Carrier Emission and Capture Kinetics in Electrothermal and Energy Tranport Models", *Microelectronics Journal*, vol. 26, No. 2/3, 1995, pp. 307-315

[5] M. Lades, G. Wachutka: "Electrothermal Analysis of SiC Power Devices Using Physically-Based Device Simulation", *Solid-State Electronics*, vol. 44, 2000, pp. 359-368.

[6] G. Wachutka: "Tailored modeling of miniaturized electro-thermomechanical systems using thermodynamic methods", in *Micromechanical Sensors, Actuators, and Systems*, Eds.: Cho, D., Peterson, J. P., Pisano, A. P., Friedrich, C., DSC-40, ASME, New York, 1992, pp. 183-197.

[7] G. Wachutka: "Physics-Based Modeling of Electromagnetic Parasitic Effects in Interconnects", *Tech. Dig. Int. Workshop on Modeling and Simulation of RF-Circuits* (Hiroshima, 2004).

[8] G. Wachutka: "Tailored Modeling: A Way to the 'Virtual Microtransducer Fab'?", *Sensors and Actuators*, vol. A47, 1995, pp. 603-612.

[9] R. Thalhammer, F. Hille, G. Wachutka: "Design and Interpretation of Laser Absorption Measurements for Power Devices", *Proc. of 1999 Int. Semiconductor Device Research Symposium* (ISDRS-99, Charlottesville, 1999), pp. 531-534.

[10] R. Thalhammer, G. Wachutka: "Virtual Optical Experiments. Part I: Modeling the Measurement Process", *Journal of the Optical Society of America A*, vol. 20, no. 4, 2003, pp. 698-706.

[11] R. Thalhammer, G. Wachutka: "Virtual Optical Experiments. Part II: Design of Experiments", *Journal of the Optical Society of America A*, vol. 20, no. 4, 2003, pp. 707-713.

[12] W. Kaindl, M. Lades, N. Kaminski, E. Niemann, G. Wachutka: "Experimental Characterization and Numerical Simulation of the Electrical Properties of Nitrogen, Aluminum and Boron in 4H/6H Devices", *Journal of Electronic Materials*, Special Issue on III-V Nitrides and SiC, Vol. 28, No.3, 1999, pp. 154-160.

[13] V. Temple, "Junction termination extension (JTE), a new technique for increasing avalanche breakdown voltage and controlling surface electric fields in p-n junctions", *IEEE Transactions on Electron Devices*, vol. 30, No. 8, 1983, pp. 954-957.

[14] U. Knipper, G. Wachutka: "Study of Time-Periodic Avalanche Breakdown Occurring in VLD Edge Termination Structures", *Proc. of the Int. Conf. on Simulation of Semiconductor Processes and Devices* (SISPAD 2007), Vienna, Austria, Sept. 25-27, 2007, pp. 189-192.

[15] U. Knipper, F. Pfirsch, T. Raker, J. Niedermeyr, G. Wachutka: "Time-Periodic Avalanche Breakdown at the Edge Termination

of Power Devices", *Proc. of the 20th International Symposium on Power Semiconductor Devices & ICs* (ISPSD 2008), Orlando, USA, May 18-22, 2008, pp. 307-310.

[16] H. Kabza, H.-J. Schulze, Y. Gerstenmaier, P. Voss, J.W.W. Schmid, F. Pfirsch, K. Platzöder: "Cosmic Radiation as a Cause for Power Device Failure and Possible Countermeasures", *Proc. of 6th Int. Symp. on Power Semiconductor Devices and ICs* (ISPSD '94), 1994, pp.9-12.

[17] W. Kaindl, G. Sölkner, H.W. Becker, J. Meijer, H.J. Schulze, G. Wachutka: "Physically Based Simulation of Strong Charge Multiplication Events in Power Devices Triggered by Incident Ions", *Proc. of 16th Int. Symp. on Power Semiconductor Devices and ICs* (ISPSD '04), 24-27 May 2004, pp. 257- 260.

[18] C. Weiß, S. Aschauer, G. Wachutka, A. Härtl, F. Hille, F. Pfirsch: "Numerical Analysis of Cosmic Radiation-induced Failures in Power Diodes", *Proc. of the European Solid-State Device Research Conference* (ESSDERC 2011), 12-16 Sept. 2011, pp. 355-358.

[19] G. Soelkner, P. Voss, W. Kaindl, G. Wachutka, K.H. Maier, H.-W. Becker: "Charge Carrier Avalanche Multiplication in High-Voltage Diodes Triggered by Ionizing Radiation", *IEEE Trans. on Nuclear Science*, vol. 47, no. 6, 2000, pp. 2365-2372.

[20] C. Weiß, G. Wachutka, A. Härtl, F. Hille, F. Pfirsch: "Predictive Physical Model of Cosmic-Radiation-Induced Failures of Power Devices", *Proc. of the 16th Int. Power Electronics and Motion Control Conference* (EPE-PEMC 2012 ECCE Europe), Novi Sad, Serbia, pp. LS2e.3-1-LS2e.3-5.

PLENARY SESSION

978-1-7281-3420-8/19 $31.00 © 2019 IEEE   30

The Fundamental Current Mechanisms in SiC Schottky Barrier Diodes: Physical Model, Experimental Verification and Implications

S. Dimitrijev, J. Nicholls, P. Tanner, and J. Han

Abstract – In this paper, we derive the equations for the current–voltage characteristics of SiC Schottky barrier diodes from the fundamental physics of thermionic emission and tunneling, as the two fundamental current mechanisms. An excellent fit between the model and the experimental data is achieved without the need for empirical fitting parameters, such as the commonly used ideality factor, and with a single set of physically meaningful parameters. This result shows that the current transport in the measured SiC Schottky diodes is not dominated by defects.

I. INTRODUCTION

Metal–semiconductor contacts and the energy barriers that they form exist in all semiconductor devices. Two fundamental current mechanisms that enable current to flow through metal–semiconductor contacts are (1) thermionic emission of carriers over the contact barrier and (2) carrier tunneling through the barrier. In spite of the fact that metal–semiconductor contacts are so common, there is a significant degree of confusion regarding the models and the parameters of the two fundamental current mechanisms [1–6]. A reason for this confusion has been the absence of experimental results that could be modeled without the need to include second-order effects. For example, carrier generation and recombination impact the carrier transport through contacts made on silicon substrates [7,8].

Silicon carbide is a wide energy-gap material, which does not suffer from carrier generation and, in the case of metal contacts to N-type SiC, there is no recombination because there are no holes in this semiconductor. Accordingly, this material provides an excellent platform for an experimental-based verification of the fundamental current mechanisms. The adequate structure for this purpose is a Schottky-barrier diode with proper edge terminations, to ensure that the measured current is through the planar region of the contact. This structure is shown in Fig. 1a.

S. Dimitrijev and J. Nicholls are with Queensland Micro- and Nanotechnology Centre, and with the School of Engineering and Built Environment, Griffith University, Nathan, Qld. 4111, Australia, E-mail: s.dimitrijev@griffith.edu.au

P. Tanner and J. Han are with Queensland Micro- and Nanotechnology Centre, Griffith University, Nathan, Qld. 4111, Australia

Commercial SiC Schottky diodes have been available for more than a decade. However, the structure of these diodes is different (Fig. 1b)—it is known as Junction Barrier Schottky (JBS) diode because it utilizes P+/P pockets to protect the semiconductor surface from the highest electric field by a dense array of reverse-biased P–N junctions. Consequently, the reverse-bias current of these diodes is dominated by the complex two-dimensional structure of the reverse-biased P–N junctions.

Recently, we have fabricated pure SiC Schottky diodes with an adequate edge termination (Fig. 1a), which are suitable for the experimental verification of the fundamental current mechanisms that we discuss in this paper. It should be mentioned that pure SiC Schottky diodes are now also available from Rohm (a large power-device manufacturer from Japan).

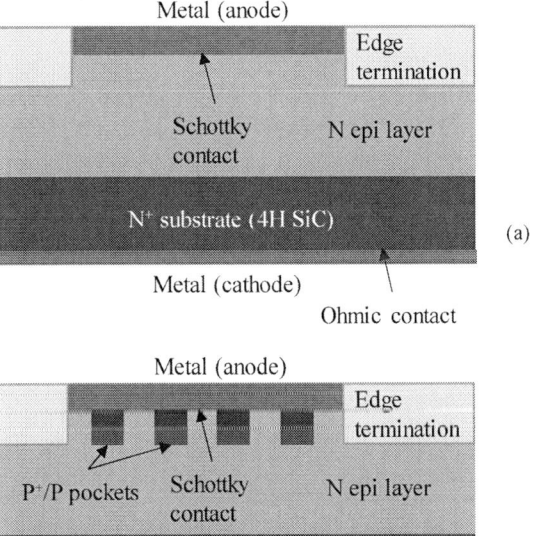

Fig. 1. Cross-sectional diagrams of a pure Schottky diode used in this paper (a) and the most commonly available junction-barrier Schottky (JBS) diode (b).

II. ENERGY BANDS

The energy-band diagram of a typical metal–SiC contact with aligned Fermi levels (E_F) is shown in Fig. 2. We consider the electrons in both metals and in the conduction band of a semiconductor as free carriers with the effective mass m^* (in addition to the generic symbol m^*, we will also use the notations m^*_m and m^*_s to distinguish between the values of the effective mass in the metal and in the semiconductor, respectively). According to the free-career model, the kinetic energy of electrons is $E_{kin} = m^* v_{th}^2/2$, where v_{th} is the thermal velocity of electrons. The condition $E_{kin} = 0$ corresponds to the bottom of the conduction band in the semiconductor (E_C), which is above the Fermi level, and to the bottom of the band in the metal, which is well below the Fermi level. In the case of aluminum, illustrated in Fig. 2, E_F is 5.6 eV above the bottom of the band, which corresponds to $v_{th} = 1.4 \times 10^6$ m/s for $m^* = m_0 = 9.1 \times 10^{-31}$ kg and for the electrons with $E_{kin} = E_F$ (note that this is 4.7% of the speed of light).

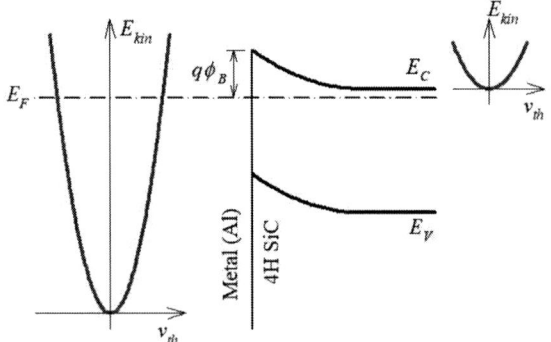

Fig. 2. Energy-band diagram of a typical metal–SiC Schottky contact with aligned Fermi levels (E_F) and with illustrations of the kinetic energies ($E_{kin} = m^* v_{th}^2/2$) of electrons in the metal and in the semiconductor. E_C and E_V are the bottom of the conduction band and the top of the valence band in SiC, respectively.

The most important parameter of a Schottky contact is the energy-barrier height, labeled by $q\phi_B$ in Fig. 2. The fundamental factor contributing to the formation of the barrier is the difference between the work function of metal and the electron affinity in the semiconductor [9]. For example, the difference between the work function of aluminum ($q\phi_m = 4.25$ eV [10]) and the electron affinity of 4H SiC ($q\chi_s = 3.6$ eV [11]) is 0.65 eV. However, this value is well below the experimentally observed barrier heights.

In real metal–semiconductor contacts, the existence of trapped charge (Q_{nit}) at distance δ from the metal surface creates an electric field, Q_{nit}/ε_s, which causes bending of the energy bands by the factor $q\delta Q_{nit}/\varepsilon_s$ [12]. The traps that are responsible for this additional component of the barrier height are usually referred to as interface traps [12];

however, we have suggested recently the use of the term "near interface traps", to highlight the fact that the charge associated with pure interface traps ($\delta = 0$) is fully compensated by the charge in the metal and, hence, does not create electric field and does not contribute to the band bending [13].

The third factor that impacts the barrier height is the image-force effect, which reduces the barrier by $q\Delta\phi_B = 2q\sqrt{qE_S/16\pi\varepsilon_s}$, where E_S is the electric field at the surface of the semiconductor and ε_s is the semiconductor permittivity [12].

The following equation summarizes the described contributions to the barrier height at a metal–semiconductor contact with aligned Fermi levels:

$$q\phi_B = q\phi_m - q\chi_s + \frac{q\delta Q_{nit}}{\varepsilon_s} - 2q\sqrt{\frac{qE_S}{16\pi\varepsilon_s}} =$$
$$q\phi_{B0} - 2q\sqrt{\frac{qE_S}{16\pi\varepsilon_s}} \qquad (1)$$

In Eq. (1), $q\phi_{B0}$ is a single material parameter, incorporating the difference between the metal work function and the electron affinity of SiC, as well as the contribution of the charge at near-interface traps when the Fermi level of the semiconductor is aligned to the Fermi level of the metal.

III. METAL-TO-SEMICONDUCTOR CURRENT

In their thermal motion inside the metal, a number of electrons hit the interface with semiconductor. If we set the x direction to be normal to the interface and label the thermal velocity in the x direction by v_{th-x}, the number of hits per unit area and unit time is

$$N_{hits} = v_{th-x}n \qquad (2)$$

where n is the concentration of electrons. Most of these electrons will bounce back, but some of them pass through the interface, either by tunneling or thermionic emission (Fig. 3). If we label the probability that an electron will traverse the interface by P and the electron charge by q, we obtain the following basic equation for the current density of electrons flowing from the metal into the semiconductor:

$$j_{m\to s} = qPN_{hits} \qquad (3)$$

In Eq. (3), both the probability P and the number of hits are variables that depend on the kinetic energy $E_{kin-x} = m^* v_{th-x}^2/2$. To account for this effect, we rewrite Eq. (3) in the following form

$$j_{m\to s} = \int_0^\infty qP(E_{kin-x})n_{hits}(E_{kin-x})dE_{kin-x} \qquad (4)$$

where $n_{hits}(E_{kin-x})$ is the number of hits per unit area, unit time, and unit energy. Taking into account the density-of-states distribution with energy, the Fermi–Dirac distribution for the probability of state occupancy, and assuming that electrons move randomly in all directions, we have derived the following equation for $n_{hits}(E_{kin-x})$ in a recently published paper [14]:

$$n_{hits}(E_{kin-x}) = \frac{4\pi m^* kT}{h^3} \ln\left[1 + \exp\left(\frac{E_F - E_{kin-x}}{kT}\right)\right] \quad (5)$$

where T is the absolute temperature, k is the Boltzmann constant, and h is the Planck constant.

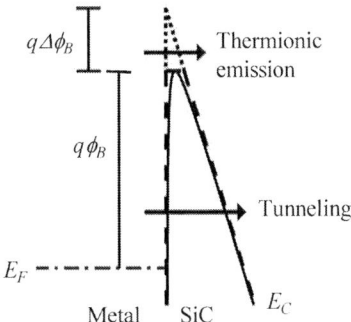

Fig. 3. Illustration of thermionic emission and tunneling as the two mechanisms of electron flow from metal to semiconductor, which dominate the reverse-bias current of Schottky diodes ($q\phi_B$ is the barrier height and $q\Delta\phi_B$ is the barrier reduction due to the image-force effect).

In Eq. (4), $P(E_{kin-x}) = 1$ for the case of thermionic emission, when $E_{kin-x} > E_F + q\phi_B$. Regarding the tunneling mechanism, we used the Wentzel–Kramers–Brillouin approximation to derive the following equation for $E_{kin-x} < E_F + q\phi_B$ [14]:

$$P(E_{kin}) = \exp\left[-\frac{4\pi\sqrt{2m_t^*}}{hqE_S}\sqrt{E_\phi}\left(\frac{2}{3}E_\phi + q\Delta\phi_B\right)\right] \quad (6)$$

where m_t^* is the tunneling effective mass, E_S is the electric field at the surface of the semiconductor, $E_\phi = E_F + q\phi_B - E_{kin-x}$, and $q\Delta\phi_B = 2q\sqrt{qE_S/16\pi\varepsilon_s}$.

To verify the model represented by Eqs. (4)–(6), we measured reverse-bias currents at different temperatures and for diodes designed for blocking voltages of 650 V and 1200 V. The results for a typical 650-V diode are shown in Fig. 4. We can see that the model fits the data well at higher temperatures, whereas the experimental current is higher at room temperature. We also observed that there is a device-to-device variation of measured currents at room temperature, which does not exist when the measurements are performed at higher temperatures. This indicates that the measured room-temperature currents include significant leakage components through defects, which become insignificant at higher temperatures.

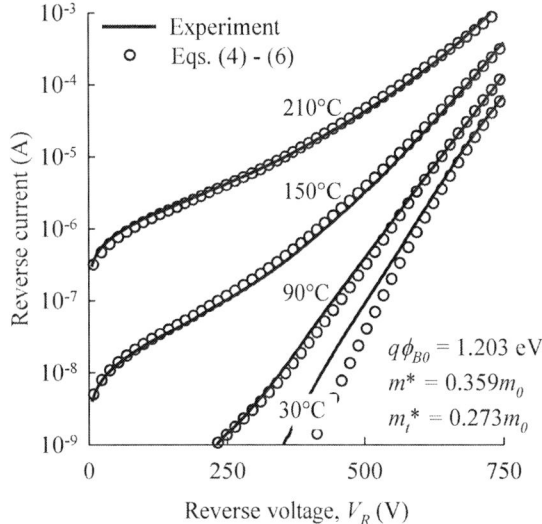

Fig. 4. Verification of the model for metal-to-semiconductor current by measured reverse-bias currents at different temperatures. Note that $q\phi_{B0} = q\phi_B + q\Delta\phi_B$.

In semiconductor theory, we frequently approximate the population of electron states by stating that the energy levels below the Fermi level are occupied, whereas those above the Fermi level are empty. With this assumption, and with reference to Figs. 2 and 3, it appears that the metal-to-semiconductor current would be dominated by tunneling of electrons with energies around the Fermi level. Given that tunneling is a temperature-independent mechanism, it is surprising to see in Fig. 4 the large temperature dependence of the reverse-bias current.

The surprise is due to the wrong assumption that the metal-to-semiconductor current is dominated by electron with energies around E_F. It is correct that the concentration of electrons well above E_F drops exponentially, but the tunneling probability increases exponentially for electrons with E_{kin-x} well above E_F. In Fig. 5, we show calculations of the function inside the integral of Eq. (4) for two different reverse-bias voltages, $V_R = 150$ V and 650 V, corresponding to the electric fields at the semiconductor surface of $E_S = 6.6 \times 10^5$ V/cm and 1.5×10^6 V/cm, respectively. We can see that the peak of the current for $V_R = 150$ V is just below the tip of the barrier, which is about 1 eV above the Fermi level. With this insight, it becomes obvious that an increase in the temperature can cause a significant current increase by increasing the concentration of electrons around the tip of the energy barrier. Figure 5b shows that the peak of the current drops toward the Fermi level at $V_R = 650$ V. This again shows that a small fraction of the metal electrons contribute to the reverse-bias current and that the current increases when the temperature increases the concentration of these electrons.

978-1-7281-3420-8/19 $31.00 © 2019 IEEE

(a) V_R=150 V (E_S = 6.6x10^5 V/cm)

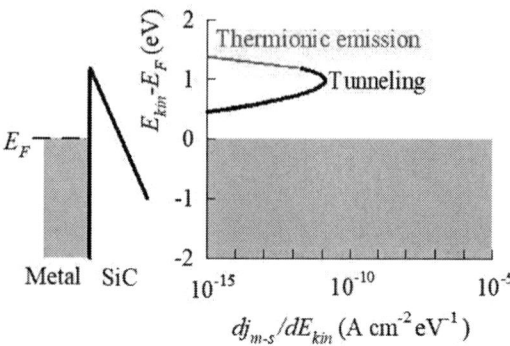

(b) V_R=650 V (E_S = 1.5x10^6 V/cm)

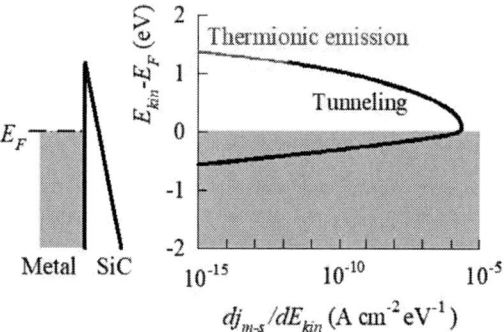

Fig. 5. Distributions of metal-to-semiconductor current at two different reverse-bias voltages: (a) 150 V and (b) 650 V.

IV. SEMICONDUCTOR-TO-METAL CURRENT

The flow of electrons from the semiconductor to metal is responsible for the forward-bias current. Figure 6 illustrates the energy bands at forward bias and shows that the dominant mechanism is thermionic emission of conduction-band electrons over the barrier (in this case, tunneling through the barrier is insignificant in comparison to the thermionic current). The kinetic energy of the electrons hitting the semiconductor–metal interface is $E_{kin-x} > 0$, which means their total energy is larger than E_C and, consequently, the probability that they will pass through the interface is $P(E_{kin-x}) = 1$. With this, the general equation for the current density of electrons flowing from the semiconductor into the metal is

$$j_{s \to m} = \int_0^\infty q n_{hits}(E_{kin-x}) dE_{kin-x} \qquad (7)$$

The derivation of the equation for the number of hits per unit area, unit time, and unit energy, $n_{hits}(E_{kin-x})$, is similar to the derivation of Eq. (5) for the electrons in metal with one important difference: As Fig. 2 shows, the bottom of the conduction band in the semiconductor is at the

energy level E_C above the bottom of the band in the metal. This means that the zero level for both the thermal velocity and the density of states, which corresponds to $E_{kin-x} = 0$, is at E_C in the case of semiconductor. Accordingly, the energy term in the Fermi–Dirac distribution is no longer just $E_{kin-x} - E_F$ but $E_{kin-x} + E_C - (E_F + qV)$, where $E_F + qV$ is the quasi-Fermi level in the semiconductor. With that, the logarithm term in Eq. (5) becomes $\ln\left[1 + \exp\left(-\frac{E_{kin-x}+E_C-E_F-qV}{kT}\right)\right]$, which can be approximated by $\exp\left(-\frac{E_{kin-x}+E_C-E_F-qV}{kT}\right)$ given that $E_{kin-x} + E_C - E_F - qV \gg kT$. Therefore, the equation for the number of hits per unit area, unit time, and unit energy becomes

$$n_{hits}(E_{kin-x}) = \frac{4\pi m^* kT}{h^3} \exp\left(-\frac{E_{kin-x}+E_C-E_F-qV}{kT}\right) \qquad (8)$$

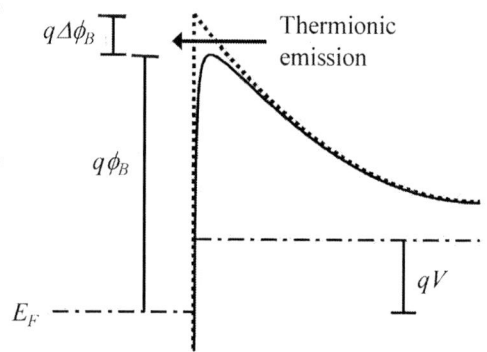

Fig. 6. Illustration of thermionic emission as the mechanism of electron flow from semiconductor to metal, which is responsible for the forward-bias current of Schottky diodes. Note that $q\phi_B = q\phi_{B0} - q\Delta\phi_B + \frac{q^2\delta D_s}{\varepsilon_s}V$ where $q\phi_{B0}$ is the barrier height without the image-force effect at $qV = 0$ (a voltage-independent parameter), qVD_s is the number of near-interface traps per unit area in the energy range between the quasi-Fermi levels, and δ is the distance of the near-interface traps from the metal surface [14].

With this equation for $n_{hits}(E_{kin-x})$, the integral in Eq. (7) can be solved, leading to the following result for the current density of electrons flowing from the semiconductor into the metal:

$$j_{s \to m} = A^* T^2 \exp\left(-\frac{q\phi_B}{kT}\right) \exp\left(\frac{qV}{kT}\right) \qquad (9)$$

where $q\phi_B = E_C - E_F$ and $A^* = \frac{4\pi q m^* k^2}{h^3}$ is the so-called Richardson constant.

According to Eq. (9), the slope of the semi-logarithmic plot, $\ln j_{s \to m}$ versus V, is equal to q/kT. However, the slope of measured forward-bias currents is always somewhat smaller than this theoretical value. It is a common practice to reduce the theoretical slope by a fitting parameter called the ideality factor, n, so that the slope becomes q/nkT. This

978-1-7281-3420-8/19 $31.00 © 2019 IEEE

discrepancy between the experimental data and the theoretical slope of q/kT is attributed to several different phenomena [3, 15–17].

The need to introduce the ideality factor is due to the inherent assumption that the barrier height $q\phi_B$ does not change with applied voltage (if we ignore the small impact through a small reduction of the electric field E_S). Referring to Eq. (1), we can see that the assumption of constant $q\phi_B$ corresponds to the implicit assumption of constant charge at the near-interface traps, Q_{nit}. The assumption that the charge at the near-interface traps does not increase as the quasi-Fermi level in the semiconductor is lifted by the applied voltage (qV in Fig. 6) is equivalent to the assumption that there are no near-interface traps with energy levels above the Fermi level of metal. However, if the density of these traps per unit area and unit energy is not zero ($D_{nit} > 0$), they will increase the barrier height as they trap additional electrons when the quasi-Fermi level is lifted by the applied voltage:

$$q\phi_B = q\phi_{B0} - 2q\sqrt{\frac{qE_S}{16\pi\varepsilon_S}} + \frac{q^2\delta D_{nit}}{\varepsilon_S}V \qquad (10)$$

Note that, in Eq. (10), qVD_{nit} is the charge per unit area trapped at the near-interface traps with energy levels between the quasi-Fermi level in the semiconductor and the Fermi level in metal (E_F). Regarding the charge at the near-interface traps with energy levels below E_F (the charge Q_{nit}), Eq. (1) shows that it is included in the term $q\phi_{B0}$.

Figure 7 shows that the fundamental equation for the thermionic emission of electrons [Eq. (9)] is an adequate model for the forward-bias current when the increase in the barrier height due to additional electron trapping [eq. (10)] is included. This means there is no need for the inclusion of the ideality factor as an empirical fitting parameter, which does not provide a direct relationship to a physical mechanism.

The experimental results in Fig. 7 show that the semi-logarithmic plots of measured currents are linear up to currents of about 100 mA. For the higher current levels, the forward voltages are higher than the values corresponding to Eqs. (9) and (10). This is the effect of serious resistance of the drift region and the substrate (r_S), which is simply accounted for by the Ohm's law:

$$V_F = V + r_S I_F \qquad (11)$$

where V_F is the measured forward voltage and I_F is the measured forward current.

V. THERMAL EQUILIBRIUM

With verified equations for the current density from metal to semiconductor (Section III) and the current density from semiconductor to metal (Section IV), it is interesting to consider the condition for thermal equilibrium, which is $I_F = 0$ for $V_F = 0$.

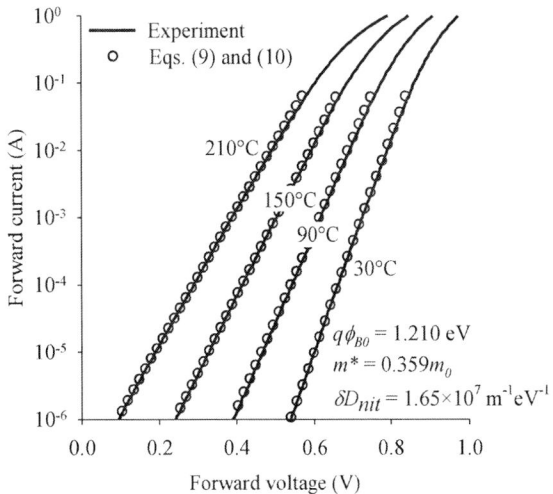

Fig. 7. Verification of the model for semiconductor-to-metal current by measured forward-bias currents at different temperatures.

Neglecting the electron tunneling from metal into semiconductor at zero bias, and approximating the logarithm term in Eq. (5) by $\exp\left(-\frac{E_{kin-x}-E_F}{kT}\right)$, Eqs. (4)–(6) can be written in the following simplified form:

$$j_{m\to s} = \frac{4\pi q m_m^* kT}{h^3} \int_{E_F+q\phi_B}^{\infty} \exp\left(-\frac{E_{kin-x}-E_F}{kT}\right) dE_{kin-x} \quad (12)$$

Note that we labeled the effective mass by m_m^* to specify that this is the mass of electrons in the metal. Solving the integral in Eq. (12), we obtain

$$j_{m\to s} = \frac{4\pi q m_m^* kT}{h^3} \exp\left(-\frac{q\phi_B}{kT}\right) \qquad (13)$$

The semiconductor-to-metal current at zero bias, obtained from Eq. (9), is

$$j_{s\to m} = \frac{4\pi q m_s^* kT}{h^3} \exp\left(-\frac{q\phi_B}{kT}\right) \qquad (14)$$

In this case, we label the effective mass by m_s^* to specify that this is the mass of electrons in the semiconductor.

The currents $j_{m\to s}$ and $j_{s\to m}$, given by Eqs. (13) and (14), respectively, would be equal and the net current at zero bias would be zero if $m_m^* = m_s^*$. However, there is no reason for this condition to be valid in different materials, such as the metal and the semiconductor in the case of Schottky diodes. To explain the consequence of this difference, assume that $m_m^* > m_s^*$. What happens then is that the excess flow of electrons from the metal into the semiconductor, when $j_{m\to s} > j_{s\to m}$, is lifting the bands to reduce the barrier height for the thermionic emission of electrons from the semiconductor and to reach the condition $j_{m\to s} = j_{s\to m}$. This band bending is equivalent to the following shift in the semiconductor quasi-Fermi level with respect to E_F [14]:

$$qV_m = kT \ln \left(\frac{m_m^*}{m_s^*} \right) \qquad (15)$$

Reducing the barrier height in Eq. (14) by this value leads to the following equation for the semiconductor-to-metal current:

$$j_{s \to m} = \frac{4\pi q m_m^* kT}{h^3} \exp \left(-\frac{q\phi_B}{kT} \right) \qquad (16)$$

This result shows that, effectively, the current from the semiconductor into metal is determined by the effective mass of electrons in the metal, in agreement with experimental results showing that the effective mass is affected by the choice of metal [14,18,19].

VI. CONCLUSION

The reverse-bias current through the metal–semiconductor contact of pure SiC Schottky diodes with an adequate edge termination is dominated by thermionic emission at lower reverse-bias voltages and by tunneling at higher voltages. An excellent fit with experimental data is achieved by a model based on the probability that an electron with a given kinetic energy will pass through the contact—either by tunneling or thermionic emission—and by a determination of the number of hits per unit area, unit time, and unit energy. An analogous model for the thermionic current of a forward-biased diode provides an excellent fit with measured forward-bias currents at different temperatures. This fit is achieved by including a barrier-increase effect due to trapping of electrons by near-interface traps with energy levels between the quasi-Fermi level in the semiconductor and the Fermi level in the metal. By including this physical mechanism, there is no need for the commonly-used empirical fitting parameter known as the ideality factor.

ACKNOWLEDGEMENT

This work was supported by SICC Material Company Ltd., China, as the industry partner in the Australian Research Council Linkage Project under Grant ARC LP 50100525.

REFERENCES

[1] W. Schottky, "Halbleitertheorie der Sperrschicht", *Naturwissenschaften*, 1938, vol. 26, pp. 843–843.

[2] H.A. Bethe, *Theory of the boundary layer of crystal rectifiers*, Cambridge, MA: Radiation Laboratory, M.I.T., 1942.

[3] F.A. Padovani and R. Stratton, "Field and thermionic-field emission in Schottky barriers", *Solid-State Electronics*, 1966, vol. 9, pp. 695–707.

[4] C.R. Crowell and V.L. Rideout, "Normalised thermionic-field (T-F) emission in metal–semiconductor (Schottky) barriers", *Solid-State Electronics*, 1969, vol. 12, pp. 89–105.

[5] E.L. Murphy and R.H. Good, Jr., "Thermionic emission, field emission, and the transition region", *Physical Review*, 1956, vol. 102, pp. 1464–1473.

[6] S.J. Fonash, "Current transport in metal semiconductor contacts—a unified approach", *Solid-State Electronics*, 1972, vol. 15, pp. 783–787.

[7] T. Arizumi and M. Hirose, "Transport properties of metal–silicon Schottky barriers", *Jap. J. Applied Physics*, 1969, vol. 8, pp. 749–754.

[8] D. Kahng, "Conduction properties of the Au–n-type–Si Schottky barrier", *Solid-State Electronics*, 1963, vol. 6, pp. 281–295.

[9] S. Dimitrijev, *Principles of Semiconductor Devices*, 2nd ed., New York: Oxford University Press, 2012.

[10] E. W. J. Mitchell and J. W. Mitchell, "The Work Functions of Copper, Silver and Aluminium", *Proceedings of the Royal Society of London. Series A, Mathematical and Physical Sciences*, 1951, vol. 210, pp. 70–84.

[11] M. Wiets, M. Weinelt, and T. Fauster, "Electronic structure of SiC (0001) surfaces studied by two-photon photoemission", *Physical Review B*, 2003, vol. 68, pp. 125321-1–125321-11.

[12] E.H. Rhoderick, *Metal-Semiconductor Contacts*, Oxford: Oxford University Press, 1978.

[13] J.R. Nicholls, S. Dimitrijev, P. Tanner, and J. Han, "The Role of Near-Interface Traps in Modulating the Barrier Height of SiC Schottky Diodes", *IEEE Trans. Electron Devices*, 2019, vol. 66, pp. 1675–1680.

[14] J.R. Nicholls, S. Dimitrijev, P. Tanner, and J. Han, "Description and Verification of the Fundamental Current Mechanisms in Silicon Carbide Schottky Barrier Diodes", *Scientific Reports*, 2019, vol. 9, pp. 3754-1–3754-9.

[15] R.T. Tung, "Electron transport at metal-semiconductor interfaces: General theory", *Physical Review B*, 1992, vol. 45, pp. 13509–13523.

[16] S.M. Sze, C.R. Crowell, and D. Kahng, "Photoelectric determination of the image force dielectric constant for hot electrons in Schottky barriers", *J. Applied Physics*, 1964, vol. 35, pp. 2534–2536.

[17] H.C. Card and E.H. Rhoderick, "Studies of tunnel MOS diodes I. Interface effects in silicon Schottky diodes", *J. Physics D*, 1971, vol. 4, pp. 1589–1601.

[18] H. Okino, N. Kameshiro, K. Konishi, A. Shima, and R.-I. Yamada, "Analysis of high reverse currents of 4H-SiC Schottky-barrier diodes", *J. Applied Physics*, 2017, vol. 122, p. 235704.

[19] N. Toyama, "Variation in the effective Richardson constant of a metal-silicon contact due to metal-film thickness", *J. Applied Physics*, 1988, vol. 63, pp. 2720–2724.

Design Techniques for Wireless Sensor Network Nodes Powered by Ambient Energy Harvesting

Z. Prijić, A. Prijić, and Lj. Vračar

Abstract — This paper outlines some techniques for the design of wireless sensor networks nodes. The nodes obtain power from small transducers that convert the energy available in their surroundings. Described examples cover thermoelectric and photovoltaic powered nodes, which are realized using off-the-shelf devices. Experimental results obtained by the practical implementation of the nodes in the network are presented. The paper also highlights the potentials of the PCB technology for improving the performances of the nodes.

I. INTRODUCTION

Micro–scale energy harvesting is a process of collecting energy from the surroundings of an electronic device and converting it to electricity necessary for its powering. Energy may come either from a natural source or as a waste from some industrial process (energy scavenging). Natural sources can be classified by origin as ambient or body, where the latter one is commonly referred to as bioenergy. Ambient sources provide radiant, thermal, and mechanical energy. By using appropriate transducers, it is possible to obtain enough electricity to supply devices designed specifically for ultra–low power applications. These devices are typically microcontrollers, sensors, and radio frequency (RF) transceivers, which are core components of wireless sensor network (WSN) nodes. Thus, a combination of energy–efficient devices with renewable energy sources on a micro–scale leads to WSN nodes which can operate autonomously for a long time, with virtually no maintenance [1–4].

The WSN node (alias mote) is an autonomous electronic system designed to sense, process, and transmit some data over the network. General architecture of the node is illustrated in Fig. 1. Transducer receives the energy from the ambient and converts it to the electricity Power management circuitry produces voltage/current levels necessary to charge the primary storage and supply the load, which consists of one or more sensors, microcontroller, and RF transceiver. It also routes excess energy, when available, to the backup storage. In many cases, the amount of harvested energy may be low or can vary considerably over time. Therefore, the design of WSN nodes relies on ultra–low power devices and their non-perpetual operation.

Sources of radiant energy are light and transmitters broadcasting in the radio spectrum where correspond-

Z. Prijić, A. Prijić, and Lj. Vračar are with Department of Microelectronics, University of Niš, Faculty of Electronic Engineering, Aleksandra Medvedeva 14, 18000 Niš, Serbia, E-mail: zoran.prijic@elfak.ni.ac.rs

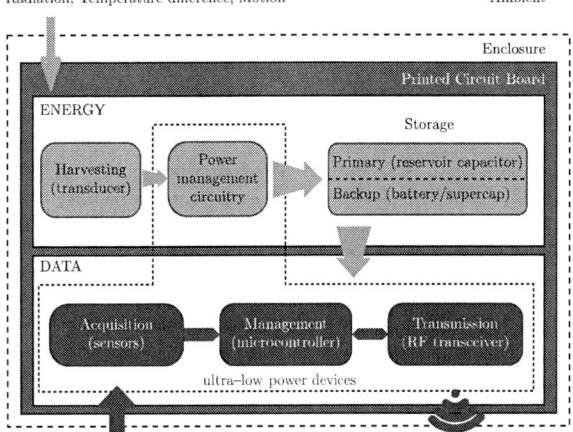

Fig. 1. Block diagram of an energy harvesting powered wireless sensor network node.

ing transducers are photovoltaic (PV) cells and antennas. Thermal sources are thermal gradient or variation, in which case energy conversion is carried out by using thermoelectric or pyroelectric devices. Mechanical sources are stress/strain, vibration or fluid flow, with the application of piezoelectric, electrostatic, electromagnetic, and various microturbine based transducers. It is also possible to use a combination of transducers in ambiances which simultaneously provide multiple energy sources (e.g. radiant and thermal) [5–9]. Typical harvested power densities for common ambient energy sources are summarized in Tab. I.

TABLE I
TYPICAL HARVESTED POWER DENSITIES FOR COMMON AMBIENT ENERGY SOURCES (BIOENERGY SOURCES EXCLUDED).

Source	Harvested power density (W/cm^2)
Light (indor/outdoor)	10µ/10m
RF	0.1µ
Thermal	10m
Mechanical (vibration/flow)	100µ/10m

Although the design of the WSN nodes follows the general principle from Fig. 1, the diversity of energy sources, as well as their availability, implies certain differences. Intensive research is being carried out to develop miniature transducers and power management circuits [10–12] specifically for energy harvesting in WSN applications. However, the use of an efficient transducer does not guarantee the harvesting efficiency of the node [13]. The topology of the power management circuitry and the way it routes energy to the storage devices should be chosen appropriately. Apart from the choice of other devices, their layout, as well as the construction of the node, also plays an important role. This paper address a few design techniques for thermoelectric and photovoltaic powered WSN nodes, along with their practical implementation using off–the–shelf transducers and devices.

II. THERMOELECTRIC–POWERED NODES

Thermoelectric powered WSN nodes are suitable for use in many environments where a temperature difference between a natural/artificial heat source and the ambient frequently occurs. Application areas span from industrial through consumer and up to environmental monitoring and wildlife tracking [14–20]. The nodes receive power from a thermoelectric generator (TEG), which consists of an array of p–n thermocouples electrically connected in series, as illustrated in Fig. 2. When

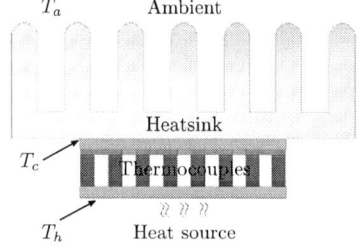

Fig. 2. Cross–section of a thermoelectric generator with heatsink attached.

a temperature difference $T_h - T_c$ exists between the hot and cold side of the TEG, a voltage appears at its external contacts due to the Seebeck effect. In an open–circuit, generated voltage is:

$$V_{TEG} = N\alpha_{pn}(T_h - T_c) , \qquad (1)$$

where N is a number of thermocouples and α_{pn} is an overall Seebeck coefficient. In a loaded circuit (Fig. 3), the current I_L produces a voltage drop on a TEG's temperature–dependent internal resistance R_{TEG}. Analytical models can accurately describe TEG as a standalone device under the assumption that T_h and T_c are constants [21]. However, the temperature difference $T_h - T_c$ is reduced due to the Peltier effect, which appears when the current I_L flows, thus affecting V_{TEG} and V_L.

To maintain the temperature difference necessary to produce usable values of V_L, the TEG requires heatsink

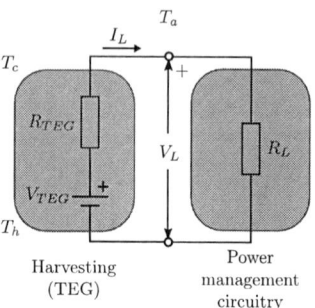

Fig. 3. The equivalent electrical circuit of a TEG with a load.

attached to the cold side, thus creating a thermally more complex system (Fig. 2). All heat transfer mechanisms (conduction, convection, and radiation) contribute to the heat flow within the system and towards the ambient. Temperature T_c depends on heatsink efficiency, which is determined by its material, geometry, and ambient conditions (natural/forced convection). It is reasonable to assume that the real heat source has an area larger than the area of the TEG, and it will also affect the thermal equilibrium of the system. The system can be described using a multiphysics approach, involving coupling between the heat transfer in solids and fluids, nonisothermal fluid flow, electric, and thermoelectric effects. It requires 3D numerical simulation resulting in the temperature distribution within the system and environment [22], as shown in Fig. 4. Simultaneous solving

Fig. 4. Temperature distribution within the system and surrounding ambient (air) under natural convection at $I_L = 18\,\text{mA}$.

of coupled equations also gives the value of load voltage, which corresponds to the temperature difference $T_h - T_c$ and accounts for the Peltier effect and variations of R_{TEG} with the load. Looking into the system, instead of the TEG alone, it is useful to express the load voltage as a function of the temperature difference between the hot side and the ambient $\Delta T_a = T_h - T_a$. A

978-1-7281-3420-8/19 $31.00 © 2019 IEEE

comparison between simulation results and experimental data is given in Fig. 5. The simulation accurately

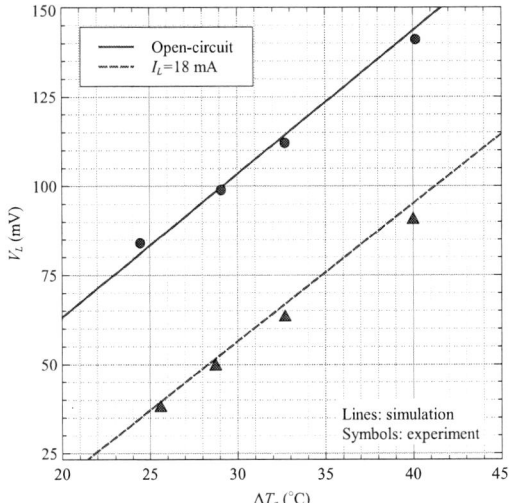

Fig. 5. Load voltage vs. ΔT_a for TEG CP08-31-06 [23] with aluminum heatsink BGA-STD-050 [24] under natural convection at $T_a = 27\,^\circ$C.

predicts the values of the generated voltage for both the open and loaded circuit conditions.

In order to design a compact WSN node, TEGs with small overall dimensions are preferred. For example, results in Fig. 5 are for the TEG which ocuppies $12 \times 12 \times 3.4\,\text{mm}^3$. Similar TEGs are built–in into WSN nodes shown in Fig. 6. The nodes were realized using

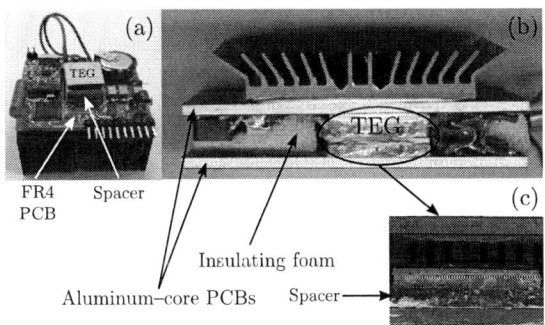

Fig. 6. Prototypes of WSN nodes realized in (a) conventional (FR4) and (b) Aluminum–core PCB technology [25]; (c) magnified view of the TEG inside.

conventional (FR4) and aluminum–core PCB technology. The advantage of the latter is because it achieves better uniformity of temperature distribution on the hot and cold side of the node [25,26]. The upper PCB effectively increases the heatsink's surface and reduces the thermal stress of the components. In both designs, an aluminum spacer was used to obtain room for components and increase the distance between the hot and

cold side of the node. It also increases the temperature difference on the TEG. The heatsink, TEG, and spacer were joined with a thermal adhesive, although a thermally conductive tape is also applicable. Space between the components was filled with thermally insulating foam to reduce undesirable heat transfer. The design of the nodes can be such that the TEG and the heatsink are out of the PCBs [27,28].

The construction of the node and layout of the components inside affect thermal equilibrium. The impact is observed by extending the study of the system from Figs. 2 and 4 to the entire node. It is possible to obtain temperature distributions inside the node under various thermal loads, as illustrated in Fig. 7. For the node

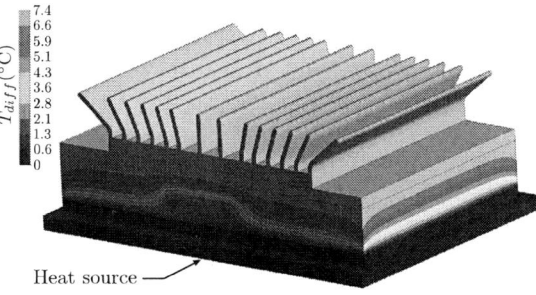

Fig. 7. Difference in temperature distributions in the WSN node from Fig. 6(b) at $T_a = 20\,^\circ$C and $T_a = 30\,^\circ$C under natural convection (numerical simulation); in both cases the heat source is adjusted to maintain $T_h = 50\,^\circ$C.

from Fig. 6(b), the value of ΔT_a necessary to produce the minimum load voltage for the start–up of the power management circuitry is about twice the value obtained for the system consisting of the TEG and heatsink only [29]. Apart from the natural convection, simulations can also give an insight into the thermal performances under forced convection by employing either heat transfer coefficients [30] or computational fluid dynamics [31]. Simulation results are useful in calibrating SPICE models based on electrothermal analogies [21,32,33].

Small TEGs provide low values of the load voltage V_L (Fig. 5), preventing linear harvesting. Therefore, a boost circuit is needed to reach the levels required for the circuitry inside the data block in Fig. 1. Additionally, to achieve maximum power transfer, R_{TEG} should be matched to the load impedance, which is realized using a Maximum Power Point Tracking (MPPT) circuitry. A variety of discrete boost and MPPT topologies are proposed in the literature [11,34–38]. Integrated solutions, tailored to various transducer types, are also available from major semiconductor manufacturers. Some of their key features are summarized in Tab. II. Note that the MPPT feature may not be applicable for nodes with small TEGs because it requires

978-1-7281-3420-8/19 $31.00 © 2019 IEEE

TABLE II
INTEGRATED POWER MANAGEMENT CIRCUITS FOR ENERGY
HARVESTING.

IC	Transducer	Startup voltage (mV)	MPPT
bq25505 [39]		100	Yes
LTC3108/9 [40]	TEG	20/30	No
LTC3105 [41]	PV Cell	250	Yes
CY39C831 [42]		350	Yes
SPV1050 [43]		75	Yes
MAX17710 [44]	Piezo PV Cell	750	No

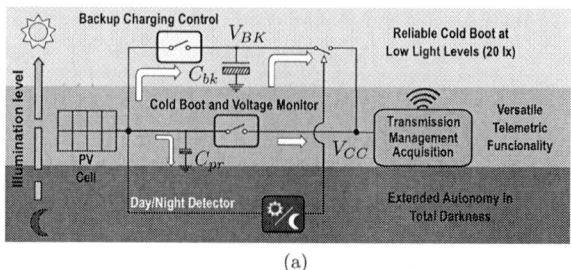

(a)

(b)

Fig. 8. Photovoltaic–powered WSN node optimized for low light
levels: (a) block diagram; (b) realized prototype (dimensions:
74 mm × 34 mm × 15 mm) [57].

higher start-up voltages and larger quiescent current. In
such a case, a boost circuit having the input impedance
close enough to R_{TEG} (a few ohms, typically) could be
a better solution [40]. By taking into account the non-
perpetual operation of the nodes, in case of approaching the critical level of stored energy, data transmission
may be suspended while acquisition can continue. By
altering power management features, it is possible to
achieve prolonged autonomy of the node in the absence
of the heat source. For example, a fully charged 200 mF
backup capacitor can supply the node from Fig. 6(b)
for 220 minutes with data logging and transmission enabled, as well as an additional 130 min with logging only
[31].

III. PHOTOVOLTAIC–POWERED WSN NODES

Photovoltaic–powered WSN nodes are also suitable
for many applications, particularly in environmental
and agriculture monitoring. The nodes utilize small solar panels, commonly referred to as PV cells, as light
energy transducers. Governing physics–based equations
and an equivalent electrical circuit of a PV cell can give
an accurate description of its current–voltage characteristics [45–47]. Experimental characterization is easily
carried out using a source-measure unit, which can also
provide a corresponding power curve [48, 49].

The main challenge in the design of PV-powered
WSN nodes is to ensure reliable cold-boot and operation
under low illumination levels. In addition to the integrated solutions from Tab. II, many other implementations rely on boost and MPPT circuits designed especially for ultra–low power consumption [50–55]. It has
been shown that such circuits can operate at light levels
down to 100 lx [56]. A further reduction of this limit is
possible without MPPT using the conceptual approach
illustrated in Fig 8a. The design utilizes high precision
voltage detectors with ultra–low current consumption
(350 nA) to control switches for primary and backup
charging circuits. An additional circuit (day/night detector) monitors the voltage level on the PV cell and de-

termines from which storage the node will receive power.
The backup storage will not be activated until the light
level falls below 20 lx. The detector also prevents complete discharging of the primary storage, thus enabling
reliable cold boot. By managing the backup storage
without using a boost converter, it is possible to achieve
prolonged autonomy of the node in total darkness. The
practical realization of the node uses two PCBs in a
stacked structure (Fig. 8b). Detailed description, including schematics and components, is given in [57].
With the fully charged 1.5 F capacitor, the node can
operate up to 81 h in total darkness, transmitting the
measured data every 200 s. In the case of data encryption, this time will be somewhat less, depending on the
encryption algorithm [58]. The node can be enclosed,
leaving only the PV cell and sensors exposed and is
suitable for indoor environmental monitoring in smart
buildings.

Figure 9 shows the WSN node designed for application in greenhouses and gardens. The node is a multi-
sensor platform, which can acquire and transmit data
about ambient and soil conditions. Passive capacitive

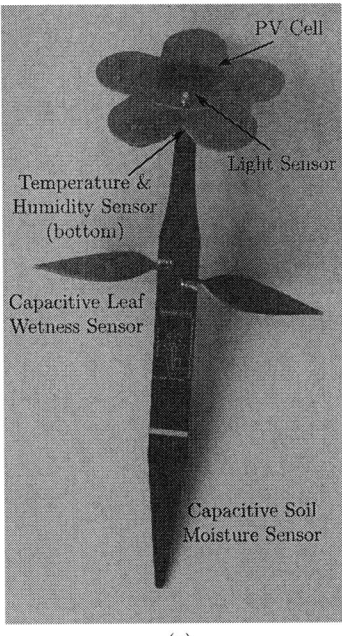

(a)

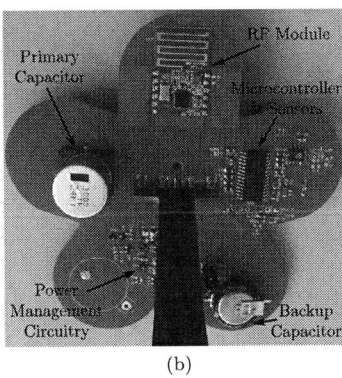

(b)

Fig. 9. Photovoltaic powered WSN node for greenhouses and gardens: (a) whole structure; (b) bottom view of the flower; overall height: 33 cm, flower diameter: 15 cm.

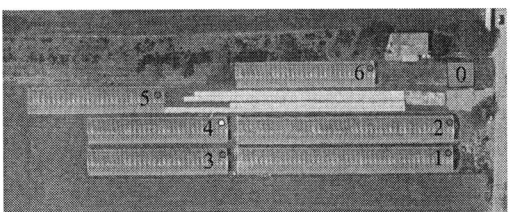

Fig. 10. Areal photography of a greenhouse farm indicating positions of the WSN members: (0) – base station; (4) – coordinator node; (1-3,5,6) – group members.

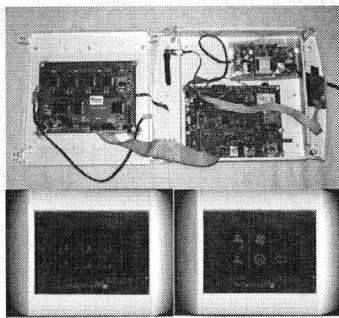

Fig. 11. Photography of the interior (top) and exterior (bottom) of the base station.

For home automation applications, a tablet or a PC equipped with a simple USB transceiver can communicate with the nodes instead of the base station. For this purpose, accompanying Windows application was developed to collect measured data. A user can also configure the nodes (Fig. 12) and tune radio link parameters. An experiment was carried out by position-

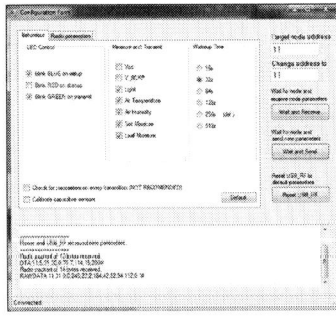

Fig. 12. Screenshot of the application, showing configuration parameters available for a WSN node.

ing a node in the corner of a terrace garden in the urban environment. Figures 13 and 14 show a part of the gathered data, which were acquired and transmitted every 512 seconds. The node can harvest enough energy to fully charge the backup capacitor at modest illuminances, which slightly exceed those of an overcast day. During the night, the backup capacitor discharges up to 3.5 V, well above the minimum of 2 V required for

sensors are realized using conventional PCB technology. The design resembles the one from Fig. 8a, without a day/night detector. A rechargeable battery may be used as backup storage, instead of a supercapacitor.

An experimental WSN consisting of 6 nodes and the base station was set up on a greenhouse farm (Fig. 10) occupying 5500 m². A multi-hop network topology was implemented with one node being the coordinator. The coordinator and base unit communicate using MiWi protocol [59]. The base unit is powered from the mains. It receives measured data from the nodes, and a user can view them on display, and initiate appropriate action (e.g. start watering system), as shown in Fig. 11. It also contains a GSM module, so the data can be transferred over the cellular network.

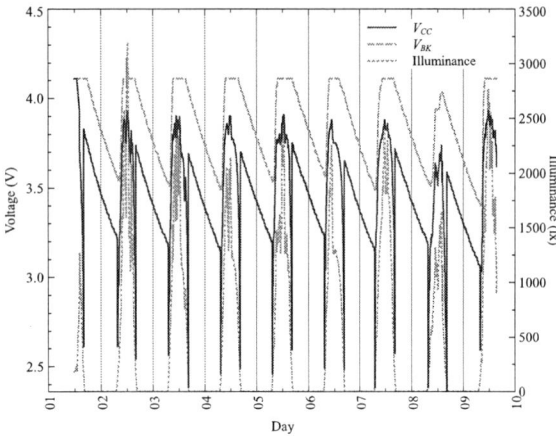

Fig. 13. Variations of the node supply V_{CC} and backup V_{BK} voltages (Fig. 8a) with the illuminance for eight days.

operation.

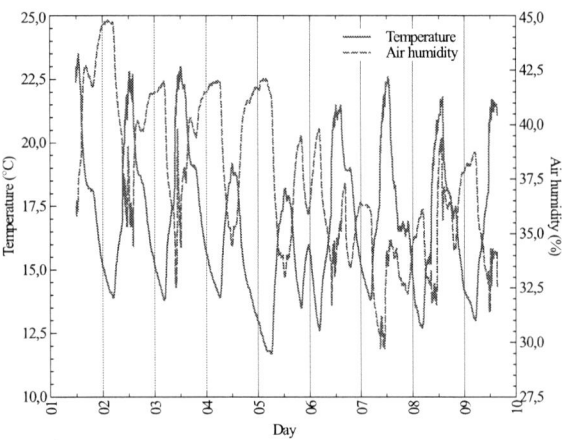

Fig. 14. The values of the temperature and humidity measured by the node for eight days from Fig. 13.

A modification of the design suitable for low light levels exploits the technique for the concentration of the solar irradiation on a PV cell. Addition of flat–panel reflectors beneficiary assists in capturing the radiation, regardless of the cell type. Figure 15 shows the practical implementation of the reflectors in WSN nodes. The reflectors are approximately the same size as a PV cell, tilted to its plane. It has been shown theoretically and experimentally that the optimum tilt angle of the reflectors is 65° [60]. The contribution of the reflectors is important at very low light levels, as illustrated in Fig 16. The time necessary to reach the startup voltage of the comparators (2.35 V in this case) is reduced significantly, which is desirable for indoor applications. Although four reflectors give better results, it is more practical to use two as shadow effects can occur. Also,

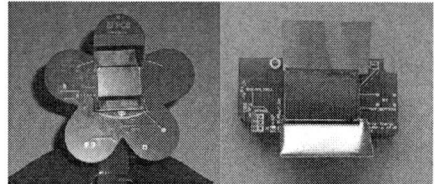

Fig. 15. Flat–panel reflectors mounted on the WSN nodes.

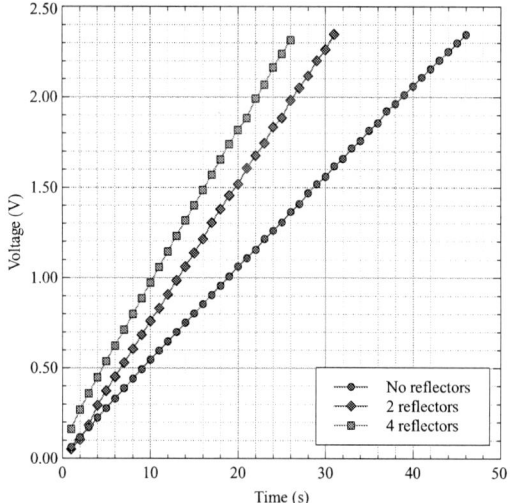

Fig. 16. Experimental charging curves of 3300 μF capacitor (C_{pr} in Fig. 8a) at 100 lx illuminance; adapted from [60].

two reflectors are much easier to mount and dismount, if necessary. It is important to note that the reflectors are produced in conventional PCB technology [60]. Because the material is almost readily available as a by-product, the cost of addition to the node is negligible.

IV. CONCLUSION

Energy harvesting is a feasible solution for powering WSN nodes. New materials, like thermoelectric nanowires, appear to be promising for increasing efficiency. Continuous dimension scaling of microelectronic circuits also results in lowering the power consumption. The room for performance improvement still exists in some less frequently exploited capabilities of existing technologies and off–the-shelf devices. Although this paper presented working design examples of compact thermoelectric and photovoltaic powered WSN nodes, additional research related primarily to the hardware reliability and embedded software optimization is needed.

ACKNOWLEDGEMENT

This work was supported in part by the Serbian Ministry of Education, Science and Technological Development under Grant TR32026 and in part by Ei PCB Factory, Niš, Serbia.

REFERENCES

[1] M. T. Penella-López and M. Gasulla-Forner, *Powering Autonomous Sensors*. Springer Netherlands, 2011.

[2] M. Belleville and C. Condemine, Eds., *Energy Autonomous Micro and Nano Systems*. John Wiley & Sons, Inc., jul 2012.

[3] S. Basagni, M. Y. Naderi, C. Petrioli, and D. Spenza, *Wireless Sensor Networks with Energy Harvesting*. John Wiley & Sons, Inc., mar 2013, pp. 701–736.

[4] N. A. Bhatti, M. H. Alizai, A. A. Syed, and L. Mottola, "Energy harvesting and wireless transfer in sensor network applications," *ACM Transactions on Sensor Networks*, vol. 12, no. 3, pp. 1–40, aug 2016.

[5] D. S. Montgomery, C. A. Hewitt, and D. L. Carroll, "Hybrid thermoelectric piezoelectric generator," *Applied Physics Letters*, vol. 108, no. 26, p. 263901, jun 2016.

[6] F. U. Khan and Izhar, "Hybrid acoustic energy harvesting using combined electromagnetic and piezoelectric conversion," *Review of Scientific Instruments*, vol. 87, no. 2, p. 025003, feb 2016.

[7] C.-C. Chen, T.-K. Chung, C.-Y. Tseng, C.-F. Hung, P.-C. Yeh, and C.-C. Cheng, "A miniature magnetic-piezoelectric thermal energy harvester," *IEEE Transactions on Magnetics*, vol. 51, p. Art. No. 9100309, July 2015.

[8] J. Estrada-López, A. Abuellil, Z. Zeng, and E. Sánchez-Sinencio, "Multiple input energy harvesting systems for autonomous IoT end-nodes," *Journal of Low Power Electronics and Applications*, vol. 8, no. 1, p. 6, mar 2018.

[9] T. Nikolić, M. Stojčev, G. Nikolić, and G. Jovanović, "Energy harvesting techniques in wireless sensor networks," *Facta Universiatis, Series: Automatic Control and Robotics*, vol. 17, pp. 117–142, 2018.

[10] P. Mullen, J. Siviter, A. Montecucco, and A. Knox, "A thermoelectric energy harvester with a cold start of 0.6°C," *Materials Today: Proceedings*, vol. 2, no. 2, pp. 823–832, 2015, 12th European Conference on Thermoelectrics.

[11] A. Montecucco and A. R. Knox, "Maximum power point tracking converter based on the open-circuit voltage method for thermoelectric generators," *IEEE Transactions on Power Electronics*, vol. 30, no. 2, pp. 828–839, feb 2015.

[12] Y.-H. Liu, Y.-H. Chiu, J.-W. Huang, and S.-C. Wang, "A novel maximum power point tracker for thermoelectric generation system," *Renewable Energy*, vol. 97, pp. 306–318, nov 2016.

[13] C. Lu, V. Raghunathan, and K. Roy, "Efficient design of micro–scale energy harvesting systems," *IEEE Journal on Emerging and Selected Topics in Circuits and Systems*, vol. 1, no. 3, pp. 254–266, 2011.

[14] J. W. Matiko, N. J. Grabham, S. P. Beeby, and M. J. Tudor, "Review of the application of energy harvesting in buildings," *Measurement Science and Technology*, vol. 25, no. 1, p. 012002, nov 2013.

[15] A. Cadei, A. Dionisi, E. Sardini, and M. Serpelloni, "Kinetic and thermal energy harvesters for implantable medical devices and biomedical autonomous sensors," *Measurement Science and Technology*, vol. 25, no. 1, p. 012003, nov 2013.

[16] X. Liu, C. Li, Y. Deng, and C. Su, "An energy-harvesting system using thermoelectric power generation for automotive application," *International Journal of Electrical Power & Energy Systems*, vol. 67, pp. 510–516, may 2015.

[17] P. Mehne, F. Lickert, E. Bäumker, M. Kroener, and P. Woias, "Energy-autonomous wireless sensor nodes for automotive applications, powered by thermoelectric energy harvesting," *Journal of Physics: Conference Series*, vol. 773, p. 012041, nov 2016.

[18] M. Nesarajah and G. Frey, "Optimized design of thermoelectric energy harvesting systems for waste heat recovery from exhaust pipes," *Applied Sciences*, vol. 7, no. 6, p. 634, jun 2017.

[19] M. Alhawari, B. Mohammad, H. Saleh, and M. Ismail, *Energy Harvesting for Self-Powered Wearable Devices (Analog Circuits and Signal Processing)*. Springer, 2017.

[20] L. Francioso and C. D. Pascali, *Thermoelectric Energy Harvesting for Powering Wearable Electronics*. Wiley-VCH Verlag GmbH & Co. KGaA, sep 2017, pp. 205–232.

[21] S. Dalola, M. Ferrari, V. Ferrari, M. Guizzetti, D. Marioli, and A. Taroni, "Characterization of thermoelectric modules for powering autonomous sensors," *IEEE Transactions on Instrumentation and Measurement*, vol. 58, no. 1, pp. 99–107, jan 2009.

[22] ANSYS Inc., "Multiphysics simulation," Ansys Inc., 2018. [Online]. Available: https://www.ansys.com/products/platform/multiphysics-simulation

[23] "Ceramic plate series CP08-31-06 thermoelectric modules," Datasheet, Laird, 2016. [Online]. Available: https://assets.lairdtech.com/home/brandworld/files/Laird-ETS-CP-Series-CP08-31-06-Data-Sheet.pdf

[24] "BGA-STD-050 heatsink," Datasheet, ABL Heatsinks, 2016. [Online]. Available: https://www.abl-heatsinks.co.uk/

[25] A. Prijić, Lj. Vračar, D. Vučković, D. Milić, and Z. Prijić, "Thermal energy harvesting wireless sensor node in aluminum core PCB technology," *IEEE Sensors Journal*, vol. 15, pp. 337–345, 2015.

[26] X. Tang, X. Wang, R. Cattley, F. Gu, and A. D. Ball, "Energy harvesting technologies for achieving self-powered wireless sensor networks in machine condition monitoring: A review," *Sensors*, vol. 18, no. 12, p. 4113, Nov. 2018.

[27] M. Guan, K. Wang, D. Xu, and W.-H. Liao, "Design and experimental investigation of a low-voltage thermoelectric energy harvesting system for wireless sensor nodes," *Energy Conversion and Management*, vol. 138, pp. 30–37, apr 2017.

[28] Y. J. Kim, H. M. Gu, C. S. Kim, H. Choi, G. Lee, S. Kim, K. K.Yi, S. G. Lee, and B. J. Cho, "High-performance self-powered wireless sensor node driven by a flexible thermoelectric generator," *Energy*, vol. 162, pp. 526–533, 2018.

[29] D. Milić, A. Prijić, Lj. Vračar, and Z. Prijić, "Characterization of commercial thermoelectric modules for application in energy harvesting wireless sensor nodes," *Applied Thermal Engineering*, vol. 121, pp. 74–82, Jul. 2017.

[30] D. Milić, A. Prijić, Lj. Vračar, and Z. Prijić, "The influence of ambient conditions on the performance of the thermoelectric wireless sensor node," *Facta Universitatis, Series: Working and Living Environmental Protection*, vol. 15, pp. 89–100, 2018.

[31] Z. Prijić, Lj.Vračar, and A. Prijić, "Design and characterization of thermoelectric energy harvesting systems for wireless sensor network nodes," in *Proc. IcETRAN 2018*, 2018, pp. 930–936.

[32] M. O. Cernaianu and A. Gontean, "High-accuracy thermoelectrical module model for energy-harvesting systems," *IET Circuits, Devices & Systems*, vol. 7, no. 3, pp. 114–123, may 2013.

[33] A. Prijić, M. Marjanović, Lj. Vračar, D. Danković, and Z. Prijić, "A steady-state SPICE modeling of the thermoelectric wireless sensor network node," in *Proc. IcETRAN 2017*, 2017, pp. MOI2.3.1–MOI2.3.6.

[34] E. A. Man, E. Schaltz, and L. Rosendahl, "Thermoelectric generator power converter system configurations: A review," in *Proceedings of the 11th European Conference on Thermoelectrics*. Springer International Publishing, 2014, pp. 151–166.

[35] A. Paraskevas and E. Koutroulis, "A simple maximum power point tracker for thermoelectric generators," *Energy Conversion and Management*, vol. 108, pp. 355–365, jan 2016.

[36] M. Guan, K. Wang, Q. Zhu, and W.-H. Liao, "A high efficiency boost converter with MPPT scheme for low voltage thermoelectric energy harvesting," *Journal of Electronic Materials*, vol. 45, no. 11, pp. 5514–5520, jul 2016.

[37] S. Twaha, J. Zhu, Y. Yan, B. Li, and K. Huang, "Performance analysis of thermoelectric generator using dc-dc converter with incremental conductance based maximum power point tracking," *Energy for Sustainable Development*, vol. 37, pp. 86–98, apr 2017.

[38] E.-J. Yoon, J.-T. Park, and C.-G. Yu, "Thermal energy harvesting circuit with maximum power point tracking control for self-powered sensor node applications," *Frontiers of Information Technology & Electronic Engineering*, vol. 19, no. 2, pp. 285–296, apr 2018.

[39] "bq25504 ultra low-power boost converter with battery management for energy harvester applications," Datasheet,

978-1-7281-3420-8/19 $31.00 © 2019 IEEE

Texas Instruments, 2015, sLUSAH0C, Rev. C. [Online]. Available: http://www.ti.com/lit/ds/symlink/bq25504.pdf

[40] "LTC3105 400mA Step-Up DC/DC Converter with Maximum Power Point Control and 250mV Start-Up," Datasheet, Linear Technology Corporation, 2010. [Online]. Available: http://www.linear.com

[41] "LTC3108 ultralow voltage step-up converter and power manager," Datasheet, Linear Technology Corporation, 2010. [Online]. Available: http://www.linear.com

[42] "CY39C831 Ultra Low Voltage Boost PMIC for Solar/Thermal Energy Harvesting," Datasheet, Cypress, Jan. 2019.

[43] "SPV1050 ultralow power energy harvester and battery charger," Datasheet, ST Microelectronics, May 2018. [Online]. Available: www.st.com

[44] "MAX17710 energy-harvesting charger and protector," Datasheet, Maxim integrated, 2012.

[45] D. Dondi, A. Bertacchini, D. Brunelli, L. Larcher, and L. Benini, "Modeling and optimization of a solar energy harvester system for self-powered wireless sensor networks," *IEEE Transactions on Industrial Electronics*, vol. 55, no. 7, pp. 2759–2766, jul 2008.

[46] H. Sharma, A. Haque, and Z. Jaffery, "Modeling and optimisation of a solar energy harvesting system for wireless sensor network nodes," *Journal of Sensor and Actuator Networks*, vol. 7, no. 3, p. 40, sep 2018.

[47] H. Sharma, A. Haque, and Z. A. Jaffery, "Solar energy harvesting wireless sensor network nodes: A survey," *Journal of Renewable and Sustainable Energy*, vol. 10, no. 2, p. 023704, mar 2018.

[48] "I–V characterization of photovoltaic cells using the Model 2450 SourceMeter source measure unit (SMU) instrument," Application Note 3224, 2012, Keithley Instruments.

[49] "IV characterizations of solar cells using the B2900A series of SMUs," Application Note 5990-6660EN, 2015, Keysight Technologies.

[50] O. Lopez-Lapena, M. T. Penella, and M. Gasulla, "A new MPPT method for low-power solar energy harvesting," *IEEE Transactions on Industrial Electronics*, vol. 57, no. 9, pp. 3129–3138, sep 2010.

[51] S. Kim, K.-S. No, and P. H. Chou, "Design and performance analysis of supercapacitor charging circuits for wireless sensor nodes," *IEEE Journal on Emerging and Selected Topics in Circuits and Systems*, vol. 1, no. 3, pp. 391–402, sep 2011.

[52] E. Dallago, A. L. Barnabei, A. Liberale, P. Malcovati, and G. Venchi, "An interface circuit for low-voltage low-current energy harvesting systems," *IEEE Transactions on Power Electronics*, vol. 30, no. 3, pp. 1411–1420, mar 2015.

[53] A. Omairi, Z. H. Ismail, K. A. Danapalasingam, and M. Ibrahim, "Power harvesting in wireless sensor networks and its adaptation with maximum power point tracking: Current technology and future directions," *IEEE Internet of Things Journal*, vol. 4, no. 6, pp. 2104–2115, dec 2017.

[54] S. Senivasan, M. Drieberg, B. S. M. Singh, P. Sebastian, and L. H. Hiung, "An MPPT micro solar energy harvester for wireless sensor networks," in *2017 IEEE 13th International Colloquium on Signal Processing & its Applications (CSPA)*. IEEE, mar 2017.

[55] B. Pozo, J. Garate, J. Araujo, and S. Ferreiro, "Photovoltaic energy harvesting system adapted for different environmental operation conditions: Analysis, modeling, simulation and selection of devices," *Sensors*, vol. 19, no. 7, p. 1578, apr 2019.

[56] A. S. Weddell, G. V. Merrett, and B. M. Al-Hashimi, "Photovoltaic sample-and-hold circuit enabling MPPT indoors for low-power systems," *IEEE Transactions on Circuits and Systems I: Regular Papers*, vol. 59, no. 6, pp. 1196–1204, jun 2012.

[57] Lj. Vračar, A. Prijić, D. Nešić, S. Đević, and Z. Prijić, "Photovoltaic energy harvesting wireless sensor node for telemetry applications optimized for low illumination levels," *Electronics*, vol. 5, no. 4, p. 26, jun 2016.

[58] Lj. Vračar, M. Stojanović, A. Stanimirović, and Z. Prijić, "Influence of encryption algorithms on power consumption in energy harvesting systems," *Journal of Sensors*, vol. 2019, 2019, article ID 8520562.

[59] *MiWi Quick Start Guide*, Microchip Technology Inc., 2019.

[60] A. Prijić, Lj. Vračar, Z. Pavlović, Lj. Kostić, and Z. Prijić, "The effect of flat panel reflectors on photovoltaic energy harvesting in wireless sensor nodes under low illumination levels," *IEEE Sensors Journal*, vol. 15, no. 12, pp. 7105–7111, dec 2015.

Hardware/Software Co-Design of Wireless LAN Transceiver: A Case Study

Z. Stamenković, K. Tittelbach-Helmrich, M. Krstić, M. Stojčev, and B. Dimitrijević

Abstract - The paper tackles an extremely important field of hardware/software co-design of wireless communication systems, which makes system and circuit designers aware of the physical implications and limitations, as well as technologists and physicists capable of coping with the system and circuit requirements in terms of power, speed, and data throughput.

I. INTRODUCTION

Mid-range and low-power wireless communications can be used to network the electronic devices like smartphones, laptops, printers, and TV sets. These wireless local area networks allow users to wirelessly connect (within close proximity) to another transceiver, that may be a router to the internet. To prevent misuse, in many cases a user authentication technique is implemented, otherwise the transceiver (router) can be freely accessed.

Wireless LAN defined by the IEEE802.11 standard [1] provides high-speed data transfer within a small region where users move from place to place. Wireless devices that access these networks are typically stationary or moving at pedestrian speeds. All wireless LANs use one of the ISM frequency bands. A license is not required to operate in these bands. WLAN can have either a star architecture, with an access point acting as central controller, or a peer-to-peer architecture, where wireless terminals self-configure into a network. These networks can reach a coverage range of a few hundred meters. We focus on WLAN since it is one of the most important technologies of computer networking, connects users to the Internet, and can be used for free in industry halls, homes, restaurants, airports, and many other places. WLAN is a technology that can help make all connections seamless and provides a high level of security and privacy.

Z. Stamenković and K. Tittelbach-Helmrich are with the IHP - Leibniz-Institut für innovative Mikroelektronik, Im Technologiepark 25, 15236 Frankfurt (Oder), Germany, E-mail: {stamenko, tittelbach}@ihp-microelectronics.com

M. Krstić is with the IHP - Leibniz-Institut für innovative Mikroelektronik, Im Technologiepark 25, 15236 Frankfurt (Oder), Germany and the University of Potsdam, Potsdam, Germany, E-mail: krstic@ihp-microelectronics.com

M. Stojčev and B. Dimitrijević are with the Faculty of Electronic Engineering, University of Niš, Aleksandra Medvedeva 14, 18000 Niš, Serbia, E-mail: {mile.stojcev, bojan.dimitrijevic} @elfak.ni.ac.rs

II. TRANSCEIVER DESIGN

WLAN is a network in which mobile devices can communicate wirelessly (over a radio link). Each mobile device is equipped with a transmitter and a receiver (in a single word, transceiver) to provide communication at mid-range distances (typically few 100 meters). The IEEE802.11 standard [1] specifies the two lowest OSI model layers of WLAN: Medium Access Control (MAC) sublayer and Physical layer (PHY). The MAC is a set of rules determining how to access the medium and send the data, while the details of transmission and reception are left to the PHY.

A. Medium Access Control Sublayer

Most WLANs operate in the infrastructure Basic Service Set (BSS) mode. In this case, all WLAN stations communicate via single wireless hops with a central entity denoted as the Access Point (AP). Normally, the AP acts as gateway to the wired communication infrastructure (Internet), so that WLAN stations can connect to peer nodes all over the world. Alternatively, WLAN nodes can form an *ad-hoc* network (independent basic service set, IBSS). This mode operates without central controller (controlling the access to the network cell, the traffic in the cell, and so on) and normally without gateway to the infrastructure.

WLAN stations sense the wireless medium before transmitting a frame using two Listen-Before-Talk (LBT) mechanisms: a Physical Carrier Sense (P-CS) and a Virtual Carrier Sense (V-CS). The first is a part of the physical layer, while the second is a MAC sublayer function. Figure 1 shows the hierarchical position of these mechanisms. If either of them indicates busy medium conditions, the station does not attempt to transmit. Once the wireless medium is idle, a station may send a frame. To avoid multiple frames being transmitted simultaneously, WLAN implements the Collision Avoidance (CA): stations need to wait for a certain random period of time before accessing the wireless medium.

P-CS is a means of the Clear Channel Assessment (CCA). It detects transmissions of similar systems: for any valid Physical Layer Convergence Protocol (PLCP) header recognized, P-CS indicates a busy wireless medium for the duration of that frame. In contrast, the Energy Detection (ED) informs on medium usage by dissimilar physical layers or pure interference. ED indicates medium usage independent of the modulation and shape of a signal.

978-1-7281-3420-8/19 $31.00 © 2019 IEEE

V-CS informs stations about ongoing or planned transmissions, e.g. expected acknowledgement frames or more fragments belonging to the same frame. All stations that are not in power-save mode, constantly monitor the wireless medium and retrieve reservation information from any frame they can decode. Frames provide the reservation information in the MAC header. If the duration field is present, stations set a Network Allocation Vector (NAV) to the respective value. The NAV works as a count-down timer. As long as the timer sets a value different than zero, V-CS indicates a busy medium.

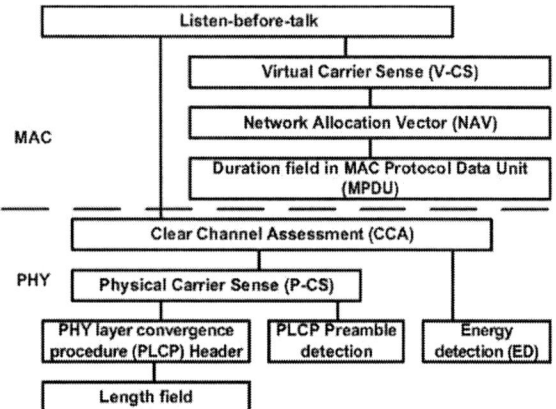

Fig. 1. Block architecture of WLAN carrier sensing.

Since the WLAN transceiver transmits and receives at the same RF channel, but the received wireless signal strength is much smaller than the emitted one, it cannot transmit and receive concurrently and cannot detect collisions. It has to rely on a collision avoidance (CA) procedure waiting for a random time period before transmitting. Two basic parameters are defined (their durations depend on the PHY layer technology):

- Short Inter-frame Space (SIFS), the minimum time gap between two transmissions on the medium.
- Time slot which defines the period needed for ED and result indication to the MAC sublayer.

The SIFS time is the amount of time to elapse between tightly coupled frame transfers, in particular between a frame and the corresponding acknowledgement frame, but also between several fragments of the same (logical) frame. The SIFS time provides sufficient time for change of the transceiver mode (the receiver of a frame switches to the transmit mode and the transmitter turns to the receive mode).

The *Distributed Coordination Function* (DCF) is a basic WLAN coordination function. It implements the Carrier Sense Multiple Access (CSMA) and Collision Avoidance (CA) mechanisms. For every frame trans-mission that fails, DCF doubles the Contention Window (CW) size and draws the new Back-off Time as a random number between 0 and CW. Afterwards, the station may try to send the frame again. After each successful frame transmission, the CW is reset to some (PHY dependent) minimum value. To detect transmission failures, the receiver acknowledges every

successfully received unicast frame. The absence of the acknowledgment frame (ACK) indicates a failure. Multicast/broadcast frames are not acknowledged.

The advantage of DCF is that it does not require a central controller to administer the channel access, nevertheless achieving an acceptable level of collisions. So it can be easily used in *ad-hoc* networks. A centrally controlled access mechanism, however, provides better efficiency, since the overhead of large waiting time due to the back-off procedure can be avoided. The WLAN standard defines a number of coordination functions. The so-called *Point Coordination Function* (PCF) was part of the original IEEE 802.11, but did not gain much popularity. Later, the IEEE802.11e amendment introduced the *Enhanced Distributed Channel Access* (EDCA) and the *Hybrid Coordination Function* (HCF). They may guarantee a certain degree of *Quality-of-Service* (QoS) for the data traffic. The basic principle is that a QoS-aware access point (QAP) assigns transmit opportunities (TXOPs) to the clients in the network, based on their individual QoS requests. This allows reducing the interframe spacing and other dead times of the data traffic in the network.

In WLAN, interference may occur at the receiver side when solely P-CS is used at the transmitting side. Depending on the network topology stations may become mutually hidden. Consider the C station being in reception range of the B station but outside P-CS range of the A station. The latter may transmit a frame to B. However, C cannot detect the transmission of A. Thus, it may also initiate a frame transmission. Depending on the distance to B, the transmission of C may significantly reduce the Signal to Noise Ratio (SNR) of the transmission from A to B. This effect is known as a hidden station problem. To avoid it, WLAN introduces an optional handshake mechanism as follows:

- Based on a manually set threshold and depending on the actual frame size, the A station sends a Request To Send (RTS) frame. Its duration field sets the NAV in stations nearby A to the duration of the intended frame exchange sequence.
- If the B station receives the frame, it replies by a Clear To Send (CTS) frame. The latter sets the NAV of stations in the surroundings of B. Thus, any station that received the RTS and/or CTS frame holds back from medium access.

B. Physical Layer

The original IEEE802.11 standard from 1999 included two physical layers operating in the unlicensed 2.4 GHz RF band: a Frequency-Hopping Spread-Spectrum (FHSS) layer and a Direct-Sequence Spread-Spectrum (DSSS) layer. Both are now completely obsolete. Later revisions (802.11a and 802.11b) added a physical layer for the 5 GHz band based on Orthogonal Frequency Division Multiplexing (OFDM) and a High-Rate Direct-Sequence Spread-Spectrum (HR/DSSS) layer at 2.4 GHz. The 802.11g amendment offers higher speed by the use of

OFDM at 2.4 GHz and backwards compatibility with 802.11b. All these modes offer data rates up to a few 10 Mbps. The 802.11n amendment adds a Multiple-Input Multiple-Output (MIMO) technology, supporting data rates up to a few 100 Mbps. The most recent amendments 802.11ac and 802.11ad specify physical layers reaching data rates of several Gbps and make the 60 GHz band available for high-speed WLAN. In this paper, we focus on the OFDM PHY for the unlicensed 5 GHz band.

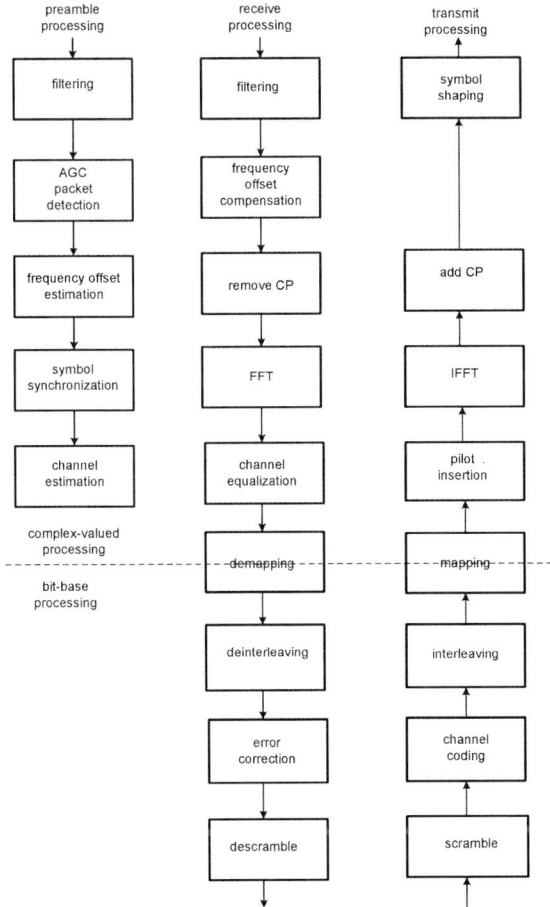

Fig. 2. OFDM digital baseband architecture.

The digital baseband (Figure 2) encodes the original digital data into a form suitable for radio transmission. The digital stream must be converted into the constellation of amplitudes and phases of the in-phase (I) and quadrature (Q) analog signal components that represents the digital signal. To increase the capacity of a radio channel, multiple baseband signals can be multiplexed over the same channel using different techniques: frequency domain multiplexing, time domain multiplexing, code division multiplexing, or a combination of those.

A PHY frame normally consists of a frame preamble, a PHY header (PLCP header, physical layer convergence procedure) and the PHY payload (PSDU, PLCP service data unit). The latter transfers the data of higher layers,

namely the MAC header, the MAC payload and usually a check sum (cyclic redundancy check) for error detection. The purpose of the preamble is to synchronize incoming frames at the receiver, i.e. to compensate differences in the RF carrier frequency (and also the baseband frequency) at both stations due to quartz crystal tolerances, and to determine the right point in time to start demodulation. Moreover, the receiver's automatic gain control (AGC) must be given some time to settle at a proper signal amplification. The PHY header usually transfers information on the modulation and coding scheme of the frame, i.e. the data rate, Forward Error Correction (FEC), etc. For variable-length frames like in WLAN, it also specifies the frame length (packet size).

The different versions of IEEE 802.11 WLAN PHY specify quite different modulation schemes [1]. They range from the comparatively simple complementary code keying in the IEEE802.11b standard to very complex MIMO schemes that use many antennas and data streams and require regular training sequences, usually as a part of the frame preamble. The OFDM modulation scheme has been specified in many variants in the IEEE 802.11a (5 GHz), 802.11g (2.4 GHz), 802.11n and 802.11ac (both 5 GHz MIMO) standards and as an option also in 802.11ad (60 GHz) standard. It is a flexible/adaptable modulation scheme that has turned out to be very robust in multi-path RF channels. Basically, the available RF band is split into a number of equally spaced sub-bands. In each sub-band, data are transferred on a RF sub-carrier (Figure 3). The IEEE802.11a standard, for instance, divides the band into 64 sub-bands, but only 52 of them are used for data transfer. The IEEE802.11n and 802.11ac define different modes that use up to 468 sub-carriers, resulting in a wide range of supported PHY data rates. Each sub-carrier may be individually modulated in BPSK, QPSK or some QAM, depending on the Signal to Noise Ratio (SNR) in that sub-band. All sub-carriers are modulated and demodulated jointly using the Fast Fourier Transform (FFT). This guarantees orthogonality of the sub-bands, i.e. they do not disturb each other. Advanced digital basebands are able to handle the numerical complexity of FFT.

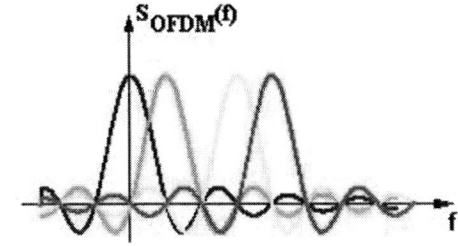

Fig. 3. OFDM modulation in frequency domain.

III. MODELING

A system level model has to capture the intent of the design and is the definition of what the design shall do. It allows designers to understand the implications of their

978-1-7281-3420-8/19 $31.00 © 2019 IEEE

choices and decisions. The model should describe system behavior and performance, define a system architecture and a framework for implementation, as well as verify all the requirements regarding performance, power consumption, cost, maintenance effort, etc. Using the developed model, the designer can consider (and compare) different design styles and do hardware/software system partitioning. There is a plenty of dedicated high-level modeling tools that support the system level design languages like SDL, SystemC, and Matlab/Simulink. These software packages include also automatic generators for hardware description language (HDL) code (for hardware implementations) and high-level programming language code (most often, C language; for software implementations). Criteria for partitioning the system into hardware and software components are usually based on the required performance (data rate and latency), functionality (simple bit operations of the data encryption, error detection and correction, and baseband processing need hardware implementations), and flexibility (reprogramming of a hardware implementation is not or hardly possible).

A. MAC Model

The first step was the creation of a detailed simulation model of the IEEE802.11a MAC protocol using the Specification and Description Language (SDL). Also, languages like SystemC would be well suited. The model was used to verify functional correctness of the initial MAC design [2]. It also served as a base for performance investigations to identify which parts of the protocol can be implemented in software (C code) and which need hardware implementation (HDL code) in order to meet the real-time requirements. The top-level structure of the model is shown in Figure 4.

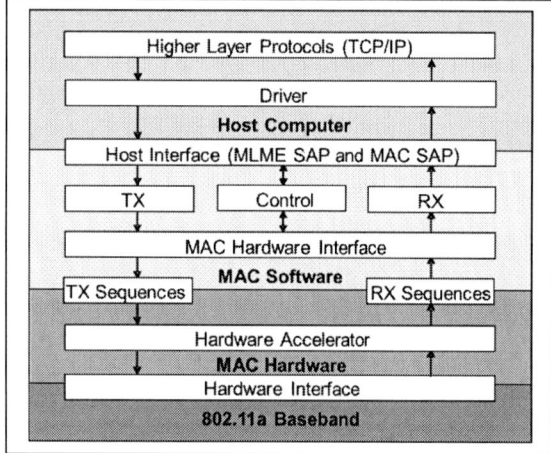

Fig. 4. Hardware/software partitioning of transceiver.

The following control blocks having low data rate, complex functionality, or potential for later extension and modification were implemented in software:
- Connection setup between access point and mobile clients including administration of power-save modes,
- Authentication/deauthentication,

- Association/disassociation, and
- Transmit and receive data queues.

The architectural blocks necessary for transport of payload data with high throughput and relatively simple functionality were implemented in hardware:
- Cyclic redundancy check (CRC) and RC4-encryption,
- Acknowledgements and retransmissions.

The hardware accelerator functions (Figure 5) include the hardware driver for the MAC-PHY interface, retrieval of Rx frame data from the physical layer byte by byte, address filtering and CRC check, and storage of the data by means of direct memory access. Moreover, the accelerator retrieves Tx frame data from a memory location, calculates and appends the check sum, and pushes the data to the physical layer. It indicates a successful reception or transmission of a frame by an interrupt, handles received and transmitted beacons with respect to timing in the WLAN cell, and extracts information on channel time allocations. The hardware accelerator also manages acknowledgments, RTS, and CTS frames. Finally, the hardware accelerator conducts the backoff procedure in the contention access period, and calculates the actual duration of a frame transmission based on its payload length and data rate.

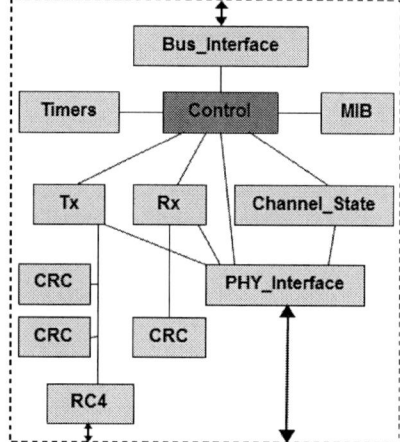

Fig. 5. MAC hardware accelerator.

In a RF-MIMO environment, additional functionality is needed:
- Determination of the optimal antenna weight coefficients for each connection by transmission and reception of the training frames or training symbols,
- Setting the appropriate weight coefficients before transmitting or receiving a frame, and
- Exchanging a short frame sequence RTS/CTS before any data frame transfer.

A single RTS frame is sent before each multicast transmission, while an RTS/CTS pair is generated before each unicast transmission. The RTS/CTS frames allow setting the optimal weights before the main frame transfer starts. They also define the duration of the following frame and acknowledgment (V-CS). The duration information from RTS/CTS frames is used to indicate the busy state until the end of frame transmission.

B. Digital Baseband Model

The digital baseband including MIMO algorithms was developed as a Matlab model composed of functions and scripts specifying the main model parameters. The most important features of the baseband Matlab model are the following:

- It implements a channel estimation, which includes the sequential transmission of 16 and 4 OFDM training symbols and applies the filtered least squares channel algorithm at both sides of the link.
- It implements a Min-MSE algorithm to obtain the beamformers from the rotated channel. The obtained rotated weights are finally corrected in order to obtain the true beamformers.
- It includes a frequency offset estimation block.
- It also corrects the quantization errors of RF weights, which are also affected by a random noise.
- The model includes a least squares estimation of the equivalent SISO channel after fixing Tx/Rx weights.
- It provides transmission rates defined in the 802.11a standard and a soft Viterbi decoder.
- The model can be used to do Monte Carlo simulations and obtain the bit and packet error rates.

A block diagram of the baseband Matlab model is shown in Figure 6. The orange blocks denote the main scripts, yellow blocks are specific MIMO operations, and green blocks represent the conventional IEEE802.11a processing steps. On the other hand, grey blocks denote simulation and measurement operations. Finally, the Matlab functions of each process step are shown in blue.

The described Matlab model uses floating-point operations to implement the baseband blocks. Floating-point values are of a large dynamic range and high precision but require more hardware resources and energy compared to fixed-point values. Therefore, the floating-point toolbox of Matlab can be helpful to obtain an upper bound on the expected performance of digital baseband, but cannot be used for generating a hardware model.

A fixed-point Matlab model was developed to emulate the baseband signals and simulate the fixed-point hardware operations. The main properties of fixed-point variables are signedness (in the proposed model they are signed) and precision (the word and fraction lengths are in bits). Other properties associated to the fixed-point operations are the round mode ('floor' or 'nearest') and the overflow mode ('wrap' or 'saturate'). All of them can easily be changed and evaluated using the developed model. This fixed-point model is bit-to-bit identical to the final hardware model.

A detailed block diagram of the digital baseband model [3] is shown in Figure 7. It is composed of two parts: baseband modules implementing the IEEE802.11a standard and baseband modules implementing new functionalities required by the MIMO analog front-end. The new functiona-lities are grouped into two main modules: channel estimator and beamforming block. These modules are active only when a MIMO training frame is detected by the Tx/Rx control block, which transfers the

signal field data to the MIMO control block in order to start the channel estimation and beamforming. The new digital baseband modules are described below.

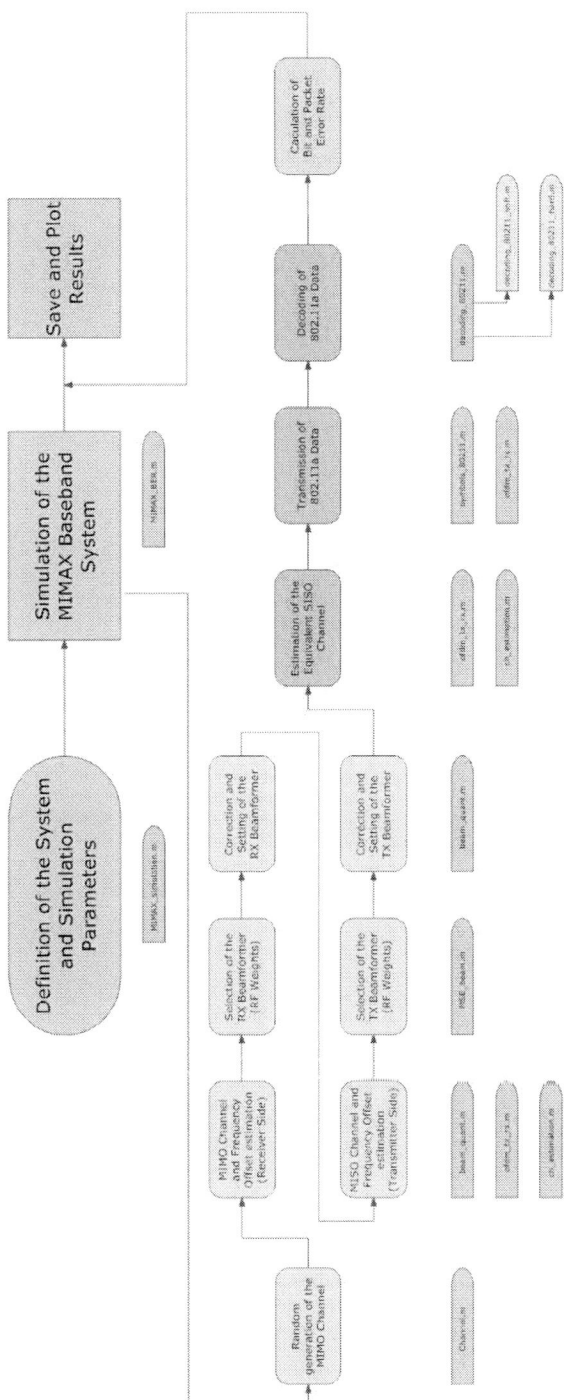

Fig. 6. Block diagram of baseband Matlab model.

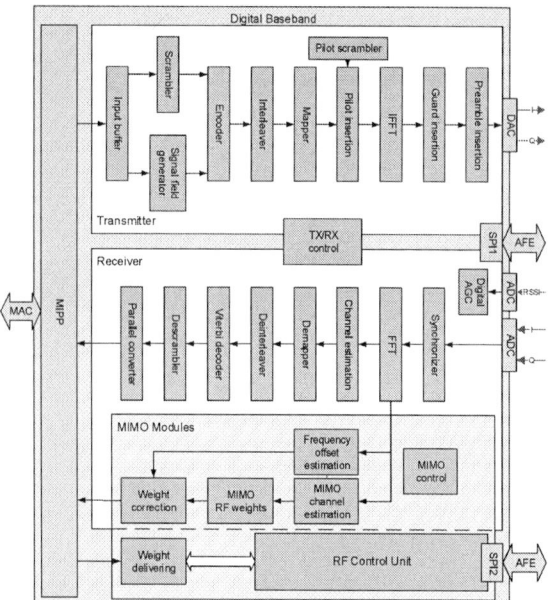

Fig. 7. OFDM MIMO baseband model.

Channel estimation: This module estimates the $n_T n_R$ MIMO channel based on the $n_T n_R$ training OFDM symbols of the received training frame. It works in the frequency domain taking the FFT signal provided by IEEE802.11a modules as input and uses a least squares estimation method.

Beamforming: It takes the estimated MIMO channel as input and computes the optimal Tx/Rx RF weights using the Max-SNR algorithm. It is the most important block in terms of complexity.

Frequency offset estimation: Due to residual frequency error at the output of the conventional IEEE802.11a synchronizer, it might be necessary to include a frequency offset estimator working in parallel with the channel estimation and RF weights modules. To estimate the frequency offset, it is necessary to transmit an additional training symbol, resulting in a training frame of $n_T n_R + 1$ training symbols.

Weight correction: This module multiplies the weights by a unitary (i.e. rotation) matrix in order to compensate the effects of the residual frequency offset and specific Tx/Rx beamformers used during training.

Weight delivery: It transfers the calculated optimal weights to the MAC (weight updating). In addition, it allows applying (from the baseband) the predefined set of weights during training (weight setting) and transferring (from the MAC) the optimal or default weights during data transmission or reception (weight uploading).

MIMO control: This module controls the signal and data flow among all MIMO blocks. It receives from the Tx/Rx control block information included in the training frame signal field (the number of Tx/Rx antennas, the number of training symbols), as well as activation and synchronization signals.

RF control unit: This is a control interface between the digital baseband and analogue front-end. It is an integrated part of the baseband.

All the MIMO blocks are activated only when a training frame is received. Therefore, they can be powered down while either processing conventional data frames or transmitting training frames. Only the MIMO control block, weight delivery block, and RF control unit remain active at any time because it must transfer and set the weights from the MAC to the RF control unit.

C. Analog Front-End Model

Figure 8 depicts the analog front-end, which is capable of performing the intended OFDM operations for the IEEE802.11a standard. The incoming standard conform signals from the antennas have a signal power of -82 to -30 dBm. They are filtered by a bandpass filter (BPF) and amplified by a low noise amplifier (LNA) before they are eventually lead into the weighting module, namely the vector modulator. This module alters phase and amplitude of the incoming RF signals in a way, so that after adding them up, the sum signal has an improved signal quality leading to higher data rates and/or higher link quality [4]. The receiver is fully compatible to IEEE802.11a and can operate as a Multiple-Input Multiple-Output (MIMO) or a Single-Input Single-Output (SISO) device.

The following down-conversion path therefore has to handle an effective dynamic range increased by approximately 12 dB. Thus, a passive down-conversion mixer is employed driven by a phase locked loop (PLL), which controls an 11 GHz voltage controlled oscillator (VCO). The last analogue stages incorporate variable gain and low pass filtering features, in order to set up optimal signal conditions for the receive analogue to digital converter (ADC). The transmit path (TX) makes use of components from the RX such as parts of the baseband (BB), variable gain amplifier (VGA), and filter. Furthermore, the same weighting module is used. The active up-conversion mixer driven by the PLL feeds its output signal to the individual TX branches by means of splitting amplifiers. The power amplifiers (PA) and antenna switch form the last part of the TX branch.

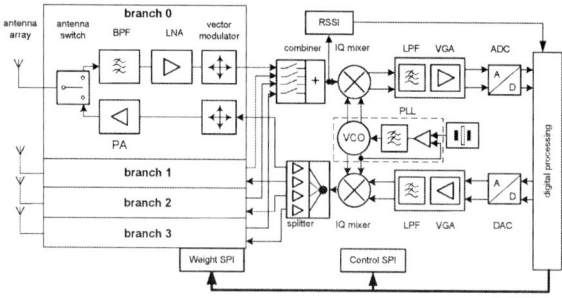

Fig. 8. Model of OFDM MIMO analog front-end.

For optimal signal processing in RF domain [5], a fully differential topology is employed for every building block. This reduces the risk and design time when assembling complex systems, since ground and supply loops are virtually eliminated.

IV. IMPLEMENTATION

The different parts of the proposed MIMO transceiver were implemented and integrated into separate subsystems and a system demonstrator for verification and testing. These subsystems comprise the MAC processor, the baseband processor with the RF control unit, and the analog transceiver (with separated transmitter and receiver chips).

A. MAC Processor

The MAC processor consists of a 32-bit MIPS 4KEp processor and a MIMO specific hardware accelerator [6]. The hardware accelerator functionality for transmit direction includes a buffer for the next frame, generation of CRC codes, and encryption. In the receive direction, the CRC check, a decryption module, a frame address filter, and generation of acknowledgements and CTS frames are implemented in hardware. The tracking channel state (busy/idle) including the back-off for sending frames, 16 timers, system time, and several interfaces are provided as hardware modules. The interfaces include a 16 bit parallel port to the physical layer, a CardBus interface to a host personal computer, a serial RS232 interface for MAC firmware download, and general purpose I/Os (GPIO). A simplified hardware architecture of the MAC processor is shown in Figure 9.

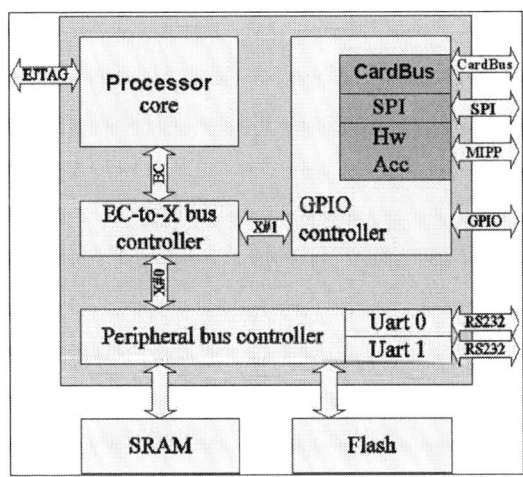

Fig. 9. Hardware components of the MAC processor.

Figure 10 represents the hardware accelerator itself. The interface to the MIPS processor core consists of data bus, address buss and some control signals. There is set of instructions for the hardware accelerator implemented in MAC software. Access to specific modules is provided by the address decoder. The status register collects any relevant information about processes in other modules and

thus allows communication with MAC software. The transmitter module provides functionality for the transmit direction and collision avoidance. The receiver fulfills its natural function-ality described earlier. The MIPP control component is a broker managing the MAC access to the PHY via the MIPP interface, which has been implemented as a 16-bit parallel port (similar to the EPP and IEEE 1284 standard). All components accessing the MAC-PHY interface are under the authority of an arbiter block.

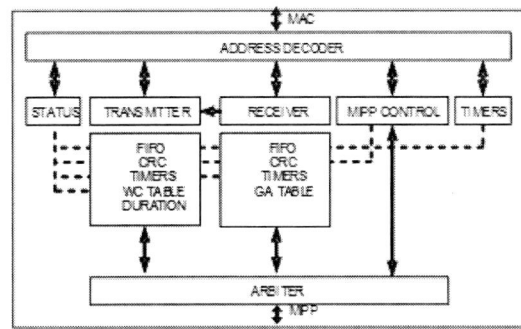

Fig. 10. Implementation of the hardware accelerator.

The MAC processor was fabricated in the IHP's 250 nm CMOS technology. The maximum operating frequency is 80 MHz. The processor including caches and hardware accelerator take most of the chip area. The hardware accelerator uses about 3 kbyte of memory (5x512 byte single-port RAM and 2x256 byte dual-port RAM). The total power dissipation of the MAC processor chip is 450 mW at a frequency of 80 MHz. It occupies a silicon area of 45 mm^2 and integrates about 1.7 million transistors.

B. Digital Baseband Processor

In order to achieve low power dissipation and optimize silicon usage, the baseband processor was divided in two principle blocks: transmitter and receiver. According to this division, the digital baseband processing includes two almost independent dataflow directions: transmit and receive. The transmitter consists of an input buffer, scrambler, signal field generator, encoder, interleaver, mapper, 64-point IFFT/FFT, as well as circuitry for pilot insertion with a pilot scrambler, guard interval insertion, and preamble insertion (see Figure 2). The IFFT/FFT is a single block used in both, receive and transmit directions in order to optimize the baseband processing. On the other hand, this solution is more complex for hardware implementation because of incomplete decoupling between the transmitter and receiver data streams. The IEEE802.11a standard defines the procedure for receiver and transmitter data processing. However, the two fundamental receiver issues (synchronization and channel estimation) are not defined by the standard. Synchronization includes the following operations: frame detection, carrier frequency offset estimation and correction (by a CORDIC processor), symbol timing estimation, extraction of the reference channel, and data reordering [7].

Channel estimation is based on a decision-directed method (and a feedback loop) with a simplified residual phase estimation and correction mechanism. Therefore, the receiver involves additional encoding, interleaving, and mapping. This concept makes use of a division unit to correct the data samples (equalizer). The receiver blocks defined by the standard are demapper, deinterleaver, descrambler, Viterbi decoder, and some buffers. In order to simplify processing of data and reduce energy consumption, the processor circuitry was divided into two domains with different clock frequencies. Computationally complex blocks of low data rates were designed to operate at 20 MHz and high data rate circuits were designed to operate at 80 MHz.

This version of the baseband processor supports up to four transmitter/receiver antennas [6]. It is reconfigurable by using the information provided in the signal field of a data frame and can operate with a smaller number of antennas. The complete reconfigurable baseband processor including conventional IEEE802.11a modules and MIMO modules (channel estimation, frequency offset estimation, and RF weights calculation/uploading) was implemented in a single FPGA chip (Figure 11). The Xilinx System Generator was used for FPGA synthesis.

Fig. 11. PCB with baseband processor FPGA.

C. Analog Front-End Transceiver

The implemented analogue front-end makes use of the channel spatial diversity and supports the transmitted and received signal weighting in the RF domain. The transceiver operates compatibly to the requirements of the described baseband and MAC processors. It is fabricated in the IHP's 250 nm SiGe BiCMOS technology. Photos of the transmitter and receiver chips are shown in Figure 12.

Because of separation of the transceiver into two chips, each of them needs its own PLL (in case of a single chip solution, a single PLL would be sufficient). Some components are identical in both chips: 50-Ω buffers, mixer core, PLL, control SPI, weight SPI, and vector modulator.

The following components are integrated in the transmitter chip: tunable filter for I and Q components with programmable gain, VCO with programmable gain mixer drivers and divide-by-2 circuit, I/Q-up-conversion mixer, 1-to-4 active signal splitter, four vector modulators for adjusting the steering vector of four branches, and four

power amplifiers to provide sufficient output power and matching properties. It occupies a silicon area of 7.5 mm² and dissipates power of 720 mW.

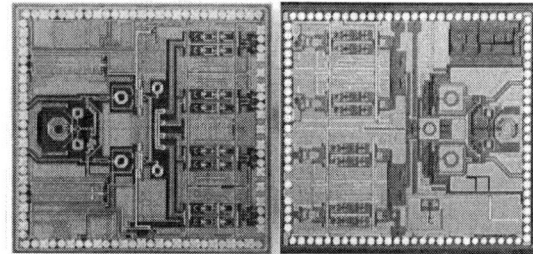

Fig. 12. Transmitter (left) and receiver (right) chips [6].

The receiver chip integrates the following components: 2 single ended LNAs, 2 differential LNAs, 4 broadband vector modulators, RF combiner, passive down-conversion mixer, voltage controlled oscillator, baseband amplifiers and aliasing filters, high-Q baseband filter, power detector circuit (RSSI), and 2 SPIs for control. It occupies a silicon area of 7.5 mm² and dissipates power of 350 mW.

V. CONCLUSION

Hardware/software co-design (standard specifications, design, modeling, and implementation) was thoroughly discussed on the example of an OFDM MIMO WLAN transceiver.

REFERENCES

[1] IEEE Standard for Local and Metropolitan Area Networks - Part 11: Wireless LAN Medium Access Control (MAC) and Physical Layer (PHY)

[2] Z. Stamenkovic, E. Miletic, M. Obrknezev, and K. Tittelbach-Helmrich, "MAC Protocol Implementation in RF-MIMO WLAN", *Proc. 16th IEEE International Conference on Electronics, Circuits, and Systems*, Yasmine Hammamet (Tunisia) 2009, pp. 303-306

[3] V. Elvira, J. Ibanez, I. Santamaria, M. Krstic, K. Tittelbach-Helmrich, and Z. Stamenkovic, "Baseband Processor for RF-MIMO WLAN", *Proc. 17th IEEE International Conference on Electronics, Circuits, and Systems*, Athens (Greece) 2010, pp. 798-801

[4] F. Gholam, J. Via, I. Santamaria, M. Wickert, and R. Eickhoff, "Simplified Architectures for Analogue Antenna Combining," *Proc. 18th ICT MobileSummit*, Santander (Spain) 2009.

[5] R. Eickhoff, R. Kraemer, I. Santamaria, and L. Gonzalez, "Developing Energy-Efficient MIMO Radios", *IEEE Vehicular Technology Magazine*, vol. 4, 2009, pp. 34-41

[6] Z. Stamenkovic, K. Tittelbach-Helmrich, M. Krstic, J. Ibanez, V. Elvira, and I. Santamaria, "MAC and Baseband Processors for RF-MIMO WLAN", *EURASIP Journal on Wireless Communications and Networking*, vol.1/207, 2011, pp.1-13

[7] M. Krstic, A. Troya, K. Maharatna, and E. Grass, "Optimized Low-Power Synchronizer Design for the IEEE 802.11a Standard", *Proc. IEEE Int. Conference on Acoustics, Speech, and Signal Processing*, Hong Kong (China) 2003, pp. 333-336

DEVICE PHYSICS, TECHNOLOGY AND CHARACTERIZATION

978-1-7281-3420-8/19 $31.00 © 2019 IEEE

On the Ballistic Transport in Si Nano-Devices

G. Golan, M. Azoulay and J. Bernstein

Abstract - The field of reliability physics for microelectronic devices is facing a significant challenge during the last decade due to the fact that the technology node (channel length and gate oxide size) has been reduced to the dimensions of charged carriers "mean free path" of silicon, which is about 15 nm at room temperature. This may involve the physical mechanism of ballistic conductance at the channel and at the gate of a MOSFET device. Recently, reliability research programs have made significant progress and developed new theoretical models and experimental methods that would fit the down scaling trend and explain the wearout mechanisms with regards to an innovative reliability physics approach. The classical, reliability common model, High Temperature Operational Life (HTOL) was found to be limited in its ability to distinguish between the dominating failure mechanisms (HCI, BTI, TDDB, EM) and the reliability physics standard methods, that are assessing the lifetime of a specific structure for just one particular mechanism at a time. More recently, a new model, named Multi Failure Mechanism, MTOL, has been introduced and posed a better understanding of the dominating failure mechanisms under various stressed conditions. Experiments were carried out on advanced technologies FPGA devices of Xilinx 45 and 28 nanometer. The insitu monitored experimental data enabled to calculate the activation energy of various degradation mechanisms, providing a more accurate and realistic prediction of the lifetime and point out on the apparent dominating wearout mechanisms. In this paper we report, for the first time, on a new phenomenon that was observed on 28 nm FPGA devices during their reliability testing. We employed the MTOL model at a temperature range of -60°C up to 160°C. The experimental results for 28 nm were compared to the 45 nm data (reported recently by Bernstein et al.) that have been recorded under identical testing conditions. From the comparison of the normalized degradation rate versus temperature, a clear deviation could be noted; the 28 nm devices have shown a distinct transition of the dominating failure mechanism at a particular temperature, whereas the 45 nm devices have not shown any transition along the entire temperature range of the test. Furthermore, the calculated values could be correlated to the recent published data, attributing the transistor channel conductance to the effect of "short channel ballistic conductance" at a lower temperature range. At higher temperatures (higher than the transition temperature), both 45 and 28 nm devices have shown similar slopes (normalized ring oscillator frequency versus temperature). To the best of our knowledge, such temperature dependence has not been reported up to now. T
his may indicate on a pronounced advantage of the lower node devices (28 nm) for operation at lower temperatures. Nevertheless, our study is ongoing to lower technology nodes (20 nm and 16 nm), which may provide additional data that will support this new hypothesis.

G. Golan, M. Azoulay, J. Bernstein are with the Department of Electrical Engineering, Faculty of Engineering, Ariel University, Ariel 40700, Israel, E-mail: gadygolan@gmail.com

I. INTRODUCTION

The common approach today for assessing microelectronics device reliability is the High Temperature Operating Life (HTOL) test which is based on the assumption that just one dominant failure mechanism is acting on the device [1, 2]. However, in practice, it is well known that multiple failure mechanisms are acting on the device simultaneously [3]. Indeed, the recently developed MTOL model predicts the reliability of electronic components by combining the Failure in Time (FIT) of multiple failure mechanisms [4]. However, the technology down scaling of microelectronic devices to the deep nanoscale dimensions (below 20 nm) is raising new reliability concerns, particularly due to the disproportion between the dimensions reduction to the lowered nominal operating current and voltage [5]. Thus, the degradation mechanisms such as Bias Temperature Instability (BTI), Hot Carrier Injection (HCI) and others, may play different roles in the wearout processes while compared to the standard reliability physics models. The most common degradation mechanisms, BTI and HCI are related to the formation of interface defects between silicon and silicon dioxide, causing the device voltage threshold shift. Here BTI is prominent along the channel and HCI is significant near the drain end [6]. However, as the channel length and the gate oxide size have become to be smaller and reached the size of the "mean free path" of carriers in silicon, it should be considered not just as the classic charge transport phenomena but also according to the "ballistic transport theory" [7]. An explicit fundamental calculation for the mean free path in silicon, results in about 12 nm (if the thermal velocity of a free charged carrier in silicon is taken as 2.5×10^{7} cm/s at room temperature, a channel mobility of 300 $cm^2/V*s$ leads to a relaxation time of 5×10^{-14} seconds and a mean-free path in the order of 12×10^{-7} cm, or 12 nm). Thus, only a few scattering events are expected to occur in a channel length of 20 to 30 nm. This points out that the properties of the carriers in these very small devices will be quite different than those of larger devices [8]. This approach is found to be in good correlation with the experimental results that we provide for the first time in this paper, in which we show the transition point (temperature) for the 28 nm devices, whereas, for the 45 nm devices no transition could be observed. It is therefore reasonable to speculate that the transition temperature may be correlated to the difference between the tested technology nodes. Yet, additional data of further lower technology nodes (20 and 16 nm) are essential in order to present an accurate calculation of the transition temperature at which one may observe the transformation of the dominating degradation mechanism. The rest of the paper is organized as follows: Section II, describes the experimental set up that we have applied for acquiring the data, this includes commercial Xilinx FPGA (45 and 28 nm), a Ring Oscillator circuit design, stressing oven (-60°C – 160°C), an insitu control plus monitoring system and data analysis procedures. Section III, provides the experimental results for the frequency degradation of the ring oscillators that were operated at several parametric ranges of temperatures, bias voltages and ring oscillator lengths (an odd number of CMOS inverters) and a discussion on the most significant and innovative results that lead to our conclusive interpretation, pointing out on an apparent ballistic conduction mechanism in the 28 nm devices at the lower temperature range. Finally, a practical impact is suggested with regards to the wearout mechanism of nano devices from the 28 nm node at low temperatures. Following the above, we continue to examine smaller devices of 20 and 16 nm in order to further verify our hypothesis and be able to extrapolate our results to future technology nodes, smaller than the mean free path of charged carriers in silicon, at the range of 5-10 nm.

II. EXPERIMENTAL METHODS AND SET-UP

The FPGA configuration that was designed for the MHTOL testing consists of the three main parts: (1) Accelerated Element (FPGA Chip); (2) Measurement system (Binary Counter); (3) Control & communication interface to a PC. The experimental setup also consists of two VHDL developments burnt into both Xilinx XC6SLX9 Spartan 6 FPGA (45 nm) on a Mojo® board, and a Zynq-7000 FPGA (28 nm) names Zybo® board, respectively. These devices consist of over 9000 logic cells (LUTs), and the developments were designed to cover the full scope of the device's components. The devices include inputs for reference signals, testing points, and a micro-USB interface. In order to allow various voltage levels, external DC power supplies delivered voltage directly to the FPGA cores. The testing boards have been subjected to high and low temperatures in standard industrial ovens. We have monitored the device temperature internally from a built in sensor and externally by using an IR camera. The detailed description of the new MHTOL experimental set up and the updated results that were employed to calculate the MHTOL matrix parameters and distinguish between the dominating failure mechanism for both 45 nm and 28 nm devices have been published recently elsewhere [9]. The testing System was synthesized and downloaded to the FPGA card. The test conditions were predefined for allowing separation and characterization of the relative contributions of the various failure mechanisms by controlling Voltage, Temperature and Frequency. Extreme core voltages and environmental temperatures, beyond the specifications, were imposed to cause failure acceleration of individual mechanisms to dominate others at each

condition, e.g. sub-zero temperatures, at a very high operating voltages, to exaggerate HCI.

For each test, the FPGA board was placed in a temperature-controlled oven, dedicated to HTOL-testing with an appropriate voltage set at the FPGA core. The board was connected to a computer via USB, and the external clock signal was fed into the chip. The tests performed for 200-500 hours, and while the device was working in the accelerated conditions - the frequencies of 140 Ring-Oscillators of different sizes were measured regularly. Initial sampling started after one working-hour in the accelerated environment, and then samples were taken, frequently at first, and in increasing intervals afterwards. The measurement data were stored in a database from which one could draw statistical information about the degradation in device performance.

Aside from the internal sampling of the data, external measurements were also taken with a 1 GHz bandwidth oscilloscope, and an eight-digit frequency counter capable of operating up to 1 GHz.

The unique testing conditions for each failure mechanism allowed us to examine its specific effect of that mechanism on the system and thus define its unique physical characteristics. A close inspection of test results yielded more precise parameters for the Acceleration Factors (AF) equations and allowed adjusting them to the device under tested.

Finally, after completing the tests, some of the experiments with different frequency, voltage and temperature conditions were chosen to construct the M-HTOL Matrix.

III. RESULTS AND DISCUSSION

The standard reliability model for failure mechanisms in semiconductor devices is the JEDEC Solid State Technology Association and it is listed in publication JEP-122G. Our previous experimental results showed unique effects of each failure mechanism in all experiments conducted. An example of this presenting the test results under high voltage (3V) and low temperature (-35°C) showed the level of degradation in performance during the experiment (%/Dec) depending on the frequency of the RO. Whereas, at low frequencies the degradation is constant, i.e. independent of frequency, which is similar to the influence of the BTI failure mechanism. In contrast, at higher frequencies, the level of degradation increases linearly with the frequency, similar to the HCI behavior (which has a strong frequency dependency). It is apparent that at high frequencies HCI is more dominant than BTI, thus this range can be referred to as HCI only. Accordingly, lower frequencies can be referred to as BTI only. Under these conditions, EM and TDDB failure mechanisms were not observed. In another experiment, the results showed clearly that the degradation remains constant and independent of frequency, demonstrating what can indicate exclusively the dominant effect of BTI failure mechanism.

978-1-7281-3420-8/19 $31.00 © 2019 IEEE

All these results verify the formulas defining the Acceleration Factors, as they are described above.

In contrast, at the present work we implement the MHTOL new method which is capable of calculating the device useful lifetime by combining the Failure in Time (FIT) of multiple failure mechanisms. A linear matrix solution combines the failure rate of each separate mechanism to calculate the actual FIT of the system based on the physics of degradation at specific operating conditions. Now that we have fully characterized the physics of failure models relating to all three mechanisms for both 45 and 28 nm FPGA's, we were able to build the Matrix Model by choosing three points, one from each mechanism, and then solve the equations against the measured FIT for each condition. The procedure for finding the results of the matrix is described in previous papers. This matrix has been then used to construct the full reliability profile whereby FIT is calculated, versus Temperature for several conditions. Interestingly, there is no apparent frequency effect, neither at high or low temperatures. The most notable difference between 45 and 28 nm is the lack of frequency effect at both low and high temperatures, leaving only one dominant failure mechanism at 28 nm. The consequence seems to be that there will be significantly improved reliability at low temperatures using 28 nm technology. Hence, there is no purpose for frequency de-rating using this technology. It is also clear that cooling the device is the major challenge and most important part of increasing the reliability of devices made using this technology. Another observation is that the voltage acceleration is much greater at 28 nm, being over 17, as compared to 3.8 for 45 nm technology. This means that the core voltage is much more sensitive and a much greater reliability advantage would be gained by lowering the voltage. The temperature effect is exactly the same for both technologies. One speculation as to the reason there are no hot carriers effects in the 28 nm technology is that the mean free path of electrons transported through the gate is much larger than the gate length. This would suggest that the electrons are transported by ballistic means by design, due to the properly strained channels. Hence, electrons are not able to accelerate to the point of causing damage due to HCI. In order to validate the findings detailed above, the cause of this phenomenon is being further researched in our laboratory. We have yet to see if this is valid for smaller size technology, but it seems that this may be an important result justifying a preference to use newer scaled technology in high-reliability applications even over older technologies that may seem to be more mature. There was a significant reliability difference between 45 and 28 nm regarding hot carriers injection (HCI) and Electromigration (EM), whereas BTI seemed to be about the same (aside from the higher voltage acceleration factor). This is probably best explained by the shorter channel alone. It seems that the stress engineering techniques successfully increase the electron and hole mobilities, such that when the temperature is reduced, the electrons may have relatively no collisions

with the oxide interface, hence completely avoiding the HCI trapping. Similarly, the current would be limited due to the inability of the electrons to move fast enough before being absorbed in the Drain. Hence, the current is automatically limited and there is also no EM resulting due to the inherent current limit of the shorter channel.

This hypothesis may apparently be confirmed by our own experimental measurements of ring oscillator (RO) frequency versus temperature shown below in figure 1.

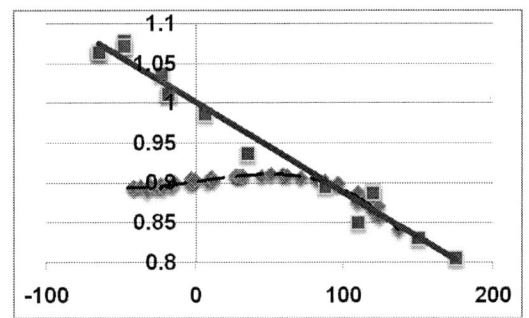

Fig. 1. Comparison of relative RO frequency versus temperature for 28 and 45 nm.

Here, we compare the relative RO frequency for the temperature range of -40°C up to 160°C. By comparing the thermal characteristics of the RO behavior, we see that the 28 nm FPGA has a maximum frequency at about 60°C which corresponds to the temperature at which the mean free path may be about the same as the channel length, 28 nm. Since the structure and materials are about the same as in 45 nm, the same mean free path would be directly in the middle of the 45 nm channel, thus providing for ample hot carriers to be trapped in the gate oxide of the longer channel devices. It is noted that such data have not been published elsewhere, yet we still need to clarify the theoretical hypothesis and further extend the experimental approach with lower technology nodes of 20 nm, 16 nm and lower. A comparison of the traditional CMOS architecture with the newly designed FINFET devices would probably add some more perspectives for understanding the short channel conductance mechanism.

REFERENCES

[1] Xilinx, Device Reliability Report, UG116 (v10.3.1), Sept. 8 2015.

[2] J.B. Bernstein, *Reliability Prediction from Burn-in Data Fit to Reliability Models*, 2014.

[3] J.B. Bernstein, et al., *Physics-of-Failure Based Handbook of Microelectronic Systems*, Reliability Information Analysis Center, Utica, NY, 2008.

[4] J.B. Bernstein, M. Gabbay, O. Delly, "Reliability matrix solution to multiple mechanism prediction", *Microelectronics Reliability*, 2014, 54, pp. 2951–2955.

[5] Ismail Saad, Khairul A. M., Nurmin Bolong, Abu Bakar A.R and Vijay K. Arora, "Computational Analysis of Ballistic

978-1-7281-3420-8/19 $31.00 © 2019 IEEE

Saturation Velocity in Low-Dimensional Nano- MOSFET", *International Journal of Simulation: Systems, Science & Technology*, 2011, Vol. 12, No.3, pp. 1-6.

[6] Yao Wang, Sorin Cotofana and Liang Fang, "A Unified Aging Model of NBTI and HCI Degradation towards Lifetime Reliability Management for Nanoscale MOSFET Circuits", *Proceedings, IEEE/ACM International Symposium on Nanoscale Architectures*, 2011, pp. 175-180.

[7] Vijay K. Arora, Desmond C. Y. Chek and Michael L. P. Tan, "The role of ballistic mobility and saturation velocity in performance evaluation of a nano CMOS circuit", *Proceedings,*

International Conference on Emerging Trends in Electronic and Photonic Devices & Systems, 2009, pp. 14-17.

[8] Peizhen Yang, W.S. Lau1, Seow Wei Lai, V.L. Lo, S.Y. Siah and L. Chan, "The Evolution of Theory on Drain Current Saturation Mechanism of MOSFETs from the Early Days to the Present Day", In: *Solid State Circuits Technologies*, Book edited by: Jacobus W. Swart, January 2010, INTECH, Croatia.

[9] Joseph B. Bernstein, Alain Bensoussan and Emmanuel Bender, "Reliability prediction with MTOL", *Microelectronics Reliability*, 2017, 68, pp. 91–97.

978-1-7281-3420-8/19 $31.00 © 2019 IEEE

Impact of γ Radiation on Charge Trapping Properties of Nanolaminated HfO$_2$/Al$_2$O$_3$ ALD Stacks

D. Spassov, A. Paskaleva, V. Davidović, S. Djorić-Veljković, S. Stanković,
N. Stojadinović, Tz. Ivanov, and T. Stanchev

Abstract – The effect of γ radiation on the charge trapping and oxide properties of MIS capacitors with nanolaminated HfO$_2$/Al$_2$O$_3$ dielectrics are presented. The irradiation with dose of 1 and 10 Mrad generates electron traps thereby substantially enhancing the memory windows of stacks. γ radiation increases the positive oxide charge of the structures, but the effect depends also on the thermal treatment of the stacks. The used doses do not deteriorate the density of interface states, leakage currents and retention characteristics.

I. INTRODUCTION

The charge trapping memories (CTM) are recently considered as one of the most promising potential replacements of the floating gate technology in the non-volatile memories as the two share the same operational principle - transfer of charge from MOSFET channel into and out of a dedicated charge storage layer (poly-Si floating gate and trap-rich dielectric in which charges are kept directly in discrete traps distributed in the dielectric band-gap, respectively). The real progress of CTM is associated with the implementation of high-k dielectrics as a charge storage medium. One of the important reliability issues for each nanoelectronic device is its radiation hardness defining the possibility of operation in a radiation environment (e.g. space navigation, radiology equipment, instrumentation for nuclear energy plants and detectors for high-energy physics experiments) where the devices are subjected to the impact of high-energy particles and/or photons which may cause generation of various electrically active defects, leakage currents, early breakdown or loss of stored information. The radiation response of high-k based MOS devices is far less understood than radiation effects in SiO$_2$ based devices. Some recent studies suggest that high-k oxides, HfO$_2$ and Al$_2$O$_3$ in particular, exhibit better radiation hardness compared to SiO$_2$ [1-3]. In most cases no severe degradation is found in regarding the leakage currents and interface state density [1,2,4] except positive charge buildup during the irradiation. An annealing effect of the radiation to the interface states density was reported [2,5]. Although an increased charge trapping in radiation treated HfO$_2$ and Al$_2$O$_3$-HfO$_2$ dielectrics has been reported [2,6], there are also studies showing a deterioration of the memory windows in charge trapping devices [7] or that memory windows are unaffected by radiation [3].

In this work we study the γ-radiation effects on the electrical characteristics and charge trapping of Al$_2$O$_3$/HfO$_2$ multilayer stacks deposited by atomic layer deposition (ALD). Recently, we have shown that these stacks have a potential for implementation as charge trapping layer in CTM devices and their charge storage ability could be tailored and enhanced by optimization of stack parameters as well as annealing processes [8,9]. The results obtained have given convincing evidence that post deposition annealing (PDA) in O$_2$ enhances substantially electron trapping in deep traps, hence charge storage ability of stacks. Therefore, the influence of oxygen annealing on radiation tolerance of Al$_2$O$_3$/HfO$_2$ stacks is also investigated.

II. EXPERIMENTAL

Nanolaminated Al$_2$O$_3$/HfO$_2$ stacks were deposited in Savannah-100 ALD system on p-type (100) B-doped (6 Ωcm) Si wafers. The stacks were build from 5 bi-layer blocks, each block containing 30 cy HfO$_2$ and 10cy Al$_2$O$_5$ sublayers. The film growth starts with Al$_2$O$_3$ deposition followed by HfO$_2$ one. Trimethylaluminum precursor and tetrakis(dimethylamido)hafnium were as precursors for Al$_2$O$_3$ and HfO$_2$ deposition, respectively and the oxidant was H$_2$O vapor. Both deposition processes were carried out at 135 °C. The overall thickness of the nanolaminated films is 26 nm determined with Woollman M2000D spectral ellipsometer. Some of the stacks received rapid thermal annealing (RTA) at 800°C in O$_2$ for 1 min. Consecutively MIS capacitors with Al top (gate) and backside contacts were defined by photolithography. The capacitors were irradiated with γ-photons from Co[60] at two total doses: 1 and 10 Mrad (Si). The irradiation was carried out without applying external voltage to capacitors. The charge trapping in the stacks was assessed from the capacitance-voltage (*C-V*) characteristics measured at 1 MHz in a dark chamber with Agilent 4980A LCR meter. The leakage currents were obtained with Keithley 236 SMU. Charge trapping properties of stacks were examined by applying

---

D. Spassov, A. Paskaleva, Tz. Ivanov and T. Stanchev are with Institute of Solid State Physics, Bulgarian Academy of Sciences, Tzarigradsko Chaussee 72, Sofia 1734, Bulgaria, E-mail: d.spassov@issp.bas.bg

V. Davidović, and S. Djorić-Veljković are with the Faculty of Electronic Engineering. University of Niš, Aleksandra Medvedeva 14, 18000 Niš, Serbia, E-mail: vojkan.davidovic@elfak.ni.ac.rs

N. Stojadinović is with the Serbian Academy of Sciences and Arts (SASA), Knez Mihailova 35, 11000 Belgrade, Serbia.

S. Stanković is with Institute of Nuclear Sciences "Vinca", University of Belgrade, Mike Petrovića 12-14, Belgrade Serbia.

978-1-7281-3420-8/19 $31.00 © 2019 IEEE

negative and positive square voltage pulses (in respect to the top electrode) with different magnitudes V_p and duration of 1 s and a subsequent C-V measurement was made to obtain the shift of the flat band voltage, V_{fb}.

III. RESULTS

A generation of positive oxide charge Q_{ox} was observed upon the irradiation. Q_{ox} of the non-irradiated stacks is positive, 0.7×10^{12} and 2×10^{12} cm^{-2} for as-grown and annealed structures respectively. The radiation induced modification of Q_{ox} is more pronounced for the as-grown samples. For them the oxide charge increases with the dose and after 10 Mrad γ exposure Q_{ox} is ~1.8×10^{12} cm^{-2}. Q_{ox} of O_2 treated films exhibit only marginal increase (ΔQ_{ox} up to 5×10^{11} cm^{-2}) without a clear dependence on the dose, suggesting that the annealed samples are less susceptive to the γ radiation damage. It should be mentioned that after irradiation Q_{ox} of as-grown samples is similar to that of annealed ones unlike the case before irradiation when the latter exhibit substantially larger Q_{ox} as compared to former ones. Irradiation does not inflict sizable changes in the slope of the C-V curves, implying that substantial interface state generation in a first approximation does not occur. In fact, the estimation of interface states density by Terman's method gives values of D_{it}^m ~ 1.5×10^{12} and 1.8×10^{12} eV^{-1}cm^{-2} at midgap before irradiation for as-grown and O_2 annealed samples. γ treatment slightly increases midgap density of states of the as-grown stacks to 1.8-2×10^{12} eV^{-1}cm^{-2} without a clear dependence of the dose. In case of O_2 treated films D_{it}^m of irradiated samples at both doses are 1.8-1.9×10^{12} eV^{-1}cm^{-2}. The initial C-V curves (measured at sweep voltage range where the charge trapping into the stacks is minimal) exhibit small hysteresis – 34 mV for the as-grown samples and ~8 mV for annealed ones. The hysteresis is attributed to the effect of slow states – traps into the dielectric in tunneling distance form Si. Irradiation with 1 Mrad decreases slightly the hysteresis of the as-grown stacks to ~29 mV, but treatment with 10 Mrad increase it slightly above the initial values (46 mV). The behavior of the O_2 annealed layers is different - a reduction of the hysteresis is observed with the dose increase: the hysteresis is 5mV and 3 mV after 1 and 10 Mrad exposure, respectively. Hence, the results corroborate the reports [1,2,4] of higher radiation hardness of HfO_2 and HfO_2/Al_2O_3 based MOS structures compared to the SiO_2-based ones. The effect of radiation mainly consists in a moderate positive oxide charge generation. The data also suggest in accordance with [1,2,4,5] that low doses of γ radiation can improve the interface properties of high-k/Si system.

In Fig. 1 evolution of C-V curve flat-band voltage at both polarities with respect to its initial position V_{fb0} versus the voltage pulse magnitude is presented in order to get more insight on the different inputs to memory window (electron trapping, hole trapping and/or electric stress-generated defects). The memory window is the difference between the curves measured at two polarities. The as-grown sample

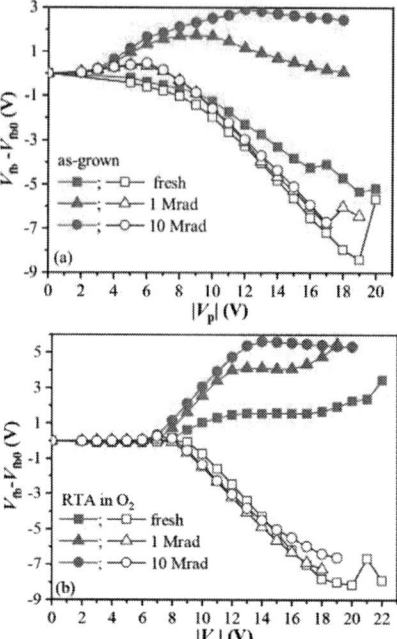

Fig. 1. Flat band voltage shifts of C-V curves as a function of the voltage pulse amplitude, before and after irradiation: (a) as-grown and (b) O_2 treated stacks. Solid symbols correspond to $+V_p$ and hollow ones to $-V_p$.

before irradiation does not reveal any electron trapping (Fig 1a). Such behavior is most likely due to existence of two competing processes – electron trapping at existent traps and stress generation of positive charge; the latter outweighs the electron trapping. As a result, net positive charge build-up is observed even when electrons are injected in the structure. In addition, the positive charge increases progressively with increasing V_p (for both positive and negative pulses) which implies that at least part of the trapped charge is related to stress-induced positively charged defects, i.e. irreversible damage. For sample after O_2 annealing (Fig. 1b) no positive charge trapping is observed up to V_p ~ -9 V; for larger pulse voltage positive charge trapping increases progressively similarly to as-deposited sample. However, unlike the as-deposited stack the annealed one shows stable electron trapping which increases with increasing $+V_p$. In other words, O_2 annealing creates electron traps which give rise to significant memory window (e.g. ΔV =9.5 V and 3.2 V at V_p =$\pm$18V is obtained for annealed and as-grown stacks, respectively). The effects of γ-radiation on the charge trapping phenomena in HfO_2/Al_2O_3 stacks could be summarized as follows (Fig. 1):

1) The positive charge trapping (hollow symbols in Fig. 1) remains nearly unaffected by irradiation. Generally, after irradiation it slightly decreases in comparison to the non-irradiated as-grown sample. For O_2 annealed sample it is slightly larger at V_p = 10-15 V. Nevertheless, the differences are small and it can be concluded that γ-radiation does not create new hole traps in both kind of stacks.

978-1-7281-3420-8/19 $31.00 © 2019 IEEE

2) It is clearly seen that radiation significantly enhances the electron trapping in both the as-deposited and O_2 annealed stacks (full symbols in Fig. 1). For as-deposited stacks after 1 Mrad irradiation, the shift due to electron trapping increases up to about $V_p = 9$ V and then it turns-around and decreases. Such behavior is most likely due to existence of two competing processes – electron trapping at radiation-induced traps and stress generation of positive charge; the latter is more pronounced at higher fields which manifests as a decrease of shift due to electron trapping at higher V_p. The irradiation with higher doses (10 Mrad) generates more electron traps, hence the turn-around effect is weaker and starts at higher V_p. The higher susceptibility of the as-deposited samples to high field stress was established in our previous work [8]. For O_2 annealed sample after initial increase of flat-band voltage shift with increasing V_p a tendency of saturation of electron trapping is observed and the saturation level increases with the radiation dose. This implies that charge trapping occurs in radiation-induced traps and are not generated by high-field stress during the measurement.

3) Thanks to the increased electron trapping both as-deposited and annealed stacks show larger memory window ΔV after irradiation. The values of ΔV measured at $V_p = 15$ V are given in Table 1. As seen, the radiation effect is more pronounced in case of the as-grown films for which due to irradiation noticeable memory windows are developed. The difference between the memory windows of as-deposited and annealed stacks diminishes with the radiation dose. This finding implies that the precursors for oxygen annealing induced traps and radiation-induced traps may be one and the same.

TABLE I
MEMORY WINDOWS OF THE AS-GROWN AND O_2 ANNEALED STACKS OBTAINED WITH $V_p=\pm 15$ V BEFORE AND AFTER IRRADIATION.

Dose	Memory window, ΔV, V	
	As-deposited	O_2 RTA
Before irradiation	1.8	6.8
1 Mrad	5.8	9.7
10 Mrad	7.7	10.62

The influence of the radiation on leakage currents is presented in Fig. 2. No leakage deterioration is detected for investigated stacks and doses in contrast to the moderate increase of J after irradiation reported in [4,6] and supporting [1] where no current deterioration is found. On the contrary, irradiation with 1 Mrad leads to lower J for both type of stacks especially in the low voltage region. The induced changes are more pronounced for the as-grown stacks, while the samples treated in O_2 seem to be weakly affected. For both stack types 10 Mrad irradiation inflicts some noise appearance at high applied V, (most noticeable under negative V). The leakage data also suggests that there is not a straightforward connection between the increase of the positive oxide charge upon γ exposure and leakage currents. This observation in combination with the overall symmetry

of J-V curves in respect to 0V indicates that current in the stacks is dominated by bulk-limited conduction mechanisms. As a first approximation the characteristics for negative V were fitted with a combination of Ohmic conduction and Poole-Frenkel (PF) mechanism:

$$J = \sigma_h E + \sigma_{PF} \exp(\frac{\sqrt{q^3 E / (\pi\varepsilon_0\varepsilon_r)}}{rkT}) \quad (1)$$

where E is the electric field, q – electron charge, ε_r is the dynamic dielectric constant, r – compensation factor, σ_h and σ_{PF} are constants. The negative applied voltage region was chosen because the Si substrate is in accumulation for $V <0$, so entire applied voltage drops across the stack, and the voltage drop in Si can be neglected. As depicted in the inset of Fig. 2a combination of Ohmic and PF conductance describes leakage of as-grown stacks very well for the pristine and 1 Mrad curves. The obtained value of ε_r ~3.6 agrees well with the refractive index of the as-grown stacks [10] and r=1.8 and 1.5 in case of fresh and 1Mrad capacitors, respectively. For the annealed layers slightly higher values of ε_r and r are obtained (ε_r =3.9-4 and r~2). The higher ε_r most likely reflects some increase of the refractive index of stacks after RTA. It should be noted, however, that the fit of the experimental J-V to (1) for annealed stacks is not as good as for the as-grown ones. J for capacitors irradiated with 10 Mrad cannot be fitted with Eq.1 and the determination of the dominant conduction mechanisms require more in depth analysis. Nevertheless,

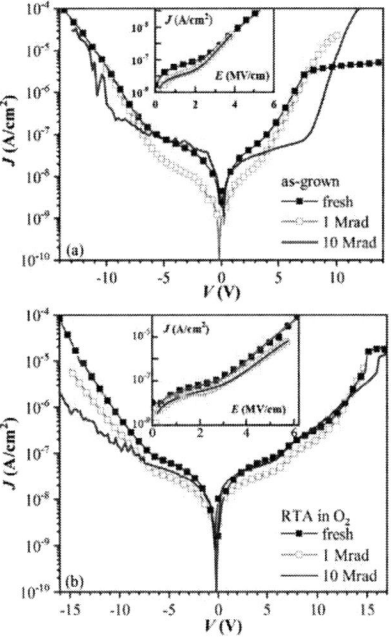

Fig. 2. Leakage currents of as-grown (a) and oxygen annealed stacks before and after irradiation. Insets: trail fit of J-E curves in negative voltage region with a combination of Ohmic and Poole-Frenkel conduction.

the results for the pristine and irradiated with 1Mrad capacitors are consistent with the finding of others [1,6] showing that PF is the main conduction mechanisms in HfO_2 and HfO_2/Al_2O_3 multilayer structures. The more precise study of the conduction in the irradiated stacks requires investigation of temperature dependence of J and will be a subject of another study.

The effect of irradiation on retention characteristics is shown in Fig. 3 for the annealed samples. As in case of the leakage currents, the used doses do not inflict shortening of the retention times. The obtained retention characteristics (V_{fb} vs. t) in case of electron trapping generally agree with $V_{fb} \sim (A - B\ln(1 + C \times t))^2$, with A, B and C - parameters, describing charge loss via PF emission dominating retention in capacitors without dedicated tunnel and blocking sublayers [11]. For the trapped holes V_{fb} seems to be proportional to $\ln(t)$ which is representative for charge loss due to tunneling processes. The different loss mechanisms might be explained with larger depth of the hole traps, making PF emission ineffective.

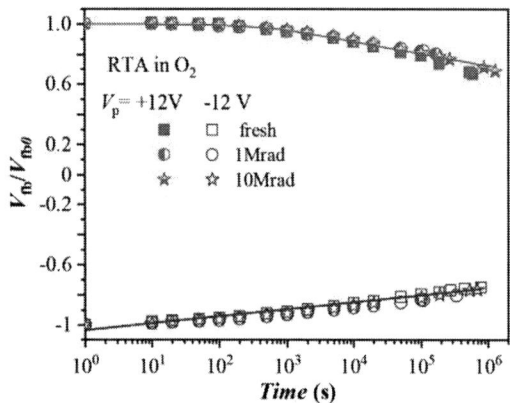

Fig. 3. Retention characteristics of pristine and irradiated O_2 treated stacks obtained with voltage pulses of ±12V and duration of 1s. Solid lines are fit to PF charge loss (after $+V_p$) and ln(t) (after $-V_p$) for electron and hole charge state, respectively.

III. Conclusion

The results clearly indicate that γ radiation significantly enhances the electron trapping in the investigated HfO_2/Al_2O_3 multilayer stacks as a result of creation of new electron traps. The trap generation, however, is not accompanied by any deterioration of the leakage current and retention characteristics of the layers implying a good radiation hardness upon γ irradiation up to 10 Mrad.

Acknowledgement

The study was conducted under a joint research project between Bulgarian and Serbian academies of sciences.

References

[1] H. García, M.B. González, M.M. Mallol ,H. Castán, S. Dueňas, F. Campabadal, M.C Acero, L. Sambuco Salomone, A. Faigón, "Electrical characterization of defects created by γ-radiation in HfO_2-based MIS structures for RRAM applications", *Journal of Electronic Materials*, 2018, vol. 47, pp. 5013–5018.

[2] S. Kaya, A. Jaksič, E. Yilmaz, "Co-60 gamma irradiation effects on electrical characteristics of HfO_2 MOSFETs and specification of basic radiation- induced degradation mechanism", *Radiation Physics and Chemistry*, 2018, vol. 149, pp. 7-13.

[3] Y.N. Xu, J.S. Bi, G.B. Xu, B. Li, K. Xi, M Liu, H.B. Wang, L. Luo, "Total ionization dose effects on charge storage capability of $Al_2O_3/HfO_2/Al_2O_3$-based charge trapping memory cell", *Chinese Physics Letters*, 2018, vol. 35, 118501.

[4] HP Zhu, Z.S. Zheng, B. Li, B.H. Li, G.P. Zhang, D.L.Li, J.T. Gao, L. Yang, Y. Cui, C.P. Liang, J.J. Luo, Z.S. Han, "Total dose effect of Al_2O_3-based metal– oxide–semiconductor structures and its mechanism under gamma-ray irradiation", *Semiconductor Science and Technology*, 2018, vol. 33, 115010.

[5] S. Maurya, "Effect of zero bias gamma ray irradiation on HfO_2 thin films", *Journal of Materials Science: Materials in Electronics*, 2016 vol. 27, pp.12796- 12802.

[6] J.M. Rafi, F. Campabadal, H. Ohyama, K. Takakura, I. Tsunoda, M. Zabala, O. Beldarrain, M.B. González, H. García, H. Castán, A. Gómez, S. Dueñas, "2 MeV electron irradiation effects on the electrical characteristics of metal–oxide–silicon capacitors with atomic layer deposited Al_2O_3, HfO_2 and nanolaminated dielectrics", *Solid-State Electronics*, 2013, vol. 79, pp. 65-74.

[7] S. Cao,·X. Ke, S. Ming, D.Wang, T. Li, B. Liu, Y. Ma, Y. Li1, Z. Yang M. Gong, M. Huang, J. Bi, Y. Xu, K. Xi, G. Xu,·S. Majumdar, "Study of γ-ray radiation influence on $SiO_2/HfO_2/Al_2O_3/HfO_2/Al_2O_3$ memory capacitor by C–V and DLTS", *Journal of Materials Science: Materials in Electronics*, 2019, vol. 30, pp 11079–11085.

[8] D. Spassov, A. Paskaleva, T.A Krajewski, E. Guziewicz, G. Luka, Tz. Ivanov, "Al_2O_3/HfO_2 Multilayer High-k Dielectric Stacks for Charge Trapping Flash Memories", *Physica Status Solidi (A)*, 2018, 1700854.

[9] D. Spassov, A. Paskaleva, T.A Krajewski, E. Guziewicz, G. Luka, "Hole and electron trapping in HfO_2/Al_2O_3 nanolaminated stacks for emerging non-volatile flash memories", *Nanotechnology*, 2018 vol. 29, 505206.

[10] D. Spassov, A. Paskaleva, E. Guziewicz, G. Luka, T.A.Krajewski, K Kopalko, A Wierzbicka, B Blagoev. "Electrical characteristics of multilayered HfO2 – Al2O3 charge trapping stacks deposited by ALD", *Journal of Physics: Conference Series*, 2016, vol. 764 01201.D

[11] K. Lehovec, A. Fedotowsky, "Charge retention of MNOS devices limited by Frenkel-Poole detrapping", *Applied Physics Letters*, 1978, vol. 32, pp. 335-338.

Determination Method for Interface State Densities Adapted to Ultrathin Dielectrics

N. Novkovski and A. Skeparovski

Abstract – In this paper modified Terman method for determination of interface state densities, adapted for nanosized dielectrics has been applied to the case of Al-gated MIS structures containing ultrathin (2.56 nm to 5.15 nm thick) SiO_2 dielectric layer.

It has been demonstrated that modified Terman method is effective in precise determination of interface states density distributions for nannosized SiO_2 dielectric SiO_2 layers. Tails towards the band edges are not present when using this method.

For all three samples a peak around 0.3 eV is observed. It corresponds to the first peak of the P_b defect. Second peak corresponding to the P_b defect cannot be clearly distinguished due to the presence of a peak related to Al gate.

I. INTRODUCTION

Determination of interface state densities in MIS structures containing ultrathin SiO_2 [1] or alternative dielectrics, presents a particular challenge, due to various effects which are limiting the straightforward application of standard methods and requiring new ones [2]. Therefore, best choice of the method and appropriate modifications convenient for novel structures [3] are to be exploited in characterization of such structures.

Several limitations have to be taken into account when designing the method of characterization, both from the side of experimental conditions and from the side of extraction of parameters and distributions. First, for MIS structures containing nanosized dielectrics, gate voltage (U_g) range for C-V measurement is to be restrained to about 1 V by absolute value, in order to avoid effects of visible degradation due to the stress, when studying fresh samples. In standard C-V measurement methods much higher voltage values are required in order to approach the accumulation capacitance value (−5 V to +5 V, or in some cases at least −3 V to +3 V). Methods of extrapolation of capacitance are therefore of crucial importance for the effective analysis of C-V characteristics. Since leakage currents substantially influence the measured values of capacitance [4], appropriate combination of frequency and voltage ranges are to be selected. Thus, it is found that high frequency (50 kHz to 1 MHz) measurements combined with low absolute voltage values (-1.5 V to +1.5 V). We previously found that serial C-R measurement mode is more convenient for characterization of such structures

N. Novkovski, and A. Skeparovski are with the Institute of Physics, Faculty of Natural Sciences and Mathematics, University "Ss. Cyril and Methodius", Arhimedova 3, 1000 Skopje, Macedonia, E-mail: nenad@iunona.pmf.ukim.edu.mk

than the dominantly used parallel one [5]. After selection of appropriate voltage range and frequency, only small correction for serial resistance is required to obtain real capacitance from the measured one [3].

In this work we study the case of Al-SiO_2-Si structures with thicknesses in the range 2.56 nm to 5.15 nm using the Modified Terman method proposed in [6]. This approach allows separating the contribution of the effect of charge quantization at the interface from the classical effect of band bending in the interfacial region, the second one being the basis for the Terman method. Thus, a correction of the result obtained by standard Terman method for the quantum charge effect is performed.

Finally, the consistency of the extrapolation method for determination of capacitance of the dielectric layer itself with the contribution of the quantum charge effect in potential drop across the semiconductor space charge layer is to be checked.

II. EXPERIMENTAL

Al-SiO_2-Si (MIS) structures used in this study were obtained as follows. The substrates were chemically cleaned by wet-chemical solution and then dipped in HF solution in order to remove the native silicon oxide. After the cleaning, thin SiO_2 layers with thicknesses of about 5 nm were grown by standard thermal oxidation in dry O_2 at 900 °C. Then, the SiO_2 layer was etched back in a diluted HF to smaller thickness down to about 2.5 nm. The thickness of the SiO_2 layer (d) was measured ellipsometrically. Three difference thickness values were obtained: 2.4 nm, 4.2 nm and 5.1 nm. MOS capacitors were formed by deposition of Al layers (300 nm thick) as top and backside contacts by sputtering. Circular gates for MIS capacitors with active areas (S) of 2.4×10^{-3} cm^2 were defined by using shadow mask. C_s-V and R_s-V curves (serial mode) were measured in the gate voltage range from −1.4 V to +1.0 V at the frequency of 100 kHz with the use of a HP 4284A LCR meter.

III. ANALYSIS

Correction method for determination of realistic capacitance from measurement results as described in [3] has been used in this work. Only results for capacitance after correction are displayed (realistic capacitance, C, Fig. 1). It is seen that in all cases displayed, saturation is far to be attained. Therefore, maximum capacitance in the measurement range cannot be used as acceptable approximation for the dielectric layer capacitance (C_i).

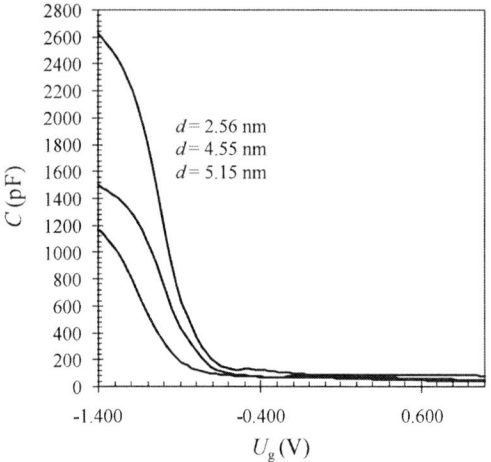

Fig. 1. *C-V* characteristics for three Al-SiO$_2$-Si structures with different thicknesses of the dielectric layer.

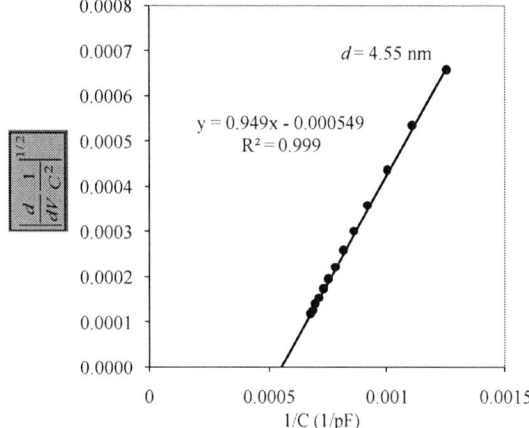

Fig. 2. Illustration of the method of extrapolation, used in this work in determination of the dielectric layer capacitance.

Capacitance (C_i) of the dielectric layer (SiO$_2$) was determined using the extrapolation method proposed by Kar [7]. Value of $1/C_i$ is obtained from the intercept of the fitted straight line of the plot of $\sqrt{\dfrac{d}{dV_g}\dfrac{1}{C^2}}$ versus $\dfrac{1}{C}$.

Illustration of the efficacy of the extrapolation method used [7] for a single case (d_{ox} = 4.55 nm) is given in Fig. 2. Exceptionally high correlation coefficient (R^2 = 0.999) has been obtained, confirming the validity of assumptions underlying in the method of extrapolation used. In addition, maximum capacitance measured is 1.50 nF, corresponding to 5.52 nm. Thickness value calculated using extrapolated capacitance is 4.55 nm that is close to the elipsometrically determined value (4.2 nm). As is seen, values obtained for the oxide thickness, using extrapolated capacitance, are close the values obtained elipsometrically for all three samples.

Central physical quantity in the determination of interface states densities in Terman method is the stretch-out of the real *C-V* curve relative to the ideal curve (ΔU_g). Details of the application of the standard and of the modified Terman method can be found in [6]. Here we describe some particular issues relevant to the present study.

First, variations of the stretch-out with oxide voltage are displayed in Fig. 3. Oxide voltage is determined using expression:

$$U_{ox,q} = U_g - U_{fb} - U_{fb} \qquad (1)$$

where U_{fb} is the flatband voltage and U_s is the theoretical voltage drop in silicon (surface potential). Subscript q for the oxide voltage indicates that no correction for quantum charge effect was done. Flatband voltage used in expression (1) is obtained by the standard method, using extrapolated value of dielectric layer capacitance.

Two distinct linear variations of ΔU_g with oxide voltage are observed for positive and negative values of the oxide voltage (gate voltages higher and lower than the flatband voltage). Only in a restrained region around the flatband condition substantial deviations from these straight lines are observed. These variations are attributed to the effect of interface states. Linear variations are accounted by the effect of quantum charge resulting in additional voltage drop proportional to the oxide voltage.

Method of determination of total interface state density over the bandgap ($D_{it,t}$) is also illustrated in Fig. 3. Briefly, $D_{it,t}$ is proportional to the voltage jump at flatband (U_{it}). The expression used is:

TABLE I

FILM THICKNESSES, PEAK POSITIONS (E), INTERFACE STATE DENSITIES (D_{IT}) AND TOTAL DENSITIES OF INTERFACE STATES (INTEGRATED D_{IT} OVER THE BANDGAP (FROM $E = E_V = 0$ eV TO $E_C = 1.12$ eV), $D_{IT,T}$)

d (nm)		E (eV)		D_{it} (10^{12} eV^{-1}cm^{-2})			$D_{it,t}$ (10^{12} cm^{-2})
Ellipsometry	Capacitance	peak 1	peak 2	peak 1	peak 2	Midgap	Integral
2.4	2.56	0.23	0.53	4.4	9.4	9.0	5.3
4.2	4.55	0.28	0.83	2.5	5.6	2.3	3.9
5.2	5.15	0.22	0.98	2.0	7.8	2.3	2.7

$$D_{it,t} = \frac{C_i}{q^2 S} U_{it}. \qquad (2)$$

where q is the unit charge and S is the surface area of the capacitor.

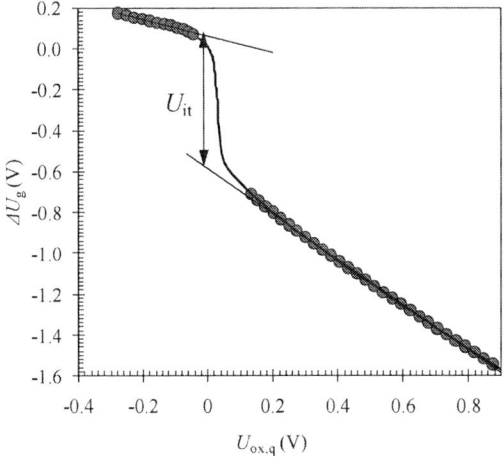

Fig. 3. Variation of the stretch-out (ΔU_g) with the oxide voltage for one of the samples used in this work.

In Fig. 3 only the measured points described with linear dependences are represented with circles, in order to show the linear trends. Part with sharp variation of ΔU_g is presented only with continuous line.

In the analysis contributions of quantization effect are considered to be described by the straight lines and are subsequently subtracted from the total to obtain solely the contribution of interface states in the variation of ΔU_g. Results obtained by this method are in agreement with the results for voltage drop in the oxide obtained by quantum mechanical calculations [8]. Therefore, these results can be treated as acceptable approximation.

Corrected stretch-out as described above has ($\delta \Delta U_g$) been subsequently used to determine distribution of interface states over the Silicon bandgap, using the expression:

$$D_{it}(U_s) = \frac{C_i}{q^2 S} \frac{d \, \delta \Delta U_g}{d U_s} \qquad . \qquad (3)$$

Results for distributions of interface state densities over Silicon bandgap for all samples are shown in Fig. 4, using the method proposed in [6]. No exponential tails are present as a result of correction for quantum charge.

Main measurement and computation results for all considered parameters are summarized in Table 1.

For all three samples a peak around 0.3 eV is observed. It corresponds to the first peak of the P_b defect

[9] in silicon dioxide (oriented silicon dangling bond pointing out of the silicon into the oxide).

The other clear peak appears to be connected to some damage created by Al metal gate. Its position varies with the oxide thickness, moving from the vicinity of the conduction band edge for the thickest film to the midgap for the thinnest film. Maximum values of D_{it} for all three films are similar (around 8×10^{12} eV^{-1}cm^{-2}).

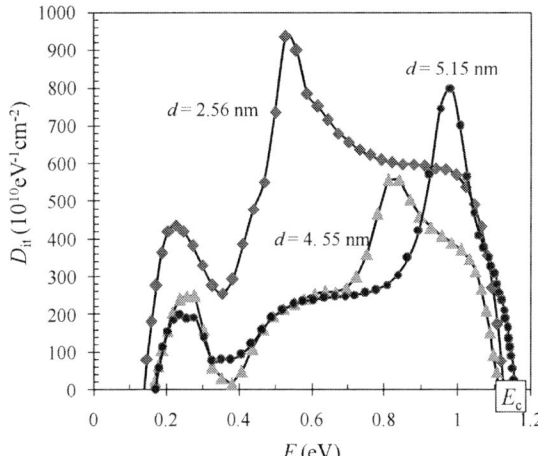

Fig. 4. Distributions of interface state densities over Silicon bandgap for all three samples.

The second peak corresponding to the P_b defect that is expected to appear near 0.9 eV is not clearly observed, due to the presence of the sharp peak attributed to the effect of the Al gate.

III. DISCUSSION

Type of plot used in Fig. 3 is particularly representative for the method used in analysis of the *C-V* curve. First, it is seen that range of the variation of U_g is from about –0.3 V to +0.9 V (span of 1.2 V). This value is about 1.2 V smaller than the range span for the gate voltage (2.4 V). The region where significant variations of ΔU_g occur is even thinner: from about –0.05 V to about +0.15 V (span of 0.2 V). This is the region of band bending, containing the information on the interface states distribution. The corresponding span in U_g is about 1.4 V. In this region surface potential varies for about 1.12 V (entire bangdap), while the oxide voltage varies for only about 0.3 V. Since the almost entire information relevant for characterization of interface states is contained in this part, restriction of the measurement range can shrink, in principle, down to 1.4 V around the flatband point.

Exponential tails typical for the standard Terman method, are not obtained here, which goes in support of the interpretation that such tails obtained by standard method are an artifact of the analysis. Two crucial arguments supporting this interpretation are given below:

First, dependences of the stretch-out from the oxide voltage observed at accumulation and inversion are strikingly linear, indicating a simple relationship between the variation of ΔU_g and the field in the oxide, such as voltage drop on a quantum capacitance with almost constant value.

Second, ΔU_g varies in a range substantially larger (1.8 V) than it is expected if is related to the surface potential, which has to be constrained in the forbidden bandgap of the semiconductor. Therefore, surface potential after attaining edge values corresponding to the top of the valence band and to the bottom of the conduction band, remains constant, while the ΔU_g varies due to the increase of the voltage on the quantum capacitance of a near interface layer in the semiconductor.

Capacitance of the insulating layer itself was extracted in this work using an independent interpolation method. This method is used only to determine extrapolation value of the capacitance and does not modify the pattern of the *C-V* curve. Therefore, it is not to be expected some interference between these two methods to modify significantly the results for the distribution of interface state densities.

IV. CONCLUSION

In conclusion, it has been demonstrated that modified Terman method allows obtaining precise results for interface state densities of ultrathin dielectrics from measurements with low gate voltages applied (gate voltage rangee from −1.4 V to +1.0 V), thus allowing avoiding visible degradation during the measurement.

Exponential tails towards the band edges are not obtained when using this method.

ACKNOWLEDGEMENT

Authors wish to express their gratitude to Albena Paskaleva and Dencho Spassov from the Institute of Solid State Physics, Bulgarian Academy of Sciences for providing the samples of outstanding quality required for this study.

REFERENCES

[1] E. I. Goldman, S. A. Levashov, V. G. Naryshkina, G. V. Chucheva, "Generation of surface electron states with a silicon–ultrathin-oxide interface under the field-induced damage of metal–oxide–semiconductor structures", *Semiconductors*, 2017, vol. 51, pp. 1136-1140.

[2] O. Rejaiba, A. F. Braña de Cal, A. Matoussi, "A comprehensive study on the interface states in the ECR-PECVD SiO_2/p-Si MOS structures analyzed by different method", *Physica E: Low-dimensional Systems and Nanostructures*, 2019, vol. 109, pp. 84-92.

[3] N. Novkovski, "Determination of interface states in metal (Ag, TiN, W)−Hf:Ta_2O_5/SiO_xN−Si structures by different compact methods", *Materials Science in Semiconductor Processing*, 2015, vol. 39, pp. 308-317.

[4] N. Novkovski, and E. Atanassova, "Frequency dependence of the effective series capacitance of metal-Ta_2O_5/SiO_2-Si structures", *Semicond. Sci. Technol.* 2007, vol. 22, pp. 533-536.

[5] N. Novkovski, and E. Atanassova, "Frequency Dependence of Characteristics of MOS Capacitors Containing Nanosized High-κ Ta_2O_5 Dielectrics", *Advances in Materials Science and Engineering*, 2017, vol. 2017, pp. 9745934-1–9745934-11.

[6] N. Novkovski, "Modification of the Terman method for determination of interface states in metal-insulator-semiconductor structures", *Journal of Physics Communications*, 2017, vol. 1, pp. 035006-1–035006-14.

[7] S. Kar, S. Rawat, S. Rakheja, and D. Reddy, "Characterization of accumulation layer capacitance for extracting data on high-κ gate dielectrics", *IEEE Transactions on Electron Devices*, 2005, vol. 52, pp. 1187-1193.

[8] J. Suné, P. Olivo, and B. Riccó, "Quantum-mechanical modeling of accumulation layers in MOS structure", *IEEE Transactions on Electron Devices* 39,1992, pp. 1732-1739.

[9] G. J. Gerardi, E. H. Poindexter, and P. J. Caplan, N. M. Johnson, "Interface traps and P_b centers in oxidized (100) silicon wafers", *Applied Physics Letters*, 1986, vol. 49, pp. 348-350.

Feasibility of applying an electrically programmable floating-gate MOS transistor in radiation dosimetry

S. Ilić, A. Jevtić, S. Stanković, and V. Davidović

Abstract— In this paper we investigated the feasibility of using a commercial programmable floating-gate MOS transistor (EPAD) as a radiation dosimeter. The results show that EPAD with zero bias have the sensitivity of 9.2 mV/Gy and low fading. EPADs with a higher initial threshold voltage show very good linearity with the radiation dose. After its annealing at 70 °C there is a visible recovery of transfer characteristics due to a parasitic parallel resistive path that occurs during irradiation. Apart from that, the threshold voltage is slightly recovered. The programming time of an EPAD increases with the absorbed dose and depends on gate biasing during irradiation.

I. Introduction

The use of p-channel MOS and VDMOS transistors as radiation dosimeters is known for decades and their characteristics have been studied in many papers [1–7]. pMOS transistors have relatively large fading, which makes them difficult for practical applications [8]. A possible solution for this problem is perhaps found in memory components such as EPROMs [9]. The idea is to find a component that can be pre-charged with electrons (programming), then erased with radiation, and re-used [10–12]. Inspired by this idea, the group of authors used a commercial programmable floating-gate device known as an EPAD (Electrically Programmable Analog Device) as a radiation dosimeter [13, 14].

We used the ALD1108E IC, which consists of four EPADs manufactured by *ALD* [15]. The EPAD is an analog IC, suitable for matched-pair balanced circuit configurations, such as current sources and current mirrors [16]. It is an exotic and impresive device, but very rare on the market [17]. The EPADs have to be programmed, which means precisely controlling the threshold voltage by storing the non-volatile charge in the floating gate of the transistor.

Without a possibility of buying a programmer, the authors designed their own programmer which was cal-

S. Ilić and A. Jevtić are undergraduate students of the Module: "Electronic Components and Microsystems" at the Faculty of Electronic Engineering, University of Niš, Aleksandra Medvedeva 14, 18000 Niš, Serbia, E-mail: ilic.stefan@elfak.rs

S. Stanković is with the Department of Radiation and Environmental Protection at the Vinča Institute of Nuclear Sciences, Belgrade, Serbia

V. Davidović is with the Department of Microelectronics, Faculty of Electronic Engineering, University of Niš, Aleksandra Medvedeva 14, 18000 Niš, Serbia

ibrated and tested [18]. That programmer can precisely adjust the threshold voltage from 1 V to 4 V. In their earlier work, the authors decapsulated and then analyzed the ALD1108E IC using an optical microscope [19]. Their intention was to understand the EPAD structure and further explain the behavior of these transistors under the effects of radiation.

II. Experimental Setup

The experiment consists of four stages: irradiation, operating mode, spontaneous recovery, and annealing. Transistors were irradiated using a ^{60}Co radiation source at the Vinča Institute of Nuclear Sciences. Operating mode, spontaneous recovery and annealing lasted 40, 36 and 168 hours, respectively.

The electrical scheme during operation mode is shown in Fig. 1. The idea for this stage of experiment was for the transistors to work in ZTC (Zero Temperature Coefficient) mode, which is about 68 μA [16].

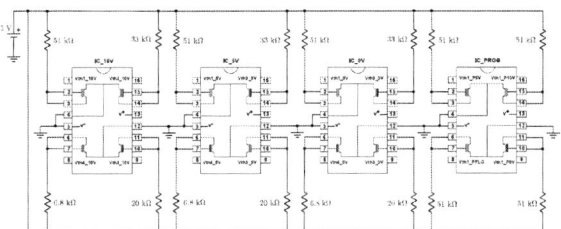

Fig. 1. The electrical scheme of the ALD1108E ICs during operation mode.

Both operating mode and spontaneous recovery were performed at room temperature, about 27 °C. Annealing was performed at 70 °C, because it is the maximum operating temperature declared by the manufacturer [16].

Four ICs, `IC_10V`, `IC_5V`, `IC_PROG` and `IC_0V`, received a total dose of 151, 149, 70 and 145 Gy, respectively. The bias on the gates during irradiation was 10 V, 5 V and 0 V.

Transistors were initially programmed at a specific V_{th}, so that each IC had transistors with a threshold voltage of 1 V, 2 V, 3 V and 4 V, respectively, except `IC_PROG`, where all transistors had a V_{th} of 1 V. This `IC_PROG` was used for examination the re-programming

978-1-7281-3420-8/19 $31.00 © 2019 IEEE

characteristics. The transistors were named depending on their initial V_{th} and the bias on the gate during irradiation (eg. Vth1_0V is the name of the transistor with 1 V initial V_{th} and 0 V bias on the gate during irradiation).

III. RESULTS AND DISCUSSION

The four stages of the experiment for the three ICs are shown in Fig. 2, 3 and 4. Three EPADs with the same initial threshold voltage are shown in Fig. 5. In order to achieve a better view of the results, all figures show annealing only for the first 96 hours, because, after that, the characteristics had not changed until the end of the experiment.

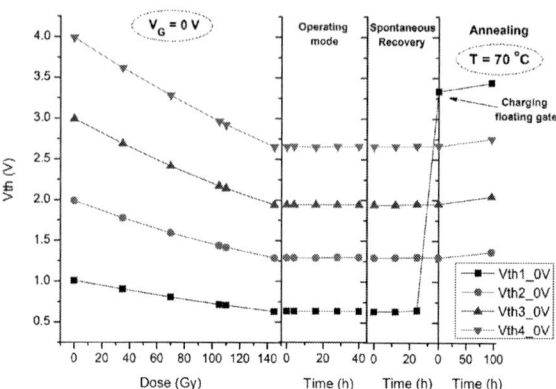

Fig. 4. The behaviour of V_{th} in EPADs during the experiment.

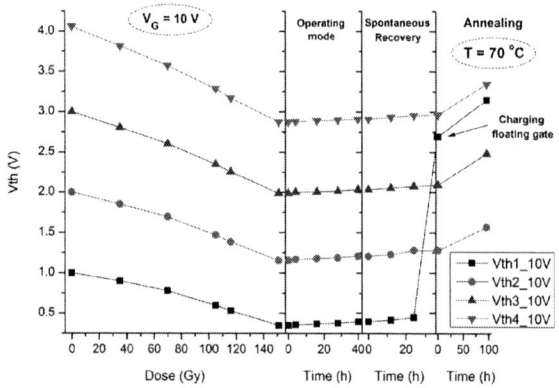

Fig. 2. The behaviour of V_{th} in EPADs during the experiment.

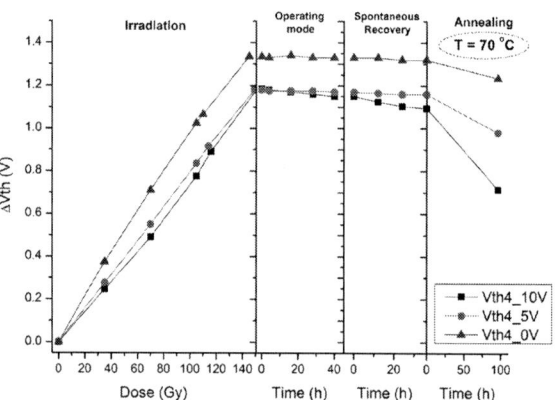

Fig. 5. The behaviour of ΔV_{th} in EPADs with the same initial V_{th} during the experiment.

by the floating gate, except for transistor whose bias was 5 V and V_{th} = 4 V, which has the best linearity of all transistors. However, linearity decreases with the increase of the absorbed dose since the floating gate discharges and reduces its electric field [11]. Also, linearity increases with the initial threshold voltage.

We can see that there is no difference between the operating mode and spontaneous recovery. This means that using EPADs after irradiation for real-time readings does not affect stability. However, annealing at 70 °C has a visible threshold voltage shift and it increases with the increase of the gate bias. That can be explained by a higher concentration of trapped charges in the oxide during irradiation when the gate is biased.

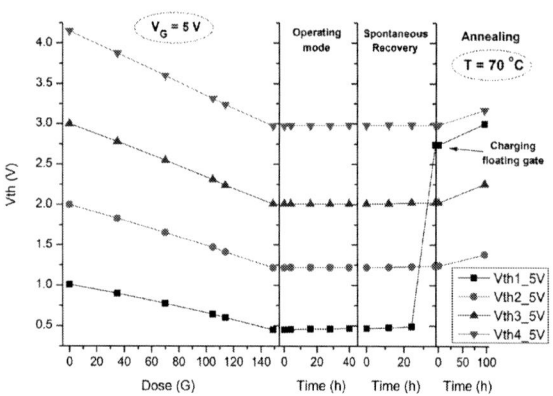

Fig. 3. The behaviour of V_{th} in EPADs during the experiment.

The linearity of the threshold voltage shift caused by the radiation depends on the superposition of the two electric fields, induced by the floating gate and control gate, available to separate radiation-generated electron-hole pairs in the oxide. By analyzing the curves during irradiation, we can point out that the 10 V and 5 V bias have stronger fields in the oxide than the one induced

The sensitivity to radiation and stability after irradiation (fading) of EPADs were calculated for the first three stages of the experiment and represented in Table I. The fading, defined as the deviation percentage of the threshold voltage during the second and third stage, was calculated using the following equation:

978-1-7281-3420-8/19 $31.00 © 2019 IEEE   68

$$f(t) = \frac{V_T(0) - V_T(t)}{V_T(0) - V_{T0}}, \qquad (1)$$

where $V_T(0)$ is the threshold voltage immediately after irradiation, $V_T(t)$ is the threshold voltage after spontaneous recovery, and V_{T0} is the pre-irradiation threshold voltage [20].

TABLE I

SENSITIVITY (mV/Gy) & FADING (%)

	10 V		5 V		0 V	
	S	F	S	F	S	F
Vth1	4.3	15.3	3.8	7.1	2.6	4.0
Vth2	5.6	14.8	5.3	3.2	4.8	1.4
Vth3	6.7	10.3	6.6	1.5	7.2	0.95
Vth4	7.8	7.6	7.9	1.3	9.2	0.75

Sensitivity increases with the threshold voltage, as it is expected. What is interesting in the results is that the sensitivity increases with the increase of the gate bias for a lower threshold voltage (Vth1 and Vth2) but decreases with the increase of the gate bias for a higher threshold voltage (Vth3 and Vth4). Fading decreases not only with an increase of the threshold voltage but also with a lower gate bias.

The most sensitive and most stable EPAD is the one with zero bias and $V_{th} = 4$ V. It is very important to emphasize that these devices have very low fading during operating mode and spontaneous recovery, which gives them great potential for radiation dosimetry applications [14]. We assume that this is because some of the radiation-generated holes are recombined with the charge stored in the floating gate during irradiation [9].

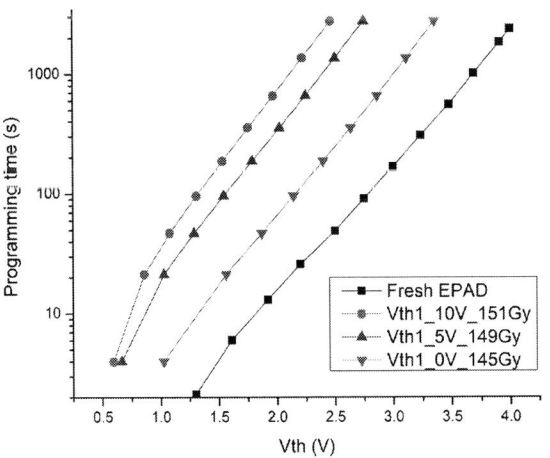

Fig. 6. Re-programming characteristics of EPADs ($V_{th} = 1$ V) after irradiation with a different voltage bias.

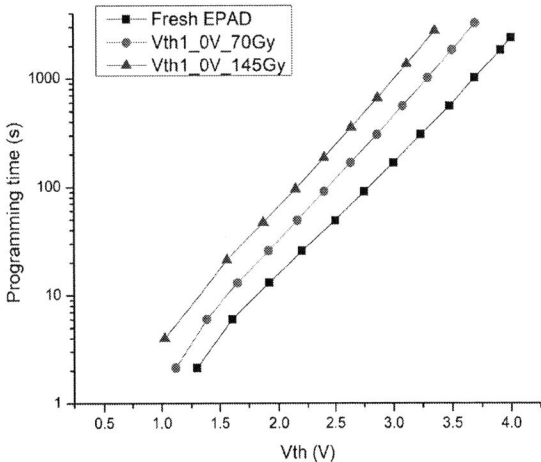

Fig. 7. Re-programming characteristics of EPADs ($V_{th} = 1$ V) with different absorbed doses.

Analyzing the re-programming curves of EPADs in Fig. 6, we can notice that the programming of EPADs becomes harder with the increase of the gate bias during irradiation. Harder programming means it takes more time to shift the V_{th} of the transistor to the desired value. As shown in Fig. 7, the programming of EPADs is harder with a higher absorbed dose. The time to shift the threshold voltage to the desired value increases with the absorbed dose. We assume that this is due to trapped charges in the oxide that compensate with the charge stored in the floating gate, thus taking more time to re-program [13].

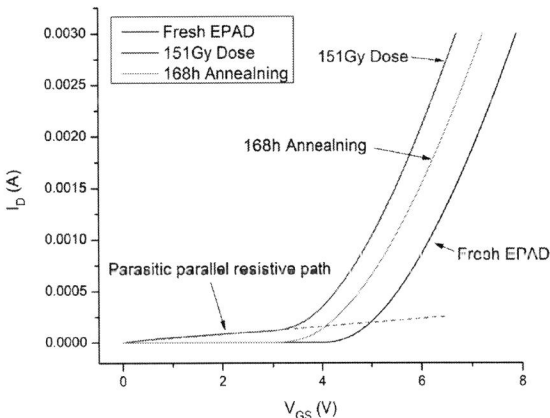

Fig. 8. Transfer characteristics of an EPAD with initial $V_{th} = 4$ V and a bias of 10 V in crucial stages.

In order to demonstrate component degradation with higher biasing (10 V), we show the transfer characteristics of an EPAD with an initial $V_{th} = 4$ V in crucial stages (Fig. 8). After the total dose of 151 Gy was received, the transfer characteristics degraded, and the parasitic parallel resistive path of approximately 25 kΩ opened in the structure. It is very interesting to notice that after annealing, the transfer characteristics had recovered.

At the end, the behavior of EPADs with higher doses of radiation should be investigated to further understand the relations between the two electric fields in the floating gate structure in order to find the possibility of extending the linearity of the threshold voltage shift dependence during irradiation.

IV. CONCLUSION

Results have shown that EPADs are sensitive to radiation. The linearity increases with the initial threshold voltage. There is also lower fading for a higher initial threshold voltage. The greatest achieved sensitivity is 9.2 mV/Gy, and the lowest fading is 0.75 % for the same transistor. The programming of EPADs is much harder for higher gate biases applied during irradiation. Also, it is harder for a higher absorbed dose. After annealing at 70 °C there is visible recovery of the transfer characteristic due to a parasitic parallel resistive path that occurs during irradiation, while threshold voltage is slightly recovered. Finally, we can conclude that EPADs have a great potential for application in radiation dosimetry and research should be continued.

ACKNOWLEDGMENT

The performed research was supported by the Serbian Ministry of Education, Science and Technological Development under Grants OI-171026 and TR-32026. The authors would like to thank the Department of Microelectronics of the Faculty of Electronic Engineering, University of Niš and the Vinča Institute of Nuclear Sciences, Belgrade, Serbia on their great assistance and cooperation. The authors also express their sincere thanks to Mr. Milorad Paunović for his help and technical assistance.

REFERENCES

[1] A. Holmes-Siedle, "The space-charge dosimeter: General principles of a new method of radiation detection," *Nuclear Instruments and Methods*, vol. 121, no. 1, pp. 169 – 179, 1974.

[2] G. Ristić, S. Golubović, and M. Pejović, "P-channel metal–oxide–semiconductor dosimeter fading dependencies on gate bias and oxide thickness," *Applied Physics Letters*, vol. 66, no. 1, pp. 88–89, 1995.

[3] Z. Savić, B. Radjenović, M. Pejović, and N. Stojadinović, "The contribution of border traps to the threshold voltage shift in pMOS dosimetric transistors," *IEEE Transactions on Nuclear Science*, vol. 42, no. 4, pp. 1445–1454, 1995.

[4] V. Davidović, D. Danković, A. Ilić, I. Manić, S. Golubović, S. Djorić-Veljković, Z. Prijić, and N. Stojadinović, "NBTI

and Irradiation Effects in P-Channel Power VDMOS Transistors," *IEEE Transactions on Nuclear Science*, vol. 63, no. 2, pp. 1268–1275, April 2016.

[5] S. Djorić-Veljković, I. Manić, V. Davidović, D. Danković, S. Golubović, and N. Stojadinović, "Annealing of radiation-induced defects in burn-in stressed power VDMOSFETs," *Nuclear Technology and Radiation Protection*, vol. 26, no. 1, pp. 18–24, 2011.

[6] D. Danković, I. Manić, V. Davidović, S. Djorić-Veljković, S. Golubović, and N. Stojadinović, "Negative bias temperature instability in n-channel power VDMOSFETs," *Microelectronics Reliability*, vol. 48, no. 8, pp. 1313 – 1317, 2008, 19th European Symposium on Reliability of Electron Devices, Failure Physics and Analysis (ESREF 2008).

[7] N. Stojadinović, S. Djorić-Veljković, V. Davidović, S. Golubović, S. Stanković, A. Prijić, Z. Prijić, I. Manić, and D. Danković, "NBTI and irradiation related degradation mechanisms in power VDMOS transistors," *Microelectronics Reliability*, vol. 88-90, pp. 135 – 141, 2018, 29th European Symposium on Reliability of Electron Devices, Failure Physics and Analysis (ESREF 2018).

[8] M. M. Pejović, M. M. Pejović, and A. B. Jakšić, "Contribution of fixed oxide traps to sensitivity of pMOS dosimeters during gamma ray irradiation and annealing at room and elevated temperature," *Sensors and Actuators A: Physical*, vol. 174, pp. 85–90, 2012.

[9] B. Lončar, P. Osmokrović, M. Stojanović, and S. Stanković, "Radioactive Reliability of Programmable Memories," *Japanese Journal of Applied Physics*, vol. 40, no. Part 1, No. 2B, pp. 1126–1129, feb 2001.

[10] J. Kassabov, N. Nedev, and N. Smirnov, "Radiation dosimeter based on floating gate MOS transistor," *Radiation Effects and Defects in Solids*, vol. 116, no. 1-2, pp. 155–158, 1991.

[11] N. G. Tarr, G. F. Mackay, K. Shortt, and I. Thomson, "A floating gate MOSFET dosimeter requiring no external bias supply," in *RADECS 97. Fourth European Conference on Radiation and its Effects on Components and Systems (Cat. No.97TH8294)*, Sep. 1997, pp. 277–281.

[12] N. G. Tarr, K. Shortt, and I. Thomson, "A sensitive, temperature-compensated, zero-bias floating gate MOSFET dosimeter," *IEEE Transactions on Nuclear Science*, vol. 51, no. 3, pp. 1277–1282, June 2004.

[13] R. Edgecock, J. Matheson, M. Weber, E. G. Villani, R. Bose, A. Khan, D. Smith, I. Adil-Smith, and A. Gabrielli, "Evaluation of commercial programmable floating gate devices as radiation dosimeters," *Journal of Instrumentation*, vol. 4, no. 02, p. P02002, 2009.

[14] R. Bose, "The development of an in-vivo dosimeter for the application in radiotherapy," Ph.D. dissertation, Citeseer, 2012.

[15] Advanced Linear Devices, Inc. [Online]. Available: http://www.aldinc.com/

[16] ALD, "Quad/dual electrically programmable analog device (epad^{TM})," 2012. [Online]. Available: http://www.aldinc.com/pdf/ALD1110E.pdf

[17] ——, "Application note ald1108," 1998. [Online]. Available: http://www.aldinc.com/pdf/AN1108.pdf

[18] A. Jevtić, S. Ilić, V. Davidović, Z. Prijić, and A. Prijić, "Characterization of electrically programmable floating gate MOS transistor," in *62nd Annual Meeting of ETRAN Society*, Palic, Serbia, 2018, pp. 304–307, in Serbian.

[19] A. Jevtić, S. Ilić, and V. Davidović, "EPADs decapsulation and analysis," in *11th Student Project Conference "IEEESTEC 2018"*, Nis, Serbia, 2018, pp. 329–333, in Serbian.

[20] G. Ristić, S. Golubović, and M. Pejović, "Sensitivity and fading of pMOS dosimeters with thick gate oxide," *Sensors and Actuators A: Physical*, vol. 51, no. 2, pp. 153 – 158, 1996.

978-1-7281-3420-8/19 $31.00 © 2019 IEEE

Heavy-Ion-Induced Single Event Burnout in SiC Schottky Diodes: Safe Operating Area

P. S. Gromova, G. G. Davydov, A. S. Tararaksin, A. S. Kolosova, D. V. Boychenko,
V. N. Vyuginov, and V. V. Luchinin

Abstract - Heavy-ion-induced single event burnout (SEB) is studied experimentally for several types of 4H-SiC Schottky power diodes with various bias voltages applied. Safe operating voltage area for each type was defined and analyzed. The comparison with Si power devices was carried out..

I. INTRODUCTION

Silicon carbide (SiC) based electronic devices seem suitable for space electronics and harsh environment [1]. There are several advantages of the SiC over conventional Si power devices (see table 1):
- lower ohmic resistance of the channel due to much dopant concentration;
- higher working voltage;
- higher power density;
- higher working temperature;
- higher working frequency.

Additionally, SiC may be characterized as radiation-tolerant process (mainly to TID effects) [2...4] due to strong bond Si-C (i.e., higher Debay temperature), high energy of charge traps generation (up to 25 eV ...35 eV) [5,6], as well as wide band-gap.

Nowadays, 4H-SiC process is mainly used as relatively inexpensive and suitable [7,8] with higher electron mobility than 6H-SiC type; power devices, fabricated on this process, are available at market.

TABLE I

SEVERAL PROPERTIES OF SI AND 4H-SIC-BASED DEVICES [9]

Device parameter	Maximum ratings	
	Si	4H-SiC
Maximum working voltage, kV	1.5	6
Breakdown electric field, MV/sm	0.4	2.7
Maximum operating temp., °C	200	> 500
Maximum current density, A/cm^2	100	2000
Thermal conductivity, W/(sm·K)	1.5	4.5

P. Gromova, G. Davydov, A. Tararaksin and D. Boychenko are with the JSC "Specialized Electronic Systems" and Centre of Extremal Applied Electronics of NRNU MEPhI, Kashirskoe shosse, 31, Moscow, Russia, E-mail: GGDavydov@mephi.ru

V. Vyuginov is with the Svetlana-Electronpribor JSC, St. Petersburg.

V. Luchinin and V.Vyuginov are with the Saint Petersburg Electrotechnical University "LETI", St. Petersburg

SiC devices for space applications must have high tolerance to high energy charged particles, primarily on SEB criterion. For Schottky diodes SEB results in irreversible increase of leakage current and functioning loss [10, 11].

A failure probability of power device depends on bias voltage during irradiation. SEB hardness of the power electronic device can be reached in safe operating area (SOA). SOA of Shottky diode corresponds to maximum reverse bias voltage, with which the devise doesn't fail under irradiation.

SOA depends on linear energy transfer (LET) value of space charge particle and tends to decrease with increasing of LET. So, higher levels of heavy ion hardness can be reached at lower reverse bias. Such "electric mitigation" (or derating) is widely used for improve sustainability to outer influence of different nature.

Several researches of Si power Schottky diodes fabricated by Diodes,inc., Vishay and so on [12] demonstrate lack of SEB with LET=60 MeV·cm^2/mg with reverse voltage derating to 50% of its appropriate maximum value. The hardness of silicon MOSFETs with breakdown voltages from 900 V to 1500 V in similar conditions are provided by derating of its reverse bias down to 20% of appropriate maximum value [13].

The purpose of this research is SOA clarification for power 4H-SiC Schottky diodes.

II. DUTS AND EXPERIMENT DESCRIPTION

A. DUT description

In most cases, layout of Schottky diode looks similar as shown in fig. 1a.

Top metal layer forms ohmic contacts to p+ regions and Schottky contacts to n- regions, so the overall device consists of interdigitated Schottky and pin diodes connected in parallel. The high-doped p+ regions reduce reverse current due to reducing field just near the Schottky contacts. The perimeter of the whole diode closed by guard rings for decrease peripheral leakage currents. Thickness of drift region is usually up to 10% of thickness of whole device. One can reach higher working voltage by increasing of drift region thickness [15] with appropriate changing of guard rings geometry.

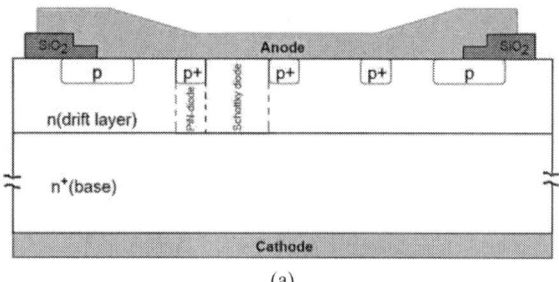

(a)

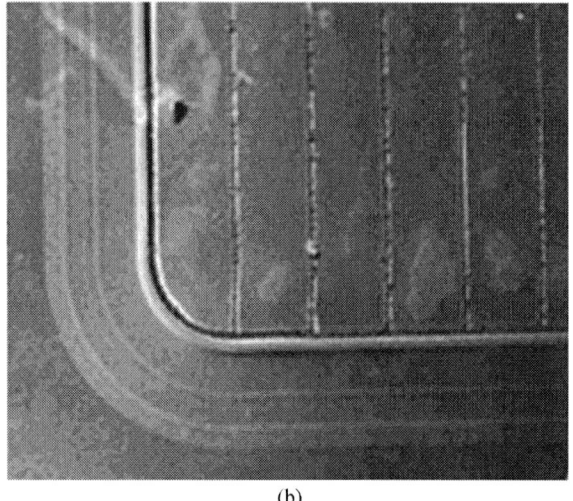

(b)

Fig. 1. Schematic of SiC Schottky Diodes structure (a) and microscope photo of DUT from B manufacturer without top metallization (b) [14, 15].

Devices-under-test (DUT) for this work are from 2 different manufacturers (A and B) and both of them have linear structure of p+ regions and floating guard rings (see fig. 1b).

DUTs manufactured by company A are with maximum reverse bias of 1.2 kV and 1.7 kV, and DUTs manufactured by B company have maximum reverse bias of 0.6 kV, 1.2 kV and 1.7 kV.

The current flow lines and equipotential lines distribution in the diode structure under bias (forward or reverse) is similar as shown in Fig. 3. The field value is maximum between A and B points (see fig. 2), so, the most heavy-ion-sensitive region (SR) is located near p-n junction. Depth of SR is usually less than 10 um.

B. Experiment description

Irradiated Schottky diodes were connected to scheme, which is shown in fig. 3. Mantigora MT-60-6P was used as high-voltage power supply [16] to provide constant and precise reverse bias VR. The digital multimeter PXIe-4071 from National instruments [17] was used to monitor the reverse current. Reverse current and forward bias were

monitored before and after irradiation. Reverse current and bias were monitored during irradiation.

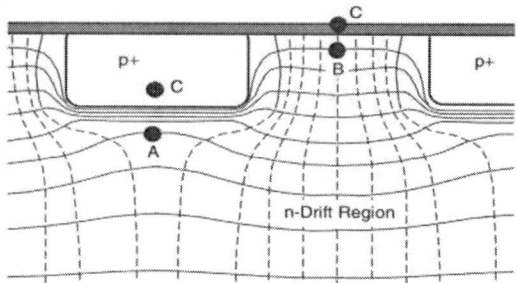

Fig. 2. Current flow (dashed) and equipotential (solid) lines in the SiC Schottky diode under bias. Note that the p+ anode and the Schottky metal, points C, are at the same potential [15]

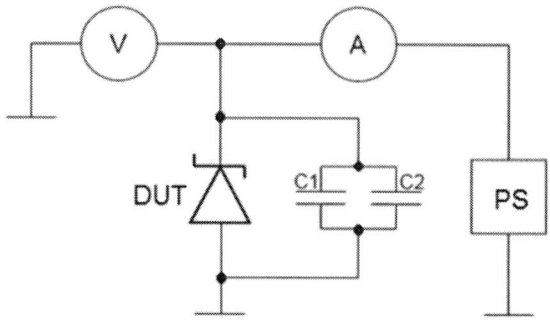

Fig. 3. The schematic of the experiment setup. PS – power supply (Mantigora MT-60-6P); A – Ampermeter (PXIe-4071 in current measuring state); V – Voltmeter (PXI-4071 in voltage measurement state); C1 = 4.5 uF, C2 = 0.22 uF.

The reverse bias value was varying from maximum working value U_{MAX} (specified in datasheet) to reduced one (to avoid SEB). Irradiation and the DUT characterization were performed at room temperature.

The part of plastic package above the crystal was removed to open the chip surface and provide sufficient penetration for the heavy ions to SR.

The heavy-ion irradiations were performed at the U-400M facility (Dubna, Russia) [18-20]. The results from Xe-, Cr-, Ar-, and Ne-ions was obtained and analyzed. Basic characteristics of the used ions are in Table 2. Cumulative fluence of each ion was about $10^6 cm^{-2}$.

TABLE II
BASIC PROPERTIES OF IONS FROM U400-M FACILITY

Ion	Energy, MeV	LET, MeV·sm²/mg	Depth, um
Xe	483.1	69	29
Kr	261.2	40	25
Ar	129.2	16	24
Ne	64.2	6	28

Experimental Results

SOA of Shottky diode for each LET value was defined as the maximum allowable reverse bias voltage UMA, at which SEB wasn't registered.

Photo of DUT surface after SEB is shown in Fig. 4. Location of damages seems to be in place of guard rings, where the electric field is the highest.

SOA for tested DUTs are shown in Fig. 5

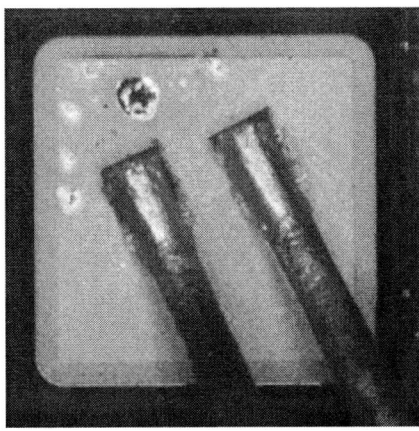

Fig. 4. Photo of DUT surface after SEB

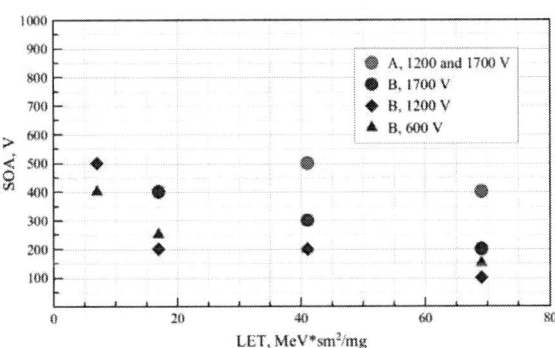

Fig. 5. SOA for several types of SiC Shottky diodes

Analyzing data in fig. 5 allow one to state the following:

1. Suitable reverse bias voltage to prevent SEB must be decreased with increasing LET value.

2. Schottky diodes, produced by A company, demonstrate higher degree of SEB hardness than devices with similar parameters produced by B.

3. For each of tested manufacturers the following trend one can observe: devices with higher working voltage will be sustain during heavy ion impact with given LET at lower ratio of U_{MA} / U_{MAX}.

For every type of DUTs, the ratio of U_{MA} / U_{MAX} was found (see fig. 6).

Devices for aerospace applications must meet the requirements of minimum ion LET (15 MeV·cm2/mg …60 MeV·cm^2/mg in most cases). If derating of 50% (like as offered in [13]) apply to tested diodes, their hardness will not exceed 10 MeV·cm^2/mg. Sustainability of all tested Schottky diodes under the impact of heavy ions with LET up to 60 MeV·cm^2/mg can be reached by reducing bias down to 10% (or less) of maximum reverse voltage from datasheet (see fig. 6).

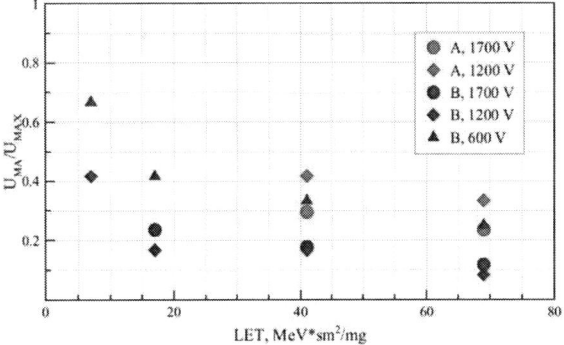

Fig. 6. SOA for several types of SiC Shottky diodes. Normalized to maximum working voltage

Conclusion

Sustainability of 4H-SiC Schottky diodes to heavy ions with LET up to 60 MeV·cm^2/mg can be provided by reducing reverse bias down to 10% (or less) of maximum reverse bias specified in datasheet. However, big specified reverse voltage and significant advantages of SiC process make SiC devices applicable in harsh environment and in aerospace equipment.

Difference in radiation behavior of SIC Schottky diodes of two different manufacturers requires performing similar radiation test for each device type to estimate possibility of usage in chosen applications.

Acknowledgement

Thank to Alexey V. Mikhailov and Alexey V. Afanasiev for provided samples and helpful discussions.

References

[1] Luchinin V.V. "Micro and Nanotechnics. Technologies of Excellence," *Nano- and Microsystem Technology*, 2016, vol. 5, pp. 259–271.

[2] Kalinina E.V. et al. "Effect of irradiation with fast neutrons on electrical characteristics of devices based on CVD 4H-SiC epitaxial layers," *Semiconductors*, 2003, vol. 37, no. 10, pp. 1260–1264.

[3] Kalinina E.V. et al. "High-dose Al-implanted 4H-SiC p+-n-n+ junctions," *Semiconductors*, 2010, vol. 44, no. 6, pp. 807–816.

[4] Lebedev A. A., Ivanov A. M., Strokan N. B. "Radiation hardness of SiC based ions detectors for influence of the relative protons," *Applied Surface Science*, 2001, vol. 184, no. 1-4, pp. 431-436.

[5] Polishchuk A. "Devices based on SiC – the present and future of power electronics," *Components & Technologies*, 2004, no. 8, pp. 3-4.

[6] A. Afanasiev. et al. "Import substitution of silicon carbide electronic components," *Strategic partnership of LETI and Svetlana PJS, Nanoindustry*, 2017, no. 8. pp. 50–60.

[7] Lebedev et al. "Wide-gap semiconductors for high-power electronics," *Semiconductors*, 1999, vol. 33, no. 9, pp. 1096–1099.

[8] Konstantinov A. O. et al. "Ionization rates and critical fields in 4H silicon carbide," *Appl. Phys. Lett.*, 1997, Vol. 71(1), pp. 90–92.

[9] G. W. Hunter et al, *High temperature electronics, communications, and supporting technologies for Venus missions*, NASA Glenn Research Center at Lewis Field, 2006.

[10] A. Javanainen, K. F. Galloway, C. Nicklaw, et al. "Heavy Ion Induced Degradation in SiC Schottky Diodes: Bias and Energy Deposition Dependence," *IEEE Trans. Nucl. Sci.*, 2017, vol. 64, no. 1, pp. 415–420.

[11] S. Kuboyama, C. Kamezawa, Y. Satoh, T. Hirao and H. Ohyama "Single-Event Burnout of Silicon Carbide Schottky Barrier Diodes Caused by High Energy Protons," *IEEE Trans. Nucl. Sci.*, 2007, vol. 54, no. 6, pp. 2379–2383.

[12] M. C. Casey et al. "Silicon Schottky Diode Safe Operating Area.Proc. of NEPP," Electronics Technology Workshop (ETW), June 13-16, 2016, https://nepp.nasa.gov.

[13] G. G. Davydov et al. "Safe Operation Area Of Trench-Gate And Low-Charge Power MOSFET," *Proc. of Conference of Radiation effects on components and systems – RADECS2018*, 16-21 September 2018, Gothenburg, Sweden.

[14] Gavrikov V, "Littelfuse Shottky diodes," *Electronics news*, 2017, no. 4, pp 4-6.

[15] T. Kimoto, J. A. Cooper, "Fundamentals of Silicon Carbide Technology," *IEEE press*, 2014, pp. 296–297.

[16] Mantigora, Series MT, Reference Manual, http://mantigora.ru/highvolt_MT.htm.

[17] National Instruments, PXI Digital Multimeter, Manual, http://www.ni.com/en-gb/support/model.pxi-4071.html.

[18] Joint Institute for Nuclear Research, U-400M facility, http://flerovlab.jinr.ru/flnr/u400m_rus.html.

[19] I. Kalagin, I. Ivanenko, G. Gulbekian, "The experimental investigation of the beam transportation efficiency through the axial injection system of the U400 cyclotron", *Proc. of Particle Accelerator Conference*, 2001, Chicago, USA. DOI 10.1109/PAC.2001.986750.

[20] V.S. Anashin, V.V. Emeliyanov, et al, "Roscosmos facilities for SEE testing at U400M FLNR JINR cyclotron", in *Proc. of 12th European Conference on Radiation and Its Effects on Components and Systems (RADECS)*, 2011, Sevilla, Spain, DOI 10.1109/RADECS.2011.6131461.

978-1-7281-3420-8/19 $31.00 © 2019 IEEE

Fabrication and Characterization of High-k Dielectrics Based Gate Stacks/MOS Capacitors for Advanced CMOS Devices

R. Vaid and R. Gupta

Abstract – This paper presents the fabrication and characterization of various sets of gate stacks: n-Si/SiO$_2$/ALD-HfO$_2$/Ti-Pt, n-Si/SiON/ALD-HfO$_2$/Ti-Pt, n-Si/SiON/ALD-ZrO$_2$/Ti-Pt and n-Si/SiON/ALD-ZrON/Ti-Pt under N$_2$ and NH$_3$ as annealing ambients. The XRD, AFM and FTIR characterizations have been performed for their structural and morphological studies; whereas the electrical characterization includes capacitance-voltage (C-V), conductance-voltage (G-V) and current-voltage (I-V) analysis. Electrical parameters such as dielectric constant (k), effective oxide thickness (EOT) and leakage current density (J) have been extracted through C-V, G-V and I-V measurements. The results suggest that SiON growth prior to HfO$_2$, ZrO$_2$ and ZrON deposition has the potential to surmount the problem of high leakage current density and interfacial traps due to sufficient amount of N$_2$ incorporated at their interface. Electrical characterization such as C-V and I-V reveals the improved results for NH$_3$ annealed ZrO$_2$ sample relative to the all other samples in terms of suppressed gate leakage current, increased dielectric constant and reduced EOT.

I. INTRODUCTION

The gate leakage current level becomes quite intolerable as the physical thickness of SiO$_2$ is reduced below 1.5 nm. In order to overcome the problem of gate leakage current, researchers have made extensive efforts for finding the alternative gate dielectrics to overcome the problem of gate leakage current. Infact, high-k dielectrics are considered as one of the best alternatives to conventional ultrathin SiO$_2$ as they can meet the scaling challenges while maintaining the gate leakage current within tolerable limit [1-3]. Moreover, the use of SiON gate dielectric having high nitrogen concentration is one of the best techniques to prevent boron penetration and gate leakage current. HfO$_2$ have remained an active area of research from the last decade due to their interesting properties like large bandgap (E$_g$ ~ 5.7 eV), high permittivity (k ~ 25), relatively large barrier height, good chemical and thermal stability on silicon. Additionally, the use of metal gate electrode with high-k gate dielectric material holds the promise of reduced leakage current and high capacitance [4-6]. Zirconium based dielectrics offer many interesting properties such as high dielectric constant (17-25), large band gap (7.8 eV), high breakdown field (7-

Dr. Rakesh Vaid is currently Professor and Head of Department of Electronics, University of Jammu, Jammu-180006 (J&K), India (Email: rakeshvaid@ieee.org).
Richa Gupta is a Research Scholar pursing Ph.D. in the Department of Electronics University of Jammu, Jammu-180006 (J&K), India (Email: 1richagupta@gmail.com).

15 MV/cm) and its excellent thermodynamically stability with Si. However, the direct deposition of Si on high-k dielectrics doesn't exhibit good results owing to the large number of defects at the interface.

To overcome these crucial issues, we preferred the pre-growth of SiON interfacial layer prior to HfO$_2$ and ZrO$_2$ deposition to refurbish the high quality interface between Si and high-k dielectrics. Additionally, the PDA of ZrO$_2$ in N$_2$ and NH$_3$ ambients for SiON/ZrO$_2$ gate stack results in good electrical properties such as thin EOT, suppressed gate leakage current and better thermal stability of the oxide layer [7-9]. Atomic layer deposition (ALD) method has been used for deposition of HfO$_2$ and ZrO$_2$ thin films owing to its various advantages such as precise thickness control, low temperature deposition, excellent adhesion, stoichiometric control of multicomponent films, excellent repeatability, low defect density and amorphous or crystalline film types depending on substrate and temperature [10]. The structural and surface morphological studies include atomic force microscopy (AFM), field emission scanning electron microscopy (FESEM) and X-ray diffraction (XRD) whereas the thickness of the grown films has been confirmed by ellipsometry and FESEM cross-sectional studies. High-frequency C-V, G-V and I-V measurements have been performed using Keithley 4200-SCS (Vega) at room temperature under light-tight conditions.

II. EXPERIMENTAL DETAILS

Single side polished (SSP) 2" n-type Si <100> substrate having resistivity of 1-5 Ω-cm were cleaned using standard RCA method. This cleaned Si wafer was then loaded into RTP-FEP centura/Gate S for interfacial layer growth of SiON ~ 5,4 nm having 5 SLM O$_2$ flow for 2 seconds at 850 °C followed by 5 SLM N$_2$ flow for 60 seconds at 1100 °C under atmospheric conditions. ALD system (Nanotech Fiji 200) was then used to deposit HfO$_2$ and ZrO$_2$ films on SiON interfacial layer at a wafer temperature of 200 °C having deposition rate of 0.95 A°/cycle. After that the wafer was split into three samples for further processing. ANNEALSYS AS-ONE 150 RTP system was used for PDA of ZrO$_2$ in N$_2$ and NH$_3$ ambient at 500 °C for 30 seconds having N$_2$ and NH$_3$ gas flow of 900 Sccm. Cauchy modeling was performed using SE800 ellipsometer to measure the thickness of HfO$_2$ and ZrO$_2$ films annealed in N$_2$ and NH$_3$ ambients, which were observed to vary from 5.2-5.3 nm and 8.2-8.3 nm respectively. Moreover, no significant change in the thickness was observed after

978-1-7281-3420-8/19 $31.00 © 2019 IEEE

annealing of ZrO_2 in N_2 and NH_3 ambients. Four target e-beam evaporator was used to deposit 100 nm thick layer having Ti ~ 20 nm and Pt ~ 80 nm over a circular area of 1.4×10^{-3} cm^2 through a shadow mask. The base and the working pressure during the deposition of Ti and Pt were kept at 7×10^{-7} torr and 3 m torr respectively. To remove the native oxide by buffered hydrofluoric acid (BHF), back side etching was performed to remove the native oxide by buffered hydrofluoric acid (BHF) followed by rinsing in de-ionized water. Al film was deposited for making back contact for n-Si/SiON/ALD-ZrO_2 gate stacks. Finally, post metallization annealing (PMA) of the fabricated gate stack structure was carried out at 420 °C for 20 minutes using forming gas (96% N_2, 4% H_2) ambient. To study the various electrical properties, C-V and I-V measurements were performed using Keithley 4200-SCS (VEGA) semiconductor parameter analyzer at room temperature

III. RESULTS AND DISCUSSION

A. Atomic Force Microscopy

Figure 1 shows the 3-D surface topography for HfO_2 film examined using AFM with a scanning area of 2×2 μm. The surface roughness of the ALD-HfO_2 film was found to be about 0.01226 nm confirming the better uniformity and smoothness.

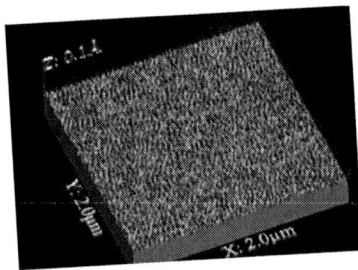

Fig. 1. 3-D AFM image of HfO_2 having 2×2 μm scanning area.

To assess the surface morphology and roughness of N_2 and NH_3 annealed ZrO_2 thin films, atomic force microscopy (AFM) was used. Figure 2 (a) and (b) shows the 2-D (top view) and 3-D AFM images of N_2 annealed ZrO_2 whereas Figure 3 (a) and (b) shows the 2-D and 3-D AFM images of NH_3 annealed ZrO_2, resulting from 2×2 μm scans. The surface roughness for NH_3 and N_2 annealed ZrO_2 film was found to be 0.49 nm and 0.13 nm respectively, which reveals the smoothness of grown films. The AFM images for both ZrO_2 and ZrON films exhibit the formation of fine grains at the surface. However, a remarkable increase in grain size has been seen in case of ZrON thin film. The increased crystallinity during ammonia annealing at 500 °C results in increase in grain size for ZrON film (Fig. 3 (a)), which basically reflects the tetragonal phase of ZrON preferred for memory applications. Based on these results, it can be concluded that NH_3 annealed ZrO_2 is quite helpful in improving the thermal stability and interfacial properties of SiON/ZrO_2 gate stack structures.

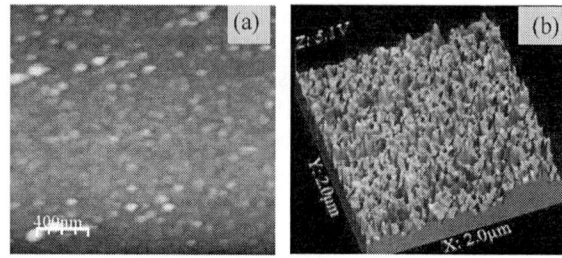

Fig. 2. (a) 2-D (top-view) and (b) 3-D View of N_2 annealed ZrO_2.

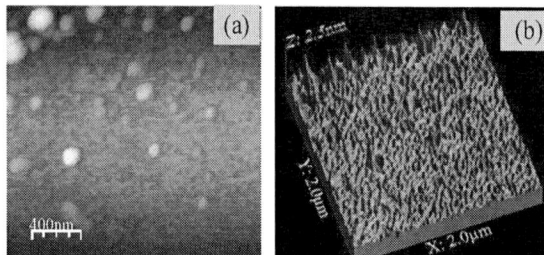

Fig. 3. (a) 2-D (top-view) and (b) 3-D View of NH_3 annealed ZrO_2.

B. XRD Analysis

XRD was carried out using Rigaku system equipped with CuKα radiation and average wavelength ~ 1.541 Å to investigate the structural analysis of RTP-grown SiON IL and degree of crystallization of ZrO_2 and ZrON thin films deposited at 500 °C. Figure 4 (a) shows the XRD pattern of

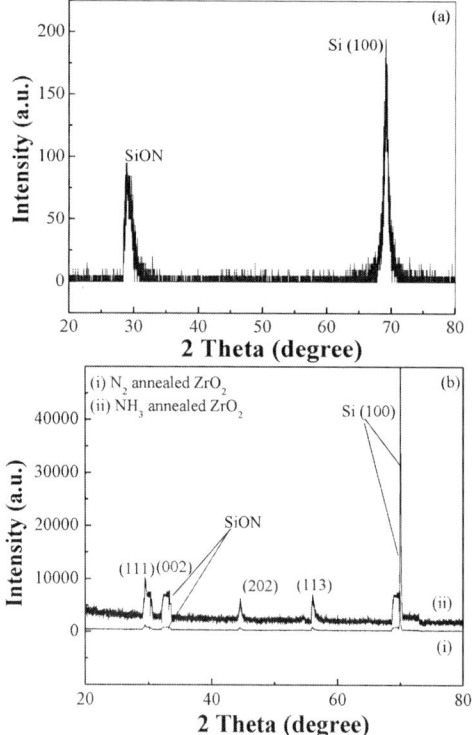

Fig. 4. (a) XRD patterns of (a) SiON IL (b) ZrO_2 films annealed in (i) N_2 and (ii) NH_3 ambients at 500 °C.

SiON IL indicating an intense peak at ~70° originating from the Si substrate (100) reflection and peak at nearly 31° confirms the presence of SiON [11-12]. Figure 4 (b) shows the XRD patterns of ZrO_2 thin film annealed in N_2 and NH_3 ambient showing the peaks corresponding to SiON interfacial layer and Si substrate (100).

After post annealing treatment of ZrO_2 films in N_2 and NH_3 ambients, it is worth to mention that crystalline peaks appeared after the improvement in degree of crystallization for NH_3 annealed ZrO_2 film. Furthermore, the tetragonal ZrO_2 phase is noticed for NH_3 and N_2 annealed ZrO_2 films which corresponds to (113), (002), (202) and (111) planes at 56.12°, 34.63°, 44.72° and 29.31° respectively. It can be suggested that ZrO_2 annealing in NH_3 is quite appreciable as the peaks are quite broader in comparison to N_2 annealed ZrO_2 and hence can replace the oxygen atom of ZrO_2 to form a thermally stable phase of ZrON.

C. Fourier Transform Infrared Spectroscopy

In order to investigate the type of elements and bonds present in the grown films, FTIR characterization has been performed. The FTIR spectrum of SiON IL was performed in absorption mode over the wave number range of 500-3500 cm^{-1} as shown by Figure 5 (a). Infrared absorption spectrum of the bare-Si wafer was used as background spectra to reflect the absorbance of grown films. Si-O-Si asymmetrical stretching is exhibited by the band assigned at 1072 cm^{-1} [13-14] whereas the band at 886 cm^{-1} reflects Si-N-Si stretching [15]. Further, the broad envelope at 3325 cm^{-1} associated with N–H stretching mode clearly indicates the presence of nitrogen [16]. The bond at 2584 cm^{-1} corresponds to Si-H stretching whereas the bond at 2918 cm^{-1} reflects the Si-OH stretching. A narrow peak at 3325.24 cm^{-1} indicates the presence of N-H stretching, which is basically the H_2 concentration in the film. However, the H_2 concentration seems to be quite low due to its bond breakage at increased deposition temperature [17] which indicates an improvement in the quality of grown SiON film.

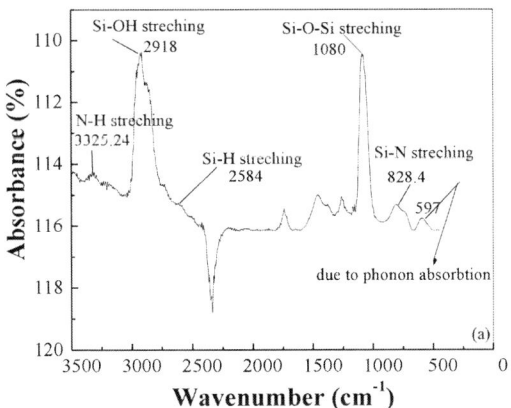

Fig. 5. (a) FTIR spectra of SiON interfacial layer.

Figure 5 (b) shows the absorbance mode FTIR spectra of the deposited HfO_2 film over the wave number range of 400-

1200 cm^{-1}. The sharp vibration bands observed between wave numbers 624-756 cm^{-1} validates the presence of HfO_2. The band at 1059.8 cm^{-1} corresponds to the peak position shifting of Hf-O towards the higher wave number series, and is somewhat reflecting its crystalline nature.

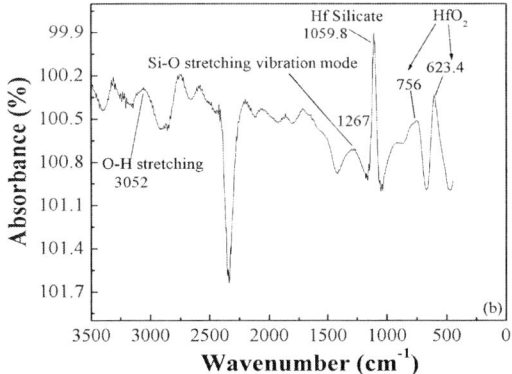

Fig. 5. (b) FTIR spectra of HfO_2 thin films.

D. C-V and I-V Measurements

Figure 6 (a) shows the capacitance-voltage characteristics of Ti-Pt/ZrO_2/SiON/n-Si and Ti-Pt/ZrON/SiON/n-Si MOS capacitors. From the graph, it is quite clear that ZrO_2 annealing in NH_3 resulted in higher capacitance as compared to ZrO_2 annealing in N_2 ambient under the same conditions.

This is mainly due to the segregation of significant amount of nitrogen incorporation into ZrO_2 thin film and also at the interface of SiON/ZrO_2 during ammonia annealing. The existence of negative effective oxide charge (Q_{eff}) in SiON/ZrO_2 and SiON/ZrON samples has shifted the flat band voltage (V_{fb}) towards positive voltage. The dielectric constant (k) was extracted from the accumulation region of C-V curve for ZrO_2 and ZrON based MOS capacitors and found to be 22.7 and 30. Here, all the calculations have been performed at high-frequency of 500 kHz. The calculated value of EOT for SiON/ZrO_2 and SiON/ZrON gate stack was 2.3 nm and 1.79 nm respectively. Figure 6 (b) shows the J-V (current density-voltage) characteristics of Ti-Pt/ZrON/SiON/n-Si and Ti-Pt/ZrO_2/SiON/n-Si MOS capacitors. Under negative bias conditions, the gate current conduction was observed mainly from the metal gate to the conduction band by the injected electrons. In reality, the conduction in the inversion region mainly occurs by the minority carrier generation from the interface states, back contact and the traps in bulk region [18]. The leakage current density was found to be 2.5×10^{-8} A/cm^2 and 2.2×10^{-9} A/cm^2 and at $V_g = \pm 1V$ for N_2 and NH_3 annealed ZrO_2 samples respectively.

IV. CONCLUSION

The physical and electrical characteristization of various types of MOS capacitors has been presented in detail. The impact of N_2 and NH_3 annealing ambients on the performance of ZrO_2 based gate stacks has been studied.

AFM study reveals smoothness of both ZrO$_2$ and ZrON films obtained after post deposition annealing of as-deposited ZrO$_2$ sample in N$_2$ and NH$_3$ ambients at 500 °C. The thicknesses of the SiON and ZrO$_2$ films were confirmed by ellipsometric measurements. XRD analysis indicates that NH$_3$ annealing can enhance the degree of crystallinity along with the formation of tetragonal phase for ZrON. Electrical characterizations namely C-V and J-V measurements indicate that NH$_3$ annealing is quite effective in increasing the capacitance, suppressing the gate leakage current density and effective oxide thickness (EOT). Apparently, the leakage current for NH$_3$ annealed ZrO$_2$ is lowest ~ 2.2×10^{-9} A/cm^2 relative to all other gate stacks. Furthermore, the EOT is reduced to about 1.79 nm for ZrON based MOS capacitor.

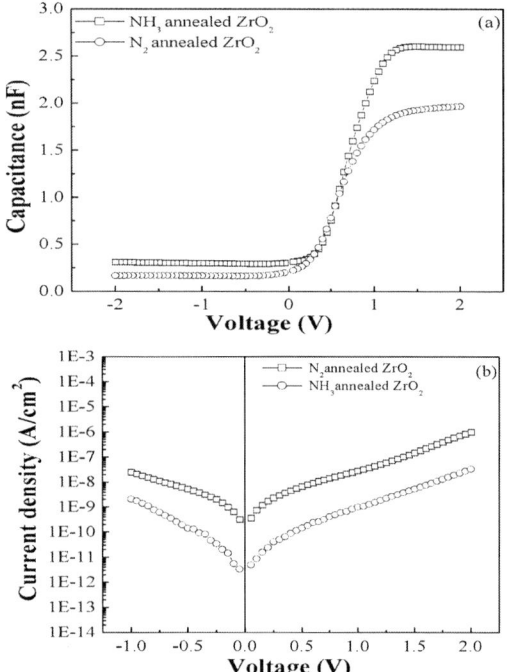

Fig. 6. (a) C-V characteristics and (b) J-V characteristics of Pt-Ti/ZrO$_2$/SiON/n-Si and Pt-Ti/ZrON/SiON/n-Si MOS capacitor.

The parameters calculated for various gate stacks at 500 KHz are shown in Table I as:

Table I

S. No.	Fabricated gate stack	Dielectric constant (k)	EOT (nm)	Leakage current density J (A/cm^2)
1.	Ti-Pt/HfO$_2$/SiO$_2$/n-Si	14	3.87	6.5×10^{-8}
2.	Ti-Pt/HfO$_2$/SiON/n-Si	14.4	3.32	2.4 × 10^{-9}
3.	Ti-Pt/ZrO$_2$/SiON/n-Si (As-deposited)	12	3.7	2.5 × 10^{-7}
4.	Ti-Pt/ZrO$_2$/SiON/n-Si (N$_2$ annealing)	23	2.3	2.5 × 10^{-8}
5.	Ti-Pt/ZrON/SiON/n-Si (NH$_3$ annealing)	30	1.79	2.2 × 10^{-9}

ACKNOWLEDGMENTS

The authors acknowledge the Centre of Excellence in Nanotechnology (CEN) at IIT Bombay under INUP scheme sponsored by DIT, MCIT and Government of India for providing us the excellent facilities of fabrication and characterization.

REFERENCES

[1] K. Nakajima, "Interface-state density of three dimensional silicon channels measured by charge pumping method", M.S. Thesis, *Tokyo Institute of Techonology*, 2012.

[2] R. G. Arns, "The other transistor: Early history of the metal-oxide semiconductor field effect transistor", *Engineering Science and Education Journal*, 7, pp. 233-240, 1998.

[3] S. E. Thompson, "A logic Nano technology featuring strained silicon", *IEEE Electron Device Letters*, 25, pp. 191-193, 2004.

[4] Y. Kim, "Conventional n-channel MOSFET devices using single layer HfO$_2$ and ZrO$_2$ as high-k gate dielectrics with poly silicon gate electrode", *IEEE International Electron Devices Meeting*, pp. 20.2.1-20.2.4, 2001.

[5] B. Cheng, M. Cao, R. Rao, A. Inani, P.V. Voorde, W. M. Greene, JMC. Stark, P. M. Zeitzoff and J. C. S. Woo, "The impact of high-k dielectrics and metal gate electrodes on sub-100 nm MOSFETs", *IEEE Transactions on Electron Devices*, 46, pp. 1537-1544, 1999.

[6] Y. K. Choi, N. Lindert, P. Xuans, S. Tang, D. Ha, E. Anderson, T. J. King, J. Bokor and C. Hu, "Sub-20 nm CMOS FinFET technologies", *International Electron Device Meeting Technical Digest*, pp. 421-424, 2001.

[7] N. Singh, "High-Performance fully depleted silicon Nano wire (diameter ≤ 5nm) gate all around CMOS devices", *IEEE Electronic Devices Letters*, 27, pp. 383-386, 2006.

[8] F. D. Agostino and D.Quercia, "Short-Channel Effects in MOSFETs", 2000.

[9] S. P. V. M. Rao, E. V. L. N. Rangacharyulu and K. L. Kishore, "Parameter Optimization of GAA Nanowire FET Using Taguchi Method", *International Journal of Engineering Research and technology*, 1, pp. 1-5, 2012.

[10] R. Gupta, D. Saikia and R. Vaid, "Argon annealed ALD-ZrO$_2$/SiON gate stack for advanced devices", *ECS Transactions*, 77, pp. 51-55, 2017.

[11] R. Gupta, R. Rajput, R. Prasher and R. Vaid, "Structural and electrical characteristics of ALD-HfO$_2$/n-Si gate stack with SiON interfacial layer for advanced CMOS technology", *Solid State Sciences*, 59, pp. 7-14, 2016.

[12] A. P. Huang, Z. F. Di, R. K. Y. Fu and P. K. Chu, "Improvement of interfacial and microstructure properties of high-k ZrO$_2$ thin films fabricated by filtered cathodic arc deposition using nitrogen incorporation", *Surface & Coatings Technology*, 201, pp. 8282-8285, 2007.

[13] H. Chakraborty and D. Misra, "Characterization of high-k gate dielectrics using MOS," *International Journal of Scientific and Research Publications*, 3, pp. 1-5, 2013.

[14] J. Robertson and R. Wallace, "High-k materials and metal gates for CMOS applications," *Materials Science and Engineering: R: Reports*, 88, pp. 1-41, 2015.

[15] J. H. Choi, Y. Mao and J. P. Chang, "Development of hafnium based high-k materials-A review", *Materials Science and Engineering R*, 72, pp. 97-136, 2011.

[16] R. Clark, "Emerging applications for high-k materials in VLSI technology", *Materials*, 7, pp. 2913-2944, 2014.

[17] N. Miyata, "Study of direct-contact HfO$_2$/Si interfaces" *Materials*, 5, pp. 512-527, 2012.

[18] R. Gupta and R. Vaid, "Effect of post deposition annealing on ALD-ZrO$_2$/SiON gate stacks for advanced CMOS Technology, *ECS Transactions*, 75 pp. 67-73, 2016.

In-depth Structure and Electrical Characteristics Study of HfOx-based Resistive Random Access Memories (ReRAMs)

B. Attarimashalkoubeh and Y. Leblebici

Abstract—**The variation issue has been known as the main hurdle in the commercialization of ReRAM technology, which encourages further studies to find a solution to resolve this issue. Less variation in switching parameters meaning better switching uniformity is only possible through precise control over formation/annihilation of filaments during the set and the reset process. In this work, we investigated more than 18 different HfO$_2$-based ReRAMs to control the variation issue, and different structural modification approaches have been studied. In the implementation of bi oxide HfO$_2$-based ReRAMs, the competency of the secondary oxide information of oxide in competition with Hf ions have been investigated and proved to affect the resistive switching significantly. Besides, the effect of the Ti buffer layer on the performance of HfO$_2$-based ReRAMs have been investigated.**

I. INTRODUCTION

Resistive Random Access Memories (ReRAMs) have been researched intensively in the last decades as one of the next-generation non-volatile memory (NVM) candidates to replace conventional charge-based memories for future computing systems. ReRAMs, with their comparatively simple structure and fabrication techniques, have been considered as one of the most promising candidates. Generally, ReRAMs consists of a metal-oxide layer sandwiched between two different metal and work based on the movement of vacancies [1]. During the set process, the oxygen vacancies are relocated and form conductive filaments (C.F.s) through the metal-oxide layer and switch ON the device (LRS), while partially/fully annihilation of the C.F.s can drift back the resistance to higher resistance states (HRS) during the so-called reset process. It is hard to control the switching mechanism of ReRAMs due to the movement of nano-scale oxygen vacancies. The variations issue in the key switching parameters has resulted from this uncontrolled formation of filaments with different geometries, which decrease the reliability of this technology and acts as a hurdle in commercializing it [2]. To solve this problem, a better understanding of the switching mechanism by having an in-depth knowledge of the interaction between layers at the interfaces is highly required [3]. In this study, several HfO$_x$-based memory devices (see Table 1) have been fabricated in

a systematic approach, and their D.C. electrical characteristics have been investigated. The Hafnium oxide material has been chosen considering its promising performance among all the other candidates (i.e., TaO$_x$, TiO$_x$, etc.). The fabrication process flow is shown in the Figure1a,b.

It has been previously reported which employing an extra metal layer has a significant effect in limiting the variation issue [4] [5] [6]. Therefore, the additional Ti metal buffer layer has been chosen carefully compatible with the adjacent metal-oxide as it is shown to be the best candidate in eliminating any additional side-effect originated from the mismatch with the metal-oxide beneath it [7]. This work provides detailed process flows for the fabrication steps and systematically investigates the resistive switching properties of more than 18 different devices. Different structural methods have been implemented to suppress the variation issue, such as bi-oxide layer (using Al$_2$O$_3$, Ta$_2$O$_5$ and TiO$_2$ as the secondary oxide together with HfO$_2$), and an additional metal buffer for the bi-oxide ReRAM. To have a better understanding and to find the optimized device, different parameters have been varied throughout this work, such as thickness and order of the oxide depositions.

II. METHOD AND MATERIALS

A. Process Flow and Fabrication

We implemented "in-VIA" design to vertically deposit the switching materials between the top and bottom electrodes (T.E., B.E.). A thin layer (5 nm) Ti has been deposited using D.C. sputtering (1000 W, 9.0 sccm Ar) on top of a 4" Si/SiO$_2$ test wafers to act as an adhesion layer for the Pt. Next, 50 nm Pt metal was deposited using D.C. sputtering (1000 W, 15.0 sccm Ar), followed with a short HMDS process and a thin coat of AZ3007 positive resist. The devices were soft baked (at 115° C to improve the adhesion of P.R. to the wafer), exposed (20 mW/Cm^2), and developed (using MF CD 26) to define the B.E. Next, we used a reactive ion etching (RIE) tool with Cl$_2$/Ar chemistry to pattern the B.E. made of Pt. A 100nm layer of *Low-Temperature Oxide* (LTO) has been deposited on the patterned B.E. using a *LPCVD* tool at 425° C, the LTO layer acts as a passivation layer and isolates T.E. and B.E. The second step of photolithography opened the VIAs within the LTO, followed by a wet etch step in buffered hydrofluoric acid (BHF) at 20° C. To deposit HfO$_2$ layer, we used an

---

B. Attarimashalkoubeh, and Y. Leblebici are with the Microelectronic System Laboratory (LSM), École polytechnique fédérale de Lausanne (EPFL), Lausanne, Switzerland, E-mail: behnoush.attarimashalkoubeh@epfl.ch

---

978-1-7281-3420-8/19 $31.00 © 2019 IEEE

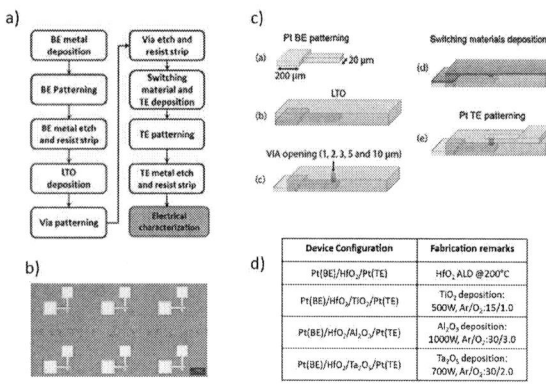

Fig. 1. a) Flow chart of fabrication steps, b) Schematic demonstration of process flow including the material deposition and the patterning steps, c)optical microscope image of a device after fabrication and d)Device configuration and fabrication details. The relatively larger pads (in yellow) are designed to provide sufficient space for the measurement steps.

Atomic layer deposition (ALD) tool with TEMAH and H_2O precursors for all memory devices at 200° C. The thickness of the deposited layer has been precisely controlled with the number of purges. Al_2O_3, TiO_2, Ta_2O_5, and metallic layer of Ti were deposited using sputtering tool. Finally, without breaking the vacuum, the Pt T.E. has been deposited and patterned as explained earlier for the B.E. Figure 1 provides more detail on the fabrication steps, with Fig.1a, schematically demonstrates the fabrication steps, Fig.1b shows an overview of the configuration of devices, and Fig.1c provides more detail for the deposition of switching materials.

To be able to attribute the switching performance of fabricated devices to their switching materials, all devices benefited from the same patterned B.E. and via; then the wafers have been diced and been used for any further fabrication steps. The design for each structure includes more than 300 different devices, providing enough samples for the analysis of device to device distribution.

B. Electrical Setup and Measurement

We utilized an Agilent B1500 parameter analyzer to study the electrical characteristics of the fabricated devices. The biases were applied to the T.E. while the B.E. was grounded during the measurement. A compliance current of 150 μA was enforced via an external transistor during forming and positive biasing of the devices. The compliance current is required to limit the flow of current in the low resistance state (LRS) and to protect the devices against hard breakdown. The devices were biased from 0 to 5 V and back to 0 V (0 $\rightarrow$ 5 $\rightarrow$ 0 V) for set and 0 to -1.5 V and back to 0 V (0 $\rightarrow$ -1.75 $\rightarrow$ 0 V) for reset. The switches of resistance from HRS to LRS (set process) in positive polarity and the LRS to HRS switch (reset process) in negative polarity exhibited the bipolar switching property regardless of the device configuration for all our fabricated devices.

III. RESULT AND DISCUSSION

For the consistency of the discussion, the paper is categorized in the following order. First, we studied the HfO_2-only ReRAMs (Pt(BE)/HfO_2/Pt(T.E.)) structures as the reference devices. Next, the implementation of bi-oxide ReRAMs are studied, this section provides an overview of the effect of the choice of the secondary oxide and the thicknesses on the electrical characterization of fabricated ReRAMs. Finally, the role of the buffer layer in improving the uniformity of electrical behavior of the devices has been studied.

A. HfO_2-alone ReRAMs

For HfO_2-alone device structure, we tried different thickness of HfO_2=1, 3, 5, 7, and 10 nm. The device was biased, as mentioned earlier. We observed the device with 1 nm of HfO_2 to be leaky (in pristine low resistance state), for HfO_2=3 nm even though the device initially appeared to be resistive, but after the forming sweep, the reset bias could not switch back the resistance to HRS, and the device failed at LRS. For a device with HfO_2=5 nm, we could successfully switch the device between HRS and LRS. Figure 2 shows the resistive switching (R.S.) behavior of the device for the first 100 cycles. The sweep plotted in red presents the first positive bias applied to our as-fabricated device (known as forming voltage) which switched the device at 4.2 V to LRS. The first negative bias (reset) could successfully bring back the device resistance state back to LRS. while the problem with the device with the HfO_2=5 nm appears to be the high variation in its switching parameters, the other devices with 7 and 10 nm also suffered from high forming voltage and unstable switching (V_f >5 V for the Pt/HfO_2(7 nm)/Pt).

B. Bi-Oxide HfO_2-based ReRAMs

The variation in the key switching parameters (V_{set}, V_{reset}, I_{HRS}, I_{HRS}) could be explained by stochasticity of the ion movements (oxygen vacancies) responsible for the set and reset operations [8]. In literature, different post-fabrication techniques have been introduced to suppress the undesired

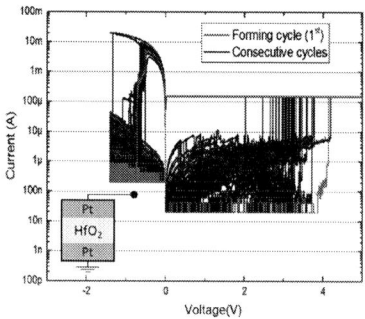

Fig. 2. The I-V characteristics of the Pt/HfO_2(5 nm)/Pt, the forming voltage is plotted in red while the next consecutive sweeps are plotted in black. The inset schematically demonstrates the biasing of the device.

variability issue of the single-layer ReRAMs, such as insertion of an external resistor in series with the ReRAM (to improve the LRS) or in parallel to improve the fluctuation in HRS states. However, there is not such a solution to stabilize the V_{set} and V_{reset} fluctuations, as it will not affect the stochasticity [9]. A promising approach to overcome this issue is the implementation of bi-oxide structures [9]. In this work, we inserted different thickness of TiO_2, Ta_2O_5 and Al_2O_5 in contact with the HfO_2(5 nm) layer. Figure 3 shows the I-V switching properties of the bi-oxide ReRAMs with 3 nm of the secondary oxide of Ta_2O_5, TiO_2, and Al_2O_3 - Pt(B.E.)/HfO_2(5 nm)/[Ta_2O_5, TiO_2, Al_2O_3](3 nm)/Pt(T.E.)- configuration. Besides the highlighted improvement in reliability of all implemented bi-oxide layer devices comparing to the HfO_2-only device choice of secondary oxide is noticeable; especially for I_{HRS}, V_{set} and the shape of reset. We assume the altered R.S. behavior among bi-oxide devices is due to their different ability in the formation of oxide in competition with the Hf in the adjacent HfO_2 layer.

The properties of formed conductive filaments in the bi-oxide devices are highly dependent on their standard Gibbs free energy of formation of oxide (ΔG) in competition with Hf in the layer beneath it. For instance, in bi-oxide device implementing Ta_2O_5 (with higher ΔG of oxidation in competition with Hf) the resulted R.S. is expected to be dissimilar to the devices employing Al_2O_3 and TiO_2 (with lower ΔG) as their secondary oxide. For Hf5Ta3, Hf has a lower ΔG comparing to Ta. Therefore it preferentially oxidized during the filament formation and injects some more oxygen vacancies in Ta_2O_5, especially near their interface. while in Hf5Ti3 and HfAl3 the situation is reversed, for instance in Hf5Al3, the Al reduces Hf in the HfO_2 layer and leaves behind oxygen vacancies, especially near the interface of the two oxides. Therefore, we can conclude that employing different secondary metal-oxide in contact with HfO_2 layer, will results in dissimilar R.S. as a result of different ΔG of oxidation in competition with the Hf, such as the utterly different reset observed for our bi-oxide devices (as is seen in Fig.3). Figure 4 presents a summary of Forming voltages (V_f) for the bi-oxide HfO_2-based ReRAMs with varied thickness of the secondary oxides. It is clear which the thickness of secondary oxide has a high impact on the R.S. of the devices, directly influencing V_f, V_{set} and V_{reset} and controlling the resistance values of HRS and LRS, for instance, the devices with higher thickness of the secondary oxide, require larger forming voltage. The high forming voltage is not desirable and generally leads to the degradation of performance; as an example, the device with Al_2O_3=7 nm failed after a few sweeps. Further device engineering is yet required to improve the uniformity of the resistive switching and lower the forming voltages. Next, to make a better conclusion of the bi-oxide devices, we reversed the order of deposition of the oxide layers (for HfO_2=5nm), Pt(B.E.)/TiO_2(3 nm)/HfO_2(5 nm)/Pt(T.E.) and Pt(B.E.)/TiO_2(5 nm)/HfO_2(5 nm)/Pt(T.E.). For devices with Pt(B.E.)/HfO_2/TiO_2/Pt(T.E.), the rupture/formation is assumed to happen in the TiO_2 layer, while for Pt(B.E.)/TiO_2/HfO_2/Pt(T.E.), the HfO_2 has this role.

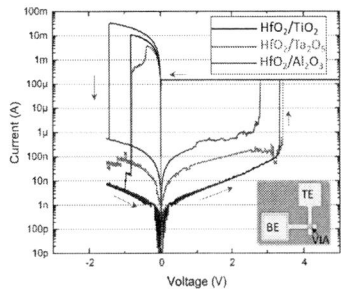

Fig. 3. I-V characterisation of bi-oxide devices with Pt/HfO_2(5 nm)/Al_2O_3(3 nm)/Pt (plotted in blue), Pt/HfO_2(5 nm)/Ta_2O_5(3 nm)/Pt (plotted in green), and Pt/HfO_2(5 nm)/TiO_2(3 nm)/Pt (plotted in black). The inset demonstrates a optical microscopic image of the devices, with B.E. and TE and the vias vertically between them.

Secondary oxide	V_f for 3nm(V)	V_f for 5nm(V)	V_f for 7nm(V)
Al_2O_3	3.4	3.9	6
TiO_2	3.9	4.1	5
Ta_2O_5	4.1	4.2	4.2

Fig. 4. The summary over forming voltages of bi-oxide ReRAMs HfO_2=5 nm, with secondary oxides for varied thicknesses.

Considering the HfO_2 layer is deposited using ALD tool, we expect higher quality rather than sputtered TiO_2 layer. Moreover, the vacuum for Pt(B.E.)/TiO_2/HfO_2/Pt(T.E.) was unavoidably broken to transfer the samples to ALD tool, which could has changed the stoichiometric of the TiO_2 layer, and reduced the creation of oxygen vacancies between the oxide layers in the interface. This assumption is confirmed by observation of higher forming voltage and lower resistance variation, especially of I_{HRS} for Pt(B.E.)/TiO_2/HfO_2/Pt(T.E.) comparing to the comparable thickness but Pt(B.E.)/HfO_2/TiO_2/Pt(T.E.).

C. Role of Buffer Layer

Another proved approach in improving the reliability of ReRAMs comprise the use of an active metal layer (buffer layer) acting as an oxygen reservoir/scavenger layer. This layer not only harvest oxygen ions from the layer beneath it (creation of oxygen vacancies), but also the oxygen ions can be stored there and later be injected back to the switching materials during the set and reset operations. In this work, we investigated the effect of insertion of Ti buffer layer (1, 3, 5 nm) on the performance of Pt/HfO_2(5 nm)/TiO_2(5 nm)/Pt ReRAMs (H5T5). The Ti=1 nm of the buffer layer resulted in relatively lower V_f (3.9 V) but was not effective in improving the variation issue observed in for the switching parameters of H5T5 device. Comparing the devices with Ti=3 and Ti=5 nm, the device with 5 nm Ti showed better switching properties with uniform R.S. properties (inset of Fig.5c). Further increase of Ti layer thickness (Ti=7 nm), led to the noticeably larger variation (comparing to the device with 5 nm Ti as buffer layer) in R.S. which eventually resulted in degradation of performance by cycling. Moreover, two devices with Pt/HfO_2(5 nm)/TiO_2(3 nm)/Pt ReRAMs (H5Ti3) and

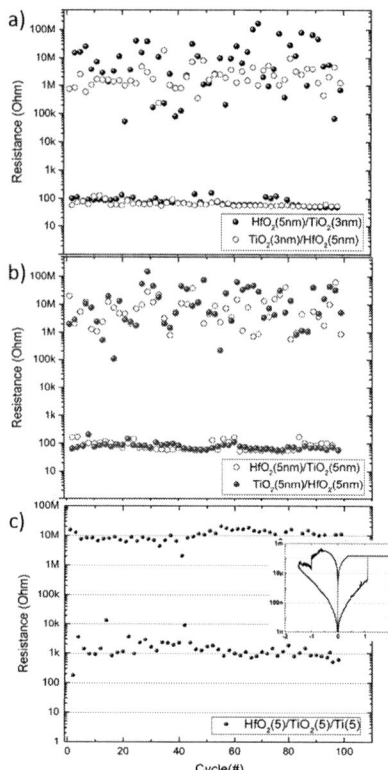

Fig. 5. Cycle to Cycle HRS/LRS distributions for devices with a) $TiO_2(3)/HfO_2(5)$ and $HfO_2(5)/TiO_2(3)$, and b)$TiO_2(5)/HfO_2(5)$ and $HfO_2(5)/TiO_2(5)$, and c)$HfO_2(5)/TiO_2(5)$ configurations.

Pt/HfO_2(5 nm)/TiO_2(3 nm)/Ti(5 nm)/Pt have been fabricated and their electrical characterization have been studied. We noticed the insertion of a 3 nm thin layer of has significantly influenced the R.S. comparing to the H5Ti3 device. The HRS for H5Ti3 has shown more resistive properties with larger variation (52 k →2.18 MΩ) while for the device with added 3 nm the distribution appeared to be significantly less resistive with a uniform switching properties (26 K→43 KΩ). Moreover, the insertion of 3 nm Ti on Hf5Ti3 device, effectively decreased the V_{set} changing from the range of 1.8-3.9 V for HfTi3 to 0.5-0.8 V for the device with extra Ti layer. This effect could be explained by the role of Ti as an active metal in creation of oxygen vacancies as the permanent defects throughout the switching materials; the more defective oxides are less resistive and required lower voltage for the formation of conductive filaments. Hence, even though the insertion of sufficiently thick buffer layer generally improves the R.S., but the stack structure of the ReRAMs should be precisely designed to reach a proper trade-off between required R.S. properties and cycle-to-cycle reliability, as the performance of the Hf5/Ti3 and H5/Ti5 with an extra Ti buffer layer appeared to be dissimilar. Among all devices with an added buffer

layer, the Pt/HfO_2(5 nm)/TiO_2(5 nm)/Ti(5 nm)/Pt showed the best switching properties with stable uniform switching among fabricated devices with uniform switching properties between HRS and LRS levels as is demonstrated in the Fig.5c.

IV. CONCLUSION

In this work, we investigated the switching properties of HfO_2-based ReRAMs with different stack structures in pursuance of improving the reliability issue observed in the HfO_2-only ReRAMs. The HfO_2-only ReRAMs appeared to suffer significantly from variation issue, resulted in the large fluctuations of their key switching parameters and degraded R.S. behavior. Three different metal oxides (TiO_2, Ta_2O_3 and Al_2O_3) introduced to the HfO_2-only devices and the observed switching properties explained by the different Gibbs free energy of oxidation. Moreover, the role of the Ti buffer layer on the device properties has been studied. As the active metal layer of Ti introduces oxygen vacancies to the switching materials of as-fabricated devices, the thickness of switching materials together with the thickness of Ti are considered to be important in defining the R.S. behavior of the ReRAMs. We observed the ReRAM with Pt/HfO_2(5 nm)/TiO_2(5 nm)/Ti(5 nm)/Pt demonstrated the best performance with sufficient on/off ratio and proper uniformity of switching parameters (V_{set}, V_{Reset}. Lowering the thickness of TiO_2 resulted in a significant reduction of on/off ratio.

REFERENCES

[1] R. Waser and M. Aono, "Nanoionics-based resistive switching memories," *Nature materials*, vol. 6, no. 11, pp. 833–840, 2007.

[2] G. Meijer, "Who wins the nonvolatile memory race?" *Science*, vol. 319, no. 5870, pp. 1625–1626, 2008.

[3] A. Paskaleva, B. Hudec, P. Jančovič, K. Fröhlich, and D. Spassov, "The influence of technology and switching parameters on resistive switching behavior of Pt/HfO2/TiN MIM structures," *Facta Universitatis, Series: Electronics and Energetics*, vol. 27, no. 4, pp. 621–630, 2014.

[4] R. Meyer, L. Schloss, J. Brewer, R. Lambertson, W. Kinney, J. Sanchez, and D. Rinerson, "Oxide dual-layer memory element for scalable non-volatile cross-point memory technology," in *Non-Volatile Memory Technology Symposium, 2008. NVMTS 2008. 9th Annual.* IEEE, 2008, pp. 1–5.

[5] H. Lee, P. Chen, T. Wu, Y. Chen, C. Wang, P. Tzeng, C. Lin, F. Chen, C. Lien, and M.-J. Tsai, "Low power and high speed bipolar switching with a thin reactive Ti buffer layer in robust HfO2 based RRAM," in *2008 IEEE International Electron Devices Meeting.* IEEE, 2008, pp. 1–4.

[6] B. Attarimashalkoubeh, J. Sandrini, E. Shahrabi, M. Barlas, and Y. Leblebici, "Effect of Hf metal layer on the switching characteristic of HfOx-based resistive random access memory," in *2016 12th Conference on Ph. D. Research in Microelectronics and Electronics (PRIME).* Ieee, 2016, pp. 1–4.

[7] C.-Y. Lin, C.-Y. Wu, C.-Y. Wu, T.-C. Lee, F.-L. Yang, C. Hu, and T.-Y. Tseng, "Effect of top electrode material on resistive switching properties of ZrO2 film memory devices," *IEEE Electron Device Letters*, vol. 28, no. 5, pp. 366–368, 2007.

[8] R. Degraeve, A. Fantini, N. Raghavan, L. Goux, S. Clima, B. Govoreanu, A. Belmonte, D. Linten, and M. Jurczak, "Causes and consequences of the stochastic aspect of filamentary rram," *Microelectronic Engineering*, vol. 147, pp. 171–175, 2015.

[9] A. Hardtdegen, C. La Torre, F. Cüppers, S. Menzel, R. Waser, and S. Hoffmann-Eifert, "Improved switching stability and the effect of an internal series resistor in HfO2/TiOx bilayer ReRAM cells," *IEEE Transactions on Electron Devices*, vol. 65, no. 8, pp. 3229–3236, 2018.

978-1-7281-3420-8/19 $31.00 © 2019 IEEE

Strong Electric Monopulses in Nonuniformly Doped Nitride Films under Negative Differential Conductivity

V. Grimalsky, S. Koshevaya, J. Escobedo-A., and J. Sanchez-S.

Abstract - The excitation of the strong nonlinear monopulses of space charge waves in the transversely non-uniform *n*-GaN and *n*-InN films is investigated theoretically. The stable numerical algorithms have been used for nonlinear 3D simulations. The monopulses of the strong electric field of durations 3 – 10 ps can be excited. The bias electric field should be chosen slightly higher than the threshold values for observing the negative differential conductivity. The doping levels should be moderate for the nitrides $10^{16}-10^{17}$ cm^{-3}. The electric monopulses of high peak values are excited from input small electric pulses. These nonlinear monopulses in the films differ from the domains of strong electric fields in the bulk semiconductors.

I. INTRODUCTION

The nitrides GaN, InN are used in the lower part of terahertz (THz) range $f = 100$ GHz – 1 THz to fabricate active and nonlinear devices. Recently it was demonstrated that *n*-InN possesses the increased values of the negative differential conductivity (NDC) [1]. The space charge waves (SCW) in *n*-GaN and *n*-InN films can be amplified due to NDC, when the magnitudes of bias electric fields are higher than the critical, or threshold, ones [2].

The linear amplification of SCW in *n*-GaN and *n*-InN films was investigated in the frequency range $f < 800$ GHz, where the non-local dependence of the electron velocity on the average electron energy was taken into account [2,3]. The typical thicknesses of the films were $2l = 0.2 - 1$ μm, their lengths were 10 – 50 μm, see Fig .1, a. Because the frequency range of amplification of SCW in the nitride films is wide and covers the lower part of THz range, it is of interest to investigate the excitation of strong nonlinear monopulses of picosecond durations without internal carrier frequency. The maximum values of the spatial increments of the linear amplification of SCW in the films achieve at the frequencies $f > 50$ GHz. Those differ from the volume crystals where the maximum increments correspond to the zero frequency [4].

In the films the influence of the boundaries of the properties of the films is principal. Namely, at the boundaries the electron mobility can be lower than in the center of the film and, moreover, NDC can be absent at the

V. Grimalsky, S. Koshevaya, J. Escobedo-A., and J. Sanchez-S. are with the Center for Investigations on Engineering and Applied Science (CIICAp), Institute for Investigations in Basic and Applied Science (IICBA), Autonomous University of State Morelos (UAEM), Av. Universidad 1001, 62209, Cuernavaca, Mor., Mexico. E-mail: v_grim@yahoo.com; svetlana@uaem.mx

boundaries, due to additional mechanisms of scattering of carriers. In the last case the non-uniform doping should be used that is increased in the center of the film $x = l$, see Fig. 1, a. The permittivities below and above the film are $\varepsilon_1 = 4$, SiO$_2$, and $\varepsilon_3 = 1$, air. The dependencies of the drift velocity and the average electron energy on the electric field used in simulations are given in Fig. 1, parts b, c [5].

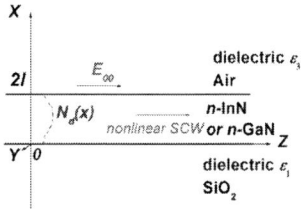

a)

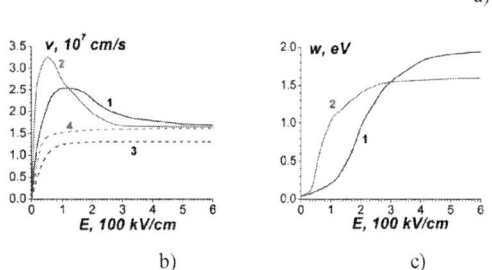

b)   c)

Fig. 1. Part a) is the geometry of the problem. $N_d(x)$ is the non-uniform doping profile. The nonlinear space charge waves are formed as strong monopulses of picosecond durations. The pulses are generally localized also along *OY* axis. Parts b), c) are dependencies of the drift velocity v and the average electron energy w on the electric field for the zinc blende *n*-GaN, curves 1, and *n*-InN, curves 2. Curves 3 and 4 are the used dependencies $v(E)$ at the boundaries of the non-uniform films GaN and InN.

II. BASIC EQUATIONS

The simplest non-local electron hydrodynamic is used jointly with the Poisson equation [2, 4, 6]. The dynamics of the electron gas is described by the total concentration n, the average velocity v, and the average electron energy w there. Below the processes are considered within the frequency range $f \leq 300$ GHz, or with the temporal scales $\geq$ 3 ps, so the diffusion-drift equation for the total electron concentration is valid:

978-1-7281-3420-8/19 $31.00 © 2019 IEEE

$$\frac{\partial n}{\partial t} + \frac{\partial j_x}{\partial x} + \frac{\partial j_y}{\partial y} + \frac{\partial j_z}{\partial z} = 0, \quad j_x = nv_x - D\frac{\partial n}{\partial x};$$

$$j_y = nv_y - D\frac{\partial n}{\partial y}; \quad j_z = nv_z - D\frac{\partial n}{\partial z}; \quad v_x = \mu(E)E_x,$$

$$v_y = \mu(E)E_y, \quad v_z = \mu(E)E_z; \quad \mu(E) \equiv \frac{v}{E}, \tag{1}$$

$$\frac{4}{3v_w(w)}\frac{\partial v}{\partial t} + v = v_d(E), \quad E \equiv (E_z^2 + E_x^2 + E_y^2)^{1/2},$$

$$E_z = E_{00} - \frac{\partial \varphi}{\partial z}, \quad E_x = -\frac{\partial \varphi}{\partial x}, \quad E_y = -\frac{\partial \varphi}{\partial y}.$$

Here μ, D are the coefficients of the electron mobility and diffusion:

$$\mu = \frac{e}{m^* v_p}; \quad D = \frac{T}{m^* v_p} \equiv \frac{\mu}{e}T. \tag{2}$$

In turn, v_p, v_w are the relaxation frequencies of momentum and energy, m^* is the effective electron mass, T is the electron temperature in energetic units, $E_{00} \equiv E_{00z}$ is the bias electric field. It is assumed that v_p, v_w, m^* are the functions of w.

The dependencies of the relaxation frequencies v_p, v_w can be calculated from Eqs. (1) in the stationary case $\partial/\partial t = 0$, taking into account the dependencies presented in Fig. 1, b, c. The relaxation frequency of the momentum is $v_p \gg v_w$; $v_w \approx 10^{12}$ s^{-1} for InN, $v_w \approx 10^{13}$ s^{-1} for GaN [2]. Therefore at the frequencies of SCW $f < 1$ GHz it is possible to neglect by the inertia of the electron gas.

The simplest model of the non-local dependence of the drift velocity is used here. In Eqs. (1) $v_d(E)$ is the stationary dependence of the drift velocity on the electric field presented in Fig. 1, b.

The equations for the dynamics of the electron gas should be added by the Poisson equation for the electric field potential. Note that the potential $\varphi = \varphi_0(x) + \widetilde{\varphi}(z,x,y,t)$ includes both the stationary part $\varphi_0(x)$ due to the non-uniform doping and the variable in time one $\widetilde{\varphi}$ due to the propagation of SCW.

$$\frac{\partial^2 \varphi}{\partial x^2} + \frac{\partial^2 \varphi}{\partial y^2} + \frac{\partial^2 \varphi}{\partial z^2} = \begin{cases} -\dfrac{e(n - N_d(x))}{\varepsilon_0 \varepsilon_2}, & 0 < x < 2l; \\ 0, & x < 0, \quad x > 2l. \end{cases} \tag{3}$$

The non-uniform doping is considered in the form:

$$N_d(x) = N_{d0}\exp(-((x - l)/x_d)^2). \tag{4}$$

Under the non-uniform doping the stationary concentration $n_0(x)$, the electric potential $\varphi_0(x)$, and the additional electric field $E_0(x)$ have been calculated from the following set of equations:

$$\frac{d^2\varphi_0(x)}{dx^2} = -\frac{e}{\varepsilon_0\varepsilon_2}(n_0(x) - N_d(x)), \; E_0(x) \equiv -\frac{d\varphi_0(x)}{dx};$$

$$n_0(x) = C \cdot \exp(-\frac{e\varphi_0(x)}{k_B T_e}), \quad C = \frac{\displaystyle\int_0^{2l} N_d(x)dx}{\displaystyle\int_0^{2l} \exp(-\frac{e\varphi_0(x)}{k_B T_e})dx}. \tag{5}$$

Eqs. (5) have been solved by the Newton method. The Poisson equation for the variable electric potential of SCW is

$$\frac{\partial^2 \widetilde{\varphi}}{\partial x^2} + \frac{\partial^2 \widetilde{\varphi}}{\partial y^2} + \frac{\partial^2 \widetilde{\varphi}}{\partial z^2} = \begin{cases} -\dfrac{e(n - n_0(x))}{\varepsilon_0 \varepsilon_2}, & 0 < x < 2l; \\ 0, & x < 0, \quad x > 2l. \end{cases} \tag{6}$$

III. LINEAR AMPLIFICATION OF SCW

To investigate the amplification of linear SCW, the solution of linearized Eqs. (1,2) is searched as the travelling waves:

$$\widetilde{n}, \widetilde{w}, \widetilde{\varphi} \sim \exp(i(\omega t - kz)). \tag{7}$$

The circular frequency $\omega \equiv 2\pi f$ is real here, whereas the longitudinal wave number is complex $k \equiv k' + ik''$. The amplification of SCW occurs when $k'' > 0$.

In the non-uniform films the dependence of the electron mobility on the coordinate x is taken as

$$\mu(x) = \mu(l) - (\mu(l) - \mu(0))\Phi(x),$$
$$\Phi(x) \equiv \exp(-(x/x_0)^2) + \exp(-((2l - x)/x_0)^2). \tag{8}$$

Here $\mu(l)$, $\mu(0)$ are the mobilities in the center and at the boundaries of the film taken from Fig. 1, b.

In Fig. 2 there are the results of the simulations of the dependencies of the spatial increments of amplification k'' on frequency of linear SCW f. In Fig. 2 the part a) is for n-GaN film. For the curves 1 – 6 the bias electric field is $E_{00} = 1.4\cdot10^5$ V/cm that corresponds to a small value of NDC. The curve 1 is for the uniform film of the thickness $2l = 0.2$ μm, the equilibrium electron concentration is $n_0 = N_d = 7\cdot10^{16}$ cm^{-3}. The curve 2 is for $2l = 0.4$ μm, $n_0 = N_d = 3.5\cdot10^{16}$ cm^{-3}. The curve 3 is for the film $2l = 0.4$ μm with non-uniform conductivity with the scale $x_0 = 0.1$ μm (see Eq. 8), $n_0 = N_d = 3.5\cdot10^{16}$ cm^{-3}. The curve 4 is for the film $2l = 0.4$ μm with non-uniform conductivity with the scale $x_0 = 0.1$ μm (see Eq. 8), $n_0 = N_d = 7\cdot10^{16}$ cm^{-3}.

In the films with the non-uniform conductivity the increments are essentially smaller than in the uniform films. The increments can be increased in the films with the non-uniform doping. The curves 5, 6 are for the nonuniformly doped films with the scales $x_d = 0.15$ μm and $x_d = 0.2$ μm. The *average* doping levels are $3.5\cdot10^{16}$ cm^{-3} there. For a comparison, the curve 7 is for a uniform film $2l = 0.4$ μm, $n_0 = N_d = 3.5\cdot10^{16}$ cm^{-3}, but the bias electric field corresponds to higher NDC $E_{00} = 1.5\cdot10^5$ V/cm.

978-1-7281-3420-8/19 $31.00 © 2019 IEEE

The analogous results have been obtained for n-InN films, see Fig. 2, b. But the bias electric fields and doping levels are smaller compared with the case of n-GaN films. The parameters are as follows. The curves 1 - 6 are for the bias field $E_{00} = 0.52 \cdot 10^5$ V/cm. The curve 1 for is for $2l = 0.2$ µm, $n_0 = N_d = 4 \cdot 10^{16}$ cm^{-3}; 2 is for $2l = 0.4$ µm, $n_0 = N_d = 2 \cdot 10^{16}$ cm^{-3}. The curves 3, 4 are for the films $2l = 0.4$ µm with the non-uniform conductivity $x_0 = 0.1$ µm, the concentrations are $2 \cdot 10^{16}$ cm^{-3} and $5 \cdot 10^{16}$ cm^{-3}. The curves 5, 6 are for the nonuniformly doped films $2l = 0.4$ µm, $x_0 = 0.1$ µm, $x_d = 0.15$ µm and $x_d = 0.2$ µm. The curve 7 is for $E_{00} = 0.55 \cdot 10^5$ V/cm, $n_0 = N_d = 2 \cdot 10^{16}$ cm^{-3}.

When the bias electric field is $E_{00} > 1.4 \cdot 10^5$ V/cm in n-GaN films the increments increase sharply; analogous situation is for n-InN films when $E_{00} > 0.52 \cdot 10^5$ V/cm.

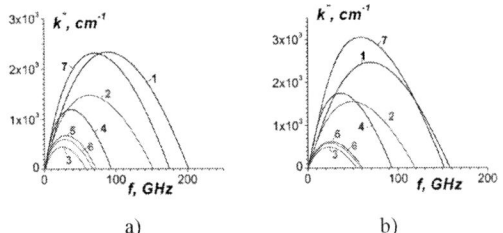

a)   b)

Fig. 2. Spatial increments of amplification of linear SCW. Prt a) is for n-GaN fims, b) is for n-InN films. Curves 1, 2, 7 are for uniform films, curves 3, 4 are for non-uniform conductivity but uniform doping, 5, 6 are for non-uniform conductivity and non-uniform doping.

IV. EXCITATION OF NONLINEAR MONOPULSES

The nonlinear dynamics of SCW nonlinear pulses has been simulated by means of the diffusion-drift equations jointly with the Poisson equation added by boundary conditions Eqs. (1), (2).

The equation for the electron concentration has been solved by the splitting with respect to physical factors [7]. The first fractional step is along OX axis, the second one is along OY, the third one is along OZ.

It is assumed the absence of the surface charge at the boundaries of the film $x = 0$, $x = 2l$. Here the electric boundary conditions are the continuity of the potential and the normal component the electric induction.

At the ends of the film the boundary conditions for the concentration are $n(z=0.x,y)=n_0(x)$ and $\partial n/\partial z(z=L_z) = 0$, the last one corresponds to the Ohmic junctions [4]. The boundary conditions $n(y=0$ or $L_y,z,x)=n_0(x)$ are used at $y = 0$ and $y = L_y$. But under simulations the film is assumed enough wide, so an influence of the boundary conditions along OY is not essential. For the variable electric potential the boundary conditions are $\tilde{\varphi} = 0$ at $z = 0$, $z = L_z$.

The Poisson equation for the electric potential $\tilde{\varphi}$ has been solved by the fast Fourier transform with respect to z and y [7], for real functions.

The initial pulses of SCW are excited by wide-band planar waveguide; the exciting field is:

$$E_z^{exc} = A \exp\left(-\left(\frac{t-t_1}{t_0}\right)^2 - \left(\frac{z-z_1}{z_0}\right)^2 - \left(\frac{y-L_y/2}{y_0}\right)^2\right). \quad (9)$$

Here z_1, z_0 are the positions of the center of the exciting element and its half-width, A is the amplitude, $A << E_{00}$. The exciting field is uniform along x, $0 < x < 2l$.

In the case when the bias electric field E_{00} corresponds to the maximum of NDC, initial strong amplification of the small pulse occurs. But then the amplified pulse deforms and as a result several powerful oscillations are formed at the output. Therefore, to create stable monopulses, the bias electric field should be chosen slightly above the threshold of NDC. Namely in this case the short monopulses are formed at the essentially nonlinear stage at the output antenna $z = z_2 \leq L_z$. Also the moderated doping levels should be applied $n_0 = 3 \cdot 10^{16} - 10^{17}$ cm^{-3} for n-GaN and $n_0 = 2 \cdot 10^{16} - 10^{17}$ cm^{-3} for n-InN films.

The typical results of numerical simulations are presented in Figs. 3-4. There are dependencies on time t of the variable part of z-component of the electric field of SCW in the center of the output antenna: $\tilde{E}_z(z = z_2, x = 2l, y = L_y/2, t) \equiv E_z - E_{00} - E_0$. The bias electric field E_{00} is marked by dot lines.

In Fig. 3 the parameters are as follows. There are n-GaN films, the threshold of NDC is $E_c = 1.25 \cdot 10^5$ V/cm. The bias electric field is $E_{00} = 1.4 \cdot 10^5$ V/cm. The parameters of the input pulse are $A = 0.5$ kV/cm, $t_1 = 10$ ps, $t_0 = 5$ ps, $z_1 = 5$ µm, $z_0 = 0.5$ µm, see Eq. 9.

In Fig. 3, part a), the results are for the uniform films. The curve 1 is for the film of the thickness $2l = 0.2$ µm, $n_0 = N_d = 7 \cdot 10^{16}$ cm^{-3}, the length is $L_z = 40$ µm, the output antenna is at $z_2 = 39$ µm. The curve 2 is for the film $2l = 0.4$ µm, $n_0 = N_d = 3.5 \cdot 10^{16}$ cm^{-3}, $L_z = 65$ µm, $z_2 = 64$ µm.

The part b) is for non-uniform films of the thicknesses $2l = 0.4$ µm, the scale of the non-uniformity of the conductivity is $x_0 = 0.1$ µm. The curve 1 is for the film with uniform doping $n_0 = N_d = 7 \cdot 10^{16}$ cm^{-3}, the length is $L_z = 50$ µm, the output antenna is at $z_2 = 49$ µm. The curves 2 and 3are for the films with the non-uniform doping with the scales $x_d = 0.15$ µm and 0.2 µm, $N_d = 3.5 \cdot 10^{16}$ cm^{-3}, $L_z = 40$ µm, $z_2 = 38$ µm and $L_z = 50$ µm, $z_2 = 49$ µm.

In Fig. 4 the results are for n-InN films. The bias electric field is $E_0 = 0.52 \cdot 10^5$ V/cm. The parameters of the input pulse are $A = 0.3$ kV/cm, $t_1 = 10$ ps, $t_0 = 5$ ps, $z_1 = 5$ µm, $z_0 = 0.5$ µm, see Eq. 9.

In Fig. 4, part a), the results are for the uniform films. The curve 1 is for the film $2l = 0.2$ µm, $n_0 = N_d = 4 \cdot 10^{16}$ cm^{-3}, $L_z = 38$ µm, $z_2 = 37$ µm. The curve 2 is for the film $2l = 0.4$ µm, $n_0 = N_d = 2 \cdot 10^{16}$ cm^{-3}, $L_z = 55$ µm, $z_2 = 54$ µm.

The part b) is for non-uniform films of the thicknesses $2l = 0.4$ µm, the scale of the non-uniformity of the conductivity is $x_0 = 0.1$ µm. The curve 1 is for the film with uniform doping $n_0 = N_d = 5 \cdot 10^{16}$ cm^{-3}, the length is $L_z = 50$ µm, the output antenna is at $z_2 = 49$ µm. The curves 2 and 3are for the films with the non-uniform doping with the scales x_d = 0.15 μm and 0.2 μm, N_d = 2·10^{16} cm^{-3}, L_z = 45 μm, z_2 = 44 μm and L_z = 70 μm, z_2 = 69 μm.

It is seen that the strong electric monopulses are formed at the output antenna. The maximum values of the output pulses exceed several times the bias electric field E_{00}. The durations of the output pulses are 3 – 10 ps. The monopulses are formed practically without any pedestal and differ from the domains in bulk semiconductors [4] that have the electric fields below the NDC threshold outside of the domain.

The results of numerical simulations are tolerant to changes of the parameters of input pulses and of the lengths of the films L_z. Under the uniform doping the shapes of the output pulses depend weakly on the transverse widths of the input pulse y_0 when $y_0 \geq 20$ μm. Under the non-uniform doping the monopulses are excited rather due to the non-uniformity of the electric field near the input end of the film $z = 0$. But the pulses at the output antenna possess the similar shape as in the case of the uniform doping.

The $\widetilde{E}_z$ component of the variable electric field is dominating in the nonlinear monopulses. This component is uniform along the thickness of the film, see Figs. 5, 6, a.

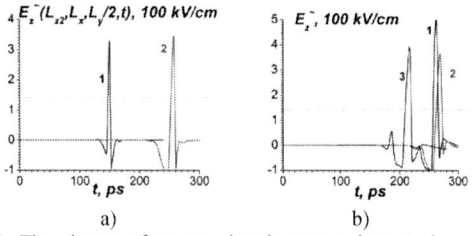

a)   b)

Fig. 3. The shapes of strong electric monopulses at the output antenna in n-GaN films. Part a) is for the uniform films; b) is for the films with non-uniform conductivity, curves 2, 3 also with the non-uniform doping.

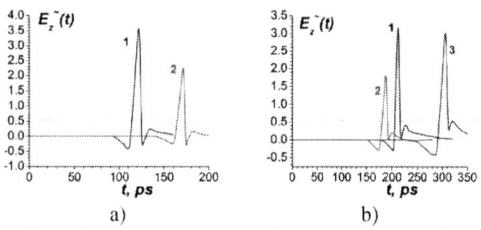

a)   b)

Fig. 4. The shapes of strong electric monopulses at the output antenna in n-InN films. Part a) is for the uniform films; b) is for the films with non-uniform conductivity, curves 2, 3 also with the non-uniform doping.

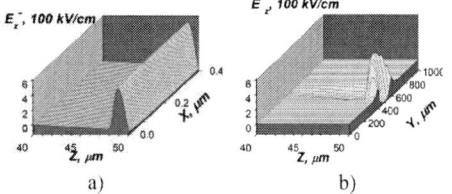

a)   b)

Fig. 5. The dependencies of $E_z\tilde{}(z,x,y=L_y/2)$ and $E_z\tilde{}(z,x=0,y)$ at t = 262 ps, non-uniform conductivity, see Fig. 3, b), curve 1.

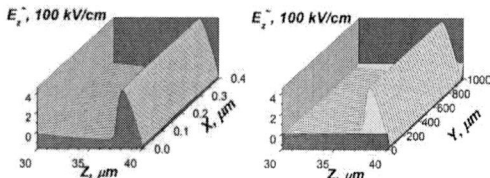

Fig. 6. The dependencies of $E_z\tilde{}(z,x,y=L_y/2)$ and $E_z\tilde{}(z,x=0,y)$ at t = 280 ps, non-uniform conductivity and non-uniform doping, see Fig. 3, b), curve 2.

V. CONCLUSIONS

The excitation of short strong monopulses of space charge waves of durations 3 – 10 picoseconds can be realized in the nitride films n-GaN и n-InN under the negative differential conductivity. The monopulses are formed from the input electric pulses of small amplitudes both in the uniform films and in films with the non-uniform conductivity and the non-uniform doping. The monopulses are realized under the amplification in the essentially nonlinear regime. The bias electric fields should be chosen slightly higher than the thresholds of the negative differential conductivity under the moderate doping levels. These monopulses differ from the domains of the strong electric fields in the bulk semiconductors.

ACKNOWLEDGEMENT

The authors are grateful to SEP-CONACyT, Mexico, for a partial support of our work.

REFERENCES

[1] P. Siddiqua, W.A. Hadi, A.K. Salhotra, M.S. Shur, and S. K. O'Leary, "Electron transport and electron energy distributions within the wurtzite and zinc-blende phases of indium nitride: Response to the application of a constant and uniform electric field", *Journ. Appl. Phys.*, 2015, vol. 117.

[2] E. Jatirian Foltides, V. Grimalsky, S. Koshevaya, and J. Escobedo-Alatorre, "Amplification of space charge waves in n-InN films of THz range", *Proc. IEEE Latin America Microwave Conference LAMC-2016*, Puerto Vallarta, Mexico, 12-14 Dec., 2016.

[3] V. Grimalsky, S. Koshevaya, M. Tecpoyotl-T., F. Diaz-A.,"Influence of nonlocality on amplification of space charge waves in n-GaN films", *J. Electromagn. Analysis & Applic. (JEMAA)*, 2011, vol. 3, no 2, pp. 33-38,

[4] S.M. Sze and Kwok N. Ng, *Physics of Semiconductor Devices*, Hobokem NJ: Wiley-Interscience, 2007.

[5] M. Levinshtein, S. Rumyantsev, and M. Shur, *Properties of Advanced Semiconductor Materials: GaN, AlN, InN*, New York: Wiley, 2001.

[6] K. Tomizawa, *Numerical Simulation of Submicron Semiconductor Devices*, Boston: Artech House Publ., 1993.

[7] W.H. Press, S.A. Teukolsky, W.T. Vetterling, and B.P. Flannery, *Numerical Recipes in Fortran*, Cambridge: Cambridge Univ. Press, 1997.

978-1-7281-3420-8/19 $31.00 © 2019 IEEE

Arrays of Bowtie Plasmonic Nanoantennas for Field Enhancement in MOEMS

M. Obradov, Z. Jakšić, I. Mladenović, D. Tanasković, and O. Jakšić

Abstract – Many micro(nano)optoelectromechanical systems (MOEMS, NOEMS) require optical (generally, electromagnetic) field localization and concentration. These include for instance photocatalytic microreactors and labs on a chip, where it is necessary to localize optical energy into a fluidic channel. Other examples are chemical and biological sensors. Plasmonics on the other hand ensures field localization down to subwavelength volumes where evanescent fields can be tailored to the shape of minuscule channels in MOEMS and NOEMS. In this work we present a possible approach to the enhancement of optical fields in MOEMS and NOEMS systems where a linear array of plasmonic bowtie structures is used to concentrate the optical field into a dielectric channel. We perform our numerical simulations using the finite element method to analyze field distributions that can be achieved by the use of the bowtie antenna and the possibility to tailor these fields. We also analyze the influence of the shape of the coupled tips of bowties to the field distribution and frequency dispersion. We conclude that arrays of plasmonic bowties could be a promising candidate for optically assisted micro and nanofluidics.

I. INTRODUCTION

Plasmonics is a rapidly expanding field in which one utilizes metal (or generally, conductor based on free electrons)-dielectric nanocomposites with subwavelength details to tailor the optical response of photonic devices. One of the principal uses of plasmonics is optical (ultraviolet, visible or infrared) field localizati-on/concentration at a subwavelength level [1]. This ensures a host of practical applications, which encompass among others ultrasensitive chemical or biological sensors, enhanced photodetectors including solar cells, photocatalytic systems including photochemical MOEMS microreactors, surface enhanced spectroscopy, optical modulators, etc. [1]. Maybe the most important goal of plasmonics is the fabrication of all-optical integrated circuits with the dimensions of VLSI and ULSI able to operate at optical frequencies [2].

One of the most effective ways to reach spatially precise and ultra-high localization of the optical fields is to use nanoantennas [3], thus ensuring a single tailorable nanofocus. A typical example of these is the plasmonic bowtie nanoantenna [4], a quasi-2D structure made of metal or another Drude-type conductor like e.g. graphene

M. Obradov, Z. Jakšić, I. Mladenović, D. Tanasković and O. Jakšić are with the Institute of Chemistry, Technology and Metallurgy, Center of Microelectronic Technologies, University of Belgrade, Njegoševa 12, Belgrade, Serbia, email: marko.obradov@nanosys.ihtm.bg.ac.rs

[5] or titanium nitride [6]. It ensures field concentration at subwavelength level using both the effect of sharp tips and the proximity effect between them [7].

In this work we consider the enhancement of optical fields in a dielectric channel using an array of plasmonic bowtie structures. The general geometry is shown in Fig. 1 and consists of an array of bowtie nanoantennas on a dielectric substrate with their tips in the direction perpendicular to a dielectric channel which is shaped in the simple form of a rectangular parallelepiped. The idea is to increase the intensity of the optical field in the channel using localization between the coupled tips of the bowties. We also consider three variations in the shapes of the bowtie tips (Fig. 2), sharp, rounded and flat.

We utilized the finite element method (Comsol multiphysics®) to calculate the response of our structures – electromagnetic field intensity in the channel and frequency dispersion of the reflection and transmission coefficients.

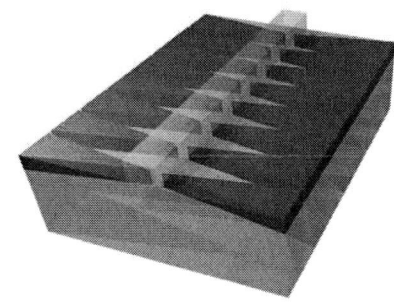

Fig. 1. General geometry of the plasmonic bowtie field enhancers; the rectangular parallelepiped on the top is our dielectric channel

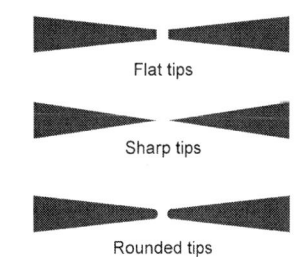

Fig. 2. Various shapes of bowtie proximal tip pairs as field enhancers

II. THEORY

For most conductors with free carrier plasma their electromagnetic properties in the optical range are well described by lossy extended Drude model. For such materials the frequency dispersion of their complex relative dielectric permittivity $\varepsilon(\omega)$ is given by [8]

$$\varepsilon(\omega) = \varepsilon_\infty - \frac{\omega_p^2}{\omega^2 + i\gamma\omega} \ , \qquad (1)$$

where ω_∞ is the asymptotic dielectric permittivity and $\gamma = 1/\tau$ is the characteristic frequency related to the damping of electron oscillations due to collisions, where τ is the relaxation time of the electron gas and plasma frequency is determined by the concentration of free carriers

$$\omega_p = \frac{ne^2}{m^*\varepsilon_0} \ , \qquad (2)$$

where n is electron concentration, e is the free electron charge ($1.6 \cdot 10^{-19}$ C), ε_0 is the dielectric permittivity of the vacuum ($8.854 \cdot 10^{-12}$ F/m), and m^* is the effective mass of electrons.

Resonant coupling between free electron plasma in conductive materials and optical radiation is the basis of the function of many types of nanoantennas, albeit not all. The structures we consider here are exclusively of plasmonic type.

Nanoantennas may be regarded as an extension of the concept of conventional radiofrequent antennas into the short frequency range where the dimensions of the antennas scale down into the nanometer domain (e.g. [9, 10]). They may function as receivers (concentrating electromagnetic waves from the far field into the near field) or emitters (scattering near field waves into the far field). Due to the full scalability of Maxwell's equations, the calculation of macroscopic antennas can be straightforwardly extended into nanodomain, naturally taking into account the difference in the dispersion properties of materials. One can find excellent reviews on nanoantennas generally in [3, 11-14].

When in the receiving mode, nanoantennas made of plasmonic materials concentrate far field radiation into a sharply defined subwavelength domain. Field enhancements in such a hotspot can easily exceed several orders of magnitude. Various practical applications of plasmonic nanoantennas are based on this, e.g. [15-18].

A plasmonic bowtie nanoantenna is a structure consisting of two approximately triangular plates made of plasmonic material with their tips facing each other, but being separated by a narrow gap. Thus such a structure combines the proximity effect that resonantly couples electromagnetic field between two triangular nanoparticles and the effect of sharp conductive tips that concentrate fields on their own. Exceptionally high field concentrations become possible in this type of structures. Bowtie nanoantennas are for instance treated in [4, 19-21].

III. RESULTS AND DISCUSSION

We examined the optical properties of our bowtie nanoantenna array shown in Fig. 1 using RF module of Comsol Multiphysics software package. The periodicity of the structure is 100 nm. The length of the entire bowtie is 300 nm and its thickness is 10 nm. The width of the both bases at the distal sides is 80 nm and the distance between the proximal tips is 10 nm. The tip width for bowties with flat tips is 10 nm and the radius of the curvature for the rounded tips is 5 nm. Bowties are made from nickel with optical parameters taken from the literature [22]. The nanoantenna array is deposited on a dielectric substrate with a refractive index $n=1.4$. The entire array is embedded within a dielectric layer with a refractive index $n=1.2$ into which a 40 nm wide channel has been built. For the sake of simplicity Fig.1 only depicts the channel directly above the bowtie tips, not the whole dielectric layer in which the channel has been dug. It is assumed that the channel is empty i.e. filled with air. Incident optical radiation arrives from the side of the substrate (backside illumination). Normal incidence is assumed.

Our numerical simulations determine the spectral and spatial properties of a TM polarized plane wave interacting with the nanoantenna array. Two parallel ports were added above and below the structure to introduce and collect electromagnetic radiation within the simulation domain. The active port is positioned below the structure for the light to enter the domain from the bottom. Floquet boundary conditions are applied to the edges of computation domain to simulate periodicity of the structure, allowing us to calculate response of the entire array using only a single bowtie. The parametric frequency sweep is used to determine the dispersive properties of the scattering parameters.

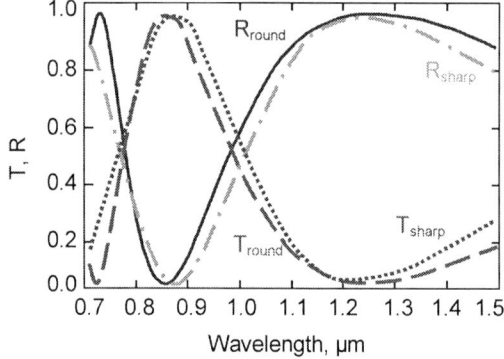

Fig. 3. Frequency dispersion of reflection coefficient for bowties with rounded (full line) and sharp tips (dash-doted line). Transmission coefficient for bowties with rounded (dashed line) and sharp tips (doted line).

978-1-7281-3420-8/19 $31.00 © 2019 IEEE

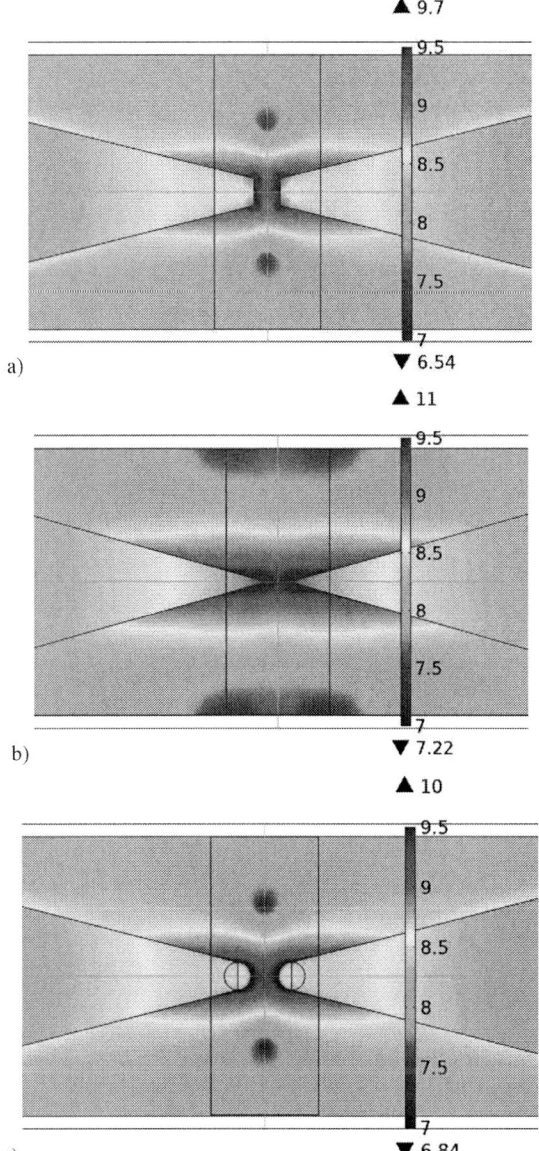

Fig. 4. Top view of field distribution in bowtie gap for various tip shapes around 850 nm wavelength: a) flat tips, b) sharp tips and c) rounded tips. Side bar is decade logarithm of field intensity.

The frequency dispersion of the reflection and the transmission coefficients for bowties with sharp and rounded tips is shown in Fig. 3. Bowties with flat tips and rounded tips have almost identical dispersive properties due to very close geometrical properties. We can readily observe a transparent band at the end of the visible and the beginning of the IR spectrum. Using nanoantennas array as a antireflective layer is particularly useful for our proposed backside illumination setup as it provides an additional degree of freedom in tailoring the spectral selectivity for potential applications.

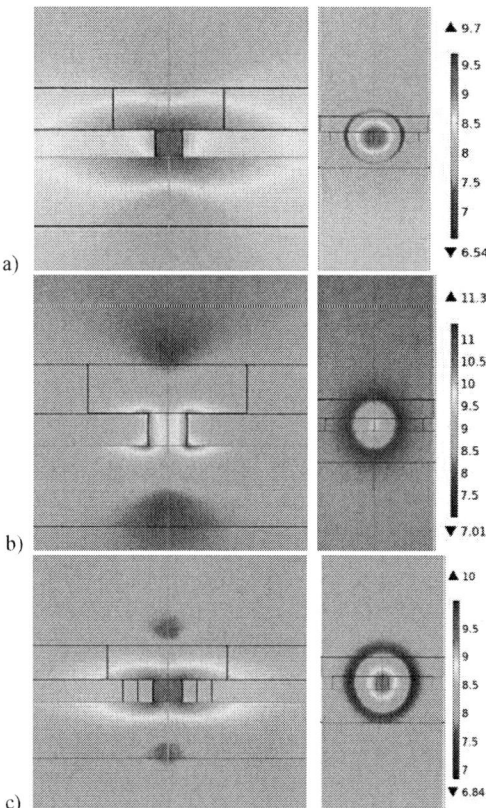

Fig. 5. Spatial field distribution in the center of the bowtie gap in plane perpendicular (left) and parallel (right) to the gap of for various tip shapes around 850 nm wavelength: a) flat tips, b) sharp tips and c) rounded tips. The side bar is decade logarithm of field intensity.

The spatial distributions of electric field intensity are shown in Fig. 4 and Fig. 5 for all three types of tips around a wavelength of 850 nm. They all exhibit high field localizations within the channel. The sharp tips have the most pronounced edge effects resulting in the highest field intensities but due to an extremely small size of the tips the near field decays fastest when moving away from the edge. The rounded and flat tip bowties cut off the near field components with the highest intensity and the highest spatial frequency due to their larger size compared to sharp tips resulting in more uniform distributions. Rounded tips

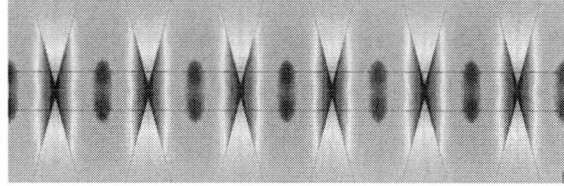

Fig. 6. Top view of field distribution along a channel filled with dielectric for a linear array of bowties with sharp tips at a wavelength of 850 nm.

localize the field in a smaller volume compared to the flat bowties. However in terms of near field intensity in the channel directly above the bowtie array all three configurations offer similar levels of enhancement because the field intensity in this region is formed by coupling of near field components that all three support. Finally, as an illustration, Fig. 6 shows the field distribution enhanced by plasmonic bowties along the dielectric-filled channel.

IV. CONCLUSION

We analyzed the use of bowtie nanoantenna arrays for enhancement of optical field in channels of MOEMS devices using FEM simulation. We also considered the influence of bowtie tip pair shape on near field enhancement. We obtained high localizations of the optical fields in the bowtie gaps leading to increased field concentration in the channel positioned directly above the gaps with solid robustness to tip shape change due to the channel positioning. The applied approach could be used for instance in microfluidics, to concentrate electromagnetic fields in the dielectric channel for the use in e.g. photocatalytic microreactors. Further applications include ultrasensitive chemical sensing, as well as a combination between photocatalytic microfluidics and *in situ* sensing in novel generation photochemical and photocatalytic reactors.

ACKNOWLEDGEMENT

This work was supported by the Serbian Ministry of Education, Science and Technological Development under Project TR32008.

REFERENCES

[1] S. A. Maier, ed., *World Scientific Handbook of Metamaterials and Plasmonics,* vols. 1-4, World Scientific, Singapore, 2018.

[2] E. Ozbay, "Plasmonics: Merging Photonics and Electronics at Nanoscale Dimensions," *Science,* 2006, vol. 311, no. 5758, pp. 189-193.

[3] L. Novotny, N. Van Hulst, "Antennas for light," *Nature Photonics,* 2011, vol. 5, no. 2, pp. 83-90.

[4] A. Kinkhabwala, Z. Yu, S. Fan, Y. Avlasevich, K. Müllen, W. E. Moerner. "Large single-molecule fluorescence enhancements produced by a bowtie nanoantenna," *Nature Photonics,* 2009, vol. 3, no. 11, art. 654.

[5] P. Avouris, M. Freitag, "Graphene photonics, plasmonics, and optoelectronics," *IEEE J. Sel. Top. Quant. Electr.,* 2014, vol. 20, no. 1, pp. 72-83.

[6] C. Hong, S. Yang, J. C. Ndukaife, "Optofluidic control using plasmonic TiN bowtie nanoantenna," *Optical Materials Express,* 2019, vol. 9, no. 3, pp. 953-964.

[7] Z. Jakšić, M. M. Smiljanić, D. Vasiljević-Radović, M. Obradov, K. Radulović, D. Tanasković, P. M. Krstajić, "Field localization control in aperture-based plasmonics by Boolean superposition of primitive forms at deep subwavelength scale," *Opt. Quant. Electron.,* 2016, vol. 48, no. 4, pp. 225.

[8] S. A. Maier, *Plasmonics: Fundamentals and Applications,* Springer Science+Business Media, New York, NY, 2007.

[9] A. Alu, N. Engheta, "Theory, modeling and features of optical nanoantennas," *IEEE T. Antenn. Propag.,* 2013, vol. 61, no. 4, pp. 1508-1517.

[10] D. Dregely, R. Taubert, J. Dorfmüller, R. Vogelgesang, K. Kern, H. Giessen, "3D optical Yagi–Uda nanoantenna array," *Nature Comm.,* 2011, vol. 2, pp. 267.1-7.

[11] P. Biagioni, J.-S. Huang, B. Hecht, "Nanoantennas for visible and infrared radiation," *Reports on Progress in Physics,* 2012, vol. 75, no. 2, pp. 024402.

[12] A. E. Krasnok, I. S. Maksymov, A. I. Denisyuk, P. A. Belov, A. E. Miroshnichenko, C. R. Simovski, Y. S. Kivshar, "Optical nanoantennas," *Physics-Uspekhi,* 2013, vol. 56, no. 6, pp. 539.

[13] M. L. Brongersma, "Plasmonics: Engineering optical nanoantennas," *Nature Photonics,* 2008, vol. 2, no. 5, pp. 270.

[14] S. V. Boriskina, H. Ghasemi, G. Chen, "Plasmonic materials for energy: From physics to applications," *Materials Today,* 2013, vol. 16, no. 10, pp. 375-386.

[15] N. Liu, M. L. Tang, M. Hentschel, H. Giessen, A. P. Alivisatos, "Nanoantenna-enhanced gas sensing in a single tailored nanofocus," *Nature Mater.,* 2011, vol. 10, no. 8, pp. 631-636.

[16] R. Adato, H. Altug, "In-situ ultra-sensitive infrared absorption spectroscopy of biomolecule interactions in real time with plasmonic nanoantennas," *Nature Comm.,* 2013, vol. 4, pp. 2154.

[17] D. Punj, M. Mivelle, S. B. Moparthi, T. S. Van Zanten, H. Rigneault, N. F. Van Hulst, M. F. García-Parajó, J. Wenger, "A plasmonic 'antenna-in-box' platform for enhanced single-molecule analysis at micromolar concentrations," *Nature Nanotech.,* 2013, vol. 8, no. 7, pp. 512.

[18] C. Simovski, D. Morits, P. Voroshilov, M. Guzhva, P. Belov, Y. Kivshar, "Enhanced efficiency of light-trapping nanoantenna arrays for thin-film solar cells," *Opt. Express,* 2013, vol. 21, no. 13, pp. A714-A725.

[19] P. J. Schuck, D. P. Fromm, A. Sundaramurthy, G. S. Kino, W. E. Moerner, "Improving the mismatch between light and nanoscale objects with gold bowtie nanoantennas," *Phys. Rev. Lett.,* 2005, vol. 94, no. 1, pp. 017402.1-4.

[20] D. P. Fromm, A. Sundaramurthy, P. J. Schuck, G. Kino, W. Moerner, "Gap-dependent optical coupling of single "bowtie" nanoantennas resonant in the visible," *Nano Lett.,* 2004, vol. 4, no. 5, pp. 957-961.

[21] N. A. Hatab, C.-H. Hsueh, A. L. Gaddis, S. T. Retterer, J.-H. Li, G. Eres, Z. Zhang, B. Gu, "Free-standing optical gold bowtie nanoantenna with variable gap size for enhanced Raman spectroscopy," *Nano Lett.,* 2010, vol. 10, no. 12, pp. 4952-4955.

[22] A. D. Rakić, A. B. Djurišić, J. M. Elazar, M. L. Majewski, "Optical properties of metallic films for vertical-cavity optoelectronic devices," *Appl. Opt.,* 1998, vol. 37, no. 22, pp. 5271-5283.

Reviewing MXenes for Plasmonic Applications: Beyond Graphene

Z. Jakšić, M. Obradov, O. Jakšić, D. Tanasković, and D. Vasiljević Radović

Abstract - MXenes are an emerging class of two-dimensional (2D) materials consisting of carbides, nitrides or carbonitrides of early transition metals. Due to the metallic type of conductivity of MXenes and their 2D structure, they are an ideal candidate to replace the conductive materials currently used in plasmonics, especially graphene. However, the use of MXenes for plasmonics and metamaterials is in its embryonal stage and practically all relevant publications appeared in the last few years.

In this contribution we first consider some methods of MXenes fabrication. We continue by analyzing the optical and electronic properties of MXenes of interest for plasmonics (especially spectral dispersion of complex relative dielectric permittivity). We proceed by reviewing some reported applications of MXenes in plasmonics, including among others conventionally designed chemical and biological sensors based on surface plasmon resonance (SPR), metasurface-based optical absorbers and MXene-based nanoantennas, as well as some possible applications in nonlinear optics.

In addition to that, we propose some novel uses of MXenes in plasmonics, including chemical and biological nanosensors, superabsorbers with a variable widths of the nanoholes, etc.

It is our opinion that the applications of MXenes in plasmonics introduced until now are only scratching the surface of a vast bulk of potential practical devices, structures and effects.

I. INTRODUCTION

Two-dimensional (2D) materials are a relatively new class of nanomaterials, with single-atom or single-molecule layered structure [1]. The first manufactured 2D material was graphene, an allotropic modification of carbon, isolated in 2004. Even today it remains the most researched and the most utilized 2D material [2]. Other similar materials include allotropes of chemical elements, like silicene, germanene, black phosphorous, stanene, etc., as well as 2D compounds like transition metal chalcogenides, graphane, germanane, borocarbonitrides, MXenes, etc. [3-5]. All of these materials have unique properties which make them usable for many applications.

MXenes are an emerging class of 2D materials consisting of carbides, nitrides or carbonitrides (the "X" part) of early transition metals (the "M" part – Ti, V, Cr,

Sc, Zr, Nb, No, Hf or Ta). They represent the largest family of 2D materials yet reported. Discovered in 2011 [6], in most cases they are foldable and moldable materials with metallic conductivity. Some MXenes are more stable than graphene, exhibit better electromagnetic properties and represent a viable alternative to this best known 2D material. In addition, there is a large number of various different MXenes and their combinations, which is convenient because it offers a choice for tailoring their properties for targeted applications.

Until now, MXenes were proposed for a vast number of practical uses, including next generation supercapacitors, Li-ion batteries and generally energy storage devices, but also electromagnetic shields, photonic and plasmonic structures and devices, optical metamaterials, highly sensitive chemical sensors, photocatalytic devices, microreactors, etc. [7].

There is an ever growing need for alternative conductive materials for nanostructured metallodielectrics in plasmonics and optical metamaterials [8]. Usual conductors like e.g. Drude metals and graphene all have their limitations. Due to the metallic type of conductivity of MXenes and their 2D structure, they are an excellent candidate to replace the currently used materials, especially graphene. However, the use of MXenes for plasmonics and metamaterials is in its embryonal stage and practically all relevant publications appeared in the last few years.

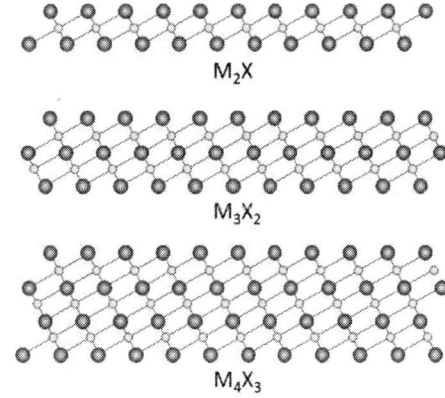

Fig. 1. MXene sheets with different numbers of layers. Top, middle and bottom are sheets with 2, 3 and 4 metal layers, respectively. Larger spheres represent transient metal, while smaller spheres are C, N or CN.

Z. Jakšić, M. Obradov, O. Jakšić, D. Tanasković, and D. Vasiljević Radović are with the Department of Microelectronic Technologies, Institute of Chemistry, Technology and Metallurgy, University of Belgrade, Njegoseva 12, 11000 Belgrade, Serbia, E-mails: jaksa, marko.obradov, olga, dragant, and dana, all at @nanosys.ihtm.bg.ac.rs

In this contribution we first consider some methods of MXenes fabrication. We continue by analyzing the properties of MXenes of interest for plasmonics (especially spectral dispersion of complex relative dielectric permittivity, and generally optical and electromagnetic properties). We proceed by reviewing some reported applications of MXenes in plasmonics, including among others conventionally designed chemical and biological sensors based on surface plasmon resonance (SPR) and metasurface-based optical absorbers, as well as some possible applications in nonlinear optics.

In addition to that, we propose some novel uses of MXenes in plasmonics, including chemical and biological nanosensors and superabsorbers with an extended bandwidth.

II. METHODS OF MXENES 2D LAYERS FABRICATION

In this Section we briefly outline the main methods for the fabrication of MXenes. The most important among them is selective etching of precursor structures to achieve delamination into monolayers.

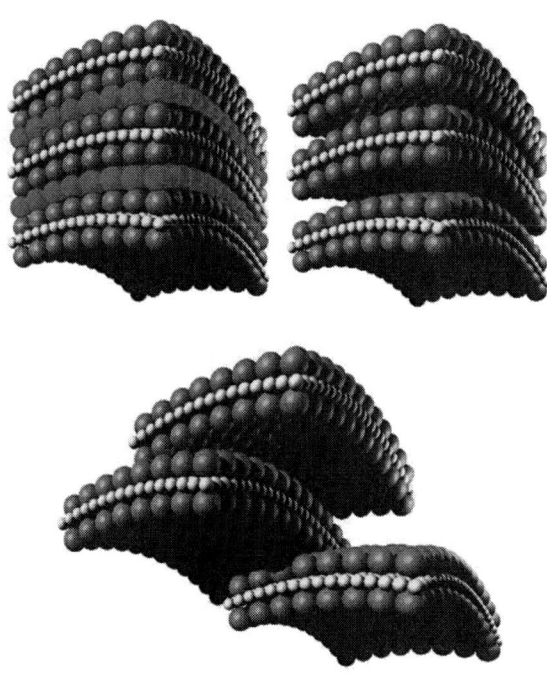

Fig. 2. Procedure of delamination of separate MXene sheets from MAX-type precursor. Top left: the MAX stage; top right: MX sheets after etching in HF ("A"-layers removed); bottom: delaminated MXene sheets. For the sake of clarity, this illustration shows the simplest M_2X structures for the case of the "bare" surface, i.e. without any termination groups.

The most common procedure (Fig. 2) is selective etching of an MAX phase precursor [9], where M stands for transition metal, X is carbide, nitride or carbonitride group and A is a sacrificial material that may be Al (actually, according to [7], the only material successfully used until now). Since the bond between M, X and A is metallic, until now no team succeeded in shearing the layers mechanically. The most often used method is etching in a hydrofluoric acid (HF), thus removing sacrificial Al and freeing MXene sheets [9].

Other methods include high-temperature etching of the MAX phase, etching from non-MAX precursors (e.g. those using a layer of Si, P, S, Ga, Ge, As, Cd, In, Sn, Tl, or Pb) [7,10]. Bottom up approach includes fabrication of MXene sheets by chemical vapor deposition (CVD) [11,12].

After producing multilayer MXene sheets, it is necessary to separate them. A method most often used to this purpose is intercalation [13]. Various organic polar substances are used for this, for instance hydrazine, isopropylamine, urea or dimethyl sulfoxide. Only a few teams tried Scotch tape exfoliation [14].

Until the moment of writing this text, no "bare" or "pristine" MXenes were produced, only with some kind of surface termination. This termination appears during the production of MXenes. During the delamination step, the "A" layer is selectively etched and replaced by surface terminations consisting of fluorine (F), hydroxyl group (OH) or oxygen (O). Thus a MXene sheet can be described as $M_{n+1}X_n(OH)_zO_yF_w$ (usually written as $M_{n+1}X_n(OH)_zT_x$ where T is the formula for surface termination).

MXenes were produced both as freestanding nanomembranes with a thickness of tens of nanometers [15] and as coatings on substrates. Numerous deposition methods were utilized to deposit delaminated flakes, including among others spin coating, drop coating, electrophoresis, layer-by-layer technique, chemical vapor deposition, etc. [16-18].

III. PROPERTIES OF MXENES OF INTEREST FOR PLASMONICS

Generally, the properties of MXenes strongly depend on the surface termination. Thus engineering of MXene surfaces has a crucial role in determining their applications. "Bare" or "pristine" $M_{n+1}X_n$ were investigated only computationally, typically using density functional theory (e.g. [19, 20]) and usually just to compare it with realistic surface terminations.

The properties of MXenes are tunable by composition, surface termination, doping or (nano)compositing with other materials. Thus they offer a tailorable and electrically conductive platform that behaves either as metal or as semiconductor and at the same time offers optical transparency in the visible [21]. Thus the first application of such material that comes to mind is obviously electromagnetic shielding [22].

978-1-7281-3420-8/19 $31.00 © 2019 IEEE

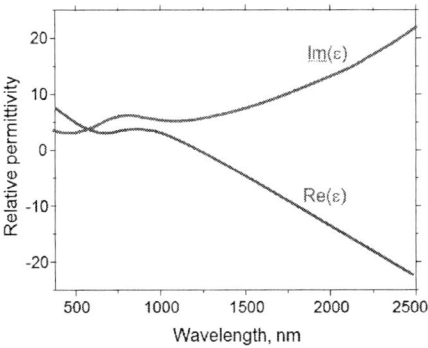

Fig. 3. Values of real and imaginary parts of relative dielectric permittivity according to [24] for a 30 nm sheets of titanium carbide (M_3X_2T).

The most important parameter of MXenes for plasmonics is their complex relative dielectric permittivity, $\varepsilon = \varepsilon_1 + i\,\varepsilon_2$. It is defined by intraband and interband transitions. The former are plasmonic by nature and are dominant in MXenes, while the influence of the latter is much weaker. Permittivity is important because it defines refractive index, absorption coefficient, transmittance, reflectance and absorptance, among others, and is directly related with electrical conductivity. Intraband transitions in MXenes are well described by Drude-Lorentz model [23]. Fig. 3 shows spectral permittivity obtained by fitting the experimental data from [24]. A Drude shape is readily observed, except in the range 500–1000 nm where the influence of interband transitions is noticeable.

IV. APPLICATIONS OF MXENES IN PLASMONICS

Since the field of MXene plasmonics is quite new, a majority of plasmonic applications of MXenes were considered only theoretically or computationally. It may be said that at this stage of development there are very few existing applications, but there is an enormous number of highly promising potential ones.

One of the most important uses is surface-enhanced Raman spectroscopy (SERS) [23,25] where MXenes are used as the conductive substrate. Another important field of use with many potentials is plasmonic sensing, albeit until now only the conventional surface plasmon resonance sensors using MXenes were proposed [26,27]. MXene-based biological sensors were also described [28].

Other types of chemical and biological MXene sensors with molecular sensitivity were also described Strong adsorption of some analytes on functionalized MXene surface leads to a marked increase of the material conductivity in the presence of adsorbed analytes like NH_3, SO_2, NO [29]. Like said before, these works were theoretical, the only experiments being done with volatile organic compounds (ethanol, methanol, acetone) [30]. We mention this type of sensors because changing conductivity

means changing permittivity so all of these designs with or even without modifications can be potentially used in plasmonics.

Nonlinear optics applications were proposed based on enhanced nonlinear (saturable) absorption aided by plasmonic mechanism when illuminated near plasma frequency. Potential nonlinear uses include optical switching, ultrafast laser applications, optical diodes (optical rectification devices) [23,31].

Another important application are broadband plasmonic metamaterial superabsorbers [24]. Highly efficient experimental absorbers were reported in ref. [32]. Applications for them are primarily found in photodetection and solar energy harvesting.

In Fig. 4 we present an MXene structure with variable widths of the nanoapertures based on a design by Aydin et al [33] that further expands the already broadband response of MXene superabsorbers.

Fig. 4. A structure of a broadband superabsorber consisting of two MXene plates with a dielectric spacer between, the top plate being textured as a mesh of crossed equilateral trapezoids which ensures a variable aperture width.

V. CONCLUSION

We reviewed potentials of MXenes for plasmonics. A bulk of the available literature points out to a conclusion that these materials are a viable alternative to graphene, not only owing to their superior properties, but also because of a much wider choice of possible compositions and types. The only hurdle appears to be the relative novelty of the field, and consequently a much smaller body of the existing research. This is a question of the current moment only, since many teams are investigating this extremely dynamic and popular field and the gap between the two is rapidly closing. Thus it is our belief that the plasmonic applications described herein are only scratching the surface of a vast bulk of potential MXene-based practical devices and structures.

ACKNOWLEDGEMENT

This work was supported by the Serbian Ministry of Education, Science and Technological Development under Project TR32008.

978-1-7281-3420-8/19 $31.00 © 2019 IEEE

REFERENCES

[1] A. Gupta, T. Sakthivel, S. Seal, "Recent development in 2D materials beyond graphene." *Progress in Mat. Sci.* vol. 73, pp. 44-126, 2015.

[2] A. K. Geim, "Graphene: status and prospects." *Science* vol. 324, pp. 1530-1534, 2009.

[3] R. Mas-Balleste, C. Gomez-Navarro, J. Gomez-Herrero, F. Zamora. "2D materials: to graphene and beyond." *Nanoscale* vol. 3, pp. 20-30, 2011.

[4] K. S. Novoselov, A. Mishchenko, A. Carvalho, A. H. Castro Neto, "2D materials and van der Waals heterostructures." *Science* vol. 353, pp. aac9439.1-11, 2016.

[5] B. Deng, R. Frisenda, C. Li, X. Chen, A. Castellanos-Gomez, F. Xia. "Progress on Black Phosphorus Photonics." *Advanced Optical Materials* vol. 6, pp. 1800365.1-15, 2018.

[6] M. Naguib, M. Kurtoglu, V. Presser, J. Lu, J. Niu, M. Heon, L.s Hultman, Y. Gogotsi, M. W. Barsoum, "Two-Dimensional Nanocrystals Produced by Exfoliation of Ti_3AlC_2," *Advanced Materials*, vol. 23, pp. 4248-4253, 2011.

[7] B. Anasori, M. R. Lukatskaya, Y. Gogotsi, "2D metal carbides and nitrides (MXenes) for energy storage." *Nature Reviews Materials* vol. 2, pp. 16098.1-17, 2017.

[8] A. Boltasseva, H. A. Atwater. "Low-loss plasmonic metamaterials." *Science* vol. 331, pp. 290-291, 2011.

[9] M. Naguib, V. N. Mochalin, M. W. Barsoum, Y. Gogotsi, "MXenes: a new family of two-dimensional materials," *Adv. Mater.* vol. 26, pp. 992–1004, 2014.

[10] J. Halim, S. Kota, M. R. Lukatskaya, M. Naguib, M. Q. Zhao, E. J. Moon, J. Pitock, J. Nanda, S. J. May, Y. Gogotsi, M.W. Barsoum, "Synthesis and characterization of 2D molybdenum carbide (MXene)," *Advanced Functional Materials*, vol. 26, pp.3118-3127, 2016.

[11] Y. Gogotsi, "Chemical Vapour Deposition: Transition Metal Carbides Go 2D," *Nature Materials*, vol. 14, pp. 1079-1080, 2015.

[12] C. Xu, L. Wang, Z. Liu, L. Chen, J. Guo, N. Kang, X.-L. Ma, H.-M. Cheng, W. Ren. "Large-area high-quality 2D ultrathin Mo 2 C superconducting crystals." *Nature materials* vol. 14, art. no. 1135, 2015.

[13] O. Mashtalir, M. Naguib, V. N. Mochalin, Y. Dall'Agnese, M. Heon, M. W. Barsoum, Y. Gogotsi. "Intercalation and delamination of layered carbides and carbonitrides." *Nature communications* vol. 4, art. no. 1716, 2013

[14] J. Xu, J. Shim, J.-H. Park, S. Lee, S. "MXene electrode for the integration of WSe2 and MoS2 field effect transistors", *Adv. Funct. Mater.* vol. 26, pp. 5328–5334, 2016.

[15] G. Liu, J. Shen, Q. Liu, G. Liu, J. Xiong, J. Yang, W. Jin, "Ultrathin two-dimensional MXene membrane for pervaporation desalination." *Journal of membrane science* vol. 548, pp. 548-558, 2018.

[16] S. Xu, G. Wei, J. Li, Y. Ji, N. Klyui, V. Izotov, W. Han, "Binder-free Ti3C2Tx MXene electrode film for supercapacitor produced by electrophoretic deposition method," *Chem. Eng. J.* vol. 317, pp. 1026-1036, 2017.

[17] H. An, T. Habib, S. Shah, H. Gao, M. Radovic, M. J. Green, J. L. Lutkenhaus, "Surface-agnostic highly stretchable and bendable conductive MXene multilayers," *Science Advances* vol. 4, art. no. eaaq0118, 2018.

[18] K. Hantanasirisakul, M.-Q. Zhao, P. Urbankowski, J. Halim, B. Anasori, S. Kota, C. E. Ren, M. W. Barsoum, Y. Gogotsi, "Fabrication of Ti_3C_2Tx MXene transparent thin films with tunable optoelectronic properties," *Advanced Electronic Materials* vol. 2, art. 1600050, 2016.

[19] M. Khazaei, A. Ranjbar, M. Arai, T. Sasaki, S. Yunoki, "Electronic properties and applications of MXenes: a theoretical review," *Journal of Materials Chemistry C,* vol. 5, pp. 2488-2503, 2017.

[20] G. R. Berdiyorov, "Optical properties of functionalized $Ti_3C_2T_2$ (T= F, O, OH) MXene: First-principles calculations," *AIP Advances* vol. 6, art. 055105, 2016.

[21] A. D. Dillon, M. J. Ghidiu, A. L. Krick, J. Griggs, S. J. May, Y. Gogotsi, M. W. Barsoum, A. T. Fafarman, "Highly conductive optical quality solution-processed films of 2D titanium carbide," *Advanced Functional Materials* vol. 26, pp. 4162-4168, 2016.

[22] F. Shahzad, M. Alhabeb, C. B. Hatter, B. Anasori, S. M. Hong, C. M. Koo, Y. Gogotsi, "Electromagnetic interference shielding with 2D transition metal carbides (MXenes)," *Science,* vol. 353, pp. 1137-1140, 2016.

[23] K. Hantanasirisakul, Y. Gogotsi, "Electronic and optical properties of 2D transition metal carbides and nitrides (MXenes)," *Advanced Materials* vol. 30, art. no. 1804779, 2018.

[24] K. Chaudhuri, M. Alhabeb, Zh. Wang, V. Shalaev, Y. Gogotsi, A. Boltasseva. "Highly broadband absorber using plasmonic titanium carbide (MXene)." *ACS Photonics* vol. 5, pp. 1115-1122, 2018.

[25] A. Sarycheva, T. Makaryan, K. Maleski, E. Satheeshkumar, A. Melikyan, H. Minassian, M. Yoshimura, Y. Gogotsi, "Two-dimensional titanium carbide (MXene) as surface-enhanced Raman scattering substrate," *J. Phys. Chem. C* vol. 121, pp. 19983-19988, 2017.

[26] Q. Ouyang, S. Zeng, L. Jiang, J. Qu, X.-Q. Dinh, J. Qian, S. He, P. Coquet, K.-T. Yong, "Two-dimensional transition metal dichalcogenide enhanced phase-sensitive plasmonic biosensors: Theoretical insight." *Journal of Physical Chemistry C* vol. 121, pp. 6282-6289, 2017.

[27] Y. Xu, Y. S. Ang, L. Wu, L. K. Ang, "High Sensitivity Surface Plasmon Resonance Sensor Based on Two-Dimensional MXene and Transition Metal Dichalcogenide: A Theoretical Study." *Nanomaterials* vol. 9, pp. 165.1-11, 2019.

[28] L. Wu, Q. You, Y. Shan, S. Gan, Y. Zhao, X. Dai, Y. Xiang, "Few-layer $Ti_3C_2T_x$ MXene: A promising surface plasmon resonance biosensing material to enhance the sensitivity," *Sens. Act. B,* vol. 277 pp. 210-215, 2018.

[29] B. Xiao, Y.-c. Li, X.-f. Yu, J.-b. Cheng, "MXenes: Reusable materials for NH_3 sensor or capturer by controlling the charge injection," *Sens. Act. B,* vol. 235 pp. 103-109. 2016

[30] E. Lee, A. VahidMohammadi, B. C. Prorok, Y. S. Yoon, M. Beidaghi, D.-J. Kim, "Room temperature gas sensing of two-dimensional titanium carbide (MXene)," *ACS applied materials & interfaces* vol. 9, pp. 37184-37190, 2017.

[31] Y. Dong, S. Chertopalov, K. Maleski, B. Anasori, L. Hu, S. Bhattacharya, A. M. Rao, Y. Gogotsi, V. N. Mochalin, R. Podila, "Saturable absorption in 2D Ti3C2 MXene thin films for passive photonic diodes," *Adv. Mat.* vol. 30, art. 1705714, 2018.

[32] W. Li, U. Guler, N. Kinsey, G. V. Naik, A. Boltasseva, J. Guan, V. M. Shalaev, A. V. Kildishev, "Refractory plasmonics with titanium nitride: broadband metamaterial absorber." *Adv. Mat.* vol. 26, pp. 7959-7965, 2014.

[33] K. Aydin, V. E. Ferry, R. M. Briggs, H. A. Atwater. "Broadband polarization-independent resonant light absorption using ultrathin plasmonic super absorbers." *Nature Communications* vol. 2, pp. 517.1-7, 2011.

Nanocrystalline Porous Nickel Ferrite Ceramics for Humidity Sensing Applications

D. L. Sekulic, Z. Z. Lazarevic and N. Z. Romcevic

Abstract - Over the last two decade, considerable interest has been received to synthesize novel ceramic nanomaterials with high humidity sensitivity characteristics for the fabrication of low-cost, fast and stable humidity sensors. This paper reports the preliminary results of humidity sensing properties of the nanocrystalline porous nickel ferrite ceramics fabricated by a conventional sintering of ultrafine nanopowders, which were successfully synthesized by soft mechanochemical processing. By using SEM and X-ray diffraction analyzes, the structural characteristics (average size of crystallites, density and porosity) of prepared $NiFe_2O_4$ ceramics were determined. At room temperature, this sensing material showed a linear response of impedance change within the wide relative humidity range from 15% to 85% at frequency of 2.5 kHz. Furthermore, the response time (adsorption process) of 26 s and the recovery time (desorption process) of 43 s have been obtained for this porous ceramic sensing material. Also, relatively small hysteresis, good repeatability and stability were also observed.

I. INTRODUCTION

Nowadays the humidity monitoring and control are attracting a great deal of attention in numerous industrial fields. Therefore, there is a constant need for development of cheaper and more stable materials with better humidity sensing properties for the fabrication of humidity sensing devices. Among various materials used for sensing elements of the humidity sensors, nanostructured ceramics based on various metal oxides offer several advantages such as high chemical, mechanical and thermal stability, as well as their porous nature that enables the rapid response dynamics and broad range of operation [1]. Recent studies show that ceramic materials which belong to the ferrospinel MFe_2O_4 (M is a divalent metal cation) type metal oxides, an important material family for electronic and information technology, are promising candidates for humidity sensor applications [2], [3].

In general, the humidity sensors based on ceramic nanomaterials can detect humidity on the principle of changes in the electrical properties of the sensing material by water vapor adsorption and by their penetration through the open pores throughout the sensing material, resulting significant changes in their electrical characteristics [4].

D.L. Sekulic is with the Department of Power, Electronic and Telecommunication Engineering, Faculty of Technical Sciences, University of Novi Sad, Trg Dositeja Obradovića 6, 21000 Novi Sad, Serbia E–mail: dalsek@uns.ac.rs

Z. Lazarevic, and N. Romcevic are with the Institute of Physics, University of Belgrade, Pregrevica 118, 11000 Belgrade, Serbia, E–mails: lzorica@ipb.ac.rs, and romcevi@ipb.ac.rs

According to the output form, ceramic humidity sensors based on the dependence of impedance of the sensing material from relative humidity (RH) are used most commonly [1], but it is also possible to realize these sensors on the principle of measuring the dependence of resistance or capacitive of the ceramic material from RH. The resistance or impedance of the resistive-type ceramic humidity sensors decreases as the RH increases, while the capacitance of capacitive-type these sensors increases with RH [5]. Taking into account the linear response, high sensitivity, wide humidity detection range and fast response, as well as low cost, resistive-type ceramic humidity sensors are becoming more prevalent than capacitive-type ones [1].

In humidity sensors based on ceramic nanomaterials, the type of conduction mechanism can be electronic or ionic [6]. In the case of electronic conduction mechanism, the molecules of water act as electron donating gas and their chemisorption increase or decrease the conductivity depending on whether the sensing material is n- or p-type semiconductor. In the case of ionic conduction mechanism, the impedance of sensing material decreases with an increase of RH due to physisorption and capillary condensation of water molecules on the surface of the ceramic material.

As a well-known multifunctional material, nickel ferrite ($NiFe_2O_4$) has received great interest in previous years due to its application potential in many electronic devices, including the ceramic humidity sensors [3]. The present study reports on humidity sensing properties of nanocrystalline porous $NiFe_2O_4$ ceramics fabricated by a conventional sintering of ultrafine powders with average particle size of 7 nm. For that purpose, the impedance of prepared nickel ferrite ceramics was measured as a function of RH at different frequencies and room temperature. We expect that the results presented in this study may offer useful guidelines to fabricate high performance humidity sensors.

II. EXPERIMENTAL DETAILS

A. Preparation of Nickel Ferrite Ceramics

Starting from mixtures of high–purity nickel(II)-hydroxide ($Ni(OH)_2$, Merck 95% purity) and hematite (α-Fe_2O_3, Merck 99% purity) as initial precursors in equimolar ratio, ultrafine $NiFe_2O_4$ nanopowders were successfully synthesized by soft mechanochemical processing in a

978-1-7281-3420-8/19 $31.00 © 2019 IEEE

planetary ball mill (Model Fritsch Pulverisette 5) after 25 hours of milling [7]. Required milling time for obtaining single phase spinel crystal structure of powders was experimentally determined by X-ray diffraction technique. Synthesized $NiFe_2O_4$ powders were pressed into circular disc-shaped pellets using a cold isostatic press. The so prepared pallets were sintered at 1100°C for 2 h (Model Lenton-UK oven) without pre-calcinations step. Heating rate was 10°C/min with nature cooling in air atmosphere.

B. Characterization

The particle size and morphology of soft mechano-chemical synthetized $NiFe_2O_4$ powders were investigated using a 200 kV transmission electron microscope (TEM, Model JEOL JEM-2100 UHR). The phase purity and crystal structure of the powder and sintered samples have been characterized by using X-ray diffractometer (XRD, Model Philips PW 1050) with Ni filtered CoKα radiation ($\lambda = 1.78897$ Å) at room temperature. The XRD data were collected in a wide range of Bragg's angles 2θ ($15° \leq 2\theta \leq 80°$) with a scanning step size of 0.05° in a 10 s per step of counting time. The microstructures of $NiFe_2O_4$ ceramics were recorded at room temperature using scanning electron microscope (SEM, Model JEOL JSM-6460LV). The sintered density of the ceramic sample was measured by applying Archimedes' principle using toluene as an immersion liquid [7].

In order to study the humidity sensing performance, the prepared porous $NiFe_2O_4$ ceramic pellet as sensing element was placed in a closed test glass chamber between two silver electrodes, which are connected to the impedance analyzer (Model HP 4194A) to measure the change in impedance with respect to relative humidity at 25°C ± 1°C. The compressed air directed into the chamber was firstly dehydrated over silica gel and $CaCl_2$ and then humidity level was varied from 15% to 85% by bubbling air through water and mixing it with dry air. In such measurement system, RH and temperature were monitored by a commercial humidity and temperature probe (Model Tecpel DTM–321).

III. RESULTS AND DISCUSSION

A. Structure Analysis

The room temperature X-ray diffraction pattern of the as-synthesized nickel ferrite nanopowders along with TEM image are presented in Fig. 1. All diffraction peaks appearing in the powder XRD spectrum confirms the formation of cubic spinel $NiFe_2O_4$ structure with the Fd3m space group. No additional secondary phases, amorphous forms or impurities were detected, indicating the high quality and single phase of the prepared nickel ferrite powder. TEM image reveals that the particle size of the synthesized $NiFe_2O_4$ powder is mostly in the range between 10 nm and 30 nm with nearly spherical shape.

Further, the obtained characteristic diffraction rings, shown in inset of Fig 1, correspond to the crystal planes of spinel structure which is in accordance with results of X-ray structure analysis.

Figure 2 shows the room temperature X-ray diffraction spectrum of $NiFe_2O_4$ ceramics fabricated by a conventional sintering of the synthesized nanopowders at 1100°C for 2 h. The position and relative intensities of all diffraction peaks match well with the reported cubic spinel nickel ferrite, and the sharp XRD peaks indicate the polycrystalline nature of sintered sample [8]. Average crystallite size of about 56 nm was determined according to the well-known Scherrer's equation using XRD data for the most intense (311) Bragg reflection [7]. In addition, the SEM results, presented in inset of Fig 2, revealed that the phase pure $NiFe_2O_4$ ceramics possess a porous structure with nano-sized grains. It can be seen that small pores are interconnected with larger pores through grain necks which is favorable for humidity sensing [9]. Based on determined sintered density of 3.93 g/cm^3 and theoretical X–ray density of 5.49 g/cm^3, the porosity of prepared nano-crystalline porous nickel ferrite ceramics was estimated to average 28.4 percent.

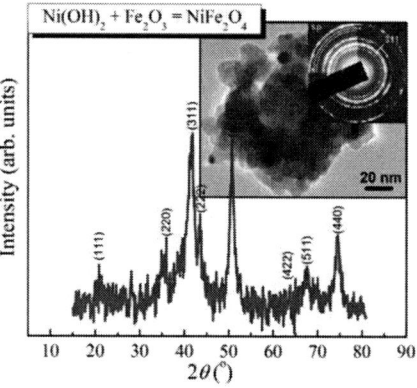

Fig. 1. XRD spectrum of the synthesized $NiFe_2O_4$ nanopowders along with TEM image and ring diffraction pattern.

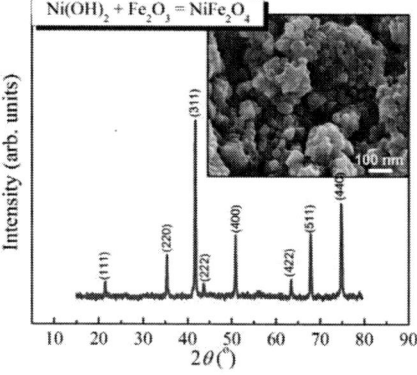

Fig. 2. XRD pattern of the nanocrystalline porous $NiFe_2O_4$ ceramics along with SEM micrograph.

B. Humidity sensing properties

The humidity sensing properties of the sensor based on porous $NiFe_2O_4$ ceramics was evaluated by determining the humidity sensitivity, hysteresis, response and recovery time, and stability under various RH values. In order to find out the optimal operating frequency, the impedance of the sensing material was measured as a function of RH at different frequencies (from 250 Hz to 25 kHz) and room temperature (25°C). In the RH range between 15% and 85%, the obtained results for several frequencies are depicted in Fig. 3. It is obvious that the impedance of the $NiFe_2O_4$-based ceramic sensor is significantly influenced by the frequency and it decreases with increasing frequency. The high humidity sensitivity and linear response in the whole RH range was observed at relatively low measurement frequency, i.e. at 2.5 kHz. At this frequency, the impedance changes by about two orders of magnitude from $4.07×10^6$ Ω to $3.25×10^4$ Ω as RH increases from 15% to 85%, exhibiting pronounced sensitivity. To quantitatively illustrate humidity sensitivity of $NiFe_2O_4$-based ceramic sensor, it is found that its sensitivity, defined as the ratio of the change in impedance against the change in RH [10], is about 57.6 kΩ/%RH. However, at frequency of 25 kHz, the impedance plot becomes flat because the direction of the applied electric field changes fast at higher frequencies and the polarization of the adsorbed water molecules cannot catch up with it [11], [12]. Therefore, the frequency of 2.5 kHz is chosen as the optimal operating frequency of the sensor based on porous $NiFe_2O_4$ ceramics for the subsequent measurements.

Humidity hysteresis, as one of the most important characteristics, is usually used to estimate the reliability of the sensor materials [13], and it is defined as the maximum difference between the adsorption and desorption curves [11]. The room temperature hysteresis characteristic of $NiFe_2O_4$-based ceramic sensor was determined by increasing the RH from 15% to 85% for water molecules absorption and then decreasing back to 15% for water molecules desorption. As can be seen from the narrow hysteresis loop curve shown in Fig. 4, the difference between impedance in the adsorption and desorption processes is obviously small under all humidity conditions indicating a good reliability of the sensing material. The maximum humidity hysteresis error [12] was calculated to be 2.36% in the range of 15-85% RH. Furthermore, the impedance of desorption process is slightly lower than that of adsorption process pointing to the fact that a relatively longer time is required to desorb the water molecules [11]. Namely, the water molecules' adsorption/desorption take place at different energy levels, and adsorption is an exothermic process, whereas desorption process needs external energy for water molecules to depart from the $NiFe_2O_4$ nanocrystals surface.

At optimal operating frequency of 2.5 kHz, the room temperature response–recovery characteristic for one cycle with RH changing from 15% and 85% is shown in Fig. 5. The response time (adsorption process) of 26 s and the recovery time (desorption process) of 43 s were observed for fabricated nanocrystalline porous $NiFe_2O_4$ ceramics as humidity sensing material. This relatively fast response and recovery behavior can be attributed to the presence of high surface area and pore volume which facilitates the exposure of accessible active sites, thus providing an easy pathway

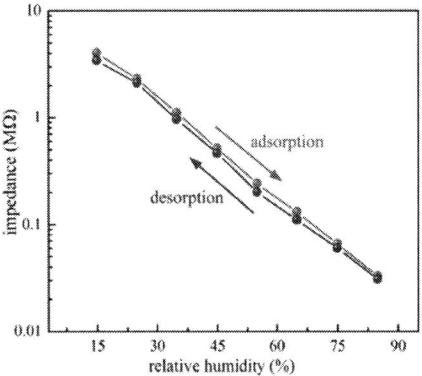

Fig. 4. Humidity hysteresis of the porous $NiFe_2O_4$ ceramics at frequency of 2.5 kHz and room temperature.

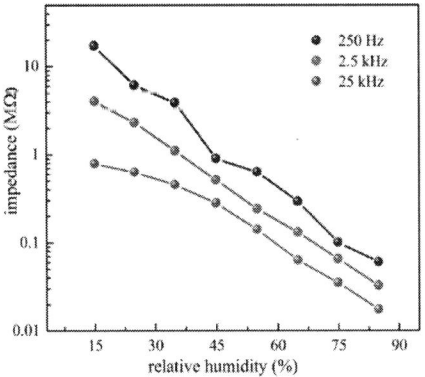

Fig. 3. Relative humidity dependence of impedance of the porous $NiFe_2O_4$ ceramics at various frequencies and room temperature.

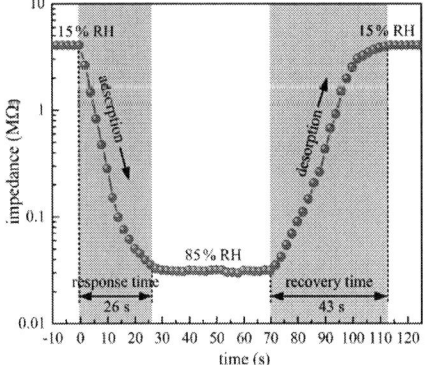

Fig. 5. Room temperature response–recovery characteristic of the porous $NiFe_2O_4$ ceramics at frequency of 2.5 kHz.

for the adsorption and desorption of water molecules on the internal and external surface of this material [13], [14].

In order to examine the stability, the nanocrystalline porous $NiFe_2O_4$ ceramics as humidity sensing material was tested repeatedly once in two days under fixed humidity levels (15%, 35%, 55%, and 75% RH) at optimal operating frequency of 2.5 kHz in a period of 10 days. As shown in Fig. 6, there are slight fluctuations in the impedances with time (less than ± 6%), which directly confirms the good stability and reliability. Therefore, porous $NiFe_2O_4$ ceramics are expected to be potential material for new high performance humidity sensors.

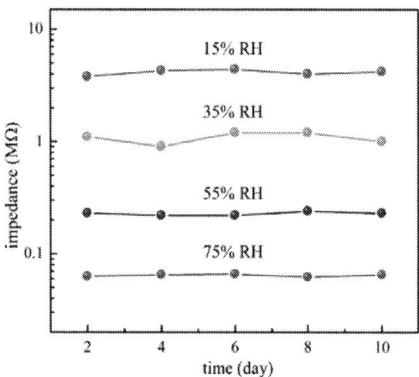

Fig. 6. Stability curves of the sensor based on porous $NiFe_2O_4$ ceramics monitored at different humidity conditions for 10 days.

IV. CONCLUSION

In this paper, it has been demonstrated that the low-cost nanocrystalline porous $NiFe_2O_4$ ceramics can be employed for designing high performance humidity sensors. The room temperature study of humidity sensing properties indicated that impedance varies about two orders of magnitude in the RH range from 15% to 85%. Furthermore, good sensing linearity, small hysteresis, relatively fast response time and recovery time were attributed to the high surface area and porous structure of as-prepared $NiFe_2O_4$ ceramics as sensing material.

ACKNOWLEDGEMENT

This research was financially supported by Ministry of Education, Science and Technological Development of the Republic of Serbia through Project No. III45003.

REFERENCES

[1] T. A. Blank, L. P. Eksperiandova, and K. N. Belikov, "Recent trends of ceramic humidity sensors development: A review", *Sensors and Actuators B*, 2016, vol. 228, pp. 416-442.

[2] X. Xu, L. Xiao, N. O. Haugen, Z. Wu, Y. Jia, W. Zhong, and J. Zou, "High humidity response property of sol–gel synthesized $ZnFe_2O_4$ films", *Materials Letters*, 2018, vol. 213, pp. 266-268.

[3] A. M. Dumitrescu, G. Lisa, A. R. Iordan, F. Tudorache, I. Petrila, A. I. Borhan, M. N. Palamaru, C. Mihailescu, L. Leontie, and C. Munteanu, "Ni ferrite highly organized as humidity as humidity sensors", *Materials Chemistry and Physics*, 2015, vol. 156, pp. 170-179.

[4] I. Petrila, and F. Tudorache, "Influence of partial substitution of Fe^{3+} with W^{3+} on the microstructure, humidity sensitivity, magnetic and electrical properties of barium hexaferrite", *Superlattices and Microstructures*, 2014, vol. 70, pp. 46–53.

[5] J. Wang, X. Wang, and X. Wang, "Study on dielectric properties of humidity sensing nanometer materials", *Sensors and Actuators B*, 2005, vol. 108, pp. 445-449.

[6] S. K. Misra, N. K. Pandey, V. Shakya, and A. Roy, "Application of undoped and Al_2O_3-doped ZnO nanomaterials as solid-state humidity sensor and its characterization studies", *IEEE Sensors Journal*, 2015, vol. 15, pp. 3582-3589.

[7] D. L. Sekulic, Z. Z. Lazarevic, M. V. Sataric, C. D. Jovalekic, and N.Z. Romcevic, "Temperature-dependent complex impedance, electrical conductivity and dielectric studies of MFe_2O_4 (M = Mn, Ni, Zn) ferrites prepared by sintering of mechanochemical synthesized nanopowders", *Journal of Materials Science: Materials in Electronics*, vol. 26, 2015, pp. 1291–1303.

[8] D. L. Sekulic, Z. Z. Lazarevic, C. D. Jovalekic , A. N. Milutinovic, and N. Z. Romcevic, "Impedance spectroscopy of nanocrystalline $MgFe_2O_4$ and $MnFe_2O_4$ ferrite ceramics: Effect of grain boundaries on the electrical properties", *Science of Sintering*, 2016, vol. 48, pp. 17-28.

[9] L. P. Babu Reddy, R. Megha, H. G. Raj Prakash, Y. T. Ravikiran, C. H. V. V. Ramanad, S. C. Vijaya Kumari, and D. Kim, "Copper ferrite-yttrium oxide (CFYO) nanocomposite as remarkable humidity sensor", *Inorganic Chemistry Communications*, 2019, vol. 99, pp. 180-188.

[10] Y. Tan, K. Yu, T. Yang, Q. Zhang, W. Cong, H. Yin, Z. Zhang, Y. Chen, and Z. Zhu,, "The combinations of hollow MoS_2 micro@nano-spheres: one-step synthesis, excellent photocatalytic and humidity sensing properties", *Journal of Materials Chemistry C*, 2014, vol. 2, pp. 5422-5430.

[11] Z. Duan, M. Xu, T. Li, Y. Zhang, and H. Zou, "Super-fast response humidity sensor based on La0.7Sr0.3MnO3 nanocrystals prepared by PVP-assisted sol-gel method", *Sensors and Actuators B*, 2018, vol. 258, pp. 527-534.

[12] D. L. Sekulic, Z. Lazarevic, C. Jovalekic, and N. Romcevic, "Characterization of yttrium orthoferrite ($YFeO_3$) nanoparticles as humidity sensor materials at room temperature", in *Proc. 30th International Conference on Microelectronics-MIEL 2017*, Nis, 2017, pp. 135-138.

[13] V. K. Tomer, S. Duhan, P. V. Adhyapak, and I. S. Mulla, "Mn-loaded mesoporous silica nanocomposite: A highly efficient humidity sensor", *Journal of the American Ceramic Society*, 2015, vol. 98, pp. 741-747.

[14] D. Bauskara, B. B. Kaleb, and P. Patil, "Synthesis and humidity sensing properties of $ZnSnO_3$ cubic crystallites", Sensors and Actuators B: Chemical, vol. 161, 2012, pp. 396- 400.

978-1-7281-3420-8/19 $31.00 © 2019 IEEE

D-mode pHEMT 0.5 um Process Characterization to Wide-Band LNA Design

D. I. Sotskov, N. A. Usachev, V. V. Elesin, A. G. Kuznetsov, K. M. Amburkin, G. V. Chukov, M. I. Titova, and N. M. Zidkov

Abstract – Results of domestic D-mode pHEMT 0.5 μm process characterization obtained during the design and testing of the single power supply wide-band low noise amplifier (LNA) are present. The simulation and test results demonstrate that designed cascode LNA has operating frequency range up to 3.5 GHz, power gain above 15 dB, noise figure below 2.2 dB, output linearity above than 17 dBm and power consumption less than 325 mW. Potential immunity of LNA to total ionizing dose and destructive single event effects exceed 300 krad and 60 MeV·cm²/mg respectively.

I. INTRODUCTION

Low noise amplifier (LNA) is important functional block in receiver paths of communication, radar or navigation systems. LNA parameters are critical to the sensitivity (noise figure (NF)), power gain (G) and input linearity (input P_{1dB}) of the receiver [1].

A depletion mode (D-mode) pseudo high-electron mobility transistor (pHEMT) process has become a good choice for LNA design because of transistor low NF, high cut-off frequency (Ft) and appropriate fabrication costs [2]. In addition, low sensitivity of pHEMT to total ionizing dose (TID) and single event effects (SEE) makes them promising for space applications [3, 4].

A disadvantage of a D-mode pHEMT based LNA is a negative bias supply requirement, which limits possible field of applications [1, 5]. Meanwhile conventional approach to single positive supply LNA design on D-mode pHEMT is based on self-biasing [1]. The major limitations to meet a wide-band performance requirements are the process design kit (PDK) model inaccuracy and poor performance in low-frequency range [1].

The purpose of this work was to introduce a domestic D-mode pHEMT 0.5 μm process [6] characterization to design a variety radiation tolerant wide-band LNAs with frequency range of 0.5-3.5 GHz.

D.I. Sotskov, N.A. Usachev, V.V. Elesin, A.G. Kuznetsov, K.M. Amburkin, G.V. Chukov, M.I. Titova, N.M. Zidkov are with National Research Nuclear University MEPhI (Moscow Engineering Physics Institute), and JSC "SPELS" Moscow, Russia, e-mails: disot@spels.ru, nausach@spels.ru, vveles@spels.ru, agkuz@spels.ru, kmamb@spels.ru, gvchuk@spels.ru, mitit@spels.ru, nmzhv@spels.ru

II. LNA DESIGN

Wide-band (multi-octave) LNAs are designed using various architectures including distributed (traveling-wave), balanced and resistive feedback configurations [7]. The resistive feedback is widely used to achieve tradeoff among several LNA performances (operated frequency range, noise figure, gain, gain flatness, linearity, VSWR, power consumption) [7, 8].

The single positive power supply cascode LNA based on resistive feedback configuration and self-biasing techniques are presented. LNA was implemented in domestic D-mode pHEMT 0.5 μm process with Ft up to 35 GHz and NFmin is 1.2 dB at 8 GHz. The LNA circuit schematic is shown in Fig. 1. It is composed of cascoded input (VT2) and output (VT1) transistors with same width of 4×150 μm [1]. A series feedback (resistor R3, capacitor C2, inductor L2) and parallel feedback (capacitor C1, resistor R1) are used to provide stability and gain flatness in wide frequency range. It is should be noted what cascode transistor with capacitance connected to the gate terminal can form a Collpits oscillator. To improve stability of the amplifier a damping resistor should be added to the gate of cascode transistor to decrease parasitic resonator quality factor [9].

LNA is designed with a single positive voltage supply 5 V. Resistive dividers implemented by R1-R2, R4-R6 and self-bias circuit R3 provides required transistors operation point. Input matching network consists of integrated spiral inductors L1 and L2. Output matching network consists of capacitor C3 and resistor R5.

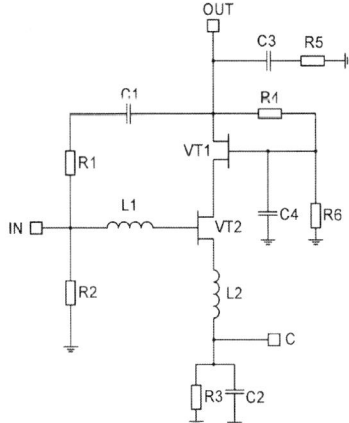

Fig. 1. Simplified circuit of the LNA.

978-1-7281-3420-8/19 $31.00 © 2019 IEEE

LNA occupies an area of 2.15×1.65 mm² and includes additional pad "C" needed for connect external bypass capacitors for enhanced performance at frequencies below 0.5 GHz.

III. LNA PERFORMANCES

A. Simulation and Measurement Setups

Simulations were performed using scalable pHEMT non-linear model and linear models for microstrip lines, inductors, T-shapes, vias and pads based on scattering (S-) parameters measurement verified up to 20 GHz and provided by foundry [6].

Measurements were performed on wafer for a significant amount (above 50) of LNA chips using specialized microwave test system, based on Cascade semi-automated probe station, vector network analyzer (VNA) and signal (spectrum) analyzer with noise figure measurement option described in [10]. The experimental setup used for LNA dies testing is shown in Fig. 2. Under test procedure S-parameters and P_{1dB} (linearity) are measured at 5 V supply, noise figure is measured at 3 V supply.

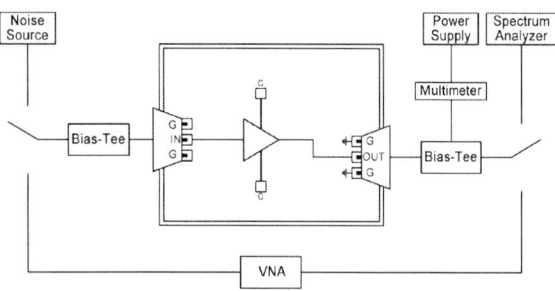

Fig. 2. Experimental setup based on the microwave probe station.

B. Simulation and Measurement Results

The simulated and measured LNA chip performances (gain, noise figure, P_{1dB} etc.) are shown in Fig. 3 and Fig. 4. The measured and simulated results are in good agreement in the frequency range of 0.5-3.5 GHz, and demonstrate that LNA has power gain above 15.3 dB, NF below 2.2 dB, output P_{1dB} above than 17 dBm (F = 1.5 GHz).

Measured performance of the presented LNA is summarized in Table I along with its counterpart from TriQuint Semi. (Qorvo Inc.).

C. Estimation of Model Accuracy

The estimation of a relative error of simulation results was carried out with following expression:

$$\delta_X = \{|Xm-Xs|/Xm\}\cdot 100\% \qquad (1)$$

where Xm and Xs are measured and simulated gain and noise figure values respectively.

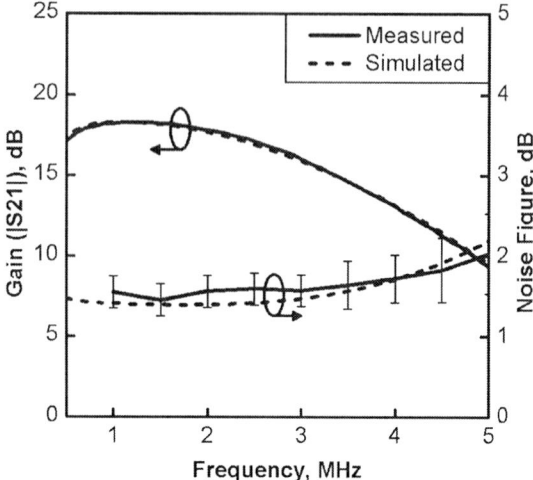

Fig. 3. Simulated and measured gain and noise figure versus frequency for the proposed LNA.

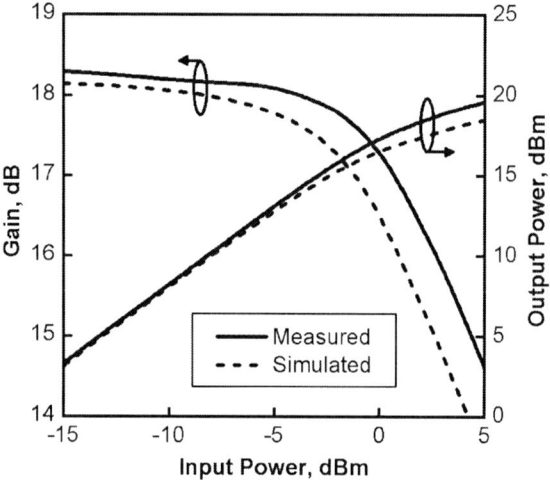

Fig. 4. Simulated and measured gain and output power versus input power for the proposed LNA.

TABLE I
MEASURED PERFORMANCE SUMMARY OF THE PROPOSED LNA

LNA	This Work	TGA5108
Technology Process	0.5 um D-pHEMT	0.5 um E/D-pHEMT
Operating Frequency, GHz	0.5…3.5	0.5…3.5
Gain, dB	15.3	15
Gain Flatness, dB	4	7
Noise Figure, dB	2.2	2.2
Output P1dB, dBm	17	20
Power Consumption, mW	325	425
Chip Size, mm²	2.15×1.65	1.49×0.85

978-1-7281-3420-8/19 $31.00 © 2019 IEEE

The analysis showed that relative gain error and noise figure error do not exceed 3 % (frequency range 0.5-3.5 GHz) and 4 % (frequency range 1-3.5 GHz) respectively which indicates a good accuracy of linear models. The relative input P_{1dB} error do not exceed 15 % (F = 1.5 GHz) for worst case, typical value is less than 5 %.

D. Estimation of Process Variation

The estimation of a coefficient of variation was carried out with following expression:

$$V_X = \{\sigma/\overline{X}\}\cdot 100\% \qquad (2)$$

where σ and $\overline{X}$ are standard deviation and mean (average) values respectively.

The measurement results analysis of the LNA chips showed that coefficient of variation do not exceed 1.3 % for gain (frequency range 0.5-3.5 GHz), 0.5 % (frequency range 1-3.5 GHz) for noise figure and 2.8 % for current consumption.

E. Low-Frequency Applications

External capacitors can be used to improve LNA performance at frequencies below 0.5 GHz. Measurement of low-frequency S-parameters and noise figure were performed at 5 V and 3V supply (VDD) respectively using special test-fixture (see Fig. 5) and microwave test system operating up to 26 GHz [10].

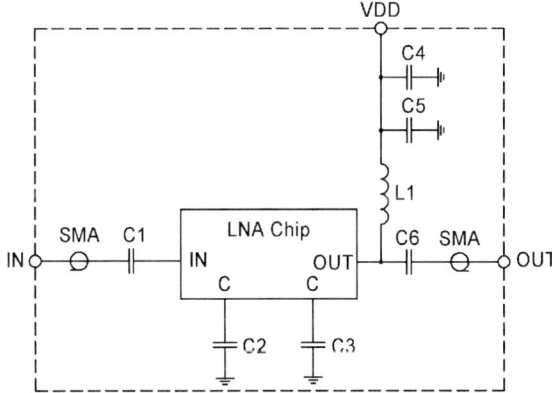

Fig. 5. Special measurement test-fixture schematic.

Measured LNA low-frequency performance with and without mounted 1 nF SMD capacitors C2 and C3 are shown in Fig. 6. Dependencies in Fig. 6 demonstrate that at frequency 100 MHz power gain and noise figure have been improved by more than 10 dB and 5 dB respectively.

F. Estimation of Radiation Tolerance

A radiation tolerance estimation was made on the base of data obtained at the SPELS and NRNU MEPHI test

center for the typical test structures: transistor and C-band two stage LNA implemented in given D-mode pHEMT 0.5 um process.

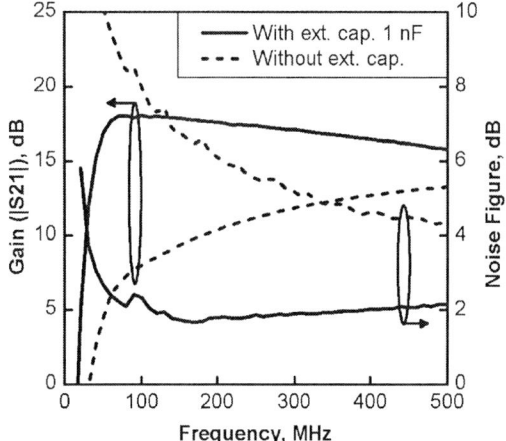

Fig. 6. Measured gain and noise figure versus frequency for the proposed LNA.

The experimental study of the test structures under TID irradiation were performed using Cs-137 "Panorama-MEPhi" irradiation facility [11]. Heavy ion irradiation performed at the facility based on U400M heavy ion cyclotron of the Joint Institute for Nuclear Research (JINR, Dubna, Russia).

According to the test results, up to an equivalent gamma dose of 300 krad no parameter degradation of the test structures is observed. Destructive SEE (SEL, SEB and other) have not been observed for heavy ions exposure with LET up to 60 MeV·cm²/mg.

IV. CONCLUSION

A domestic D-mode pHEMT 0.5 µm process was characterized to design a variety of single power supply wide-band LNAs with operating frequency range up to 3.5 GHz and above, power gain above 15 dB, noise figure below 2.2 dB, output linearity above than 17 dBm.

It is should be noted that investigated process is a good candidate for radiation tolerant LNA and control circuits (switches, attenuators, phase shifters) design and provides relatively high process design kit (PDK) models accuracy, low process variations and immunity to total ionizing dose and destructive single event effects up to 300 krad and 60 MeV·cm²/mg respectively.

ACKNOWLEDGEMENT

This work was supported in accordance with agreement between Ministry of Education and Science of the Russian Federation and National Research Nuclear University MEPhI № 8.2373.2017/4.6.

REFERENCES

[1] G. Gonzalez. *Microwave Transistor Amplifiers: Analysis and Design.* 2nd ed. Pearson. 1996. P. 528.

[2] G.D. Vendelin, A.M. Pavio, U.L. Rohde. *Microwave Circuit Design Using Linear and Nonlinear Techniques.* John Wiley & Sons Ltd, 2005, p. 1058.

[3] D.V. Gromov, V.V. Elesin, S.A. Polevich, et al. "Ionizing-Radiation Response of the GaAs/(Al, Ga)As PHEMT: A Comparison of Gamma- and X-ray Results", *Russian Microelectronics*, 2004, Vol. 33, No. 2, pp. 111-115.

[4] Chukov G.V., Elesin V.V., Nazarova G.N., Nikiforov A.Y., Boychenko D.V., Telets V.A., Kuznetsov A.G., Amburkin K.M., "SEE testing results for RF and microwave ICs" *2014 IEEE Radiation Effects Data Workshop*, 2014, pp. 233-235.

[5] H.-C. Chiu et al., "Enhancement- and Depletion-Mode InGaP/InGaAs pHEMTs on 6-Inch GaAs Substrate", *Proceedings Asia-Pacific Microwave Conference*, 2005, pp. 1-4.

[6] Fazylkhanov O.R., Pushnitsa I.S., Strelnikov S.I., Kalyakin M.A., Filaretov A.H. "Process Design Kit verification - methodology and practice", 2017, *Crimico*, pp. 143-149.

[7] I.J. Bahl. *Fundamentals of RF and Microwave Transistor Amplifiers.* John Wiley & Sons. 2009. p. 671.

[8] G. Wang, J. Liu et al., "The Design of Broadband LNA with Active Biasing based on Negative Technique", *Journal of Microelectronics, Electronic Components and Materials*, 2018, vol. 48, no. 2, pp. 115 – 120.

[9] Bagher Afshar, Ali M. Niknejad. "X/Ku Band CMOS LNA Design Techniques", *IEEE Custom Integrated Circuits Conference*, 2006, pp. 389-392.

[10] D.I. Sotskov, V.V. Elesin, K.M. Amburkin, G.N. Nazarova, N.A. Usachev, A.Y. Nikiforov. "Design and Testing Issues of a High-Speed SOI CMOS Dual-Modulus Prescaler for Radiation Tolerant Frequency Synthesizers", *Proc. 30th Int. conf. on microelectronics (MIEL 2017)*, 2017, pp.329-332.

[11] A.S. Artamonov, A.A. Sangalov, A.Y. Nikiforov, V.A. Telets, D.V. Boychenko. "The New Gamma Irradiation Facility at the National Research Nuclear University MEPhI", *IEEE Radiation Effects Data Workshop*, 2014, pp. 258–261.

Practical Evaluation of Optocouplers TID-Hardness Research Method Using an X-ray Unit

S. V. Novikov, M. E. Cherniak, R. K. Mozhaev, and D. V. Boychenko

Abstract – The article describes an optoisolators radiation hardness testing method to the effects of absorbed dose using an x-ray unit. Technological issues and features of the research are highlighted.

I. INTRODUCTION

Opto-isolators (also known as optocouplers) are optoelectronic devices used to protect low-voltage circuit parts from overloads by providing electrical isolation from the high-power section of an electrical circuit.

Devices for galvanic isolation of electrical circuits are widely used in complex electronic equipment, including space applications [1], such as communications satellites. Such application imposes certain requirements on radiation hardness of this class of devices to the effects of ionizing radiation. To verify the possibility of a chip usage, it is necessary to perform radiation-hardness tests.

II. OPTOCOUPLERS CLASSIFICATION

An optocoupler circuit includes an input part that converts an electrical signal into a light one; an output part that receives the light signal and converts it into an electrical signal and an optical channel connecting the input and output parts of the optocoupler. An infrared LED is usually used as the input part of an optocoupler.

Depending on the structure of the output elements, the following classes of optocouplers can be distinguished:

- Optocouplers with an output phototransistor (Figure 1a). This class is the most common. Devices of this type are able to provide galvanic isolation while maintaining control currents ranging from 1 to 50 mA. Another variation of this circuit is a combination of a photodetector diode and a conventional output bipolar transistor.

- Optocouplers with an output photodiode. They differ from the previous class by lower transmission coefficient, but higher response-speed.

- Optoelectronic solid-state relays (opto-relay), in which, a switching element, is usually a complementary pair of MOSFETs. They are efficient secondary switches with galvanic isolation between the control inputs and the load and are intended for switching the load in AC and DC circuits (Fig. 1b).

Sergey V. Novikov, Maksim E. Cherniak, Roman K. Mozhaev, and Dmitry V. Boychenko are with the National Research Nuclear University (NRNU) "MEPhI" and JSC "SPELS", Kashirskoe shosse 31, Moscow, Russia, E-mail: svnov@spels.ru

- Digital optocouplers (Figures 1c and 1d) allow organizing electrical protection in systems operating at frequencies of the order of 10 MHz. Depending on a type of the input part, digital optocouplers with a digital input and an input LED can be distinguished.

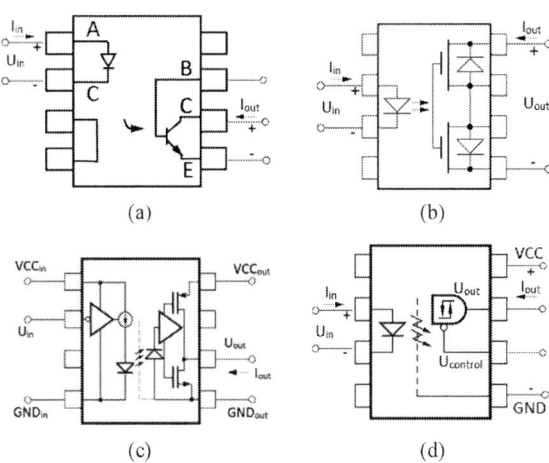

Fig. 1. Block diagrams of optocouplers with an output phototransistor (a), high-power optoelectronic switches (b), digital optocouplers with an input LED (c) and a digital input (d).

There are two options for the layout of the optocoupler circuit elements. The first option is when the output switching part is located inside the case separately from the light emitter and photodiode. The second option - the output transistor is located under the LED, in this case, irradiation on an x-ray source will lead to uneven illumination and distorted test results.

III. DEVICE UNDER TEST, ITS PARAMETERS AND CHARACTERISTICS

As a subject of the research, an opto-relay with an output pair of MOSFETs was selected. According to the results of [2], the most noticeable performance deterioration of such opto-relays is a leakage current degradation. Therefore, the comparative characteristic of the reference parameters is based on leakage current parameter values, depending on the total ionizing dose (TID).

To identify the radiation-sensitive element, opto-relay radiation tests were carried out, having a photodiode and a complimentary pair of MOSFET transistors as the output

978-1-7281-3420-8/19 $31.00 © 2019 IEEE

part, while the photodiode is located under the LED, separately from the transistors (figure 2a).

Some modifications of the test samples have been made before tests. The first sample group – without the top-cover and the compound (figure 2b). The second one - without the top-cover, the compound and the LED (Figure 2c).

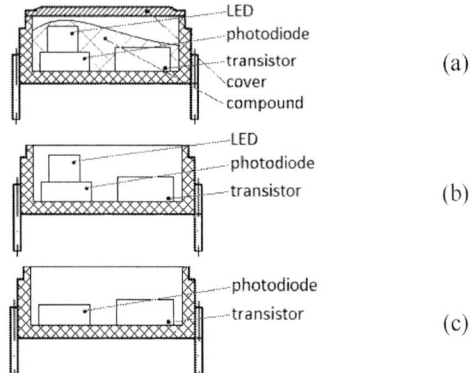

Fig. 2. Opto-isolators configurations used in the research. Common opto-relay (a), opto-relay without cover and compound (b), opto-relay without cover, compound and LED (c).

In this research, the samples without top cover, compound and LED, specially prepared by the manufacturer, were used. But if this is not possible to get a bare device, the compound on the chip can be removed using chemical and mechanical methods. The result of the preparation of samples for the research is shown in figure 3.

Fig. 3. Optocoupler sample with a removed compound.

IV. EXPERIMENTAL TECHNIQUE

Optocouplers Total Ionizing Dose (TID) radiation tests can be carried out using an electron accelerator LINAC operating in bremsstrahlung mode or an isotope source facility [3, 4].

Gamma radiation of such facilities is highly penetrative which causes the uniform ionization of all circuit elements [5]. One should take into account the possibility of radiation damage to measuring equipment [6, 7].

An alternative solution is to use equipment that reproduces TID effects - an X-ray source. X-ray unit advantages are the simplicity of operation and low cost of tests. A significant drawback is the fact that the X-rays penetrating ability is rather low as compared to gamma. This means that the irradiated circuit elements of the opto-relay cannot be covered with compounds or other circuit elements, otherwise, uneven exposure of the components will occur, which will affect test results.

A. Test facilities.

This research was performed using LINAC "U-31/33", operating in electron braking mode, and X-ray complex "RIK-0401". The parameters of the test equipment are given in Table I.

TABLE I
CHARACTERISTICS OF TEST FACILITIES

The name of the unit	X-ray simulation complex "RIK-0401"	LINAC "U-31/33"
Type of radiation source	X-ray characteristic and braking radiation	braking radiation
Maximum quantum energy	45 keV	2,1 MeV
Average quantum energy	10 keV	0,5 MeV
Tube current	105 uA	-
Maximum exposure dose rate	-	1 Gy/s

B. Experiment description

The devices under test are three types of optocouplers with an output MOSFET, differing in the power of the switched load.

The first stage of the experiment was LINAC irradiation of each of the three types of optocouplers. Tests were performed for the modified optocouplers without the upper lid and the compound (figure 2b), and also for the ones without the upper cover, the compound and the LED (figure 2c). During irradiation, the optocouplers operate in active mode without current flowing at the input ($I_{IN} = 0$ mA), and the maximum operating voltage is applied to the output terminals ($U_{out} = U_{max}$ V).

The second stage of the experiment was the repetition of the tests on the x-ray unit.

Table II shows the Exposure Dose Rate (EDR) for each test.

The results of the experiments are presented in Fig. 4, the data analysis is given in Table III.

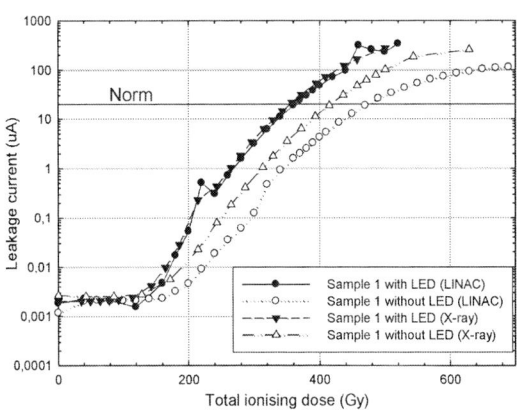

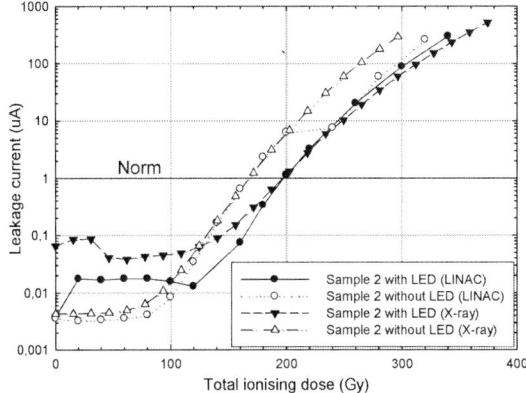

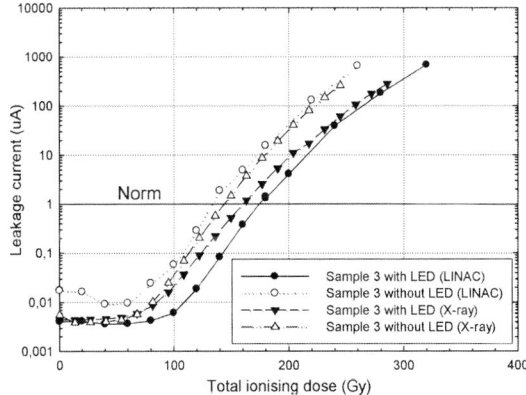

Fig. 4. Graphs of the leakage current level vs. total ionization dose for the three types of optocouplers.

TABLE II
EXPOSURE DOSE RATE

	EDR at LINAC, Gy/s	EDR at X-ray unit, Gy/s
Sample 1	0,31	0,14
Sample 2	0,33	0,39
Sample 3	0,33	0,34

TABLE III
THE SCATTER OF THE TEST RESULTS FOR THE OPTOCOUPLER WITH LED AND WITHOUT LED

	K_{LINAC}, %	K_{X-ray}, %
Sample 1	26,7	16,2
Sample 2	11,8	18,2
Sample 3	28,6	9,5

K_{LINAC} − the ratio of the difference between the levels of resistance to the dose effects of an optocoupler with LED and without LED to the average level of resistance to dose effects when tested on an LINAC.

K_{X-ray} − the ratio of the difference between the levels of resistance to the dose effects of an optocoupler with LED and without LED to the average level of resistance to dose effects when tested on an X-ray simulation complex.

C. Discussion

The degradation dynamic of the leakage current for samples with LED and without LED is the same for the tests on a LINAC and an x-ray unit. The obtained levels of TID hardness for different modifications of samples differ by less than the dosimetry error (TABLE III). During the tests on the x-ray unit, the temperature of the optocoupler can be easily set to higher or lower. It is also possible to select the most sensitive area of the device, using the method of masking parts of the object with lead plates, which will protect separate elements from x-ray radiation.

V. CONCLUSION

A research technique for optocouplers TID hardness evaluation using both LINAC and X-ray tester was developed. The obtained test results are in agreement for two configurations and demonstrate that the main radiation-induced mechanism of opto-relay parameters degradation is the degradation of the output transistors. At the same time, the degradation of the photodiode has practically no effect on the parameters of the optocoupler.

Consequently, TID hardness assurance tests of optocouplers can be carried out using an x-ray unit, though it is important that test components are manufactured in a configuration without a compound. The comparative gamma-radiation tests are used for the X-ray dosimetry results verification.

REFERENCES

[1] R. Reed, "Guideline for Ground Radiation Testing and Using Optocouplers in the Space Radiation Environment", electronic file available at http://radhome.gsfc.nasa.gov/radhome/papers, 2002.

[2] A.S. Epstein and P.A.Trimmer, "Radiation Damage and Annealing Effects in Photon Coupled Isolators", *IEEE Transactions on Nuclear Science*, 1972, vol. NS-19, p. 391.

[3] W.J YU, "Bias Dependence of Total Dose Effect of Partially Depleted SOI MOSFET", *High Energy Physics and Nuclear Physics*, 2007, vol.31, no. 9, pp. 819-822.

[4] A.H. Johnston and T.F. Miyahira, "Hardness Assurance Methods for Radiation Degradation of Optocouplers", *IEEE Transactions on Nuclear Science*, 2005, vol. 52, no. 6, pp. 2649-2656.

[5] E. V. Petrova, N. A. Komarova, M.E. Cherniak, A.V. Ulanova, and A.Y. Nikiforov, "Hardware/software solution for optocouplers with output MOSFET transistors based on National Instruments PXI-platform", in *Proc. 2016 International Siberian Conference on Control and Communications,* SIBCON 2016 National Research University "Higher School of Economics" Moscow; Russian Federation; May 12 -14, 2016, article number 7491764.

[6] S. Shmakov, M. Cherniak, A. Boruzdina, P. K. Skorobogatov, and A. Y. Nikiforov, "Automated data processing system for oscilloscope measurements during an environmental test", *MATEC Web of Conferences*, 2016, vol. 79, article number 01068.

[7] O. A. Kalashnikov, P. V. Nekrasov, A. Y. Nikiforov, V. A. Telets, G. V. Chukov, and V.V. Elesin, "System-on-chip: Specifics of radiation behavior and estimation of radiation hardness," *Russian Microelectronics*, 2016, vol. 45, no. 1, pp. 33-40.

Proton Accelerator's Direct Ionization Single Event Upset Test Procedure

A. O. Akhmetov, G. S. Sorokoumov, A. A. Smolin, D. V. Bobrovsky,
D. V. Boychenko, A. Y. Nikiforov, and A. E. Shemyakov

Abstract - The paper presents single event upset (SEU) experimental results in Spartan-6 FPGA due to direct and indirect proton ionization. High energy proton beam and aluminum foils were used to decrease proton energy down to 1...20 MeV to observe proton direct ionization upsets.

I. INTRODUCTION

Nowdays, there are a lot of critical ICs for space applications that use sub-90nm process and low energy protons are able to produce enough energy in sensitive volume to cause single event effects. Many experimental results are published on modern CMOS and SOI SRAM proton direct ionization [1,2] but there is much less information about direct ionization influence to proton rate calculation. Authors in [2] presented the publications analysis on direct ionization experimental data and suggest proton rate calculation method taking into account direct ionization [3]. Maximum ratio between direct and indirect ionization SEU rate according [3] is about 14 times for 1 g/cm^2 aluminum shielding. But space equipment in general has much more shielding than 1 g/cm^2 and it is not clear if direct ionization should be considered for proton rate calculation. If the difference between direct and indirect proton rate calculation is less than for example ten percent then it is not necessary to perform the complex and expensive IC's direct ionization experiment.

The main aim of the paper is to look for proton direct ionization upsets in modern 45 nm CMOS FPGA using high energy proton beam (about 30 MeV) and aluminum foils to reduce energy down to 1 MeV.

II. DEVICE UNDER TEST

Proton direct ionization is critical for sub-90nm process [7] and 45nm CMOS FPGA was chosen for the experiment. XC6SLX4 Xilinx FPGA was used to obtain direct and indirect ionization. Spartan-6 FPGAs store the customized configuration data in SRAM-type internal latches. The number of configuration bits is 2722512 for XC6SLX4. Spartan-6 has a lot of memory types but in our experiment

A.O. Akhmetov, G.S. Sorokoumov, A.A. Smolin, D.V. Bobrovsky, A.Y.Nikiforov and D.V. Boychenko are with Specialized Electronic Systems (SPELS) and National Research Nuclear University (NRNU) "MEPHI", Moscow, Russia (e-mail: ahmet@spels.ru).
A.E. Shemyakov is with P.N. Lebedev Physical Institute of the Russian Academy of Sciences, Protvino, Russia

only configuration memory was tested because SEU in configuration memory is much more important than the other memory SEU types. DUT has copper metallization. Internal DUT structure is shown in figure 1. DUT package consists of SiO$_2$ and carbon (spectroscopy is shown in figure 2) and has 0.45 mm thickness above the chip.

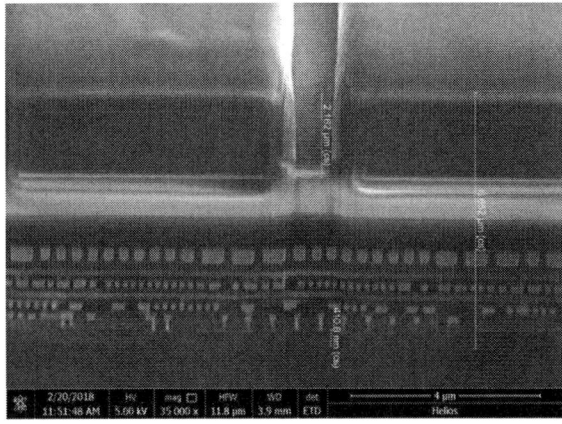

Fig. 1. Internal XC6SLX4 structure.

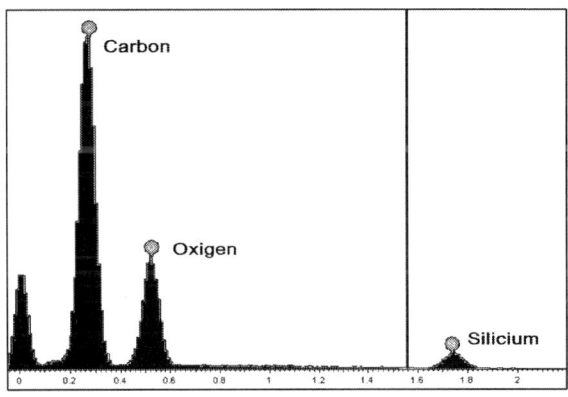

Fig. 2. XC6SLX4 package spectroscopy.

III. TEST FACILITY

DUT was irradiated using proton accelerator with adjustable energy at the medical center in Protvino (Moscow area, Russia). Proton energy range in the experiment was from 30 MeV to 230 MeV. Brief information about facility is published in [4]. All tests were performed in air because facility does not have vacuum chamber.

The difference between description in [4] and our experiment was the wider energy range down to 30 MeV.

IV. EXPERIMENTAL RESULTS

SEU measurement was performed using automated system [5,6,8,9,11].

DUT had been programmed with test configuration before irradiation and configuration memory was verified for upsets after irradiation. DUT at the test facility is shown in figure 3.

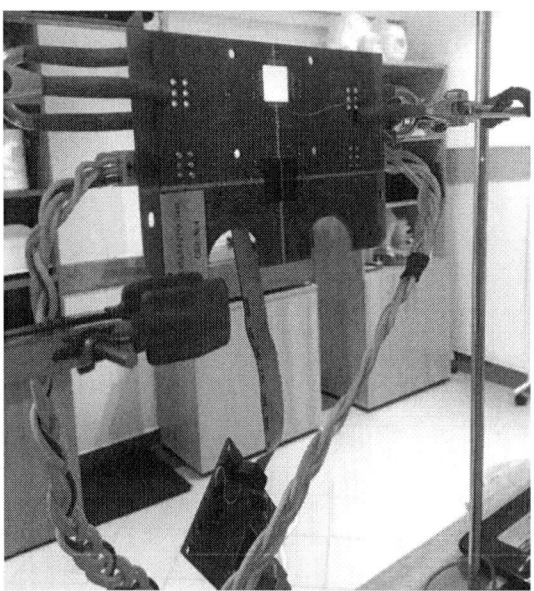

Fig. 3. XC6SLX4 centered in proton beam at the proton accelerator.

At the first experiment stage irradiation was done with proton energies 30, 50, 70, 100 and 230 MeV to measure indirect cross-section. Proton fluence was $6 \cdot 10^{10}$ proton/cm^2 for each energy and a number of SEUs was about 1000 (statistical error 3...5%). Indirect ionization configuration memory SEU cross section is shown in figure 4.

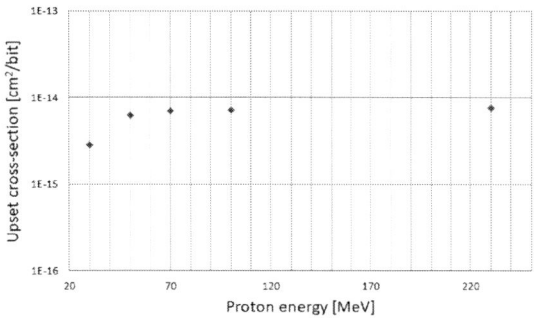

Fig. 4. XC6SLX4 indirect SEU cross section.

At the second stage proton energy was set down to 30 MeV and a set of aluminum foils was used to reduce energy in sensitive volume up to 0 MeV (all protons stopped in foils and the package). Proton fluence was $1.2 \cdot 10^{10}$ proton/cm^2 for each foils set and a number of SEUs was about 50 (statistical error no more than 15%). Fluence was decreased to minimize irradiation time because it is about 10 times longer for 30 MeV than for 230 MeV. The aluminum thickness was from 3000 to 3600 um and foils located near the DUT package.

«Direct ionization» configuration memory SEU cross section is shown in figure 5.

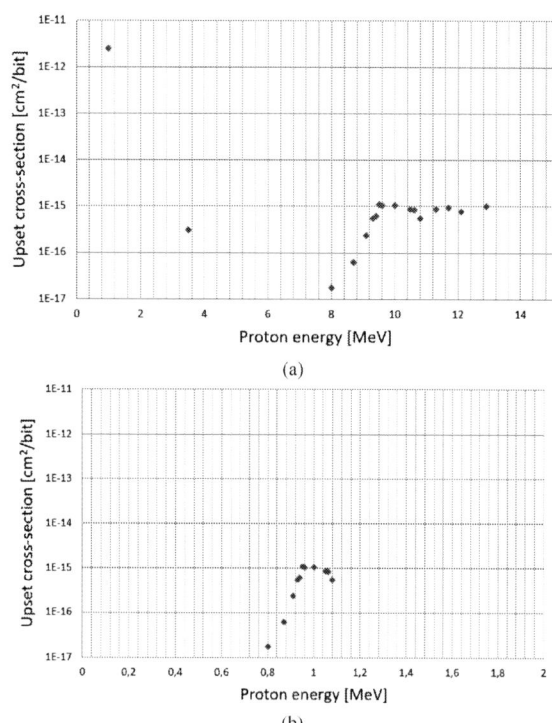

Fig. 5. XC6SLX4 «direct SEU» cross section: two GEANT4 modelling for different package materials and proton straggling parameters: (a) – upset cross section near "1 MeV" was corrected to the fluence reduced by the package; (b) – modelling with another package parameters.

V. DISCUSSION

SRIM and GEANT4 simulation was performed to determine proton energy in sensitive volume. Experimental results shown in figure 5 (a and b) has a cross-section peak at 1 MeV proton energy. But proton fraction with 1 MeV energy is about 1% in full proton energy spectrum when protons interact with aluminum and package above the chip. Experimental cross section was multiplied to about 100 based on GEANT4 simulation for figure 5a and it is close to results in [2]. Figure 5b shows that with another simulation parameters (package material and proton straggling) one can obtain result that is absolutely contrary to figure 5a.

Based on GEANT4 simulation one can conclude that it is possible to reduce proton energy up to 3 times without taking into account proton spectrum and fluence changes after shielding. In this case fluence and spectrum error is close to initial flux and fluence error measurement (10…20%). If energy is reduced more than 3 times it should be noted that energy spectrum and fluence are changed dramatically. Fluence in this case should be corrected based on modeling of proton interaction with shielding.

Proton flux shielding dependence for 2000 km circular orbit is shown in figure 6 and fraction in upset rate due to indirect and direct ionization is shown in figure 7. It should be noted that heavy ion upset rate [10] is not considered in figure 7. One can see that proton flux at low energies is dramatically decrease and upset rate fraction due to direct ionization is not so significant for shielding thicknesses close to real spacecrafts (about 1 g/cm^2 and over) as compared to outer space without shielding conditions. If we take into account heavy ion upset rate (it has been done for example in [1]) the fraction of direct ionization to the total rate would be insignificant.

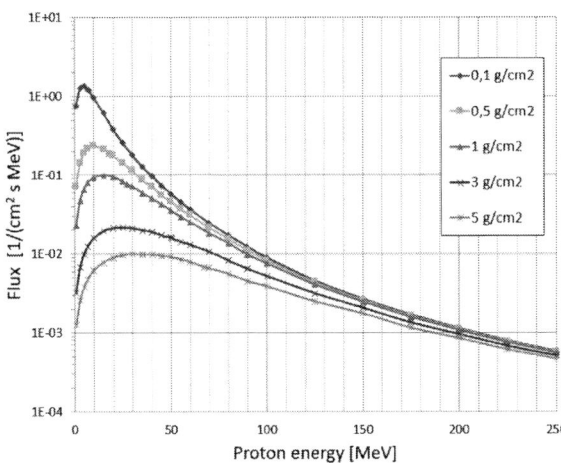

Fig. 6. Proton flux for circular orbit 2000 km 60° shielding dependence.

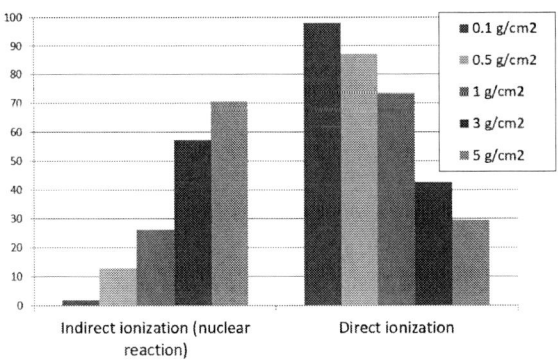

Fig. 7. Fraction [%] in upset rate (protons only) due to indirect (left side) and direct (right side) proton ionization.

VI. CONCLUSION

Proton direct ionization for modern IC's like RAM, FPGA or microprocessor can increase proton upset rate in about 10 times, but for real IC's shielding conditions of more than 3…5 g/cm^2 the upset rate increasing is not so essential and is not more than 2 times.

It is possible to reduce proton energy up to 3 times without taking into account proton spectrum and fluence changes after shielding. If energy is reduced at more than 3 times then fluence should be corrected based on modeling of proton interaction with shielding.

Based on experimental results and previous papers analysis one can conclude that there are only two ways to observe proton direct ionization:

- the first one is based on original method presented in [1]. Based on the method one can calculate proton direct ionization SEU rate without knowledge about proton direct SEU cross-section versus proton energy;
- the second one is based on direct experiment inside vacuum chamber with decapsulated IC. It is complex experiment and there are very few accelerators suitable for that irradiation.

Proton direct ionization experiments in air (with or without degraders) has a lot of test condition variations due to energy straggling. It is very difficult to perform fluence and proton energy precise calculation in sensitive volume.

ACKNOWLEDGMENT

Authors wish to thank Anatoliy Bazhan of the CJSC PROTOM, Protvino, Russia, for the proton irradiation.

Also authors wish to acknowledge Armen Sogoyan and Alexander Chumakov from NRNU MEPhI - SPELS, Moscow, Russia for their valuable advices.

REFERENCES

[1] N.A. Dodds, M.J. Martinez, P.E. Dodd, M.R. Shaneyfelt, F.W. Sexton, J.D. Black, D.S. Lee, S.E. Swanson, B.L. Bhuva, K.M. Warren, R.A. Reed, J. Trippe, B.D. Sierawski, R.A. Weller, N. Mahatme, N.J. Gaspard, T. Assis, R. Austin, S.L. Weeden-Wright, L. W. Massengill, G. Swift, M. Wirthlin, M. Cannon, R. Liu, L. Chen, A. T. Kelly, P. W. Marshall, M. Trinczek, E. W. Blackmore, S. J. Wen, R. Wong, B. Narasimham, J. A. Pellish, and H. Puchner, 'The Contribution of Low-Energy Protons to the Total On-Orbit SEU Rate', IEEE Transactions on Nuclear Science, vol. 62, no. 6, pp. 2440–2451, Dec. 2015.

[2] J. Guillermin, N. Sukhaseum, A. Privat, P. Pourrouquet, T. Cardaire, N. Chatry, F. Bezerra and R. Ecoffet, "Assessment of the direct ionization contribution to the proton SEU rate", RADECS 2016 proceedings

[3] N. Sukhaseum, A. Samaras, L. Gouyet, P. Pourrouquet, N. Chatry, F. Bezerra, R. Ecoffet and E. Lorfèvre, 'A Calculation Method for Proton Direct Ionization Induced SEU Rate from

Experimental Data: Application to a Commercial 45nm FPGA', 2014, NSREC 2014 Proceedings – [PB-5]

[4] A.O. Akhmetov, A.V. Yanenko, A.I. Bazhan "Proton accelerator with adjustable energy for ICs radiation test" (2013) Proceedings of the European Conference on Radiation and its Effects on Components and Systems, RADECS, № 6937364

[5] L.A. Tambara, A.O. Akhmetov, D.V. Bobrovsky, F.L. Kastensmidt "On the characterization of embedded memories of Zynq-7000 all programmable SoC under single event upsets induced by heavy ions and protons" (2015) Proceedings of the European Conference on Radiation and its Effects on Components and Systems, RADECS, 2015-December, № 7365643

[6] A.E. Rudenkov, A.O. Akhmetov, D.V. Bobrovsky, A.I. Chumakov, A.N. Schepanov "Automated measuring system for MIL-STD-1553 integrated circuits functional and parametric control" (2017) 2017 International Siberian Conference on Control and Communications, SIBCON 2017 - Proceedings, № 7998546

[7] A.Y. Nikiforov, D.V. Boychenko, V.A. Telets, A.A. Smolin, V.V. Elesin, A. V. Ulanova, and L.N. Kessarinskiy "Basic trends in electronic components product range development:

Radiation hardness aspects," in Proc. 30th Int. Conf. on Microelectronics, MIEL 2017; Nis, Serbia, October 2017, pp. 45-48.

[8] D.V. Bobrovsky, A.A. Pechenkin, A.A. Novikov, A.I. Chumakov , N.V.Ryasnoy, and Y.V. Churilin, "Flip-chip ICs SEE testing technique," in Proc. 30th Int. Conf. on Microelectronics, MIEL 2017; Nis, Serbia, October 2017, pp. 309-311.

[9] A.O. Akhmetov, D.V. Bobrovskiy, A.S.Tararaksin, A.G. Petrov, L.N. Kessarinskiy, D.V. Boychenko, A.I. Chumakov, A. Rousset, and C. Chatry "IC SEE comparative studies at UCL and JINR heavy ion accelerators," in 2016 IEEE Radiation Effects Data Workshop, REDW 2016, Portland; United States,3 April 2017, article number 7891720

[10] A.V. Sogoyan, A.I. Chumakov, A.A. Smolin, A.V. Ulanova, and A.B. Boruzdina "A simple analytical model of single-event upsets in bulk CMOS," Nuclear Instruments and Methods in Physics Research, Section B: Beam Interactions with Materials and Atoms, vol. 400, 1 June 2017, pp. 31-36, 2017.

[11] V.A. Marfin, P.V. Nekrasov, O.A. Kalashnikov, and A.Y. Nikiforov, "Functional testing of digital signal processors in radiation experiments," Russian Microelectronics, vol. 46, no. 3, pp. 149-154, 2017.

DEVICE PHYSICS, TECHNOLOGY AND CHARACTERIZATION – POSTER SESSION

978-1-7281-3420-8/19 $31.00 © 2019 IEEE

Reliability of Various Type of Gas-filled Surge Arresters Under DC Discharge

E. Živanović, S. Veljković, M. Živković, and M. Pejović

Abstract – The delay response of Littelfuse gas-filled surge arresters was determined using the time delay method. It is a qualitative method, which can give neither the number densities of ions nor neutral active states in the glow and afterglow, but it can be used for qualitative observation of ions and neutral active states decay in the afterglow to such low concentrations where other methods cannot be applied. It has also enabled the estimation of recombination and de-excitation times of the mentioned particles due to their recombination on the tube walls, electrodes and in gas. Experimental data of mean values of breakdown voltage of these components were obtained for voltage increase rates from 1 to 10 Vs^{-1} analysed with discretized dynamic method presented in this paper. Besides, an influence of overvoltage values on the dependence of electrical breakdown voltage vs. relaxation time was also been investigated.

I. INTRODUCTION

Efficient overvoltage protection of electronic components and systems is of great importance for their proper operation [1], [2]. The gas-filled surge arresters are non-linear components used in overvoltage protection. In literature they are known as surge voltage protectors or gas discharge tubes. Gas-filled surge arresters (GFSA) have several advantages compared to the over-voltage protection components. Their main advantages are the ability to conduct high currents (>5 kA), low intrinsic capacity (<1 pF), high insulating resistance (>1 GΩ) and low resistance in conducting regime (~0.1 Ω) [3]-[5]. They can be used for overvoltage protection in the range from 70 to 1200 V. It is most often used in protective circuit in telecommunications as well as high voltage engineering. The major drawbacks in GFSAs application is their delay response and cut off delay upon voltage disconnection as well as relatively large deviation in activation voltage, which goes up to 20% with respect to values usually found in datasheets [6].

The operating principle of these components is based on the electrical breakdown in gases. Based on the type of applied voltage on the component, the electrical breakdown in gas can be static or dynamic.

The breakdown is dynamic, and it is characterized by dynamic breakdown voltage when the rate of voltage change is equal to or greater than the velocity of the elementary processes during the gas breakdown.

However, the static DC breakdown occurs when breakdown voltage has an unchangeable value, and it is the static breakdown voltage. Its value is of a great importance to define for many gas-filled devices. Due to statistical nature of breakdown process, when the voltage applied on the gas component, breakdown cannot appear instantly. The time elapses between the moment of application of voltage higher than breakdown voltage and the moment when the gas-filled surge arresters current starts to flow is called the electrical breakdown time delay t_d. Delay response of gas-filled surge arresters is usually referred as electrical breakdown time delay. The electrical breakdown time delay consists of the statistical time delay t_s and the formative time t_f, i.e. $t_d = t_s + t_f$ [7]. Statistical time delay is the time interval between the moment of operating voltage application and the appearance of a free electron which initiates the breakdown. Formative time is the time taken from the end of the statistical time delay to the onset of breakdown, characterized by the collapse of the applied voltage as a self-maintained glow [7].

The various parameters have an influence on electrical breakdown time delay, but the most important is the relaxation time τ which represents the time interval between two successive measurements when there is no voltage on the used component [8]. This dependence, $\bar{t}_d = f(\tau)$, is called the memory curve.

II. EXPERIMENT DETAILS

The samples used in the experiment are the gas-filled surge arrester of the manufacturer Littelfuse. One type of used gas-filled surge arrester is designed to operate at a voltage of 120 V AC (in this paper marked as LF1) and the other at 230 V AC voltage (designated as LF2). The technical characteristics of the gas arrester LF1 are as follows: the breakdown voltage in the DC mode varies from 230 V to 340 V, the insulating resistance is 10 GΩ at 100 V, and the capacitance is 1.5 pF [6]. The following technical characteristics for the gas arrester LF2 are known: the breakdown voltage in the DC mode varies from 184 V to 276 V, the insulating resistance is 1 GΩ at 50 V, and the capacitance is 1.5 pF [6]. These characteristics are shown in detail in Table I. The cross section of these components is shown in Figure 1.

E. Živanović, S. Veljković, M. Živković and M. Pejović are with the Department of Microelectronics, Faculty of Electronic Engineering, University of Niš, Aleksandra Medvedeva 14, 18000 Niš, Serbia, E-mail: emilija.zivanovic@elfak.ni.ac.rs

TABLE I
LITTELFUSE GAS-FILLED SURGE ARRESTERS SPECIFICATIONS AT 25°C

Device	Device Specifications (at 25°C)						
	Operating voltage AC (V)	Breakdown voltage in the DC mode (V)	DC Breakdown - typical (V)	Insulation Resistance	Capacitance (pF)	Arc Voltage -on state Voltage (V)	Nominal Impulse Discharge Current (x10@8/20μs) (kA)
LF1	120	230 to 340	285	10 GΩ at 100 V	1.5	~ 25	5
LF2	230	184 to 276	230	1 GΩ at 50 V	1.5	~10	5

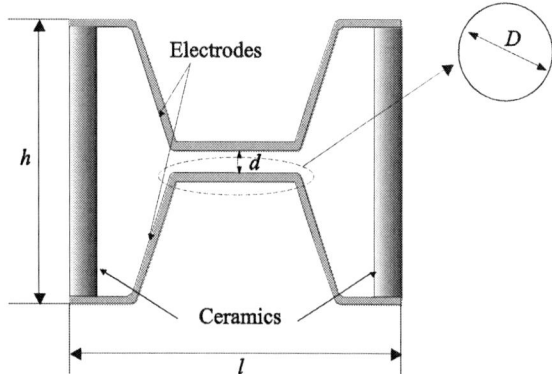

Fig. 1. Gas-filled surge arresters cross section.

The experimental investigations of DC breakdown voltage U_b for small voltage increase rates and electrical breakdown time delay t_d of commercial gas-filled surge arresters are presented in this paper. In addition, this paper examines the influence of the relaxation time and overvoltage on the delay response of gas-filled surge arresters based on the

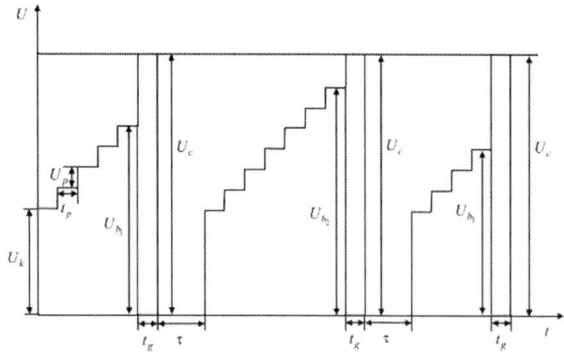

Fig. 2. Measuring cycles for U_s determination: U_{b1}, U_{b2}, breakdown voltages, τ – relaxation time.

memory curve. It should be noted that GFSA used in our experiment are commercial components and the manufacturer does not give the gas type specification.

The static breakdown voltage U_s can be defined as minimal value of breakdown voltage U_b for which breakdown probability still has finite value. It was estimated using the dynamic discretized method [9]. This method is based on voltage increase in steps, with strictly defined duration voltage step, until breakdown. Voltage step and its duration can be arbitrarily chosen. In order to determine the static breakdown voltage U_s, as deterministic variable, the mean value of the dynamic breakdown voltage $\overline{U}_b$ as a function of voltage increase rate k were extrapolated until the intersection with $\overline{U}_b$ axis using linear fit. The intersection point for $k = 0$ correspond to U_s value. Namely, this method is based on the determination of $\overline{U}_b = f(k)$ dependence, where $\overline{U}_b$ is the mean value of measured dynamic breakdown voltages and $k = U_p/t_p$ is the voltage increasing rate, where U_p is the increasing voltage step, t_p is duration of step, i.e. the time interval between successive voltage steps. The breakdown measurement cycle is shown in Figure 2.

It can be seen from Figure 2, the voltages U_{b1}, U_{b2} and U_{b3} values are different because of stochastic nature of breakdown process. It requires to perform a lot of measurements of breakdown voltage and to use the mean value. In this experiment $\overline{U}_b$ was determined based on a hundred measurements of breakdown voltage.

After breakdown, the glow current flows through the gas during the glow time t_g, which is achieved by maintaining a constant value of voltage U_c. When gas relaxation interval τ finished, the voltage U_k is again reached. The whole procedure is repeated the desired number of times.

It should be noted that the rise of breakdown voltage in the steps is the main feature of this method because it enables more precisely breakdown voltage determination than earlier used method [10]. Before any research and testing of gas electronic components, it is necessary to determine the value of static breakdown voltage.

978-1-7281-3420-8/19 $31.00 © 2019 IEEE

The results shown in Figure 3 relates to the estimation of the static breakdown voltage for two early mentioned gas-filled surge arresters. Obtained result has shown that the linear function gives a good agreement with experimental results. In this case the estimated value for the static breakdown voltage is 284 V and 238 V for gas-filled surge arresters LF1 and LF2, respectively. Comparing obtained values with the technical documentation confirms the adequacy of the method used to estimate the value of static breakdown voltage. The breakdown voltage measurements as well as the electrical breakdown time delay measurements were performed with the electronic systems which block diagram and principle of operation in detail are shown [9].

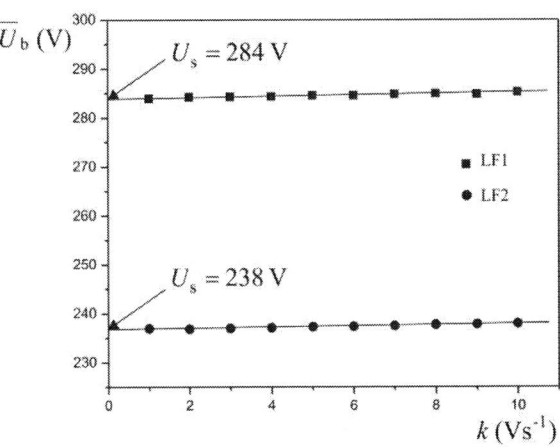

Fig. 3. Mean value of breakdown voltage as a function of voltage increase rate for gas-filled surge arresters.

III. RESULTS AND DISCUSSION

A family of dependences of gas components responses for different relaxation times and overvoltage is shown. The overvoltage is most often expressed in percentages and is defined as $(\Delta U / U_s) \cdot 100\%$, where ΔU is the difference in operating voltage U_w ($U_w > U_s$) and static breakdown voltage, i.e. $\Delta U = U_w - U_s$. The aim of this study was to examine the effect of the applied voltage at the electric breakdown time delay is the function of relaxation time. In this way, there is a possibility to determine the overvoltage on which the corresponding component will work reliably. It can be noticed from Figure 4 that the first component marked as LF1 works reliably for overvoltage higher than 20%. While the second one designated as LF2 is reliable for overvoltage higher than 30% (see Figure 5).

In addition, in order to established overvoltage values below which components are not reliable, the time delay method allows evaluating the delay response of these components. Namely, the memory curves shown in Figures 4 and 5 indicate to values of GFSA, i.e. approximate to 10 μs, as well as in the range of 10 to 30 μs for LF1 and LF2, respectively.

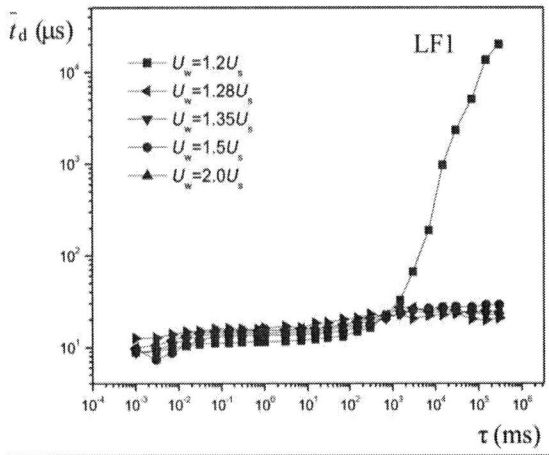

Fig. 4. Mean value of electrical breakdown time delay as a function of relaxation time for different values of overvoltage for gas-filled surge arresters LF1.

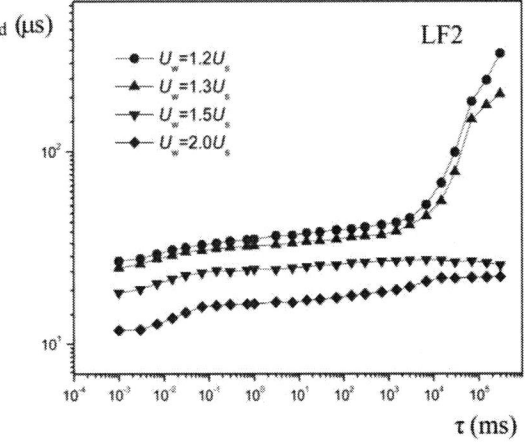

Fig. 5. Mean value of electrical breakdown time delay as a function of relaxation time for different values of overvoltage for gas-filled surge arresters LF2.

The fact that the manufacturer does not give the specification of used gas type in GFSA, disables the further detailed analysis of possible processes which are responsible for breakdown appearance as well as the gas relaxation. What appears to be a possibility is to compare the obtain results with earlier obtained results [11]-[13] which relate on the separation of mechanisms responsible for the breakdown initiation in gases.

IV. CONCLUSION

On the basis of the above mention consideration, the following could be concluded. The static breakdown voltage, as deterministic variable, can be precisely estimated using the dynamic discretized method, i.e. fitting the dependence of mean value of the electrical breakdown voltage as a function of the voltage increase rate. Estimated values of static breakdown voltage for used GFSAs in this manner are in good agreement with the nominal values as can be seen from Table I. Their tolerance range for Littelfuse gas-filled surge arresters, LF1 and LF2 is about 1% and 3%, respectively.

The experimental data of electrical breakdown time delay are used for the delay response of GFSAs. The analysis whose results are shown in Figure 4 indicates the following. As it can be seen for overvoltages greater than 20% this type of gas arrester LF1 works reliably. However, for overvoltages less than 20% also works reliably, but for relaxation times up to 1 s. In addition, the results show that the delay response of this component is about 10 μs is for all voltage values applied to the sample.

An analysis of the results obtained for the LF2 sample, shown in Figure 5, indicates that this type of gas arrester is reliable up to the applied voltages greater 30% than the static breakdown voltage, for relaxation times up to 300 s, as long as the experimental investigation was carried out. For applied voltages of 30% lower than static breakdown voltage, and confirmed by 20%, this type of arrester works reliably for relaxation times up to 8 s. In addition, the results indicate that the value of overvoltage slightly affects the delay response, which increases with the growth in the applied voltage, in the range of 10 to 30 μs.

Our further research will be reflected in the possibility of continuing the research with the aim of comparing the obtained results by performing additional analysis of the gas-filled surge arresters' samples produced by EPCOS. In addition, we investigate mechanisms that lead to an increase in the electrical breakdown time delay after a certain value of overvoltage, thereby affecting the reliability of the components themselves.

ACKNOWLEDGEMENT

This work has been supported by the Ministry of Education, Science and Technological Development of Republic of Serbia under the contract no. 32026.

REFERENCES

[1] K. Stanković, M. Vujisić and E. Dolićanin, "Reliability of semiconductors and gas-filled diodes for over-voltage protection exposed to ionizing radiation", *Nuclear Techol. Radiat. Prot.*, 2009, vol. 24, pp. 132-137.

[2] B. Lončar, S. Stanković, A. Vasić, P. Osmokrović, "The influence of gamma and X-radiation on pre breakdown current and resistance of commercial gas-filled surge arrsters", *Nuclear Technol. Radiat. Prot.*, 2005, vol. 20, pp. 59-63.

[3] M. M. Pejović and M. M. Pejović, "Investigations of breakdown voltage and time delay of gas-filled surge arresters", *J. Phys. D: Appl. Phys.*, 2006, vol. 39, pp. 4417-4422.

[4] M. M. Pejović, K. Stanković, I. Fetahović, M. M. Pejović, "Processes in insulating gas induced by electrical breakdown responsible for commercial gas-filled surge arresters delay response", *Vacuum*, 2017, vol. 137, pp. 85-91.

[5] P. Osmokrović, B. Lončar and S. Stanković, "Investigations of the optimal method for improvement of the protective characteristics of gas-filled surge arresters built in radioactive sources", *IEEE Trans. Plasma Sci.*, 2002, vol. 30, pp. 1876-1880.

[6] https://www.littelfuse.com/~/media/electronics/product_catalogs/littelfuse_gdt_catalog.pdf.pdf

[7] J. M. Meek and J. D. Craggs, *Electrical breakdown of gases*, New York, USA: Wiley, 1987.

[8] M. M. Pejović, G. S. Ristić and J. P. Karamarković, "Electrical breakdown in low pressure gases", *J. Phys. D: Appl. Phys.*, 2002, vol. 35, pp. R91-R103.

[9] M. M. Pejović, "Digital system for vacuum and gas-filled devices testing", *Rev. Sci. Instrum.*, 2005, vol. 76, p. 015102.

[10] M. M. Pejović, Č. S. Milosavljević and M. M. Pejović, "The estimation of static breakdown voltage for gas-filled tubes at low pressures using dynamic method", *IEEE Trans. Plasma Sci.*, 2003, vol. 31, pp. 776-781.

[11] E. N. Živanović, "Investigation of the effect of additional electrons originating from the ultraviolet radiation on the nitrogen memory effect" *Facta Universitatis, Series: Electronics and Energetics*, 2015, vol. 28, pp. 423-437.

[12] M. M. Pejović, N. T. Nesić, M. M. Pejović and E. N. Živanović, "Afterglow processes responsible for memory effect in nitrogen", *Journal of Applied Physics*, 2012, vol. 112, p. 013301 (10pp).

[13] E. N. Živanović, "Influence of combined gas and vacuum breakdown mechanisms on memory effect in nitrogen", *Vacuum*, 2014, vol. 107, pp. 62-67.

978-1-7281-3420-8/19 $31.00 © 2019 IEEE

Effect of Rare-Earth Ions on Electrical Properties of BaTiO₃ Ceramics

V. Paunović, M. Đorđević, V. Mitić, and Z. Prijić

Abstract - The effect of rare earth ions on dielectric permittivity, dielectric losses and microstructure of barium titanate ceramics doped with Ho, Er and Yb at the concentrations of 0.05 wt% to 1.0 wt% was investigated. The doped BaTiO₃ were obtained by solid-state sintering reaction at a temperature of 1320°C in air atmosphere.

The electrical properties were measured in the temperature interval from 30°C to 180°C at different frequencies. The results of electrical investigation showed that samples doped with a low additives concentration, with a spherical and large grains, have high dielectric constant value. ε_r at Curie temperature is 4250, 3828 and 3011 for 0.05 wt% Ho, Er and Yb respectively. By increasing dopant concentration and decreasing grain size, the dielectric permittivity value decreases. All investigated samples have low values of dielectric losses ranged from 0.01 to 0.09. Also, samples with a low concentration of additives follow Curie-Weiss's law. The Curie temperature of doped samples is slightly lower than Curie temperature of the undoped ceramics and it is 126-128 °C. The Curie constant C for all examined samples increases with increasing dopant concentration, so that the highest values were measured for the samples doped with 1.0 wt% of the additive.

I. INTRODUCTION

Owing to its outstanding electrical properties, ceramic with perovskite structure attracts significant scientific and commercial interest that results in various industrial applications. Among them are piezoelectrics, capacitors, PTC-thermistors, and dielectric resonators for wireless communication [1] - [4].

Dielectric characteristics depend to a large extent on the ceramics composition, the methods of synthesis starting powder, the sintering methods and the obtained microstructure [5] - [7].

Dielectric or semiconducting properties of BaTiO₃ ceramics are due to dopants that can occupy the places of the Ba^{2+} or Ti^{4+} ions, depending on the ions radii that is being embedded. Since the ions radii in trivalent cations of rare earths such as Yb^{3+}, Er^{3+}, Dy^{3+} are in size between ion radii Ba^{2+} or Ti^{4+} ions, dopant ions can occupy A or B positions (A=Ba and B=Ti) in perovskite BaTiO₃ ceramic structure. At low concentrations of additive, Ba^{2+} ions are replaced and solid solutions form. At higher concentrations of the additive (1.0 wt %), Ba^{2+} or Ti^{4+} ions may be substituted [8] - [10].

V. Paunović, M. Đorđević, V. Mitić, and Z. Prijić are with the Department of Microelectronics, Faculty of Electronic Engineering, University of Niš, Aleksandra Medvedeva 14, 18000 Niš, Serbia, E-mail: vesna.paunovic@elfak.ni.ac.rs

As the applied dopants influence the temperature of the phase transformation and the Curie constant, it is best seen through the dependence: dielectric permittivity vs. temperature. For BaTiO₃ doped ceramics, the phase transformation can have a very sharp transition from the ferroelectric to the paraelectric region, but the diffusion phase transition (relaxor ceramics) can also occur [11]-[13].

To examine the behavior of BaTiO₃ in the paraelectric phase, besides the Curie-Weiss law, the modified Curie-Weiss law was also used, which describes the deviations from the linearity ε_r=f(T) due to diffuse phase transformation.

In this paper, a relationship among type and concentration of rare earth ions, microstructure and dielectric properties of BaTiO₃ ceramics is examined.

II. EXPERIMENTAL

BaTiO₃ ceramics doped with rare earth ions was obtained by conventional solid-state sintering. Dopant concentration ranged from 0.05 to 1.0 wt% Ho, Er or Yb. The starting powders were, after calcination, mixed in isopropyl alcohol, dried and pressed into pellets under a pressure of 120 MPa and then sintered in air atmosphere at a temperature of 1320°C for four hours.

The microstructure of samples was investigated by scanning electron microscope, JEOL, SEM-5300 equipped by EDS (Energy Dispersive Spectrometer) system. The electrical characteristic were measured using LCR meter Agilent 4284A, in temperature range from 30 to 180°C and in frequency range from 20 Hz to 1MHz.

The block diagram that explain preparation of BaTiO₃ samples and their electrical characterization is given in Fig.1.

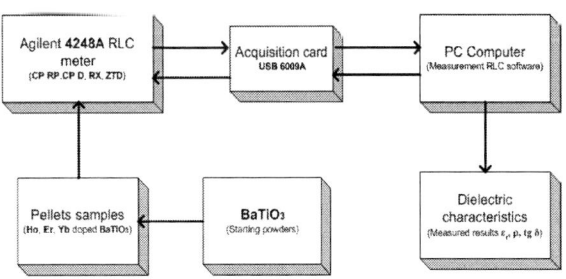

Fig. 1. Block diagram that explain preparation of BaTiO₃ samples and their electrical characterization.

III. RESULTS AND DISCUSSION

A. Microstructure characteristics

The relative density of doped samples is ranged from 82% to 90% of theoretical density (TG), with the highest density being measured for Er doped samples.

SEM investigation (Fig. 2) indicates that the microstructural characteristics of the samples doped with Er_2O_3 and Yb_2O_3 are quite similar. For the low content of additive for both series of samples, are characteristic polygonal grains and abnormal grain growth with average grain size from 20-40 µm for Er doped (Fig. 2a) and from 15-30µm for Yb doped samples (Fig. 2c). As the concentration of the additive increases, the grain size decreases. For samples doped with 1.0wt% of additive the grain size is from 5 to 20 µm (Fig.2b and 2d).

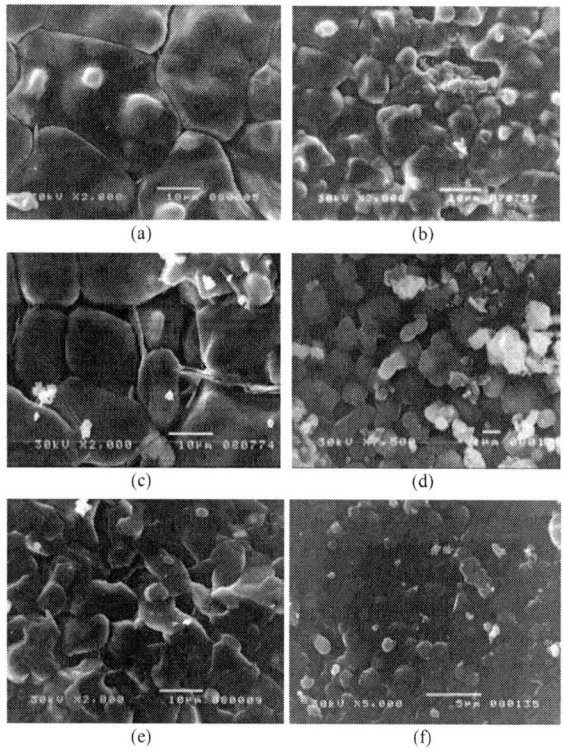

(a)   (b)

(c)   (d)

(e)   (f)

Fig. 2. SEM images of doped $BaTiO_3$ ceramics a) 0.05 wt% Er, b) 1.0 wt% Er, c) 0.05 wt% Yb, d) 1.0 wt% Yb, e) 0.05 wt% Ho, f) 1.0 wt% Ho.

For Ho doped $BaTiO_3$ ceramic are characteristic spherical grains. For lower Ho concentrations (0.05wt% Ho) there is a characteristic abnormal grain growth and grain size ranging from 10-30µm (Fig. 2e). With the increase in the concentration of additive, the mean grain size decreases, so that the specimens doped with 1.0 wt% Ho have a fine-grained and relatively uniform microstructure with a mean grain size from 2 to 5 µm (Fig. 2f).

B. Electrical characteristics

The influence of the additive type and the resulting microstructure on the dielectric permittivity can be observed over the dependence of ε_r on the temperature and frequency (Fig. 3 and Fig. 4). The temperature range in which the dielectric permittivity was investigated was from 30 to 180° C and the frequency range from 20Hz to 1MHz. Based on the curves of dielectric permittivity vs. temperature, it can be seen that the highest dielectric permittivity values at Curie temperature ε_r = 4250 show samples with additive concentration of 0.05wt% Ho. The lowest dielectric permittivity values at the Curie temperature have samples with the highest concentration of additives (1.0 wt %) and it is ε_r = 1167. It is expected that the dielectric permittivity have lower values in samples with a higher concentration of additives, due to the inhomogeneous distribution of additives and lower density of the samples.

The highest value of the dielectric permittivity (ε_r = 1349) at room temperature has 0.0Er/BaTiO3 samples. The lowest values (ε_r = 545) were measured in samples doped with 1.0 wt% Ho.

For the Ho and Yb doped samples are characteristic lower values of the permittivity at room temperature, which are the consequence of a lower sintering density on the one hand and, on the other hand, the appearance of the domain structure in abnormal grains.

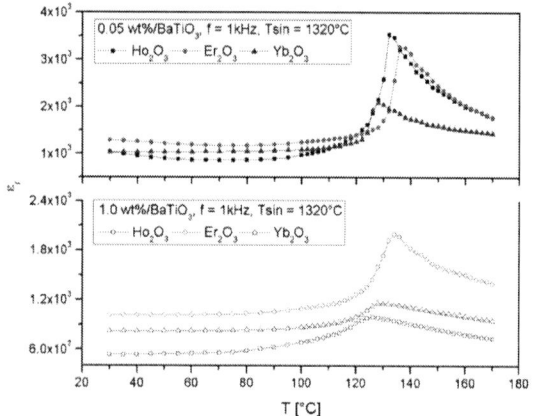

Fig. 3. Dielectric permittivity of 0.05wt% and 1.0wt% doped samples as a function of temperature at 1 kHz.

The dependence of the dielectric permittivity on the temperature shows that the dielectric permittivity above the Curie temperature (T_C), in the paraelectric region, follows the CurieWeiss law. For all tested series of the samples, a sharp transition from the ferroelectric to the paraelectric phase at the Curie temperature is characteristic. With the increase in dopant concentration, the dielectric permittivity decreases and its change in a wide temperature interval, from room to Curie temperature, is considerably lower. The Curie temperature (T_C), for all samples it ranges from 126°C to 128°C and is lower in relation to the Curie temperature for undoped $BaTiO_3$ ceramics where T_C=132°C.

978-1-7281-3420-8/19 $31.00 © 2019 IEEE   118

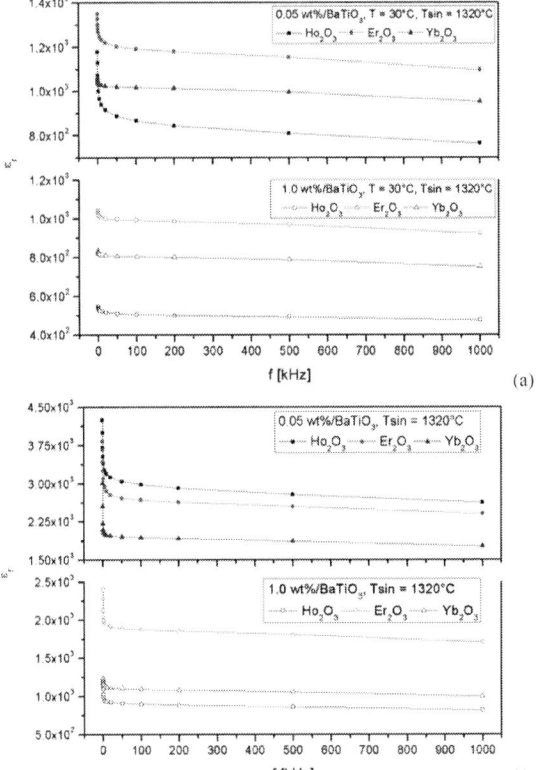

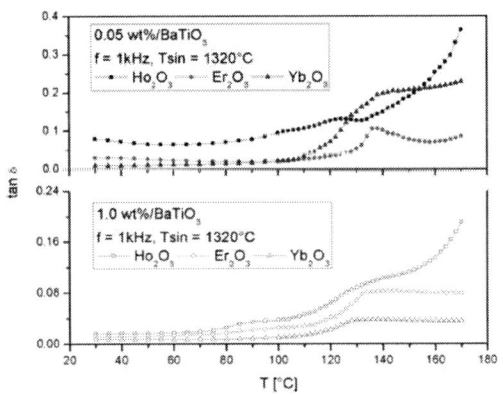

Fig. 4. Dielectric permittivity of 0.05wt% and 1.0wt% doped samples as a function of frequency at a) room temperature and b) Curie temperature.

Dielectric permittivity vs. frequency dependence at room temperature and Curie temperature is given in Fig. 4. It has been observed that the investigated samples have higher dielectric permittivity values at lower frequencies. Dielectric permittivity for all samples after initial high values at lower frequencies decreases with increasing frequency and achieves a constant value for f> 20kHz.

Fig. 5. Dielectric losses of 0.05wt% and 1.0wt% doped samples as a function of temperature at 1 kHz.

Fig. 5 shown the dependence of the dielectric losses on temperature measured at a frequency of 1 kHz for Ho, Er and Yb doped samples. The dielectric loss (tan δ) values at room temperature are ranging from 0.01 to 0.09. For all investigated samples it is characteristic that tan δ increases with increasing temperature. The smallest increase in tan δ is for samples with a concentration of additives of 1wt%, and the highest increase in samples with a concentration of additive of 0.05wt%. Similar dependencies have been obtained for other concentration of additive and other frequencies.

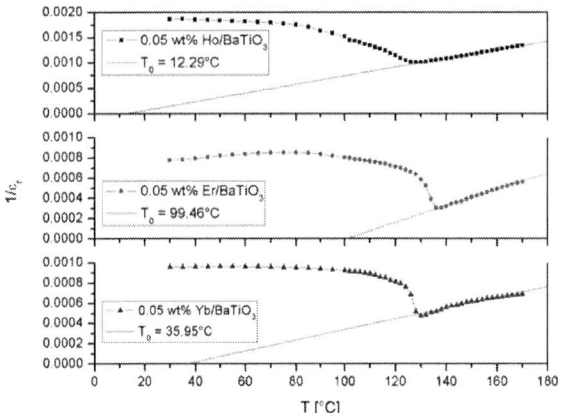

Fig. 6. Reciprocal value of ε_r in function of temperature for 0.05wt% Ho, Er and Yb doped samples.

By fitting the curves $1/\varepsilon_r$ vs. temperature (Fig. 6), the Curie constant (C) and the Curie-Weiss temperature (T_0) for the doped samples were calculated and their values are given in Table 1. The Curie-Weiss temperature T_0 has a lower value than Tc for all series of samples.

TABLE I
DIELECTRIC PROPERTIES FOR DOPED $BaTiO_3$ CERAMICS

Add. (wt%)	ε_r 300K	$\varepsilon_{r\,max}$	Tc [°C]	T_0 [°C]	$C\,10^5$ [K]	γ
0.05Ho	1177	4250	128	12.2	1.442	1.01
0.1Ho	807	2965	128	93.4	2.153	1.02
0.5Ho	683	1882	126	29.3	5.901	1.03
1.0Ho	545	1163	128	94.7	5.911	1.07
0.05Er	1349	3828	128	99.4	1.983	1.07
0.1 Er	1117	2015	128	57.4	2.106	1.11
0.5Er	1066	2103	128	64.8	2.946	1.17
1.0Er	1045	2405	128	46.6	3.047	1.20
0.05Yb	1059	3011	128	35.9	3.017	1.05
0.1 Yb	1070	2396	128	69.6	3.401	1.08
0.5Yb	1288	2603	126	70.3	3.875	1.11
1.0Yb	833	1237	128	45.1	3.997	1.14

978-1-7281-3420-8/19 $31.00 © 2019 IEEE 119

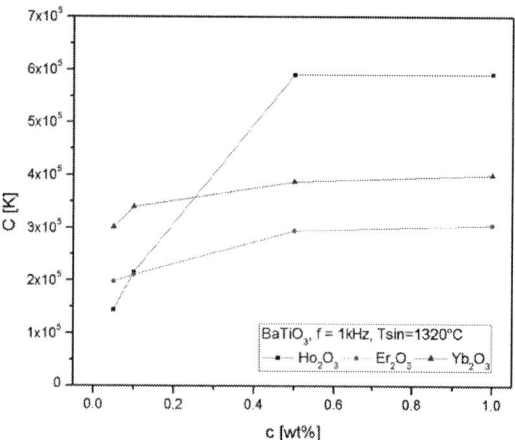

Fig. 7. Curie constant C as a function of additive content measured at f=1 kHz.

Curie constant C (Fig. 7) for all series of tested samples increases with increasing dopant concentration, so that the highest values have samples doped with 1.0wt% of additive (C=5.9·10^5 K, 3.04·10^5 K and 3.9·10^5 K for 1.0wt% Ho, Er and Yb doped samples respectively). This change is in correlation with the change in sample density and reduction of grain size with increasing additive concentration.

The value of the critical non-linearity exponent γ ranged from 1.0-1.12 for lower additive concentrations and up to 1.20 for concentrations of 1.0 wt% of additive (Table 1). Such results are consistent with experimental data and a sharp phase transition from the ferroelectric to the paraelectric phase for all samples.

IV. CONCLUSION

Comparative investigations of rare earth ions influence on the microstructure and dielectric characteristics of Ho, Er and Yb doped BaTiO$_3$ ceramics are given in this paper. At lower concentrations of additives, all ceramics are characterized by abnormal grain growth and grain size ranging from 20-40 μm. For higher additive concentration (1.0 wt %) the grain size decreased and ranged from 2-20 μm for all sample series. The highest dielectric constant values at room temperature and the largest change in dielectric permittivity with temperature show the samples with the lowest additive concentration. The dielectric permittivity at 1 kHz and room temperature for 0.05Er/BaTiO$_3$ is 1349, for 0.05Yb/BaTiO$_3$ is 1059 and for 0.05 Ho/BaTiO$_3$ is 1177. The Curie temperature of the doped samples is 126-128°C, independent of the concentration of the additive. The Curie constant C increases with increasing additive concentration so that the highest values have samples doped with 1.0 wt% of additives. The value of the critical non-linearity exponent γ ranged from 1.0-1.12 for lower additive concentrations and up to 1.20 for concentrations of 1.0 wt% of additive.

ACKNOWLEDGEMENT

The authors gratefully acknowledge the financial support of Serbian Ministry of Education, Science and Technological Development. This research is a part of the Projects OI-172057, TR-32026, and TR-33035.

REFERENCES

[1] S.F. Wang, G.O. Dayton: "Dielectric Properties of Fine-grained Barium Titanate Based X7R Materials", *Journal of the American Ceramic Society*, 1999, Vol. 82, No. 10, pp. 2677 – 2682.

[2] C. Pithan, D. Hennings, R. Waser: "Progress in the Synthesis of Nanocrystalline BaTiO$_3$ Powders for MLCC", *International Journal of Applied Ceramic Technology*, 2005, Vol. 2, No. 1, pp. 1 – 14.

[3] D.H. Kuo, C.H. Wang, W.P. Tsai, "Donor and Acceptor Cosubstituted BaTiO$_3$ for Nonreducible Multilayer Ceramic Capacitors", *Ceramics International*, 2006, Vol. 32, pp.1-5.

[4] A.K.Yadav, C.Gautam, "Dielectric Behavior of Perovskite Glass Ceramics", *J. Mater Sci: Materials in Electronics*, 2014, Vol. 25, pp. 5165-5187.

[5] M.S. Alkathy, A.Hezam, K.S.D. Manoja, J. Wang, C. Cheng, K. Byrappa, K.C. James Raju, "Effect of sintering temperature on structural, electrical, and ferroelectric properties of lanthanum and sodium co-substituted barium titanate ceramics", *Journal of Alloys and Compounds*, 2018, Vol. 762, pp 49-61.

[6] Lj. Zivkovic, V. Paunovic, N. Stamenkov, M. Miljkovic, "The Effect of Secondary Abnormal Grain Growth on the Dielectric Properties of La/Mn Co-Doped BaTiO$_3$ Ceramics", *Science of Sintering*, 2006, Vol.38, pp. 273-281.

[7] V. Paunović, Z.Prijić, M.Djordjević, V. Mitić, "Enhanced dielectric properties in La modified barium titanate ceramics", *Facta Universitatis, Series: Electronics and Energetics*, 2019, Vol. 32, No 2, pp. 179 – 193.

[8] H. Kishi, N. Kohzu, J. Sugino, H. Ohasato, Y. Iguchi, T. Okuda: "The Effect of Rare-earth (La, Sm, Dy, Ho and Er) and Mg on the Microstructure in BaTiO$_3$", *Journal of the European Ceramic Society*, 1999, Vol. 19, No. 6-7, pp. 1043 – 1046.

[9] V. Paunović, Lj. Živković, V. Mitić: "Influence of Rare-earth Additives (La, Sm and Dy) on the Microstructure and Dielectric Properties of Doped BaTiO3 Ceramics", *Science of Sintering*, 2010, Vol. 42, No. 1, pp. 69 – 79.

[10] K.J. Park, C.H. Kim, Y.J. Yoon, S.M. Song, "Doping Behaviors of Dysprosium, Yttrium and Holmium in BaTiO$_3$ ceramics", *J.E.Ceram.Soc.*, 2009. Vol. 29, pp. 1735-1741.

[11] R. Zhang, J.F. Li, D. Viehland, "Effect of Aliovalent Substituents on the Ferroelectric Properties of Modified Barium Titanate Ceramics: Relaxor Ferroelectric Behavior", *J.Am.Ceram.Soc.* 2004, Vol.87, pp. 864-870.

[12] Y. Wang, K.Miao, W.Wang, Y.Qin, "Fabrication of lanthanum doped BaTiO3 fine-grained ceramics with a high dielectric constant and temperature-stable dielectric properties using hydro-phase method at atmospheric pressure", *J.E.Ceram.Soc.*,2017, Vol. 37, pp. 2385–2390.

[13] V. Paunović, V.V.Mitić, Lj. Kocić, "Dielectric characteristic of donor-acceptor modified BaTiO$_3$ ceramics", *Ceramics International*, 2016, Vol. 42, pp.11692–11699.

Generation of Harmonics of Terahertz Radiation in Paraelectrics in a Wide Temperature Range

V. Grimalsky, S. Koshevaya, J. Escobedo-A., and Y. Gomez-B.

Abstract – It is investigated theoretically the generation of higher electromagnetic harmonics of the lower part of terahertz range in nonlinear paraelectrics like SrTiO₃. The efficient generation of harmonics is possible when the bias electric field is applied to the crystal to get the quadratic nonlinearity. The initial focusing of the pump first harmonic makes possible to realize the selective generation of proper higher harmonics. In the paraelectric SrTiO₃ the efficient harmonic generation occurs in the wide range of temperatures 50 – 200 K.

I. INTRODUCTION

The nonlinear dielectrics and semiconductors can be used as volume nonlinear materials in the terahertz (THz) range 0.1 – 30 THz. The ferroelectrics in the non-polar phase are utilized as the nonlinear dielectrics, so-called paraelectrics like SrTiO₃, KTaO₃ [1,2]. The crystalline paraelectrics possess high values of the permittivity ≥500, see Fig. 1, a, high electrodynamic nonlinearity, and low losses at relatively low temperatures [2,3]. The crystalline SrTiO₃ possesses high cubic nonlinearity and low losses in the lower part of THz range 0.5 – 2.5 THz at moderately low temperatures 50 – 200 K [3]. In a distinction from the microwave range, there exists the frequency dispersion in THz range when the frequency is near the soft mode frequency, i.e. the lowest frequency of the optical modes of the lattice oscillations [4].

The nonlinear electromagnetic (EM) field dynamics is determined mainly by the soft mode dynamics [4]. In crystalline SrTiO₃ the soft mode frequency decreases with the decrease of the temperature [3,4], it varies from $f_T \approx 2.5$ THz at $T = 200$ K down till $f_T \approx 0.6$ THz at $T = 50$ K [3], see Fig. 1, b. Due to high values of the permittivity, in the lower part of THz range the EM wavelengths are quite small there < 10 μm, so the crystal lengths can be chosen ≤ 1 cm.

Due to the electrodynamic dielectric nonlinearity the frequency multiplication in the volume SrTiO₃ in microwave and THz ranges takes place. In this paper the frequency multiplication in THz range in bounded crystalline SrTiO₃ is investigated theoretically. The wide range of temperatures $T = 60$ K – 200 K is considered.

V. Grimalsky, S. Koshevaya, J. Escobedo-A., and Y. Gomez-B. are with the Center for Investigations on Engineering and Applied Science (CIICAp), Institute for Investigations in Basic and Applied Science (IICBA), Autonomous University of State Morelos (UAEM), Av. Universidad 1001, 62209, Cuernavaca, Mor., Mexico. E-mail: v_grim@yahoo.com; svetlana@uaem.mx

Because the nonlinearity increases with the increase of $\varepsilon(0)$, i.e. at lower temperatures, the input intensities of EM wave can be lower there. In turn, at higher temperatures the soft mode frequency increases, so a possible frequency range for the harmonics generation can be wider.

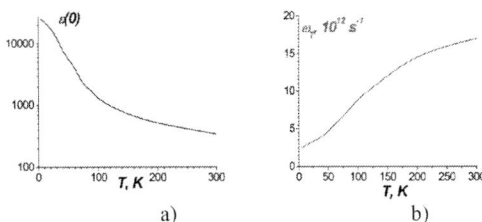

Fig. 1. The used parameters of paraelectric SrTiO₃. Part a) is the dependence of the low-frequency permittivity $\varepsilon(0)$ on the temperature T. Part b) is the dependence of the soft mode frequency $\omega_T \equiv 2\pi f_T$ on T.

II. BASIC EQUATIONS

The nonlinear propagation of EM waves with the electric field component $E_x = E$ is considered along OZ axis within the crystalline paraelectric SrTiO₃. The dielectric nonlinearity in SrTiO₃ is due to the nonlinear properties of the lattice polarization due to the optical soft mode $\varepsilon_0 P$, where ε_0 is the electric constant, SI units. Below the two-dimensional case is considered, and the basic equations that describe the nonlinear EM wave propagation in paraelectric crystalline SrTiO₃ are [3,4]:

$$\frac{\partial^2 P}{\partial t^2} + \gamma \frac{\partial P}{\partial t} + \omega_T{}^2 (1 + \frac{P^2}{P_0{}^2})P = \varepsilon(0)\omega_T{}^2 E;$$

$$\Delta E \equiv \frac{\partial^2 E}{\partial y^2} + \frac{\partial^2 E}{\partial z^2} = \frac{1}{c^2}\frac{\partial^2 P}{\partial t^2}; \qquad (1)$$

$$D \equiv \varepsilon_0(E + P + \varepsilon_\infty E) \approx \varepsilon_0 P.$$

The inequality is used $\varepsilon_\infty \leq 10 << \varepsilon(0)$, where ε_∞ is high frequency permittivity. Here $P_0 \equiv \varepsilon(0)E_0$, the parameter E_0 characterizes the value of nonlinearity. At the temperature $T = 80$ K it is $E_0 = 60$ kV/cm. At lower temperatures the nonlinearity increases, $E_0 \sim \varepsilon(0)^{-3.2}$. The dissipation γ is minimal at $T = 80$ K $\gamma = 2 \cdot 10^{11}$ s⁻¹, it slightly increases at another temperatures. At $T = 110$ K SrTiO₃ possesses the structural transition, so the vicinity of this temperature is not considered here.

Due to the central symmetry of paraelectrics, the cubic nonlinearity is present in Eqs. (1), and the odd harmonics with the numbers 3, 5, 7,… can be excited. The efficiency of such a frequency multiplication is low, so it is rather better to consider the case when the bias electric field E_s is present, the dominating nonlinearity becomes quadratic, and the effective generation of higher EM harmonics occurs in SrTiO$_3$:

$$E = E_s + \widetilde{E}, \quad P = P_s + \widetilde{P}; \quad P_s \cdot (1 + \frac{P_s^2}{P_0^2}) = \varepsilon(0) \cdot E_s. \quad (2)$$

Here P_s is the constant polarization that corresponds to the bias field E_s. But an influence of the cubic nonlinearity can be essential at relatively higher temperatures $T > 100$ K, so the basic equations for slowly varying amplitudes of EM harmonics include both quadratic and cubic nonlinear terms. Eqs. (1) are reduced to the equation for the variable part of the polarization:

$$\Delta(\frac{\partial^2 \widetilde{P}}{\partial t^2} + \gamma \frac{\partial \widetilde{P}}{\partial t} + \omega_{Ts}^2 \widetilde{P}) - \frac{\varepsilon(0)\omega_T^2}{c^2} \frac{\partial^2 \widetilde{P}}{\partial t^2} =$$

$$= -3 \frac{P_s}{P_0^2} \omega_T^2 \Delta(\widetilde{P}^2) - \frac{1}{P_0^2} \omega_T^2 \Delta(\widetilde{P}^3);$$

$$\frac{\partial^2 \widetilde{E}}{\partial y^2} + \frac{\partial^2 \widetilde{E}}{\partial z^2} = \frac{1}{c^2} \frac{\partial^2 \widetilde{P}}{\partial t^2}; \quad (3)$$

$$\omega_{Ts}^2 \equiv \omega_T^2 \cdot (1 + 3\frac{P_s^2}{P_0^2}).$$

Below the nonlinearity is assumed as moderate: $|\widetilde{E}| < E_s < E_0$. $(|\widetilde{P}| < P_s < P_0)$. The stationary case of the harmonic generation is investigated. The method of slowly varying amplitudes is applied [5] here.

The solution for E is searched as

$$\widetilde{E} = \frac{1}{2} \sum_{j=1,2,3...} A_j(z,y) e^{i(\omega_j t - jk_1 z)} + c.c.; \quad (4)$$

The structure of the basic set of equations for slowly varying amplitudes $A_j(z,y)$ of the variable electric field is presented in Eqs. (5). Here z, y are the longitudinal and the transverse coordinates, ω_j are frequencies of harmonics, k_j are their wave numbers; $N^{(2)}_{jm} \sim E_s$, $N^{(3)}_{jmp}$ are the coefficients of the quadratic and cubic nonlinearity, $\Gamma_j \sim \omega_j^2$ are dissipation coefficients. The phase mismatches between the first harmonic and higher ones are $\Delta k_j \equiv k_j - jk_1$.

The following relations should be satisfied: $|\partial A_j / \partial z| \ll k_j |A_j|$ [5]. The coefficients of nonlinearity possess complex structure and are not given here.

$$\frac{\partial A_j}{\partial z} + \frac{i}{2k_j} \frac{\partial^2 A_j}{\partial y^2} + i(k_j - jk_1)A_j + \Gamma_j A_j -$$

$$- 3i(N^{(3)}_{jjj}|A_j|^2 + 2\sum_{m \neq j} N^{(3)}_{mmj}|A_m|^2)A_j =$$

$$= i \sum_{m<j} N^{(2)}_{jm} A_{j-m} A_m + 2i \sum_{m>j} N^{(2)}_{jm} A_{m-j}^* A_m +$$

$$+ i \sum_{\substack{m,p, \\ j > m+p}} N^{(3)}_{jmp} A_{j-m-p} A_m A_p + \quad (4)$$

$$+ 3i \sum_{\substack{m,p; \\ j<m+p; \\ m,p \neq j}} N^{(3)}_{jmp} A_{m+p-j}^* A_m A_p +$$

$$+ 3i \sum_{m,p} N^{(3)}_{jmp} A_{m+p+j} A_m^* A_p^*.$$

The following notations are used:

$$\omega_j \equiv j\omega_1; \quad k_j \equiv \text{Re}(k(\omega_j)) > 0;$$

$$k(\omega_j) = \frac{\omega_T}{c} (\frac{\varepsilon(0)\omega_j^2}{\omega_{Ts}^2 - \omega(\omega - i\gamma)})^{1/2} \equiv k_j - i\Gamma_j. \quad (5)$$

There exists the self-action of EM waves due to the cubic nonlinearity that limits the efficiency of the generation of higher harmonics in the case of absence of the bias electric field E_s.

The layer of SrTiO$_3$ has a finite width L_y along OY axis. Eqs. (5) are added by the boundary conditions:

$$A_1(z=0) = A_{10} \exp(-(\frac{y - L_y/2}{y_0})^6)M(y);$$

$$A_j(z=0) = 0, \quad j > 1; \quad A_j(y=0, L_y) = 0, \quad j \geq 1 \quad (6)$$

Here A_{10} is the maximum amplitude of the input pulse; $F_t(t)$ is a dependence of the long input pulse on t, which is close to rectangular; the parameter y_0 determines the width of the input pulse along the transverse coordinate y. The length of the crystal is L_z. The function $M(y)$ is due to possible focusing of the input beam, as discussed below. Because the permittivities of the lateral dielectrics at $y < 0$ and $y > L_y$ are much smaller than the permittivity of SrTiO$_3$, the electric field does not penetrate into these dielectrics. But the nonlinear dynamics occurs within the crystal SrTiO$_3$ and does not depend on the width L_y, but depends on the width of the EM beam, so an influence of the lateral boundaries is not essential. It is considered the case without reflections at the input $z = 0$ and the output $z = L_z$ of the crystal.

The dynamics of the frequency multiplication has been investigated under the focusing of the input wave. It is assumed that the input pulse is excited by the circular antenna of the radius of curvature R, as seen from Fig. 2. In this case at the input EM wave obtains the additional phase shift:

978-1-7281-3420-8/19 $31.00 © 2019 IEEE

$$-k_1\Delta(y) = -k_1(R - (z_F{}^2 + (y - L_y/2)^2)^{1/2}), \quad (7)$$

where z_F is the distance from the input to the focus point, $z_F = (R^2 - (L_y/2)^2)^{1/2}$, k_1 is the wave number of the first harmonic. Correspondingly, in the plane $z = 0$ the input amplitude, Eqs. (6), obtains the multiplier $M(y) = \exp(-ik_1\Delta(y))$.

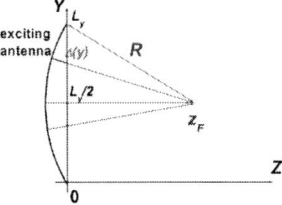

Fig. 2. The focusing of the input wave by the circular antenna. Here R is the radius of curvature of the antenna, z_F is the focus point. The phase shift is due to the variable distances $\Delta(y)$.

III. GENERATION OF HIGHER HARMONICS

The effective generation of higher harmonics of high numbers can be realized under relatively small bias electric fields. The frequency of the input EM wave should be chosen 10 - 50 times lower than the soft mode frequency. When the focusing of the input wave at the first harmonic occurs due to the circular antenna, it is possible to excite the higher harmonics with an increased efficiency. There is a possibility to select the proper number of the EM harmonic 3, 4,...,12, when the value of the bias electric field E_s, the radius of the focusing antenna R, and the length of the crystal L_z are chosen properly. The input amplitudes are essentially smaller there, when compared with the case of the self-action due to the pure cubic nonlinearity. When the generation of higher harmonics takes place under different temperatures, there is a possibility to cover the lower part of THz range from 0.3 THz at lower temperatures up till 2.5 THz at higher temperatures.

The typical values of the bias electric field E_s and the input amplitudes for the first harmonic are $E_s/E_0 = 0.2 - 0.5$, $A_{10}/E_0 = 0.02 - 0.1$, and therefore $P_s \approx \varepsilon(0)E_s$. At higher bias fields the saturations of the polarization P_s and of the quadratic nonlinearity occur. Moreover, at high bias fields the nonlinear dissipation takes place [3], which is not considered here. At smaller bias fields $E_s/E_0 < 0.2$ the quadratic nonlinearity is small and the efficiency of generation of higher harmonics is low, even when the focusing is applied. The typical values of the frequencies of the first harmonics are $\omega_1 = 10^{11} - 6 \cdot 10^{11} \text{s}^{-1}$. Higher harmonics are in the lower part of THz range.

The typical results of numerical simulations are presented in Figs. 3-7. The main attention is paid to the distributions of the harmonics in the center of the nonlinear crystal $y = L_y/2$. In Figures the values of the variable electric field $\tilde{E}$ are related to the characteristic nonlinear electric field E_0.

The used parameters of the crystal $SrTiO_3$ are given in Table 1.

TABLE I
USED PARAMETERS OF CRYSTALLINE $SrTiO_3$

T, K	$\varepsilon(0)$	ω_T, 10^{12} s^{-1}	γ, 10^{11} s^{-1}	E_0, kV/cm
60	4000	5	2.5	21
80	2000	6	2	60
150	700	12	3	290
200	500	15	4	480

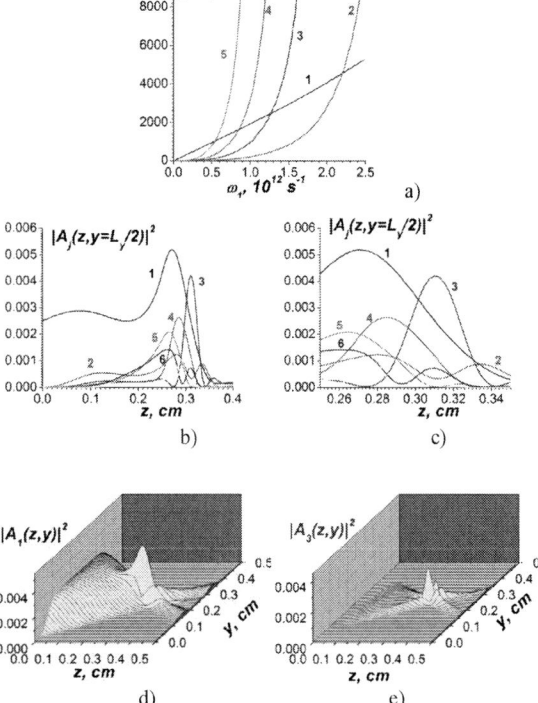

Fig. 3. The generation of higher harmonics at the temperature $T = 60$ K. Part a) is the dispersion relation for the first harmonic, curve 1, and the mismatches for higher harmonics, curves, 2 -5. Parts b), c) are distributions of intensities of harmonics in the center of the crystal, general and detailed views. The curves $1 - 6$ correspond to the proper numbers of harmonics. Parts d), e) are the spatial distributions of the intensities of the first and the third harmonics within the crystal.

In Fig. 3 the parameters are as follows. The temperature is $T = 60$ K, the bias electric field is $E_s/E_0 = 0.3$; the frequency of the first harmonic is $\omega_1 = 2\cdot10^{11}$ s^{-1}, the amplitude of the first harmonics is $A_{10}/E_0 = 0.05$, the initial width is $y_0 = 0.2$ cm, the curvature radius of the exciting antenna is $R = 0.35$ cm. It is seen that the third harmonic is dominating, but also higher harmonics with the numbers 4,5,6 possess relatively high intensity levels within the crystal. Moreover, they have the peak levels at the different distances z.

In Fig. 4 the temperature is $T = 60$ K, $E_s/E_0 = 0.3$; $\omega_1 = 4\cdot10^{11}$ s^{-1}, the amplitude of the first harmonics is $A/E_0 =$

978-1-7281-3420-8/19 $31.00 © 2019 IEEE

0.05, the initial width is $y_0 = 0.08$ cm, the curvature radius of the exciting antenna is $R = 0.15$ cm. Here the second harmonic is dominating.

In Fig. 5 the temperature is $T = 80$ K, $E_s/E_0 = 0.3$; $\omega_1 = 2 \cdot 10^{11}$ s^{-1}, the amplitude of the first harmonics is $A/E_0 = 0.05$, the initial width is $y_0 = 0.2$ cm, the curvature radius of the exciting antenna is $R = 0.35$ cm.

In Fig. 6 the temperature is $T = 150$ K, $E_s/E_0 = 0.3$; $\omega_1 = 5 \cdot 10^{11}$ s^{-1}, the amplitude of the first harmonics is $A/E_0 = 0.05$, the initial width is $y_0 = 0.2$ cm, the curvature radius of the exciting antenna is $R = 0.4$ cm.

In Fig. 7 the temperature is $T = 200$ K, $E_s/E_0 = 0.3$; $\omega_1 = 6 \cdot 10^{11}$ s^{-1}, the amplitude of the first harmonics is $A/E_0 = 0.05$, the initial width is $y_0 = 0.2$ cm, the curvature radius of the exciting antenna is $R = 0.4$ cm.

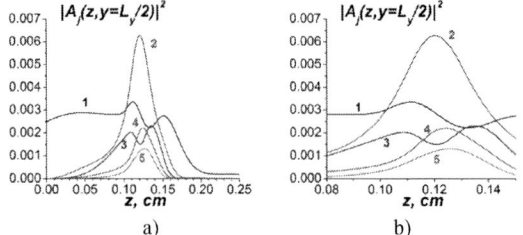

a)   b)

Fig. 4. The generation of higher harmonics at the temperature $T = 60$ K. The frequency of the first harmonic, its width, and the radius of the input antenna differ from Fig. 3.

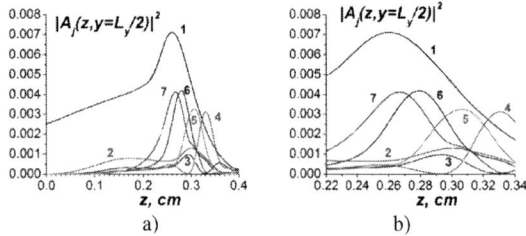

a)   b)

Fig. 5. The generation of higher harmonics at the temperature $T = 80$ K. Parts a), b) are distribution of intensities of harmonics in the center of the crystal, general and detailed views.

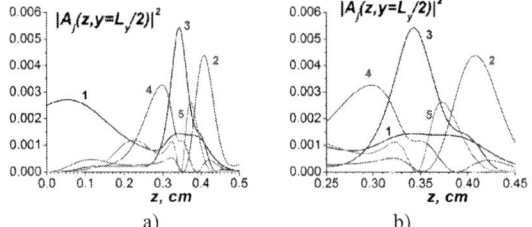

a)   b)

Fig. 6. The generation of higher harmonics at the temperature $T = 150$ K. Parts a), b) are distribution of intensities of harmonics in the center of the crystal, general and detailed views.

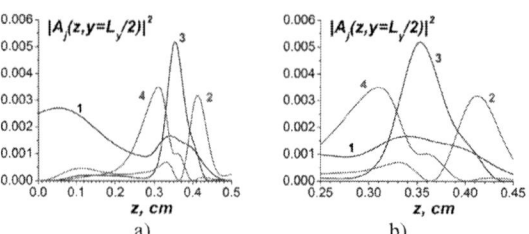

a)   b)

Fig. 7. The generation of higher harmonics at the temperature $T = 200$ K. Parts a), b) are distribution of intensities of harmonics in the center of the crystal, general and detailed views.

It is seen that at higher temperatures it is possible to get higher EM frequencies due to the frequency multiplication. But, because the nonlinearity is smaller, i.e. the parameter E_0 is higher, it is necessary to use the higher input levels of the intensity of the first EM harmonics.

IV. Conclusions

The paraelectric crystals like SrTiO$_3$ can be used for the generation of higher harmonics of the lower part of terahertz range below the soft mode frequency under moderate levels of cooling $T = 50 - 200$ K. The moderate bias electric fields should be applied, to provide the quadratic nonlinearity, as well the focusing of the pump first harmonic. In the case of the initial focusing it is possible to select the efficiency of the generation of proper harmonics within the crystal. Under lower temperatures $T \sim 60$ K there are higher nonlinearity levels, whereas at higher temperatures $T > 120$ K the soft mode frequency is higher and the wider frequency range can be used for the frequency multiplication.

Acknowledgement

The authors are grateful to SEP-CONACyT, Mexico, for a partial support of our work.

References

[1] M., Perenzoni and D.J. Paul, Eds., *Physics and Applications of Terahertz Radiation*, New York: Springer, 2014.

[2] S. Gevorgian, *Ferroelectrics in Microwave Devices, Circuits and Systems*, New York: Springer, 2009.

[3] O.G. Vendik, Ed., *Ferroelectrics in Microwave Technology*. Moscow, Sov. Radio (in Russ.), 1979.

[4] M.E. Lines and A.M. Glass, *Principles and Applications of Ferroelectric and Related Materials*, Oxford: Clarendon Press, 1977.

[5] D.L. Mills, *Nonlinear Optics. Basic Concepts*, New York: Springer, 1998.

978-1-7281-3420-8/19 $31.00 © 2019 IEEE

Study of Nanoporous Anodic Aluminum Oxide as a Template Filled with Piezoelectric Materials

T. Tsanev, M. Aleksandrova and V. Videkov

Abstract – This paper is devoted to method of nano-structuring of piezoelectric potassium niobate ($KNbO_3$) material in order to improve its voltage generating ability for potential application as energy harvesting. The main goal is geometrical structuring of the piezoelectric material to obtain higher piezoelectric voltage per unit volume in comparison with thin non-structured film from the same material. This is possible due to template properties of porous anodic aluminum oxide (AAO). It was found that $KNbO_3$ grown by sputtering in the nanopores of AAO generates an effective piezoelectric voltage of 410 mV from volume of 0.001 cm^3 at mass load of 40 g and frequency of 50 Hz. For comparison, non-structured films with thickness lower than 200 nm and area resulting in a volume of 0.00006 cm^3 the generated voltage was 1.78 mV, which means that this type of nanostructuring leading to two orders of magnitude increase of the surface-to-volume ration enhances the piezoelectric response 2.3 times.

I. INTRODUCTION

Piezoelectric nanostructures have attracted attention as possible candidates for thin film energy harvesting and sensing applications [1]. Usually, due to the small thickness of the piezoelectric film (in nanometer range), the generated charge and therefore the produced voltage are not sufficient for reliable power supply. This imposes increase of the specific functional area without losing portability. Suitable example is related to zinc oxide nanowires, which however are most often grown by chemical processes, non-conventional for the microelectronic industry [2].

One of the most effective ways to increase the piezoelectric yield is stress concentration. Nanostructuring is effective tool to fabricate multiple formations serving as stress concentrators like nanowires, nanorods and nanoneedles [3]. These nanostructures have been realized in some oxides, like ZnO [4]. Structuring piezoelectric materials like PVDF and PZT with this type of technology is also achievable [5]. Structuring properties of those oxide materials are good base to improve piezoelectric parameters of nanogenerators [6]. Possible template for growing nanowires by conventional, non-chemical methods is anodization of aluminum for preparation of an array of Porous Anodic Aluminum Oxide (PAAO). Such template assisted nanostructures are grown on aluminum substrates. Their parameters can be adjusted by controlling the

T. Tsanev, M. Aleksandrova and V. Videkov, are with the Department of Microelectronics, Technical University of Sofia, 8 St. Kliment Ohridksi Blvd, 1000 Sofia, Bulgaria, E-mail: zartsanev@tu-sofia.bg, m_aleksandrova@tu-sofa.bg, videkov@tu-sofia.bg

duration of anodization at constant temperature, voltage and electrolyte concentration, or constant voltage, process time and electrolyte concentration with a temperature change [7]. There is a possibility to use PAAO matrix for plasma and atomic layer deposition. [8], [9]. The matrix itself can be built as a separate template or as a carrying substrate. In the latter case, it can be free standing or placed on another substrate with vacuum deposited aluminum on it. Various electrolytes may be used to modify the matrix parameters [10]. The anodization rates are usually established as a function of the time and of the temperature of anodic solution. This is realized, in order to determine the depth and the diameter of the pores. At certain process parameters, it is supposed that faster growing speed of the oxide will be achieved, as well as layer with different diameter of the nanopores will be obtained. By such nanostructuring it is expected enhancement of the transducers' piezoelectric performance (energy harvesting or sensing devices). This gives a good base for optimizing the parameters of the future nanoelectromechanical (NEMS) piezoelectric devices.

In this paper we present the results for filling of AAO with lead-free piezoelectric material. The degree of penetration of the potassium niobate particles was compared for nanopores prepared at different modes of anodization. By the authors knowledge study of the piezoelectric performance of filled by sputtering AAO with piezoelectric material has not been done yet.

II. EXPERIMENTAL SECTION

At the beginning of experiment it is realized a preparation before anodizing of four aluminum substrate with purity if 99.3% and dimensions 10 mm x 40 mm x 0.1 mm. All substrates were cleaned in solution of 40% NaOH for 30 sec at 30° C. Any left natural oxide formation was removed by solution of 40g/l CrO_3 90 ml/l H_3PO_4 at temperature of 75°C for 20 min.

PAAO layers with variety of thicknesses (pores diameters) were produced in electrolyte solution of 5% H_3PO_4 without overheating or degradation of the substrates and oxide layer, at process conditions that were recommended elsewhere [11]: Us = 150V (necessary voltage for this of electrolyte solution); I = from 0.7 A/dm^2 to 2 2A/dm^2, (function of solution temperature and dimensions of anodized area) and temperature off electrolyte Ta = from 13°C to 15°C, (±0.1°C) t = from 10 min to 120 min. Variation of the process parameters affects pore diameters,

978-1-7281-3420-8/19 $31.00 © 2019 IEEE

wall thickness, layer thickens and geometrical order of oxide structure. The setup for growing PAAO is shown in Fig. 1.

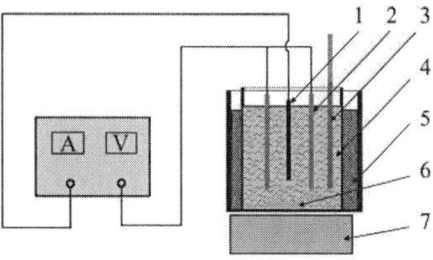

1 Aluminium foil Anode; 5 Cooling Water;
2 Steel Catode; 6 Magnetic Capsule;
3 Temperature control; 7 Magnetic Stirrer;
4 Electrolyte; 8 DC power supply

Fig. 1. Setup for aluminum anodization process.

Obtained structures of PAAO – cantilever, membrane type after anodization are shown in Fig. 2.

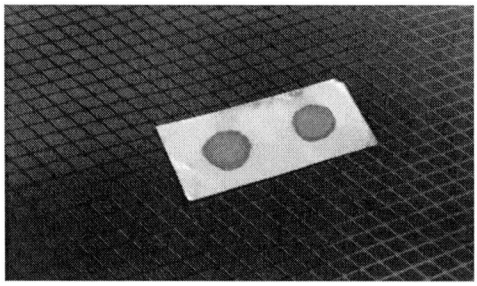

Fig. 2. Membranes of Porous Anodic Aluminum Oxide.

Possibility for using anodic nanostructure with two sided open pores, as a template for filling with piezoelectric material $KNbO_3$ is investigated. Vacuum radiofrequency sputtering is used for coating the Al_2O_3 (Fig. 3).

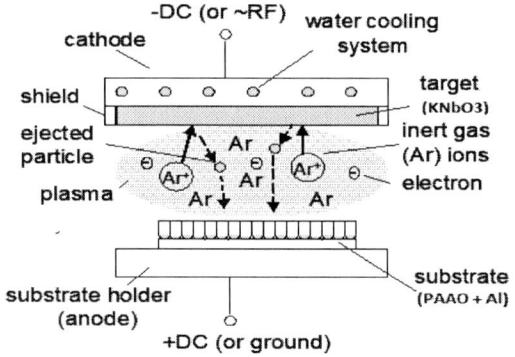

Fig. 3. Vacuum radiofrequency sputtering of $KNbO_3$ on substrate of Porous Anodic Aluminum Oxide. [12]

For comparing the results before and after nanostructuring of the $KNbO_3$, we use our previous research

[10] for non-structured films from the same material with thicknesses below 200 nm. The charge generation ability for both types of structures was investigated. Surface morphology and cross-section views are monitored by scanning electron microscopy (SEM) combined with Energy Dispersive X-ray (EDX) microanalysis for chemical elements distribution in depth in the pores. Electrical signals generated from the structures are recorded by digital oscilloscope.

The non-structured sample has surface area of $3cm^2$ which mean that volume is around 0.00006 cm^3 for film with thickens of 200 nm. The thickness of the structured piezoelectric film is up to 2 µm and around 0.5 µm overlay, for each side of the substrate, which means that the total thickness is up to 5 µm. With those parameters it was calculated functional volumes of structured and non-structured $KNbO_3$ on the substrate (Fig.4).

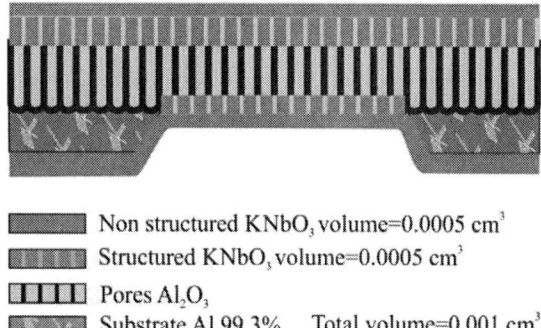

Non structured $KNbO_3$ volume=0.0005 cm^3
Structured $KNbO_3$ volume=0.0005 cm^3
Pores Al_2O_3
Substrate Al 99.3% Total volume=0.001 cm^3

Fig. 4. Schematic cross-section view of the layers in the nanostructured sample.

For measurement of the generated voltage from the non-structured film, a simple structure with two aluminum electrodes deposed by vacuum thermal evaporation was prepared. For measurement of the piezoelectric response of the PAAO filled with potassium niobate, poly(3,4-ethylenedioxythiophene) polystyrene sulfonate (PEDOT:PSS) conductive polymer was one side casted on the filled AAO, following by drying for 20 min at 120°C for removing of the water solution content. Afterward, silver adhesive paste was supplied for contact with the measurement probes. To generate voltage from all samples it is used free standing cantilever method, reported elsewhere [12]. The vibrations created equivalent mass weight load to 40 g to avoid breaking of the PAAO fine structures. The frequency of vibration was relatively low, namely 50 Hz.

III. RESULTS AND DISCUSSION

As a result it was obtained PAAO structures with total thickness of 18 µm with penetration of $KNbO_3$ from two sides of the substrate, as it is shown from the SEM image of the cross-section Fig.5. The brighter zones near the top sides of the pores represent the niobium accumulation, which is indication about the distribution of the material on the inner walls of the pores.

978-1-7281-3420-8/19 $31.00 © 2019 IEEE

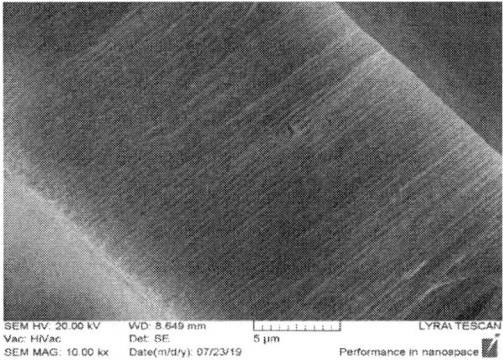

Fig. 5. SEM image (cross-section view) of the porous structure in oxide layer shows the penetration of KNbO$_3$ material.

Additionally, EDX analysis provided elemental concentration along the length of the AAO tube. The results clearly showed that the inner sides are poorly coated, but the piezoelectric material is strongly concentrated near the opening of the pores, tending to close them. The elemental analysis showed that penetration level of piezoelectric material is up to 2.5 µm form the two sides of the inner walls of the pores, as is shown in Fig.6.

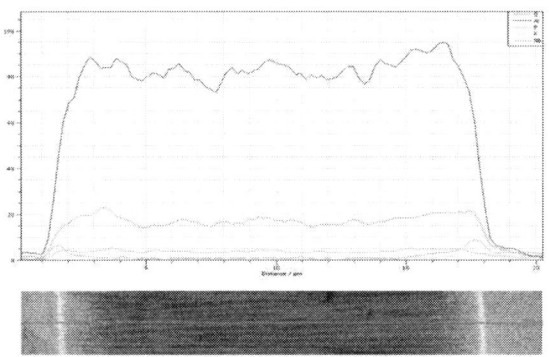

Fig.6. Elemental analysis (cross-sectional view) of the AAO filled with piezoelectric material.

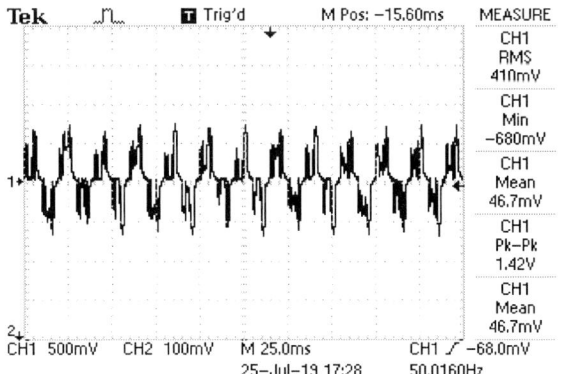

Fig. 7. Piezoelectric voltage from structured by AAO template film (<5000 nm) of KNbO$_3$ measured by oscilloscope.

The magnitude and shape of the generated voltage from the nanostructured KNbO$_3$ film is shown in Fig. 7 (maximum 710 mV).

For comparison the results from the surface morphology and the corresponding voltage from dense and flat (non-structured) KNbO$_3$ are shown below.

The SEM result shows fine granular, uniformly distributed coating on the substrate with average roughness lower than 5% from the films thickness (Fig. 8).

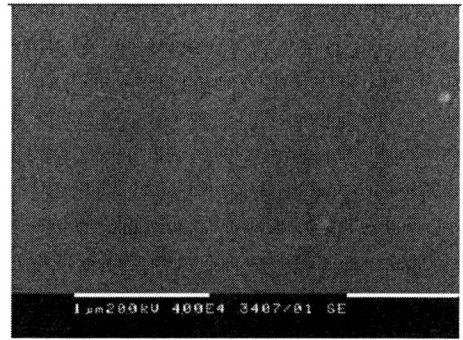

Fig.8. SEM image of the surface of sputtered, non-structured thin film from KNbO$_3$.

The generated piezoelectric voltage from the non-structured KNbO$_3$ film is shown in Fig. 9. (maximum 11.88 mV).

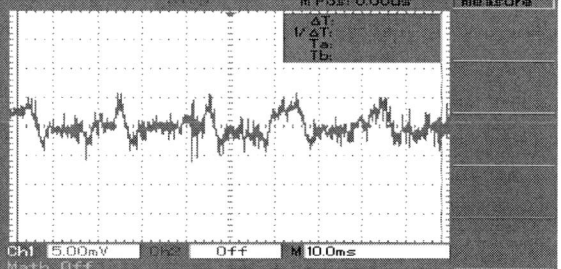

Fig. 9. Piezoelectric voltage from non-structured thin (<200 nm) film of KNbO$_3$.

The results for structured film of KNbO$_3$ show that structured piezoelectric layer has not completely filled the nanopores and nanotubes in the volume of the PAAO matrix. However, it was obtained higher generated voltage, due to the increased surface to volume ratio of the piezoelectric material for the nanostructured coating as compared to the non-structured.

IV. CONCLUSION

In this paper the possibility of using AAO as a template for nanostructuring of piezoelectric materials for energy harvesting applications was studied. The results showed that the nanostructuring by anodization of aluminum and sputtering of the piezoelectric material could be successful in terms of piezoelectric voltage increase. However, some challenges should be still overcome, like the poor coverage in depth of the AAO. Generated voltage from this structure is relatively high when it is considered the small amount of piezoelectric material. It is expected to achieve even higher voltage when pores of the PAAO matrix are fully filled. For this purpose, our future work will be directed to decrease the thickness of the nanostructured AAO and to increase the accuracy in control of the process parameters (especially temperature), to obtain nanopores completely filed with $KNbO_3$. The presented results give a good base for future researches for structuring of piezoelectric materials.

ACKNOWLEDGEMENT

The authors acknowledge BNSF's grant КП06-Н27/1. Part of this study is related to PhD project for thesis development 192ПД0006-03 at NIS, TU-Sofia.

REFERENCES

[1] X. Li, M. Sun, X. Wei, Ch. Shan, and Q. Chen, "1D Piezoelectric Material Based Nanogenerators:Methods, Materials and Property Optimization", *Nanomaterials,* 2018, vol.8, 188.

[2] N. Ahmed, A. Aled, R. L. Cameron, P. P. Thierry, and G. G. Maffeis, "Investigation of the growth parameters of hydrothermal ZnO nanowires for scale up applications", *J. Saudi Chem. Soc.,* 2018, vol. 22, pp. 538-545.

[3] Z. Wang, X. Pan, Y. He, Y. Hu, H.Gu, and Y. Wang, "Piezoelectric Nanowires in Energy Harvesting Applications",

Advances in Materials Science and Engineering, 2015, vol. 2015, article ID 165631

[4] J. Briscoe, S. Dunn, *"Nanostructured Piezoelectric Energy Harvesters"*, Springer Briefs in Materials, 2014.

[5] I. Dakua, and N. Afzulpurkar, "Piezoelectric Energy Generation and Harvesting at the Nano-Scale: Materials and Devices", *Nanomaterials and Nanotechnology,* 2013, vol. 3, pp. 1-16.

[6] Z. L. Wang, "Piezoelectric Nanostructures: From Growth Phenomena to Electric Nanogenerators", *MRS Bulletin,* 2007, vol. 32, pp. 109-116.

[7] R. Abdel-Karim, and S. M. El-Raghy, *"Fabrication of Nanoporous Alumina"*, Chapter in book Nanofabrication using Nanomaterials, One Central Press (OCP), UK, 2016.

[8] J. M. Mánuel at all, *Engineering of III-Nitride Semiconductors on Low Temperature Co-fired Ceramics*, Scientific Reports (2018) 8:6879 | DOI: 10.1038/s41598-018-25416-6

[9] S Blagoev, at all, *Atomic layer deposition of ZnO:Al on PAA substrates,* INERA Conference: Vapor Phase Technologies for Metal Oxide and Carbon Nanostructures, Journal of Physics: Conference Series 764 (2016) 012004 doi:10.1088/1742-6596/764/1/012004

[10] T. Kikuchi, D. Nakajima, O. Nishinaga, S. Natsui, and R. O. Suzuki, "Porous Aluminum Oxide Formed by Anodizing in Various Electrolyte Species", *Current nanoscience,* vol. 11, pp. 560-571

[11] F. Nasirpouri, *"Electrodeposition of Nanostructured Materials"*, Springer, Switzrland, 2017.

[12] M. P. Aleksandrova, T. D. Tsanev, G. H. Dobrikov, "Study of piezoelectric behavior of sputtered $KNbO_3$ nanocoatings for flexible energy harvesting", *International Conference of Alternative Energy Sources, Materials & Technologies (AESMT'19),* Sofia, Bulgaria, 2019.

An AlSiN Nanocomposite Film with Improved Mechanical Parameters for Multifunctional Applications

L. Kolaklieva, R. Kakanakov, T. Cholakova, H. Bahchedzhiev, and V. Chitanov

Abstract - An AlSiN film was deposited by cathodic arc evaporation. The structure and composition of the film were investigated by XRD, EDS and XPS analyses. The film had a nanocomposite structure consisting of (AlSi)N nanograins incorporated in a Si_3N_4 matrix. The film hardness determined by nanoindentation measurements was found to be 46 GPa. The AlSiN nanocomposite demonstrated improved elastic strain to failure (H/E=0.10), very good resistance to plastic deformation (H^3/E^2=0.44) and elastic recovery of 67%, which indicated improved toughness. The nanocomposite AlSiN film exhibited a low friction coefficient of 0.9 against the diamond indenter and enhanced wear resistance with wear rate of 7.56×10^{-10} mm^3/ (N.GPa.m).

I. INTRODUCTION

Aluminum nitride films doped with Si have been subjected to research interest since many years because of their potential for multifunctional applications. Initially, the studies were focused on the optical properties and applications in optoelectronics such as field emission devices [1], UV light emitters [2], and optical coatings with refractive index tailored by change of the Si content [3]. Pélisson et al. [4], [5] investigated the effect of Si content on the microstructure and mechanical properties of transparent AlSiN coatings. Maximum hardness of 30 GPa at 10% of Si content was reported [4]. It was shown that the coating transmission in the UV–Visible spectra is not affected by the Si amount in the interval from 2.5 at.% to 18 at.% [5]. Recently, the mechanical properties of AlSiN films were investigated toward their application as solar selective coatings [6] and coatings for surface protection in the machining industry [7]-[10]. The studies reported on dependence of microstructure, composition, optical and mechanical properties on the Si content. Additionally, several studies have shown that the use of an AlSiN film in multilayer structures or as an interlayer in diamond coatings improves their mechanical and tribological behavior [11]-[12]. The reported AlSiN films exhibited high resistance to plastic deformation, high temperature hardness, high oxidation resistance and better elastic strain to failure, but their hardness did not exceed 35 GPa. However, in many applications, the AlSiN films should combine the superhardness (> 40 GPa) with higher

L. Kolaklieva, R. Kakanakov, T. Cholakova, H. Bahchedzhiev, V. Chitanov are with the Central Laboratory of Applied Physics, Bulgarian Academy of Sciences, 61 St. Petersburg Blvd., 4000 Plovdiv, Bulgaria, E-mail: ohmic@mbox.digsys.bg

elasticity, which results in better toughness. Two ratios, H/E (elastic strain to failure) and H^3/E^2 (resistance to plastic deformation) have been used as ranking parameters correlated with the film toughness [13]. Hence, higher hardness and lower elastic modulus are, better the toughness is. Usually, the films with enhanced hardness exhibit a higher elastic modulus [14], which makes difficult to satisfy the requirements for enhanced toughness.

This paper presents the results from investigation of the developed AlSiN film with superhardness and improved elasticity, i.e. improved toughness.

II. EXPERIMENTAL DETAILS

A. Film deposition

The AlSiN film was deposited onto 5 mm - thick disks (coupons) of high-speed stainless steel (HSS) with a diameter of 10 mm for mechanical and tribological measurements and on 10x10 mm square stainless steel DIN 1.4541 substrates (SS) for morphology, composition and structure investigations. Prior to deposition, the substrates were ultrasonically cleaned in an alkaline solution, rinsed in de-ionized water and dried at 130 °C. The film was deposited by cathodic arc evaporation from lateral rotating LARC® cathodes in Platit $\pi 80^+$ equipment. Cleaning treatments by Ar discharge and by Cr ions at a bias voltage of 1000 V, were performed prior to film deposition. Deposition was performed in a nitrogen (99.9999 %) atmosphere at a pressure in the range of 9×10^{-1} Pa to 4 Pa. Firstly, a contact layer composed of Cr adhesion and CrN transition layers were deposited from a Cr (99.99 wt. %) cathode, followed by a CrAlSiN gradient layer. Next, the main structure of AlSiN was formed by evaporation from an AlSi alloy (82 at.% Al, 18 at.% Si) cathode. The AlSiN layer was deposited at a current of 160 A of the Al+Si cathode, and a bias voltage of –90 V. The deposition was performed at a constant temperature of 500 °C. After deposition, two-hour annealing at 525 °C in a nitrogen ambience was performed.

B. Characterization methods

The mechanical properties were investigated by nanohardness measurements using Compact Platform CPX (MHT/NHT) equipment (CSM Instruments, Anton Paar). Nanoindentation was performed with a diamond Berkovich indenter within the loading interval of 5 mN – 500 mN.

The Oliver&Pharr method was used to determine the nanohardness, elastic modulus, penetration depth, and stiffness from the load – displacement curves [15]. Coating thickness was measured using the ball-erosion method [16]. Surface observation and composition analysis were performed on a JEOL JSM 6390 electron microscope equipped with an INCA Oxford EDS energy dispersive detector. Powder X-ray diffraction patterns were collected within the range from 5.3 to 80° 2θ with a constant step of 0.02° 2θ on a Bruker D8 Advance diffractometer with Cu Kα radiation and a LynxEye detector. Phase identification was performed with Diffracplus EVA using ICDD-PDF2 Database. The mean crystallite size was determined with the Topas-4.2 software. The XPS measurements were carried out on AXIS Supra electron spectrometer (Kratos Analitycal Ltd.) using AlK$_\alpha$ radiation with photon energy of 1486.6 eV. The energy calibration was performed by normalizing the C1s line of adsorbed adventitious hydrocarbons to 284.6 eV. The data-processing software ESCApe^{TM} of Kratos Analytical Ltd. was used for the binding energy (BE) determination and deconvolution of the peaks. The adhesion was characterized by scratch tests. A spherical Rockwell indenter with radius of 200 μm and a cone angle of 120° was used at a normal force progressively increasing from 1 N to 30 N and constant scratch velocity of 0.5 N/min. Wear test was performed by the Rockwell indenter at load of 5 N on distance of 30 mm.

III. RESULTS AND DISCUSSION

The thickness as determined from the image of the abraded craters formed in the film and the substrate was 5.0 μm. The nanohardness (H), elastic modulus (E) and elastic recovery (W$_e$) were evaluated from the loading-unloading curves measured in nanoindentation tests. The measurements were performed on HSS coupon specimens with a hardness of 13 GPa. Poisson's ratio was assumed to be 0.25. The dependence of the nanohardness and elastic modulus on the indentation depth is presented in Fig. 1. Maximum nanohardness of 46 GPa was determined at the indentation depth of 221 nm, which is within the 10% interval of the film thickness. Consequently, the substrate influence could be considered negligible and the obtained superhardness can be accepted intrinsic for the film. With the increase of the indentation depth, the hardness decreases slowly to 27 GPa at 1200 nm indentation depth. It is due to the increased effect of the low-hard substrate (13 GPa). The elastic modulus has similar indentation-depth dependence like the film hardness. Once the indentation depth has increased, the elastic modulus decreases due to the increased contribution of the substrate having a lower elastic modulus of 285 GPa. An elastic modulus of 468 GPa corresponding to the maximum hardness, was calculated for the AlSiN film. As a physical property of the material, elastic modulus is influenced by the structure and composition of the coatings. Hence, varying the composition and structure could be adjusted to obtain a superhard film with a low elastic modulus.

The AlSiN coating exhibit H/E and H^3/E^2 ratios of 0.10 and 0.44, respectively. This result indicates the high elastic strain prior to failure and enhanced ability to absorb energy at deformation before fracture. The elastic recovery of 67% was determined, which implies high elasticity and improved toughness.

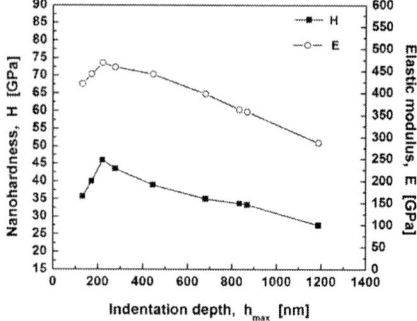

Fig. 1. Dependence of the nanohardness and elastic modulus on the indentation depth of the AlSiN film.

The composition of the AlSiN film was determined by EDS analyses in several points on the surface. Fig. 2 presents the element content in two marked points. The results allow to consider the film stoichiometric with regard to the N/(Al+Si+Cr) ratio. The Si content of 8.6 at.% is higher than the solubility limit of Si in the AlN lattice, which facilitates the grain size decrease and formation of an amorphous SiN$_x$ phase in the film [7]. The presence of a negligible Cr amount in the AlSiN film is caused by diffusion from the gradient CrAlSiN layer and the Cr/CrN contact layer, which process occurs at high deposition and post-annealing temperatures.

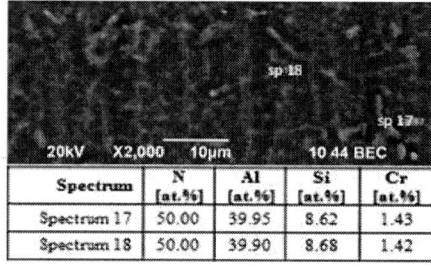

Spectrum	N [at.%]	Al [at.%]	Si [at.%]	Cr [at.%]
Spectrum 17	50.00	39.95	8.62	1.43
Spectrum 18	50.00	39.90	8.68	1.42

Fig. 2. A SEM image and element composition of AlSiN film.

The chemical bonding state and chemical composition were identified by XPS analysis. Fig. 3 illustrates the fitted Cr2p, Al2s, Si2p and N1s high-resolution photoelectron spectra of the AlSiN film. The peaks centered at 575.5 eV and 584.8 eV of the deconvoluted Cr2p spectra were recognized as Cr2p$_{3/2}$ and Cr2p$_{1/2}$, respectively, with a peak splitting of 9.3 eV. They correspond to Cr-N bonds indicating the presence of a CrN phase [17]. The Cr2p$_{3/2}$ peak at 576.8 eV could be attributed to Cr-N with assignment of Cr$_2$N phase. However, an overlap of Cr$_2$N and Cr$_2$O$_3$ peaks is possible [18]. The peak centered at

978-1-7281-3420-8/19 $31.00 © 2019 IEEE

579.0`eV might be assigned for CrO_3. The maximum of the binding energy of Al2p was determined at 74.0 eV. This position coincides with the Cr3s peak, because of that it is difficult to distinguish their contribution in the spectrum. Therefore, the Al2s spectrum was taken for quantification. The core level of the Al2s peak was detected at 118.7 eV. The Al2p and Al2s peak positions correspond to the Al-N bond in the AlN phase [19]. The main peak at 396.6 eV of the deconvoluted N1s peak was recognized as nitrogen bonded to aluminum (Al-N) [10]. The component at 397.6 eV could be assigned to nitrogen in Si-N [10]. The low intensive peak at 395.4 eV is most probably due to N-N defects in AlN [20]. In the Si2p spectrum, the binding energy for the Si2p level appeared at 101.5 eV, which is in good agreement with stoichiometric Si_3N_4 [21]. Thus, the XPS analysis revealed three types of bonds, Al-N and Si-N, determining the composition of the AlSiN film, and very small amount of Cr-N, which contribution is negligible=

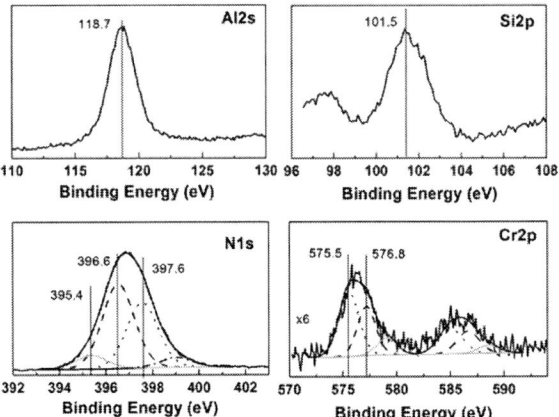

Fig. 3. XPS Cr2p, Al2s, N1s and Si2p spectrta of the AlSIN film.

XRD patterns of the AlSiN coatings are presented in Fig. 4. The peaks of the substrate (marked SS) correspond to a cubic unit cell (S.G. Fm-3m). XRD spectrum exhibit peaks from different lattice planes, indicating that the film is polycrystalline with random grain orientation. The AlSiN film comprises of one crystalline phase. The detected diffraction peaks are slightly shifted to the larger 2θ angles than that of the standard bulk material. This result suggests that Si atoms substitute for the Al atoms in the AlN lattice and the formation of the (AlSi)N solid solution is possible [4]. A mean crystallite size of 5 nm was determined in the AlSiN film. No Si-N phase was detected in the film, which indicates that SiN_x might exist in the film in amorphous state. Hence, the XRD results showed that the AlSiN film has a nanocomposite structure which is composed of two phases, nanocrystalline and amorphous. During deposition, the amount of silicon reaching the film exceeds its solubility limit in the AlN [4] lattice and forms the amorphous Si_3N_4 phase, which segregates along the (AlSi)N crystallite boundaries. The formed Si_3N_4 phase interrupts the continuous growth of nitrides and causes

refinement of the film structure. Due to the presence of an amorphous Si_3N_4 matrix, the dislocation and crack propagation are prevented and the grain boundary sliding is hampered. Thus, solid solution and nanocomposite hardening occur, which explains the observed enhancement of the film hardness.

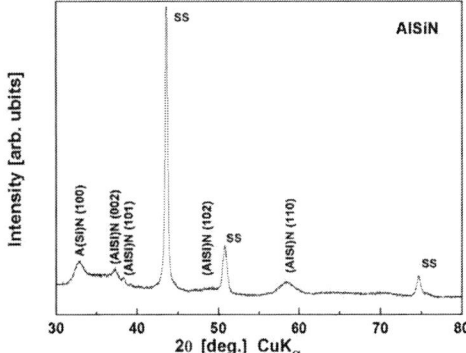

Fig.4. XRD pattern of the AlSiN film.

The performed scratch test revealed excellent adhesion strength of the developed AlSiN film to the substrate (Fig. 5). No critical loads were observed, which implies the film ability to tolerate the deformations. The signal of the friction force (F_t) increased smoothly in the depth. The undulating F_t line in some sections is due to the rough surface evidenced by the prescan profile (R_p) and residual depth (R_d). No cracks, track edge chipping and delamination were obtained in the scratch track (insert in Fig. 5) at progressively increased load F_n from 0.1 N to 30 N, indicating the good adhesion to the substrate. A quite low coefficient of friction μ= 0. was 09 measured.

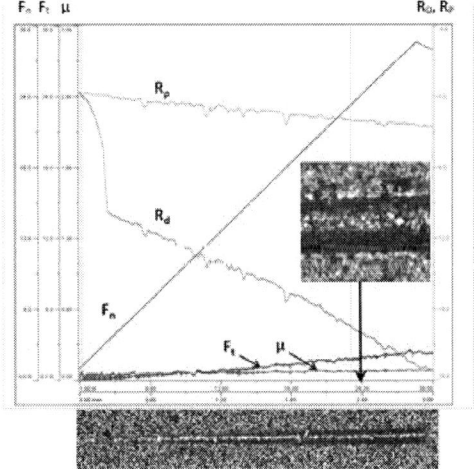

Fig. 5. A plot of the scratch parameters and an image of the corresponded scratch track of the AlSiN film.

The wear rate K was evaluated from the formula $K = V / (F_n \cdot L)$, where V is the worn volume, F_n – the applied normal force and L - the length of the worn track. The wear rate K of 3.48×10^{-08} mm^3/(N.m) was determined at F_n=5 N, a 30 mm worn tract and a worn volume of 5.22×10^{-03} mm^3. Specific wear rate K/H of 7.56×10^{-10} mm^3/(N.GPa.m) was calculated for the maximum hardness of 46 GPa.

III. CONCLUSION

The results of this study reveal that the developed AlSiN film has a nanocomposite structure consisting of an amorphous Si_3N_4 matrix, in which nanocrystallites with a size of 5 nm are incorporated. The nanocrystals are dominantly composed of (AlSi)N solid solution. This structure causes enhanced film hardness of 46 GPa and reduced elastic modulus of 468 GPa. The increased elasticity relates to the better elastic recovery of 67%. The AlSiN film exhibits very good elastic strain prior to failure (H/E=0.10) and resistance to plastic deformation (H^3/E^2=0.44). This result implies improved resistance to mechanical degradation and failure under higher stresses, and improved ability to absorb energy at deformation before fracture. Besides the enhanced hardness and improved toughness, the AlSiN film possess a low friction coefficient of 0.09 and very low wear rate of 7.56×10^{-10} mm^3/(N.GPa.m).

Consequently, the developed AlSiN film is suitable for application in solar cells, micro- and nanomechanical systems, optical coatings requiring high wear resistance, as well as machining industry as a protective coating improving the hardness and toughness of the tools.

ACKNOWLEDGEMENT

This work was supported by the European Regional Development Fund within the OP "Science and Education for Smart Growth 2014 - 2020", Project CoC "Smart Mechatronic, Eco- And Energy Saving Systems And Technologies", № BG05M2OP001-1.002-0023

REFERENCES

[1] Y. Taniyasu, M. Kasu, and T. Makimoto, "Field emission properties of heavily Si-doped AlN in triode-type display structure", *Applied Physics Letters*, 2004, vol. 84, pp. 2115 pp. 2117.

[2] M. Hermann, F. Furtmayr, A. Bergmaier, G. Dollinger, M. Stutzmann, M. Eickhoff, "Highly Si-doped AlN grown by plasma-assisted molecular-beam epitaxy", *Appl. Phys. Lett.*, 2005, vol. 86, 192108, pp. 1-3.

[3] A. Bendavid, P.J. Martin, H. Takikawa, "The properties of nanocomposite aluminium–silicon based thin filmsdeposited by filtered arc deposition", *Thin Solid Films*, 2002, vols. 420-421, pp. 83-88.

[4] A. Pélisson, M. Parlinska-Wojtan, H.J. Hug, J. Patscheider, "Microstructure and mechanical properties of Al–Si–N transparent hard coatings deposited by magnetron sputtering", *Surface & Coatings Technology*, 2007, vol. 202, pp. 884–889.

[5] A. Pélisson-Schecker, H. Hug, J. Patscheider, "Morphology, microstructure evolution and optical properties of Al–Si–N nanocomposite coatings", *Surface & Coatings Technology*, 2014, vol. 257, pp. 114–120.

[6] L. Rebouta, A. Sousa, M. Andritschky, F. Cerqueira, C.J. Tavaresa, P. Santillib, K. Pischowba, "Solar selective absorbing coatings based on AlSiN/AlSiON/AlSiO$_x$layers", *Applied Surface Science*, 2015, vol. 356, pp. 203–212.

[7] A.Schecker, H. Hug, J. Patscheider, "Morphology, microstructure evolution and optical properties of Al–Si–N n anocomposite coatings", *Surface & Coatings Technology*, 2014, vol. 257, pp. 114–120.

[8] J. Musil, G. Remnev, V. Legostaev, V. Uglov, A. Lebedynskiy, A. Lauk, J. Procházka, S. Haviar, E. Smolyanskiy, "Flexible hard Al-Si-N films for high temperature operation", *Surface & Coatings Technology*, 2016, vol. 307, pp. 1112–1118.

[9] T. Nguyen a, T. Nguyen, G. Nguyen, V. Le, "Effect of the Si content on structure and mechanical properties in Al$_{1-x}$Si$_x$N materials", *Vacuum*, 2016, vol. 129, pp. 1-8.

[10] X Jiang, F.C. Yang, W.C. Chen, J.W. Lee, C.L. Chang, "Effect of nitrogen-argon flow ratio on the microstructural and mechanical properties of AlSiN thin films prepared by high power impulse magnetron sputtering", *Surface & Coatings Technology*, 2017, vol. 320, pp. 138–145.

[11] V. Le, T. Nguyen, S. Kim, K. Pham, "Effect of the Si content on the structure, mechanical and tribological properties of CrN/AlSiN thin films", *Surface & Coatings Technology*, 2013, vol. 218, pp. 87–92.

[12] A. Gaydaychuk, S. Zenkin, S. Linnik, "Influence of Al-Si-N interlayer on residual stress of CVD diamond coatings", *Surface & Coatings Technology*, 2019, vol. 357, pp. 348-352.

[13] A. Leynad and A. Matthews, "On the significance of the H/E ratio in wear control: a nanocomposite coating approach to optimised tribological behavior", *Wear*, 2000, vol. 246, pp. 1–11.

[14] R. Ritchie, "The conflicts between strength and toughness", *Nat. Mater.*, 2011, vol. 10, pp. 817-822.

[15] W.C. Oliver and G.M. Pharr, "An improved technique for determining hardness and elastic modulus using load and displacement sensing indentation experiments", *J. Mater. Res.*, 1992, vol. 7, pp. 1564-1583

[16] D. Mikičić, A. Kunosić, M. Zlatanović, "Contact Force Determination in Abrasive Wear Test", *Tribology in industry*, 200, vol. 527, pp. 34.

[17] Y. Wang, S. Zhang, J. Lee, W. Lew, D. Sun, B. Li, "Toward hard yet tough CrAlSiN coatings via compositional grading", *Surface & Coatings Technology*, 2013, vol. 231, pp. 346–352.

[18] A. Conde, G. Cristóbal, T. Tate, J. Damborene, "Surface analysis of electrochemically stripped CrN coatings", *Surface & Coatings Technology*, 2006, vol. 201, pp. 3588–3595.

[19] NIST, "X-ray Photoelectron Spectroscopy Database, National Institute of Standards and Technology", Gaithersburg, 2012 (http://srdata.nist.gov/xps/).

[20] L. Rosenberger,R. Baird, E. McCullen,G.AunernG.Shreve, "XPS analysis of aluminum nitride films deposited by plasma source molecular beam epitaxy", *Surface and Interface Analysis*, 2008, vol. 40, pp. 1254–1261.

[21] S. Tan, X. Zhang, X. Wu, F. Fang, J. Jiang, "Effect of substrate bias and temperature on magnetron sputtered CrSiN films", *Applied Surface Science*, 2011, vol. 257, pp. 1850–1853.

Artificial Neural Network for Composite Hardness Modeling of Cu/Si Systems Fabricated Using Various Electrodeposition Parameters

I. Mladenović, J. Lamovec, V. Jović, M. Obradov, K. Radulović,
D. Vasiljević Radović, and V. Radojević

Abstract –Copper coatings are produced on silicon wafer by electrodeposition (ED) for various cathode current densities. The resulting composite systems consist of 10 μm monolayered copper films electrodeposited from sulphate bath on Si wafers with sputtered layers of Cr/Au. Hardness measurements were performed to evaluate properties of the composites. The composite hardness (H_c) was characterized using Vickers microindentation test. Then, an artificial neural network (ANN) model was used to study the relationship between the parameters of metallic composite and their hardness. Two experimental values: applied load during indentation test and current density during the ED process were used as the inputs to the neural network. Finally, the results of the composite hardness (experimental and predicted) were used to estimate the film hardness (H_f) of copper for each variations of the current density. This article shows that ANN is an useful tool in modeling composite hardness change with variation of experimental parameters predicting hardness change of composite Si/Cu with average error of 6 %. Using created ANN model it is possible to predict microhardness of Cu film for current density or indentation load for which we do not have experimental data.

I. INTRODUCTION

Artificial neural network (ANN) is a numerical model designed to simulate information processing of a human brain. They are used in complex non-linear systems using the preexisting empirical data to learn about the system. As such ANNs are used for assessment, prediction, decision making and diagnostics [1,2]. The neural network consists of simple processors, called neurons. Each neuron has inputs and generates output signals that are sent to other neurons in the network as inputs via the interconnections. ANN approach is used in many fields of chemical and material engineering such as: prediction of yield strength, tensile strength and elongation of cast alloys [3], for estimation on of the deposition rate of copper-tin during electroplating, hardness predictions of nickle-CBN composites [2], evaluating the change of wood hardness during heat treatment [4], etc.

I. Mladenović, J. Lamovec, V. Jović, M. Obradov, K.Radulović, and D. Vasiljević Radović are with the Department of Microelectronic Technologies, Scientific Institution Institute of Chemistry, Technology and Metallurgy, University of Belgrade, Njegoševa 12, 11000 Beograd, Serbia, E-mail: ivana@nanosys.ihtm.bg.ac.rs
V. Radojević is with the Department of Materials Engineering, Faculty of Technology and Metallurgy, UB, Serbia.

Electrodeposited copper films are used in fabrication of micro-electro-mechanical (MEMS) devices for a wide range of applications [5]. Mechanical properties of electrodeposited copper films on silicon substrates in electronic devices heavily influence the lifetime of the devices [6]. As such it is especially important to analyze the hardness of the composite systems which depends on several factors, such as the microstructure and hardness of the film and of the substrate, thickness of the film etc. [7].

Microindentation is one of the best known methods for the evaluation of mechanical properties of films and coatings. In cases where the thickness of the film is small, the substrate hardness affects the hardness of the film, such a measured hardness is called composite hardness.

Based on experimental measurements, the database was created. Using that database we created neural network model that we used to predict composite hardness of our Cu/Si system. The measurements and predicted values of composite hardness were used to calculate the hardness of the copper film.

II. ARTIFICIAL NEURAL NETWORK (ANN)

In this study, a proposed ANN model was designed using the Matlab Neural Network Toolbox and using a multi-layer perception (MLP) model for prediction. The MLP architecture consists of an input layer, one or more hidden layers, and an output layer [4]. The input layer consists of two input nodes: applied load during indentation measurement and applied current density during ED process. The hidden layer utilizes three neurons, and the output layer consists of one output node: composite hardness of the Cu/Si systems.

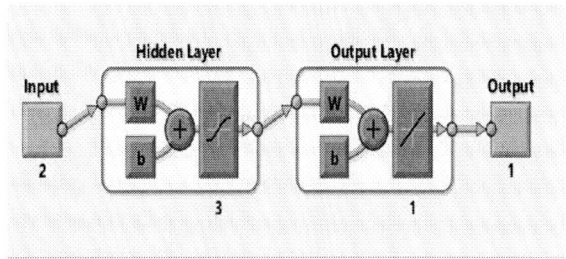

Fig. 1. Block diagram of ANN used in this study.

ANN block diagram used in this study is given in Fig. 1. The hidden layer uses a hyperbolic tangent sigmoid transfer function and the training algorithm is the Levenberg-Marquardt backpropagation.

The database-containing 60 indentation hardness measurements of the fabricated Cu/Si composites was randomly divided into three groups: 42 data points (70% of the total data) used for the ANN training process, 9 data points (15 % of the total data) for validation group, and 9 data points (15 % of all data) for the testing process.

The network performance can be estimated through the error of deviation between actual and predicted values. The mean absolute percentage error (MAPE), the mean square error (MSE) and determination coefficient (R^2) were utilized to evaluate the performance of the ANN. The errors were calculated using the following formulas:

$$MAPE = \frac{1}{N}\left\{\sum_{i=1}^{N}\left[\left|\frac{Hc_i - Hcp_i}{Hc_i}\right|\right] \times 100\right\} \quad (1)$$

$$MSE = \frac{1}{N}\sum_{i=1}^{N}(Hc_i - Hcp_i)^2 \quad (2)$$

$$R^2 = 1 - \frac{\sum_{i=1}^{N}(Hc_i - Hcp_i)^2}{\sum_{i=1}^{N}(Hc_i - Hc')^2} \quad (3)$$

where Hc_i represent the experimental output, Hcp_i represent the predict output, N represent the total number of samples and Hc' represents the mean of predicted outputs.

III. HARDNESS DETERMINATION OF COPPER FILMS USING THE WORK-OF-INDENTATION MODEL

Mathematical models used to calculate the hardness of the thin film from the measured composite hardness depends on the type of composite systems. The composite hardness model of Korsunsky [8] was chosen and applied to experimental and prediction data in order to calculate the copper film hardness. According to this model correlation between composite hardness H_c, film hardness H_f, and substrate hardness H_s is given as:

$$Hc = Hs + \left[\frac{1}{1 + k' \cdot (d^2/t)}\right] \cdot (Hf - Hs) \quad (4)$$

$$k' = \frac{k}{49 \cdot t} \quad (5)$$

where k is a dimensionless materials parameter related to the composite response mode to indentation, d is the indent diagonal and t is the thickness of the film.

Prior deposition process, the substrate hardness was first determined, experimentally. We used Proporcional

Specimen Resistance (PSR) model [9]. The calculated hardness of the substrate was 14.65 GPa, measured for the first 5 load points.

IV. EXPERIMENTAL PROCEDURE

For these experiments substrate of Si wafers (4 inch, (100) orientation) was chosen and prepared. The wafer was cut in parts about 1cm wide, standard cleaning and drying procedures. The plating base on the silicon wafers were sputtered layers of 10 nm Cr and 100 nm Au.

Copper films were electrodeposited from a 100 ml sulphate bath [6]. Electrochemical deposition was carried out using direct current galvanostatic mode with the 5 current density values (10, 33.33, 50, 66.67 and 100 mA/cm^2). Based on the platting surface, current density, and duration of deposition process, thickness of copper deposits was 10 μm.

The mechanical properties of the composite systems Si/Cr/Au/Cu were characterized using Vickers microhardness tester "Leitz, Kleinharteprufer DURIMET I" with loads ranging from 2.4515 N down to 0.04903 N. Indentation was done at room temperature. The dwell time was 25 s. The average values of impression diagonals d (in m), were calculated from several independent measurements on every specimen for different applied loads, P (in N). The composite hardness, Hc (in GPa), was calculated using the formula (6).

$$Hc = \frac{0.01854 \cdot P}{d^2} \quad (6)$$

Topographic examination was done by the metallographic microscope "Carl Zeiss Epival Interphako".

V. RESULT AND DISCUSSION

In this section, experimental results were compared to prediction values for composite hardness from the ANN model as shown in Fig. 2. Large disagreements between experimental and predicted values were noted for low and high loads. At low loads, the Vicker's diagonal size is small and difficult to read. These errors are known as indentation size errors or load errors and must be included in the assessment of hardness as correlation coefficients. The estimated mean absolute square errors for the first two load points are over 10 %. Another critical area is at the end of the composite region, when using a load over 1.5 N. Here, the effect of the substrate hardness becomes significant and the composite hardness increases. The important factor is the depth of the penetration of the top of an indenter. The substrate starts to contribute the measured hardness at the penetration depth 0.07-0.20 times the coating thickness [10].

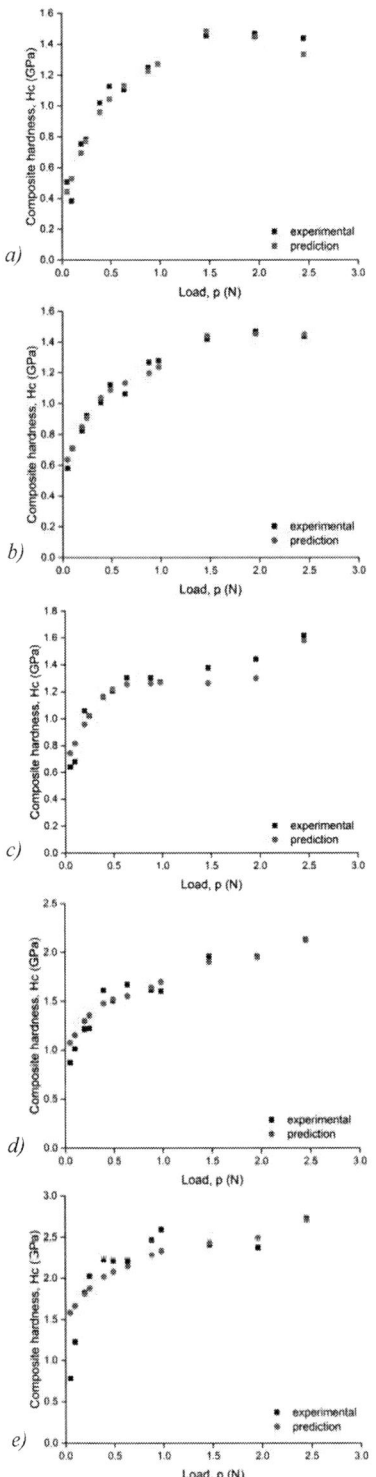

Fig. 2.Comparasion of composite hardness values (experimental and prediction) depending on the applied indentation load for different current density: a) 10 ; b) 33.33; c) 50; d) 66.67; e) 100 mA/cm².

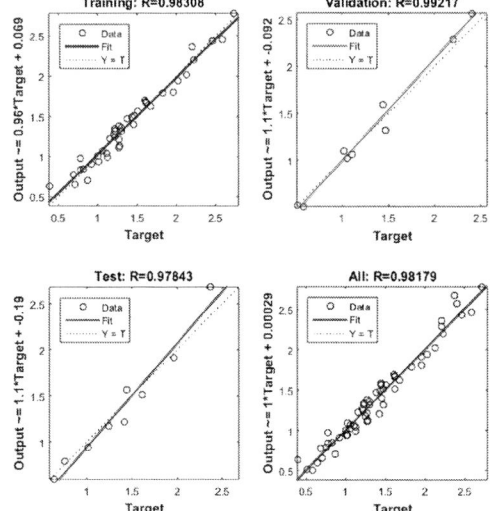

Fig. 3.ANN regression for microhardness modeling.

TABLE I
RESULTS OF THE CRITERIA USED IN PREDICTION COMPOSITE
HARDNESS CHANGE

data	samples	MSE	R
training	42	9.489e-3	9.831e-1
validation	9	1.023e-2	9.922e-1
testing	9	2.010e-2	9.784e-1

Predictive ability of the models was evaluated using performance indicators (2) and (3): MSE and R^2 for training, validation and testing data as shown in Table I. Ideal values are MSE=0 and R^2=1. The plot on Fig.3 shows a regression between network outputs and network targets. If the training were perfect, the network outputs and the targets would be equal. The R-values were found as 0.983 for training, 0.922 for validation and 0.978 for testing. With the result above, it is possible to say that the proposed model was well trained and showed an acceptable accuracy in predicting the composite hardness change with variations of current density and applied load.

TABLE II
RESULTS OF THE FILM HARDNESS CHANGE FOR EXPERIMENTAL AND
PREDICTION VALUES

sample	experimental		prediction	
j (mA/cm2)	H_f (GPa)	k'	H_f (GPa)	k'
10	0.605	0.133	0.687	0.122
33.33	0.739	0.109	0.852	0.099
50	0.852	0.105	0.921	0.110
66.67	1.056	0.212	0.988	0.204
100	1.591	0.312	1.457	0.386

To estimate the film hardness independently of the substrate Korsunsky model was applied in order to determine absolute hardness of the films. Fitted results are shown in Table II. The increase in the hardness of the film with increasing current density is evident for each sample. The predicted results of the film hardness and experimental values are close.

The next step is predicting the data on which the network was not trained. Three new values of current density were selected (15, 65 and 85 mA/cm^2). The results of prediction film hardness according ANN model are given in Table III. In Fig. 4. prediction of composite hardness for two current densities that are outside the range of experimental measurement are shown.

TABLE III
RESULTS OF THE FILM HARDNESS CHANGE FOR PREDICTION VALUES

sample	Prediction	
j (mA/cm^2)	H$_f$ (GPa)	k'
15	0.724	0.116
65	0.895	0.189
85	0.904	0.403

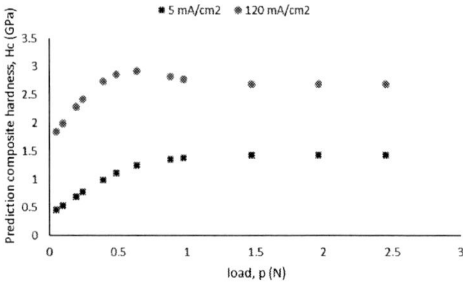

Fig. 4. Predicted composite hardness according to ANN model for current densities of 5 and 120 mA/cm^2.

It can be seen that predicted results follow experimentally established dependencies on current density and load values and are in line with results presented in Fig.2 and Table 2. In the same way, the system's hardness can be predicted for any load point that has not been experimentally performed.

VI. CONCLUSION

Composite systems of electrochemically deposited Cu films on Si (100) substrates were prepared and investigated.

An Artificial Neural Network model was developed and tested for predicting composite hardness of Cu/Si systems using total of 60 experimentally obtained data records. In this article, the focus was on modeling the effects of current density and indentation load on composite and film hardness via ANN predictions.

We have shown excellent consistency and good agreement between ANN predicted results and experimental

measurements. Added advantage is that ANN is constantly learning improving with each iteration. Predicting the composite hardness or film hardness over the ANN model is useful when we want to define the properties of a material in advance or to evaluate feasibility of a situation not experimentally performed.

Our future research will focus on more complex ANNs ideally bypassing the need for any analytic approximations in determining the thin film hardness.

ACKNOWLEDGEMENT

This work was funded by Ministry of Education, science and Technological Development of Republic of Serbia through the orijects TR 32008 and TR 34011.

REFERENCES

[1] D. M. Habashy, H. S. Mohamed, E. F. M. El-Zaidia, "A simulated neural system (ANNs) for micro-hardness of nano-crystalline titanium dioxide", *Physica B: Condensed Matter,* 2019, vol. 556, pp.183-189.

[2] T. L. Frango, K. Ramanathan, G. N. K. RameshBapu, P. Marimuthu, "Artificial Neural Network (ANN) modeling for predicting hardness of Ni-CBN composite coatings", *International Journal of Advanced Engineering Technology,* 2016, vol. 7, no. 2, pp.1234-1237.

[3] M.S. Ozerdem, S. Kolukisa, "Artificial neural network approach to predict the mechanical properties of Cu-Sn-Pb-Zn-Ni cast alloys", *Materials and design,* 2009, vol.30, no. 2, pp.764-769.

[4] T. H. V. Nguyen, T. T. Nguyen, X. Ji, K. T. L Do, M. Guo, " Using Artificial Neural Networks (ANN) for Modeling Predicting Hardness Change of Wood during Heat Tretment", *IOP Conf.Series: Materials Science and Engeneering,* 2018, 394, 032044.

[5] A. Maciossek, B. Lochel, H. J. Quenzer, B. Wagner, S. Schulze, J. Noetzel, "Galvanoplating and sacrificial layers for surface micromachining", *Microelectronic Engineering,* 1995, vol. 27, no.1, pp. 503-508.

[6] N. D. Nikolić, Z. Rakocević, K. I. Popov, "Reflection and structural analyses of mirror-bright metal coatings", *Journal of Solid State Electrochemistry,* 2004, vol. 8, no.8, pp. 526-531.

[7] J. Lamovec, V. Jović, M. Vorkapić, B. Popović, V. Radojević, R. Aleksić, "Microhardness analysis of thin metallic multilayer composite films on copper substrates", *Journal of Mining and Metallurgy, Section B: Metallurgy,* 2011, vol.47, no.1 , pp. 53-61.

[8] A. M. Korsunsky, M. R. McGurk, S. J. Bull, T.F. Page, "On the hardness of coated systems", *Surf.&Coat.Technol.,* 1998, vol.1, no.99, pp. 171-183.

[9] H.Li, R.C. Bradt, "The indentation load/size effect and the measurement of vitreous silica", *Journal of non-crystalline solids,* 1992, vol. 146, no.1, pp. 197-212.

[10] J. Lamovec, V. Jovic, D. Randjelovic, R. Aleksic, V. Radojevic, "Analysis of the composite and film hardness of electrodeposited nickel coatings on different substrates", *Thin solid films,* 2008, vol.516, no.23, pp. 8646-8654.

[11] A. Augustin, K. R. Udupa, K. U. Bhat, "Effect of coating current density on the wettability of electrodeposited copper thin film on aluminium substrate", *Engineering and Material Sciences,* 2016, vol. 8, pp. 472-474.

Modeling Errors of the MISFET-Based Sensors' Characteristics

B. Podlepetsky and N. Samotaev

Abstract - The accuracy of circuits', electrical and electro-physical models of MIS transistor sensors' elements, taking into account the errors of simplifying assumptions, approximations, extrapolations and experimental dispersions' characteristics in determining the parameters of the models and measured physical quantity is estimated.

I. INTRODUCTION

It took more than 30 years after the patented idea [1] to create real field-effect transistors with a metal-dielectric-semiconductor structure (MISFET) [2], [3]. MISFETs are used in electronic devices and systems as elements of integrated circuits (ICs), as well as sensitive elements (SE) of sensors of temperature, pressure, power of light radiation, ionizing radiation doses, humidity, concentrations of ions in liquids, concentrations of gas molecules [4]-[10]. Therefore, MISFETs can be considered as a multifunctional SE of different physical quantities X. Technological and circuits' CAD systems (e.g. TCAD, SPICE, VERILOG-A) based on physical, electrical and engineering-physical models are used for the design of MISFETs-based ICs and sensors [11]-[22].

Many models have been developed to describe experimental characteristics of MISFET and integrated circuits on their basis. In an effort to create general models, take into account the influence of different factors and to improve the accuracy of modeling, many researchers include additional parameters in the model, which can not always be determined experimentally [14], [18]-[21]. This, in fact, complicates the models themselves, reduces their reliability and accuracy, complicates their practical use. Simplified mathematical models with a small number of parameters do not always have a physical meaning, but are accurate enough to approximate experimental dependences with an uncomplicated extraction procedure of the adjusted parameters [11], [16], [22]. All models are approximate, some models contradict each other, and some complement the previous analogues.

In most models, MISFETs are considered as elements of circuits. However, when MISFETs are SE of sensors, the direct application of the known models of MISFETs and CAD for the design of sensors and devices becomes

B. Podlepetsky and N. Samotaev ares with the Department of Micro- and Nanoelectronics, National Research Nuclear University MEPhI (Moscow Engineering Physics Institute), 31 Kashirskoe shosse, 115409 Moscow, Russia, Email: bipod45@gmail.com, NNSamotaev@mephi.ru

difficult or impossible without taking into account the technological features of the SE and the operating conditions of their application. The models obtained on the basis of approximations of experimental data are used to predict the performance of MISFET-based integrated sensors and measuring devices. The accuracy of the prediction depends on the errors of the models.

The motivation of this work is to estimate the accuracy of circuits', electrical and electro-physical models of MISFET-based SE of the sensors, taking into account the errors due to simplifying assumptions, approximations, extrapolations and dispersions of experimental characteristics.

II. THE SOURCE DATE FOR MODELING

As SEs are usually used n-channel MISFET, having transconductance b at the same geometric parameters greater, than p-channel transistors. The length L and width w of the channel and the thickness of the sub-gate dielectric d of MISFET-SE are relatively large compared to MISFETs of ICs (Fig. 1).

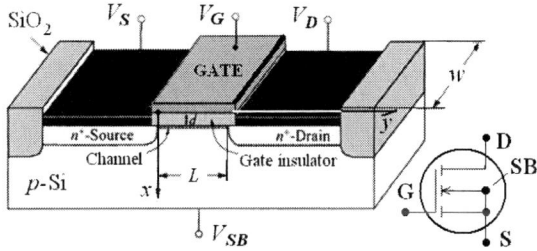

Fig.1. A structure fragment and a designation of n-channel MISFET- based SE.

The specific electrical capacity $C_0 = (\varepsilon_0\varepsilon)/d$ and parameter $a = (2q \cdot \varepsilon_S\varepsilon_0 N_A)^{0.5}/C_0$ are used in MISFET performance models as technological parameters . N_A, ε and ε_s are acceptors' concentration, permittivity of SiO_2 and Si. Values q and ε_0 are electron charge and vacuum dielectric constant.

Errors of technological parameters are determined by technological standards or errors of measuring instruments, the relative values of which are usually in the range from 5% to 10%.

The dependence of the sensor's output signal Y on value of measured physical quantity X, influencing factor parameter z and time t can be presented as

978-1-7281-3420-8/19 $31.00 © 2019 IEEE

$$Y(X, z, t) = Y_0(Y_{00}, z, t) + \Delta Y(X, z) \qquad (1)$$

Values of Y_{00}; Y_0; ΔY are respectively the primary value of Y when $t = 0$; the initial value of Y when $X = 0$; the changes in Y under the influence of the measured X and z (conversion function).

Different types of sensors have different models of conversion functions $\Delta Y(X)$, which are determined by the approximation of the average values of experimental data $\Delta Y_{ai}(X_i)$. Example of the experimental functions $Y(X)$ of the hydrogen sensor is shown in Fig. 2.

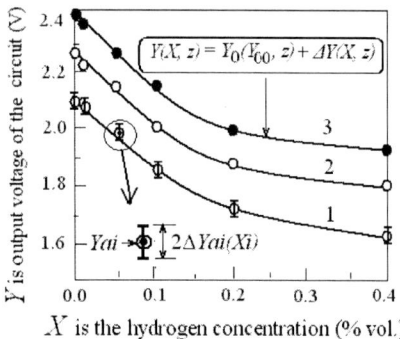

Fig. 2. Example of the experimental functions $Y(X)$ of the hydrogen sensor for the circuit 2 (Fig. 3) at V_D is 1 V and drain currents I_D: 0.05 mA (1); 0.1 mA (2); 0.3 mA (3). Symbols are experimental points, lines are approximation dependences (results from [22]).

The error's band $[2\Delta Y_{ai}(X_i)]$ of values $Y_{ai}(X_i)$ is determined by the parameter $\theta_{Yi}(X_i)$. The approximation error is estimated by deviations of values $\Delta Y_{ai}(X_i)$ from the corresponding model values $\Delta Y(X)$.

$$\Delta Y_{ai} = \frac{1}{N} \cdot \sum_{n=1}^{N} |\Delta Y_{ni}| \qquad (2)$$

$$\theta_{Yi} = \frac{1}{N} \cdot \sum_{n=1}^{N} \left| \frac{\Delta Y_{ni} - \Delta Y_{ai}}{\Delta Y_{ai}} \right| \qquad (3)$$

Indexes i, n and N are the serial number values X_i, the measurements' number X_{ni} and number of the measurements of X_i respectively.

III. CIRCUITS' MODELS OF MISFET BASED SENSORS

To measure value X by using MISFET-sensor, transistor is embedded in device's measuring circuits. Typically, MISFETs are applied in analogue circuits, in which the informative parameter is the output voltage V. The informative parameter of the MISFET may be threshold voltage V_T. Typical circuits are shown in Fig. 2, where parameter $U(X) = V_G - V_T$.

The general sensitivity S being equal to $|dV/dX|$ is $|S_C \times S_T|$, where the circuit's sensitivity S_C is $|dV/dV_T|$, and

threshold voltage's sensitivity of MISFET S_T is dV_T/dX. Of practical interest for sensors are the errors in determining values of X. The maximum absolute $\Delta(X)$ and the relative δX errors can be presented as

$$\Delta(X) = |\Delta(V)/S|; \quad \delta X = |\Delta(V)/(S \cdot X)| \times 100\% \qquad (4)$$

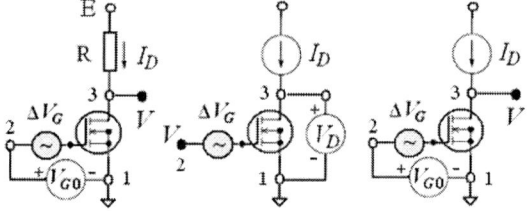

Circuit 1
$V = E - R \cdot I_D[U(X)]$,
$S_C = bRU(X)$

Circuit 2
$V = V_{G0} + \Delta V_G(X)$,
$S_C = 1.0$

Circuit 3
$V = U(X) - [U^2 - I_D]^{0.5}$,
$S_C < 0.4$

Fig. 3. The volt-metric circuits of MISFET-based sensors.

Value of $\Delta(V)$ is the maximum absolute error of voltage V:

$$\Delta(V) = \left| \begin{matrix} \Delta V_V + \theta_V(X) \cdot \Delta V(X) + \\ + \Delta V_p(X, \{p_k\}) + \Delta V_z(X, \{z\}) \end{matrix} \right| \qquad (5)$$

Value of ΔV_V is absolute voltage measurement error; the dispersion factor $\theta_V(X)$; $\Delta V_p(X, \{p_k\})$ and $\Delta V_z(X, \{z\})$ are additional errors associated with fluctuations of electrical parameters p_k of sensor circuit and influence of external factors z respectively. Since usually the values $\Delta(V)$ and S depend on X, the errors ΔX and δX will also depend on X.

Typically use either the circuit 1, when the parameters of the function $\Delta V(X)$ are known, and we can set the optimal gate bias, or the circuit 2, when we need to investigate the characteristics of new types of MISFETs' sensors. Examples of values $\Delta(X)$ and δX vs. ΔX are shown in Fig. 4.

IV. ELECTRO-PHYSICAL MODELS OF MISFET

Electro-physical models of MISFET associate the electrical parameters of a transistor with physical parameters that are sensitive to the value of X.

The dependences of gate voltage V_G on the surface potential φ_s are the basis of many electrical models of MISFET.

$$V_G(\varphi_s) = \varphi_s + \varphi_{ms} - [Q_s(\varphi_s) + Q_{ss}(\varphi_s) + Q_{te}]/C_0 \qquad (6)$$

$$Q_s(\varphi_s) = -a \cdot C_0 \cdot \{\varphi_s + \varphi_T \cdot \exp[(\varphi_s - 2\varphi_{s0})/\varphi_T]\}^{0.5} \qquad (7)$$

Values of φ_{ms}, Q_s, Q_{ss} and Q_{te} are respectively the potential of metal-semiconductor work functions difference, charge densities in semiconductor, in interface semiconductor-dielectric and in dielectric [21].

978-1-7281-3420-8/19 $31.00 © 2019 IEEE

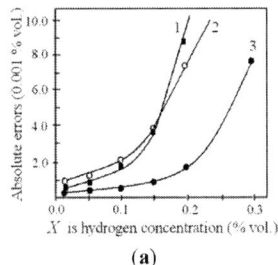 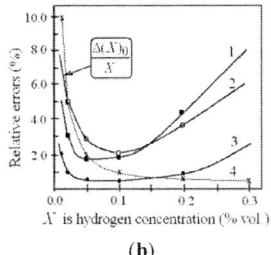

(a)   **(b)**

Fig. 4. The errors $\Delta(X)$ (a) and δX (b) vs. ΔX of hydrogen sensor: 1 – X_0 is zero; 2 – after H_2 expositions by doses 25 (% vol. × min); 3 – the errors due only to $\Delta V_V = 1$ mV; 4 – $\Delta(X)_0$ is 0.001% vol.

The degree of influence of these parameters on the sensitivity to X is presented in Table 1. Values of T, D, C and C_I are respectively temperature, dose of ionizing radiation, concentrations of gas molecules and ions in the solution. Thermal potential $\varphi_T = kT/q$ and $\varphi_{s0} = \varphi_T \ln(N_A/n_i)$, k is Boltzmann constant, n_i is intrinsic carrier concentration. Signs "+", "–" and "?" correspond to sensitivity to X, absence of sensitivity to X and if sensitivity to X the studies on the issue.

TABLE I
THE DEGREE OF PARAMETERS' INFLUENCE ON SENSITIVITY TO X

X	φ_{ms}	Q_s	Q_{ss}	Q_{te}	z	Δb	ΔV_T
T	–	+	+	?	D	?	
D	–	–	+	+	T	+	
C	+	–	–	+	T; D	–	+
C_I	φ_e +	–	–	–		–	

Note that in ion-sensitive transistors, in which the role of the gate is played by the fluid electrolyte and the reference electrode, the sensitivity of the sensor is determined by only the electrode potential φ_e. In other types of sensors sensitivity depends on two parameters, a separate contribution of which deserves special research. For this purpose, analytical models (8) at $\varphi_s \in [\varphi_{s0}, 2\varphi_{s0}]$ and (9) at $\varphi_s \in [\varphi_{s1}, \varphi_{s2}]$, based on simplifying assumptions and piecewise linear approximations of nonlinear functions, are proposed. Examples of approximations $V_{GA}(\varphi_s)$ of model (6) for ionizing radiation dose MISFET sensor are shown in Fig. 5:

$$V_{GA}(\varphi_s) = V_G(\varphi_{s0}) + (\varphi_s - \varphi_{s0}) \cdot (V_T - V_{G1})/\varphi_{s0} \quad (8)$$

$$V_{GA} = V_1 + (\varphi_s - \varphi_{s1}) \cdot (V_2 - V_1)/(\varphi_{s2} - \varphi_{s1}) \quad (9)$$

The approximations' errors $\Delta(V_G)$ being equal to $|V_G - V_{GA}|$ for (8) and (9) are:

$$\Delta(V_G)_1 = a \cdot [\varphi_s^{0.5} - 0.414 \quad \varphi_s \cdot (\varphi_{s0})^{-0.5} - 0.586 \; (\varphi_{s0})^{0.5}] \quad (10)$$

$$\Delta(V_G)_2 = a \cdot [f_s - f_{s1} - (\varphi_s - \varphi_{s1}) \cdot (f_{s2} - f_{s1})/(\varphi_{s2} - \varphi_{s1})] \quad (11)$$

$$f_k = \{\varphi_k + \varphi_T \exp[(\varphi_k - 2\varphi_{s0})/\varphi_T]\}^{0.5}] \;, \quad k \in \{s; \; s1; \; s2\}.$$

Maximum values of $\Delta(V_G)_m$ depend on temperature and parameter a. For this example, values of $\Delta(V_G)_{1m}$ and $\Delta(V_G)_{2m}$ are respectively 12 mV and 0.3 V.

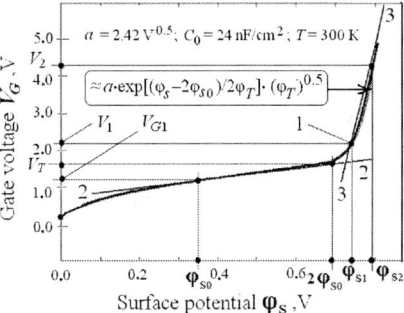

Fig. 5. The exact dependence of $V_G(\varphi_s)$ is shown by the "fat" line 1, piecewise linear approximations are shown by straight lines 2 and 3 [23].

The operating points and ranges of ΔV_G or of the drain currents ΔI_D of MISFET sensors are in the region of strong inversion (at $U > 0$). The values of maximum changes ΔV_G are in the range (0.2–1) V (for chemical sensors) and (1–3) V (for physical sensors). In linear approximation, these changes should be covered by the selected range $(V_2 - V_1)$. Since the approximation error increases with increasing $(V_2 - V_1)$, it is better to use a nonlinear approximation for physical sensors, which also allows obtaining an analytical dependence $\varphi_s(V_G)$.

Approximating conversion functions of the threshold voltage $\Delta V_T(X)$ for different sensor types are presented in Table 2. All model parameters are determined experimentally.

TABLE II
APPROXIMATING FUNCTIONS $\Delta V_T(X)$ FOR DIFFERENT SENSORS

X	$\Delta V_T(X)$
T	$S \cdot \Delta T$; $(S < 0)$
D	$\Delta V_{sM} \cdot [1 - \exp(-k_2 D)] - \Delta V_{iM} [1 - \exp(-k_1 \cdot D)]$
C	$\Delta V_{TM} \cdot [1 - \exp(-kC)]$ or $\Delta V_{IM} \cdot \ln(C/C_0)$
C_I	$\Delta \varphi_e = \pm(RT) \ln(f \cdot C_I)/(nF)$

V. ELECTRICAL MODELS OF MISFET

Electrical models of n - channel MISFET (transfer CVCs) at a constant voltage V_D before and after exposure to the measured value X are shown in Fig.6. The drain current is indicated in relative unit $i_D = (I_D/I_{D0})$. The current I_{D0} corresponds to the threshold voltage V_T, which for all sensors under the action of X is reduced by $\Delta V_t(X)$ being equal to $\Delta Q_{te}/C_0$. The simplified 3-component electrical model (12) is the most widespread.

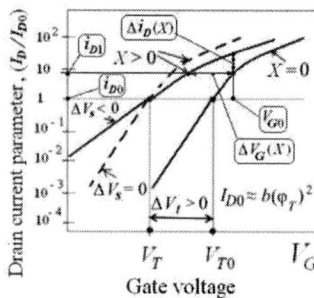

Fig. 6. Deformation of CVC under the action of X.

$$i_D(U) = \begin{bmatrix} \exp\{U/[(c+\gamma)\cdot\varphi_T]\} \ at \ i_D < 1; \\ 1 + 0.5(U/\varphi_T)^2 \quad at \ i_D \in [1; i_{D1}]; \\ V_D \cdot (U - 0.5V_D)/(\varphi_T)^2 \ at \ i_D > i_{D1} \end{bmatrix} \quad (12)$$

$I_{D0} \approx b \cdot (\varphi_T)^2$; $i_{D1} = 0.5(V_D/\varphi_T)^2$; $b(D_{ss}) = [\mu_n(D_{ss})\cdot wC_0]/L$; $c = 1 + 0.414a\cdot(\varphi_{s0})^{-0.5}$. Value of $\Delta V_s(X) = \Delta Q_{ss}/C_0 < 0$ and $\Delta V_s \approx q \cdot D_{ss}(X) \cdot [\varphi_{s0}(T) - \varphi_s]/C_0$, where D_{ss} is energy density of surface states at Si-SiO$_2$ interface. Typically, the operating points of the sensors are in a mode of strong inversion ($i_D > i_{D1}$), in which the deviation of the model (12) from the accurate models [12]-[20] can be up to 20%.

VI. CONCLUSION

The general estimations of the errors of circuits', electro-physical and electrical models of MISFETs as a sensitive element of different sensors are given. Sensors based on MISFET have common disadvantages: large variation of parameters of responses (up to 15 %), thermal and temporal instability of the parameters, restricted deadlines. As they are used in the fields of measurement, where the relative errors are within (10 – 20)% (ionizing radiation dose, concentrations of ions and gas molecules), it is possible to apply simplified models MISFETs to predict errors of the characteristics of devices based on them.

ACKNOWLEDGEMENT

This work was supported by the Russian Science Foundation (Grant Agreement 18-79-10230 of 08.08.2018).

REFERENCES

[1] J. E. Lilienfeld, "Method and Apparatus for Controlling Electrical Currents", *U.S. Patent № 1,745, 175* (1930).

[2] Kahng, Dawon, "Electric Field Controlled Semiconductor Device," *U. S. Patent No. 3,102,230* (Filed 31 May 31, 1960, issued August 27, 1963).

[3] C. T. Sah, "A new semiconductor triode, the surface-potential controlled transistor", *Proceedings of the IRE*, 1961, vol. 49, No.11, pp. 1623-1634.

[4] P. Bergveld, "Development, operation and application of the ion-sensitive field-effect transistor as a tool for lectrophysiology", *IEEE Trans. Biomed. Eng.*, 1972, vol. BME-19, pp. 342-351.

[5] A. Sibbald, "Recent advances in field-effect chemical microsensors", *J. Mol. Electron.*, 1986, vol. 2, pp. 51-83.

[6] I. Lundström, S. Shivaraman, C. Svensson, and L. Lundkvist, "A hydrogen-sensitive MOS field-effect transistor", *Appl. Phys. Lett.*, 1975, vol. 26, pp. 55–57.

[7] I. Lundström, M.Armgarth, A.Spetz, and F.Winquist, "Gas sensors based on catalytic metal-gate field-effect devices", *Sensors and actuators*, 1986, vol.3-4, pp. 399-421.

[8] I. Lundström, H. Sundgren, F. Winquist, M. Eriksson, C. Krants-Rülcker, and A. Lloyd-Spets, "Twenty-five years of field effect gas sensor research in Linköping", *Sensors and Actuators B*, 2007, vol.121, pp. 247-262.

[9] B.I. Podlepetsky, "Sensitive elements of sensors based on transistors with structure electrode – insulator – semiconductor", *Sensors and Systems*, 2010, no. 3, pp. 66-77, (in Russian).

[10] M.M. Pejovic, "P-channel MOSFET as a sensor and dosimeter of ionizing radiation", *Facta Universitatis. Series: Electronics and Energetics*, 2016, vol.29, no. 4, 509-541.

[11] J. R. Brews, "A charge sheet model of the MOSFET", *Solid-State Electron.*, 1978, vol. 21, pp. 345–355.

[12] G. Baccarani, M. Rudan, and G. Spadini, "Analytical i.g.f.e.t model including drift and diffusion", *IEE J. Solid-State and Electron Devices*, 1978, vol. 2, pp. 62-68.

[13] F. Van de Wiele, "A long channel MOSFET model", *Solid-State Electron.*, 1979, vol. 22, pp. 991-997.

[14] J. R. Brews, "Physics of MOS transistor", in *Silicon Integrated Circuits , Part A*, Ed. D. Kahng, Applied Solid-State Science Series, Academic Press, New York, 1981.

[15] P. P. Guebels and F. Van de Wiele, "A small geometry MOSFET models for CAD applications", *Solid-State Electron.*, 1983, vol.26, pp. 267-263.

[16] J. Park, P. K. Ko, and C. Hu, "A charge sheet capacitance model of short channel MOSFET's for SPICE", *IEEE Trans. Computer-Aided Design*, 1991, vol. CAD-10, pp. 376-389.

[17] A. R. Boothryod, S. W. Tarasewicz, and C. Slaby, "MISNAN-A physically based continuous MOSFET model for CAD applications", *IEEE Trans. Computer-Aided Design*, 1991, vol.CAD-10, pp. 1512-1529.

[18] N. Arora, "MOSFET DC Model", in: *MOSFET Models for VLSI Circuit Simulation. Computational Microelectronics.* pp. 230-324, Vienna: Springer, 1993.

[19] G. Gildenblat, L. Xin, W. Wu, W. Hailing, A. Jha, R. van Langevelde, G. D. J. Smit, A. J. Scholten, and D. B. M. Klaassen, "PSP: An advanced surface-potential-based MOSFET model for circuit simulation", *IEEE Trans. Electron Devices*, 2006, vol. 53, no. 9, pp. 1979-1993.

[20] V.V. Denisenko Compact models of MOSFETs for SPICE in micro-and nanoelectronics. - M.: FIZMATLIT, 2010, 408 pp. (In Russian).

[21] I. S. Esqueda, H. J. Barnaby, and M. P. King, "Compact modeling of Total Ionizing Dose and Aging Effects in MOS Technologies", *IEEE Trans. Nucl. Sci.*, 2015, vol. 62, 1501-1515.

[22] B. Podlepetsky, "Integrated Hydrogen Sensors Based on MIS Transistor Sensitive Elements: Modeling of Characteristics", Autom. Remote Control., 2015, vol. 76, pp. 535-547.

[23] B. Podlepetsky, A. Bakerenkov, and Yu. Sukhoroslova, "Radiation sensitivity modeling technique of sensors' MIS-transistor elements", *Automation and Remote Control*, 2018, vol. 79, 180-189.

Flexible Oxide-Polymeric Composites for Piezoelectric Energy Harvesting

M. Aleksandrova, G. Kolev, Y. Vucheva, I. Pandiev, and K. Denishev

Abstract - In this study is presented technology for fabrication of a piezoelectric element as an alternative energy source (energy harvesting). The elements are produced on flexible polyethylene naphthalate (PEN) substrates and consist of novel lead-free nanocomposite [Ga-doped ZnO (GZO) - polyvinylidene fluoride (PVDF)]. The oxide film is deposited by vacuum radiofrequency (RF) sputtering and PVDF is pulverized by spray coating system. Aluminum and gold metal coatings are investigated as electrodes for optimal extraction of the generated electric energy from the piezoelectric coating. It was showed that PVDF spray deposition reduces the surface roughness of GZO film with 1.4 %. Piezoelectric response is measured at different applied dynamic loads with the two types of electrodes, as well as for oxide-only film. It was found that PVDF based composite leads to improved interface conditions for electrode coating, such as low parasitic capacitances. The highest obtained piezoelectric voltage is ~ 586 mV at 40 g mass weight load with frequency of 50 Hz for gold coated GZO+PVDF. This voltage is 41% higher and more stable in the time sweep in comparison with the case at PVDF-free piezoelectric film, and 29% higher than the composite element, but with aluminum electrode. The interface capacitance is 3 orders of magnitude lower (nF vs μF) and the contact resistance is 15 times smaller (Ω vs $k\Omega$) when the interfaces are with gold, which optimizes the electric energy collection and enhances the energy harvesting performance.

I. INTRODUCTION

Flexible electronics and flexible energy harvesting technology have attracted attention recently as emerging alternative to the conventional silicon based power supplied technology for sensing. Nowadays, devices related to health monitoring, or scavenging of waste mechanical energy should be compact, lightweight, sensitive to weak activating stimuli and battery-less if possible. They use motion or small displacement in order to convert force or pressure change into electrical energy due to the piezoelectric effect. That's why such elements, used for example as self-supplied detectors of blood pressure, breathing capacity or simply utilizing the surrounding

M. Aleksandrova, G. Kolev, Y. Vucheva, and K.. Denishev are with the Department of Microelectronics, Faculty of Electronic Engineering and Technologies, Technical University of Sofia, Kliment Ohridski blvd 8, bl. 1, 1000 Sofia, Bulgaria, E-mail: m_aleksandrova@tu-sofia.bg,georgi_klv@abv.bg, dani_30@abv.bg, khd@tu-sofia.bg.
I. Pandiev is with the Department of Electronics, Faculty of Electronic Engineering and Technologies, Technical University of Sofia, Kliment Ohridski blvd 8, bl. 1, 1000 Sofia, Bulgaria, E-mail: ipandiev@tu-sofia.bg.

vibrational energy, rely on thin films grown on plastic substrates [1]. Because of the small thickness and activating volume these elements are inefficient (only few percent of conversion efficiency). Generated voltage is in the range of few hundred of millivolts from area of ~16 cm^2, but the current is usually in nanoampere range, which makes the produced power practically unuseful [2]. A lot of complex compounds have been doped to additionally increase the electrical conductivity and piezoelectric coefficients, and to decrease the loss dissipation factor of the piezoelectric films. For example, $Na_{0.5}Bi_{0.5}TiO_3$ has been doped by Er^{3+} ions [3], PZT type ceramics have been doped by Nb and Li [4], lead zirconate titanate (PZT) – by Sm [5] and even graphene with boron composite has been used to improve the piezoelectric behavior [6]. Additionally, the composites between variety of polymers and piezoelectric oxides have been explored to enhance the mechanical performance of the structures, because of the favorable elasticity modulus of the polymeric materials [7]. After establishing the proper composition of the piezoelectric material, metal electrode layers with specific work function, adhesion, mechanical stability and technological compatibility should be applied at the opposite poles of the samples in order to collect effectively the generated charges [8]. In our previous study it was found that GZO films exhibited higher piezoelectric current than undoped ZnO [9]. It was demonstrated that the surface roughness could be reduced to 6% from the total film thickness by control of the plasma power, but it cannot be reduced further, which influenced negatively the output signal. Thus, decrease of the surface roughness is necessary in order to improve the contact parameters of the structure and to achieve stronger and smoother piezoelectric signal. Additionally, the electrode effect on the energy generation from flexible harvesting element with piezoelectric oxide-polymer composite has not been studied.

In this paper, novel composite between ZnO doped by Gallium (GZO) and piezoelectric polymer polyvinylidene fluoride (PVDF) has been prepared and deposited on a flexible substrate. Their piezoelectric response has been studied at different metal electrodes. Effect of the PVDF insertion on the piezoelectric oxide film uniformity and interface capacitances has been investigated. The voltage-mass load plots have been analyzed for all cases. Conclusions about the performance of the flexible oxide-polymeric composites have been made in terms of energy harvesting applications.

II. Experimental Section

Polyethylene naphthalate (PEN) substrates were rinsed in isopropyl alcohol and dried by pressured air. In order to compare the electrode effect on the output voltage and piezoelectric element's overall performance, two metal electrodes (aluminum and gold) were deposited on both sides of the film. The electrode size effect on the electrical output of the samples was not investigated, so all electrode areas are the same – 6 cm^2. Bottom and top gold (Au) electrodes with thickness of 90 nm were deposited by vacuum DC sputtering at plasma current 30 mA and sputtering argon pressure 1.10^{-2} Torr. Bottom and top aluminum electrodes with thickness of 200 nm were deposited by vacuum thermal evaporation at basic pressure of 10^{-5} Torr and deposition rate of ~20 nm/s. The bottom gold electrodes were patterned by conventional photolithography with etching solution of potassium iodide. The bottom aluminum electrodes were patterned by inverse photolithography (lift-off) with sacrificial layer of photoresist. Top electrodes (Au and Al) were patterned by shadow mask. In this way four identical active areas are formed on a single substrate in one technological cycle. Ga-doped ZnO (GZO) film with thickness of ~90 nm was RF sputtered at plasma voltage of 1.1 kV and argon pressure of $2.5.10^{-2}$ Torr. Polyvinylidene fluoride (PVDF) solution was pulverized with setup for fine aerosol formation at substrate temperature of 80°C. Surface roughness of the samples before and after PVDF deposition onto GZO film was monitored by atomic force microscope (AFM) MFP-3D, Asylum Research, Oxford Instruments. Cantilever type samples were prepared and mounted on laboratory-made stand for piezoelectric voltage measurements. In this way, enough length is provided for single point fixing and free standing opposite end of the sample. The lab-made shaker driven by sine-wave generator excites the base clamp, transferring in this way the vibrations to the bonded sample (fig. 1a).

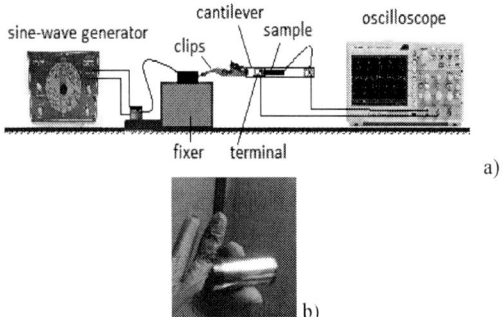

Fig. 1. a) Principle of piezoelectric voltage generation; b) photo of the fabricated flexible piezoelectric sample (energy harvester).

The frequency of the generated vibrations was 50 Hz and the maximal equivalent mass weight applied on the piezoelectric cantilever sample was 40 g. Oscillograms of the piezoelectric voltage were recorded by Tektronix TDS 1012B oscilloscope with activated noise rejection filter function. The samples' capacitances were measured by LCR-819 Gwinstek. Photo of a prepared sample is shown in Fig. 1b.

III. Results and Discussion

AFM scanning was conducted for GZO coating with thickness of 90 nm, obtained at previously reported optimized sputtering conditions [9]. Fig. 2a shows the 3D surface topography of this film, which exhibits relatively high RMS roughness of 5.2 nm for scanned area 40 x 40 μm. This is 5.8% from the total film thickness at maximum allowable 5% [10]. For comparison, AFM 3D surface topography of ~836 nm composite coating GZO+PVDF was also presented. RMS roughness is lower than the maximum allowable - 36 nm, which is 4.4% from the total film thickness (Fig. 2b). This analysis shows that application of polymeric coating by solution process could make the surface smoother, decreasing its roughness with 1.4 % in comparison with the case of polymer-free oxide film. This is a precondition for uniform contact with the electrode, which could make the device electrically reliable and generating noise-free signals. In addition, the PVDF contributes with its piezoelectric charge, which gains the piezoelectric voltage.

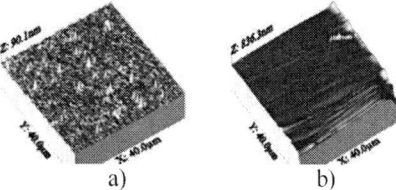

Fig. 2. AFM 3D surface topography of a) GZO coating and b) composite coating GZO+PVDF.

Effect of the metal electrode on the piezoelectric structure performance was explored for single layer energy harvesting elements with symmetrical configuration of the top and bottom electrodes from gold and aluminum. Fig. 3 shows the voltage-mass weight plot of the both type of samples (GZO and GZO+PVDF) for the cases of Au and Al contacts. As can be seen, the composition GZO-PVDF exhibited twice stronger piezoelectric response in comparison with pure GZO. Thus, the output voltage for GZO/Au based structure was 256 mV versus 414.6 mV for GZO-PVDF/Al. Gold electrode is favorable for the piezoelectric charge collection at the composite film, gaining the piezoelectric voltage with 41.5 % to 586.7 mV in comparison with the structure, using aluminum electrode. This behavior could be ascribed to the low temperature deposition process of the gold electrode, which doesn't affect thermally the interface zone at the piezoelectric film [11]. In the same time, the hot Al particles produced during evaporation could partially damage the temperature sensitive piezoelectric polymer constituent at the surface, forming the interface zone with unspecified compositional and electrical properties. In addition, the work function of the gold is supposed to be energetically suitable for the piezoelectric composite due to similar values of the work function and piezoelectric material band energy [12], suggesting low height of the interfacial energy barriers. It is believed that the work function of the metal electrodes also play a role for the energy losses together with the surface state of the piezoelectric film. Typical values of the work function for gold and aluminum are 5.1 eV and 4.06 eV [13]. Based on the electrical measurements, it is evident that the generated output voltage is lower for the metal with the lower work function. This assumption, as well as the data from the AFM, were well supported by the contact capacitances measurement, presented in Table 1.

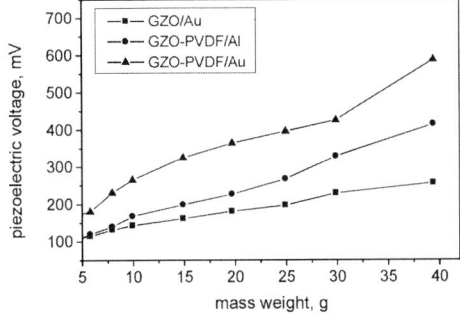

Fig. 3. Comparison of the piezoelectric voltage at different mechanical loads for different sample compositions with aluminum and gold electrodes.

TABLE I
INTERFACE CAPACITANCE OF THE SAMPLES

f, Hz, $U_{bias}=1V$	C_{GZO}, F	C_{Al}, F	C_{Au}, F
30	$13,57.10^{-6}$	12.10^{-9}	$5,5.10^{-9}$
50	$9,08.10^{-6}$	$2,6.10^{-9}$	$1,4.10^{-9}$
1000	$1,78.10^{-6}$	$1,3.10^{-9}$	$0,55.10^{-9}$

The selected low frequencies are often met at the harvesting elements activated from biological processes and 1 kHz is standard frequency. As is shown, the junction capacitance is the highest for the pure GZO, having also the highest roughness of 5.8% from the total film thickness. The interface capacitance significantly decreased for composite film GZO + PVDF and aluminum electrode with three orders of magnitude due to the smoother surface, as compared with the GZO. The interface capacitance further decreased (almost twice) due to replacement of the aluminum with gold. The results for the C-f relations are in good agreement with the reported in the literature [14]. It can be noted that the rate of increase of the voltage output is different in the different ranges of the mass-load conditions – the curve is non-linear up to 15 g, corresponding to charges activation losses, following by broad range of loads (up to 30g) with linear response and

ending with small deviations from the linearity, because of charges accumulation at the interface, affecting the internal electric field distribution [15].

Because the composite samples with gold electrodes exhibited superior piezoelectric behavior over those with aluminum contacts, they were selected as demonstrators of the signal generating abilities. Oscillograms of GZO (Fig.4a) and GZO+PVDF composite films (Fig. 4b) sandwiched between gold electrodes at vibration load of 30 g were measured at frequency of 50 Hz to show the shape of the generated signals.

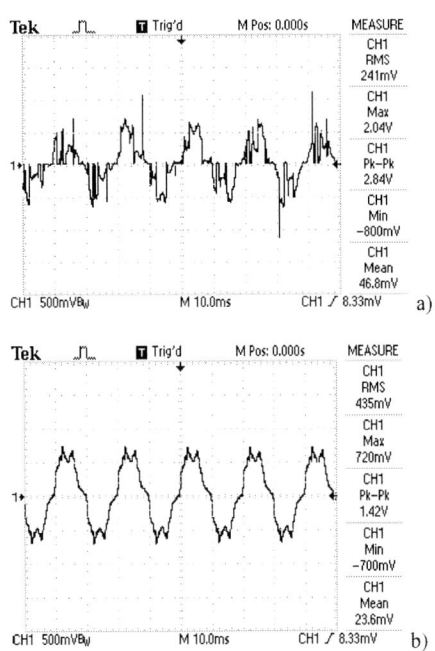

Fig. 4 Energy harvesting response at low frequency and low mechanical loading of samples with gold electrodes and piezoelectric a) GZO film; b) GZO+PVDF film.

It can be seen that the former sample generates piezoelectric voltage with effective value of 241 mV with significant noise level and harmonic distortions. The smoother and thicker piezoelectric composite film provides stable contact conditions, resulting in greater generated voltage (rms value of 435 mV) with smaller non-linear distortions of the signal.

The interface capacitance affects the rise and fall times of the generated piezoelectric voltage when sharp, square-shaped signal is applied as a poling voltage (fig. 5). The structure is polarized with time delay dependent on the interface states and contact quality between the piezoelectric film and the metal electrode. Although this characteristic is more important for the force sensor applications rather than for energy harvesting this time constant is very informative and useful for determination of the parasitic contact resistance, where significant part of the electrical energy is usually lost in form of voltage drop.

978-1-7281-3420-8/19 $31.00 © 2019 IEEE

It could be seen that for the film exhibiting higher roughness (GZO/Au) this time constant at input signal with frequency of 50 Hz is approximately 10 ms (fig. 5a).

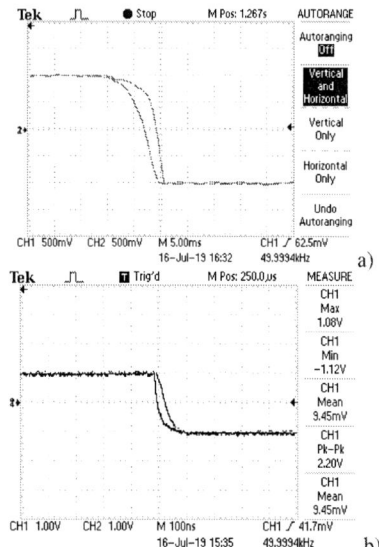

Fig. 5. Piezoelectric response at squared-wave signal and time delay of the piezoelectric voltages, according to the used piezoelectric film a) GZO/Au; b) GZO+PVDF/Au.

Taking into account that this time is a product of the contact resistance and interfacial capacitance, and knowing this capacitance from Table 1 ($9,08.10^{-6}$ F), it was calculated that the contact resistance is 1.1 kΩ. For comparison, the time constant at the structure with the smoother composite GZO+PVDF film is 100 ns. Considering capacitance of $1,4.10^{-9}$ F, for the contact resistance was obtained 15 times smaller value – namely ~71.4 Ω. This value suggests energetically favorable conditions for charges collection and it is the reason for the highest measured piezoelectric voltage, referring the voltage – mass weight plot.

IV. CONCLUSIONS

In this research, the feasibility of oxide-polymer composite on flexible substrate with two different metal electrodes for energy harvesting applications is addressed. It was observed that the output voltage is strongly dependent on the electrode work function and surface roughness of the piezoelectric film. The voltage outputs variation for the both elements with gold and aluminum electrodes were investigated at different input mass loads. The voltage output follows near to linear dependence, which extends the possibility of the elements to act as a pressure or force sensor. Future work will be related to increase of the thin films stability against stronger mass load without cracking after greater number of bending cycles (>100) and optimization of the ratio GZO:PVDF.

ACKNOWLEDGEMENT

This work was financially supported by the Bulgarian National Science Fund, grant No DH 07/13.

REFERENCES

[1] G. Hwang, M. Byun, C. Jeong, and K. Lee, "Flexible piezo-electric thin-film energy harvesters and nanosensors for bio-medical applications", *Adv Healthc. Mater.* 2015, vol. 4, pp. 646-658.

[2] C. Zhang et.al., "Fully Rollable Lead-Free Poly (vinylidenefluoride)-Niobate-Based Nanogenerator with Ultra-Flexible Nano-Network Electrodes", *ACS Nano*, 2018 vol. 12, pp. 4803-4811.

[3] L. Luo, P. Du, W. Li, W. Tao, and H. Chen, "Effects of Er doping site and concentration on piezoelectric, ferroelectric, and optical properties of ferroelectric $Na_{0.5}Bi_{0.5}TiO_3$", *J. Appl. Phys.*, 2013, vol. 114, 124104.

[4] C. Tănăsoiu, E. Dimitriu, and C. Miclea, Effect of Nb, Li doping on structure and piezoelectric properties of PZT type ceramics, *J. Europ. Ceram. Soc.*, 1999, vol. 19, pp. 1187-90.

[5] Sh. B. Seshadri et.al., "Unexpectedly high piezoelectricity of Sm-doped lead zirconate titanate in the Curie point region", *Scientific Reports*, 2018, vol.8, Article number: 4120.

[6] Y. Badali, S. Koçyiğit, A. Aytimur, Ş. Altındal, and İ. Uslu, "Synthesis of boron and rare earth stabilized graphene doped polyvinylidene fluoride (PVDF) nanocomposite piezoelectric materials", *Polymer Composites*, 2019, in press.

[7] S. Lee, B. Yeom, Y. Kim, and J. Cho, "Layer-by-layer assembly for ultrathin energy-harvesting films: Piezoelectric and triboelectric nanocomposite films", *Nano Energy*, 2019, vol. 56, pp. 1-15.

[8] S. Park, Y. Kim, H. Jung, J.Y. Park, N. Lee, and Y. Seo, "Energy harvesting efficiency of piezoelectric polymer film with graphene and metal electrodes", *Sci. Rep.*, 2017, vol. 7, 17290.

[9] M. Aleksandrova, T. Tsanev, G. Dobrikov, G. Kolev, M. Sophocleous, J. Georgiou, and K. Denishev, "Sputtering of Ga-doped ZnO nanocoatings on silicon for piezoelectric transducers", *8th International Scientific Conference "Techsys 2019"*, Plovdiv, Bulgaria, 2019.

[10] D. Lehmann, F. Seidel, and D. R. T. Zahn, "Thin films with high surface roughness: thickness and dielectric function analysis using spectroscopic ellipsometry", *Springerplus*, 2014, vol. 3, pp. 82-89.

[11] A. Koochekzadeh, E. Alamdari, A. Barzegar, and A. Salardini, "Thermal Effects of Platinum Bottom Electrodes on PZT Sputtered Thin Films Used in MEMS Devices", *Key Eng. Mat.*, 2010, vol. 437, pp. 598-602.

[12] A. Varpula, S. Laakso, T. Havia, J. Kyynäräinen, and M. Prunnila, "Contacting mode operation of work function energy harvester", *J. Phys.:Conf. Ser.*, 2014, vol. 557, 012010.

[13] Klaus D. Sattler, *Fundamentals of Picoscience*, New York, CRC Press, 2013.

[14] Z. Abas, H. Kim, L. Zhai, J.Kim, and J. Kim, "Electrode effects of a cellulose-based electro-active paper energy harvester", *Smart Mater. Struct.* 2014, vol. 23, 074003.

[15] R. Pranchov, *Materials Science*, Sofia, TU-Sofia Publishing House, 2011.

Pt Resistive Film Sensors

Z. Stanimirović and I. Stanimirović

Abstract – Temperature sensors are being widely used in various aspects of our lives. One of the most important areas of their application is environmental monitoring. In air velocity measurements resistive temperature detectors are often used. For this reason, as an initial step in development of novel thermo-anemometer that will allow 3D wind speed measurements, a series of Pt resistive film sensors that combine custom made thin-film material and thick-film technology was realized. In this paper these sensors will be presented from design and manufacturing stage to evaluation of sensor performances.

I. INTRODUCTION

Temperature is one of the mostly measured environmental parameters. Researchers are continuously making efforts to improve ways of temperature sensing and consequentially temperature sensors are present everywhere. They are present in our homes, workplaces, different means of transportation, wide variety of electrical devices. They are being used in various applications that include areas such as industry, medicine, biology, agriculture etc. There are several different contact and non-contact types of temperature sensors available on the market and the most commonly used ones are thermistors, thermocouples, and resistance temperature detectors. Resistive temperature detectors or RTDs are electrical resistance temperature sensors formed using films like platinum. They are probably the most popular type of temperature sensors [1-4]. They exhibit more linear and more accurate behavior over wide temperature ranges than widely used thermocouples. Platinum RTDs come in a number of construction forms and are often used because of their properties such as accuracy and stability. In environmental monitoring they are known for their usage in air velocity measurements. Thermo-anemometry is based on two RTDs that are being installed at the end of the probe - one measuring the ambient temperature as a reference and the second continuously maintaining a constant temperature above the ambient one. The current required to maintain the second RTD's temperature is related to the air velocity and, based on the current usage, the air velocity is being calculated. Having in mind that thermo-anemometers have the lack of moving parts they are being replaced by expensive complex ultrasonic anemometers that measure wind speed in three directions. As an initial step in development of novel thermo-anemometer that will allow 3D wind speed

Z. Stanimirović, and I. Stanimirović are with Institute for Telecommunications and Electronics IRITEL a.d. Belgrade, Batajnički put 23, 11080 Belgrade, Republic of Serbia, E-mail: zdravkos@iritel.com

measurements, a series of custom designed RTDs have been realized. These Pt resistive film sensors combine custom made thin-film material and thick-film technology. In this paper we would like to present results obtained during all steps of Pt resistive film sensor development – from design and manufacturing stage to evaluation of sensor performances.

II. TECHNOLOGICAL PROCESS

A number of Pt resistive film sensors with different resistances were formed (Fig. 1) using a combination of custom made thin-film Pt composition, commercially available ceramic substrates and thick-film compositions and standard thick-film processes. 96% alumina ceramic substrates have been used because of the fact that they do not react chemically with Pt composition regardless of the temperature. Also, thermal expansion coefficients of alumina and Pt thin-film composition are similar and therefore stress and strain between Pt film and substrate are minimal. Resistors of several different forms were realized on $50.8\times50.8\times0.65$ mm^3 Kyocera alumina substrates. Resistor forms and dimensions can be seen in Fig. 2.

Fig. 1. Pt resistive film sensors.

Standard thick-film process was performed using 200 mash screens with 10-12 μm thick emulsion. Thick-film Pd/Ag conductive composition from DuPont QM series designated QM21 was screen printed, leveled at a room temperature (21°C) for 15 min, dried in an infra-red conveyor dryer in 10 min cycle at 150 °C and fired in a conveyor belt furnace in 30 cycle with peak temperature of 850 °C held for 10 min. Then, custom made Pt thin-film composition with sheet resistivity of 0,3 Ω/sq was deposited using thick-film screen printing process. Special care was taken of keeping Pt composition free of contamination in order to provide stable resistor performances. Pt composition was printed in twelve different forms, four of which were serpentines that will be evaluated in this paper. Four layers of the composition were screen printed and each layer was leveled and dried and all four dry layers were simultaneously fired. Fired Pt thin-film resistive sensors were less than 10 μm thick. They were of relatively simple construction and the chosen production process provided high repeatability of their characteristics. Sensors presented in Fig. 1 covered surfaces of 16 mm^2, 8 mm^2, 5.32 mm^2 and 4 mm^2, respectively. All serpentine lines were 4 mm wide but their lengths varied as given in Table I.

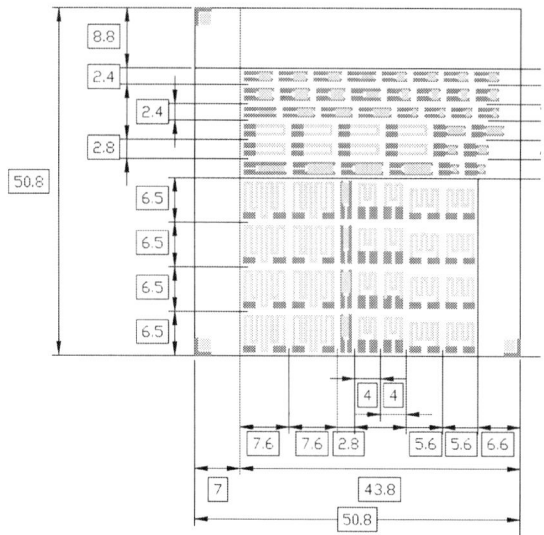

Fig. 2. Dimensions and forms of Pt resistive film sensors formed on a single alumina substrate.

III. EXPERIMENTAL RESULTS

Formed Pt film sensors were heated to temperatures ranging from 0 to 150 °C. Temperature measurements were performed using Fluke Ti95 infrared camera (Fig. 3). For Pt sensor resistance measurements, a concept of computer-based data acquisition (DAQ) and National Instruments software package LabVIEW were used. Experimental setup is shown in Fig. 3 where NI-9264 is a voltage output module, NI-9205 is a voltage input module and R_l=99,7 kΩ is a wire wound resistor.

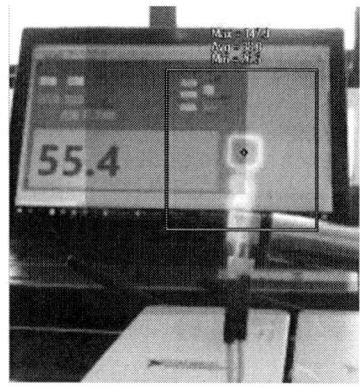

Fig. 3. Temperature measurements performed using infrared camera.

For temperatures between 0 and 850°C resistance change of P$_t$ resistive sensor can be given as:

$$R_t = R_0(1 + A_t t + B_t t^2). \tag{1}$$

However, for small temperature ranges such as 0 to 150 °C, the resistance change can be considered linear and can be described as:

$$R_t = R_0(1 + \alpha t), \tag{2}$$

where R_0 is resistance of the resistor at 0°C, t is temperature given in °C and α is resistance's temperature coefficient of the resistive material used.

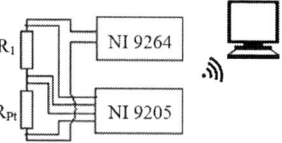

Fig. 4. Experimental setup (R_{Pt} – Pt resistive sensor test sample, R_l – wire wound resistor).

Special caution was exercised during sensor excitation because the current through the Pt film sensor causes self-heating that changes the temperature of the sensor and appears as a measurement error. The resulting temperature measuring error can be given as:

$$\Delta T = PS \tag{3}$$

where P is the power loss and S the self-heating coefficient given in K/mW. The self-heating coefficients depend on the particular composition used. Because self-heating is dependent on the thermal contact between the Pt sensor and the surrounding medium, efficient heat transfer to the environment allows higher measured currents to be used. Standard recommended current for Pt resistors with resistances ≤100 Ω is 1 mA. Therefore, special attention was paid to the signal conditioning circuitry so that self-heating was negligible. For custom made Pt thin-film composition used in the experiment, minimum currents applied to the sensor ranged from 0.3 mA to 0.5 mA and corresponding power dissipations were between 2.5 µW to 5 µW.

Having in mind the most common disadvantages of Pt resistive sensors such as relatively low sensitivity, low resistance, effect of self-heating and non-linear characteristics, a number of measurements was performed. Characteristic values for each serpentine form are given in Table I.

Fig. 5 shows resistance vs. temperature characteristics for four Pt resistive film sensors whose parameters are given in Table I. Measured resistance values are presented with symbols, while lines depict designed resistance values. The resistance proved to be almost linear with temperature. The higher the temperature was, the larger was the resistance. As the consequence of the technological process designed and achieved resistance values differed. Higher resistances were in better agreement with designed values than lower resistances due to serpentine line width and resistive film thickness consistency. However, values of temperature coefficient of resistance were in accordance with expected values. The values of the temperature coefficient of resistance, α, fell between 0.00453 ppm/°C and 0.00564 ppm/°C. The largest values of α were found for sensors with longest serpentines.

TABLE I
PARAMETERS FOR Pt FILM SENSORS (l-SERPENTINE LENGTH, Rd − DESIGNED RESISTANCE VALUE, Rm − MEASURED RESISTANCE VALUE, α − TEMPERATURE COEFFICIENT OF RESISTANCE)

	l (mm)	$R_{d\ at\ 0°C}$ (Ω)	$R_{m\ at\ 0°C}$ (Ω)	α (ppm/°C)	Sensitivity (Ω/°C)
R_1	10	9	9.34	0.00453	0.04
R_2	13.3	11.2	11.57	0.00359	0.04
R_3	20	15	14.9	0.00504	0.08
R_4	40	30	30.29	0.00564	0.17

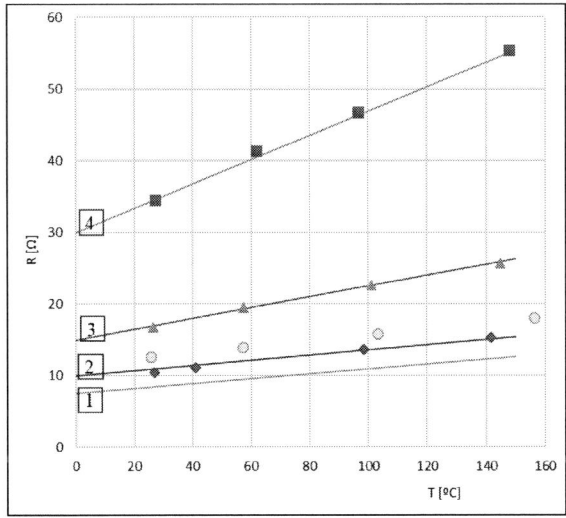

Fig. 5. Resistance vs. temperature characteristics for four Pt resistive film sensors with different geometries (Measured values: ◊ - R₁, ○ - R₂, ∧ - R₃, ⊓ - R₄; designed values: 1 - R₁, 2 - R₂, 3 - R₃, 4 - R₄).

When thermal sensing is in question, the amount to which the resistance changes as temperature changes is an important parameter known as sensitivity. It depends on resistance as well as on temperature coefficient of resistance. Referring the Table I, it can be seen that higher resistance will increase sensitivity. Increase in sensitivity with length of the resistor for resistors presented in Table I is shown in Fig. 6.

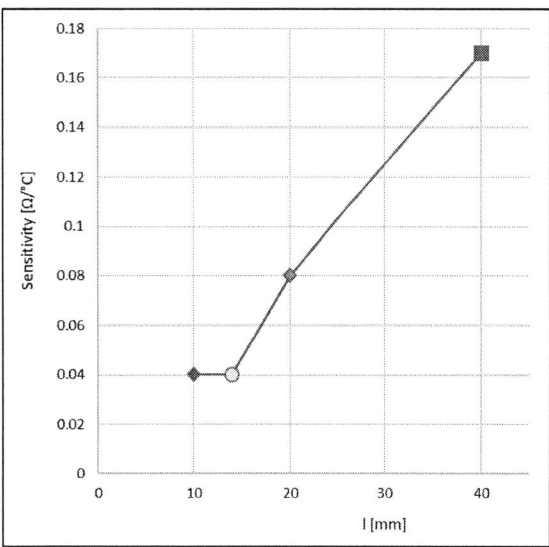

Fig. 6. Sensitivity vs. resistor length for four Pt resistive film sensors with different geometries (◊ - R₁, ○ - R₂, ∆ - R₃, □ - R₄).

IV. CONCLUSION

Presented Pt resistive film sensors with their simple construction, good repeatability of characteristics and adequate range of temperatures will be used in thermo-anemometry. Their advantages also include chemical stability, relative ease of manufacture, excellent reproducibility of their electrical characteristics and interchangeability of sensing resistors. The main characteristics of realized Pt film sensors are presented in table II.

TABLE II
MAIN CHARACTERISTICS OF REALIZED Pt FILM SENSORS

Substrate material	96% alumina ceramics
Conductive material	Commercially available Pd/Ag conductive composition
Resistive material	Custom made Pt thin-film composition
Sheet resistance	0,3 Ω/sq
Technology	Thick-film
Thickness of the fired Pt resistive film	< 10 μm
Type of sensors	Serpentine resistors
Resistor lengths	10 mm - 40 mm
Designed resistor values	9 Ω - 30 Ω at 0 °C
Measured resistance values	linear with temperature
Temperature coefficient of resistance	0.00453 ppm/°C - 0.00564 ppm/°C
Sensitivity	0.04 Ω/°C - 0.17 Ω/°C

ACKNOWLEDGEMENT

This research was partially funded by Ministry for Education, Science and Technological Development of the Republic of Serbia under contracts III44003 and III45007.

REFERENCES

[1] B.P. Nagaraju and K.J. Rathanraj, "Precision Temperature Measurement Using Resistance Temperature Detector", *Asian Review of Mechanical Engineering*, 2013, ISSN 2249 – 6289, Vol. 2 No. 1, pp.1-6.

[2] F. Mailly, A. Giani, R. Bonnot, P. Temple-Boyer, F. Pascal-Delannoy, A. Foucaran, A. Boyer, "Anemometer with hot platinum thin film", *Sensors and Actuators A: Physical*, 2001,Vol. 94, Iss. 1–2, pp. 32-38.

[3] K.G. Kreider, D.C Ripple and W.A Kimes, "Thin-Film Resistance Thermometers on Silicon Wafers", *Measurement Science and Technology*, 2009, Volume 20, Number 4.

[4] N. Hambali, S. Saat, M.A. Ahmad, M.S. Ramli, M.A. Ishak, "Computer-based System for Calibration of Temperature Transmitter using RTD", *3rd International Conference on Information Management, Innovation Management and Industrial Engineering, 2010*, pp. 332-336.

High-Voltage Pulse Trimming of Thick-Film Resistors – Some Modelling Aspects

I. Stanimirović and Z. Stanimirović

Abstract – Novel applications of thick-film technology require precision trimming of thick-film resistors that are being used as both resistive and sensing elements. High-voltage pulse trimming provides high precision trimming leaving the resistor body undamaged and resistor geometry intact. This paper introduces an adapted statistical bimodal bond conductance approach to analysis of high-voltage pulse trimmed thick-film resistors. Application of high-voltage pulses causes transitions from contact conductance to tunneling conductance and vice versa in resistors whose disordered structure is being presented using 3D planar random resistor network.

I. INTRODUCTION

Thick-film technology that has been in use for past few decades is experiencing a renascence due to various novel implementations. They include ceramic and hybrid micro-electro-mechanical systems as well as various sensing applications. This comeback can be attributed to the thick-film technology's ability to integrate sensor and actuator elements as well as signal processing electronic circuitry on the same substrate. Thick-film resistors are being used as both sensing and resistive elements and, due to new demands, they are facing stricter requirements that include significant reduction in dimensions, tighter resistance tolerances and increasing need for buried thick resistive films. Therefore, standard processes of adjusting resistance values of thick-film resistors such as air-abrasive and laser trimming had to be replaced by less invasive process - high-voltage pulse trimming method.

Several papers have been published dealing with different aspects of this trimming method that include electrical stability characterization, investigation of primary parameters and structural properties of thick resistive films along with change in low-frequency noise characteristics of these resistors [1-5]. In [6], theoretical statistical description of voltage discharge effects on disordered composites was presented based on random resistor network model. Since the resistance change and related physical mechanisms due to high-voltage pulse trimming strongly depend on the nature of resistors, this paper introduces standard percolation approach presented in [6] adapted to high-voltage pulse trimming of thick resistive films.

I. Stanimirović, and Z. Stanimirović are with Institute for Telecommunications and Electronics IRITEL a.d. Belgrade, Batajnički put 23, 11080 Belgrade, Republic of Serbia. E-mail: zdravkos@iritel.com

II. BIMODAL BOND CONDUCTANCE DISTRIBUTION

In [6] a generalized statistical description of high-voltage pulse trimming process was presented. It was based on a random resistor network model that introduced 3-dimesional cubic lattice whose bonds had random conductance values g. Bimodal distribution of bond conductances was given as [6]:

$$\rho_0(g) = p_0 \delta(g - g_1) + (1 - p_0)\delta(g - g_2), \quad (1)$$

where $g=g_1$ conductance had a fraction $0 \geq p_0 \geq 1$ of bonds, and $g=g_2$ conductance had a remaining fraction $1-p_0$. Under the influence of high-voltage trimming process, the bond conductances change their values. According to [6] bond with conductance g_1 changes into g_2 with probability W_{12}. Also, a g_2 bond becomes g_1 bond with probability W_{21} where probabilities W_{12} and W_{21} depend on the microstructure of the composition. After n subsequently applied high-voltage pulses resistor conduction becomes [6]:

$$G_n = \frac{(2p_n - 1)(g_1 - g_2)}{2} + \\ + \sqrt{\frac{(2p_n-1)^2(g_1-g_2)^2}{4}} + g_1 g_2 \,, \quad (2)$$

where:

$$p_n = p_0(1 - W_{12} - W_{21})^n + \\ + \frac{W_{21}[1-(1-W_{12}-W_{21})^n]}{W_{12}+W_{21}}. \quad (3)$$

One of the most important results presented in [6] is that asymptotic value of fraction of bonds with conductance g_1 designated as p:

$$p = \lim_{n \to \infty} p_n = \frac{1}{1 + \frac{W_{12}}{W_{21}}} \quad (4)$$

depends on W_{12}/W_{21} ratio and can be greater or smaller than the initial value p_0. Also, model presented in [6] gives the condition under which the final conductance may increase or decrease depending on probabilities of transitions from conductance g_1 to conductance g_2 and conductance g_2 to conductance g_1. For $g_1 > g_2$:

$$p > p_0 \to G_n > G_0, \\ p < p_0 \to G_n < G_0. \quad (5)$$

The model would be more realistic if the fact that local resistance depends on the state of the resistor network would be taken into account. Microscopic processes have different causes depending on the system in question. In polysilicon, local joule heating can enhance structural changes in grain boundaries. Electromigration or breakthrough are also some of microscopic processes that can be introduced into the trimming process analysis. All of these occurrences depend on the local characteristics of the resistor network.

III. ADAPTED MODEL

Value of the parameter p that determines what fraction of contacts has certain conductance value and conductance values g_1 and g_2 depend on a model that is being used to describe the disordered structure. This allows multiple approaches to interpretation of obtained trimming results.

Probability of the conductance transition processes depends on the nature of the process in question and according to [6] it is based on boundary processes. The nature of the boundary depends on the very nature of the process in question and local parameters that stimulate the process such as temperature, applied electrical field, etc.

Instead of the bimodal bond conductance distribution where during high-voltage pulse trimming conductance transitions g_1 to g_2 and g_2 to g_1 occur, in thick resistive films multimodal conductance can be introduced. In case of boundary circumstances, it can be assumed that transitions g_{12} to g_2 as well as g_{21} to g_1 may occur during the trimming process.

IV. DISCUSSION

If we present disordered structure of thick resistive film as 3D planar random resistor network then it can be assumed that charge transport takes place via chains of conducting particles. In one chain neighboring conducting particles can be sintered or separated by thin glass layers. Therefore, present conducting mechanisms are conducting through conducting particles and sintered contacts and tunneling through glass barriers. In that case conductances g_1 and g_2 are contact conductance g_c and tunneling conductance g_t respectively. When high-voltage trimming process of thick resistive films is in question, the boundary case is being taken into account. The trimming process induces transitions of the contact of two conduction particles to metal-insulator-metal cell and vice versa. Then the fraction of contacts between neighboring conducting particles prior to the trimming process can be given as:

$$ p_i = {N_{ci}}/{N}, \qquad (6) $$

where N is the total number of bonds in the resistive film and N_{ci} is the total number of contacts between neighboring conducting particles in the resistive film prior to high-

voltage pulse resistor trimming. If N_{bi} designates the total number of barrier resistances in the thick-film resistor prior to trimming then:

$$ 1 - p_i = {N_{bi}}/{N}. \qquad (7) $$

In that case the initial conductance of the resistor becomes:

$$ G_i = \frac{M^2}{N_{ci}\frac{\rho}{\pi a} + N_{bi}\frac{h^2 s}{q^2 A (2mq\Phi_B)^{1/2}} exp\left[\left(-\frac{32\pi^2 mqs^2\Phi_B}{h^2}\right)^{1/2}\right]} \qquad (8) $$

where M is the number of parallel conducting chains in the resistive film that take part in the conduction process, ρ is specific contact resistance, a is radius of the barrier cross section, q and m are electron charge and its effective mass respectively, h is Planck's constant, A is the tunneling area, Φ_B is the potential barrier height with respect to the Fermi energy and s is the potential barrier width.

After n subsequent high-voltage pulses applied to the resistor the fraction of contacts of neighboring conducting particles in trimmed resistive film becomes:

$$ p_{tn} = {N_{ctn}}/{N}, \qquad (9) $$

and accordingly, the total number of barrier resistances becomes:

$$ 1 - p_{tn} = {N_{btn}}/{N}. \qquad (10) $$

Then the final conductance of the thick-film resistor after the impact of n subsequent high-voltage pulses becomes:

$$ G_{tn} = \frac{M^2}{N_{ctn}\frac{\rho}{\pi a} + N_{btn}\frac{h^2 s}{q^2 A (2mq\Phi_B)^{1/2}} exp\left[\left(\frac{32\pi^2 mqs^2\Phi_B}{h^2}\right)^{1/2}\right]}. \qquad (11) $$

Fractions of contacts between neighboring conducting particles p_{tn} for low-ohmic and high-ohmic thick film resistors, each trimmed by 10 sequential high-voltage pulses, were calculated. Obtained results are shown in Fig. 1 and Fig. 2. Structural parameters used in calculations were taken from [7].

Experimental results showed that resistance of the low-ohmic resistor increases with the applied trimming process. In Fig. 1 it can be seen that p_{tn} decreases with the increase in resistance. This occurance is due to the decreasing number of contacts between neigbouring conducting paricles randomly distributed throughout the resistor network meaning that dominant transition for low-ohmic resistors is $g_c \rightarrow g_t$.

When high-ohmic resistor is in question experimental results showed that resistance decreases with the applied trimming process. In Fig. 2 it can be seen that p_{tn} increases with the decrease in resistance. Increasing number of contacts causes this occurance and therefore dominant transition for high-ohmic resistors is $g_t \rightarrow g_c$.

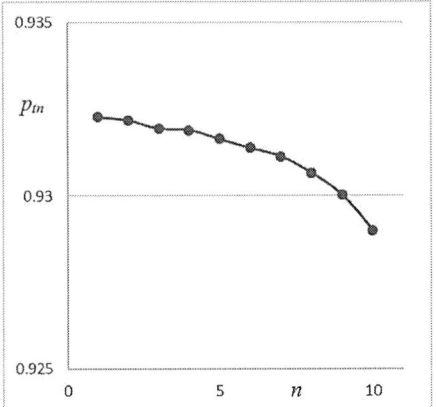

Fig. 1. Fraction of contacts between neighboring conducting particles p_{tn} for low-ohmic thick film composition (sheet resistance: 1 kΩ/sq)

Results obtained by using bimodal bond conductance distribution are not in collision with previously published experimetal investigations related to high-voltage pulse trimming of thick resistive films [3-4].

In low-ohmic resistors with sheet resistances ≤ 1 kΩ/sq dominant conducting mechanism is metallic conduction through conducting particles and direct contacts between neighboring conducting particles in conducting chains of random resistor network. The trimming process can induce changes in contact area between resistive and conductive material, cause migrations of the conductive material, cause defects due to presence or migration of defects and impurities present in thick resistive film. All these occurrences may result in resistance increase and accordingly in decrease in number of contacts between neighboring conducting particles.

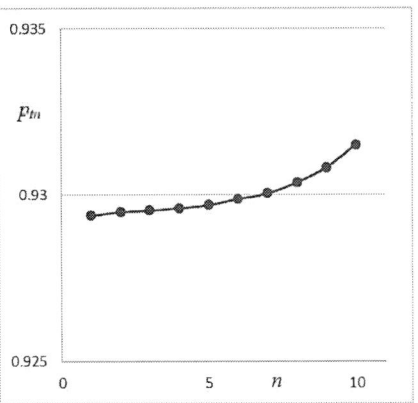

Fig. 2. Fraction of contacts between neighboring conducting particles p_{tn} for high-ohmic thick film composition (sheet resistance: 100 kΩ/sq)

In high-ohmic resistors with sheet resistances ≥ 100 kΩ/sq dominant conducting mechanism is tunneling through thin glass barriers present between the most of the neighboring conducting particles due to the high volume fraction of insulating phase in the resistor. The trimming process affects concentration of traps present in insulating barriers causing resistance decrease and increase in number of contacts between neighboring conducting particles.

Although it is a generalized approcah, bimodal bond conductance distribution and choice of values p_i, g_c and g_t that describe the disordered structure confirm validity of the interpretation of experimental results for resistance changes caused by high-voltage trimming process. However, thick resistive films are specific composite systems where multimodal conductance can be introduced instead of the bimodal bond conductance distribution where during high-voltage pulse trimming $g_c \rightarrow g_t$ and $g_t \rightarrow g_c$ conductance transitions occur. In case of boundary circumstances, transition $g_{ct} \rightarrow g_t$ may occur due to the glass intrusion between neighboring conducting particles during glass phase melting caused by the trimming process. Also, $g_{tc} \rightarrow g_c$ transition may occur due to conducting channel formation through the insulating layer caused by breakthrough and electromigration. It should be pointed out that both the bimodal and the multimodal bond conductance distribution take into account only the change of two technological parameters - numbers of contacts and barriers in conducting chains that form random resistor network subjected to high-voltage pulse trimming.

V. CONCLUSION

In this paper one particular percolation approach [6] based on bimodal bond conduction distribution adapted to thick-film resistor high-voltage pulse trimming is presented. The key contributions of this analysis are:

- In bimodal bond conductance distribution when thick resistive film is in question two conductances g_1 and g_2 are contact conductance g_c and tunneling conductance g_t.
- In low-ohmic resistors with sheet resistances ≤ 1 kΩ/sq dominant conducting mechanism is metallic conduction and the consequence of the performed trimming process is the increase in resistance. This occurance is due to the decreasing number of contacts p_{tn} between neigbouring conducting paricles randomly distributed throughout the resistor network meaning that the dominant transition for low-ohmic resistors is $g_c \rightarrow g_t$
- In high-ohmic resistors with sheet resistances ≥ 100 kΩ/sq dominant conducting mechanism is tunneling through thin glass barriers and the trimming process leads to resistance decrease and increase in number of contacts p_{tn} between neighboring conducting particles. For these reasons the dominant transition for high-ohmic resistors is $g_t \rightarrow g_c$.
- In thick resistive films multimodal conductance can be introduced instead of the bimodal bond conductance

distribution due to of boundary circumstances such as conducting channel formation through the insulating layer or glass intrusion between neighboring conducting particles.

- The bimodal and the multimodal bond conductance distribution take into account only the change of two technological parameters as the consequence of the trimming process. These parameters are numbers of contacts and barriers in conducting chains that form 3D planar random resistor network.

- Other technological parameters that can be affected by high-voltage pulse trimming but are not included in this analysis are number of traps in one conducting chain and number of conducting chains in the resistor. Both of these parameters can be altered due to the trimming process but their contribution to conducting processes is minor and does not strongly affect the validity of presented statistical approach.

ACKNOWLEDGEMENT

This research was partially funded by Ministry for Education, Science and Technological Development of the Republic of Serbia under contracts III44003 and III45007.

REFERENCES

[1] A. Dziedzic, A. Kolek, W. Ehrhardt, H. Thust, "Advanced electrical stability characterization of untrimmed and variously trimmed thick-film and LTCC resistors", *Microelectronics Reliability*, 46, 2006, pp. 352-359.

[2] B. Rambabu and Y. Srinivasa Rao, "Multiple High Voltage Pulse Stressing of Polymer Thick Film Resistors", *Active and Passive Electronic Components*, Vol. 2014, Article ID 319213, 5 pages

[3] I. Stanimirović, Z. Stanimirović, "Influence of HVP trimming on primary parameters of thick resistive films", *J Mater Sci: Mater Electron*, Vol. 28, 2017, pp.8002-8010

[4] Z. Stanimirović, Ivanka Stanimirović, "Effects of High Voltage Pulse Trimming on Structural Properties of Thick-Film Resistors", *Science of Sintering*, 49, 2017, pp. 91-98

[5] I. Stanimirović, Z. Stanimirović, "Low-Frequency Noise in Thick-Film Resistors Due to Laser and High-Voltage Pulse Trimming", *Proc. of 41st International Spring Seminar on Electronics Technology ISSE 2018*, May 16-20, 2018, Zlatibor, Serbia.

[6] C. Grimaldi, T. Maeder, P. Ryser and S. Strassler, "A random resistor network model of voltage trimming", *J.Phys.D: Appl.Phys.*, 37, 2004, pp. 2170-2174.

[7] M.M. Jevtić, Z. Stanimirović, I. Stanimirović, "Evaluation of thick-film structural parameters based on noise index measurements", *Microelectronics Reliability*, 41, 2001, pp. 59-66.

SnO$_2$-Pd as a Gate Material for the Capacitor Type Gas Sensor

N. Samotaev, K. Oblov, A. Litvinov, and M. Etrekova

Abstract - The article describes the result of the use SnO$_2$-Pd thin films as a gate for structure measured ppb range of NO$_2$ gas by the capacitive method. The technological aspects of fabrication SnO$_2$-Pd gate and one comparison by metrological parameters with the classical Pd gate field effect sensor are discussed. The use of SnO$_2$-Pd material allows improvement in sensitivity of NO$_2$ by an order of magnitude compare the classical Pd based gate field effect sensors.

I. INTRODUCTION

Using of SnO$_2$ material for fabrication field effect gas sensors based on Schottky diode effect firstly describing in work [1], somewhat later, a study of the sensitivity of field-effect sensors to NO$_2$ began [2]. Over time, structurally and technologically gas sensors based on the field effect were divided into capacitor type and transistor type sensors. Each type of sensor has its advantages and disadvantages. For the transistor type, this is a small crystal size (about 1x1 mm) and high compatibility with microelectronic technology, for a capacitor type, this is a high sensitivity and the possibility of using exotic materials and technological methods for their deposition. Advantages in one place turn into disadvantages in another. For condenser type it is a large crystal size (4x4 mm) [3] and as a result high thermal inertia. For the transistor type, insufficient sensitivity entails the use of high operating temperatures (close to 500 ° C) and SiC technology [4]. Our work is designed to improve the gas-sensitive properties of condenser-type sensors through the use of technological methods and the speed of response to gas through methodological methods (mathematical calculation and gas sampling method) and adapt it to working in explosives detection applications.

Most explosives are nitro-containing substances, trace of which possible detect only in sub-ppb concentrations by NO$_2$ conversion technique [5]. Materials based on SnO$_2$ are widely used to measure NO$_2$ by resistive type MOX sensors [6-9], but MOX sensors margin of stable sensitivity is limited by sub-ppm range [10]. The further increase in sensitivity to NO$_2$ is possible in the technological combination of well proven material used nowadays in MOX sensors and high sensitivity of field-effect sensors.

N. Samotaev, K. Oblov, A. Litvinov, M. Etrekova are with Department of Microelectronics, Institute of Nanoengineering in Electronics, Spintronics and Photonics, National Research Nuclear University MEPhI (Moscow Engineering Physics Institute), Kashirskoe highway 31, 115409 Moscow, Russian Federation, E-mail: nnsamotaev@mephi.ru. Fax: +74993242111

II. MATERIALS AND METHODS

The scheme of capacity sensor developed in described experiment are present on figure 1. That are classical metall – insulator - semiconductor (MIS structure) condencer with Pd based material with sub-layer of SnO$_2$.

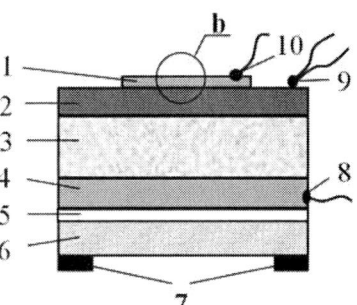

Fig. 1. Capacitor sensor: 1 - SnO$_2$/Pd gate; 2 - SiO$_2$ film; 3 - Si substrate; 4 -Al electrode; 5 - insulator; 6 - heater; 7, 8, 10 - the electric contacts; 9 - thermistor.

The MIS structures presented in this work were fabricated using the method of pulsed laser deposition of thin films. With this method of evaporation, the vaporized substance consists of neutral atoms and ions, and their energies can be tens and hundreds of electron volts [11]. Such high energy values allow atoms and ions to penetrate deep enough into the substrate. Therefore, in this case, the metal-dielectric interface is a layer with a variable stoichiometric composition — from the pure material of the metal gate to the pure material of the dielectric. Morphological studies of the palladium film showed that the film deposited on the dielectric layer using the method of pulsed laser deposition consists of nanocrystals with sizes from 10 to 100 nm and contains numerous pores of different sizes. Therefore, when diffused into the film, gas molecules easily reach the region of the metal-dielectric interface, bypassing diffusion through the bulk of palladium crystallites. This statement can be confirmed by SEM photo of Pd-SnO$_2$ gate present on figure 2.

A. MIS structures fabrication technology

In [12] was shown that the characteristics of field-effect sensors strongly depend on the composition metal-dielectric transition layer, which is determined by the materials of metal gate and insulator dielectric and methods of their deposition. For a new capacity type sensor manufactured n-type silicon substrate with a 0.1 μm SiO_2 layer thickness was used. Layer of SnO_2 is additionally deposited through the shadow mask on the SiO_2 film by the method of magnetron sputtering. This method of deposition allows you to form films with a high effective surface area. The Pd film with a thickness of 100 nm was coated SnO_2 film through the shadow mask by pulse laser deposition method.

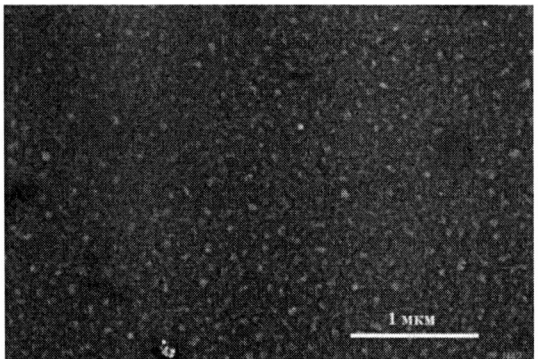

Fig. 2. SEM picture of the catalytic Pd gate sediment by pulsed laser deposition technique on SnO_2 material.

The Pd-SnO_2-SiO_2-Si structure was basis of capacitor type gas sensor which photo present on Fig. 3 and cross-section scheme of structure on Fig. 1, respectively. In parallel with the SnO_2-Pd based sensor, a similar series of capacity type sensors without SnO_2 sub-layer, only with a 100 nm Pd gate, were made in order to compare changes of gas-sensitive characteristics with Pd-SnO_2-SiO_2-Si structure. For MIS structure Pd-SiO_2-Si were used the identical geometrical dimensions, thickness of materials and substrate.

Fig. 3. Capacitor sensor's photo assembled with film resistive heater and thermistor. Package type is TO-8 (11 mm in diameter).

The optimal geometrical dimensions of the fabricated MIS structures and thickness of the Pd gate material were selected on the basis of the preliminary experiments described in the work [13], where the concentration of NO_2 gas was studied during experiments with conversion of TNT explosives. Indeed, according to [14], the vapors pressure of TNT at room temperature does not exceed 5...7 ppb (60 μg/m³, or 60 pg of 1 cm³) and need to be orientated on 1...10 ppb NO_2 range detection.

B. Testing gas sensitive properties of fabricated MIS structures

During the study, the dynamic characteristics of the sensor samples were measured when a step pulse of NO_2 concentration was applied. According to the sensitivity mechanism model [9], the response time of the MIS sensor is largely determined by the diffusion rate of gas molecules through the pores of the metal gate and in the transition layer. The diffusion coefficient increases with increasing temperature. Therefore, increasing the temperature of the sensor reduces its response time to chlorine concentration. But with increasing temperature, the sensitivity of the sensor will decrease due to the degeneration of its Volt-Farad (C-V) characteristics. Fig. 4 shows the C-V characteristics at different temperatures. It can be seen that as the temperature increases, the slope of the linear part of the C-V characteristic decreases significantly, which leads to a decrease in the sensitivity of the sensor to zero at high temperatures. This contradiction can be resolved with the help of pulsed heating of the sensor discussion of whom will be presented below.

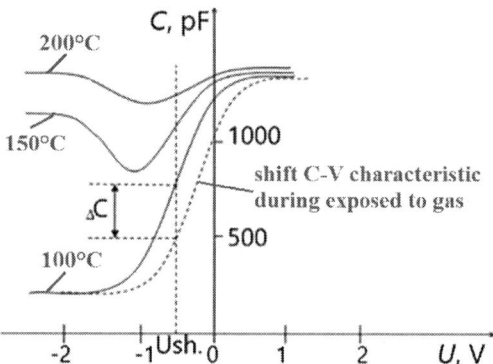

Fig. 4. Scheme of capacitance-voltage characteristic for Pd gate sensor at different heating temperatures and shift C-V characteristic during exposed to gas.

The capacitance-voltage characteristics for SnO_2-Pd and Pd base sensors at temperatures of 170°C, 140°C and 100°C were studied and presented on Fig. 5. Also sensitivity to NO_2 at different temperatures presented on Fig. 6 and Table. 1. On Fig. 5 and Fig. 6 it can be seen that the slope of the characteristic decreases with increasing temperature, therefore, the sensitivity of the sensor should decrease. Explanation of effect is given on Fig. 4. The Fig. 6 shows the time responding capacity sensor with SnO_2-Pd gate to 108 ppb NO_2 at different temperature and it is possible to see decreasing response time and sensitivity during increase of the working temperature of sensor. This behavior is standard for the field effect sensor.

978-1-7281-3420-8/19 $31.00 © 2019 IEEE   154

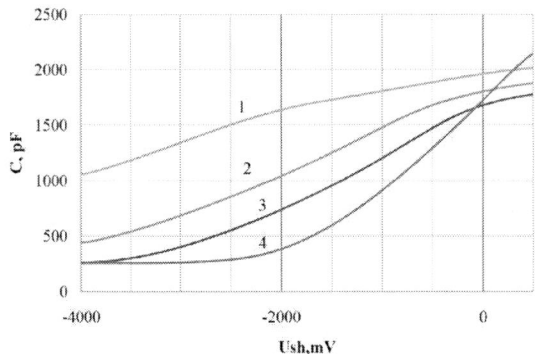

Fig. 5. The capacitance-voltage characteristics for capacity sensor at temperatures: 1 - 170°C, 2 - 140°C, 3 - 100°C, 4 - pure Pd gate at 100°C.

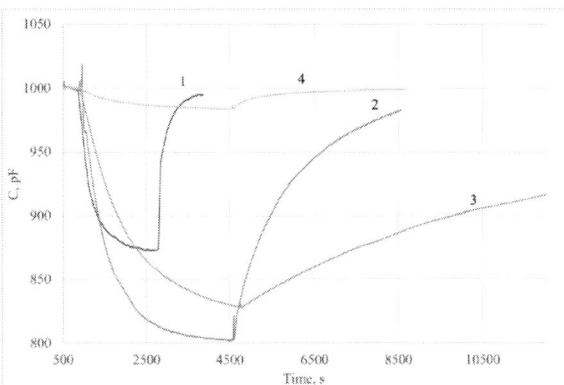

Fig. 6. Responses of the capacitor sensor to 108 ppb NO_2 concentration at temperatures: 1 - 170°C, 2 - 140°C, 3 - 100°C, 4 - pure Pd gate at 100°C.

TABLE I
THE RESPONSE OF CAPACITIVE SENSORS TO 108 PPB NO_2
CONCENTRATION AT DIFFERENT TEMPERATURES.

T, °C	Gate structure	
	Pd-SnO$_2$	Pd
	Response, pF	
100	172	20
140	195	-
170	130	-

The Fig. 6 shows the time response of MIS stucture with gate SnO$_2$-Pd to 108 ppb NO_2 (curve number 2). For complete of response to gas step pulse need time about one hour. That time are not enough for applications for the detection of explosives, where as a rule no more than 5 s are required for analysis [5].

C. Methods of improving selectivity and response time MIS structures

A significant improvement in speed of MIS sensor response is possible through the use of mathematical methods for processing the signal that occurs when a gas concentration appears [15]. Since the MIS sensor signal represents charge changes in the capacitor structure caused by the diffusion of polar gases (NO_2 in our case), the shape of the response curve will always be constant. It means that is possible to judge the current gas concentration by the derivative of the MIS sensor signal (the rate of increase of the signal). Consequently, it is possible to increase the sensor response time to ≈ 10 seconds, which are necessary for recognizing a trend in a change in the rate of signal build-up.

High sensitivity to NO_2 was obtained in the laboratory. The absolute error of such measurements is small and allows you to measure the concentration of NO_2 at the level of 1 ppb. However, using a MIS sensor as a sensitive element in a gas analyzer in real (atmospheric) conditions, the measurement error may increase by an order of magnitude due to the influence of uncontrolled external conditions. Tests have shown that in real (atmospheric) conditions, using a gas analyzer with a MIS sensor, one can measure NO_2 concentrations ranging from 10 ppb. The selectivity of the gas analytical system can be improved by using a two-channel measuring system with filtration of NO_2, presented in [16] for concentrations of about 0.1 ppb of H_2S gas.

Combining the two methods presented above will allow to create a device for detecting NO_2 with good selectivity and speed, although it must be borne in mind that gas diffusion processes at sub-ppb concentrations tend to increase the reaction time due to gas adsorption on surfaces of gas way inside of gas-analytical systems. The expected response time of the gas analysis system is of the order of a minute, and another minute will be required to clean the gas path in a two-channel system with NO_2 gas filtration.

In any case, gas MIS sensors have advantages, since their response is simple physical adsorption, which does not produce chemical conversion initial gas atmosphere to other chemical compounds, for example, as it happens in semiconductor MOX sensors. It means there are no errors and no disturbance is introduced into the analyzed atmosphere and it is possible to change the impulse mode of operation without fear.

III. CONCLUSION

The using of the new type SnO$_2$-Pd gate in capacity type sensor gives possible to raise sensitivity by almost ten times but disadvantage for such approach is increasing response and relaxation times of the sensor. Possible approach for improving response and relaxation times is using pulse temperature mode increasing diffusion rate through the gate present in work [17]. Extrapolation of present measurement results suggests stable detection NO_2 concentrations by the Pd-SnO$_2$-SiO$_2$-Si structure in the region of 1 ppb and less.

ACKNOWLEDGEMENT

This work was supported by the Russian Science Foundation (Grant Agreement 18-79-10230 of 08.08.2018).

REFERENCES

[1] M.S. Shivaraman, I. Lundström, C. Svensson, and H. Hammarsten, "Hydrogen sensitivity of palladium-thin-oxide-silicon schottky barriers", *Electronics Letters*, 1976, vol. 12 (18), pp. 483-484.

[2] J. Fogelberg, H. Dannetun, I. Lundström, and L.-G. Petersson, "A hydrogen sensitive palladium metal-oxide-semiconductor device as sensor for dissociating NO in H_2-atmospheres" *Vacuum*, 1990, vol. 41 (1-3), pp. 705-708.

[3] I.N. Nikolaev, E.V. Emelin, "Portable NO2 gas analyzer in the concentration range 0.02-2 ppm based on a MDS-sensor" *Measurement Techniques*, 2004, vol. 47 (11), pp. 1113-1115.

[4] M. Andersson, R. Pearce, and A.L. Spetz, "New generation SiC based field effect transistor gas sensors, *Sensors and Actuators, B: Chemical*, 2013, vol. 179, pp. 95-106.

[5] J.W. Gardner, J Yinon, "Electronic noses & sensors for the detection of explosives" Springer, 2004.

[6] <https://www.sgxsensortech.com/products-services/industrial-safety/metal-oxide-sensors/>

[7] K. Grossmann, U. Weimar, and N. Barsan, "Semiconducting Metal Oxides Based Gas Sensors" *Semiconductors and Semimetals*, 2013, vol. 88, pp. 261-282.

[8] K. Oto, A. Shinobe, M. Manabe, H. Kakuuchi, Y. Yoshida, and T. Nakahara, "New semiconductor type gas sensor for air quality control in automobile cabin" *Sensors and Actuators, B: Chemical*, 2001, vol. 77 (1-2), pp. 525-528.

[9] J. Zhang, S. Wang, Y. Wang, Y. Wang, B. Zhu, H. Xia, X. Guo, S. Zhang, W. Huang, and S. Wu, "NO2 sensing performance of SnO2 hollow-sphere sensor" *Sensors and Actuators, B: Chemical*, 2009, vol. 135 (2), pp. 610-617.

[10] G.D. Khuspe, R.D. Sakhare, S.T. Navale, M.A. Chougule, Y.D. Kolekar, R.N. Mulik, R.C. Pawar, C.S. Lee, and V.B. Patil, "Nanostructured SnO2 thin films for NO2 gas sensing applications" *Ceramics International*, 2013, vol. 39 (8), pp. 8673-8679.

[11] E.V. Zhovannik, V.S. Kulikauskas, and I.N. Nikolaev, "Structure of Pd-(III)Si intermediate region during laser deposition of palladium" *Fizika i Khimiya Obrabotki Materialov*, 1998, vol. 5, pp. 48-52.

[12] E.V. Zhovannik, I.N. Nikolaev, D.G. Stavkin, V.M. Shevlyuga, R.M. Imamov, and A.A. Lomov, "Studies of transient region between laser-deposited palladium and (111)Si", *Kristallografiya*, 1996, vol. 41 (5), pp. 935-939.

[13] A.V. Litvinov, N.N. Samotaev, M.O. Etrekova, and A.A. Mikhailov, "The detection of nitro compounds by using MIS-sensor" *IOP Conference Series: Materials Science and Engineering*, 2019, vol. 498 (1), article 012020.

[14] J.C. Qxley, J.L. Smith, W. Luo, and J. Brady, "Determining the vapor pressure of diacetone diperoxide (DADP) and hexamethylene triperoxide diamine (HMTD)" *Propellants Explosives Pyrotechnics*, 2009, vol. 34(6), pp. 539-543.

[15] I.N. Nikolaev, V.V. Krashevskaya, "A fast resistive sensor for explosive hydrogen concentrations" *Measurement Techniques*, 2004, vol. 47 (3), pp. 304-307.

[16] L.N. Kalinina, A.V. Litvinov, I.N. Nikolaev, and N.N. Samotaev, "MIS - Field effect sensors for low concentration of H2S for environmental monitoring" *Procedia Engineering*, 2010, vol. 5, pp. 1216-1219.

[17] N. Samotaev, A. Litvinov, and M. Etrekova. "Improving Detection Chlorine by Field Effect Gas Sensor with Using Temperature Pulse Mode", in *Proc. 17th International Meeting on Chemical Sensors IMCS 2018*, Vienna, Austria, 2018, pp. 474-475.

978-1-7281-3420-8/19 $31.00 © 2019 IEEE

Rapid Prototyping of MOX Gas Sensors in Form-Factor of SMD Packages

N. Samotaev, K. Oblov, A. Ivanova, A. Gorshkova, and B. Podlepetsky

Abstract - this work discusses the design of flexible laser micromilling technology for fast prototyping metal oxide based (MOX) gas sensors in SMD packages as a alternative to traditional silicon clean-room technologies. By laser micromilling technology possible to fabricate custom Micro Electro Mechanical System (MEMS) microhotplate platform and also SMD packages for MOX sensor, that gives complete solution for integration one in devices using IoT conception.

I. INTRODUCTION

Now day's state of art in electronic components has orientation on conception of Internet of things (IoT). A special practical problem of introducing the IoT, it is noted that it is necessary to ensure maximum autonomy and low cost of measuring instruments – sensors of different physical parameters. One of interested parameters for measuring by IoT instruments is concentration of hazardous gases – explosive and toxic. In the feed of gas concentration measurements now the leading sensor technologies is MOX sensors. The total MOX sensor design is a combination of three argent parts. The microhotplate responsible for level of power consumption and parameters of temperature cycling mode of MOX gas sensor. The chemical composition of MOX gas sensitive layer is responsible for sensitivity to target gases. The package is responsible for the variety of fields where MOX gas sensor can be used [1]. Elaboration of MOX gas sensors in last decades is concentrated on technologies which allow to obtain a low-cost product for mass application such as sensors for monitoring of indoor air quality (IAQ) [2] or breath test application [3]. Using MOX sensors in form-factor as surface mounted devices (SMD package) gives easily integration into a variety of consumer wearable, smartphones, tablets, industrial ventilation devices and other IoT applications. Technological solutions which posses to obtain MOX sensors in SMD form-factor are presented by the leading producers of MOX sensors (SGX, AMS, Bosch, Sensirion, Figaro, ect.) and include the combined usage of silicon MEMS microhotplate [4-6] SMD package made of different materials such as plastic [4], metal-plastic [5] and metal-ceramic [6].

N. Samotaev, K. Oblov, A. Ivanova, A. Gorshkova, B. Podlepetsky are with Department of Microelectronics, Institute of Nanoengineering in Electronics, Spintronics and Photonics, National Research Nuclear University MEPhI (Moscow Engineering Physics Institute), Kashirskoe highway 31, 115409 Moscow, Russian Federation, E-mail: nnsamotaev@mephi.ru. Fax: +74993242111

II. MATERIALS AND METHODS

The main idea of our developed technological flow based on laser micromilling is wide flexibility in developing of MEMS and SMD structures. Using equipments only widely presented on the market and refusing of technological steps needs a clean rooms support. Only semi custom 3D printing type software is especially developed product for laser micromilling system need for successful development and production of MEMS and SMD structure during our experiments. Software is need for translation CNC code to 4-axis laser micromilling setup and online measuring of geometrical parameters of MEMS and SMD structure for corrections micromilling procedure during automatic production. Print screen of developed software for using in ceramic micro milling process presen on Fig. 1.

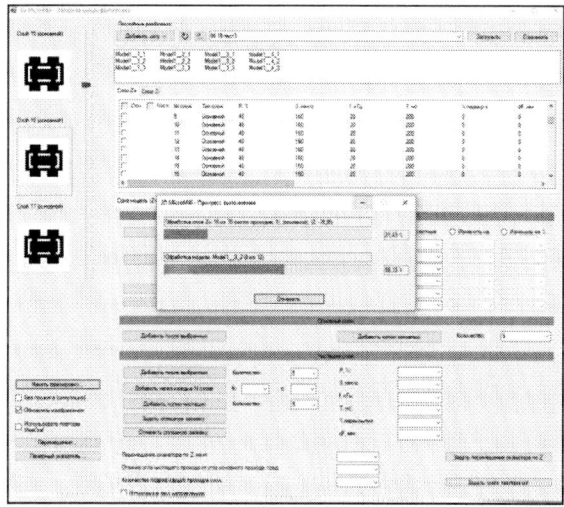

Fig. 1. Print screen of developed software for using in ceramic micromilling process. The print screen shows the process of fabrication the bottom part of the SMD package (left top images).

A. Technology flow for ceramic MOX sensor fabrication

During our work, an Ytterbium pulsing 20W fiber laser with a wavelength of 1.064 μm and tunable pulse duration from 50 to 200 ns is used (present on fig. 2). This laser emitter is installed on the four-coordinate portal complex, which allows the laser scanner to be moved across wilde field. The processing of ceramic substrates is carried out in a snap-in fixed in a rotational device, which allows processing flat substrates on both sides, cylindrical substrates over the entire surface area. Currently, fiber markers are most often used in industry for marking various types of products and are not intended for 3D laser milling, despite the fact that the technical capabilities of any laser marker allow it.

Fig. 2. Photo of 4-axis laser micro milling setup used in work. On right bottom angle screen with software present of fig.1. On left bottom corner is processing of ceramic substrates is carried out in a snap-in fixed in a rotational device during process of laser micro milling.

Fabrication of MOX sensors includes the following main steps:
• 3D modeling MEMS microhotplate with both bottom and top parts of the SMD package (Fig.3 and Fig.4,a) in 3D CAD programs with output file in STL format and also 2D modeling of MEMS and SMD parts metallization topology in DXF format;
• Optionally MEMS microhotplate parameters can be simulated in software is used to design products and semiconductors, as well as to create simulations that test a product's durability, temperature distribution, fluid movements, and electromagnetic properties, which allows to predict approximate thermal characteristics of the fabricated MOX sensor (example of modeling is presented in the paper [7]);
• Carry out processing of ceramics by using 4-axis laser facility (Fig. 2) for monolithic ceramics laser micromilling and 3D models of bottom and top parts of the SMD package and MEMS microhotplate. As a blank for processing, we used substrates with a thickness of 500 μm 99.9% alumina oxide ceramics;
• Platinum metallization deposition according with 2D model of topology, metallization annealing paying attention to specification on aerosol-jet or screen-print materials using for metalization (Fig. 5,a and b); In our case, the thick-film platinum based material already used in the work [8] was deposited;
• Optionally the metallization can be processed (local annealing or ablation) by laser according with 2D model of

metallization topology; In our case, this was not done, but it was done when applying films in the work [9].
• MOX gas sensitive layer deposition and annealing on the MEMS middle of microhotplate (Fig.6). In our case, the material based on SnO_2 [10] was done by a droap coating;
• Assembling separate parts of sensor into one SMD package and adhesion with special glass (Fig.4 b and Fig.5,c). In our case, the material for fairing was platinum consists glass [11] which used for metallization (present on Fig 5,a and b as black past on contact pads of SMD package)

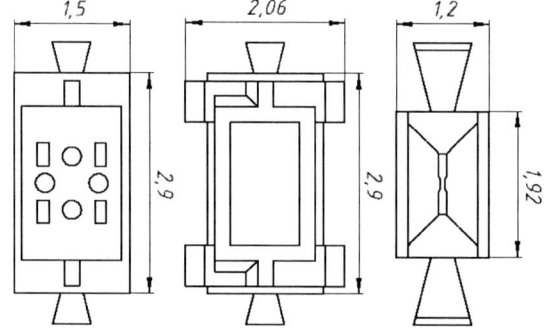

Fig. 3. Sketch with dimension in mm for top part of SOT-23 package, bottom part of SOT-23 package and frame with MEMS microhotplate (from left to right).

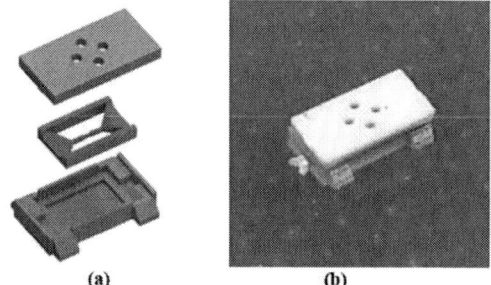

(a)   (b)

Fig. 4. 3D model of SOT-23 SMD package using for SMD package fabrication; (b) The SOT-23 assembled SMD package after fabrication by using laser micro milling technique.

(a)   (b)   (c)

Fig. 5. (a) Deposition Pt paste on bottom part of SOT-23 SMD package; (b) Deposition Pt paste on top side of MEMS microhotplate part inside SOT-23 SMD package. (c) The SOT-23 assembling SMD package before firing Pt-paste.

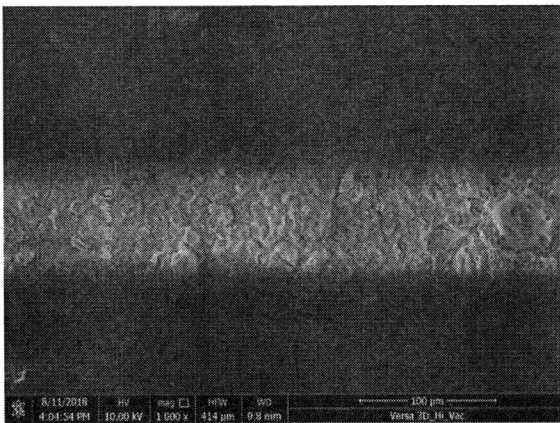

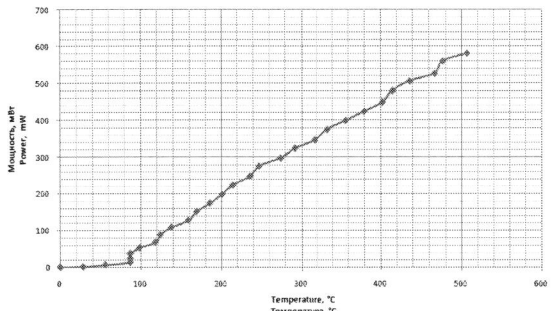

Fig. 8. Measured dependence power consumption of MOX gas sensor in SOT-23 SMD package versus working temperature of MEMS microhotplate which present on Fig.7.

Fig. 6. SEM-photo central section of the MEMS microhotplate immediately after laser micro milling, the scale of the linear size is presented in the lower right corner of the image. That part of MEMS microhoplate used for droap coating deposition of gas sensetive material based on SnO_2.

B. Tests fabricated MEMS microhotplate in SMD package

After fabrication MOX gas sensor in SOT-23 package were soldered to PCB for providing thermal characteristics measurement. Additionally the big central holein top part of SMD package is cut by a laser for visual observation of the temperature of microhotplate by using an IR camera. The view of SOT-23 SMD package soldered package present on Fig. 7.

Power consumption of fabricated MOX gas sensor in SOT-23 SMD package is ~500m W@450 ^{0}C (that working temperature is used for stable detection of methane gas by SnO_2 based materials [12]). This value is comparable to what is currently available for serial sensors on the market, for example TGS 816 (825 mW) [13] or TGS 2611(280 mW) [14] produced by Figaro company. But this temperature-power consumption working regime is not the limit for the described above technology rapid prototyping of ceramic MEMS sensors. Using described tech flow, experiments were carried out to fabricate a possible minimum size of MEMS microhotplate. The minimum size of the manufactured microhotplate with 250 mW power consumption at 450°C by track width was 30 µm and 20 µm thickness in SMD SOT-23 package type (3.0x1.4x1.0 mm with max dissipating power at 20°C is 350 mW) were achieved. Tests of MEMS microhotplate describes in work [15].

III. CONCLUSION

By laser micromilling technology possible to fabricate custom MEMS microhotplate platform and also SMD package for MOX sensor, that gives complete solution for integration in devices using IoT conception (smart phones, tablets, industrial ventilation devices and etc.). The 3D design and fabrication of MEMS microhotplates and packages products occurs simultaneously that give opportunity for ultra-fast time making unique solutions for MOX sensors without looking at standard solutions (primarily the package type) and on your own requirements (number of microhotplates, hot spot size on microhotplates, diameter holes in package cap and etc.). The cheapness of the equipment for the manufacture of MOX sensors, the lack of strict requirements for production facilities and the resistance of the using ceramic to high temperatures and harsh operating conditions make the technology offered in the our work attractive for small and medium-sized production of MOX sensors in SMD packages.

Rapid Prototyped MOX Gas Sensor in SOT-23 form-factor of SMD package in described work have same power consumption like wide use commercial sensors and

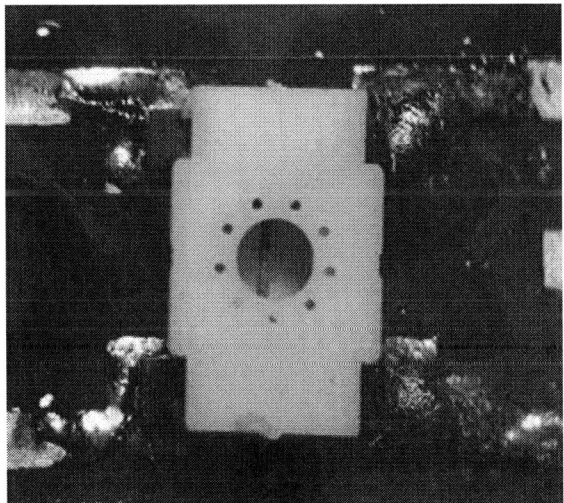

Fig. 7. MOX gas sensor in SOT-23 SMD package soldered to PCB for thermal characteristics measurement. The big central holein top part of SMD package is cut by a laser for visual observation of the temperature of microhotplate by using an IR camera.

Plot with neasured dependence power consumption of MOX gas sensor in SOT-23 SMD package versus working temperature of MEMS microhotplate present on Fig.8.

is fully completed sensor device. Also useful advantages of fully ceramics MOX sensor (MEMS microhotplate and SMD package) is long term stability against harsh environmental conditions including extreme temperature and acid or alkaline gases produced by industrial and natural catastrophes.

ACKNOWLEDGEMENT

This research was sponsored by the Federal Ministry of Education and Research (BMBF) in Germany founding No. 02P15B520, the Israel Innovation Authority and the Ministry of Science and Higher Education of the Russian Federation founding with unique identifier RFMEFI58718X0054.

REFERENCES

[1] I. Simon, N. Bârsan, M. Bauer, and U. Weimar, "Micromachined metal oxide gas sensors: opportunities to improve sensor performance", *Sensors and Actuators, B: Chemical*, 2001, vol. 73 (1), pp. 1-26.

[2] N. Yamazoe, and K. Shimanoe, "New perspectives of gas sensor technology", *Sensors and Actuators, B: Chemical*, 2009, vol. 138 (1), pp. 100-107.

[3] J.-W. Yoon, and J.-H. Lee, "Toward breath analysis on a chip for disease diagnosis using semiconductor-based chemiresistors: Recent progress and future perspectives", *Lab on a Chip*, 2017, vol. 17 (21), pp. 3537-3557.

[4] Environmental Sensors. Available online: https://www.sensirion.com/en/environmental-sensors/gas-sensors/ (accessed on 30 July 2019).

[5] Bosch Sensortec. Available online: https://www.bosch-sensortec.com/bst/products/all_products/bme680 (accessed on 30 July 2019).

[6] Figaro sensor TGS8100. Available online: http://www.figarosensor.com/products/entry/tgs8100.html (accessed on 30 July 2019).

[7] N. Samotaev, K. Oblov, A. Gorshkova, A. Ivanova and M. Etrekova, "Rapid Prototyping of MOX Gas Sensors in Form-Factor of SMD Packages", *Proceedings MDPI* 2019, vol. 14, p. 52.

[8] A.A. Vasiliev, A.S. Lipilin, A.M. Mozalev, A.S. Lagutin, A.V. Pisliakov, N.P. Zaretskiy, N.N. Samotaev, and A.V. Sokolov, "Gas sensors based on MEMS structures made of ceramic ZrO2/Y2O3 material" *Proceedings of SPIE - The International Society for Optical Engineering*, 2011, vol. 8066, article 80660N.

[9] K. Oblov, A. Ivanova, S. Soloviev, N. Samotaev, A. Pisliakov, A. Sokolov, and A. Vasiliev, "Fabrication of metallization by laser sintering of micro powders", *Physics Procedia*, 2015, vol. 72, pp. 470-474.

[10] A.A. Vasiliev, R.G. Pavelko, S.Yu. Gogish-Klushin, D.Yu. Kharitonov, O.S. Gogish-Klushina, A.V. Sokolov, and N.N. Samotaev, "Alumina MEMS platform for impulse semiconductor and IR optic gas sensors", *Transducers and Eurosensors'2007 - 4th International Conference on Solid-State Sensors, Actuators and Microsystems*, 2007, article 4300563, pp. 2035-2037.

[11] J.H. Kim, J.S. Sung, Y.M. Son, A.A. Vasiliev, V.V. Malyshev, E.A. Koltypin, A.V. Eryshkin, D.Yu. Godovski, A.V. Pisyakov, and S.S. Yakimov, "Propane/butane semiconductor gas sensor with low power consumption", *Sensors and Actuators, B: Chemical*, 1997, vol. 44 (1-3), pp. 452-457.

[12] A.A. Vasiliev, A.V. Pisliakov, A.V. Sokolov, O.V. Polovko, N.N. Samotaev, W. Kujawski, A. Rozicka, V. Guarnieri, and L. Lorencelli, "Gas sensor system for the determination of methane in water" *Procedia Engineering*, 2014, vol. 87, pp. 1445-1448.

[13] Figaro. TGS 816 Product Information. Available online: https://www.figaro.co.jp/en/product/docs/tgs816_product_in fomation_rev01.pdf (accessed on 30 July 2019).

[14] Figaro. TGS 2611-C00 Product Information. Available online: https://www.figaro.co.jp/en/product/docs/tgs2611-c00_product%20information%28en%29_rev00.pdf (accessed on 30 July 2019).

[15] N. Samotaev, K. Oblov, and A. Ivanova, "Laser micromilling technology as a key for rapid prototyping SMD ceramic MEMS devices", *MATEC Web of Conferences*, 2018, vol. 207, article number 04003.

Analysis of Intrinsic Stochastic Fluctuations of the Time Response of Adsorption-Based Microfluidic Bio/Chemical Sensors: the Case of Bianalyte Mixtures

I. Jokić, Z. Djurić, K. Radulović, M. Frantlović, and P. M. Krstajić

Abstract – Real-time in situ operation of bio/chemical sensors assumes detection of chemical substances or biological specimens in samples of complex composition. Since sensor selectivity cannot be ideal, adsorption of particles other than target particles inevitably occur on the sensing surface. That affects the sensor response and its intrinsic fluctuations which are caused by stochastic fluctuations of the numbers of adsorbed particles of all the adsorbing substances. In microfluidic sensors, such response fluctuations are a result of coupled adsorption, desorption and mass transfer (convection and diffusion) processes of analyte particles. Analysis of these fluctuations is important because they constitute the adsorption-desorption noise, which limits the sensing performance. In this work we perform the analysis of fluctuations by using a stochastic model of sensor response after the steady state is reached, in the case of two-analyte adsorption, considering mass transfer processes. The results enable estimation of the ultimate sensing performance of adsorption-based microfluidic bio/chemical sensors of different sensing areas, operating in bianalyte mixture environments.

I. INTRODUCTION

Microfluidic bio/chemical sensors are highly sensitive devices for detection of biological specimens or chemical substances [1]-[3]. A significant advantage of these devices compared to conventional laboratory equipment is their capability of real-time in-situ operation. However, such applications assume the use of native samples taken from the environment or living organisms, which are of complex composition.

In adsorption-based devices sensor response is determined by the number of adsorbed particles of a certain species on the sensing surface. In microfluidic sensors this number is determined both by adsorption-desorption (AD) processes and mass transfer (MT) processes (convection and diffusion of particles toward the sensing surface and away from it). In the case of native (complex) samples, not only particles of the target substance, but also those of other substances existing in the sample can be adsorbed,

due to the non-ideal sensor selectivity. Therefore, it is necessary to take into account AD processes and mass transfer of all the adsorbing substances in the analysis of sensor response. This is also true for the analysis of intrinsic response fluctuations that originate from the inevitable fluctuations of the number of adsorbed particles, which determine the AD noise and the sensor minimal detectable concentration.

Analysis of AD fluctuations of sensor response is performed by using stochastic models. The models that take into account the coupling between AD and MT processes are derived for the case of an ideally selective sensor, in which only the target substance is adsorbed on the sensing surface [4], [5]. Recently a stochastic model that takes into account MT was presented for adsorption of two substances, and it was used for the analysis of the expected number of adsorbed particles of both species, which determines the sensor response kinetics [6]. However, until now, the analysis of fluctuations was not performed by using a stochastic model of sensor response in the case of two-analyte adsorption, and considering MT.

In this work we first give a review of the sensor response stochastic model in the case of coupling of adsorption-desorption and mass transfer of substances present in a bianalyte mixture. The model enables investigation of the influence of various parameters on sensor response fluctuations. We use the steady-state form of the model for the analysis of variances and the covariance of the stochastic numbers of adsorbed particles depending on the sensing surface area, which determine the intrinsic fluctuations of sensor response. This analysis is useful for estimation of the ultimate noise performance and the minimal detectable signal of adsorption-based sensors (e.g. plasmonic, surface/bulk acoustic wave, micro/nano-cantilever, nanowire FET sensor) of different sensing areas.

II. MODELING OF SENSOR RESPONSE

The mathematical model that describes the stochastic AD processes on the sensing surface, coupled with mass transfer in the sensor microfluidic chamber for the case of a bianalyte mixture is obtained starting from the Master equation for the bivariate gain-loss processes and the definitions of the first and the second moments of a random vector variable $\mathbf{N}=[N_1 \ N_2]$, where N_1 and N_2 are the stochastic numbers of adsorbed particles of the two

---

I. Jokić, K. Radulović, M. Frantlović, and P. M. Krstajić are with the ICTM – Center of Microelectronic Technologies, University of Belgrade, Njegoševa 12, 11000 Belgrade, Serbia, E-mail: ijokic@nanosys.ihtm.bg.ac.rs

Z. Djurić is with the Institute of Technical Sciences of SASA, Serbian Academy of Sciences and Arts, Knez Mihailova 35, 11000 Belgrade, Serbia, E-mail: zoran.djuric@itn.sanu.ac.rs

analytes. The model is in the form of five differential equations

$$\frac{d<N_i>}{dt} = <A_i(N_1, N_2)> - <D_i(N_i)> \quad (1)$$

$$\frac{d\sigma_i^2}{dt} = <A_i(N_1, N_2)> + <D_i(N_i)>$$
$$+ 2<(N_i - <N_i>)[A_i(N_1, N_2) - D_i(N_i)]> \quad (2)$$

$$\frac{d\sigma_{12}}{dt} = <(N_1 - <N_1>)[A_2(N_1, N_2) - D_2(N_2)]>$$
$$+ <(N_2 - <N_2>)[A_1(N_1, N_2) - D_1(N_1)]> \quad (3)$$

Here $<N_i>$ is the expected value and σ_i^2 is the variance of the number of adsorbed particles of the type i ($i=1$ or 2), $\sigma_{12} = \sigma_{21}$ is the covariance of the random variables N_1 and N_2, while A_i and D_i are the probabilities of increase and decrease of the number of adsorbed particles of the ith substance by one in unit time, respectively. Each of Eqs. (1) and (2) replaces two equations of that system, of which one is obtained for $i=1$, and the other one for $i=2$. The system of Eqs. (1)-(3) is obtained by assuming that at any given moment a change of the number of adsorbed particles is possible for one substance only, by +1 or -1. The transition probabilities in unit time, A_i and D_i, are determined by using the two-compartment model [7] to approximate the temporally and spatially dependent concentrations of substances in reaction chambers of microfluidic sensors. Such concentrations are a result of the coupling of AD processes, convection, and diffusion of analyte particles of two substances. By assuming 1:1 competitive binding of analyte particles to adsorption sites on the sensing surface, and the Langmuir adsorption, the two-compartment model yields the following expressions for the probabilities of transition between the adjacent states of the random vector variable **N** [6]

$$A_i = k_{ai} \frac{C_i + \frac{k_{di} N_i}{k_{mi} A}}{1 + \frac{k_{ai}(N_m - N_1 - N_2)}{k_{mi} A}} (N_m - N_1 - N_2) , \; D_i = k_{di} N_i$$

where C_i is the concentration of the ith analyte in the sample entering the microfluidic chamber, k_{ai} and k_{di} are its adsorption and desorption rate constants, k_{mi} is the mass transfer coefficient of the analyte i [7], N_m is the number of surface binding sites, and A is the sensing surface area (the binding sites surface density, which is the parameter used in the numerical simulations, equals N_m/A).

Eqs. (1)-(3) become the system of equations for the first two moments after the Taylor expansion of bivariate functions $A_i(N_1, N_2)$, centered at the expected values. In the steady state the system is

$$A_{ie} - D_{ie} + \frac{1}{2}\left[\frac{\partial^2 A_i}{\partial N_1^2}\sigma_{1e}^2 + 2\frac{\partial^2 A_i}{\partial N_1 \partial N_2}\sigma_{12e} + \frac{\partial^2 A_i}{\partial N_2^2}\sigma_{22}^2\right] = 0 \quad (4)$$

$$A_{ie} + D_{ie} + \left[2\left(\frac{\partial A_i}{\partial N_i} - \frac{\partial D_i}{\partial N_i}\right) + \frac{1}{2}\frac{\partial^2 A_i}{\partial N_i^2}\right]\sigma_{ie}^2$$
$$+ \left(2\frac{\partial A_i}{\partial N_j} + \frac{\partial^2 A_i}{\partial N_1 \partial N_2}\right)\sigma_{12e} + \frac{1}{2}\frac{\partial^2 A_j}{\partial N_j^2}\sigma_{je}^2 = 0 \quad (5)$$

$$\frac{\partial A_2}{\partial N_1}\sigma_{1e}^2 + \left(\frac{\partial A_1}{\partial N_1} - \frac{\partial D_1}{\partial N_1} + \frac{\partial A_2}{\partial N_2} - \frac{\partial D_2}{\partial N_2}\right)\sigma_{12e} + \frac{\partial A_1}{\partial N_2}\sigma_{2e}^2 = 0 \quad (6)$$

where $i=1$ or 2, $j=1$ or 2, $i \neq j$ in Eq. (5), $A_{ie} = A_i(<N_{1e}>, <N_{2e}>)$, $D_{ie} = D_i(<N_{ie}>)$, and all the partial derivatives are calculated for $N_1 = <N_{1e}>$ and $N_2 = <N_{2e}>$. By solving this system we obtain the steady-state expected values, variances and the covariance of stochastic numbers of adsorbed particles ($<N_{1e}>$, $<N_{2e}>$, σ_{1e}^2, σ_{2e}^2, and σ_{12e}), which are a measure of steady-state intrinsic fluctuations of sensor response.

As the sensor response due to adsorption of two substances is $r = m_1 N_1 + m_2 N_2$ (m_1 and m_2 are the weight factors that represent the average contribution of a single adsorbed particle of the first and second analyte to the sensor response), the steady-state variance of sensor response is given by the expression

$$<(\Delta r)_e^2> = m_1^2 \sigma_{1e}^2 + m_2^2 \sigma_{2e}^2 + 2m_1 m_2 \sigma_{12e} \quad (7)$$

The intrinsic response fluctuations in steady state due to the coupling of AD and MT processes, known as AD noise, are expressed as the square root of $<(\Delta r)_e^2>$, and they are completely determined by the solutions of the system of equations (4)-(6).

III. ANALYSIS OF SENSOR STOCHASTIC RESPONSE

The presented mathematical model is used for the analysis of the steady-state expected values, variances and the covariance of the numbers of adsorbed molecules of two proteins on the surface of a microfluidic biosensor. The parameter values are given in Table I. The results given in Figs. 1 and 2 show the dependencies of these quantities on the sensing surface area in the range from 10^{-13} m^2 to 10^{-11} m^2.

TABLE I
PARAMETER VALUES

Parameter	Analyte 1	Analyte 2
Concentration	$6 \cdot 10^{17}$ 1/m^3	$1.2 \cdot 10^{18}$ 1/m^3
Adsorption rate constant	$1.3 \cdot 10^{-19}$ m^3/s	$1.3 \cdot 10^{-20}$ m^3/s
Desorption rate constant	0.08 s	0.02 s
Mass transfer coefficient	$2 \cdot 10^{-5}$ m/s	$2 \cdot 10^{-5}$ m/s
Binding sites density	$6 \cdot 10^{15}$ 1/m^2	

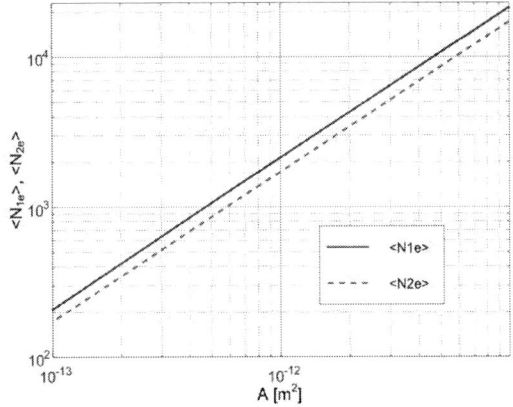

Fig. 1 Expected values of the numbers of adsorbed particles of two proteins depending on the sensing surface area. Two proteins are competitively adsorbed.

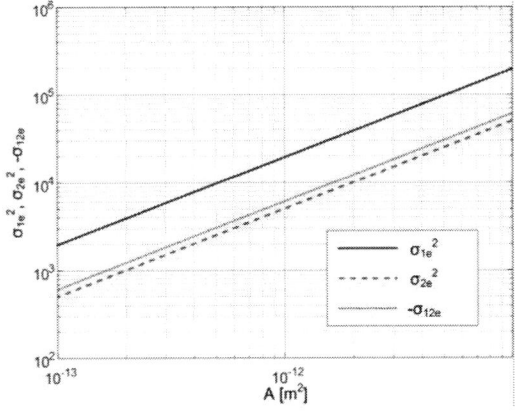

Fig. 2 Variances and covariance of the numbers of adsorbed particles of two competitive proteins depending on the sensing surface area.

As it can be seen in Fig. 1, there is a continuous increase of the expected values of the numbers of adsorbed particles of two proteins as the surface area increases. Also, it can be seen that these two values are close to each other although the analyte 2 has a lower affinity (defined by the affinity constant $K_a = k_{ai}/k_{di}$) for adsorption sites. Therefore, the competitive (in most cases unwanted) adsorption of another substance can significantly affect the expected value of the steady-state sensor response, which equals $<r_e> = m_1<N_{1e}> + m_2<N_{2e}>$, although this influence also depends on the values of the weight factors m_1 and m_2.

Variances and covariance of N_1 and N_2 also increase in sensors of larger sensing areas, as shown in Fig. 2. The covariance is negative, so the diagram shows its absolute value. It is of the same order of magnitude as the variances of the numbers of adsorbed particles. Based on Eq. (7) it can be concluded that, due to the negative covariance value, depending on the values of parameters m_1 and m_2,

the steady-state variance of sensor response (and thus the sensor's AD noise) in the case of competitive adsorption can be higher or lower than in the case of single substance adsorption (adsorption of only one substance corresponds to the sensor of ideal selectivity).

The diagrams in Figs. 1 and 2 are general, as they show the statistical parameters that describe stochastic fluctuating numbers of adsorbed particles, so they are valid for any type of adsorption-based sensor (surface plasmon resonance sensors, bulk and surface acoustic wave sensors, microcantilever sensors etc.), regardless of its measurement parameter (refraction index, acoustic wave frequency, deflection or oscillation frequency of a cantilever). When the values of the parameters m_1 and m_2 are taken into account for a given type of sensor, the presented diagrams enable the estimation of the influence of competitive adsorption on the steady-state expected value of sensor response, as well as its fluctuation, i.e. on the sensor's AD noise that determines the sensor's ultimate performance.

IV. CONCLUSION

In this paper the analysis is presented of the steady-state expected values, variances and the covariance of the numbers of adsorbed molecules of two proteins on the surface of a microfluidic biosensor, depending on the sensing surface area. Intrinsic fluctuations of the numbers of adsorbed particles are caused by the stochastic nature of adsorption-desorption and mass transfer processes of two analytes, so we used a stochastic model of sensor response that takes into account the coupling of these processes. Since the results and conclusions stemming from the analysis refer to the numbers of adsorbed particles, they are general in the sense that they are valid for any type of adsorption-based sensor (surface plasmon resonance sensors, bulk and surface acoustic wave sensors, microcantilever sensors etc.) whose response is governed by the number of adsorbed particles on the sensing surface.

The results enable estimation of the quantitative influence of competitive adsorption on the expected value of the steady-state sensor response and its intrinsic fluctuation, i.e. the adsorption-desorption noise and the ultimate performance of sensors with different sensing surface areas that operate in various bianalyte mixture environments. Depending on the type of substances and the values of parameters that relate the number of adsorbed particles and the time response of a given sensor, the response fluctuations can be higher or even lower than in the case of single substance adsorption on the sensing surface.

ACKNOWLEDGEMENT

This work has been funded by the Serbian Ministry of Education, Science and Technological Development (Project TR 32008) and by the Serbian Academy of Sciences and Arts (Project F-150).

REFERENCES

[1] E. K. Sackmann, A. L. Fulton, and D. J. Beebe, "The present and future role of microfluidics in biomedical research", *Nature*, 2014, vol. 507, pp. 181-189.

[2] G. Luka, A. Ahmadi, H. Najjaran, E. Alocilja, M. DeRosa, K. Wolthers, A. Malki, H. Aziz, A. Althani, and M. Hoorfar, "Microfluidics integrated biosensors: A leading technology towards lab-on-a-chip and sensing applications", *Sensors*, 2015, vol. 15, pp. 30011–30031.

[3] K.-K. Liu, R.-G. Wu, Y.-J. Chuang, H. S. Khoo, S.-H. Huang, and F.-G. Tseng, "Microfluidic systems for biosensing", *Sensors*, 2010, vol. 10, pp. 6623-6661.

[4] I. Jokić, Z. Djurić, K. Radulović, and M. Frantlović, "Stochastic time response of adsorption-based micro/nano-biosensors with a fluidic reaction chamber: the influence of

mass transfer", in *Proc. 30th International Conference on Microelectronics MIEL 2017*, Niš, Serbia, 2017, pp. 127-130.

[5] G. Tulzer, and C. Heitzinger, "Fluctuations due to association and dissociation processes at nanowire-biosensor surfaces and their optimal design", *Nanotechnology*, 2015, vol. 26, pp. 025502 1-9.

[6] I. Jokić, Z. Djurić, K. Radulović, and M. Frantlović, "Analysis of stochastic time response of microfluidic biosensors in the case of competitive adsorption of two analytes", *MDPI Proceedings*, 2018, vol. 2, pp. 991 1-5.

[7] D. G. Myszka, X. He, M. Dembo, T. A. Morton, and B. Goldstein, "Extending the Range of rate constants available from BIACORE: interpreting mass transport-influenced binding data", *Biophysical Journal*, 1998, vol. 75, pp. 583-594.

978-1-7281-3420-8/19 $31.00 © 2019 IEEE

Modeling Noise and Stability of Affinity-Based MEMS, NEMS and NOEMS Sensors of Ternary Gas Mixtures

O. Jakšić, I. Jokić, Z. Jakšić, M. Obradov, D. Tanasković, D. Randjelović, and D. Vasiljević Radović

Abstract – We address noise and stability of adsorption-based sensing of ternary gas mixtures in affinity-based MEMS, NEMS and NOEMS sensors. We investigate mechanisms of such chemical sensing in diverse industrial environments where ternary gas mixtures are of importance. In all existing sensing devices signal fluctuations determine their ultimate performance, and in affinity-based nanodevices the prevailing noise is caused by adsorption and desorption of different species at the sensor surface. We consider analytically and numerically detection of three-component gas mixtures. We present results obtained by applying the conventional method for assessing stability of chemical reactions to three component monolayer adsorption. Gas molecules modeling as spherical particles is suitable for applying any model regarding population dynamics, but taking molecular dimension and orientation into account is more realistic. Noise analysis in time and frequency domain is performed for two different classes of ultrahigh sensitivity adsorption based gas sensors: (nano) plasmonic sensors (refractometric devices) and MEMS/NEMS resonator-based devices.

I. INTRODUCTION

Affinity-based devices belong among the most sensitive known chemical and biological sensors. This is especially valid for the resonance-based sensors. The analyte adsorbed to the surface of such devices changes the resonant frequency and this resonance shift is measured. The resonance can be electromagnetic, like in various plasmonic and metamaterial sensors. The adsorbed species modify the refractive index, while plasmonic effects cause extreme localizations of optical fields, leading to ultra-high sensitivities that may even ensure single molecule detection [1]. Another large group of ultrasensitive chemical sensors is based on mechanical resonance in MEMS and NEMS (Micro- or NanoElectroMechanical Systems) [2]. The resonating parts can be vibrating micro and nanocantilevers, and the adsorbed mass changes their resonant frequency.

Sensing of ternary gas mixtures is of importance in diverse industrial fields. One such field is calibration of ternary gas mixtures [3]: Anaerobic Growth Mixture (0.5-10% CO_2, 0.5-10% H in N_2) and Medical Laser Mixture

Olga Jakšić, Ivana Jokić, Zoran Jakšić, Marko Obradov, Dragan Tanasković, Danijela Randjelović, and Dana Vasiljević Radović are with the Department of Microelectronic Technologies, Institute of Chemistry, Technology and Metallurgy, University of Belgrade, Njegoseva 12, 11000 Belgrade, Serbia, E-mails: olga, ijokic, jaksa, marko.obradov, danijela, dana, all at @nanosys.ihtm.bg.ac.rs

(4.5-9% CO_2, 13.5-15% N_2 in He) or ternary biological atmosphere gas mixtures [4]. Another field are gas-filled radiation detectors containing a conducting gas, an insulating gas, and a third gas with a low ionization energy increasing the net number of electrons producing a current [5].

Methods currently used for ternary gas sensing may suffer from hardware issues, as for instance problems related to bandwidth filtering in absorption based measurement systems [6] or software issues (like e.g. those related to complex processing of data acquired from triplets of microresonators in QCM arrays [7]). Systems for monitoring gas composition either require expensive equipment or have limitations related to simpler setups, so true simultaneous gas detection seems to be a promising solution.

In all existing sensing devices, regardless of their type, signal fluctuations and noise determine their ultimate performance. In affinity devices the main mechanism of their function and simultaneously a fundamental source of noise are the processes of adsorption and desorption.

In this contribution we present our results of modeling noise and stability in time and frequency domain for various three-gas mixtures. Different ways of modeling the adsorption process are discussed (Langmuir, Lagergren or the second order reaction model) alongside with criteria for modeling the sorption rate constants. After choosing the proper reaction model based on the established criteria, noise analysis in both time and frequency domain is performed using stochastic analysis for both of the above mentioned classes of affinity based gas sensors – plasmonic devices and devices based on MEMS/NEMS resonators. Issues regarding how to distinguish oscillations buried in noise and noise-mimicking oscillations are also discussed.

II. THEORY

Figure 1 schematically shows competitive adsorption of three-component gas mixture. Two kinds of molecular gases are shown as an example (O_2 and H_2) mixed with monatomic gas (He). Free floating species are shown, as well as the same species bound to gold atoms functioning as adsorption sites. A molecule structure is obtained using the chemical structure information available in the PubChem Substance and Compound database [8] through the unique chemical structure identifier CID. Then, molecular dimensions and projected surface areas were obtained using Marvin 5.9.3, 2012, ChemAxon [9]

978-1-7281-3420-8/19 $31.00 © 2019 IEEE

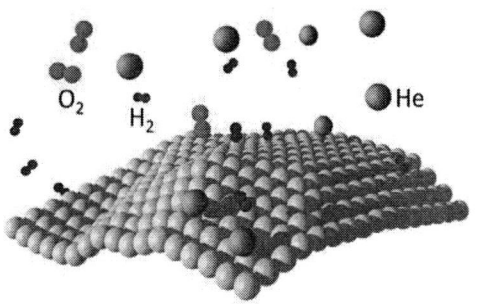

Fig. 1. Schematic presentation of competitive adsorption of three-component gas mixture (molecules of oxygen and hydrogen and monatomic helium) on a gold surface.

The stoichiometric equation valid for monolayer adsorption of ternary mixtures on the solid surface is

$$A_{g,i} + A_f \underset{k_{di}}{\overset{k_{ai}}{\rightleftarrows}} A_{a,i}, \quad i = 1, 2, 3 \tag{1}$$

where a molecule in gas phase is A_g, a free adsorption site is A_f and an adsorbed molecule is A_a, k_a is the adsorption rate constant and k_d is the desorption rate constant.

The set of deterministic differential rate equations governing the process kinetics (the time evolutions of the numbers of adsorbed molecules of each species, N_{ai}) can be formed in different ways. In [10] a detailed analysis and modeling are presented. Here we use an equation set taking into account the molecular size, so that each species i has a different surface density of adsorbed molecules, σ_i, and each one can populate the whole surface binding to a different number of adsorption sites, M_{bi}.

The linearization criteria are given in [11]. One of them is fulfilled when the number of molecules in the gas phase (N_{0i}-N_{ai}) can be considered constant.

$$\frac{dN_{ai}}{dt} = k_{ai}(N_{0i} - N_{ai})\left(M_{bi} - \sum_{j=1}^{r}\frac{\sigma_i N_{aj}}{\sigma_j}\right)$$
$$-k_{di}N_{ai}, \quad i = 1, 2, 3 \tag{2}$$

We estimate the refractive index change, measured by plasmonic sensors, by applying the effective medium theory [12]. Assuming that the refractive indices of the analytes in the mixture, n_{Ai}, and of the environment, n_e, are known, the refractive index change is proportional to the weighted sum of numbers of adsorbed molecules

$$RIC = \sum_{i=1}^{r} N_{ai}\frac{(n_{Ai} - n_e)}{M_{bi}} = \sum_{i=1}^{r} w_i N_{ai} \tag{3}$$

The same premises used for establishing the deterministic set (1) are also used for addressing the intrinsic fluctuations of refractive index change caused by the random approach of molecules to the surface and their random residential time, for the adsorption process is genuinely stochastic in nature. The stochastic analysis in time domain shows that expectation and dispersion (variance) of the refractive index change in linear systems are also weighted sums [13].

$$E\{RIC\} = \sum_{i=1}^{3} w_i E\{N_{ai}\}, \quad D\{RIC\} = \sum_{i=1}^{3} w_i D\{N_{ai}\} \tag{4}$$

The stochastic analysis in frequency domain, informative on random behavior in equilibrium, performed by applying the Langevin method (by adding the stochastic zero mean white noise term to every equation of the deterministic set), shows a more complex dependence [14]. Written in the matrix form, the power spectral density of the refractive index change caused by the fluctuations is

$$PSD_{RIC}\cdot(j\omega) =$$
$$\mathbf{\Theta}^T(j\omega\mathbf{I} + \mathbf{M})^{-1}\ \mathbf{GG}^{*T}\ \left[(-j\omega\mathbf{I} + \mathbf{M})^{-1}\right]^T\mathbf{\Theta} \tag{5}$$

where $\mathbf{\Theta}$ is column matrix of weighting coefficients w_i, $\mathbf{M}$ is the quadrature matrix of coefficients multiplying N_{ai} in the deterministic equation set (2) adapted so that $k_{ai}(N_{0i}-N_{ai})$ are considered constant and $\mathbf{G}$ is the diagonal matrix of powers of added independent noise terms, where each term in the diagonal corresponds to one of the species in the mixture and represents quadruple of the equilibrium desorption rate (or adsorption rate, since in equilibrium they are equal) [15].

Equilibrium fluctuations in adsorption with nonlinear, second order kinetics have been addressed analytically, by employing linearization [16]. In equilibrium, perturbations from the stationary state are small enough so that Taylor expansion is valid. After applying Taylor expansion on the equation set (2), one obtains a set of linear differential equations suitable for the procedure based on the Langevin method that led to equation (5), ($\mathbf{M}$ and $\mathbf{G}$ matrices refer now to the obtained linearized equation set).

However, far from equilibrium, for the stochastic analysis in time domain, we do not use linearization of nonlinear systems, but rather the method based on stochastic simulation algorithms (SSA) [17]. Nonlinear systems affected by fluctuations, apart from noisy response, may exhibit possible instabilities [18]. There are multiple methods to assess the stability of rate processes [18]. Here we use the linear stability analysis to find out if the stationary solutions remain stable after a small perturbation. The equation matrix would be

$$\frac{\partial \mathbf{N}}{\partial t} = f(\mathbf{N}, \mu) \tag{6}$$

If we represent $\mathbf{N}$ as a sum of stationary vector $\mathbf{N_s}$ and perturbation $\mathbf{n}$, after a Taylor expansion around $\mathbf{N_s}$ we have

$$\frac{\partial \mathbf{n}}{\partial t} = f(\mathbf{N}_s + \mathbf{n}, \mu) - f(\mathbf{N}_s, \mu) = \mathbf{M}_\alpha \cdot \mathbf{n} \tag{7}$$

$\mathbf{M}_\alpha$ is the Jacobian matrix with elements $(\partial f_i/\partial N_j)$ calculated at stationary state. The solutions to (7) have a form $\mathbf{n} = \mathbf{u}\exp(\lambda t)$ where $\mathbf{u}$ is eigenvector λ (a set of Jacobian eigenvalues). They determine whether the perturbation rises or diminishes over time. Thus the steady state remains stable if the Jacobian trace is negative, $\mathrm{tr}(\mathbf{M}_\alpha) < 0$ and the Jacobian determinant is positive, $\det(\mathbf{M}_\alpha) > 0$.

Analogously, the outlined methodology used for refractometric sensing is valid for any resonating structures whose operation is affected by adsorption. A weighted sum of the numbers of adsorbed molecules, where the number of the adsorbed molecules of each species is weighted with the weight of its molecules, M_{ai}, represents the overall adsorbed mass and all variables dependent on mass can be further used according to laws of mathematical statistics.

III. RESULTS AND DISCUSSION

The setup for our numerical experiments is based on our in-house MATLAB code developed for the analysis of MOEMS and NOEMS in the case of monolayer adsorption. The code is freely available online [19].

Figure 2 represents time evolutions of the mean refractive index change due to adsorption of the anaerobic growth mixture on two different plasmonic materials: graphene, Fig. 2a), and gold, Fig. 2b). Along with the time evolution of the mean refractive index change of the mixture, Figs 2a, b demonstrate the time evolutions of individual components at a pressure equal to their partial pressure in the mixture. Typical presence of knees in the log diagrams of their responses is desirable for simultaneous detection of analytes in the mixture. The process kinetics and the knee positions depend on various parameters, some of which cannot be changed for each species independently in order to have a simultaneously ensured selectivity. However, desorption energy is the parameter with the greatest impact on the kinetics of the processes in the mixture because its influence on the expressions for the rate constants is exponential, and because it is analyte-specific.

We see in Figs. 2, 3 that graphene is more selective than gold. Desorption energy is not the only parameter that differs in these two numerical experiments. Another one is the surface density of adsorption sites. Analytes cannot cover the surface more densely than the adsorption centers allow. H_2 on graphene is packed in its native density, and on gold in the density defined by gold atoms.

The knee position may be appropriate for qualitative analysis of the mixture composition in some experiments. In Fig. 2a) the presence of hydrogen is recognizable, but quantitative composition is hard to deduce. Fig. 4 shows that in experiment shown in Fig. 2a) equilibrium coverages are not in accord with the actual amounts of components in the mixture. Fig. 4a that shows greater equilibrium coverage of CO_2 corresponds to the adsorption of the composition with less CO_2 (set of lines with lighter shade in Fig. 2a). The overall coverage and the overall response are greater. However, the refractive index change depends on the indices of separate analytes. The complex interplay between the analyte – surface specific parameters like refractive index, desorption energy and surface density determines the component footprint in the final response. Fig. 5 shows the power spectral density of refractive index change due to adsorption of three separate gases and their ternary mixture for the same parameters as in Fig. 4.

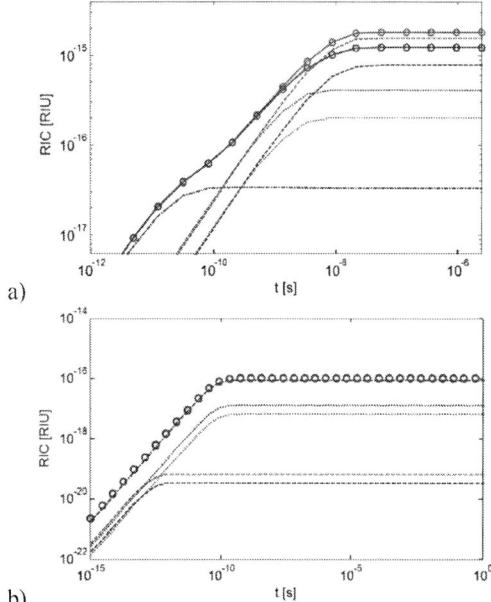

a)

b)

Fig. 2. Time evolution of the mean value of refractive index change due to adsorption of anaerobic growth mixture at 300 K, overall pressure of 10^5 Pa, on two different plasmonic surfaces of 1 mm^2: a) graphene and b) gold. Lighter lines refer to CO_2 5%, H_2 10 % and N_2, darker lines refer to CO_2 10%, H_2 5 % and N_2. Dotted lines refer to time evolutions of CO_2 as if it were alone, dashed lines refer to time evolutions of H_2, dash-dot lines refer to N_2 and o symbols refer to time evolution of the mixtures.

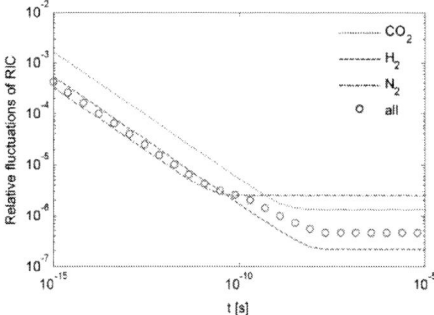

Fig. 3. Relative fluctuations of refractive index change due to adsorption of separate gases and their ternary mixture for anaerobic growth mixture of CO_2 (5%), H_2 (10%) and N_2 at room temperature, overall pressure 10^5 Pa on the plasmonic surface of graphene with area of 1 mm^2.

We analyze the dynamics of fluctuations in equilibrium. We use the frequency response for the estimation of dominance of intrinsic noise sources. In some applications a-d fluctuations may bury other intrinsic noise sources. On the other hand, there are applications where a-d noise is buried. Theoretical results imply that the process is stable, there are no bifurcation phenomena in ternary adsorption, but oscillations may occur in numerical and lab experiments due to numerical error propagation or, just like in a monocomponent case, due to oscillations of some system parameter.

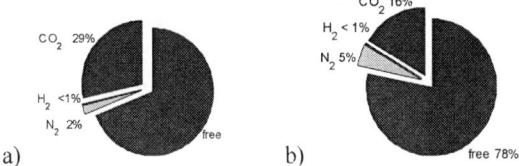

a)   b)

Fig. 4. Equilibrium surface coverage of two different compositions of the anaerobic growth mixture on graphene surface of $1mm^2$, at 300 K, overall pressure of 10^5 Pa. a) 5% of CO_2 and 10% of H_2 in the mixture b) 10% of CO_2 and 5% of H_2.

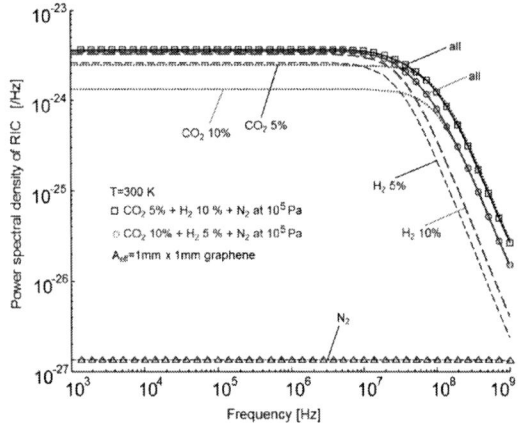

Fig. 5. Power spectral density of refractive index change due to adsorption of three separate gases and their ternary mixture for two different compositions of anaerobic growth mixture, one being CO_2 5% , H_2 10% and another CO_2 10%, H_2 5%.

IV. CONCLUSION

Noise and stability of adsorption-based sensing of ternary gas mixtures are addressed in order to contribute to the research related to the simultaneous multicomponent gas sensing in diverse fields where ternary gas sensing is important. This paper gives complete instructions for the deterministic and stochastic analysis of ternary adsorption, whether modeled as a linear Lagergren/Langmuir process or as a process with second order kinetics, in time and frequency domain, suitable for investigation of noise and response of related affinity-based plasmonic sensors and resonant structures whose operation is adsorption dependent. Additionally, the paper addresses the stability of ternary adsorption. We believe this is a useful analytical and numerical tool for testing new plasmonic materials and new micro and nanostructures.

ACKNOWLEDGEMENT

This work was supported by the Serbian Ministry of Education, Science and Technological Development under Project TR32008.

REFERENCES

[1] D. K. Lim, K. S. Jeon, H. M. Kim, J. M. Nam, Y. D. Suh, "Nanogap-engineerable Raman-active nanodumbbells for single-molecule detection," *Nature Mater*, 2010, vol. 9, pp. 60–67.

[2] M. Li, H. X. Tang, M. L. Roukes. "Ultra-sensitive NEMS-based cantilevers for sensing, scanned probe and very high-frequency applications." *Nature nanotech*. 2, 2007, no. 2, 114.

[3] F. Rossi, O. Kylián, M. Hasiwa, "Decontamination of surfaces by low pressure plasma discharges," *Plasma Processes and Polymers*, 2006, vol. 3, pp. 431-442.

[4] http://www.chemixcalibrationgases.com/biologicalatmosphere-growth-mixture.html, retrieved on March 15, 2019.

[5] L. G. Christophorou, S. R. Hunter. "Ternary gas mixture for diffuse discharge switch." U.S. Pat. 4,751,428, 1988.

[6] J. K. Sell, B. Jakoby, "A simple mid-infrared measurement system based on a tunable filter for the analysis of ternary gas mixtures," *Measur. Sci. Tech*, 2013, vol. 24, art. 084006.

[7] B. Mumyakmaz, A. Özmen, M. A. Ebeoğlu, C. Taşaltın, "Predicting gas concentrations of ternary gas mixtures for a predefined 3D sample space," *Sens. Act. B*, 2008, vol. 128, pp. 594–602.

[8] S. Kim, J. Chen, T. Cheng, A. Gindulyte, J. He, S. He, Q. Li, B. A. Shoemaker, P.A. Thiessen, B. Yu, L. Zaslavsky, J. Zhang, E. E. Bolton. "PubChem 2019 update: improved access to chemical data", *Nucleic Acids Res*, 2019, vol. 47, no. D1, pp. D1102-1109.

[9] http://www.chemaxon.com, retrieved on July 15, 2019.

[10] O. Jakšić, D. Randjelović, Z. Jakšić, Ž. Čupić, Lj. Kolar-Anić, "Plasmonic Sensors in Multi-Analyte environment: rate constants and transient analysis", *Chem. Eng. Research and Design*, 2014, vol. 92, pp. 91-101.

[11] O. Jakšić, I. Jokić, Z. Jakšić, Ž. Čupić, Lj. Kolar-Anić. 2014. "Adsorption-induced fluctuations and noise in plasmonic metamaterial devices." *Physica Scripta* (T162). 014047, 2014.

[12] Z. Jakšić, O. Jakšić, Z. Djurić, C. Kment, "A consideration of the use of metamaterials for sensing applications: field fluctuations and ultimate performance", *J. Opt. A: Pure Appl. Opt*, 2007, vol. 9, pp. S377–S384.

[13] O. Jakšić, Z. Jakšić, Ž. Čupić, D. Randjelović, Lj. Kolar-Anić, "Fluctuations in Transient Response of Adsorption-Based Plasmonic Sensors", *Sens. Act. B: Chemical*, 2014, vol. 190, pp. 419–428.

[14] O. Jakšić, Z. Jakšić, J. Matović, "Adsorption-Desorption Noise in Plasmonic Chemical/Biological Sensors for Multiple Analyte Environment", *Microsys. Tech.*, 2010, vol. 16, no. 5, pp. 735-743.

[15] C. Gardiner, *Stochastic methods: a handbook for the natural and social sciences*, 4th ed., Springer, 2009.

[16] I. Jokić, O. Jakšić, "A second-order nonlinear model of monolayer adsorption in refractometric chemical sensors and biosensors case of equilibrium fluctuations," *Opt. Quantum Electron.*, 2016, vol. 48, no. 353, pp. 1-7.

[17] D. T. Gillespie, A. Hellander, L. R Petzold, "Perspective: Stochastic Algorithms for Chemical Kinetics." *J. Chem. Phys.* 2013, vol. 138 (17) art. 170901.

[18] O. Jakšić, Z. Jakšić, M. Rašljić, Lj. Kolar-Anić, "On Oscillations and Noise in Multicomponent Adsorption: The Nature of Multiple Stationary States", *Adv. in Math. Phys.*, 2019, vol. 2019, Art. 7687643.

[19] O. Jakšić. *ADmoND: MathWorks Matlab Package for Simulation of Monolayer Adsorption Processes in Nano Devices*, Mendeley Data, 2018.

[20] M. S. Mehand, B. Srinivasan, G. De Crescenzo, "Optimizing Multiple Analyte Injections in Surface Plasmon Resonance Biosensors with Analytes Having Different Refractive Index Increments." *Scientific Reports*, 2015, vol. 5, art. 15855.

Effects of Endspoiling on Microchannel Plate Performance

A. Stanković, I. Zlatković, R. Nikolov, B. Brindić, and D. Pantić

Abstract - Microchannel plate (MCP) as a key part of image intensifiers is an electron multiplier device. It consists of several million very thin, conductive glass channels (4-25 micrometers in diameter) fused together. Each channel functions as an independent electron multiplier according to the principle of secondary emission of an electron. During the final stage of manufacturing an MCP, an electrode is evaporated onto the input and output surface of the MCP [1]. During evaporation of the output electrode, the coating penetrates each pore by a depth of about 2 pore diameters which leads to removing the secondary ·emission characteristic from the end of the MCP pore. Its effect on gain and spatial resolution will be depicted in this paper. As well as the influence of deposited layer thickness on surface resistance.

I. INTRODUCTION

Microchannel plates (MCP) are widely used as electron multipliers in image intensifiers for night vision devices. They have many advantages: compactness, high gain, good timing properties due to the low channel length, excellent amplitude distribution and two-dimensional imaging when is used in conjunction with phosphor screen. Night-vision image intensifiers are used for low-light imaging: military, law enforcement, hunting wildlife, border surveillance, navigation, and entertainment. There are three main components of image intensifiers: photocathode, microchannel plate, and phosphor screen. Photocathode converts incident photons into photoelectrons that are amplified by the microchannel plate and converted back into photons by the phosphor screen. When the multiplied cloud of electrons from MCP strike the screen, tens of thousands of photons will be generated for every single photon that was initially converted by the photocathode. The characteristics of each component, as well as the careful consideration of its impact on each other, will contribute to the overall quality of the resulting image.

A. Stanković, I. Zlatković, and D. Pantić are with the Department of Microelectronics, Faculty of Electronic Engineering, University of Niš, Aleksandra Medvedeva 14, 18000 Niš, Serbia, E-mail: aleksandra.stankovic@elfak.rs, ivanzlatkovic90@hotmail.com, dragan.pantic@elfak.ni.ac.rs.

R. Nikolov is with Photon Optronics, Bulevar Svetog Cara Konstantina 80-82, 18000 Niš, Serbia, E-mail: rade.nikolov@photonoptronics.rs

Branislav Brindić is with Sova HD, Bulevar Svetog Cara Konstantina 80-82, 18000 Niš, Serbia.

II. FUNCTIONAL PRINCIPLE OF MICROCHANNEL PLATE (MCP)

Each microchannel represents a continuous electronic multiplier, in which multiplication is performed in the presence of a strong electric field. An electron that enters the channel collides with the channel wall. This collision leads to the release of new electrons, which are moving through the channel along the parabolic paths and re-collide with the channel wall creating new secondary electrons. The result of these repeated collisions is an avalanche of electrons giving an output electron gain of around 10^3 for a typical single MCP at an applied operating voltage of 800 V. After passing one cascade of the electrons through the channel, it takes a certain amount of time to recover (in order of nanosecond) in order to be able to register a new signal. In figure 1 the principle of microchannel plate is shown.

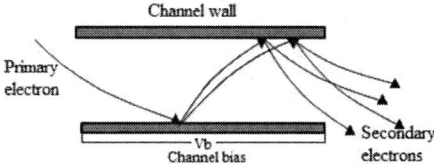

Fig. 1. Functional principle of MCP

III. EVAPORATION PROCESS

The evaporation process is the process of applying a thin film of NiCr or Cr to the input and output side of the MCP. This film represents the contact electrodes to which the voltage is applied.

• Cr is used as a layer because it has good adhesion on the glass if a good vacuum is reached. Also, it's good for the contact layer because it is robust, so MCP can be used again.

• NiCr is good in terms of layer resistance. But it's not very stable, it's sensitive to chemical treatment as well as to ultrasonic warming, it can be easily scratched and the part of the evaporated layer can be peeled off.

The thickness of the electrodes is controlled to have a surface resistance of 100 to 200Ω between the MCP edges. In general, the electrodes are evaporated to uniformly penetrate into the channels. The penetration depth of the output side of MCP, known as end spoiling (Fig.2),

978-1-7281-3420-8/19 $31.00 © 2019 IEEE

significantly affects the angular and energy distributions of the output electrons, it is usually controlled to be deeper in order to collimate the output electrons. The input electrode penetration increases the probability that the first impact of an incident electron occurs on the inner glass surface, and not on the input electrode. [2]

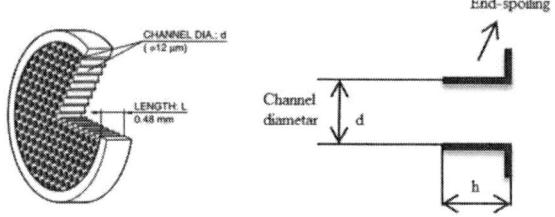

Fig. 2. Structure of MCP (left) and schematic view of Cr penetration into the channel (right)

To create electrical contact, chromium (Cr) layer of specified thickness was deposited by evaporation on the input and output surfaces of the MCP using a SELENE 500 evaporation device. The MCPs are loaded in a rotating planetary fixture, rotating around their axis and around the center of the circle (carousel). So, as the microchannel plate is rotated in a high vacuum, the metallic material is evaporated from the source onto the microchannel plate. MCP rotates about an axis which is parallel with the axis of the MCPs channels, the metallic material (Cr) from the source coats not only onto the input surface of the microchannel plate but also into the channels themselves, depositing Cr in range of 0.3-2.2 pore diameters into the MCP. The distance into the channels to which the metallic electrode material will coat is dependent upon the angulation of the source with respect to the axis of the channels.

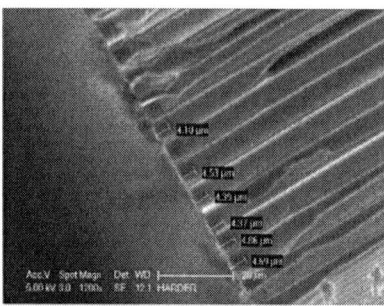

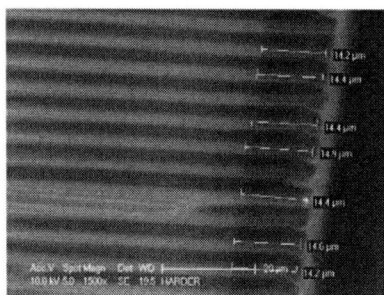

Fig. 3. SEM micrographs of input (up) and output (down) penetration of Cr into the channels of MCP.

The angulation of the channels relative to the evaporation source is held constant to produce a uniform depth of penetration of the metallic electrode coating into the channels.

Figure 3 is an electron microscopic view of a cross-sectioned MCP in the region of input (first picture) and output (second picture) electrode. In the picture, the darker part represents penetration of the metallic electrode that is measured to be around 4μm on the input and around 14μm on the output of MCP.

IV. EXPERIMENTS AND DISCUSSION

The first experiment was performed with different thicknesses of the deposition layer. Measurements of surface resistance are done after evaporation and they are shown in table 1.

TABLE1
RESULTS OF EXPERIMENTS

Parameters	MCP1	MCP2	MCP3	MCP4
Film	Single Cr	Single Cr	Single Cr	Single Cr
Input Angle	32.5°	32.5°	32.5°	32.5°
Output Angle	8.54°	8.54°	8.54°	8.54°
Thickness on input (Å)	1000	1500	2200	2400
Thickness on output (Å)	1000	1500	1400	2400
Surface resistance on input (Ω)	200	190	160	150
Surface resistance on output (Ω)	92	50	53	30
Input penetration	0.6d	0.62d	0.71d	0.66d
Output penetration	2.5d	1.8d	2.4d	1.74d

The goal of the experiment was to examine the influence of this parameter on surface resistance of MCP. The presumption was that with smaller thickness, resistance will be greater, and vice versa. To confirm that, the thickness was set from 1000Å to 2400Å. Input penetration depth h_{in}= (0.3-0.7)*d, output penetration depth h_{out}= (1.7-2.4)*d, where d is diameter of channel. All other parameters were the same: input angle 32.5°, output angle 8.54° and the rate of chromium deposition is 5 Angstroms per second in a high vacuum. As expected, surface resistance was greater with lower thickness of the deposited layer.

A. Influence of End Spoiling on Spatial Resolution

Since each channel of the MCP serves as an independent electron multiplier, MCP resolution is determined by the channel diameter (d) and pitch (center-to-center distance). When the output from MCP is observed with a phosphor screen, the resolution also depends on the MCP electrode depth penetrating into the channels, the MCP-screen gap, and the accelerating voltage. The typical spatial resolution of an MCP composed of 10 µm diameter channels, which is observed with a phosphor screen, is about 40 lp/mm. [3] The unit for spatial resolution is line pairs per millimeter (lp/mm).

In the next experiment, four MCPs are chosen and they've been evaporated in order to examine the influence of deposited chromium layer on the resolution. First, a layer of 1400Å is deposited, for which resistance of approximately 52Ω is obtained. Then, additional 500Å and 1300Å on the first two MCPs, and 1000Å and 750Å on last two, to get final resistance of approximately 24Ω. In table 2 measured surface resistances are given for different layer thicknesses of deposited chromium. After each evaporation process, MCPs are tested. MCP testing represents a check of electrical parameters of the microchannel plate. As well as electronic image which includes multy-to-multy (represents the difference in contrast between the multy-fibers), black spots, FPN (fixed pattern noise- appears as a hexagonal structure on an electronic image), emission. The best difference of layer thickness in the electronic image is seen on MCP 3. It is found that in saturation, FPN is worse for the layer thickness of 3200 Å which is the only indication that the resolution is better, that is shown in figure 4.

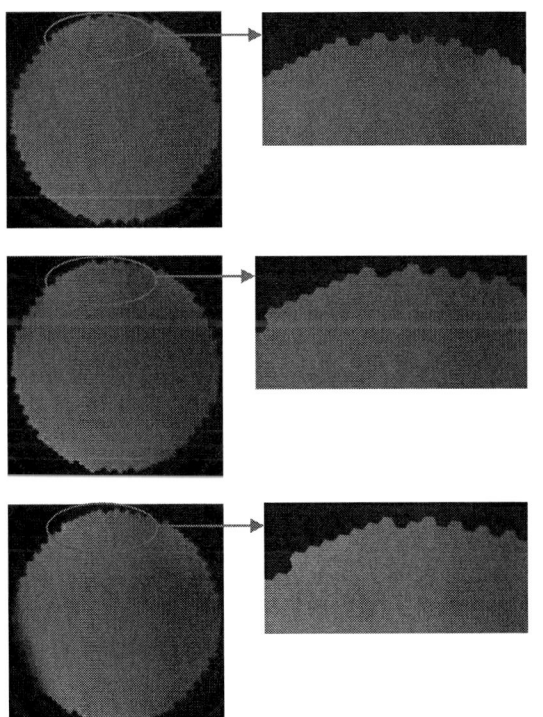

Fig. 4. MCP3 in saturation with different thickness of Cr-layer (1400Å, 2400 Å and 3200 Å, respectively)

Next, measurements of resolution are done in a tube with MCP 18-6 under following conditions: voltage on MCP – 600V, the distance between MCP and screen variate from 0.43 to 0.66 mm; results are shown in Table 3. In the first column, MCP output resistance of the metallization layer is given, and in the third column measured resolution. From the previous said we know that for thicker metallization layer we get lower resistance. For correct comparison modules with very close distances between MCP and screen are chosen (highlighted rows). We can see that thinner Cr-layer showed the lower resolution (64 lp/mm) in compare with thick Cr-layer (68 lp/mm).

TABLE 2
MEASURED SURFACE RESISTANCE FOR DIFFERENT LAYER THICKNESS

Layer thickness	Resistance of MCP1	Resistance of MCP2	Resistance of MCP3	Resistance of MCP4
1400 Å	52 Ω	51.5 Ω	53 Ω	52 Ω
1900 Å	44 Ω	42.5 Ω	/	/
2400 Å	/	/	34 Ω	35 Ω
3200 Å	23 Ω	24 Ω	25 Ω	24 Ω

TABLE3
RESULTS OF MEASURED RESOLUTION IN TUBE

MCP output resistance (Ω)	MCP-screen gap (mm)	Resolution (lp/mm)	SNR
50	0.61-0.63	57	28.2
50	0.59-0.62	57	28.2
50	0.43-0.49	64	29.9
50	0.55-0.59	64	26.2
20-30	0.55-0.57	68	27.2
20-30	0.57-0.6	62	29.8
20-30	0.59-0.66	57	28.7

MCPs with thicker metallization from output work as a lens and collimate the output electron beam. It allows improvement of resolution for minimum one element (from 2 to 4 lp/mm), and the general contrast of the image improves as well. The results are consistent with the literature [4, 5, 6].

B. Influence of End Spoiling on Gain

The gain of an MCP, g, is given by the following equation:

$$g = \exp (G*(L/D)), \qquad (1)$$

where G is the secondary emission characteristics of the channel called gain factor (G~0.23) and L/D is the length-to-diameter ratio of the channel. Generally, L/D is designed to be around 40, which produces a gain of 10^4 with an applied voltage of 1kV. The gain of MCP is mostly affected by the number of the collision of electrons inside the channel and the secondary electron yield per collision. The yield of a secondary electron depends on the L/D ratio and bias angle of MCP. Channels of MCP's are set at a small bias angle to the major axis of the MCP in order to increase the chances of electrons colliding with the channel walls. In correlation with that, metallic coating material, such as chromium, has secondary emission of less than unity, which means that it can produce no more than 1 electron per collision as it is absorbed. Approximately 20% of electrons that impinge the input surface of MCP are lost to electrode coating within channels, that means that the information they carry will be lost so it can't be amplified and contribute to the output of the MCP, leading to decrease in gain. [7] Also, input penetration contributes to a better signal-to-noise ratio (SNR).

V. CONCLUSION

The electron beams emerging from the channels are spread before reaching the phosphor screen, and they reduce the resolution of MCP. End spoiling technique of the output electrode is commonly used to upgrade the resolution of the output section [8]. MCP with thicker metallization from output works as a lens and collimates the output electron beam. It improves resolution for minimum one element (from 2 to 4 lp/mm), general contrast of the image improves as well. However, the endspoiling factor causes a decrease in the gain, as well as input penetration of metallic coating. The typical value of penetration (h) ranges from 1.7d to 2.4d on the output side of MCP and 0.3d to 0.7d on the input. It is found that the best penetration depth on the input side is 0.5d and on the output side 2.4d. Endspoiling is just one of the parameters of MCP that have an influence on gain and resolution. The design and selection of various geometrical and electrical parameters of MCPs must be carefully considered based on specific applications of MCP and requirements for the image intensifier tube.

ACKNOWLEDGMENT

We are grateful to Photon Optronics and Sova HD corporations that have created conditions for performing experiments and measuring characteristics of the microchannel plate.

REFERENCES

[1] J. L. Wiza, "Microchannel plate detectors," *Nuc. Instrum. Methods*, 1979, vol. 162, pp. 587-601.

[2] Shymanska, "Spatial Resolution of Infrared Imaging Systems", *International Journal of Applied Physics and Mathematics*, 2016, vol. 6, pp. 207-217.

[3] "MCP Assembly", Hamamatsu. (n.d.). Retrieved from: https://www.hamamatsu.com/resources/pdf/etd/MCP_TMCP 0002E.pdf

[4] L. Chen, X. Wang, T. Jinshou1, T. Zhao, L. Chunliang, L. Hulin, W. Yonglin, S. Xiaofeng, W. Xing, S. Jianning, S. Shuguang, C. Ping, T. Liping, H. Dandan, G. Lehui, "The Gain and Time Characteristics of Microchannel Plates in Various Channel Geometries", *IEEE Transactions on Nuclear Science*, 2017, vol. 64, issue 4, pp. 1080-1086.

[5] T. H. Hoenderken, C. W. Hagen, J. E. Barth, P. Kruit, "Influence of the microchannel plate and anode gap parameters on the spatial resolution of an image intensifier", *Journal of Vacuum Science & Technology B: Microelectronics and Nanometer Structures*, 2001, Vol. 19, No. 3, p. 843.

[6] G. J. Price G. W. Fraser, "Calculation of the output charge cloud from a microchannel plate", *Nucl. Instrum. Methods Phys. Res. A Accel. Spectrom. Detect. Assoc. Equip*, 2001, vol. 474, pp. 188-196.

[7] R. L. Pierle, M. Gilpin, H. G. Parish, P. Lin, U.S. Patent US5776538A, 1998.

[8] L. Chen, X. Wang, T. Jinshou1, T. Zhao, L. Chunliang, L. Hulin, W. Yonglin, S. Xiaofeng, W. Xing, S. Jianning, S. Shuguang, C. Ping, T. Liping, H. Dandan, G. Lehui, "Simulation of the effects of coated material SEY property on output electron energy distribution and gain of microchannel plates", *Nuclear Instr.*, 2016, Vol. 840, pp. 133-138.

Electrical Characteristics of $Ag_{10}(As_{40}S_{30}Se_{30})_{90}$ as Resistive Switching Material for Potential Application in Memory Devices

K. O. Čajko, D. L. Sekulić, D. M. Petrović, T. B. Ivetić, and S. R. Lukić–Petrović

Abstract - Complex four–component amorphous chalcogenide glass $Ag_{10}(As_{40}S_{30}Se_{30})_{90}$ was synthesized with the melt quenching technique in cascade regime of heating. In this paper, we are focused on the study of electrical characteristics of semiconductor glass $Ag_{10}(As_{40}S_{30}Se_{30})_{90}$ as resistive switching material for potential application in memory devices. Experimental results revealed that bipolar switching mechanism is present in prepared planar shaped $Ag/Ag_{10}(As_{40}S_{30}Se_{30})_{90}/Ag$ glass sample, which is confirmed during multiple cycles at ambient temperature. The observed resistive switching mechanism was attributed to the formation and rupture of the conducting filament. Further, the difference in R_{OFF}/R_{ON} state indicates large storage potential.

I. INTRODUCTION

In view of the physical limitations of typical Si–based flash memory devices, resistive random access memory (RRAM) has been considered as a potential candidate for next generation non–volatile memory device with the advantages of its higher integration density and faster storage speed [1]. Basically, RRAM can be switched under an external electric field and exhibits two resistance switches between a high resistance state (OFF state) and a low resistance state (ON state), which is equivalent to the "erasing" process and "writing" process, respectively [2]. Such a typical resistive switching memory cell consists of a capacitor like structure, in which a switching medium layer (insulating or semiconducting material) is sandwiched between two conductive electrodes [3]. Switching medium layer is a place for information storage.

Besides the choice of storage medium material and selection of electrodes, a switching mechanism in a formed sandwich planar shaped material is also dependent on the operation mode [4]. In the literature [5], it is stated that the resistive switching effect can be observed in two different operation modes depending on the polarity of the applied voltage. Namely, unipolar and bipolar switching modes are distinguished regarding on the same or opposite bias polarities, respectively, depending on the different switching mechanisms and material systems. Further, the explanation of the resistive switching mechanism is based on the creation and partial destruction of conductive filament (CF) [5]. This CF behaves in the form of a bridge when connecting both electrodes corresponds to the state needed for the storage of the information. In numerous studies, investigations of metal filaments grow were carried out on planar shaped chalcogenide glass such as $Ag/Ag–As_2S_3/Au$ [6], Ag_2S [7], $Ag/Ag–Ge–Se/Ni$ [8,9], etc.

In recent years, resistive switching properties have been observed in various materials, including chalcogenide semiconductors, but the switching mechanism has not yet been fully clarified. The fact that when an appropriate voltage is applied the resistance of chalcogenide materials changes by several orders of magnitude, and that the material can be transferred to a conducting state, can be considered as one of the features of this class of materials and allows them to find application and use in different non–volatile memory devices. Advantage of non–volatile devices is enough low temperature i.e. room temperature [10]. Also, it is well known that by doping the chalcogenide glass with metals it changes its properties. Inter alia, base matrix of the glass doped with silver could improve the electrical properties of the material or it could influence the reduction of the resistance.

In this study, preliminary results of resistive switching characteristics of amorphous $Ag_{10}(As_{40}S_{30}Se_{30})_{90}$ chalcogenide for potential application in next generation non–volatile memory devices are presented. Examination of electrical resistivity of the synthesized material and room temperature current–voltage (*I–V*) characteristics of planar shaped $Ag/Ag_{10}(As_{40}S_{30}Se_{30})_{90}/Ag$ were performed with the aim to provide valuable information from the viewpoint of potential application of this functional material in the field of memory devices.

II. EXPERIMENTAL PART

Semiconducting glassy $Ag_{10}(As_{40}S_{30}Se_{30})_{90}$ material was prepared with the melt quenching technique in cascade regime of heating from pure 5N elemental components. Weighted components were vacuum sealed in the ampoule and put in the rocking furnace for 48 hours. Special quartz ampoules that were previously cleaned and prepared were

---

K.O. Čajko, D.M. Petrović, T.B. Ivetić and S.R. Lukić–Petrović are with the Department of Physics, Faculty of Sciences, University of Novi Sad, Trg Dositeja Obradovića 4, 21000 Novi Sad, Serbia, E–mail:kristina.cajko@df.uns.ac.rs,   dragoslav.petrovic@df.uns.ac.rs, tamara.ivetic@df.uns.ac.rs, svetlana@df.uns.ac.rs.

D.L. Sekulić is with the Department of Power, Electronic and Telecommunication Engineering, Faculty of Technical Sciences, University of Novi Sad, Trg Dositeja Obradovića 6, 21000 Novi Sad, Serbia E–mail: dalsek@uns.ac.rs

---

978-1-7281-3420-8/19 $31.00 © 2019 IEEE

used for the synthesis that can withstand high temperatures. Length of the ampoule was 25 cm with internal diameter 1.3 cm and thickness of the wall being 2 mm. The technological map consisted of six plateaus and two heating rates with the maximum temperature of the synthesis being 950 ^{0}C. Heating of the elemental components began at room temperature, which automatically increased up to 250 ^{0}C at a rate of 2 ^{0}C/min. After reaching the determined temperature, the ampoule was maintained on the first plateau for the next 4 hours. In the second heating stage, the oven temperature at the same rate increased up to 350 ^{0}C where the ampoule was kept for 5 hours. In the third and fourth part of the cascade heating process, temperatures were reached up to 450 ^{0}C and 550 ^{0}C, respectively, at the same heating rate 2 ^{0}C/min as in the previous stages. The lengths of temperature plateaus at these temperatures are the same and last for 4 hours. The heating rate in the last two stages of the experiment was 1 ^{0}C/min, where the fifth temperature plateau was 850 ^{0}C, and the ampoule was kept at this temperature for 8 hours. The last sixth temperature plateau was at a temperature of 950 ^{0}C. The ampoule with the melt was held at that temperature for at least 12 hours. It was then taken out from the oven and air–quenched. After quenching it was placed in alumina powder (Al_2O_3) and left approximately 24 hours until complete cooling. This ensures fast cooling at room temperature and reduction of stress in the newly formed glass. The heat treatment profile with the map can be found in our recent study [11]. After a complete synthesis process, the prepared sample was taken out by breaking the ampoules. Images of the ampoule and synthesized glass are presented in the Fig.1.a. Visually it can be noted that the synthesized material posses the outer surface that is perfectly smooth with a glossy shine and the color is distinctly metallic gray. It is characterized by shell–like fracture, which is one of the main features of amorphous materials.

Fig. 1. (a) Images of the ampoule and synthesized $Ag_{10}(As_{40}S_{30}Se_{30})_{90}$ glass with its fractures and surface; (b) Prepared Ag/Ag_{10} $(As_{40}S_{30}Se_{30})_{90}/Ag$ in planar shape.

The resistivity of synthesized $Ag_{10}(As_{40}S_{30}Se_{30})_{90}$ glass was measured by two probe method as a function of temperature to confirm its semiconductor nature. Measurement was performed in the temperature range from 298 K to 423 K.

In order to examine the resistive switching behaviour, the room temperature current–voltage (*I–V*) characteristics planar shaped $Ag/Ag_{10}(As_{40}S_{30}Se_{30})_{90}/Ag$ were measured by

a computer–controlled Source Meter Keithley 2410 with two electrodes. In this measurement setup, a personal computer with in–house built software implemented by LabVIEW was used for acquisition of measured data. Preparation of the sample consisted of a forming a plan parallel plate by polishing with abrasives of different grain size up to thickness 0.975 mm, where electrodes were applied in the form of silver paste. Electrodes were deposited on larger areas ~ 13.6 mm^2 of the sample forming the top and bottom electrodes and left for 48 hours to cure. Such prepared planar shaped $Ag/Ag_{10}(As_{40}S_{30}Se_{30})_{90}/Ag$ for the *I–V* measurements is presented in Fig. 1.b.

III. RESULTS AND DISCUSSION

The variation of obtained values of electrical resistivity (ρ) in the temperature range from room temperature to 423 K is shown in Fig. 2. It is evident that ρ exponentially decreases with the temperature indicating that the conduction is a thermally activated process. Such behaviour is quite common for amorphous chalcogenides and suggests that prepared $Ag_{10}(As_{40}S_{30}Se_{30})_{90}$ glass possesses a negative temperature coefficient of resistance (NTCR) behaviour usually shown by semiconductor materials [12]. This result is in accordance with the available data of the resistance of chalcogenide glasses from the $Ag_x(As_{40}S_{30}Se_{30})_{100-x}$ system, $1 \leq x \leq 5$ at. % Ag [13].

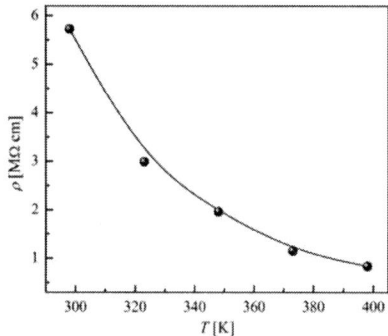

Fig. 2. Temperature dependence of electrical resistivity of prepared $Ag_{10}(As_{40}S_{30}Se_{30})_{90}$.

Fig. 3. shows the room temperature *I–V* curve of prepared planar shaped $Ag/Ag_{10}(As_{40}S_{30}Se_{30})_{90}/Ag$ by sweeping the DC voltage in the sequence of $0V \rightarrow 11V \rightarrow 0V \rightarrow -11V \rightarrow 0V$. The sweeping voltage step was 0.1V. It is obvious that bipolar resistive switching behavior with good reproducibility in this structure is present. Namely, when voltage ranging from 0 V to 11 V was applied to the Ag electrode, the current increased gradually reaching its maximum value at about 6.8 V, indicating a switch from high resistance (OFF state) to low resistance (ON state), which is called a SET process. Afterward, as the applied voltage swept from 0 V to −11 V, the current decreased reaching its minimum value at about −7.1 V and transition from low resistance (ON state) to high resistance (OFF state) appeared, regarded as RESET process. The arrows in Fig. 3 indicate the sweeping directions of the applied voltage.

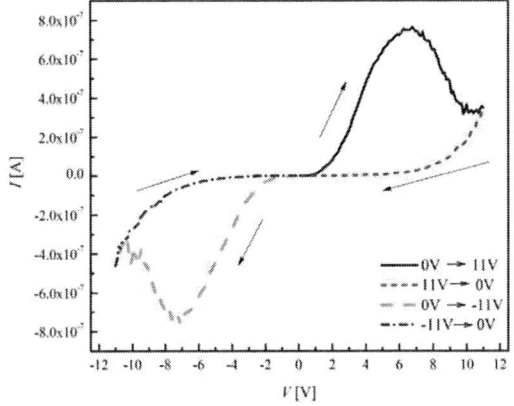

Fig. 3. Room temperature current–voltage characteristic of planar shaped $Ag/Ag_{10}(As_{40}S_{30}Se_{30})_{90}/Ag$. Arrows denote the direction of the voltage sweep.

In the Fig. 4 is given the I–V curve in logarithmic scale. It can be observed from the graphic that the obtained I–V curve shows a relatively symmetric feature.

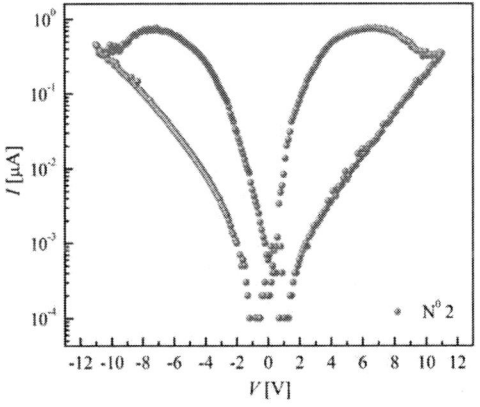

Fig. 4. I–V curve in logarithmic scale during a bipolar voltage sweep.

Fitting the linear parts of the I–V curve given in Fig. 3 can be estimated the values of resistance in OFF and ON states, denoted as R_{OFF} and R_{ON}, respectively. It was determined that the ratio R_{OFF}/R_{ON} is about two orders of magnitude. This ratio in OFF and ON states provides a memory window in prepared planar shaped $Ag/Ag_{10}(As_{40}S_{30}Se_{30})_{90}/Ag$ which demonstrates the good potential for constructing a device for non–volatile memory applications [3]. In the literature [14], it is usually stated that the difference between these states R_{OFF}/R_{ON} can range

from 10 to 10^3, further large ratios are desirable for good storage information providing better differentiation.

Finally, recording of the I–V curve was repeated three times where it remained its pattern showing stable reproducible response as can be seen in Fig. 5.

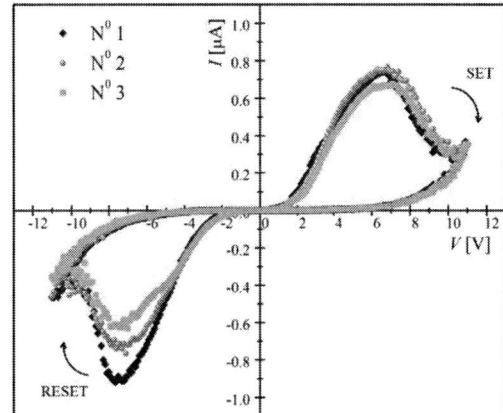

Fig. 5. Repeatability of the current–voltage characteristic of planar shaped $Ag/Ag_{10}(As_{40}S_{30}Se_{30})_{90}/Ag$ obtained during three measurement cycles.

Establishment of SET ("writing") and RESET ("erasing") processes could be explained by the formation and degradation of the so–called conductive filament in silver ion material [15]. Namely, in the literature it is pointed out that the ionic conduction in chalcogenide glasses is of importance for the current bridging random access memories, whereby the ON/OFF changes are interpreted by formation of special pathway and later its degradation if opposite voltage polarity is applied between the two electrodes consisting of metal [15,16]. It is also indicated that such effects appear in systems containing metal, some chalcogen element and an additive, which in the case of the synthesized $Ag_{10}(As_{40}S_{30}Se_{30})_{90}$ sample is completely satisfied (Ag – metal, S and Se – chalcogen elements and As has a roll of additive). According to this fact, it can be assumed that the observed behaviour of this chalcogenide glass and its transformation between OFF and ON states are the result of the auto-diffusion and migration of silver ions due to the applied electric field. The assumption that there are silver ions, which would carry out the aforementioned transformation, is based on the fact that in our previous study [17] the phase separation in this sample, i.e. presence of silver–rich regions was indicated. Namely, the existence of phase separation in $Ag_{10}(As_{40}S_{30}Se_{30})_{90}$ is shown through performed DSC measurements [17]. Furthermore, results related to the investigation of the optical band gap for the samples with $x \leq 5$ at. % Ag indicated the possibility of phase separation for compositions with a higher silver content than $x = 5$ at. % Ag [11]. This also supports the existence of filaments responsible for the mentioned effect.

IV. CONCLUSION

In this paper, the temperature dependence of electrical resistivity of the synthesized $Ag_{10}(As_{40}S_{30}Se_{30})_{90}$ glass confirmed its NTCR characteristics. Preliminary results of a study of the resistive switching effect in a planar shaped chalcogenide glass with Ag electrodes for potential application in the formation of switching components and their application in the memory devices are given. Analysis of the recorded room temperature current–voltage curve revealed that this chalcogenide material possesses a bipolar switching mechanism, showing its stable reproducible response. Also, by fitting the obtained current–voltage curve the resistance value in ON (set) and OFF (reset) state was estimated. The results showed that the proposed planar configuration displays a large resistance window, namely $R_{OFF}/R_{ON} \approx 10^2$. The writing and erasing processes were explained through the creation of a conductive pathway by silver ions and its rupture, respectively. Obtained results in this study indicate that this functional material possesses a wide range of application in non–volatile memory devices.

ACKNOWLEDGEMENT

This research was financially supported by the Provincial Secretariat for Higher Education and Scientific Research of the Autonomous Province of Vojvodina through Projects: "Optimization of metal content in chalcogenide matrix as a basis for application in electronic components" (142-451-3441/2018-02) and "Properties and electrical characteristics of doped amorphous chalcogenide materials and nanostructured ceramics" (142-451-2080/2019-01), and by the Ministry of Education, Science and Technological Development of the Republic of Serbia through Project No. ON171022.

REFERENCES

[1] Y. Yu, F. Yang, S. Mao, S. Zhu, Y. Jia, L. Yuan, M. Salmen and B. Sun, "Effect of anodic oxidation time on resistive switching memory behavior based on amorphous TiO2 thin films device", *Chem. Phys. Lett.*, 2018, 706, pp.477–482.

[2] D. Zhou, F. Chen, S. Han, W. Hu, Z. Zang, Z. Hu, S. Li and X. Tang, "Resistive switching characteristics of AgInZnS nanoparticles", *Ceram. Int.*, 2018, 44, S152– S155.

[3] P. Han, B. Sun, S. Cheng, F. Yu, B. Jiao and Q. Wu, "An optoelectronic resistive switching memory behavior of Ag/α–SnWO₄/FTO device", *J. Alloys Compd.*, 2016, 681, pp.516–521.

[4] F. Pan, S. Gao, C. Chen, C. Song and F. Zeng, "Recent progress in resistive random access memories: Materials,

switching mechanisms, and performance", *Materials Science and Engineering R*, 2014, 83, pp.1–59.

[5] Y. Li, S. Long, Y. Liu, C. Hu, J. Teng, Q. Liu, H. Lv, J. Suñé and M. Liu, "Conductance Quantization in Resistive Random Access Memory", *Nanoscale Research Letters*, 2015, 10:420

[6] Y. Hirose, H. Hirose, "Polarity dependent memory switching and behavior of Ag dendrite in Ag photodoped amorphous As_2S_3 films", *J. Appl. Phys.*, 1976, vol. 47, No.6, pp.2767–2772.

[7] M. Morales–Masis, S.J. van der Molen, W.T. Fu, M.B. Hesselberth and J.M. van Ruitenbeek, "Conductance switching in Ag_2S devices fabricated by in situ sulfurization", *Nanotechnology*, 2009, 20, 095710 (6pp).

[8] M.N. Kozicki, M. Mitkova, "Mass transport in chalcogenide electrolyte films–materials and applications", *J. Non–Cryst. Solids*, 2006, 352, pp.567–577.

[9] M.N. Kozicki, C. Ratnakumar, M. Mitkova, "Electrodeposit Formation in Solid Electrolytes" in: *Proc. of Non–Volatile Memory Technology Symposium*, 2006, pp.111–115.

[10] M. Frumar, T. Wagner, K. Shimakawa and B. Frumarova, "Crystalline and Amorphous Chalcogenides, High–Tech Materials with Structural Disorder and Many Important Applications", in *Nanomaterials and Nanoarchitectures, A Complex Review of Current Hot Topics and their Applications, NATO Science for Peace and Security Series - C: Environmental Security*, M. Bardosova and T. Wagner, Eds., pp. 151–238, Springer, 2013.

[11] K.O. Čajko, S.R. Lukić–Petrović and D.D. Štrbac, "Absorption Edge and Optical Band Gap of Ag–$As_{40}S_{30}Se_{30}$ Amorphous Samples", *Acta Phys. Pol. A*, 2015, vol. 127, No. 4, pp.1286–1288.

[12] S. Cui, D. Le Coq, C. Boussard–Pledel and B. Bureau, "Electrical and optical investigations in Te–Ge–Ag and Te–Ge–AgI chalcogenide glasses", *J. Alloys Compd.*, 2015, 639, pp.173–179.

[13] K.O. Čajko, D.L. Sekulić, S. Lukić–Petrović, M.V. Šiljegović and D.M. Petrović, "Temperature–dependent electrical properties and impedance response of amorphous $Ag_x(As_{40}S_{30}Se_{30})_{100-x}$ chalcogenide glasses", *J. Mater. Sci: Mater Electron*, 2017, 28, pp.120–128.

[14] Y. Gonzalez–Velo, H. J. Barnaby and M. N. Kozicki, "Review of radiation effects on ReRAM devices and technology", *Semicond. Sci. Technol.*, 2017, 32, 083002 (44pp).

[15] B. Zhang, P. Kutalek, P. Knotek, L. Hromadko, J. M. Macak and T. Wagner, "Investigation of the resistive switching in Ag_xAsS_2 layer by conductive AFM", *Appl. Surf. Sci.*, 2016, 382, pp.336–340.

[16] V. Sousa, "Chalcogenide materials and their application to Non–Volatile Memories", *Microelectronic Engineering*, 2011, 88, pp.807–813.

[17] K.O. Čajko, S.R. Lukić–Petrović, M. V. Šiljegović, G.R. Štrbac, D.M. Petrović, "Specificity of thermally induced crystallization in the glasses of Ag–As–S–Se system", in *Proc. 10ʰ International Scientific Conference Contemporary Materials 2017*, Banja Luka, 2018, pp.171–182.

Modelling of ΔV_T in NBT Stressed P-Channel Power VDMOSFETs

N. Mitrović, D. Danković, Z. Prijić, and N. Stojadinović

Abstract – Negative bias temperature instabilities in commercial IRF9520 p-channel power VDMOSFETs were studied in order to design an analytical model for this effect. A modelling framework is proposed, which tends to be in line with earlier obtained experimental data. Since the pulsed voltage stressing caused generally lower shifts as compared to static stressing performed at the same temperature with equal stress voltage magnitude, different kind of models are considered for static and pulsed stressing. Differences between static and pulsed NBT stress depend on both stress duty cycle and frequency, and the differences become more significant as the duty cycle decreases and frequency increases. Because of that, described model intent to apply to specific stress signal, but also to be adaptable to different types of stress signal. Modelling of threshold voltage shifts induced by pulsed negative bias temperature stress has been done on the bases of experimental results.

I. INTRODUCTION

Negative bias temperature instabilities (NBTI) have been found to occur mostly in p-channel MOSFETs operated at elevated temperatures (100 - 250°C) under negative gate oxide fields in the range 2 - 6 MV/cm [1-3]. NBTI are manifested as the decrease in device transconductance and absolute drain current and the increase in device threshold voltage (V_T) and absolute "off" current. NBTI have been identified as a critical limiting factor that determines device lifetime.

In this paper, we will analyze the threshold voltage instabilities observed in p-channel power VDMOS (Vertical Double Diffused MOS) transistors subjected to NBT stressing. The main part of this research is the threshold voltage modelling based on the experimental data.

II. EXPERIMENTAL RESULTS

Devices used in this study were commercial p-channel power VDMOSFETs IRF9520 [4]. Degradation of power MOS devices under various stresses (irradiation, high field, temperature, and even hot carrier) has been subject of extensive research [5-9], but very few authors seem to have addressed the NBTI in these devices. Practical realization of the NBT stress and measurement setup is explained in

N. Mitrović, D. Danković, Z. Prijić, N. Stojadinović, are with the Faculty of Electronic Engineering, University of Niš, Aleksandra Medvedeva 14, 18000 Niš, Serbia, E-mail: nikola.mitrovic.nis@gmail.com

N. Stojadinović is with the Serbian Academy of Sciences and Arts (SASA), Knez Mihailova 35, 11000 Beograd, Sebia.

details in [5]. To provide suitable results for ΔV_T modelling, we have done experiments on different groups of devices subjected to pulsed NBT stressing under typical conditions (-45 V, 175°C) for 24 hours, with different on-times and off-times. In this paper, accent is on the stress signal with DTC = 50%, where both pulse on- and pulse off-time are 50 μs, and on static stress signal.

The results for these stress signals are shown in Fig. 1. As can be seen, NBT stress-induced threshold voltage shifts are most significant in the case of static stress.

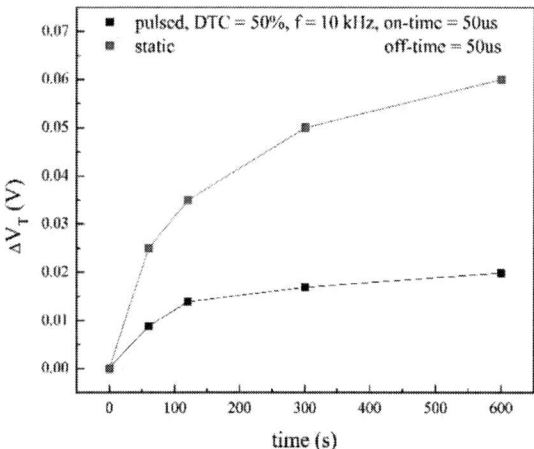

Fig. 1. Experimental results for threshold voltage during the static and pulsed NBT stress in p-channel power VDMOSFETs IRF9520.

Our previous results have pointed to the existence of characteristic time constant (25 μs) related to the recoverable and permanent components of stress-induced degradation [10]. Namely, it was found that 25 μs off-time of the pulsed stress voltage was sufficient to completely remove the recoverable component of degradation created during the preceding pulse on-time. This investigation is in line with literature data [11, 12]. Fig. 2 shows graphical illustration of threshold voltage shifts during the static and pulsed NBT stress.

III. MODELS FOR PULSED AND STATIC NBTI

There have been several efforts in developing analytical model for NBTI [13]. In the interest to fully model the specific characteristic, micro-level signal must be analyzed first. Experiments have shown that micro-level signal has specific form displayed in Fig. 3.

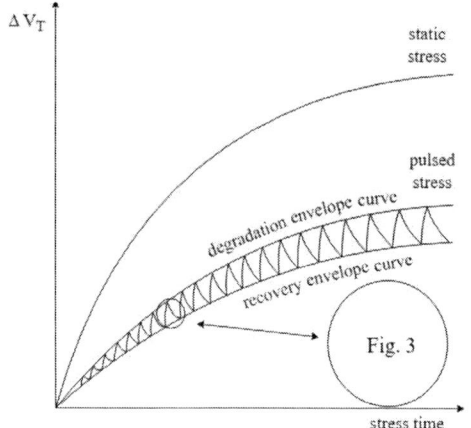

Fig. 2. Graphical illustration of NBTI.

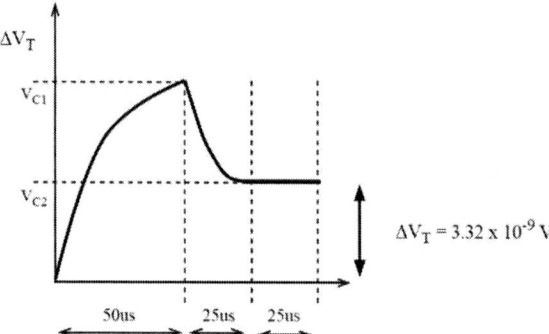

Fig. 3. Graphical illustration of micro-level signal during the pulsed NBT stress.

Rising part of the curve can be presented as charging of capacitor (responds to first 50 µs of period), while descending part of the curve must be divided in two parts, decreasing part (recoverable component) and constant part (permanent component) [10]. Decreasing part can be presented as discharging of capacitor, while at constant part, discharging must be stopped. How experimental results show that recoverable component has specific duration (25 µs), other 50 µs of the micro-level signal period must be split in half, 25 µs for recoverable component and 25 µs for permanent component. That means that the modeling should be done in three parts. In order to perform modeling an electrical circuit is designed, as shown in Fig 4.

Pulsed voltage is applied to capacitor through resistor R_I, and an ideal diode D_I, so that capacitor C is charging. Capacitor is discharging through resistor R_D and second ideal diode D_2. Ideal diodes are connected solely because of their directional properties. Diode D_2 ensures that pulsed voltage will completely be used to charge the capacitor, while diode D_I ensures that capacitor will be discharged through resistor R_D. Resistor R_I is used to limit the charge rate of the capacitor, and resistor R_D is used to limit the discharge rate of the capacitor. Also, very important part of

the circuit is voltage controlled switch S_I, connected to the discharging part of the circuit. Since discharging of the capacitor should be allowed only for 25 µs (in this concrete case, from 50 µs to 75 µs), controlling signal for voltage controlled switch should be set high only during that time through one period. That way, discharging can be enabled only for appropriate interval. Voltage on capacitor, V_C, corresponds to ΔV_T produced by this defect. It is important to note that for the circuit to operate on different conditions, pulsed voltage for the stress and the controlling voltage for the switch must be synchronized.

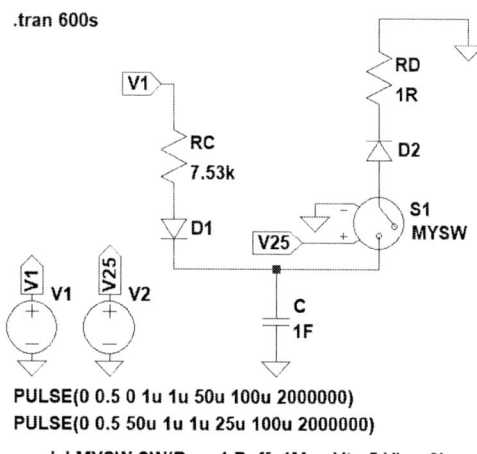

Fig. 4. Circuit for pulsed NBT stress modelling.

To determine the values of pointed resistors and capacitor, experimental results must be further analyzed first. Fig. 1 shows that for 600 seconds, ΔV_T will rise from 0 V to 0.0199 V [14]. Since the frequency of the applied signal is 10 kHz, period is 100 µs. In 600 seconds, 6×10^6 pulses are applied, and therefore, 6×10^6 micro-level signal are produced. Since ΔV_T for described amount of pulses is experimentally measured, increment of each micro-level signal, ΔV_T, can be calculated. As previously noted, for chosen pulsed signal, only permanent component of off-stress-induced degradation contributes to the ΔV_T growth, so, calculated $\Delta V_T = 3.32 \times 10^{-9}$ V presents remainder between V_{C1} (voltage until which a capacitor is charged during a period) and V_{C2} (voltage until which a capacitor is discharged during a period). With appropriate equations, circuit can be fully described.

Voltage to which capacitor is charged during charge zone presents voltage from which capacitor starts to discharge in discharge zone. Therefore, equation that describes the circuit transforms to:

$$V_{C2} = V_F \left(1 - e^{-\frac{t_c}{R_C\,C}} \right) \left(e^{-\frac{t_d}{R_D\,C}} \right) \qquad (1)$$

In Eq. 1 unknown values are values of resistance of resistors R_D and R_C, and value of capacitance of capacitor C. In order to try to calculate unknown values, some assumptions must be made. So, we have to choose R_D and C, in order to calculate resistor R_C from Eq. 1. In order to further simplify calculation for unknown elements, referent resistance of R_D will be set to 1 Ω, and referent capacitance of capacitor will be set to 1 F. With these assumptions, calculation is further proceeded.

$$3.32 \cdot 10^{-9} = 5 \cdot 10^{-1} \cdot \left(1 - e^{-\frac{5 \cdot 10^{-5}\,s}{R_C \cdot 1}} \right) \cdot \left(e^{-\frac{2.5 \cdot 10^{-5}\,s}{1 \cdot 1}} \right) \quad (2)$$

In Eq. 2, only unknown value is R_C. Calculation is easily continued, leading to:

$$R_C = 7.527 \text{ k}\Omega \quad (3)$$

Using the calculated values of the elements of electrical circuit micro-level simulation is performed and experimental results are shown in Fig. 5.

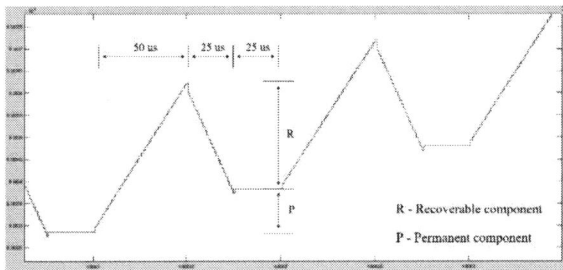

Fig. 5. Micro-level simulation results.

As can be seen in Fig. 5 simulation outputs match experimental results in terms of the shape of the curve. Therefore, a complete simulation can be done for entire time range of 600 s. Comparison between experimental and simulation results is given in Fig. 6.

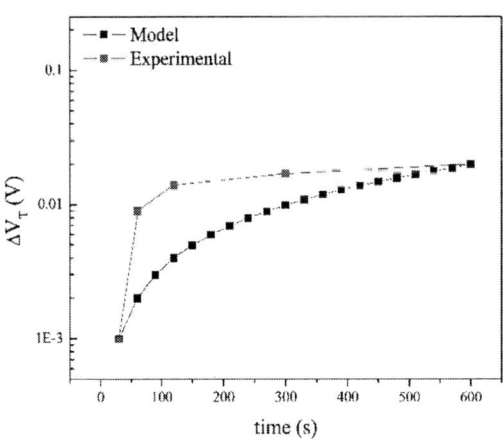

Fig. 6. Comparison between experimental and modelled results of pulsed NBT stress given in lin-log scale.

Considering that the value of exponent of the curve is mostly diverse in the starting part of the curve, and that for modelling a constant rising was predicted, largest deviation is noticed in the beginning. With time, how exponent of the curve tends to stabilize, deviation is significantly smaller.

While for pulsed stress degradation both charging and discharging must be taken into account, for static stress, circuit is rather simplified, mostly because of absence of discharging part of the circuit (Fig. 7). That fact is also in line with experimental measurements that points out that ΔV_T growth is faster with static stress [10, 14].

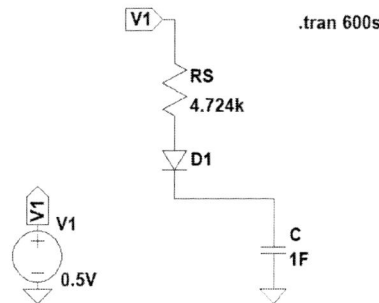

Fig. 7. Circuit for static NBT stress modelling.

Equations for the model are slightly changed. Since only the charging of the capacitor should be assumed, to describe the circuit, standard capacitor equation is used:

$$V_{C1} = V_F \left(1 - e^{-\frac{t_c}{R_S C}} \right) \quad (4)$$

Since that capacitor value is already marked as $C = 1$ F, only unknown value in Eq. 4 is resistor R_S. Taking in consideration experimental results, and known values of stress signals, solving of the equations leads to solution:

$$R_S = 4.724 \text{ k}\Omega \quad (5)$$

Simulation results obtained using this circuit are shown in Fig. 8. As already mention, in the starting part of

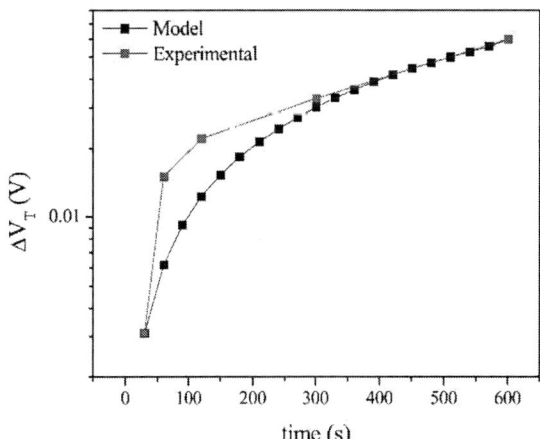

Fig. 8. Comparison between experimental and modelled results of static NBT stress given in lin-log scale.

the curve, due to instability of growth, deviation is detected. However, for the largest part of the period, results provide noticeably better match, that tend to continue for further measuring.

IV. CONCLUSION

Described model is designed based on experimental measurements, suggesting that the modelling needs further development. For both static and specific pulsed stress of VDMOSFET, model provides similar curves, but it needs to be detailly improved to obtain a perfect match. It would be good, to develop a model based on experimental results for randomly selected values of frequency and duty cycle that would match the recovery curve.

ACKNOWLEDGEMENT

This work has been supported by the Ministry of Education, Science and Technological Development of the Republic of Serbia under the Grants no. OI-171026 and TR-32026, and by Serbian Academy of Sciences and Arts (SASA) under the Grant no. F-148.

REFERENCES

[1] J.H. Stathis, S. Zafar, "The negative bias temperature instability in MOS devices: A Review", *Microelectronics Reliability*, 2006, vol. 46, pp. 270-286.

[2] D.K. Schroder and J.A. Babcock, "Negative bias temperature instability: Road to cross in deep submicron silicon semiconductor manufacturing", *Journal of Applied Physics*, 2003, vol. 94, pp. 1-18.

[3] S. Ogawa, M. Shimaya, and N. Shiono, "Interface-trap generation at ultrathin SiO₂ (4-6 nm)-Si interfaces during negative-bias temperature aging", *Journal of Applied Physics*, 1995, vol. 77, pp. 1137-1148.

[4] IRF9520N Data sheet, Int. Rectifier 1998, www.irf.com

[5] I. Manić, D. Danković, A. Prijić, Z. Prijić, N. Stojadinović, "Measurement of NBTI Degradation in p-channel Power VDMOSFETs", *Informacije MIDEM, Journal of Microelectronics, Electronics Components and Materials*, 2014, vol. 44, no. 4, pp. 280-287.

[6] N. Stojadinović, I. Manić, V. Davidović, D. Danković, S. Djorić-Veljković, S. Golubović, S. Dimitrijev, "Electrical stressing effects in commercial power VDMOSFETs", *IEE Proceedings: Circuits, Devices and Systems*, 2006, vol. 153, no. 3, pp. 281-288.

[7] N. Stojadinović, S. Golubović, V. Davidović, S. Djorić-Veljković, S. Dimitrijev, "Modelling radiation-induced mobility degradation in MOSFETs", *Physica Status Solidi (A)*, 1998, vol. 169, no.1, pp. 63-66.

[8] M.M. Pejović, "P-channel MOSFET as a Sensor and Dosimeter of Ionizing Radiation", *Facta Universitatis, Series: Electronics and Energetics*, 2016, vol. 29, no. 4, pp. 509-541.

[9] V. Davidović, D. Danković, S. Golubović, S. Djorić-Veljković, I. Manić, Z. Prijić, A. Prijić, N. Stojadinović, S. Stanković, "NBT stress and radiation related degradation and underlying mechanisms in power VDMOSFETs", *Facta Universitatis Series: Electronics and Energetics*, 2018, vol. 31, no. 3, pp. 367-388.

[10] D. Danković, N. Stojadinović, Z. Prijić, I. Manić and A. Prijić, "Recoverable and Permanent Components of VT Shift in Pulsed NBT Stressed P-Channel Power VDMOSFETs", in *Proc. 29th International Conference on Microelectronics (MIEL 2014)*, Belgrade (Serbia), May 2014, pp. 297-300.

[11] T. Nigam, "Pulse-stress dependence of NBTI degradation and its impact on circuits", *IEEE Transactions on Device and Materials Reliability*, 2008, vol. 8, no. 1, pp. 72 -78.

[12] M.-F. Li, D. Huang, C. Shen, T. Yang, W. J. Liu, and Z. Liu, "Understand NBTI Mechanism by Developing Novel Measurement Techniques", *IEEE Transactions on Device and Materials Reliability*, 2008, vol. 8, no. 1, pp. 62 -71.

[13] M.A. Alam, "A Critical Examination of the Mechanics of Dynamic NBTI for PMOSFETs", in *Technical Digest of the IEDM2003*, Washington, DC (USA), 2003, pp. 345-348.

[14] D. Danković, I. Manić, N. Stojadinović, Z. Prijić, S. Djorić-Veljković, V. Davidović, A. Prijić, A. Paskaleva, D. Spassov, and S. Golubović, "Modelling of Threshold Voltage Shift in Pulsed NBT Stressed P-Channel Power VDMOSFETs", in *Proc. 30th International Conference on Microelectronics (MIEL 2017)*, Nis (Serbia), October 2014, pp. 297-300.

Comparison of Radiation Characteristics of HfO$_2$ and SiO$_2$ Incorporated in MOS Capacitor in Field of Gamma and X Radiation

S. Stanković, D. Nikolić, N. Kržanović, L. Nadjdjerdj, and V. Davidović

Abstract – The paper presents the application of a numerical method for the determination of the absorbed dose rate of gamma and X radiation in the dielectric thin layer of hafnium dioxide (HfO$_2$) or SiO$_2$, which is located in the structure of the MOS capacitor. Considering the radiation characteristics of the selected dielectrics, it can be concluded that there are advantages of HfO$_2$ over SiO$_2$ in the radiation field with high-energy X-rays. Similar radiation effects should be expected for the interaction of dielectric material with gamma ray photons originating from a Co-60 source, both in the dielectric layer with SiO$_2$ and in that with HfO$_2$. It can be concluded that for the same radiation absorbed dose into the MOS capacitor with HfO$_2$, there are a greater number of generated electron-hole pairs, in which case the value of effective trapping efficiency is smaller than if SiO$_2$ was used.

I. INTRODUCTION

In recent years, modern semiconductor MOS structures were designed with a silicon dioxide (SiO2) as the selected material. The advancement of new techniques for detection of ionizing radiation requires the development of electronic components, which in their structure have new materials with improved radiation characteristics. In this paper we will consider the possibility of using hafnium dioxide (HfO$_2$) as a new material that could be used in semiconductor technology instead of silicon dioxide (SiO$_2$). Hafnium dioxide (HfO$_2$) has been considered to be one of reasonable alternative solutions due to a suitable high dielectric constant (k > 20), the relatively large band gap (5.68 eV) and thermal stability with Si substrate [1-4].

In order to investigate the basic structural and electrical characteristics of hafnium dioxide, it has been shown that it is suitable to use a metal oxide semiconductor (MOS) capacitor. In one inspirational study [5], the HfO$_2$ thin film have been deposited on p-type (1 0 0) silicon wafer using RF magnetron sputtering technique. The Al/HfO$_2$/Si MOS capacitors have been fabricated and after the post metallization, the high frequency CV measurements

S. Stanković, D. Nikolić, N. Kržanović and L. Nadjdjerdj are with the Institute of nuclear sciences "VINČA", University of Belgrade, Mike Petrovića Alasa 12-14, 11351 Belgrade, Serbia, E-mail: srbas@vin.bg.ac.rs

V. Davidović is with the Faculty of Electronic Engineering, University of Niš, Aleksandra Medvedeva 14, 18000 Niš, Serbia, E-mail: vojkan.davidovic@elfak.ni.ac.rs

were conducted suggesting that the flat/band voltage shifts towards negative potential due to positive trap charge.

The annealed HfO$_2$ films showed very low leakage current about 10^{-7} A/cm^2 at 1 V and the dielectric constant, equivalent oxide thickness , barrier height, effective charge carriers and interface trap density are determined 22.47, 1.64 nm, 1.28 eV, $9.3 \cdot 10^9$, $9.25 \cdot 10^{11}$ cm^{-2} eV^{-1} respectively [5]. In another study, electrical measurements of MOS capacitor with Al/HfO$_2$/Si gate stacks indicate that 300 ^{0}C annealed HfO$_2$ sample demonstrates improved electrical performance [6]. A low interface-state density of $2.82 \cdot 10^{12}$ cm^{-2} eV^{-1}, small border trap density of $1.28 \cdot 10^{11}$ cm^{-2} and low leakage current of $3.65 \cdot 10^{-5}$ A/cm^2 at applied substrate voltage of 2 V have been obtained.

The researchers experimentally studied the behavior of MOS capacitor with HfO$_2$ in the gamma radiation field and observed that the midgap and flatband voltage shifts show that radiation does not cause significant variation, especially at low doses [7-9]. The leakage current through the hafnium oxide increases with the increase of total dose, and this can be attributed to the ionic irradiation induced electron-hole pairs and their transport through the film of HfO$_2$. In doing so, the interface barrier height decreases with the increase of total dose, which can be attributed to the oxygen vacancy generation and charge trapping in hafnium oxide [10]. To understand the physical significance of trapped charge densities, it is necessary to do its estimation with an effective trapping efficiency for MOS devices [9,11]. Since the MOS capacitor was found to be changing in the field of radioactive radiation in its characteristics of different types, the question was raised how, after the experiments, the results should be explained using appropriate models of physical electronics. Considering that there is a dependence in the change of effective trapping efficiency on the total absorbed dose, this paper conducted a study comparing the radiation characteristics for SiO$_2$ and HfO$_2$, that were used in a numerical method based on a physical model of X transport and gamma radiation through a thin layer of dielectric oxide in a MOS capacitor.

II. NUMERICAL METHOD

The paper presents the application of a numerical method for the determination of the absorbed dose of gamma and X radiation in the dielectric thin layer of hafnium dioxide, which is located in the structure of the MOS capacitor (Fig.1). The relation on the basis of the numerically calculated absorbed dose of radiation is obtained by using the theory of the physical transport of photons in a thin layer of dielectric. In doing so, it is necessary to know the spatial dependence of the photon flux of gamma or X-ray in a volume of the dielectric, as well as the values of the total mass attenuation coeficient (TMAtC) and total energy absorbed mass coeficient (TMEnAbC) for SiO_2 or HfO_2 as a radiation characteristics of the material from which is made a dielectric.

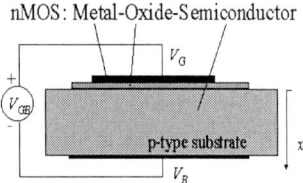

Fig.1. MOS capacitor standard structure (sketch of the model).

The attenuation of the particle flux Φ when gamma and/or X photons pass through the target dielectric (HfO_2 or SiO_2), which has a thickness of x, is given by:

$$\Phi = \Phi_0 \cdot \exp[-(\mu_m / \rho) \cdot \rho x] \quad (1)$$

where Φ_0 is incident photon flux (photons/cm^{2}s), (μ_m/ρ) is total mass attenuation coefficient (cm^2/g) calculated by XCOM program [12], ρ is the density (g/cm^3) of the target material. Total mass attenuation coefficients for SiO_2 and HfO_2 are calculated and presented at Fig. 2 and Fig. 3, respectively.

The XCOM program provides total mass attenuation coefficients and cross sections and as well as partial cross sections for the following processes: incoherent scattering, coherent scattering, photoelectric absorption, and pair production in the field of the atomic nucleus and in the field of the atomic electrons. For compounds, the quantities tabulated are partial and total mass interaction coefficients, which are equal to the product of the corresponding cross sections times the number of target molecules per unit mass of the material. The reciprocals of these interaction coefficients are the mean free paths between scatterings, between photo-electric absorption events, or between pair production events. The sum of the interaction coefficients for the individual processes is equal to the total attenuation coefficient. Total attenuation coefficients without the contribution from coherent scattering are also given, because they are often used in gamma and/or X-ray transport calculations.

The energy absorbed by thin film of dielectric as target material is expressed over the values of total mass energy absorbed coefficient (μ_{en}/ρ) that are determined and stored in data base NISTIR 5632 [13]. If the absorber is chemical compound or mixture, its total energy absorbed coefficient can be approximately evaluated from the coefficients for the constituent elements according to the weighted average:

$$\mu_{en} / \rho = \sum_i w_i \cdot (\mu_{en} / \rho)_i \quad (2)$$

where w_i is proportion by weight of the i-th constituent.

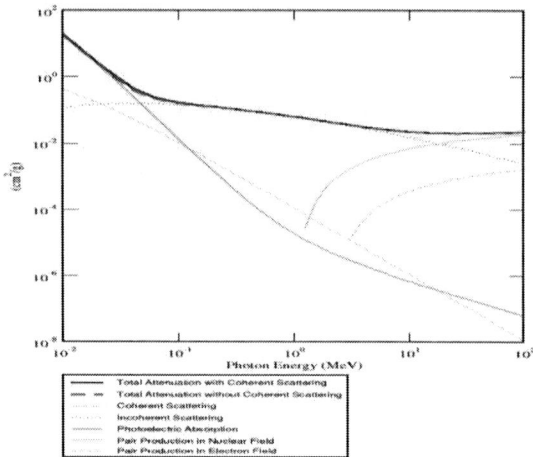

Fig. 2. Total mass attenuation coefficient (cm^2/g) for SiO_2 in photon energy range from 0.01 MeV to 100 MeV.

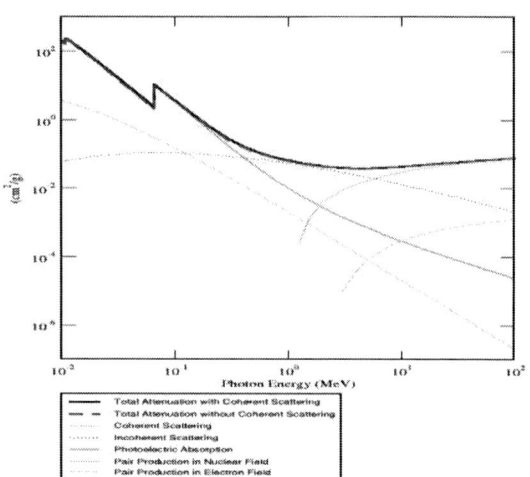

Fig. 3. Total mass attenuation coefficient (cm^2/g) for HfO_2 in photon energy range from 0.01 MeV to 100 MeV.

The absorbed dose rate D_R in a thin film of dielectric is given by:

$$D_R = \int_0^{d_{OX}} E \cdot \Phi_0 \cdot \exp\left[-(\mu_m / \rho) \cdot \rho x\right] \cdot \frac{(\mu_{en} / \rho)}{d_{ox}} dx \quad (3)$$

where d_{ox} is the thickness of dielectric as irradiated material.

When the thickness d_{ox} very small, the absorbed dose rate can be approximately given by:

$$D_R \cong E \cdot \Phi_0 \cdot \frac{\mu_{en}}{\rho} \quad (4)$$

where Φ_0 is incident photon flux with energy E(MeV).

978-1-7281-3420-8/19 $31.00 © 2019 IEEE   182

Based on the radiation characteristics and the absorbed dose rate, it is possible to perform the estimation of effective trapping efficiency for MOS device based on the relation [9,11]:

$$f_{ot} = (-\Delta V_{mg}\varepsilon_{ox})/(qk_g f_y t_{eq} t_{phys} D_R \Delta\tau) \quad (5)$$

where are ΔV_{mg} is the midgap voltage shift, ε_{ox} is the dielectric constant of thin layer of oxide, -q is electronic charge, k_g is the number of electron-hole pairs generated per unit dose, f_y is the charge yield, t_{eq} is the equivalent oxide thickness, t_{phys} is the physical thickness of the alternative dielectric and D_R is the total absorbed dose rate, $\Delta\tau$ is the time interval of irradiation.

III. RESULTS AND DISCUSSION

Taking into account the settings of the applied physical model for the transport of photons of X and / or gamma radiation through thin layers of dielectric oxide from the MOS capacitor, which were incorporated into the numerical method, the corresponding calculations were approached. After preparing the baseline data, the XCOM program and the NISTIR5632 database were used to determine the values of the radiation characteristics μ_m/ρ and μ_{en}/ρ. The results of the calculation for the ratio of absorbed dose rate and incident photon flux with energy of 1.25 MeV (gamma rays from ^{60}Co source) and 20 MeV (X-rays) are shown in Table 1 for SiO$_2$ and in Table 2 for HfO$_2$.

TABLE I

RESULTS OF CALCULATION FOR RADIATION CHARACTERISTICS OF SIO$_2$ FOR TMATC (M_M/P), TMENABC (M_{EN}/P) AND ABSORBED DOSE RATE TO INCIDENT PHOTON FLUX RATIO FOR GAMMA IRRADIATION CO-60 (1.25 MEV) AND HIGH ENERGY X RAYS (20 MEV)

Type of Irradiation	Gamma rays Co-60 (1.25 MeV)	X-rays (20 MeV)
μ_m/ρ (cm/g)	0.05693	0.02036
μ_{en}/ρ (cm/g)	0.02661	0.01546
D_R/Φ_0 (MeVcm^{2}s^{-1})	0.03326	0.30911

TABLE II

RESULTS OF CALCULATION FOR RADIATION CHARACTERISTICS OF HFO$_2$ FOR TMATC (M_M/P), TMENABC (M_{EN}/P) AND ABSORBED DOSE RATE TO INCIDENT PHOTON FLUX RATIO FOR GAMMA IRRADIATION CO-60 (1.25 MEV) AND HIGH ENERGY X RAYS (20 MEV)

Type of Irradiation	Gamma rays Co-60 (1.25 MeV)	X-rays (20 MeV)
μ_m/ρ (cm/g)	0.05534	0.05184
μ_{en}/ρ (cm/g)	0.02702	0.03128
D_R/Φ_0 (MeVcm^{2}s^{-1})	0.03378	0.62561

When comparing the radiation characteristics for HfO$_2$ and SiO$_2$, it can be concluded that the gamma irradiation Co-60 (1250 keV) values for TMAtC and TMEnAbC have similar values, while for high energy X rays of 20 MeV, the values of both total mass coefficients for HfO$_2$ are greater than those for SiO$_2$ by more than twice. In the next step, it is also obtained through the approximate relation (4) that for high-energy X-rays the value of the absorbed dose rate to incident photon flux ratio for HfO$_2$ is greater than that for SiO$_2$ by more than 200%.

In order to clarify the results obtained, it should be recalled that the most sensitive part of the MOS virgin to the effect of ionizing radiation is a layer of dielectric oxide, in which electron-hole pairs are generated when photons pass through the material medium. In addition, for HfO$_2$ it is estimated that the value of the average energy required creating an electron-hole pair is 15 eV, while for SiO$_2$ it is 18 eV [11]. This means that for the same incident photon flux of high-energy X-rays, the sensitivity of the MOS capacitor with HfO2 is higher, and in a shorter time interval an equal amount of absorbed radiation dose is obtained than if SiO2 was used.

After proper analysis, the researchers found that the typical value of effective trapping efficiency for HfO$_2$ is less than three times that of SiO$_2$, since they have a lower defect density [11]. However, in relation (5) it can be observed that effective trapping efficiency also depends on the parameters that can be affected by ionizing radiation. Therefore, it is important to emphasize that for the same absorbed radiation dose, effective trapping efficiency has a lower value for HfO$_2$ than SiO$_2$, since in this case the number of electron-cavity pairs generated is higher in HfO$_2$.

IV. CONCLUSION

Our research effort has been focused on demonstrating the importance of the numerical method, which is based on a chosen physical model of photon transport through thin layers of dielectrics from MOS devices. Considering the radiation characteristics of the selected dielectrics, it can be concluded that there are advantages of HfO$_2$ over SiO$_2$ in the radiation field with high-energy X-rays. The MOS capacitor with HfO$_2$ is more sensitive to high-energy photons and this has been shown by calculating the ratio of absorbed dose rate to flux of incident photons. Similar radiation effects should be expected for the interaction of dielectric material with gamma ray photons originating from a Co-60 source, both in the dielectric layer with SiO$_2$ and in that with HfO$_2$. By studying the relation that defines effective trapping efficiency in the field of ionizing radiation, it can be concluded that for the same radiation absorbed dose into the MOS capacitor with HfO$_2$, there are a greater number of generated electron-hole pairs, in which case the value of effective trapping efficiency is smaller than if SiO$_2$ was used. This research has led to the general conclusion that high-k materials such as HfO$_2$ can become an alternative to SiO$_2$ in MOS devices because they have special advantages in terms of radiation characteristics that are of great importance for gamma and X-ray detection.

978-1-7281-3420-8/19 $31.00 © 2019 IEEE

ACKNOWLEDGEMENT

This work was provided under financial supported by the Ministry of Education and Science, Republic of Serbia.

REFERENCES

[1] G.D. Wilk, R.M. Wallace, "Electrical properties of hafnium silicate gate dielectrics deposited directly on silicon", *Appl. Phys. Lett.*, 1999, vol.74(19), pp. 2854-2856.

[2] M. Gutowski, J.E. Jaffe, C.L. Liu, M. Stoker, R.I. Hegde, R.S. Rai, P.J. Tobin, "Thermodynamic stability of high-k dielectric metal oxides ZrO2 and HfO2 in contact with Si and SiO2", *Appl. Phys. Lett.*, 2002, vol.80(11), pp. 1897-1899.

[3] I. Kang, B.H. Lee, W.J. Qi, Y. Jeon, R. Nieh, S. Gopalan, K. Onishi, J.C. Lee, "Electrical characteristics of highly reliable ultrathin hafnium oxide gate dielectric", *IEEE Elect. Dev. Lett.*, 2000, vol. 21, pp. 181-183.

[4] T. Tan, Z. Liu, H. Lu, et al. "Band structure and valence-band offset of HfO_2 thin film on Si substrate from photoemission spectroscopy", *Appl. Phys. A*, 2009, vol. 97, pp. 475-479.

[5] P.M. Trimali, A.G. Khairnar, B.N. Joshi, A.M. Mahajan, "Structural and electrical characteristic of RF-sputtered HfO2 high-k based MOS capacitors", *Solid-State Elect.*, 2011, vol.62, pp. 44-47.

[6] J. Gao, G. He, J.W. Zhang, B. Deng, Y.M. Liu, "Annealing temperature modulated interfacial chemistry and electrical characteristics of sputtering-derived HfO2/Si gate stack", *Jrnl. Alloys and Compounds*, 2015, vol.647, pp. 322-330.

[7] A.Y. Kang, P.M. Lenahan, J.F. Conley, "The radiation response of the high dielectric-constant hafnium oxide/silicon system, IEEE Trans. Nucl. Sci., 2002, vol.49(6), pp. 2636-2642.

[8] J.R. Schwank, M.R. Shaneyfelt, D.M. Fletwood, J.A. Felix, P.E. Dodd, P. Paillet, V. Ferlet-Cavrois, "Radiation effects in MOS oxides", *IEEE Trans. Nucl. Sci.*, 2008, vol.55(4), pp.1833-1853.

[9] F.B. Ergin, R. Turan, S.T. Shishiyanu, E. Yilmaz, "Effects of gamma radiation on HfO2 based MOS capacitor", *Nucl.Instr.Meth. in Phys.Research B*, 2010, vol.268, pp.1482-1485.

[10] M. Ding, Y. Cheng, X. Liu, X. Li, "Total dose response of hafnium oxide based metal-oxide-semiconductor structure under gamma-ray Irradiation", *IEEE Trans. Diel. Electr. Insulation*, 2014, vol.21(4), pp.1792-1800.

[11] J.A. Felix, D.M. Fleetwood, R.D. Schrimpf, J.G. Hong, G. Lucovsky, J.R. Schwank, M.R. Shaneyfelt, "Total-dose radiation response of hafnium-silicate capacitors", *IEEE Trans. Nucl. Sci.*, 2002, vol.49(6), pp.3191-3196.

[12] XCOM Photon Cross Section Data Base, *NBSIR 87-3597*, NIST's Standard Reference Data Program, 2010.

[13] J.H. Hubbell, S.M. Seltzer, *NISTIR 5632*, 1995, NIST, USA.

Reduced Low Dose Rate Sensitivity (RLDRS) in Bipolar Devices

V. Pershenkov, A. Bakerenkov, A. Rodin, V. Felitsyn, V. Telets, and V. Belyakov

Abstract - Possible physical mechanism of reduced low dose rate sensitivity in bipolar devices is described. The reduced sensitivity can be connected with a specific position of effective Fermi level relatively acceptor and donor radiation-induced interface traps.

I. INTRODUCTION

Several types of bipolar devices demonstrate enhanced degradation during low dose rate irradiation in comparison with irradiation at high dose rate for the same total dose level. These devices are known as ELDRS-susceptible (Enhanced Low Dose Rate Sensitivity) [1]. Since the physical mechanism of the ELDRS effect [2] is connected with suppressing of the accumulation of the radiation defects at high dose rate (HDR) we can consider the effect as Reduces High Dose Rate Sensitivity (RHDRS) [3]. Nevertheless the term ELDRS is used for these devices in literature.

Recent research of bipolar technology shows [4], that reducing degradation with decreasing of the dose rate is an inherent property of devices, which are not susceptible to ELDRS effect. We can consider these devices as ELDRS-free [5]. The decreasing of the radiation degradation at low dose rate (LDR) irradiation, as the rule, is not considered during hardness assurance tests of devices for space applications. It can lead to significant underestimation of the operation life time of these devices in real space environment.

Since devices, which demonstrate reduced degradation at low dose rate irradiation are usually considered as ELDRS-free, it will be useful to single out them into separate class using term RLDRS (Reduced Low Dose Rate Sensitivity).

The purpose of this work is to describe possible physical mechanism of RLDRS effect using the conversion model [6,7], which enables to estimate numerically the degradation of electrical parameters of bipolar devices during specified space mission.

A. S. Bakerenkov is with the National Research Nuclear University MEPhI (Moscow Engineering Physics Institute), Moscow, Russian Federation (corresponding author to provide phone: +7 499 324 01 84; fax: +7 (499) 324 21 11; e-mail: AS_Bakerenkov@list.ru).

V.S. Pershenkov, A.S. Rodin, V.A. Felitsyn, V.V. Belyakov and V.A. Telets are with the National Research Nuclear University MEPhI (Moscow Engineering Physics Institute), Moscow, Russian Federation (e-mail: ASRodin@mephi.ru).

In this work the conversion model is shortly described. The qualitative and quantitative models of the RLDRS effect are considered.

II. CONVERSION MODEL

The model is based on the assumptions that the ELDRS effect in bipolar devices is directly connected with increasing of the surface recombination current due to interface trap buildup at SiO_2/Si interface near emitter junction. We suppose, that the interface trap buildup can be described by H-e model [8,9]. According to the model the interface trap buildup is connected with positive oxide trapped charge conversion due to interaction with substrate electrons and not with the action of hydrogen ions only. Nevertheless the H-e model is not in conflict with the most popular hydrogen model [10]. The H-e model takes into account the contribution of substrate electrons to interface trap buildup process [11].

In [6,7] is supposed that interface trap buildup is connected with a conversion of rechargeable part of trapped positive charge located opposite the silicon forbidden gap [12]. An interaction of thermally excited rechargeable positive charge Q_{ot} and tunneling substrate electrons leads to interface trap buildup (Fig. 1,a). The positive charge Q_{ot} can be neutralized by hole emission to silicon valence band (Fig. 1b).

The conversion mechanism of trapped positive charge or E'_γ center to interface state can be following. According to the model [13], positively charged Si atom in the E'_γ center is sp2-hybridized (planar configuration), while neutral Si^0 atom is sp3- hybridized (tetrahedral configuration). The electron energy levels in this defect strongly depend on the distance between Si^+ and Si^0 atoms [14,15]. When this distance is large, these levels are located in the oxide close to the Si midgap [14]. When these two atoms are bonded and the distance between them is small, the energy levels of bonding electrons are close to the edges of SiO_2 bandgap. The first electron capture to the E'_γ center changes defect configuration [13-14] which results in the electron energy levels shift from Si midgap toward the Si valence and conduction bands, the distance between Si atoms being of intermediate value. The electrons in this transformed defect are expected to be in the intermediate sp2-sp3 configuration and could form diffusion orbital (it is the orbital which extends toward neighbour Si further than ordinal sp3 one) [15]. This defect configuration may be assumed stable and electron energy levels are proposed to remain unchanged when one of the electrons is removed.

As is supposed in [6,7] there are two types of oxide traps that are located in the oxide interface opposite the silicon forbidden gap: shallow traps with a short time of conversion responsible for the degradation at high dose rates (HDR), and deep traps that determine the excess base current at greater times of irradiation or LDR. The duration of HDR irradiation process is relatively short, that not enough to convert all radiation-induced positive charge to interface traps. Therefore at long-time LDR irradiation we can observe the increasing of the degradation.

III. PHYSICAL MODEL OF RLDRS

A. The qualitative physical model

The surface recombination current, which is responsible for a radiation degradation of the base current, depends on the concentration of interface traps and the surface potential on interface screen oxide-base region along emitter junction perimeter.

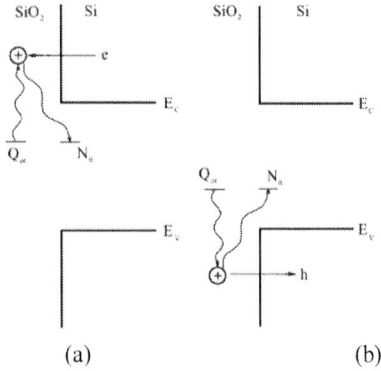

Fig. 1. Conversion of rechargeable oxide charge Q_{ot} to interface traps N_{it} : capture of an electron e (a); emission of a hole h (b), E_c and E_v are energy levels of Si conduction and valence band.

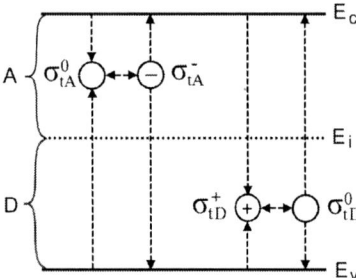

Fig. 2. Acceptor-like (A) and donor-like (D) interface traps.

Using well known assumption, we consider that in the top half the interface traps act as acceptors, while in the bottom half they act as donors (Fig.2). The empty acceptor-like traps are neutral, the filled acceptor-like traps are negatively charged. The empty donor-like traps are positively

charged and the filled donor-like traps are neutral. The capture cross section of a neutral trap is approximately 10^{-15} cm^{-2} the order of atomic dimensions. The capture cross section of a charged trap is one-two orders greater (10^{-14} - 10^{-13} cm^{-2}) due to columbic interaction with injected to base minority carriers. The charge state of trap depend on its position relatively Fermi level on the surface.

In this work we suppose that interface traps in the top and bottom half of the silicon forbidden gap occupy some effective mono levels E_{tA} and E_{tD}. Fig.3 shows energy location of these traps in the forbidden gap for p-base region of npn transistor (the real possible distribution of these traps in the forbidden gap is shown by dotted lines). The acceptor-like traps in the top half of the forbidden gap always is empty for any location of Fermi level in p-base region. The capture cross section of the neutral acceptor-like traps σ_{tA}^{0} equals approximately 10^{-15} cm^{-2}.

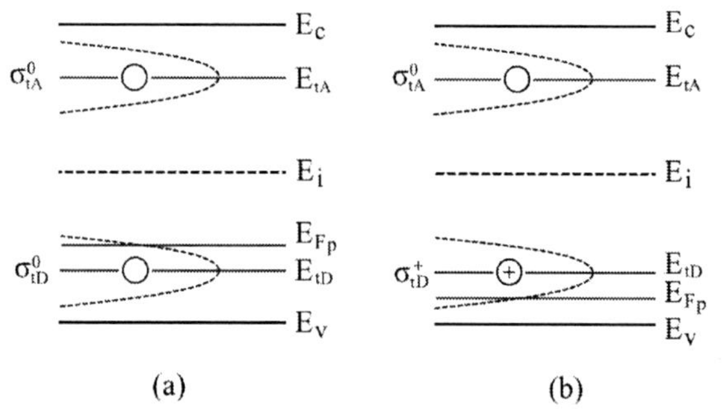

Fig. 3. The energy location of the accepter-like E_{tA} and donor-like E_{tD} traps relatively Fermi level E_{FP} in the forbidden gap for p-base region of npn transistor: Fermi level E_{FP} is located above the effective mono levels of the donor-like traps E_{tD} (a); E_{FP} is located below level E_{tD} (b). The dotted lines show the real possible distribution of interface traps in the forbidden gap

The capture cross section of the donor-like traps strongly depends on their position relatively Fermi level E_{FP}. In Fig.3,a Fermi level is located above the effective mono levels of the donor-like traps E_{tD}, but in Fig.3,b it lies below level E_{tD}. The charge state of the donor-like traps depends on their position relatively Fermi level. The filled donor-like traps below Fermi level are neutral and their capture cross section corresponds neutral traps σ_{tD}^{0} (approximately 10^{-15} cm^{-2}) (Fig. 3a). The empty donor-like traps above Fermi level are positively charged (Fig. 3b) and their capture cross section essentially increases. The capture cross section of the positively charged traps may be equal 10^{-14} - 10^{-13} cm^{-2}.

We suppose that RLDRS and ELDRS devices differ due to a feature of the Fermi level position in base region relatively position of the radiation-induced surface traps in the silicon forbidden gap.

The case of Fig.3a is feature for RLDRS devices. In that case the acceptor-like and donor-like traps are neutral. The recombination rate of injected from emitter electrons connects with their capture on traps with relatively small capture cross section (10^{-15} cm^{-2}). For this reason at low dose rate irradiation the excess base current is relatively small in spite of all trapped oxide charges are converted to interface traps during long time irradiation. The increasing of dose rate (the reducing of the irradiation time) leads to increasing of the non converted trapped positive charge and the increasing of the excess base current (Fig.4a). Qualitatively it is explained by that: a greater positive charge attracts the injected electrons to the surface that leads to increasing of the recombination rate. Quantitatively it is described by the increasing of the surface potential on interface screen oxide-base region (will be considered in full version of paper). The saturation of the excess base current degradation at the high dose rate (Fig.4a) connects with a contribution of the conversion of the shallow traps [6,7], when the conversion of deep traps becomes insignificant.

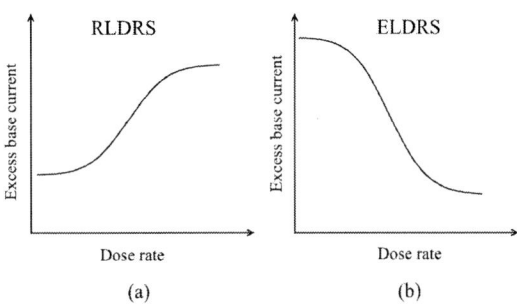

Fig. 4. The excess base current versus dose rate for RLDRS (a) and ELDRS (b) devices.

The case of Fig.3b is a feature of ELDRS devices. In that case the acceptor-like traps are neutral while the donor-like traps are positively charged. The recombination rate of injected from emitter electrons connects with their capture on the positively charged traps with relatively large capture cross section (10^{-14} - 10^{-13} cm^{-2}). Therefore the excess base current is relatively large. According to conversion model [6,7] the increasing of dose rate (the reducing of the irradiation time) leads to decreasing of the interface trap concentration and the excess base current reduces (Fig.4b). The increasing of the trapped positive charge yield at HDR, like RLDRS devices, has not significant effect since the recombination is connected essentially with a capture of electrons on positively charged traps. Besides the increasing of the dose rate leads to reducing of the effect of the positively charged interface traps. This issue will be considered in full version of paper.

The position of Fermi level in p-base depends on a specific feature of the manufacturing process: a doping level in p-base region and a value of the initial positive trapped charge in screen oxide above base. The case of Fig.3a is characterized by a low level of p-base doping or a large value of the initial positive trapped charge in screen oxide. This positive charge shifts the energy level of the donor-like traps E_{tD} below Fermi level E_{FP} (Fig. 5a).

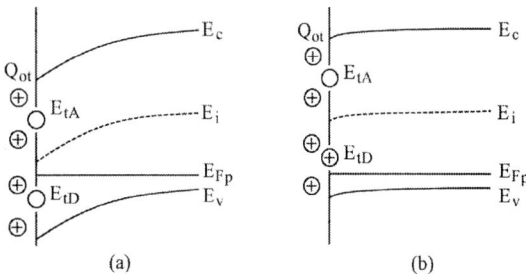

Fig. 5. Fermi level E_{FP} location relatively the position of donor-like traps E_{tD} on surface: (a) for case Fig.3, a (feature for RLDRS devices); (b) for case Fig.3, b (feature for ELDRS devices).

The case shown in Fig. 3b is realized when the p-base region is strongly doped or the screen oxide has a small technological positive charge. It corresponds the position of the donor-like traps E_{tD} above Fermi level E_{FP} (Fig. 5b). The initial features corresponding Fig. 3a or Fig. 3,b can be used for classification RLDRS and ELDRS devices.

An effect of the positive charge of the donor-like traps will be discussed in full version of paper. A similar mechanism can be described for n-base region of pnp bipolar structures using acceptor-like traps in top half of the silicon forbidden gap.

B. The quantitatively model

We suppose that the accumulation and annealing of the positive oxide trapped charge Q_{ot} are described by the following equation:

$$\frac{dQ_{ot}}{dt} = K_{ot}P - \frac{Q_{ot}}{\tau_D} - K_{RICN}Q_{ot}P, \qquad (1)$$

where Q_{ot} is the oxide trapped charge; where K_{ot} is a coefficient characterizing the accumulation of trapped charge; P is dose rate; τ_D is the conversion time of deep traps; K_{RICN} is coefficient connected with RICN (Radiation Induced Charge Neutralization) effect [16].

First term in the right-hand side of (1) represents the trapped charge accumulation in thick oxide by dispersion transport of radiation-induced holes. Second term in the right-hand side of (1) is responsible for the neutralization of deep trap charge by electrons from substrate. Third term in the right-hand side of (1) characterizes the annealing of the positive charge by radiation-induced electrons (RICN effect).

The interface trap buildup N_{it} can be expressed as follows:

$$\frac{dN_{it}}{dt} = \frac{1}{q}\frac{Q_{ot}}{\tau_D} + \frac{1}{q}K_{RICN}Q_{ot}P. \qquad (2)$$

First and seconds terms in the right-hand side of (2) represent the interface traps buildup through the conversion of trapped charge by the substrate electrons and radiation-induced electrons.

The total concentration of interface traps equals the sum of donor-like the acceptor-like and traps

$$N_{it} = N_{tA} + N_{tD}, \qquad (3)$$

where N_{tA} is the concentration of the acceptor-like and traps in the top half of the forbidden gap; N_{tD} is the concentration of the donor-like and traps in the bottom half of the forbidden gap.

The excess base current equals the sum of a recombination on the acceptor-like and donor-like traps:

$$\Delta I_b = (v_{th}\sigma_{tA}N_{tA} + v_{th}\sigma_{tD}N_{tD})\frac{p_s n_s}{p_s + n_s + 2n_i}, \qquad (4)$$

where p_s and n_s are the surface concentration of holes and electrons; v_{th} – thermal velocity.

The relationships (1) – (4) is allowed to a quantitatively estimate RLDRS effect. It can be done by numerical modeling. The disadvantage of this approach connects with necessity of using a lot of unknown fitting parameters. For simplicity it is possible to use the approach described in [6,7].

IV. CONCLUSION

The possible physical mechanism of a reduced low dose rate sensitivity (RLDRS) and enhanced low dose rate sensitivity (ELDRS) in bipolar devices can be connected with specific position of Fermi level in base region relatively radiation-induced interface traps in forbidden gap. Acceptor- and donor-like interface traps may be neutral or charged according to their position relatively Fermi level. The capture cross section of a charged trap is one-two orders greater than a neutral trap due to columbic interaction with injected to base minority carriers. For RLDRS devices interface traps are neutral while for ELDRS devices they are charged. As the result the effect of low dose rate irradiation is quite different for these devices.

The qualitative physical and quantitatively models of RLDRS are presented. The quantitatively model involves the fitting parameters extraction technique that allows using it for numerical estimation of radiation degradation for arbitrary dose rate, total dose and temperature.

REFERENCES

[1] R.L. Pease, R.D. Schrimpf, D.M. Fleetwood, "ELDRS in bipolar linear circuits: A review", *IEEE Transactions on Nuclear Science*, 2009, vol. 56, Issue 4, pp. 1894-1908.

[2] D.M. Fleetwood, S. L. Kosier, R. N. Nowlin, R. D. Schrimpf, R. A. Reber, Jr., M. DeLaus, P. S. Winokur, A. Wei, W. E. Combs and R. L. Pease, "Physical Mechanisms Contributing to Enhanced Bipolar Gain Degradation at Low Dose Rates", *IEEE Trans. Nucl. Sci.*, 1994, vol. NS-41, no. 6, pp.1871-1883.

[3] H.P. Hjalmarson, R.L. Pease, S.C. Witczak, M.R. Shaneyfelt, J.R. Schwank, A.H. Edwards, C.E. Hembree, T.R. Mattson, "Mechanisms for radiation dose-rate sensitivity of bipolar transistors", *IEEE Trans. Nucl. Sci.*, 2003, vol. NS-50, no.6, pp.1901-1909.

[4] K. Kruckmeyer, L. McGee, B. Brown, D. Hughart, "Low dose rate test results of National Semiconductor s ELDRS-free bipolar amplifier LM124 and comparators LM139 and LM193", *Proc. IEEE Radiation Effect Data Workshop Record*, 2008, pp.110-117.

[5] J. Boch, A. Michez, M. Rousselet, S. Dhombres, A.D. Touboul, J.-R. Vaille, L. Dusseau, E. Lorfevre, N. Charty, N. Sukhaseum, F. Saigne, "Dose rate switching technique on ELDRS-free bipolar devices", *IEEE Trans. Nucl. Sci.*, 2016, vol. 63, no.4, pp.2065-2071.

[6] V.S. Pershenkov, D.V. Savchenkov, A.S. Bakerenkov, V.N. Ulimov, A.Y. Nikiforov, A.I. Chumakov, A.A. Romanenko, "The conversion model of low dose rate effect in bipolar transistors", *RADECS Proceeding*, 2009, pp. 286-393.

[7] V.S. Pershenkov, D.V. Savchenkov, A.S. Bakerenkov, V.N. Ulimov, "Conversion model of enhanced low dose rate sensitivity in bipolar ICs", *Russian Microelectronics*, 2010, vol. 39, no. 2, pp. 91-99.

[8] A.V. Sogoyan, S. V. Cherepko, V. S. Pershenkov, V. I. Rogov, V. N. Ulinov, V. V. Emelianov, "Thermal- and Radiation-Induced Interface Traps in MOS Devices", *RADECS Proceeding*, 1997, pp. 69-72.

[9] V. Sogoyan, S. V. Cherepko, V. S. Pershenkov, "The hydrogenic-electron model of accumulation of surface states on the oxide-semiconductor interface under the effects of ionizing radiation", *Russian Microelectronics*, 2014, vol. 43, no. 2, pp. 162-164.

[10] F. B. McLean, "A Framework for Understanding Radiation-Induced Interface States in MOS SiO$_2$ Structures," *IEEE Trans. Nucl. Sci.*, 1980, vol. NS-27, no.6, pp. 1651-1657.

[11] S.K. Lai, "Interface trap generation in silicon dioxide when electrons are captured by trapped holes", *Journal of Applied Physics*, 1983, vol. 54, pp. 2540-2546.

[12] V. V. Emelianov, A. V. Sogoyan, O. V. Meshurov, V. N. Ulimov, V. S. Pershenkov, "Modeling the Field and Thermal Dependence of Radiation-Induced Charge Annealing in MOS Devices", *IEEE Trans. Nucl. Sci.*, 1996, vol. 43, no. 6, pp. 2572-2578.

[13] K.L.Yip and W.B.Fowler, "Electronic structure of E-centers in SiO", *Phys. Rev. B*, 1975, vol. 1(6), pp. 2427-2338.

[14] E.P. Reilly and J. Roberston, "Theory of Defects in Vitreous Silicon Dioxide", *Phys. Rev. B*, 1981, Vol. 27(6), pp. 3780.

[15] W.H.Flygare, *Molecular Structure and Dynamics,* Prentice-Hall Inc., Englewood Cliffs, NJ, 1978.

[16] D.M. Fleetwood, "Radiation-induced charge neutralization and interface-trap buildup in metal-oxide-semiconductor devices", *Journal of Applied Physics*, 1990, vol. 67, Issue 1, pp. 580-583..

Blocking of Impacts of Single Ionizing Particles by CMOS C-Element in Two-Phase Systems

V. Ya. Stenin, Yu. V. Katunin and K. A. Petrov

Abstract - The work presents the TCAD simulation of the 65 nm bulk CMOS C-element as resistant to the single-event transients. The charge collection from a track of a single nuclear particle simulates in impacted on drain regions of the transistors, which leads to the error pulses in the output of 2-phase inverters and C-element. The TCAD simulation used the tracks along the normal to the chip. The linear energy transfer from a particle to the track is 60 MeV·cm2/mg.

I. INTRODUCTION

The C-elements used in asynchronous logic [1], synchronous rad-hard designs [2] and for redundant self-correcting flip-flops [3]. In publications, it is found under the names keeper-less C-element [1], tristate transmission gate [4]. The effect of the single nuclear particles on combinational logic elements is an error pulse [5] that cause the temporary changes logical state (single event transient – SET) of such elements. Increased robustness to the effects of single nuclear particles depends on the both schemes and topology design. TCAD simulation of a charge collection from nuclear particle tracks is a virtual experimental base for the study of elements of microelectronic systems. The results obtained in this study allow us to evaluate the capabilities of silicon 65-nm CMOS bulk technology as the basis for the design of the elements of high-performance microprocessor systems designed for space applications. The purpose of the work is to study the C-element in the chain of two-phase inverters, when single ionizing particles affect to transistors of the C-element.

II. ELEMENTS OF TWO-PHASE SYSTEMS

Fig. 1 presents the chain of two 2-phase inverters, the C-element and the 2-phase converter based on the two standard inverters. The 2-phase logic makes it possible to avoid data failure, when the impact on only one of the phases [6]. The part of 2-phase logic is new C-element [7] based on the trigger with spacing of transistors into two groups (Spaced Transistor Groups DICE - STG DICE) so that charge collection from the track of single particle by only one of the transistor groups does not result failure.

Yu. V. Katunin, and K. A. Petrov are with the Scientific Research Institute of System Analysis, Russian Academy of Sciences, Nakhimovsky pr. 36-1, 117218 Moscow, Russia, E-mail: katunin@cs.niisi.ras.ru, petrovk@cs.niisi.ras.ru.

V. Ya. Stenin is with the National Research Nuclear University MEPhI (Moscow Engineering Physics Institute), Kashirskoe sh. 31, 115409 Moscow, Russia, E-mail: vystenin@mephi.ru

In this work, we study C-element with 2-phase inputs. If the inputs are in-phase, the C-element transfers this state on the output. If the inputs are antiphase, the C-element stores the last common-mode state on the output.

The transistors on Fig. 1 are indicated by two digits as $N_{1.1}$ or $P_{2.1}$, where the first digit means the number of the transistor in the each element, and the second digit – the number of the of this element in the chain. The digital designations of the nodes voltages are similar: $V1.1$ and $V2.1$ – the first and second outputs of the first inverter and so on; V_{OUT} – the output of the C-element.

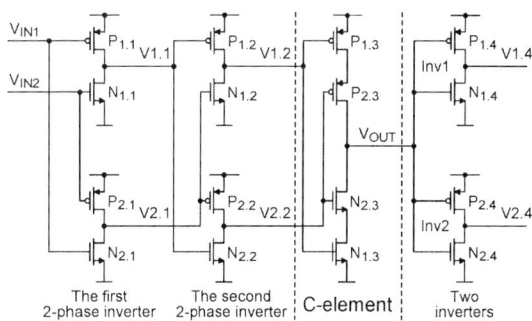

Fig. 1. The chain of two 2-phase inverters, the C-element and the 2-phase converter on standard inverters.

Fig. 2 shows the layout of the topology of the simulated chain. The asterisks indicate location of the input points of the tracks. The points 1nC, 2nC, 2n correspond to the charge collection at NMOSFETs and the points 1pC, 2pC, 2p corresponding PMOSFETs.

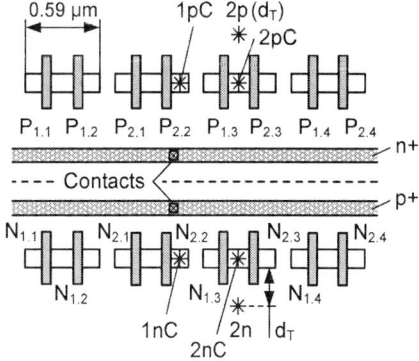

Fig. 2. The layout of the topology of the chain with 2-phase inverters, asterisks indicate input points of the particle tracks, d_T – the distance from the track input point 2n to the point 2nC.

978-1-7281-3420-8/19 $31.00 © 2019 IEEE

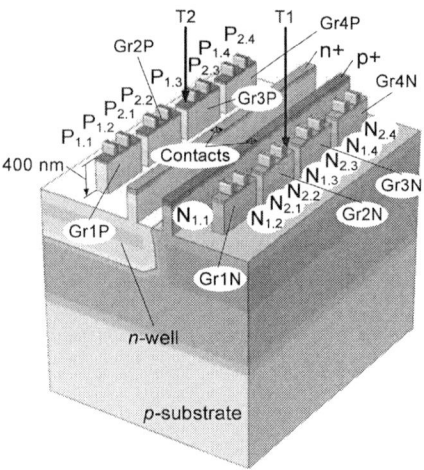

Fig. 3. 3D TCAD physical structure of the chain, the directions of tracks are at the normal to the surface (as tracks T1 and T2), n+ and p+ areas – guard strips.

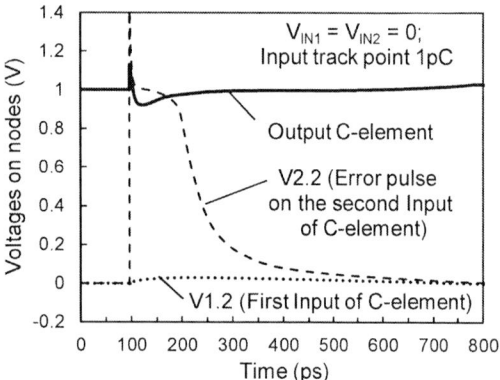

Fig. 4. The voltages on nodes of the C-element at the charge collection from the track with LET = 60 MeV·cm2/mg during the storage mode, the input track point 1pC.

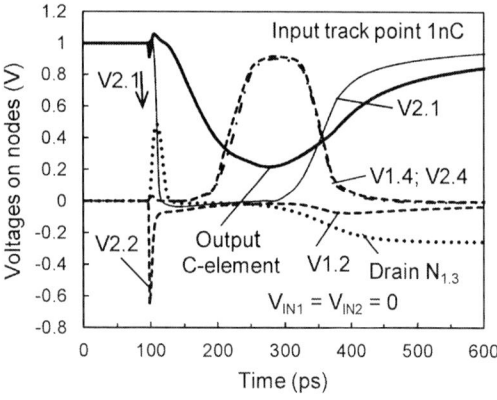

Fig. 5. The voltages on nodes of the C-element at the charge collection from the track with LET = 60 MeV·cm2/mg during the storage mode, the input track point 1nC (Fig. 2).

III. METHODOLOGY OF SIMULATION USING TCAD

Fig. 3 presents 3D TCAD device physical model of the chain. The chain consists of transistors groups surrounded a shallow trench isolation (400 nm depth) which remoted in Fig. 3 for make visual the model. The "parallelepipeds" in Fig. 3 contain the eight groups (Gr1N–Gr4N and Gr1P–Gr4P) of two transistors each including of 2-phase inverters, the C-element and the 2-phase converter. The groups Gr1N, Gr2N, Gr1P and Gr2P combine the pairs of transistors of 2-phase inverters, which are simultaneously in different states (closed and open). The groups Gr3N, Gr3P combine pairs of the transistors of the C-element.

The 3-D device structure comprises of transistors developed taking into account the models presented in the work [8]. The effect of a particle strike simulated as a collecting of charge carriers generated at the tracks with linear energy transfer (LET) from a particle to the track 60 MeV×cm2/mg. This corresponds to the maximum of the LET range, which often used for the experimental study of the fault stability of CMOS elements under the influence of heavy charged particles [5]. The results obtained using Sentaurus Device simulator at the temperature of 25°C and the supply voltage of 1.0 V for 65 nm CMOS bulk structure. The channel width of each CMOS transistors are 150 nm. This work presented results of the simulation a charge collection from particle tracks for the C-element without simultaneously a switching transient.

IV. SIMULATION OF CHARGE COLLECTION

A. Input track points 1pC and 1nC in groups Gr2N and Gr2P

The index "C" of the input track points in Fig. 2 means tracks at the drain regions of transistors. C-element on the two-phase inverter stores of the logical state on the capacity of the output node.

Fig. 4 shows the dependencies of voltages on nodes in the case the inputs of the chain are $V_{IN1} = V_{IN2} = 0$, and the track passes through the point 1pC at the group Gr2P (Fig. 2) with LET = 60 MeV·cm2/mg. In the group Gr2P, the transistor $P_{2.1}$ is open but the transistor $P_{2.2}$ is closed. The transistor $P_{2.2}$ collects a charge through its drain area and the error pulse of the positive polarity with the amplitude of 1 V is formed at its drain (Fig. 4) and at the second input of the C-element. The two inputs of the C-element comprise to antiphase. In this case, C-element stores the last common-mode state at its output, passing into a high resistive state.

Fig. 5 presents the case, when the error pulse at the output of the C-element is formed as a result of charge collection only by transistors $N_{1.3}$ and $N_{2.3}$ of the C-element that are in the Gr3N group near the Gr2N group. In this case, the track passes through the 1nC point (the drain of the open transistor $N_{2.2}$) in the group Gr2N (Figs. 2 and 3). The result is the small error pulse of the negative polarity (the amplitude of 0.78 V) at the output of the C-element (Fig. 5). When the output of C-element reaches the pulse amplitude value, the output node begins to restore to the state "1".

B. Input track point 2nC in group Gr2N of C-element

C-element consists of two cascode-connected transistor pairs; each pair forms its own structural group Gr3N and Gr3P (Fig. 3). The source of the transistor $N_{1.3}$ in the group Gr3N connected to the common bus; the drains of transistors $N_{2.3}$ and $P_{2.3}$ connected to the output of the C-element (Fig. 1).

Fig. 6 shows the dependencies of the voltages at the nodes of the second two-phase inverter and C-element (V1.2, V2.2, V_{OUT}) and at the nodes of the standard inverters (V1.4, V2.4) during the charge collecting from the track at LET = 60 MeV·cm2/mg. The track point 2nC is near the regions of drains and sources of the transistors $N_{1.3}$ and $N_{2.3}$ in the group Gr3N of the C-element. The charge collected immediately after the occurrence of the track moves the transistor $N_{1.3}$ at 100 ps (Fig. 6) in the mode of inverse offset. In this case, the rapid removal of electrons from the track through transistors $N_{1.3}$ и $N_{2.3}$, that accompanied by the formation the jump of a voltage at the front of the error pulse exceeding the amplitude value of the error pulse on the output of the C-element.

In the antiphase with the error pulse, the voltages V1.4 and V2.4 at the outputs of the two traditional inverters are synchronously changed. After the time 500 ps the charge collection from the track is completed and the output of C-element returns to the initial state "1".

C. Input track points outside the groups Gr3N and Gr3P of C-element

If the input track point located on the insulating layer of silicon oxide in the region of the NMOS transistors, charge forms in the silicon layer under the insulating layer. The error pulse on the output of the C-element occurs in case the charge is diffusing in the sufficient quantity from a track up to the group of transistors Gr3N collecting a charge. Fig. 7 shows the dependencies of the node voltages of the C-element for the case when the input point of the track 2n was at the distance d_T = 300 nm to the group Gr3N (Fig. 2). The diffusion process of non-basic charge carriers causes delay and an inertia of the voltage change at the output node of the C-element in Fig. 7 in comparison with the case when the input point of the track 2nC passes through the group of transistors Gr3N (Fig. 6).

The delay at the transition of the transistor $N_{1.3}$ until the inverse offset mode can be 10-30 ps relative to the moment of the occasion of the track, and for the transition of the transistor $N_{2.3}$ until inverse offset mode – delay 30-100 ps. At distances of 100-300 nm from the input track point 2n to the Gr3N group boundary, the voltage between the drain and the source of the transistor $N_{1.3}$ in inverse offset mode can reach $V_{DRAIN.N1.3}$ = -(0.40–0.45) V (Fig. 7).

The distance d_T has the variable value from the track input point 2n to the point 2nC, either from the point 2p to the point 2pC (Fig. 2). For simulation of the charge collection generated in the silicon under the insulating oxide layer were used the five input track points. Their distances are d_T = 0.1, 0.2, 0.3, 0.5 and 0.65 microns from

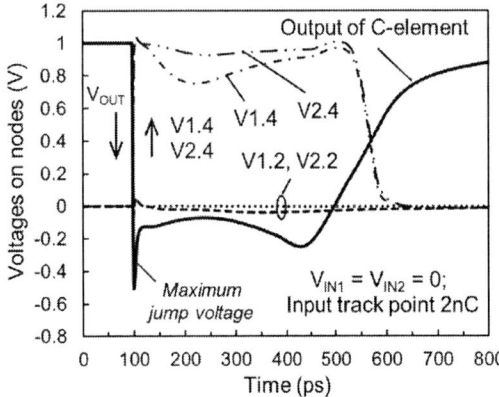

Fig. 6. The nodes voltages of the C-element in the time during a charge collection from the track with LET = 60 MeV·cm2/mg, the collection of a charge through NMOS transistors $N_{1.3}$, $N_{2.3}$ of the C-element, the input track point 2nC (Fig. 2).

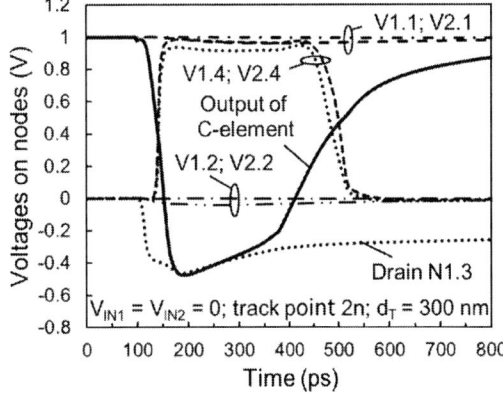

Fig. 7. The nodes voltages of the second two-phase inverter, C-element and two standard inverters during the charge collection from the track at LET = 60 MeV·cm2/mg, the input track point 2n (Fig. 2) at the distance of 300 nm to the point 2nC.

the track point input up to the border of the group Gr3N and the five input track points up to the group Gr3P characterized the same distances.

Dependencies of the amplitude $V_{A\,ERR}(d_T)$ and the duration $t_{ERR}(d_T)$ of the error pulse at the output of the C-element as functions of the distance of the input track point d_T are shown in Fig. 8. Dependencies simulated with the voltages at the main inputs of the chain V_{IN1} = V_{IN2} = 0 for the NMOS transistors in the group Gr3N, and at V_{IN1} = V_{IN2} = 1 V for the PMOS transistors in the group Gr3P and at LET = 60 MeV·cm2/mg. The duration of the pulses in Fig. 8 determined at 0.3 V up from the base of the pulse.

The coordinates of the 2nC and 2pC points correspond to the zero distance d_T = 0. The solid lines on Fig. 8 represent the dependencies of the amplitude and duration of the error pulses for the region of NMOS transistors; the point line represents the dependence of the maximum jump at front of the error pulse as on Fig. 6. The dashed lines represent the dependencies on Fig. 8 for the region of PMOS transistors.

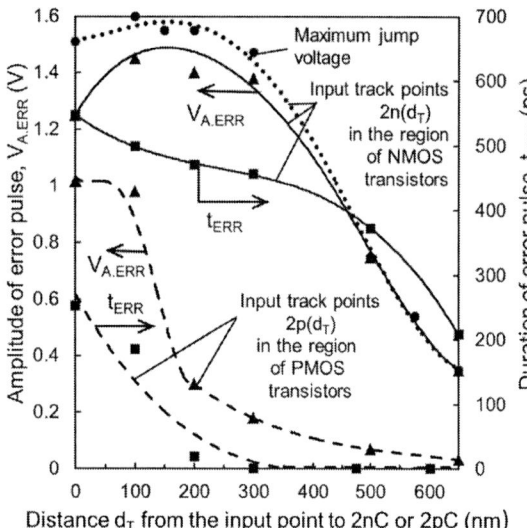

Fig. 8. The amplitudes and durations of the error pulses at the output of C-element for tracks in the silicon layer under the insulating layer outside the transistors groups in dependencies of the distance d_T (Fig. 2) from the input track point 2n (or 2p) to the point 2nC (or 2pC) for the tracks with LET = 60 MeV·cm2/mg.

For the region of NMOS transistors the pulses of error on the output of the C-element are with amplitudes of 1 V or more, if the distance d_T from the input track point 2n to the border of the group Gr3N does not exceed 450 nm (the solid lines on Fig. 8 at LET = 60 MeV·cm2/mg). In this case, the durations of the error pulses are 350-550 ps.

For the case the input track points in the region of PMOS transistors, the amplitude of the error pulse at the output of the C-element is 1.0 V when the input track points 2p displacement until 200 nm from the border of the group Gr3P (the dashed lines on Fig. 8). In this case, the durations of the error pulses are 200–250 ps. The data on Fig. 8 confirm the insignificance of pulse noises generated the track with the input point displacement more 100 nm from the border of the group Gr3P.

V. THE MAIN RESULTS OF SIMULATION

1) The switching delay of the C-element into the high resistive state and storing the output logical state is 20-25 ps in case of violation of the in-phase inputs signals, regardless of the duration transients in previous 2-phase inverters.

2) The duration of storage of the voltage at the capacitance of the output node limited by the values of the output current of the C-element discharging the capacity of the output node, and is 10-20 ns.

3) The propagation delay of C-element for common-mode input signals is 25-40 ps in the case of the 65 nm CMOS bulk technology. When the change two in-phase signals coincides with the start of a charge collection by the 2-phase inverter or C-element, the switching delay of the C-element is increased till 120-160 ps due to the combining of propagation transients and the transients during of the charge collection from a track under an impact of a single particle.

4) In the charge collection directly by the transistors of the C-element be formed an error pulse at its output. The amplitude and the duration of an error pulse depends on the input track point and are maximum for the points through the drains N- or PMOS transistors of C-element (2nC or 2pC). The transient state of the C-element can be up to 500 ps.

5) Error pulses with amplitudes of about 1 V and a duration of 400-500 ps can be formed at linear energy transfer 60 MeV·cm2/mg at tracks only with input track points spaced up to 400-500 nm from the NMOS transistors.

VI. CONCLUSION

The results of the study allow us to assess the reliability CMOS circuits so as 2-phase inverters with C-elements while a charge collecting from the track of single nuclear particles. C-elements according to the CMOS 65 nm bulk technology reliably block error impulses arising in the chain of 2-phase inverters under the action of single particles with energy transfer to the track 60 MeV·cm2/mg.

REFERENCES

[1] D. E. Muller, and W. S. Bartky, "A theory of asynchronous circuits," in *Proc. of International Symposium on the Theory of Switching,* Cambridge, M.A.: Harvard University Press, 1959, pp. 204-243.

[2] P. Hao, S. Chen, P. Huang, J. Chen, and B. Liang, "A Novel SET Mitigation Technique for Clock Distribution Networks," *IEEE Trans. on Device and Mater. Rel.*, 2018, vol. 18, pp. 1-8.

[3] C. Ramamurthy, A. Gujja, V. Vashishtha, S. Chellappa, and L. T. Clark, "Muller C-element Self-corrected Triple Modular Redundant Logic with Multithreading and Low Power Modes," in IEEE Xplore (Conference Section, RADECS-3017), e-book, 2019, pp.184-187.

[4] R. J. Baker, *CMOS Circuit Design, Layout, and Simulation,* Hoboken, New Jersey: John Wiley & Sons, Inc., 2010.

[5] *Soft errors in Modern Electronic Systems* / M. Nicolaidis, Ed. New York: Springer, 2011.

[6] Yu. V. Katunin, and V. Ya. Stenin, "Simulation of Single Event Effects in STG DICE Memory Cells," *Russ. Microelectronics*, 2018, vol. 47, pp. 20-33.

[7] Yu. V. Katunin, and V. Ya. Stenin, "Logical C-element on STG DICE Trigger for Asynchronous Digital Devices Resistant to Single Nuclear Particles," *Russ. Microelectronics*, 2019, vol. 48, pp. 143-156.

[8] R. Garg, S. P. Khatri, *Analysis and Design of Resilient VLSI Circuits: Mitigating Soft Errors and Process Variations*, New York: Springer, 2010.

Nonparametric Statistical Analysis of Radiation Hardness Threshold Variation in CMOS IC Wafer Lots Series with the Aim of Process Monitoring

Yu. I. Bogdanov, N. A. Bogdanova, D. V. Fastovets, Y. M. Moskovskaya, A. V. Sogojan, and A.Y. Nikiforov

Abstract - Nonparametric statistical criteria usage has been considered for estimating radiation hardness threshold variation of CMOS IC wafer lots series. It gives one the possibility to assess every newly manufactured IC lot by a small sample size to determine whether test sample and the whole lot statistically belong to the previously tested general reference group of samples or not. The approach allows us to minimize the overirradiation factor and obtain a statistically reliable radiation test result even for small-size test samples.

I. INTRODUCTION

IC process statistical monitoring is usually based on the assumption that the wafers parameters distribution is normal. However, in the case of statistical radiation hardness monitoring during production process, IC's parameter distributions are not always normal. In this case non-parametric statistical evaluation criteria can provide a significant effect [1].

The usage of non-parametric statistical criteria for IC radiation harness evaluating is considered under the following condition: the accumulation of IC lots quantity leads to a change in the conformity degree between the group under consideration and the basic - reference group.

The advantage of non-parametric statistical criteria as compared to commonly used parametric criteria is a freedom from assumptions about the particular statistical distribution type. This property of non-parametric evaluation criteria is very important as the type of statistical distribution is usually unknown beforehand.

It is proposed in this paper to determine the exact quantitative statistical criteria for CMOS IC accessory to the basic reference group by using all possible permutations arising in comparison of two statistical samples.

The usage of computer technologies provides high-precision reliable results that cannot be obtained from any statistical tables.

Yu.I. Bogdanov, Y.M. Moskovskaya, A.V. Sogojan, A.Y. Nikiforov, are with NRNU MEPhI and JSC SPELS, Moscow, Russia, e-mail: ymmos@spels.ru.
Yu.I. Bogdanov, N.A. Bogdanova, D.V. Fastovets are with Valiev Institute of Physics and Techonology of Russian Academy of Sciences and National Research University of Electronic Technology (MIET), Russia, Moscow, e-mail: bogdanov_yurii@inbox.ru

In case of large samples sizes (it is impossible to obtain explicit enumeration of all permutations using modern computers or supercomputers) we use the Monte-Carlo statistical test method. This method allows us to construct approximate criteria that ensure high precision of estimations, sufficient for any practical tasks. It should be noted that the standard approach based on the approximate asymptotic estimation usage often leads to an insufficiently low accuracy of radiation hardness evaluation results.

II. THEORETICAL ANALYSIS OF NON-PARAMETRIC STATISTICAL CRITERIA USAGE

Statistically, we check the hypothesis of the equality of two probability distributions. Alternative hypotheses are constructed to estimate the shift and scale parameters. We consider various non-parametric tests including the Lehmann-Rosenblatt test, Mann-Whitney test and Wilcoxon test, as well as the generalized Wilcoxon-Siegel-Tukey test [1-3].

Let us consider the Lehmann-Rosenblatt test using the following statistics T:

$$T = \frac{1}{nm(n+m)} \left[n \sum_{i=1}^{n} (r_i - i)^2 + m \sum_{j=1}^{m} (s_j - j)^2 \right] - \frac{4nm-1}{6(n+m)}. \quad (1)$$

Here, we consider two samples of size n and m respectively. The r_i, $i = 1,...,n$ parameters are the indexes of the first sample elements, and s_j, $j = 1,...,m$ are the indexes of the second sample elements in the general ordered sample obtained from the combined sample. To obtain the exact statistical test, it is necessary to investigate all permutations. The total number of all permutations is

$$N = \frac{(n+m)!}{n!m!}.$$

The following example illustrates the situation where the difference between the test group of lots and base set of lots increases statistically significantly as the number of lots accumulates. Let A and B – be two lots with three CMOS IC samples in each. Total ionizing dose hardness threshold have been previously determined for each sample in conventional radiation test. Figure 1 corresponds to the lot

A. The Lehmann-Rosenblatt test is performed in the $n = 3$ case (test lot *A* with 3 samples) and $m = 18$ (basic lot *B* with 18 samples). The total number of permutations is $N = 1330$.

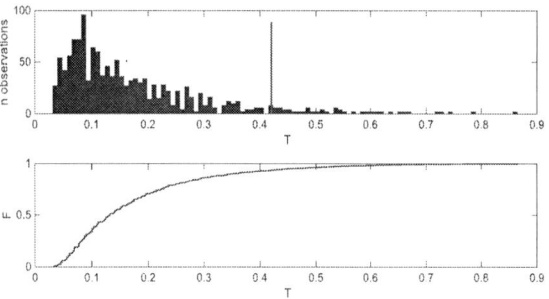

Fig 1. The Lehmann-Rosenblatt test for n=3; m=18 (lot *A*). The top figure is T parameter distribution histogram. The bottom figure is a cumulative distribution of T parameter. The red vertical line corresponds to the parameter value T=0.420634920634921 calculated by the formula (1) for two considered samples. The significance level corresponds to the total probability to the right of this line.

Among the total number 1330 of possible permutations, and by means of explicit enumeration, we found out that 88 permutations have had the statistical value that was not less than the obtained value $T = 0.420634920634921$ from formula (1). Therefore, the significance level (p-value) is:

$$p = \frac{88}{1330} = 0.0661654135338346.$$

The obtained value exceeds the selected significance level $\alpha = 0.05$. Thus, it indicates the absence of statistically significant difference in radiation hardness thresholds between the test lot and the basic lot.

A similar numerical experiment for B test lot showed that among the total number 1330 of possible permutations, 122 permutations have the statistical value are not less than the obtained value $T = 0.365079365079365$ using formula (1). Therefore, the significance level (p-value) is:

$$p = \frac{122}{1330} = 0.0917293233082707.$$

The obtained value exceeds the selected significance level $\alpha = 0.05$ again. It is indicating the absence of statistically significant difference in radiation hardness between the test lot and the basic lot.

Finally, let us consider the Lehmann-Rosenblatt test for the combination of considered lots *A* and *B*. Now $n = 6$ and $m = 18$. Note that the total number of possible permutations has become much higher $N = 134596$. Among these 134596 possible permutations, 1750 permutations have the statistical value are not less than the obtained value $T = 0.673611111111111$ using formula (1). Therefore, the significance level (p-value) is:

$$p = \frac{1750}{134596} = 0.0130018722696068.$$

The obtained value is significantly below the selected significance level $\alpha = 0.05$. It indicates the statistically significant difference in radiation hardness between the *A* and *B* lots and the basic lot.

Note that the radiation hardness of the test *A* and *B* lots is statistically significantly higher than the radiation hardness of the basic lot - some process correction have been done before *A* and *B* lots that improved radiation hardness.

It is important to note that it is possible to take into account all samples that passed the statistical test and add them to the original basic sample. Such approach is justified when the technological level doesn't degenerate over time. Thus, the basic sample will be expanded. It will improve the evaluation of statistical significance level while considering further lots of ICs. Since the *A* lot has no statistically significant differences with the base lot, we will add all wafers of this lot to the base lot. Thus, the basic lot will consist of 18 + 3 = 21 samples. Let us compare this combined lot with lot *B* (3 samples). As a result, we obtain that among of 2024 possible permutations, 390 permutations have statistical value are not less than the obtained value 390 $T = 0.2589285714285714$ using formula (1). Therefore, the significance level is:

$$p = \frac{390}{2024} = 0.19268774703557312.$$

The obtained value is significantly below the selected significance level $\alpha = 0.05$. It indicates the statistically significant difference in radiation hardness between the *B* lot and new base lot obtained by joining base and *A* lots.

III. PRACTICAL ASPECTS OF NON-PARAMETRIC STATISTICAL CRITERIA USAGE TO MONITORING LOTS OF IC WAFERS IN MASS PRODUCTION

An important practical example of non-parametric statistical criteria application is the implementation of CMOS IC wafers lots radiation hardness monitoring during mass production [4-6]. The proposed approach allows us to make a rational choice of the overirradiation factor (K_L) which is used to guarantee radiation hardness in case of small-sizes samples according to Russian national radiation test regulations.

The radiation overirradiation factor K_L is the ratio of the increased total ionizing dose (TID) level X_L in radiation test to the TID level X_S, specified in the requirements, i.e.

$$K_L = \frac{X_L}{X_S} \qquad (2)$$

K_L can be determined by the following formula:

$$K_L = \frac{M - z_P \sigma}{M - z_R \sigma} = \frac{1 - z_P V}{1 - z_R V} \qquad (3)$$

Here, z_R and z_P are the quantiles of the normal distribution, corresponding to the probability levels R and P, respectively. The parameter R is the probability of preserving the working ability.

The parameter P is the lower limit of the confidence interval, determined from the relation:

$$P = (1-\gamma)^{1/n}, \qquad (2)$$

where γ is the confidence level, and n is the sample size.
Parameters M and σ are the estimated mean and standard deviation of the normal population, $V = \sigma/M$ is the coefficient of variation. The calculation method is illustrated in Figure 2.

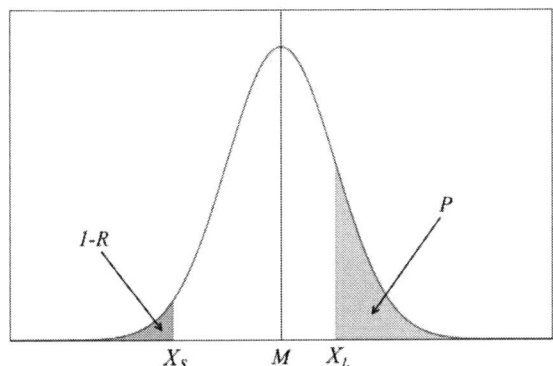

Fig. 2 Graphic illustration of the method for calculating the overirradiation factor K_L. The tail probability to the right of the level X_L is equal to P, the tail probability to the left of the level X_S is equal to $1-R$.

The considered statistical monitoring corresponds to the following parameters: $R = 0.95$, $\gamma = 0.9$, $V = 0.25$.

Thus, for a regular sample (3 samples), according to the formula (3) we obtain $K_L = 1.74$. This implies the necessity of significant overdose of each sample with a positive test result to confirm the specified TID hardness threshold requirements. This approach assumes that the production process has a higher stability than standard TID hardness level of product.

Since the task of ensuring stability of a product from each lot is essentially equivalent to the problem of ensuring the stability of the production process, then the TID hardness should be described by the statistical regulation rules. Therefore, it can be assumed that each following sample of the production lot is a part of the general parent group of all previously tested samples from previously released lots. If such general sample group satisfies the conditions of homogeneity then each following lot increases the total size of the general sample and K_L factor can be reduced by increasing the number of samples up to the level close to the case $K_L = 1$ when the general sample size is more than 40. It is important to note, that we take into consideration only lots which satisfy TID hardness

requirements. Thus, the lower bound is strictly defined and only the variation above the boundary is considered.

In the previously considered example using non-parametric statistical estimation, it is quantitatively justified that the tested sample (with only 3 samples from A lot) is a part of the general parent group with the sample size 21. This allows us to accept the value for this number of samples $K_L = 1.16$, i.e. statistically reliable test results are provided with a significant reduction of the test overirradiation degree.

If according to non-parametric statistical estimation during lots monitoring it is established that the considered sample doesn't belong to the general sample of previously manufactured and tested products, then it means that some changes (declared or undeclared) must have occurred in the technological process. Therefore, the current process differs from the previous one (it may become "better"). Thus, in order to guarantee the statistical reliability of test results it is necessary to obtain the positive testing result with the usage of a proper (high) overirradiation factor (for example $K_L = 1.74$ for a sample size 3) and thus start formation of a new basic sample and further enlarge it with next lot samples.

If the tested samples obtained from several serial lots according to the results of non-parametric statistical estimation don't belong to the initial general sample (as in the previous example of the combined sample $A+B$) but do belong to the new basic sample (provided by higher hardness level, for example), then it is necessary to perform several corrective actions to characterize the modified process (for example, recertification of the technological process, changing product parameters standards and etc.).

IV. CONCLUSIONS

To summarize, the usage of non-parametric criteria with intensive usage of the computer technologies allows one to solve problems of the radiation hardness variation monitoring of CMOS IC wafers lots during mass production with high precision and statistical reliability. In particular, using the estimation results, it is possible to perform a rational choice of the overirradiation factor (K_L) which is used to guarantee statistically reliable radiation test results even with small-sizes samples.

REFERENCES

[1] L.N. Bolshev and N.V Smirnov, *Tables of Mathematical Statistics (in Russian)*, Nauka, Moscow, 1983.

[2] E. L. Lehmann, *Nonparametrics: Statistical Methods Based on Ranks*, Holden-Day, San Francisco,1975.

[3] S.Siegel, and J. W.Tukey, *J. Amer. Statist. Ass.*, 1960, 55, pp 429–444.

[4] Y.M. Moskovskaya, A.Y.Nikiforov, D.V.Bobrovskiy, A.V.Ulanova, A.A. Zhukov. "Process parameters variations influence on CMOS IC's hardness to total ionizing dose", *IEEE 30th International Conference on Microelectronics (MIEL)*, 2017.

[5] Z. Stamenkovic and N. Stojadinovic, "New Defect Size Distribution Function for Estimation of Chip Critical Area in Integrated Circuit Yield Models", *Electronics Letters*, 1992, vol. 28, pp. 528-530.

[6] Yu. Moskovskaya, "Common methodological approachto evaluation of radiation resistanceof gate arrays and semicustom very large scale IC based on them", *Nanoindustry*, 2017, vol. 1, pp. 50-59.

Radiation Hardness Evaluation of LEDs Based on InGaN, GaN and AlInGaP Heterostructures

D. S. Ukolov, N. A. Chirkov, R. K. Mozhaev, and A. A. Pechenkin

Abstract – The radiation hardness results of light emitting diodes (LED) in green, blue and red regions of the spectrum and in white, based on InGaN, GaN and AlInGaP structures are presented. The technical aspects of monitoring parameters during exposure are described, and LEDs response to various radiation exposures are given.

I. INTRODUCTION

Nowadays there is a large amount of LED types, which differ among themselves both in light-emitting region structure and technical parameters, as well as in the variety of designs and light collimating methods [1].

Semiconductor light emitters, in particular LEDs, have proven to be a suitable replacement for gas-discharge emitters or lightbulbs [2]. LEDs are used in all areas of illuminating engineering, as light source devices and as display devices. When using light-emitting diodes in space applications or in places with high background radiation [3], they impose certain requirements on the ability of device to withstand total ionizing dose and single event effects in outer space. LEDs must perform their functions and keep parameter values within specified limits under and after radiation exposure.

II. DEVICE UNDER TEST. PARAMETERS AND CHARACTERISTICS

The paper compares radiation hardness of semiconductor emitters (Table 1) with different types of structures, radiation spectrum and structures to neutrons, protons and gamma radiation. As have been mentioned before, the radiation sensitive parameters are luminous flux and luminous intensity. These parameters depend on the light-emitting structure and the method of collimating the optical radiation.

Therefore, in order to determine and compare radiation hardness, it is important not only to test samples with different radiating structure, but with different construction as well.

LEDs #1,2,3 were manufactured using InGaN / GaN processes and have a protective layer of a compound. A luminophore is deposited on the radiating surfaces of white light indicators. When exposed to ionizing radiation, the

Denis S. Ukolov, Nikolay A. Chirkov, Roman K. Mozhaev, Alexander A. Pechenkin are with the National Research Nuclear University (NRNU) "MEPhI" and JSC "SPELS", Kashirskoe shosse 31, Moscow, Russia, E-mail: dsukol@spels.ru

optically transparent compound can darken and degrade the radiative characteristics of the luminophore.

TABLE I
TYPES OF LEDs UNDER TEST

Structure	Wavelength, nm	Housing / construction	LED number
InGaN/GaN	485	SOT-23	1
	565		2
	White		3
GaN	560	5 LEDs in 1	4
	680		5
InGaN/Al$_2$O$_3$	White	SMD1608	6
		SMD3528	7
		SMD3535	8
		type emitter	9
AlInGaP/ Al$_2$O$_3$	590	SMD1608,	10
		SMD3528	11
		SMD3535	12

Semiconductor LED indicators #4,5 are based on GaN epitaxial structures. Light-emitting elements have a difference in the concentration and type of dopants of the conductivity layer. For red light emitting elements, the n-type conductivity layer is doped with Zn and O; for green light-emitting elements, the p-type layer was doped with N. These indicators have a protective polymeric optically transparent material with the addition of the corresponding dye.

LEDs # 6-9 of white color are based on two types of InGaN/Al2O3 epitaxial structures. The SMD3535 type emitters has a protective layer of optically transparent polymer that forms the lens. The LEDs in the SMD1608 and SMD3528 package do not have lens, and the crystal is covered with a protective layer of compound.

LEDs #10-12 have yellow hue, are based on AlInGaP/Al2O3 structures and use SMD1608, SMD3528 and SMD3535 packages.

III. EXPERIMENTAL PLAN

The total ionizing dose (TID) research was carried out using a Co^{60} isotope source and an electron accelerator operating in bremsstrahlung mode with the maximum energy of 2.2 MeV. Light parameters were controlled before and after exposure using the «Ophir» photosensor. Irradiation of light-emitting diodes was carried out both at

978-1-7281-3420-8/19 $31.00 © 2019 IEEE

normal (T = +25°C), low (T = - 60°C) and high temperatures (T = +85°C).

Research of resistance to structural damage were carried out on a pulsed fast neutron reactor with an energy of 14 MeV in the unbiased mode of device operation and on a proton accelerator with a proton beam energy of 200 MeV with periodic monitoring, with a fluence pitch of $2 \cdot 10^{10}$ p/cm², optical parameters.

IV. RESULTS AND DISCUSSION

During the LEDs No. 4 TID hardness research, a change in the optical power of the emitters was observed.

Differences in radiation hardness were discovered when exposed to TID irradiation. It was found out that radiation hardness depends (Figure 1 and 2) on the dopant's concentration and the type of dopants in the conductivity layer. The measurements also demonstrated that the protective optical compound, made with the addition of a dye, the color of which corresponds to the emission color of the LED, changed its optical properties.

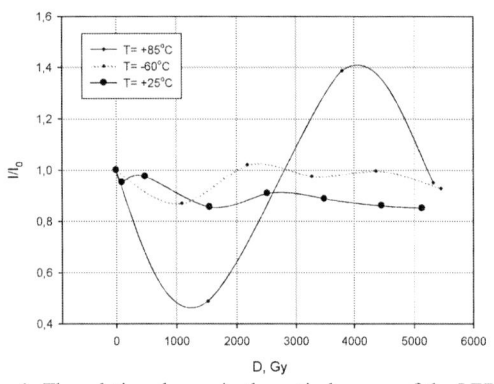

Fig. 1. The relative change in the optical power of the LED №5 depending on the absorbed dose at different ambient temperatures.

For LED №3, optical power degradation was also noticed (Figure 3). The most severe degradation of optical power was observed for samples irradiated at room temperature.

For all types of heterostructures, increase in optical power depending on the absorbed dose occurs after irradiation. Presumably, this effect is associated with radiation-stimulated annealing of defects and mechanical stresses in the heterostructure layers introduced at the growth stage and/or at the diode manufacturing stage [3].

The parameters monitoring during structural damage test with a pulsed fast neutron reactor showed that after exposure to neutrons with a fluence of $\Phi = 2,3 \cdot 10^{12}$ n/cm², the parameters were degraded [4] according to Table 1. LED No. 4 demonstrated the lowest hardness level.

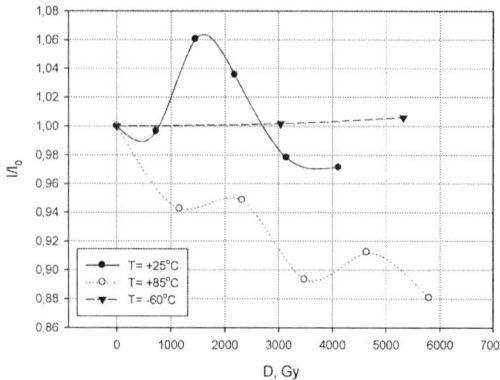

Fig. 2. The relative change in the optical power of the LED №4 depending on the absorbed dose at different ambient temperatures.

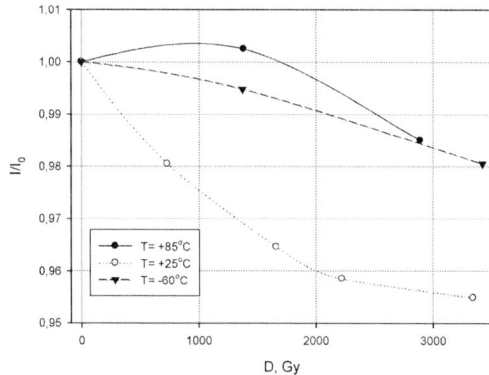

Fig. 3. The relative change in the optical power of the LED №3 depending on the absorbed dose at different ambient temperatures.

To validate hardness levels to structural damage, an additional testing using a proton accelerator with proton beam energy of 200 MeV was carried out with periodic monitoring of optical parameters [5]. The dependence of the light intensity of semiconductor emitters on the fluence of protons was obtained. Degradation of optical parameters observed during irradiation with fluence up to $2 \cdot 10^{10}$ p/is shown in Figure 4.

TABLE II
THE RELATIVE CHANGE IN THE OPTICAL POWER OF LEDs №4,5 DEPENDING ON THE NEUTRON FLUENCE.

Φ, n/cm²	I/I_0 LED №4	I/I_0 LED №5
0	1	1
2,3E+12	0,023	0,623
1,1E+13	0,024	0,229

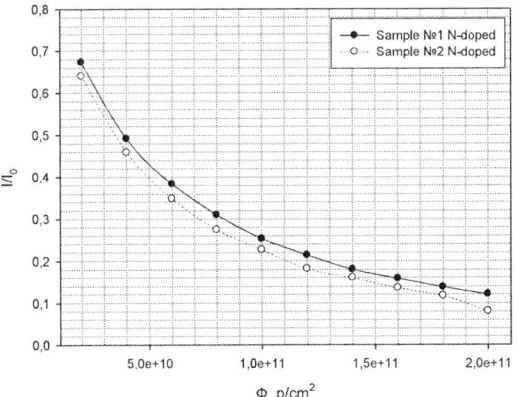

Fig. 4. The relative change in the optical power of the LEDs №4. as a function of the proton fluence.

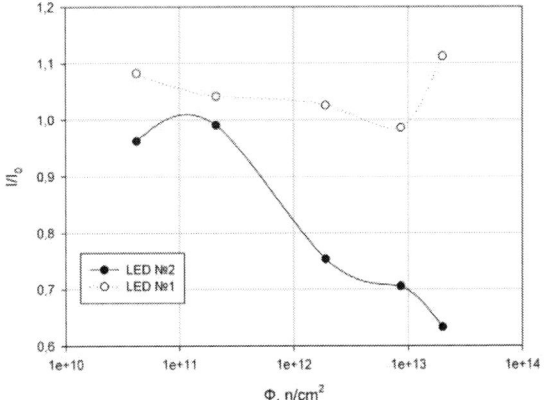

Fig. 5. The relative change in optical power from the neutron fluence of LED № 1 and LED № 2.

The measurements carried out for green and blue LEDs in SOT-23 package showed that green LEDs are less resistant to structural damage effects (Figure 5).

The difference between light emitting structures of LED No. 1 and LED No. 2 is concern with concentrations. The LED №1 has lower concentration of In, than the LED №2.

Since the epitaxial structures used in these LEDs differ only in the concentration of dopants, this difference in hardness levels is due to the increased molar mass of In in samples with a green emission color.

Despite the high concentration of penetrating dislocations in InGaN / GaN epitaxial films, which are caused by the discrepancy between these layers and sapphire substrates, semiconductor emitters on these structures have a high efficiency of radiative recombination. One of the problems when growing InGaN layers with high in content is its evaporation from the surface, as a result of which dislocations are formed. When irradiated with gamma rays, protons or neutrons, radiation-induced relaxation of defects occurs and the subsequent restructuring of the initial simple defects in heterostructures into more complex ones, which leads to decrease in LEDs efficiency. Figure 6 and figure 7 show the dependence of the optical power degradation on

neutron fluence for InGaN/Al2O3 and AlInGaP/Al2O3 LED structures respectively. LEDs №8,9 demonstrate a similar degradation dynamic, in contrast to LEDs №6,7. LEDs No. 6.7 show a stepwise change in optical power, which occurs due to radiation-stimulated relaxation of defects in the light-emitting structure of the LED.

This difference is most likely due to the structural design of the LEDs [4].

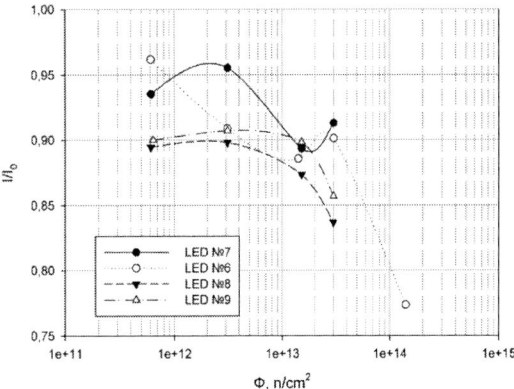

Fig. 6. Relative change in optical power from neutron fluence of LEDs LED № 6-9.

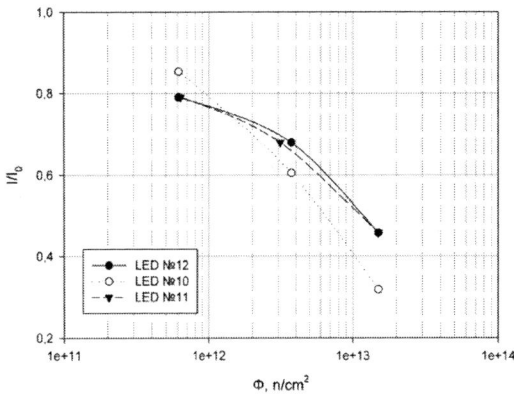

Fig. 7. Relative change in optical power from neutron fluence of LEDs LED № 10-12.

IV. Conclusion

Radiation hardness research and comparative analysis of semiconductor indicators show that LED structures are sensitive to ionizing radiation. An essentional parameters degradation is observed when exposed to protons and neutrons.

It is noted that green luminescence color LEDs based on N-doped GaN heterostructure exhibit lower radiation resistance to structural damage, and TID radiation, compared to structures based on GaN heterostructure with Zn and O doping.

Low resistance to structural damage is also shown by green LEDs based on the InGaN/GaN herostructure with an increased molar content of In.

978-1-7281-3420-8/19 $31.00 © 2019 IEEE

Therefore, the impurity type and its concentration affect the resistance to neutron and proton radiation.

For AlInGaP/Al$_2$O$_3$ LED structure, the effect of neutrons with a fluence of the order of $\Phi = \cdot 10^{13}$ n / cm^2 is critical. Optical power falls by more than 50%.

Light-emitting structures based on InGaN / Al2O3 have the best radiation resistance to structural damage. Optical power does not fall below 75% when exposed to $\Phi = \cdot 10^{14}$ n/cm^2.

LEDs with different package types sensitivity to structural damage show that protective compounds have little effect on change of photometric parameters during exposure to radiation.

REFERENCES

[1] E. Fred Schubert. *Light-Emitting Diodes* (3rd Edition). – E. Fred Schubert, 2018.

[2] H.N. Becker, A.H. Johnston, "Proton damage in LEDs with wavelengths above the silicon wavelength cutoff", *Nuclear Science IEEE Transactions on*, 2004, vol. 51, no. 6, pp. 3558-3563.

[3] Orlova, K.N. "Resistance of LEDs Based on AlGaInP Heterostructures to Irradiation by Fast Neutrons", *J. Chem. Chem. Eng.*, 2013, No7, pp. 409-413.

[4] R.K. Mozhaev, M.E. Cherniak, A. V. Ulanova, and A. Y. Nikiforov, "Estimation technique for LED sensitivity to structural damage based on minority carrier's lifetime measurements," *in Proc. 30th Int. Conf. on Microelectronics*, MIEL 2017; Nis, Serbia, October 2017, pp. 161-164.

[5] M.E. Cherniak, A. A. Pechenkin, R.K. Mozhaev, A. V. Ulanova, and A. Y. Nikiforov, "Automated measurement system for optoelectronic devices based on National Instruments PXI-platform," *in Proc. 2017 International Siberian Conference on Control and Communications*, SIBCON 2017, S. Seifullin Kazakh Agrotechnical University, Astana; Kazakhstan; June 29 -30, 2017, article number 7998545

Simulation of Errors Impulses from Single Ionizing Particles in CMOS Triple Majority Gates

Yu. V. Katunin, V. Ya. Stenin and A. G. Prozorova

Abstract - This work presents the TCAD simulation of the 65-nm bulk CMOS Triple Majority Gate as resistant to the single-event transients. The charge collection from the track of a single nuclear particle simulates as impacted at the every transistors of the gate which leads to the noise pulse of the output of this gate. The TCAD simulation used the tracks along the normal to the chip surface with input track points at the each transistor. The linear energy transfer to the track is 60 MeV·cm2/mg.

I. INTRODUCTION

The effect of the single nuclear particles on combinational logic elements is an error pulse [1] that causes the temporary changes of logical state (single event transient – SET). Increased robustness to the effects of single particles depends on schemes and the topological design. At design rules less 100 nm in CMOS VLSI the effect of charge sharing between adjacent elements becomes noticeable [2]. This effect can be useful for the design of the elements topology with the compensation of the error pulse using quenching [3]. This technique was used in the matching logic of the translation lookaside buffers [4].

In our work, we present the TCAD simulation results of a charge collection by the NAND-based Triple Majority Gate (TMG), whose topology was optimized to use the quenching effect. This TMG is most commonly used due to the acceptable sensitivity to SET [5] and the occupied area.

II. VOTING IN TRIPLE MAJORITY LOGIC

In TMR (triple modular redundancy), three identical logic phases are used to compute the same set of specified Boolean function. When there are failures, the output signals of the phases may be different. The TMG are used to voting for the correct output signal. The majority gates itself are sensitive to SET. Fig. 1 shows the shows the scheme of the TMG element based on the NAND elements only.

Fig. 2 presents the transistors version of the TMG scheme based on the NAND elements only. Fig. 3 shows the standard layout of the topology of this TMG element. We propose the new version of the layout of the topology of

Yu. V. Katunin, and A. G. Prozorova are with the Scientific Research Institute of System Analysis, Russian Academy of Sciences, Nakhimovsky pr. 36-1, 117218 Moscow, Russia, E-mail: katunin@cs.niisi.ras.ru

V. Ya. Stenin is with the National Research Nuclear University MEPhI (Moscow Engineering Physics Institute), Kashirskoe sh. 31, 115409 Moscow, Russia, E-mail: vystenin@mephi.ru

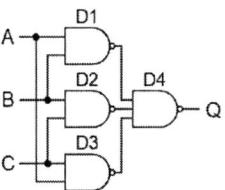

Fig. 1. Triple Majority Gate on the NAND gates only.

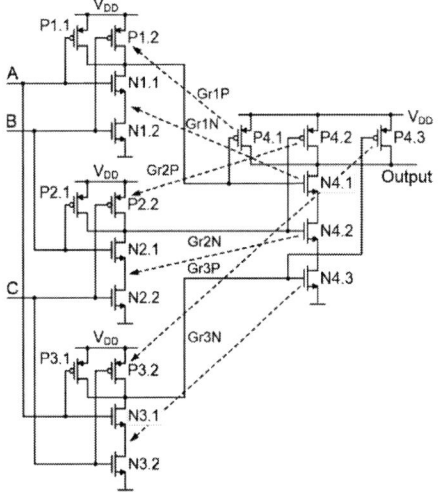

Fig. 2. The scheme of the Triple Majority Gate with the transistors of three 2NAND gates and the one 3NAND gate.

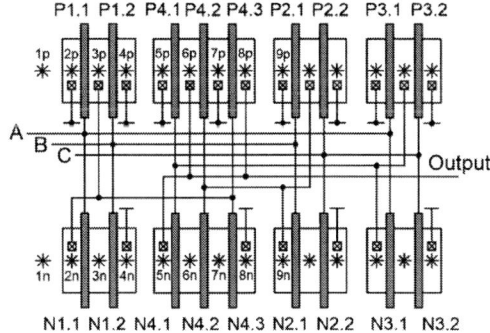

Fig. 3. The standard layout of the topology of the TMG element.

this element with the error pulses correction. Arrows on Fig. 2 indicate the charge coupling of the transistors needed for the coorection of the error pulse.

978-1-7281-3420-8/19 $31.00 © 2019 IEEE

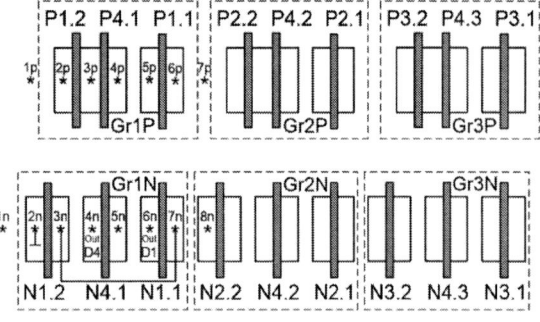

Fig. 4. The layout of the topology of the proposed TMG element with the alternation of transistors of NAND gates for the correction of the error pulses in the each groups of transistors.

Fig. 4 demonstrates the layout of the topology of the TMG element with the alternation of transistors of NAND gates. All transistors located in the six groups, three same groups of NMOS transistors Gr1N–Gr3N and three same groups of PMOS transistors Gr1P–Gr3P. In each group one transistor, relating to the output gate D4, is surrounded by two transistors, incoming to one of the input gates D1–D3. The charge collection by closed transistors of the input gate, located in the same group, can cause a noise pulse (SET) at its output. For that reason the transistor of the output gate D4, located in the same group, will become closed and sensitive to a charge collection. Joint charge collection by transistors of logic gates, that are adjacent in the chain, will contribute to correct the noise pulse at the output of the gate D4 in cases when it is not logically masked by the gate D4.

III. METHODOLOGY OF SIMULATION

Fig. 5 presents 3D TCAD device physics model of the proposed Triple Majority Gate (TMG) with the logic gates D1, D2, D3 and D4 on Fig. 1. The design of the element consists of the transistor's groups with a shallow trench isolation. All transistors correspond with the six transistors groups Gr1N–Gr3N and Gr1P–Gr3P. Each group includes

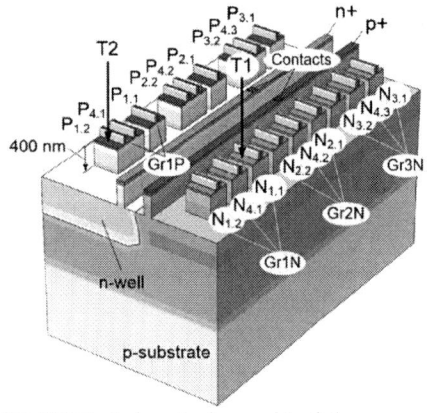

Fig. 5. 3D TCAD device physics model of the proposed Triple Majority Gate. Tracks directions are along the normal to the surface (T1 and T2); n+ and p+ areas – guard strips with contacts.

two transistors of the 2NAND element and one transistor of the same conductivity type of the 3NAND element, the gate of which connected to the output of 2NAND element. The shallow trench isolation of the transistors (400 nm depth) does not show on Fig. 5.

Single event transients at the elements during 3-D TCAD simulation depends on the linear energy transfer to the track and the direction of the track. The results obtained using the Sentaurus Device simulator at the temperature of 25°C and the supply voltage of 1.0 V for the 65 nm CMOS bulk structure. The widths of the N- and PMOS transistors are 400 nm. The 3-D device was comprised of transistors developed taking into account the models presented in the work [6]. TCAD simulation is preferable to SPICE simulation, despite the more duration of calculations, which is about 28 hours. The required accuracy of the quenching effects evaluation depends on the topology-related nature of a charge collection by sensitive volumes of logic elements that are adjacent in the chain. It justifies the use of TCAD.

IV. SIMULATION RESULTS

The linear energy transfer from a particle to the track is 60 MeV·cm2/mg. The track directions are along the normal to the surface of the chip. The input track points are along

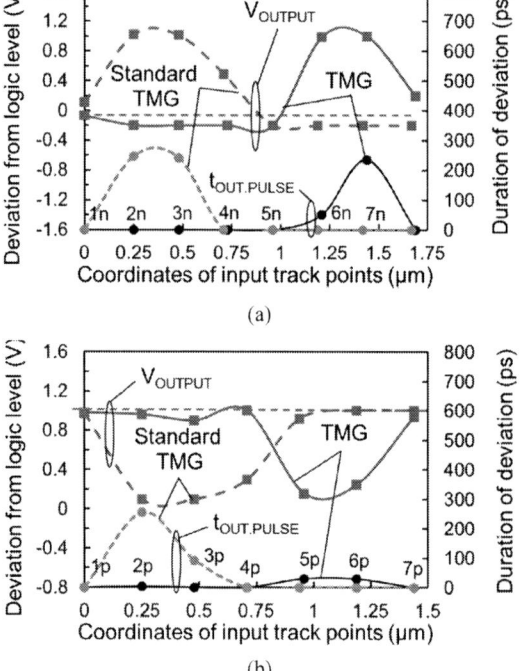

Fig. 6. The voltage deviations on the nodes from logic level and error pulse durations for the proposed TMG and for the standard TMG variant (Figs. 3, 4) without corrections as the functions of the track points input coordinates when TMG inputs: (a) A = B = C = 0; (b) A = B = 1, C = 0. Solid lines - proposed TMG, dashed lines - standard TMG.

the line drains of transistors (Fig. 5: tracks T1 and T2). The step of the displacement of input track points is 0.25 μm. The voltage deviations from logic level of the output node V_{OUTPUT} and the durations of the impulses of errors $t_{OUT.PULSE}$ presented in Figs. 6 and 7 for two triple majority gates (the proposed TMG and the standard TMG topologies). These are functions of the track points input coordinates. The dependencies of the element with the TMG topology presented by the solid lines. The curves of the element with the standard topology are by the dashed lines.

A. Corrections of error pulses in proposed TMG

Fig. 6 shows the dependencies with corrections of the nodes voltage deviations and the durations of the of errors in the transistors group Gr1N for the TMG element with the proposed topology, when A = B = C = 0 (Fig. 6a), and in the transistors group Gr1P, when A = B = 1, C = 0 (Fig. 6b). Fig. 6 shows also the dependencies without corrections of the nodes voltage deviations and the durations of the impulses of errors in the cases of the standard TMG element, when A = B = C = 0 (the dashed lines in Fig. 6a) and when A = B = 1, C = 0 (the dashed lines in Fig. 6b).

The dependencies with corrections of the impulses of errors (solid lines in Fig. 6a) have not practically the voltage deviations on nodes in the group Gr1N of TMG for the input track points 1n–5n. The maximum error pulses on the output of the TMG element with correction are when the track passes through the point 6n (drain N1.1) and the point 7n (drain N1.2). This is the result of a charge collecting by serial-connecting transistors N1.1 and N1.2 (Fig. 2).

The dependencies with corrections (solid lines in Fig. 6b) have not practically the voltage deviations on nodes in the group Gr1P for the input track points 1p–4p (Fig. 4). This is the case of the parallel connection of transistors P1.1, P1.2 in the 2NAND gate D1 (Fig. 2). The maximum noise pulses have the amplitudes 0.8–0.9 V but the small durations 100–200 ps for the input track points 5p and 6p (Fig. 6b).

B. Cases not correction of error pulses

Fig. 7 shows the dependencies without corrections of the impulses of errors for the TMG element with the proposed topology. Fig. 7a presents the dependences of voltages on the nodes of the 3NAND gate D4 (Fig. 2) during charge collection for the input track point 4p with LET = 60 MeV·cm2/mg for the transistors in the group Gr1P, when A = B = C = 0. The deviations on the nodes from logic level are 0.6–1.4 V with the durations 250–300 ps. In this case we have not any corrections of the nodes voltage deviations.

Fig. 7b shows the dependencies for the group Gr1N of TMG when A = B = 1, C = 0. In these cases the 3NAND element D4 (Fig. 2) has the maximum false output signals $V_{D4.OUT}$ = 1.1 V with the durations 330 ps for the input track point 4n. The significant deviations are when the charge is collected by the closed NMOS transistor N4.1 of the 3NAND element D4 (the track points 3n–5n of

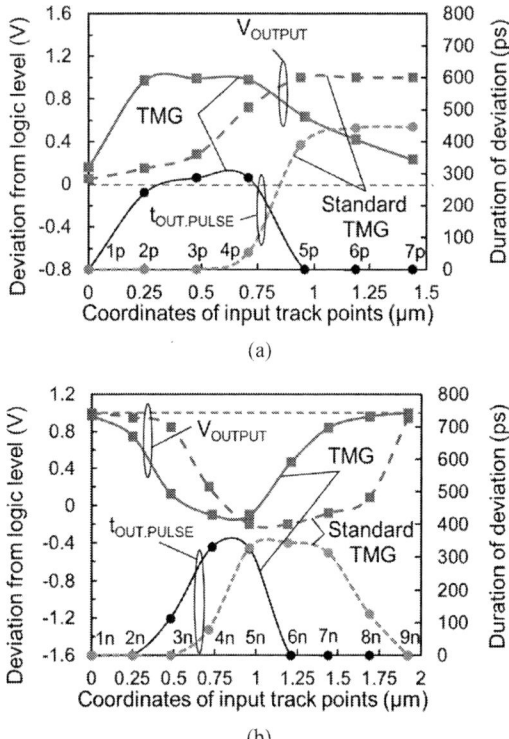

Fig. 7. The voltage deviations on the nodes from logic level and error pulse durations in case of non corrections of error pulses for two TMG variants: proposed TMG and standard (Figs. 3, 4) as the functions of the track points input coordinates when TMG inputs: (a) A = B = C = 0; (b) A = B = 1, C = 0. Solid lines - proposed TMG, dashed lines- standard TMG.

dependencies in Fig. 7b). In this case, we have the serial connection of transistors N4.1, N4.2 and N4.3 in 3NAND element D4. The deviations from the logic level at the output of the TMG element have values of 0.6–1.1 V with the duration 100–325 ps. The proposed TMG element and the TMG element with the standard topology show the same dependencies of the output error pulses on Fig. 7.

C. Examples of Correction in the Group Gr1N

Fig. 8 shows dependencies for the impact to the NMOS transistors N4.1 and N1.1 of NAND elements D1 and D4, when the TMG inputs are A = B = C = 0. These two NMOS transistors in the Gr1N group begin to collect the charge from the track (the input point 6n on Fig. 4), which reduces the voltage at the output 2NAND D1 and on the gate of the NMOS transistor N4.1 of D4, located in the group Gr1N (Fig. 4) for the error correction. The closed transistor N4.1 begins to collect a charge, the amplitude of the error pulse at the output D4 reduces (Fig. 8). After 50 ps, the voltage at the drain of the transistor N4.1 becomes less 0.7 V, which keeps the output logic of the TMG element nearly "0". This is the case of the correction (minimizing) the amplitude and the duration of the error pulse.

978-1-7281-3420-8/19 $31.00 © 2019 IEEE   203

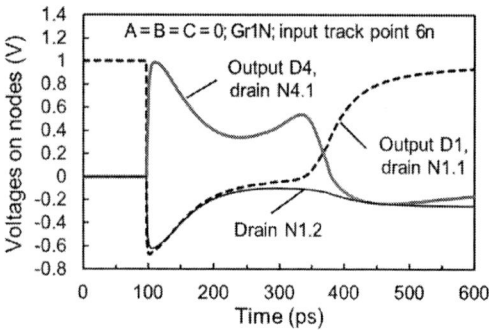

Fig. 8. The nodes voltages of the proposed TMG element (Fig. 4) in the group Gr1N during charge collection from the track with LET = 60 MeV·cm2/mg; the input track point 6n (Fig. 6a), A = B = C = 0.

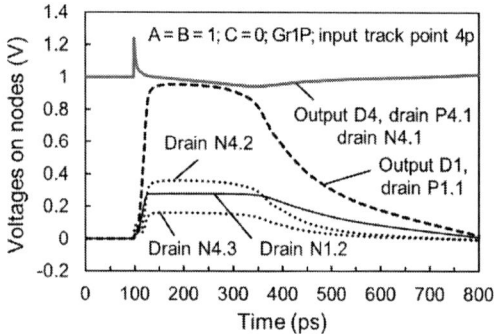

Fig. 9. The nodes voltages of the proposed TMG element (Fig. 4) in the group Gr1P during charge collection from the track with LET = 60 MeV·cm2/mg; the input track point 4p (Fig. 6b), A = B = 1; C = 0.

D. Examples of Correction in the Group Gr1P

Fig. 9 shows curves for the impact to the drain of the open PMOS transistors P4.1 of the gate D4 (the track input point is 4p in Fig. 4), when the TMG inputs are A = B = 1, C = 0. The transistor P4.1 is in the open state and passes to the inverse offset mode in the starting of the charge collecting. This fixes the output voltage of the TMG element on the stable level logical "1". The closed PMOS transistors P1.2 and P1.1 collect a charge from track, forming a pulse on the gate node of the transistor P4.1. This pulse does not close the transistor P4.1 and the TMG element keeps the stable output level logic "1" without an error pulse. This is the case of the minimizing the amplitude and the duration of the error pulse.

V. ANALYSES OF SIMULATION RESULTS

The layout of the transistors groups is based on the combining two or three transistors with interconnection drains and sources. For example, the transistors of D1 and D4 gates combined into groups Gr1N and Gr1P, collect simultaneously a charge from the track of a particle. These are the transistors N1.2, N4.1, N1.1 in the group Gr1N and the transistors P1.2, P4.1, P1.1 in the group Gr1P.

If the output transistors N4.1 or P4.1 of D4 gate are open in the initial mode of the TMG element, they can keeping the level of output TMG signal compensating and blocking changes.

The output signal keeps the correct value if open (or closed) transistors in the initial moment of a charge collecting from the track go into the inverse offset mode and if this is blocking the voltage change at the output node.

The transition groups from state with the noise pulses in the state without any noise pulses is associated with the correction effect that is missing in TMG elements of the traditional topology without alternation of transistors of the logical elements D1, D2, D3 and the element D4. In TMG elements of the traditional topology without alternation of their transistors the significant noise pulses occur for all input signals of the TMG item.

VI. CONCLUSION

The CMOS Triple Majority Gate on the NAND gates using topology of quenching provides mainly the best characteristics in modelling impacts of single ionizing particles. This element has less transistors than the traditional TMG element on AND–OR gates and may be the basis for design the triple modular redundancy logic in 28–65 nm CMOS hardened systems resistant to single nuclear particles.

ACKNOWLEDGEMENT

The research is supported in the framework of the State of Russian Federation assignment, project 0065-2019-0008.

REFERENCES

[1] *Soft errors in Modern Electronic Systems* / M. Nicolaidis, Ed. New York: Springer, 2011, pp. 27–54.

[2] N. N. Mahatme, S. Jagannathan, T. D. Loveless, L. W. Massengill, B. L. Bhuva, S.-J. Wen, and R. Wong, "Comparison of combinational and sequential error rates for a deep submicron process," *IEEE Trans. Nucl. Sci.*, 2011, vol. 58, pp. 2719–2725.

[3] N. M. Atkinson, A. F. Witulski, W. T. Holman, J. R. Ahlbin, B. L. Bhuva, and L. W. Massengill, "Layout technique for single-event transient mitigation via pulse quenching," *IEEE Trans. Nucl. Sci.*, 2011, vol. 58, pp. 885–890.

[4] Yu. V. Katunin, and V. Ya. Stenin, "TCAD Simulation of the 65-nm CMOS Logical Elements of the Decoders with Single-Event Transients Compensation," in *Proc. of 2018 Workshop on Electronic and Networking Technologies (MWENT)*, Moscow, 2018, pp. 1–6.

[5] I. A. Danilov, M. S. Gorbunov, and A. A. Antonov, "SET Tolerance of 65 nm CMOS Majority Voters: A Comparative Study," *IEEE Trans. Nucl. Sci.*, 2014, vol. 61, pp. 1597–1602.

[6] R. Garg, S. P. Khatri, *Analysis and design of resilient VLSI circuits: mitigating soft errors and process variations*, New York: Springer, 2010, pp. 194–205.

978-1-7281-3420-8/19 $31.00 © 2019 IEEE

Influence of Temperature over Impedance of Different Inkjet Printed Patterns and Substrates

B. Nikolova, G. Nikolov, E. Gieva, I. Ruskova, and M. Mladenov

Abstract - Inkjet printing technology turns out to be a fast-growing technology that, due to its lower cost, is becoming more and more widely used in the production of sensors. The electrical properties and parameters of such sensors are dependent on the type of ink, its composition, the type of the substrate, and the number of printed layers. In this paper, we study the influence of the shape, the type of the substrate and the temperature on the impedance of the printed structures.

I. INTRODUCTION

Printing electronics is a set of technologies that literally "print" functional electronic layers (dielectric, conductor and semiconductor) to create various electronic components such as sensor elements and even whole electronic circuits. The general requirement for all printing technologies is that printing materials should be in the form of inks, which can consist of both organic substances and metallic nanoparticles such as silver, copper or aluminium. The ink we use is with silver nanoparticles [1, 2].

The determination of the capacitance of the layer structures (Interdigital capacitor) is very complicated. On the one hand, it is difficult to determine the overlapping area of the more complex forms. Also the electric field is non-homogenous and passes through different environments - air and dielectric substrate. The most frequent modelling of such capacitors is through finite element method (FEM) and numerical simulations using programs such as COMSOL [3, 4]. A major problem with the physical modelling of Inkjet print structures is that there is a lack of information about the physical and electrical parameters of the resulting layers.

The use of paper and polymer backing in Inkjet printing technology has as its main advantage low cost but also a substantial disadvantage – lack of information about their electrical parameters. In publication [5] we gave results of measurements of dielectric permeability and dielectric losses of five substrates types In this study, we

B. Nikolova is with the Department of Technology and Management of Communication Systems, Faculty of Telecommunication, Technical University of Sofia, Sofia, Bulgaria, E-mail: bnikol@tu-sofia.bg

G. Nikolov and M. Mladenov are with the Department of Electronics, Faculty of Electronic Engineering and Technologies, Technical University of Sofia, Sofia, Bulgaria, E-mail: gnikolov@tu-sofia.bg

E. Gieva and I. Ruskova are with the Department of Microelectronics, Faculty of Electronic Engineering and Technologies, Technical University of Sofia, Sofia, Bulgaria, E-mail: gieva@ecad.tu-sofia.bg; ruskova@ecad.tu-sofia.bg

have established that these properties strongly depend on the temperature and the humidity of the environment.

On the other hand, in publication [6] we have described and proposed a method for sheet resistance measurement of conductive layers, made by Inkjet printing technology. The results show, that the thickness of the printed layers strongly depends on the substrate's type. That is why the measured sheet resistances are with different value.

It is very easy to determine the different electrical parameters of the studied structures, such as capacitance, resistance, inductance, etc., from the impedance characteristics. The obtained results directly can be used to determine the transfer function of temperature sensors realized by printed structures. They will also be useful for other sensor applications of the printed structures.

From the above, it follows that it is good to study the impedance parameters on different structures on different substrates. It will be also very interesting to study these characteristics as a function of the temperature change.

II. MATERIALS AND STRUCTURES

Fig. 1 presents a classical capacitive form, its dimensions and photos of the printed structure on the three substrates. Similarly, Fig. 2 and 3 show the dimensions and photos of the rest two structures. Their shape and area are different from the classical one, in order the change in the capacitance and impedance to be investigated.

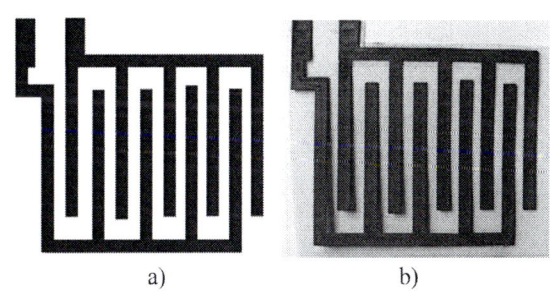

a)   b)

Fig. 1. a) Type of form 1, b) Substrate PET.

In this article, we use the Epson® Stylus C88 + printer and the Metalon® JS-B25P Inkjet conductive ink [3]. We have chosen three different shapes to measure impedance characteristics. In order to determine the impact of the substrate, we chose the structures to be printed on three different substrates: Bond uncoated paper with a grammage of 80 g/m2 and thickness d = 0.1 mm, Photopaper

(90 g/m2, d = 0.120 mm) and Novel PET film (d = 0.1 mm). In Figures 1 to 3 we have shown the shape of our three capacitive structures and pictures of the printed structures on the various substrates.

After the structures have been printed, they are not directly sintered. We perform several sample measurements from 25 °C to 100 °C to investigate the repeatability of the results and whether different effects are observed, also whether sintering influences the samples in subsequent measurements at temperature change measurements. The influence of temperature on samples is of great importance for their possible application as sensors.

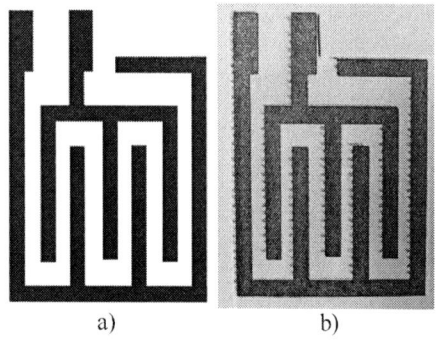

Fig. 2. a) Type of form 2, b) Substrate Paper

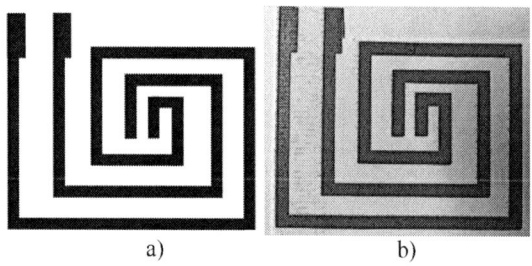

Fig. 3. a) Type of form 3, b) Substrate Photopaper.

III. EXPERIMENTAL SETUP

The main objective of the study is to assess the impact of temperature on the impedance of printed capacitive structures (Fig. 1 – 3). To obtain these features, LCR Bridge Hameg 8118 is used and a specially designed and implemented test fixture. Its basic technical characteristics are: measurement functions - L, C, R, |Z|, X, |Y|, G, B, D, Q, Θ, Δ, M, N; frequency range from 20Hz to 200kHz; basic accuracy 0.05%. The device can be controlled by Universal Serial Bus (USB) or RS-232 and can be programmed using LabVIEW.

In order to investigate and analyze the influence of temperature, the printed structures are placed in a thermal camera. The thermocouple is used to measure and monitor the temperature accurately.

The measurements are performed at frequency of 1 kHz, value of the voltage 1V and temperature 25 °C.

IV. MEASUREMENTS AND RESULTS

After printing the samples (shown in Figures 1 - 3), we connect them to a test attachment and place them in a thermal chamber. We monitor the change of temperature and measure capacity and impedance at 5 °C. Figures 4 and 5 show the graphs of change of capacity and impedance, depending on the change of temperature, for sample 1.

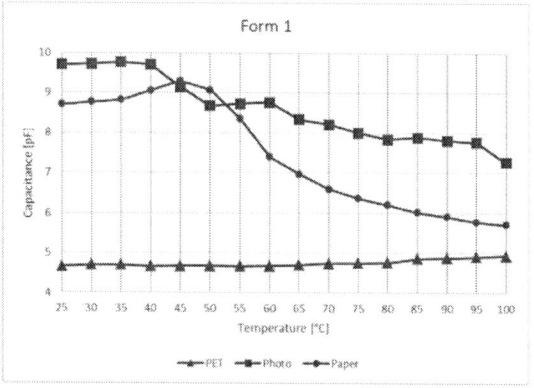

Fig. 4. Capacity dependence on temperature change for sample 1

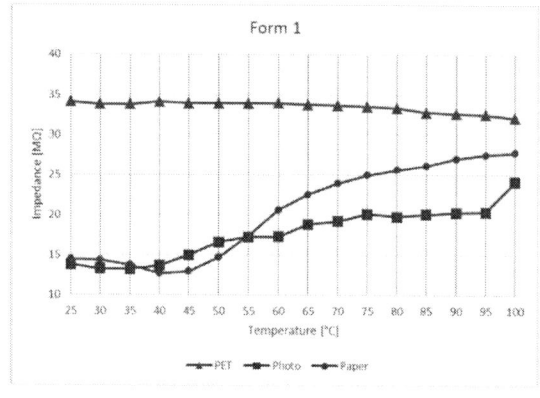

Fig. 5. The dependence of the impedance on changing the temperature for a sample 1

The particularity in this form is that the greatest change is the capacity and impedance of a structure made on paper, and the smallest of the PET substrate. An interesting result is obtained with the structure realized on photo paper, in which, with increasing temperature, the capacity decreases monotonically - from about 10 pF to about 7 pF. This means that the structure thus obtained can be used as a temperature sensor. Similarly, Figures 6 and 7 present the graphs of capacity and impedance dependence on temperature for sample 2.

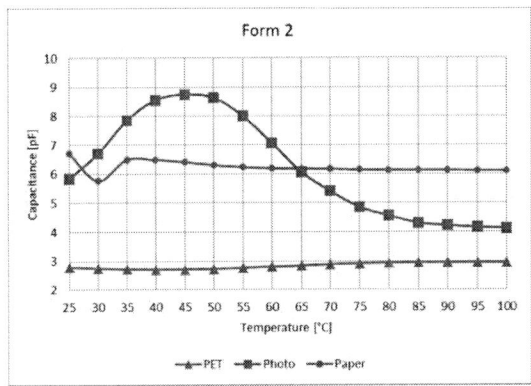

Fig. 6. Capacity dependence on temperature change for sample 2

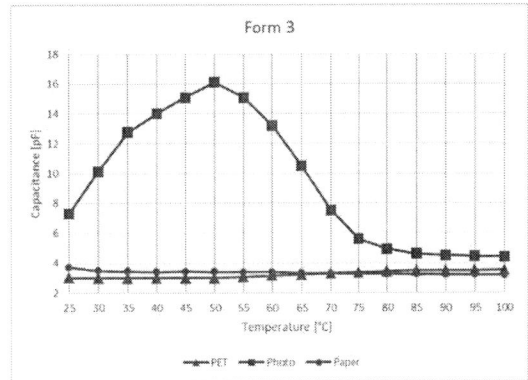

Fig. 8. Capacity dependence on temperature change for sample 3

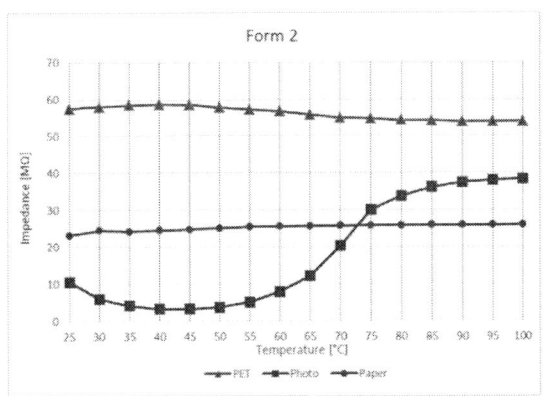

Fig. 7. Impedance vs. temperature for a sample 2

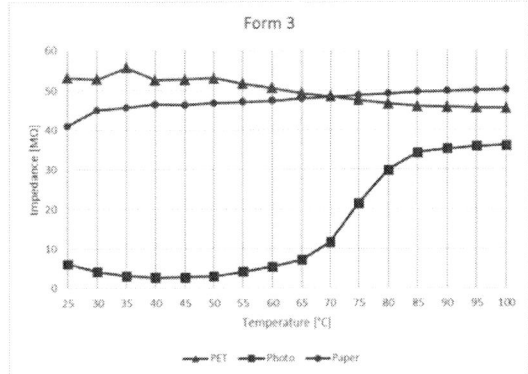

Fig. 9. Impedance vs. temperature for a sample 3

The main difference with form 1 is that the strongest temperature dependence in this form occurs when it is printed on photo paper. The particular case is that dependency is not monotonous - first increases, then decreases. This means that if this structure is to be used for a temperature sensor, this can only be done for a certain temperature range, for example from 50 °C to 100 °C.

Similarly, for sample 3, Figures 8 and 9 show the graphs of capacity and impedance variation depending on temperature. Here again it turns out that the most severe temperature dependence is this form, printed on photo paper, this structure has the highest capacity - 16 pF at 50 °C.

Based on the results presented in the above graphs, the following conclusions can be drawn:

1. Regardless of the substrate used and the temperature change form, all the studied structures exhibit a totally capacitive character - the increase of the capacitance leads to a reduction of the impedance and vice versa.

2. Again, the conclusion is drawn that the substrates used have a very strong influence on the parameters of the printed structures. Furthermore, it is seen that the same pattern of samples on different substrates does not show the same trend in temperature variation of capacitance and impedance.

3. Generally, the use of a PET substrate results in a structure with the smallest capacity (less than 5 pF), which is also least affected by the temperature. This means that this substrate can be used over a wide temperature range as an insulator.

4. Generally, the greatest change in temperature and impedance in structures is realized on photo paper. This means that they can be used as temperature sensors but in a limited temperature range. It would also be advisable to examine these structures at least once for temperature changes to establish repeatability of the obtained dependencies.

5. The greatest capacity is obtained when the structure is realized on photo paper, and the smallest - on a PET substrate. This indicates that essentially the capacitance is determined by dielectric properties of the substrates and the results obtained correspond to the [5] measured dielectric permeability of the various flexible substrates.

In order to illustrate the difference between the capacity of the three substrates in Figure 10, the graphs of the temperature change capacity of a structure realized on a PET substrate are shown.

It can be seen that the largest capacity has form 1 (about 5 pF), and the smallest form is 3 (less than 3 pF). In general, this difference is not very significant, since the dimensions of all three shapes are of one order. Also, the capacity for all

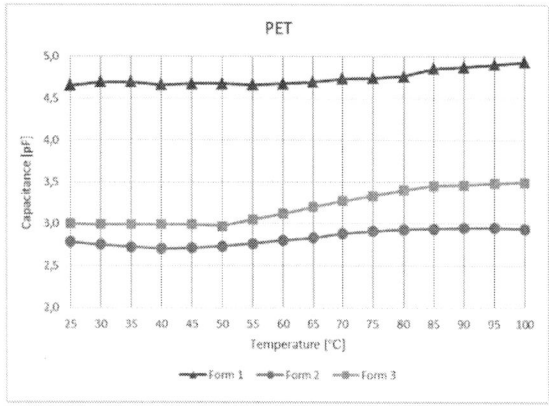

Fig. 10. Capacity dependence on temperature for substrate PET

forms almost does not change with increasing temperature. The curves are almost straight lines and the values do not differ greatly.

Tables 1 and 2 give specific values for the capacity measured in [pF] and the impedance measured in [MΩ] for each of the three samples on the three types of substrates only for three temperatures - room, average and maximum.

TABLE I

THE CAPACITY OF THE STUDIED SAMPLES, MEASURED IN [pF]

	Substrate	Temperature °C		
		25	50	100
Form 1	PET	4,65	4,67	4,93
	Photo paper	9,70	8,76	7,24
	Paper	8,70	9,07	5,71
Form 2	PET	2,78	2,73	2,94
	Photo paper	5,83	8,63	4,12
	Paper	6,71	6,30	6,10
Form 3	PET	3,00	2,97	3,49
	Photo paper	7,29	16,10	4,39
	Paper	3,70	3,39	3,16

V. CONCLUSION

Experimental studies of the effect of temperature on the impedance properties of different types of interdigital capacitive structures were performed in the presented paper. The structures are printed on Inkjet technology on three different flexible substrates: Bond uncoated paper, Photopaper and PET film.

From the results obtained and the analyses made, it can be seen that the substrates has a strong influence on the capacity and impedance of the samples. The influence of temperature also depends strongly on the type of the pad as well as on the shape of the capacitive structure. For a PET substrate, stability of the capacity values that are not highly dependent on the temperature change is observed. Research and analysis can be used to design sensor structures for specific applications.

TABLE II

THE IMPEDANCE OF THE STUDIED SAMPLES, MEASURED IN [MΩ]

	Substrate	Temperature °C		
		25	50	100
Form 1	PET	34,12	33,90	32,10
	Photo paper	13,76	16,57	24,11
	Paper	14,46	14,69	27,78
Form 2	PET	57,30	57,80	54,04
	Photo paper	10,60	3,78	38,40
	Paper	23,27	25,20	26,04
Form 3	PET	53,16	53,10	45,47
	Photo paper	6,17	3,10	36,17
	Paper	41,00	46,80	50,20

ACKNOWLEDGEMENT

This paper is carry out in the frame of the project DM17-9/20.12.2017 "Research, design and modeling of flexible nanocomposite sensors based on inkjet technology".

REFERENCES

[1] C. Gaspar, J. Olkkonen, S. Passoja and M. Smolander, "Paper as Active Layer in Inkjet-Printed Capacitive Humidity Sensors", *Sensors,* 2017, vol. 17(7), p. 1464.

[2] G. Mattana and D. Briand, "Recent advances in printed sensors on foil", *Materials Today,* 2015, vol. 19, no. 2, pp. 88-99.

[3] G. González, E. S. Kolosovas-Machuca, E. López-Luna, H. Hernández-Arriaga and F. Javier González, "Design and Fabrication of Interdigital Nanocapacitors Coated with HfO₂", *Sensors,* 2015.

[4] X. Hu and W. Yang, „Planar capacitive sensors – designs and applications", *Sensor Review,* 2010, vol. 30, no. 1, pp. 24–39.

[5] B. Nikolova, G. Nikolov, E. Gieva, and I. Ruskova, „Dielectric Properties Measurement of Flexible Substrates", in Proc. *IEEE XXVII International Scientific Conference ET2018,* 2018.

[6] E. Gieva. G, Nikolov and B. Nikolova, "Sheet Resistance Measurement of Inkjet Printed Layers", *42nd International Spring Seminar on Electronics Technology ISSE2019,* 2019.

[7] https://www.novacentrix.com/products/inkjet-starter-kit

978-1-7281-3420-8/19 $31.00 © 2019 IEEE

Virtual System for Measurement of Inkjet Printed Resistive and Capacitive Structures

G. Nikolov, E. Gieva, B. Nikolova, and I. Ruskova

Abstract - The presented article describes the method and sequence in creating a virtual measurement system using a modern LabVIEW-controlled measuring instrument. We have implemented a virtual environment for measuring different electrical characteristics of printed resistive and capacitive sensor structures through Inkjet technology. The determination of different parameters of the printed structures makes it possible to characterize them and to determine their applications.

I. INTRODUCTION

Inkjet printed layers have a fast development over the recent years in the manufacturing of cost effective sensors. The base electrical properties of printed sensor structure strongly depend on the composition of the inks, type of substrates and number of prints. Therefore, in order to obtain electrical parameters of printed functional layers over different flexible substrates it is needed to apply specific measurement techniques [1].

These specific measurement methods, respectively, require a large number of diverse measuring devices with different ranges and functional capabilities. Many of the publications review the printing methods, the types of structures, the ways of preparing the inks and the results of the sensor structures [4, 5, 6]. However, in a very small part [7], attention is paid to the measuring instruments and control software used for the study of structures.

The present paper examines the characteristics of printed capacitive and resistive structures on different types of flexible substrates. The focus of the article is on the approach of creating a virtual measurement system consisting of modern measuring instruments that are controlled by a graphical programming language.

G. Nikolov is with the Department of Electronics, Faculty of Electronic Engineering and Technologies, Technical University of Sofia, Sofia, Bulgaria, E-mail: gnikolov@tu-sofia.bg

E. Gieva and I. Ruskova are with the Department of Microelectronics, Faculty of Electronic Engineering and Technologies, Technical University of Sofia, Sofia, Bulgaria, E-mail: gieva@ecad.tu-sofia.bg; ruskova@ecad.tu-sofia.bg;

B. Nikolova is with the Department of Technology and Management of Communication Systems, Faculty of Telecommunication, Technical University of Sofia, Sofia, Bulgaria, E-mail: bnikol@tu-sofia.bg

II. MATERIALS, METHODS AND VIRTUAL SYSTEM ARCHITECTURE

We have chosen four different structures that we printed on different substrates using the Epson® Stylus C88 + printer. We printed three layers of each structure with Metalon® JS-B25P Inkjet conductive ink and then processed them at 100 °C for 60 minutes [3]. Once the structures have been prepared, their characteristics will be determined by the created virtual measurement environment. We use LabVIEW [2] and modern Hameg 8112 RLC Bridge and Rigol DM3068 to create the virtual environment. This article aims to measure the resistance and capacity of the printed three capacitive and resistive shapes. One of the classical interdigital capacitive shape is shown in Figure 1, and a classical resistive meander type is shown in Figure 2.

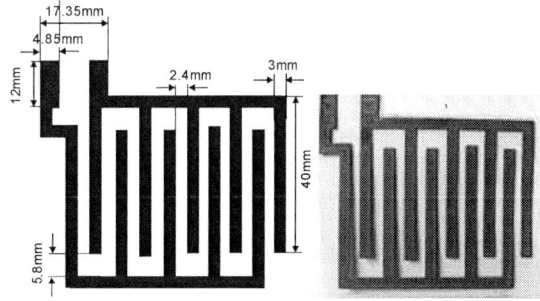

Fig. 1. Interdigital capacitive structure.

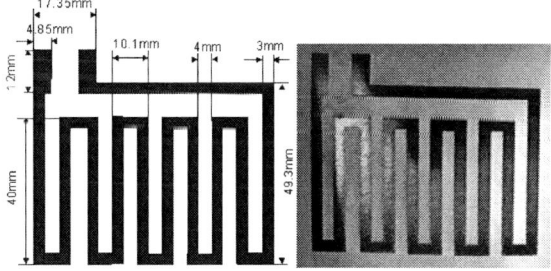

Fig. 2. Meander structure form of resistivity.

The block diagram of the hardware part of the designed and implemented virtual measurement system is shown in Fig. 3. Only two measuring devices are shown in the diagram, but with the approach used, their number can be significantly increased.

The first is the Hameg 8112 RLC Meter. This device allows measurement of impedance and capacity in the frequency range from 20Hz to 200kHz. It is controlled by a computer via a USB interface and is only available from one PC. The second one is the Rigol DM3068 digital multi meter which has a resolution of 6 $^{1/2}$ digits and is driven by the LXI (LAN extension for Instrumentation) standard with instrument drivers. For the operation of LXI-based devices, LAN Switch and DHCP Server are required. Once an IP address is associated with such a device, it becomes accessible for management from any computer connected to the Internet.

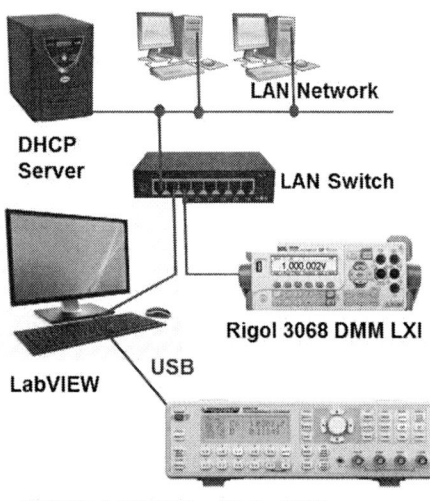

Fig. 3. Architecture of designed virtual system.

The technical specification of the measuring instruments used for the functions and ranges in which they are used are given in Table 1.

It is noticeable that the measurement of capacity with the multi meter is associated with a big error. This is due to the fact that DMM uses a method for measuring capacity with applied constant current into the capacitance, and measure the voltage changing rate.

For the purpose of this study, a special test fixture has been designed and manufactured to allow the measuring equipment to access the printed structures. The plugin allows a 4-wire connection of the structures to the devices, as can be

seen from Fig. 4. Decreasing the contact resistance is achieved by tightening the contact areas with screws.

TABLE I
SELECTED TECHNICAL SPECIFICATION OF THE MEASURING
INSTRUMENTS

Function	Range	Accuracy
Rigol DM3068		
Resistance	200.0000Ω	0.010% + 8mΩ
Capacitance	2.000nF	2% + 50pF
Hameg 8112 LCR Bridge Meter		
Impedance	100 MΩ	0.5% + 5mΩ
Impedance	100 Ω	0.1% + 1mΩ
Capacitance	100pF	0.2% + 0.1pF

Fig. 4. Implemented Test fixture.

A block diagram created in the programming environment LabVIEW for controlling the system multi meter RIGOL DM 3068 is shown in Fig. 5. For the preparation of the code, a classic design pattern has been used with sequential performance of instrument driver functions and multiple reads of the data through the loop. The Configuration Measurement feature is polymorphic and allows configuration of the Ethernet, USB, GPIB or RS232 interfaces. On the Fig. 5. the drivers are configured for Ethernet control (LXI).

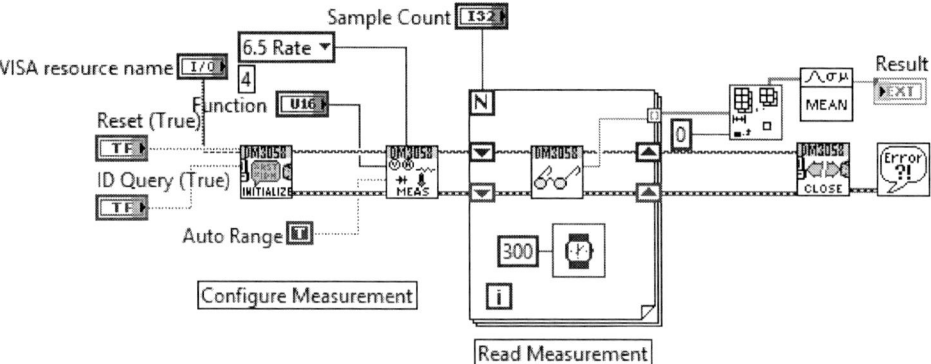

Fig. 5. Block diagram to control Rigol DM3068 via Ethernet.

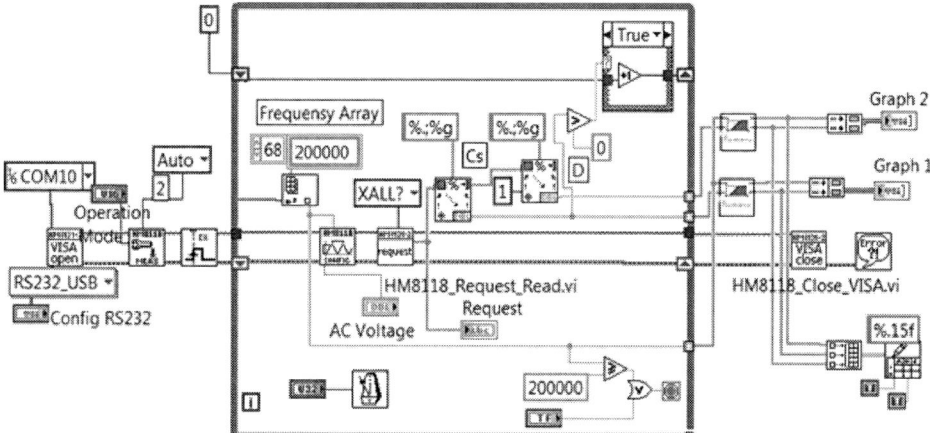

Fig. 6. Block diagram to control HM 8112 LCR via USB.

Similarly, a graphical program application was created for the Hameg 8112 LCR shown in Fig. 6, but this device is driven by a USB interface. The device is designed in such a way that the USB interface works as a virtual port (in this case COM10). This hardware disadvantage of the measurement device causes frequent generation of invalid inertial results, which greatly impedes the graphical representation of the results and the omission of values for some frequencies. This inconvenience is avoided by software. At the top of the program code presented in Fig. 6 through the *Shift registers* and the *Case structure* there is provided a method by which each invalid report (negative value) or string "Error" is not allowed to increase the frequency but repeats the measurement at the same frequency until receiving of a valid result. Frequency range is performed in the *While loop*, with fixed frequencies available in the Frequency Array with 69 values. When reaching a frequency of 200kHz, the cycle stops its performing. After the cycle, two subVIs are used to eliminate zero results and visualize the measurement process by XY Graph (Graph 1 and Graph 2 in Figure 6)

IV. EXPERIMENTAL RESULTS

The front panel of the created virtual system in the active 4-Wire Resistance measurement mode with Rigol DM3068 is shown in Figure 7. The figure shows the attached IP address of the device according to the LXI standard. To the structure of the meander of Fig. 2, the wires are connected by a 4-Wire method, 25 measurements are performed and an average value is induced.

Similarly, Fig. 8 shows the front panel of the system, but when measuring the capacity of the interdigital structure of Fig. 1. As mentioned, measurement of low-capacity multi meters is associated with a large error, which also reveals the large difference in the values measured by Rigol DM 3068 at DC current and HM 8118 LCR at alternating voltage.

Results in graphical view obtained from the virtual system in impedance and frequency domain measurement are shown in Figures 9, 10 and 11.

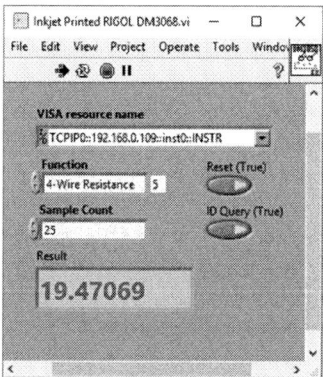

Fig. 7. Front panel of Virtual system measuring active resistance of silver Meander structure printed over PET.

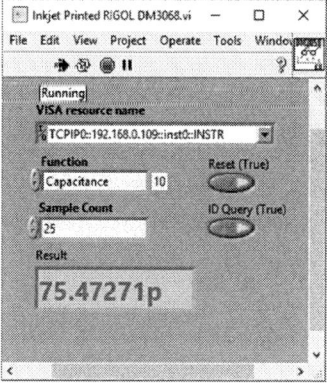

Fig. 8. Front panel of Virtual system measuring capacitance of silver Interdigital structure printed over PET.

978-1-7281-3420-8/19 $31.00 © 2019 IEEE

Fig. 9 and 10 show the frequency dependencies of the impedance and the capacity of the Interdigital capacitive structure of Fig. 1 for three different PET substrates, Bond Paper and Photopaper. It is apparent the strong frequency dependence of the structure, the most stable in terms of capacity is the structure on PET.

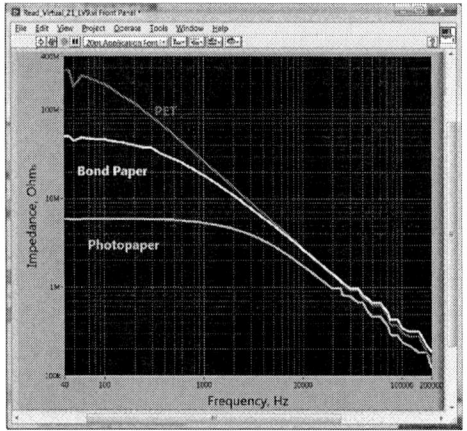

Fig. 9. Front panel of Virtual system measuring impedance vs. frequency.

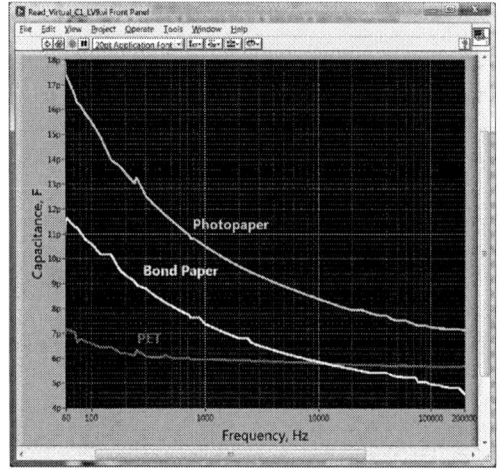

Fig. 10. Front panel of Virtual system measuring capacitance vs. frequency.

Figure 11 shows the frequency response of the meander impedance of Fig. 2. Here we can emphasize the extremely stable impedance behaviour of the structure in the range up to 200 kHz. Another thing that impresses is that the impedance of the structure on Bond Paper is on the order of magnitude larger than that on Photopaper and PET.

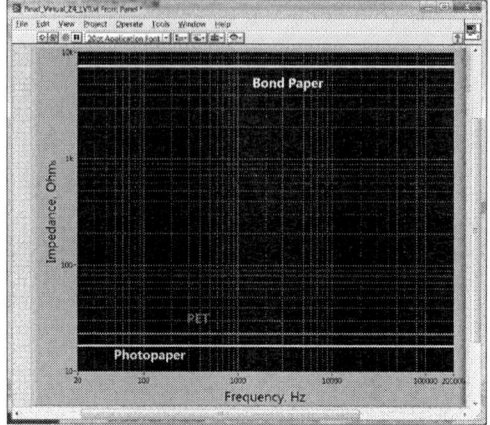

Fig. 11. Front panel of virtual system measuring impedance vs. frequency for resistive structure.

V. Conclusion

A virtual environment has been created to measure the electrical characteristics of resistive and capacitive sensor structures. The sensor structures were printed with Inkjet technology. The environment was created with the LabVIEW software program. The article focuses mainly on the approach of creating virtual measuring systems suitable for the study of printed sensor structures, without a detailed analysis of the properties of the structures. The suggested approach would be useful for researchers and engineers in the field of measuring technology and sensors.

Acknowledgement

This paper is carried out in the frame of the project DM17-9/20.12.2017 "Research, design and modelling of flexible nanocomposite sensors based on inkjet technology".

References

[1] A. Arazna, K. Janeczek, K. Futera, "Mechanical and thermal reliability of conductive circuits inkjet printed on flexible substrates", *Circuit World*, vol. 43, issue. 1, pp.9-12, 2017.

[2] http://www.ni.com/en-rs/shop/labview/labview-details.html.

[3] https://www.novacentrix.com/products/inkjet-starter-kits.

[4] M. Hartwig, R. Zichner and Y. Joseph, *Inkjet-Printed Wireless Chemiresistive Sensors—A Review,* Chemosensors, 2018.

[5] S. Zuk, A. Pietrikova, "Capacitive Sensors Realized on Flexible Substrates", *Electroscope*, Ročník, 2017.

[6] B. Andò, S. Baglio, A. R. Bulsara, T. Emery, V. Marletta and A. Pistorio, "Low-Cost Inkjet Printing Technology for the Rapid Prototyping of Transducers", *Sensors*, 2017.

[7] D. Barmpakos, A. Segkos, C. Tsamis and G. Kaltsas, "A Disposable Inkjet-Printed Humidity and Temperature Sensor Fabricated on Paper", *MDPI Proceedings*, 2018.

Investigation on Body Potential in Cylindrical Gate-All- Around MOSFET

M. Kessi, A. Benfdila, A. Lakhelef, L. Belhimer, and M. Djouder

Abstract - Cylindrical gate all around (CGAA) MOSFET has been studied and the body potential has been simulated and studied. The center and surface potentials models are studied and a comparison is made between the two potential behaviours for a CGAA MOSFET. Moreover, the channel doping concentration and the band diagrams are obtained using the finite element numerical method by solving poisson's equation in the cylindrical coordinate system.

I. INTRODUCTION

As MOSFET$_S$ continue to get smaller between source and drain has led to an impending power crisis and reduces the capability of gate electrode to control the body potential distribution in the channel [1]. The short-circuit from source and drain is facing serious problems, like Drain Induced Barrier Lowering (DIBL) and the threshold voltage roll off [2]. As a result the off state current increases and the On-Off current ration are degraded [3]. Several designs structures (MuGFET$_s$) have been developed [4], all of them are targeting the enhancement of the better channel control due to the action of multiple gate electrodes surrounding the channel. Cylindrical Gate-All-Around MOSFET$_S$ have been regarded as a promising technology for sub-10-nm CMOS devices, because has provide the best short channel device performance , better gate controllability, suppressed floating-body, improved transport property and CMOS compatibility, compared with other non-classical device structures[5-6], because do not have corner effects due to the circular cross-section and cylindrical body. Moreover, the center and surface potential of the CGAA structure has been investigated and the comparison is made with them.

II. DEVICE STRUCTURE AND PARAMETERS

The schematic diagram of the Cylindrical GAA (CGAA) MOSFET structures used for simulation is shown in Fig. 1. The radial directions are assumed to be along radius and lateral direction along z-axis of the cylinder. The details of device physical parameters used in the structure are shown in Table 1.

The authors are within the Micro and Nanotechnology Research Group (MNRG), Faculty of Electrical Engineering and Computer Sciences, University M. Mammeri, Tizi-Ouzou, UMMTO DZ 15000, Algeria. E-mail: mohamed.kessi@ummto.dz and abenfdil@ictp.it

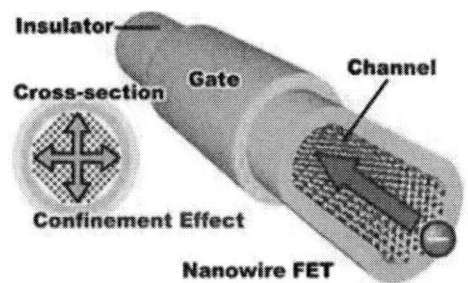

Fig. 1. Schematic structure of Cylindrical Gate-All-Around (CGAA) MOSFET$_S$. (After [4])

TABLE I
VALUES OF VARIOUS PARAMETERS USED IN SIMULATION.

Symbol	Parameter	Value
N_a	Impurity doping in the channel	10^{16}cm^{-3}
N_d	Impurity doping in source and drain	10^{20}cm^{-3}
R	Channel radius	5nm
t_{si}	Silicon film thickness	10nm
L_S , L_D	Length of source and drain	5nm
t_{ox}	Oxide thickness	2nm
L	Channel length	30nm
ε_0	Permittivity of vacuum	$8.8*10^{12}$F/m
ε_{si}	Permittivity of silicon	$11.85* \varepsilon_0$
ε_{ox}	Permittivity of oxide	$3.9* \varepsilon_0$
T	Absolute temperature in Kelvin	300K
Φ_M	Metal work function	4.6eV

III. RESULTS AND DISCUSSION

In this section, we are presented the results obtained from numerical simulation of the potential distribution Φ (r, z) in the source-channel-drain region. The center and surface potential has been obtained by solving the following 2-D asymmetric Poisson's equation in cylindrical coordinate system.

The doping profile in the drain / channel / source regions is presented in Fig. 2. The electron density is higher in the source and drain regions and the hole density is higher in the channel.

The doping profile is very important because it gives us information on the desired current levels and the direction of the electric field vectors in the device.

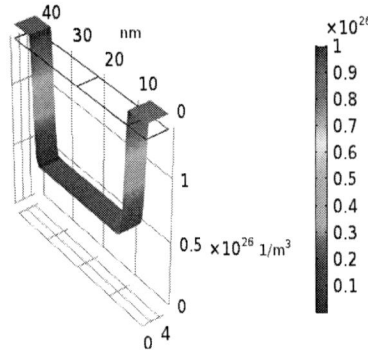

Fig. 2. Doping concentration of Si CGAA MOSFET$_S$.

Fig. 3. illustrates the energy band diagrams in the radial directions along the tunneling FET body.

The band diagram provides information on the levels of conduction, Fermi and valence band energy as a function of the length of the channel.

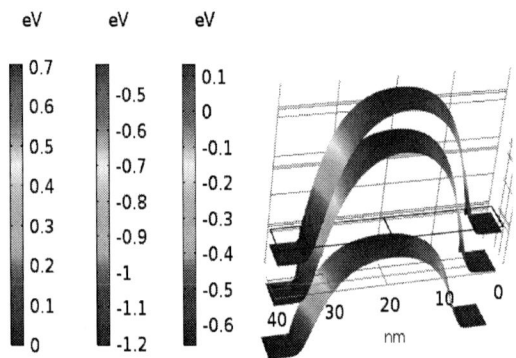

Fig. 3. Band diagram in the both (Si) (z axis) of a CGAA

In order to study the effect of the parameter variations on the basic characteristics described above, we have considered the profile described above in all simulations.

A. Influence of the gate voltage (V_{GS}) Variations

Fig. 4. and Fig. 5. show the variation of the surface and center potentials along the source-channel-drain for different gate-to-source voltages of the CGAA MOSFET. When the gate-to-source voltage is reduced, the controllability of the gate voltage on the channel becomes better compared with the influence on the source and drain ends.

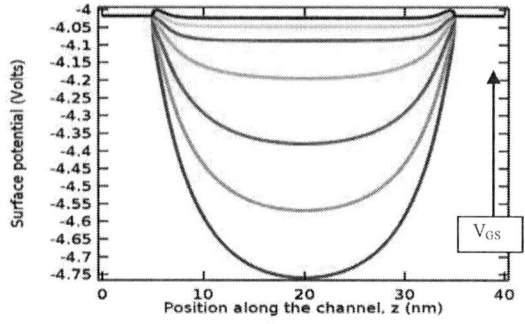

Fig. 4. Surface potential along the channel length at various gate voltages V_{GS} (start: -0.2 V, step: 0.2V, stop: 1V), V_{DS}=0V.

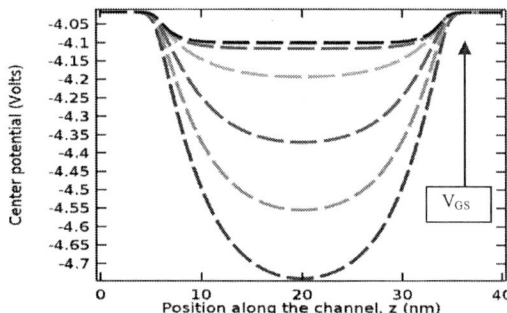

Fig. 5. Center potential along the channel length at various gate voltages (start: -0.2 V, step: 0.2V, stop: 1V) and V_{DS}=0V.

B. Influence of the oxide thickness (t_{OX}) Variations

Fig. 6. shows the variation of the center potential along the channel for different oxide thicknesses. When the oxide thickness is reduced, the controllability of the gate over the channel potential increases. The downscaling of the tox reduces SCEs.

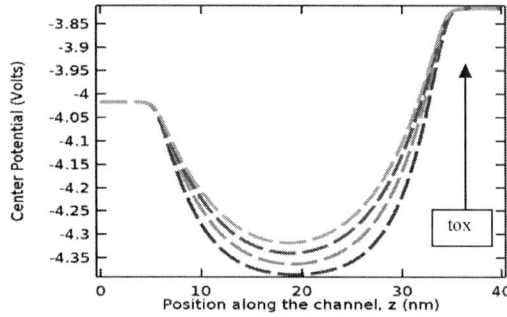

Fig. 6. Center potential along the channel length at various oxide thicknesses tox (start: 1nm, step: 0.1nm, stop:4nm), V_{DS}=0.2V.

978-1-7281-3420-8/19 $31.00 © 2019 IEEE

C. Influence of Channel thickness (t_{Si}) Variations

Fig. 7. presents the variation of the center potential along the channel for different radius of silicon body thickness (tsi) of CGAA MOSFET. When the (tsi) is reduced, the controllability of the gate over the channel becomes important in comparison with the effect exerted by the source/drain.

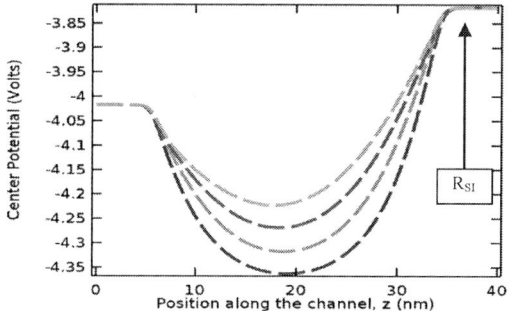

Fig. 7. Center potential along the channel length at various silicon thicknesses (start: 5nm, step: 2.5, stop: 12.5nm), $V_{DS}=0.2V$

D. Influence Gate Works function (Φ_M) Variations

Fig. 8. illustrates the variation of the center potential along the channel for different values of gate work function (φ_M). When (φ_M) augments, the controllability of the gate over the channel potential increases.

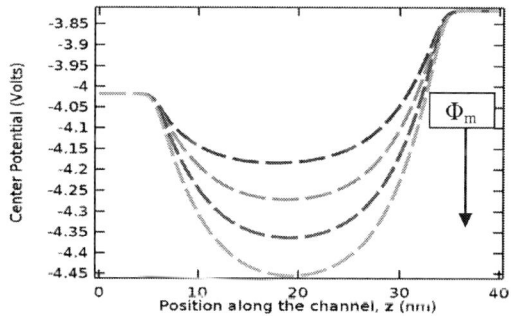

Fig. 8. Center potential along the channel length at various gate work function Φ_m (start: 4.4eV, step: 0.1eV, stop: 4.7eV), $V_{DS}=0.2V$ and $V_{GS}=0.2V$.

E. Influence of the Drain voltage (V_{DS}) Variations

Fig. 9. reveals the center potential along the channel length at different values of the drain voltage. The presence of a DIBL effect can be observed, since the minimum point of central potential indicates an upward movement with the increase of the drain voltage.

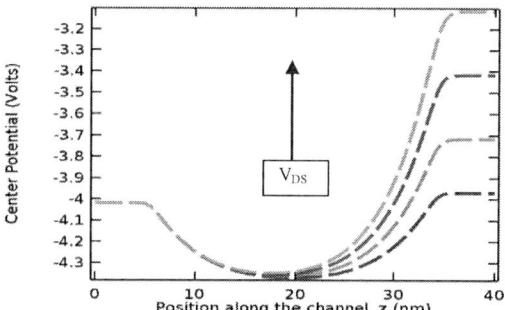

Fig. 9. Center potential along the channel length at various drain voltages V_{GS} (start: 0.05 V, step: 0.3V, stop:0.95V) and $V_{GS}=0.2V$.

F. Comparison of the center and surface potential

Fig. 10. and Fig. 11. display the comparison of the surface and center potential along the source-channel-drain at various gate to source voltage of CGAA MOSFET with two value of drain voltage ($V_{DS}=0V$ and $V_{DS}=0.9V$). It can be noted that source channel barrier height at channel center is lower than that of the surface. The presence of DIBL effect can be easily observed from Fig. 11. The threshold voltage depends on the drain bias, gate length, body and oxide thickness, gate work function, doping profile, etc.

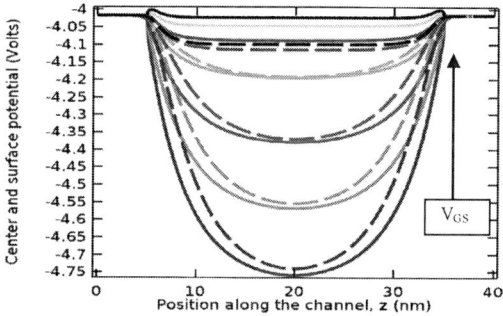

Fig. 10. Comparison of the center and surface potential along the channel at various gate voltage V_{GS} (start:-0.2, step:0.2V, stop:1V) and $V_{DS}=0V$.

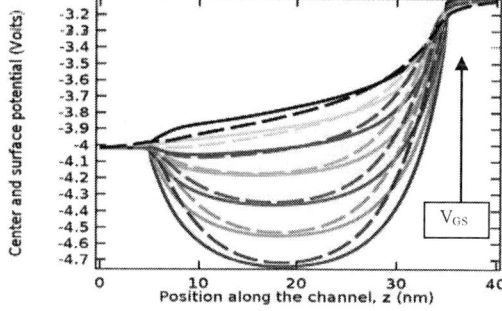

Fig. 11. Comparison of the center and surface potential along the channel at various gate voltages V_{GS} (start:-0.2, step: 0.2V, stop:1V) and $V_{DS}=0.9V$.

IV. CONCLUSION

In this paper, we have studied the potential distribution in the body of the Cylindrical Gate-All-Around (CGAA) MOSFET$_s$, using the finite element, numerical method by solving the Poisson's equation in the cylindrical coordinate system. The results show excellent accuracy comportment of the potential in the channel of CGAA.

REFERENCES

[1] U. A. Maduagwu and V. M. Srivastava "Performance of Potential Distribution of CSDG MOSFET Using Evanescent Mode Analysis", *International Journal of Emerging Technologies in Learning (iJET)*, 2019, 7(4), pp. 5649-5653.

[2] T. K. Sachdeva, S.K. Aggarwal, A. K. Kushwaha, "Design, Analysis and Simulation of 30 nm Cylindrical Gate all around MOSFET," *IEEE Trans. IJARCCE*, 2016, vol. 5, Issue 10.

[3] M. Kessi; A. Benfdila; A. Lakhlef "Investigation on cylindrical gate-all-around (GAA) tunnel FETS scaling", *Proc. 30th International Conference On Microelectronics (MIEL 2017)*, Niš, Serbia, October, 9th-11th, 2017.

[4] B. Buvaneswari "A Survey on Multi Gate MOSFETS", *International Journal of Innovative Research in Science, Engineering and Technology*, 2014, Vol. 3, Special Issue 3.

[5] M. Kessi, A. Benfdila, A. Lakhlef, "FinFET versus GAAFET Performances and Perspectives", European Material Research Society (EMRS) Spring Meeting. At: Nice, France, May 2019.

[6] B. Jena, K. P. Pradhan, P. K. Sahu, S. Dash, G. P. Mishra, S. K. Mohapatra "Investigation Oncylindrical Gate Allaround (GAA) To Nanowire MOSFET for Circuit application", *Facta Universitatis Series: Electronics and Energetics*, 2015, Vol. 28, No. 4, pp. 637-643.

CIRCUIT AND SYSTEM DESIGN AND TESTING

978-1-7281-3420-8/19 $31.00 © 2019 IEEE

SAR ADC Architecture with Fully Passive Noise Shaping

D. Osipov, A. Gusev, V. Shumikhin, and St. Paul

Abstract— **A new fully passive noise-shaping architecture for successive approximation register (SAR) analog-to-digital converters (ADCs) was proposed. A first-order noise transfer function (NTF) with zero located nearly at one can be achieved. The additional pole increases the efficiency of noise shaping to further 3 dB. So, the use of higher over sampling ratios (OSR) and increased effective number of bits (ENOB) is possible.**

The architecture was applied to the design of a 9.8-bit ENOB SAR ADC in a 65 nm complementary metal-oxide semiconductor (CMOS) of United Microelectronics Corporation (UMC) with OSR equal to 10. A 6-bit capacitive DAC was used. The proposed architecture provides 3.8 additional bits in ENOB. The equalent input bandwitdth is equal to 200 kHz with the sampling rate equal to 4 MS/s.

I. INTRODUCTION

The SAR ADCs are most commonly used types of analog to digital converters (ADCs) nowadays. The most benefits of this architecture are the low power consumption and easy scaling for newer technology nodes. A lot of effort is performed to further decrease the power of this ADC. The typical SAR ADC can be split in three parts: comparator, capacitive DAC and digital logic. All these parts consume much more power if the resolution of SAR ADC increases. Furthermore, the DACs capacitors are binary weighted, so the increase of resolution to one bit requires the increase of the area twice.

Several ideas were introduces to lower the power consumption of the DAC, for example, the advanced switching algorithms [1] or the modification of the reference voltage circuit [2]. In this paper, we introduce another approach to decrease the DACs power and area. When using noise shaping ADC requires less resolution in the DAC to achieve more effective number of bits (ENOB) in the ADCs output.

So, the idea of noise shaping applied to the SAR ADCs quantization noise is interesting. Theoretically, with the first order noise shaping and the over sampling ratio (OSR) of 10, the four-bit increase in the resolution of the ADC can be achieved. In case of 10 bit ADC, it means that only 6 bit DAC with much lower area is

D. Osipov, A. Gusev and St. Paul are with the Institute of Electrodynamics and Microelectronics, University of Bremen, Otto-Hahn Allee 1, 28359 Bremen, Germany, E-mail: {osipov, agusev, steffen.paul}@uni-bremen.de

V. Shumikhin is with the ASIC Lab, National Research Nuclear University MEPhI, Kashirskoe shosse, 31, 115409, Moscow, Russia E-mail: vvshumikhin@mephi.ru

needed. Furthermore, such DAC can be switched much faster without the use of advanced switching methods for high speed ADCs like merged sampling [3].

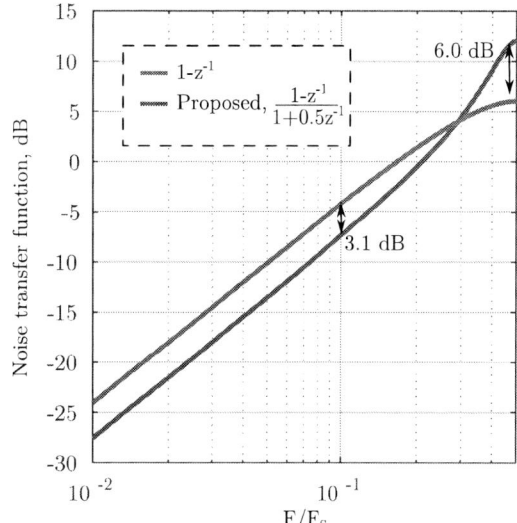

Fig. 1. The proposed noise shaping transfer function (NTF) compared with the simple first order NTF.

Recently several first order noise shaping schemes were published [4]. The most of them does not achieve the zero at one if fully passive schema is used, which decreases the efficiency of noise shaping with higher OSRs. Other schemes, which provide more efficient NTF, utilize the operational amplifiers [5,6]. The ADCs architectures with higher order NTFs also were published. However, the higher noise shaping order usually means the sufficient rise of ADCs area. For example, in [7] the third order scheme is proposed, which requires the 3 times increase of overall capacitance of the converter.

The comparison of the ideal first order NTF (typical SAR NS schemes have zero at $z^{-1} < 0.75$, so they are less efficient) and the one proposed in this paper is shown in Fig. 1. The proposed noise shaping schema provides the NTF zero, which is close to one. So, the higher OSRs can be used. It has an additional pole also, which increases the noise shaping efficiency for 3 dB.

The rest of the paper is organized as follows. In the section I the proposed noise shaping architecture is described. In the section II we show the implementation of the proposed noise shaping architecture in 65 nm CMOS technology. The simulation results are given in section III. Finally, the section IV concludes the paper.

II. Proposed Architecture

The proposed architecture is shown in Fig. 2. Unlike the classical SAR ADC, in the proposed architecture the comparator is connected to the DAC through attenuation capacitor C_r, which also is particularly used for the storage of DACs residue voltage (it provides the pole in the NTF). The capacitor $2^M C_0$ is used to store the quantization error of previous sample.

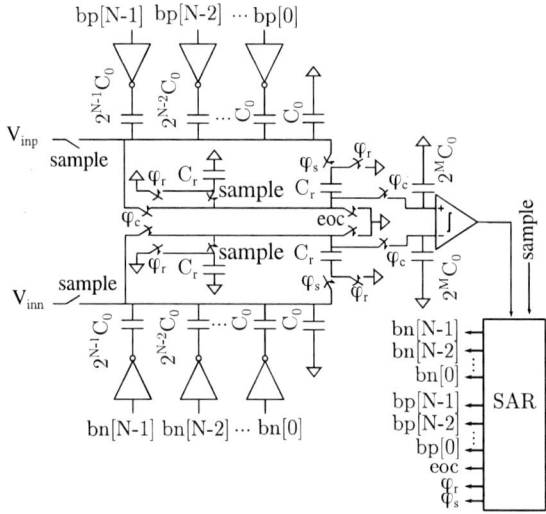

Fig. 2. Proposed NS SAR architecture.

The simplified functional diagram of the proposed SAR ADC architecture is shown in Fig. 3. Initially, the input signal $x(n)$ is sampled on the top plates of the capacitive DAC. At the same time, the residue voltage of previous sample (voltage on the capacitive DAC after the end of conversion) saved on the capacitor C_r is divided by two. Without this step the pole will be equal to one and the circuit will become unstable. The capacitor $2^M C_0$, with the previous quantization error divided by $\alpha = (1/2^N C_0 + 1/2^M C_0 + 1/C_r)$, is disconnected from the DAC. After the end of sampling phase, the DAC is connected to the quantization error storage capacitor $2^M C_0$ through the capacitor C_r. The charge redistribution occurs, which can be described as:

$$\begin{cases} \Delta q = 2^N C_0 (x(n) - k(n)) \\ \Delta q = C_r (k(n) - g(n)) + 0.5 C_r V_{res}(n-1) \\ \Delta q = 2^M C_0 (g(n) - e(n-1)/\alpha) \end{cases} \quad (1)$$

where $k(n)$ and $g(n)$ are the voltages on the DAC and quantization error storage capacitor $2^M C_0$, respectively, after charge redistribution, $e(n-1)$ is the quan-

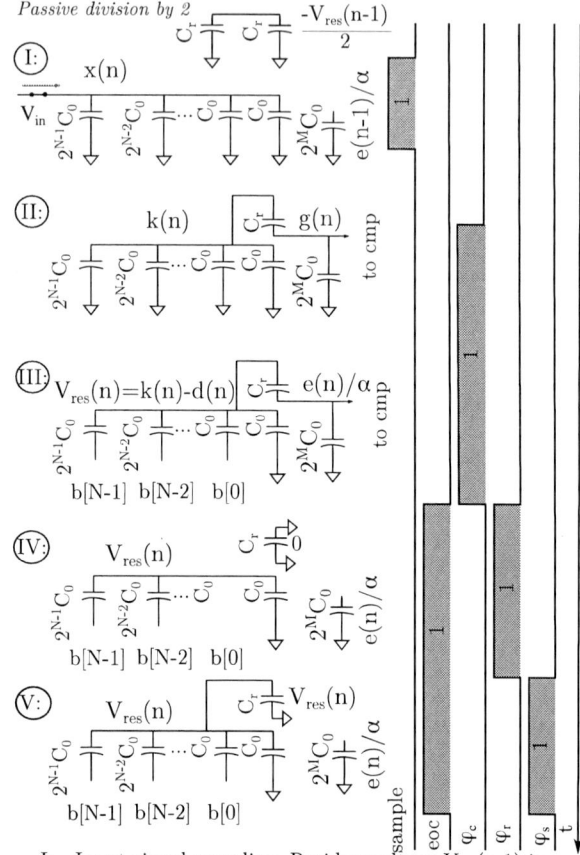

I: Input signal sampling. Residue voltage $V_{res}(n-1)$ is divided by two.

II: Charge redistribuiton.

III: Conversion (the quantization error of previous sample is stored in storage capacitor $2^M C_0$).

IV: Reset of residue storage capacitor C_r.

IV: Storage of the residue voltage $V_{res}(n)$.

Fig. 3. Functional diagram of the proposed ADC architecture.

tization error of previous sample. After the charge redistribution the conversion begins. As the capacitor $2^M C_0$ is connected directly to comparator, after the end of conversion, its voltage will be equal to $e(n)/\alpha$. The voltage at the DAC will be equal to

$$V_{res}(n) = g(n) - d(n), \quad (2)$$

where $d(n)$ represents the digital DAC input (ADCs output).

After the end of conversion the new value of $V_{res}(n)$ can be saved on capacitor C_r (if $C_r << 2^N C_0$).

From (1)-(2) the following equation can be written for the determination of output digital code $D(z)$:

978-1-7281-3420-8/19 $31.00 © 2019 IEEE

$$D(z)\left(1+\left(0.5+\frac{1}{2^N\alpha}\right)z^{-1}\right)=X(z)(1+0.5z^{-1})$$
$$+\,E(z)\left(1-\frac{\frac{1}{2^N C_0}+\frac{1}{C_r}}{\alpha}z^{-1}-\frac{\frac{0.5}{2^N C_0}}{\alpha}z^{-2}\right)\quad(3)$$

If $1/2^N \ll 1$, this equation can be simplified to:

$$D(z)=X(z)+\frac{1-\frac{\frac{1}{2^N C_0}+\frac{1}{C_r}}{\alpha}z^{-1}}{1+0.5z^{-1}}E(z)\quad(4)$$

If $N,M\gg 1$, the equation (4) can be rewritten as:

$$D(z)\approx X(z)+\frac{(1-z^{-1})}{(1+0.5z^{-1})}E(z).\quad(5)$$

So, the circuit will perform the first order noise shaping. The pole gives additional 3 dB noise shaping in the input frequency band.

For circuit implementation we selected the M=N-1 and $C_r=C_0$. So, the exact form of the digital output is:

$$D(z)\approx X(z)+\frac{(1-0.943z^{-1})}{(1+0.5z^{-1})}E(z).\quad(6)$$

With the selected OSR ratio of 10, the deviation of the zero from 1 can be neglected. The increase of M can give the arbitrary approximation of the zero to one. However, the increase of M has an undesirable effect: the increase of area of the ADC.

III. Circuit implementation in 65 nm CMOS of UMC

In order to verify the proposed architecture, we designed the SAR ADC in 65 nm UMC technology. The minimum size 51.16 fF metal-insulator-metal (MIM) capacitors available in the technology were used as C_0. All switches are made as single nmos transistors except of input sampling switch, where bootstrapped switch architecture is used. Single stage strong arm comparator is adopted in the schematic. The common-mode voltage on the input of the comparator sequentially decreases from $V_{ref}/2$ devided by the attenuation multiplier to zero, so one p-type input differential pair can be used. The comparator schematic is shown in Fig. 4.

The SAR logic needs only slightly modification to provide additional control signals. The modified SAR logic is shown in Fig. 5. The SAR ADC is fully asynchronous, the clock signal is generated with help of comparators output signals and a delay circuit. It can be seen that only 6 additional combinational logic blocks and one delay are used to generate the signals for noise shaping.

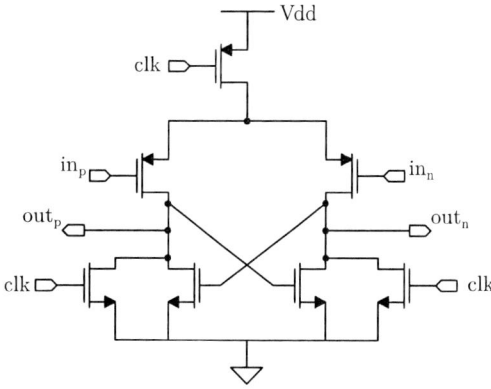

Fig. 4. Simple one stage comparator used in this design.

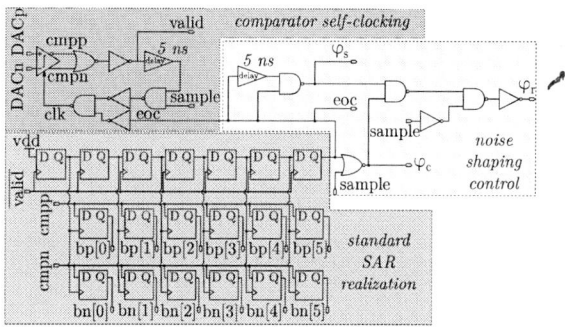

Fig. 5. Control logic for the proposed NS SAR architecture.

IV. Simulation Results

The sampling rate of the ADC was set to 4 MS/s. The simulated output spectrum for a 194.458 kHz input signal is shown in Fig. 6. SAR ADC achieves the ENOB of 9.8 bits in 200 kHz bandwidth. The whole power consumption of the proposed ADC implemented in 65 nm CMOS equals to 7.99 μW. The power distribution of single blocks is shown in Fig. 7. It can be seen that the implementation of the proposed noise shaping scheme does not affect the power distribution of SAR ADC. The achieved FoM equals to 22.43 fJ/conv.-step (for the effective Nyquist frequency of 400 kS/s).

The comparison with the state of the art noise shaping schemes is given in table I.

V. Conclusion

The new passive noise shaping scheme for SAR ADCs was proposed. It provides the first order noise shaping with the zero located at one, this means that higher OSRs can be used. The pole located at 0.5 provides additional 3 dB of noise shaping among the whole input frequencies range. The noise shaping can save the area of SAR ADC, while providing the high resolution. In our implementation example, we achieved 10 bit res-

TABLE I
State of the art Noise Shaping SAR ADCs.

	[4]	[8]	[9]	This work
Architecture				
NTF zero	0.75	0.5	0.65	**1**
Need of OTA	No	Yes	Yes	No
Comparator modification	Yes	Yes	Yes	**No**
Input attenuation	No	Yes	No	No
Number of unit capacitors, C0	2^N	2^{N+1}	2^N	$3(2^{N-1})$
Circuit Performance				
Technology, nm	130	65	65	65
Bandwidth, kHz	125	6250	11000	200
DAC size, bit	10	8	8	**6**
ENOB, bit	12	10	9.35	9.8
Additional bits	2	2	1.35	**3.8**
OSR	8	4	4	10
FoM, fJ/conv.-step	59.6	14.8	35.8	22.43
FoMs, dB	167	165.19	163.31	**166.0**
Verification	Meas.	Meas.	Meas.	Sim
Year	2016	2015	2012	2019

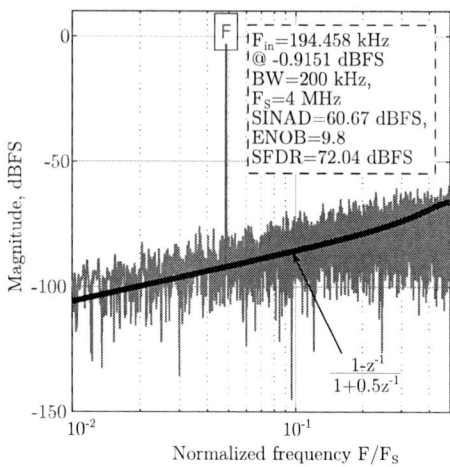

Fig. 6. Simulated output spectrum of the proposed ADC for a 194.458 kHz input signal sampled at 4 MS/s.

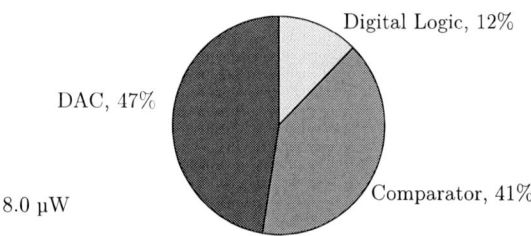

Fig. 7. Power distribution of different parts of the proposed ADC.

olution using 6 bit DAC and OSR ratio equal to 10. Because of zero location near to one even higher OSRs can be used.

The proposed architecture was implemented in the 65 nm UMC technology. The efficient input signal bandwidth is equal to 200 kHz, the achieved FoM is equal to 22.43 fJ/conv.-step. The classical SAR ADC architecture should be modified very slightly, near the same control logic can be used with only 5 additional combinational logic blocks and one delay. The proposed architecture also does not require the modification of comparator.

Acknowledgement

This work was supported by German Research Foundation (DFG), Project number: 389481053

References

[1] D. Osipov and S. Paul, "Two advanced energy-back sar adc architectures with 99.21 and 99.37% reduction in switching energy," *Analog Integrated Circuits and Signal Processing*, vol. 87, no. 1, pp. 81–91, 2016.

[3] X. Wang, X. Zhou, and Q. Li, "A energy-efficient high speed segmented prequantize and bypass dac for sar adcs," in *2014 IEEE 57th International Midwest Symposium on Circuits and Systems (MWSCAS)*, Aug 2014, pp. 97–100.

[2] D. Osipov and S. Paul, "Low power sar adc with two-step switching scheme in 65 nm standard cmos process," in *2017 IEEE 30th International Conference on Microelectronics (MIEL)*, Oct 2017, pp. 209–212.

[4] W. Guo and N. Sun, "A 12b-enob 61w noise-shaping sar adc with a passive integrator," in *ESSCIRC Conference 2016: 42nd European Solid-State Circuits Conference*, Sep. 2016, pp. 405–408.

[5] A. Matsuzawa and M. Miyahara, "Sar+$\Delta\Sigma$ adcs with open-loop integrator using dynamic amplifier," *IEICE Electronics Express*, vol. 15, no. 6, pp. 20 182 002–20 182 002, 2018.

[6] M. Shahghasemi, R. Inanlou, and M. Yavari, "An error-feedback noise-shaping sar adc in 90 nm cmos," *Analog Integrated Circuits and Signal Processing*, vol. 81, no. 3, pp. 805–814, Dec 2014. [Online]. Available: https://doi.org/10.1007/s10470-014-0434-6

[7] P. Payandehnia, H. Mirzaie, H. Maghami, J. Muhlestein, and G. C. Temes, "Fully passive third-order noise shaping sar adc," *Electronics Letters*, vol. 53, no. 8, pp. 528–530, 2017.

[8] Z. Chen, M. Miyahara, and A. Matsuzawa, "A 9.35-enob, 14.8 fj/conv.-step fully-passive noise-shaping sar adc," *IEICE Transactions on Electronics*, vol. 99, no. 8, pp. 963–973, 2016.

[9] J. A. Fredenburg and M. P. Flynn, "A 90-ms/s 11-mhz-bandwidth 62-db sndr noise-shaping sar adc," *IEEE Journal of Solid-State Circuits*, vol. 47, no. 12, pp. 2898–2904, Dec 2012.

Improving Magnitude Response of Comb Two-Stage Structure Using Simple Multiplierless Filters

G. Jovanovic Dolecek

Abstract - This paper presents novel two-stage comb-based decimation structure for even decimation factors. First stage and second stages are decimated by $M/2$ and 2, respectively, where M is the decimation factor. In the second stage is the cascade of multiplierless corrector filter and sharpened cascade of the corrector and comb. Filters in second stage improve magnitude characteristic of comb in odd folding bands and decrease comb passband droop. In the first stage is the cascade of comb and multiplierless modified filter. This modified filter improves alias rejection not only in even folding bands, but also in odd folding bands. Method is illustrated with two examples and the corresponding structure is presented. Finally, the method is compared with similar methods in literature.

I. INTRODUCTION

Decimation is the process of decreasing sampling rate by an integer, called decimation factor. Decimation has applications, for example in Sigma-Delta Analog/Digital Converters (SD ADC), communications, subband coding among others, [1]. The process of decimation introduces aliasing, which may deteriorate decimated signal. Consequently, aliasing must be eliminated by a filter, called decimation filter.

The most simple decimation filter is a comb filter which has all its coefficients equal to unity. Comb attenuates aliasing, which occur in the bands around comb zeros, called folding bands. However, the attenuation in folding bands is not sufficient in many applications, and must be increased. Different methods were proposed to increase aliasing rejection in comb folding bands, [2-6].

Additionally, comb magnitude characteristic is not flat in the passband. Different methods were proposed for the compensation for the comb passband droop, [7]-[10].

However, only few methods simultaneously improve both, comb passband and stopband, [11]-[14].

A two-stage comb structure for even decimation factors was proposed in [12]. The decimation factors for the first and second stage are $M/2$ and 2, where M is a decimation factor.

In the second stage is placed a multiplierless corrector filter, for the defined number of the cascaded combs K, $K=1,\ldots,5$.

The filter provides an increased attenuation in all odd folding bands, and a decreased passband droop, in comparison with a comb filter.

G. Jovanovic Dolecek is with the Department of Electronics, Institute INAOE, Emrique Erro 1, 72740 Tonantzintla, Puebla, Mexico, E-mail: gordana@ieee.org

In [13] was proposed to increase aliasing rejection also in even folding bands, while in [14] was proposed only one corrector filter, independent on the comb parameter K. However, like in Method in [12], aliasing rejection in even comb folding bands is not improved.

The goal of this proposal is to improve comb magnitude characteristic also in even folding bands, using a simple corrector filter and sharpening technique, while keeping a low comb passband droop. Additionally, unlike in Method in [12] the corrector should be independent on the comb parameter K.

II. DESCRIPTION OF METHOD

A. *Two-Stage Structure*

We consider here even decimation factor M, $M=(M/2)$ 2. First stage is decimated by $M/2$, while the second stage is decimated by 2. Comb filters in the first and second stages, are:

$$H_1(z) = \frac{1}{M/2}\frac{1-z^{-M/2}}{1-z^{-1}}; H_2(z) = (1+z^{-1})/2 . \quad (1)$$

B. *Choice of Corrector and Sharpening Polynomial*

Like in [14] we pick up here the simplest corrector filter in [12], used for $K=1$, denoted as $C(z)$:

$$C(z) = -3 + 2z^{-1} + 16z^{-2} + 16z^{-3} + 2z^{-4} - 3z^{-5} . \quad (2)$$

The corrector filter (2) is placed in the second stage. In order that this corrector also works for values of $K>1$, i.e. for $K=1,\ldots,5$, the sharpening technique is applied for the cascade of $C(z)$ and comb filter $H_2(z)$. We propose to apply the simplest sharpening polynomial [15]:

$$Sh\{X\} = 2X - X^2 = X(2-X) , \quad (3)$$

where X is the sharpened filter. To this end, from (3) the sharpened cascade of corrector (2) and comb $H_2(z)$ is given as:

$$Sh\{C(z)H_2(z)\} = C(z)H_2(z)[2 - C(z)H_2(z)] . \quad (4)$$

Note that sharpening (3) should be applied for filter of odd length. The corrector (2) has an even length. However, the cascade of $C(z)$ and comb $H_2(z)$ is odd. Sharpened filter (4) is cascaded with the corrector $C(z)$. Therefore the filter at second stage, $H_{22}(z)$ is given as:

$$H_{22}(z) = C(z)Sh\{C(z)H_2(z)\} = C^2(z)H_2(z)[2 - C(z)H_2(z)] \qquad (5)$$

C. Filters in First Stage

In the first stage is introduced the modified comb:

$$G_1(z) = 1 + 3/2z^{-1} + z^{-2}. \qquad (6)$$

The filter (6) is expanded by 2:

$$G_1(z^2) = 1 + 3/2z^{-2} + z^{-4}. \qquad (7)$$

The cascade of filters (6) and (7),

$$G(z) = G_1(z)G_1(z^2), \qquad (8)$$

introduces zeros in comb folding bands and thus improve aliasing rejection in those bands. The effect of modified comb (8) on comb zeros is shown in following example.

Example 1:

We consider comb filters for two values of M: 8 and 12, and $K=1$. In both cases, the modified comb is cascaded with the comb. Fig. 1, in the first row, shows pole-zero plots of comb filters, while in the second row are shown pole-zero plots of cascade comb and modified comb.

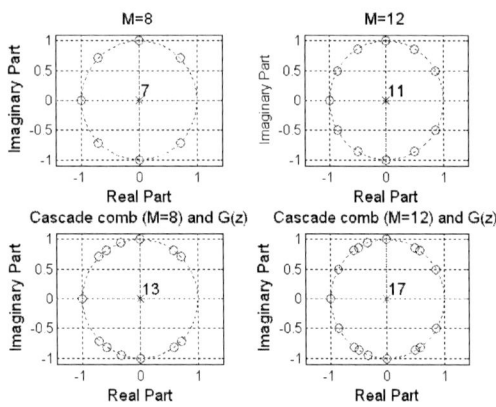

Fig. 1. Pole-zero plots in Example 1.

Note that the additional zeros are introduced in comb folding bands for both values of M. Modified comb (8) is cascaded with filter $H_1(z)$ in the first stage:

$$H_{11}(z) = H_1^K(z)G(z), \qquad (9)$$

where $G(z)$ is given in (8), and $H_1(z)$ in (1).

D. Transfer Function

Transfer function of the proposed filter is given as:

$$H_p(z) = H_{11}(z)H_{22}(z^{M_1}), \qquad (10)$$

where $H_{11}(z)$ and $H_{22}(z)$ are given in (9) and (5), respectively. The method is illustrated in the following example.

Example 2:

We consider $M=14$ and $K=4$. The magnitude responses are compared with the corresponding comb filter in Fig. 2. The passband zooms are also shown. Note that the aliasing rejection is improved in all folding bands except in the second one, and that the passband droop is significantly decreased.

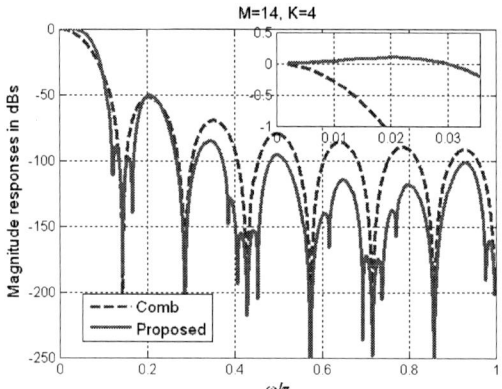

Fig. 2. Magnitude responses in Example 2.

In next example is demonstrated that the same designed parameters can be used for different values of M and K.

Example 3:

We consider now $M=20$ and $K=5$. The magnitude characteristics of the proposed filter and corresponding comb are contrasted in Fig. 3. Note again that the proposed filter has better magnitude characteristic in both, passband and folding bands.

Principal features of the proposed method are:

- The proposed structure is a multiplierless two-stage structure.
- The modified comb in first stage does not depend on comb parameters M and K.

- The corrector filter in second stage does not depend on comb parameters M and K.
- Magnitude response is improved in folding bands, while keeping a low passband deviation.
- The method is convenient for even decimation factors.

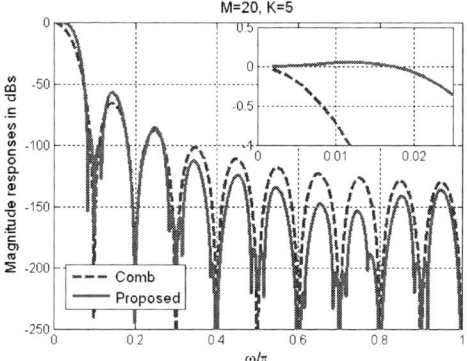

Fig. 3. Magnitude responses in Example 3.

E. Structure

Two-stage structure is shown in Fig. 4a. More detailed structures of first and second stages are given in Figs. 4b and 4c.

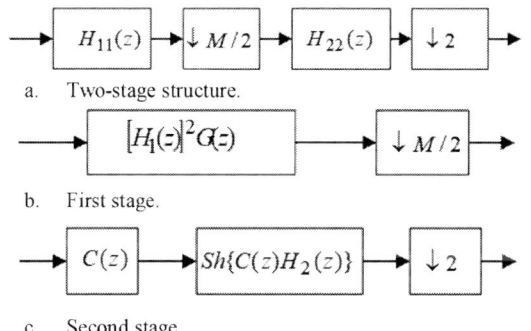

a. Two-stage structure.

b. First stage.

c. Second stage.

Fig. 4. Structures.

Filters in the first stage can be moved to lower rate using the polyphase decomposition:

$$H_{11}(z) = \sum_{k=0}^{N-1} z^{-k} H_{1,k}(z^{M/2}), \qquad (11)$$

where $H_{11}(z)$ is given in (1) and $H_{1,k}(z^{M/2})$ are the polyphase components.

III. COMPARISONS

In this section we compared the proposed method with the methods in literature based on corrector filters.

A. Comparison with method in [14]

For a sake of comparison is chosen decimation factor $M=14$ and $K=4$. The magnitude responses along with the passband zooms are given in Fig. 5.

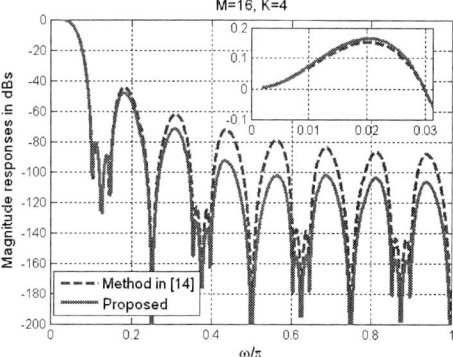

Fig. 5. Comparison with Method in [14].

Both methods have similar complexity. However, the proposed method provides better aliasing rejection, while the passband characteristics are similar.

B. Comparison with Method in [12]

We consider comb parameters: $M=18$ and $K=3$. Since $K=3$, in the second stage, in Method [12], is used the corrector filter $C_3(z)$ with coefficients:

$$[1,1,-5,-5,12,24,12,-5,-5,1,1].$$

This corrector is cascaded with K combs in the second stage. In the first stage is the cascade of K combs. The proposed method is slightly more complex due to sharpening at second stage and modified comb filter in first stage. However, in the proposed method it is used the simplest corrector from [12] for all values of K, $K=1,\dots5$. The overall magnitude responses and the passband zooms are presented in Fig. 6. The proposed method exhibits better aliasing rejection, while the passband characteristics are similar.

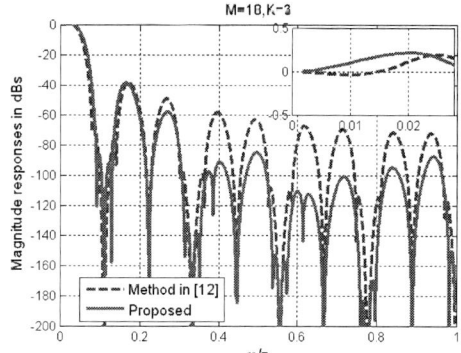

Fig. 6. Comparisons with Method in [12].

IV. CONCLUSION

The novel two-stage comb-based decimation structure is presented in this paper. Method is based on corrector filters introduced in [12]. Unlike to method in [12], in the proposed structure, is used only one corrector for all values of comb parameter K. This is achieved by applying the sharpening technique for the cascade of comb and corrector at second stage. Additionally, the modified combs are inserted in the first stage. As result, the comb aliasing rejection is improved in comparison with methods from literature based on corrector filters. However, the passband characteristics are similar.

REFERENCES

[1] G. Jovanovic Dolecek (Ed), "Advances in Multirate Systems", *Springer Int. Publishing*, NY, 2018.

[2] M. Laddomada, "Comb-based decimation filters for Sigma-Delta A/D converters: Novel schemes and comparisons", *IEEE Trans. on Signal Processing.*, vol. 55, No. 5, May 2007, pp. 1769-1779.

[3] B. P. Stošić, and V. D. Pavlović, "Design of new selective CIC filter functions with passband-droop compensation", *Electronics Letters*, vol. 52, No. 2, 2016, pp. 115-117.

[4] J. O. Coleman, "Chebyshev stopbands for CIC decimation filters and CIC-implemented array tapers in 1D and 2D", *IEEE Trans. on Circ. and Syst. I*, vol. 59, No.12, December 2012, pp. 2956-2968.

[5] G. Jovanovic Dolecek, R. Garcia Baez, and M. Laddomada, "Design of efficient multiplierless modified cosine-based comb decimation filters: Analysis and implementation", *IEEE Trans. on Circ. and Syst. I*, vol.36, No.5, May 2017, pp.2031-2049.

[6] B. P. Stošić, and V. D. Pavlović, "Design of selective CIC filter functions", *AEU-International Journal of Electronics and Communications*, vol. 68, No. 12, 2014, pp. 1231-1233.

[7] G. Jovanovic Dolecek., R. Garcia Baez., G. Molina Salgado, and J de la Rosa, "Novel multiplierless wideband compensator with high compensation capability", *Circuits, Systems and Signal Processing*, Springer, vol.36, No.5, 2017, pp. 2031–2049.

[8] G. Molnar, and M. Vucic, "Closed-form design of CIC compensators based on maximally flat error criterion", *IEEE Trans. Circ. Syst. II: Express Briefs*, vol. 58, 2011, pp. 926-930.

[9] G. Jovanovic Dolecek, and S. K. Mitra, "Simple method for compensation of CIC decimation filter", *Electronics Letters*, vol. 44, No. 19, September 2008, pp. 1162-1163.

[10] G. Molnar, A. Dudarin, and M. Vucic, "Design of multiplierless CIC compensators based on maxinun passband deviation", in *Proc. Int, Conference MIPRO 2017*, Opatija, Croatia, May 2017.

[11] A. Y. Kwentus, Z. Jiang, and A. N. Willson, Jr., "Application of filter sharpening to cascaded integrator-comb decimation filters", *IEEE Trans. on Sig. Processing*, vol.45, February 1997, pp.457-467.

[12] G. Jovanovic Dolecek and A. Fernandez, "Novel droop-compensated comb decimation filter with improved alias rejections", *Int. Journal for Elect. and Comm. (AEUE)*, vol.67, May 2013, pp.387-396.

[13] G. Jovanovic Dolecek, and L. Sepulveda, "Improving magnitude response in two-stage corrector comb structure," in *Proc. 2017 International Conf. ICCDCS*, 2017, pp.73-76.

[14] G. Molina, Salgado, G. Jovanovic Dolecek, J. de la Rosa, "Novel two-stage comb decimator with improved frequency characteristic", in *Proc .of IEEE Conference LASCAS*, 2015, pp.1-4.

[15] J. F. Kaiser, R.W. Hamming, "Sharpening the response of symmetric non-recursive filter", *IEEE Transactions on Acoustic, Speech, and Signal Processing*, vol. 25, 1977, pp. 415–422.

Hardware Implementation of Selected Statistical Quantities for Applications in Automotive V2I Communication System

M. Banach, R. Długosz, and T. Talaśka

Abstract— In this work we propose a concept of a transistor level implementation of a simplified iterative methods for computing several basic statistical quantities, such as the mean and the variance. The motivation behind the presented work is the realization of a calibration algorithm for determining the positions of the V2I (vehicle-to-infrastructure) communication devices in novel automotive applications. Such devices, mounted in fixed points of the road and urban infrastructure (RSU – road side equipment) will be used to support autonomous vehicles moving in urban and suburban environments. The role of the calibration procedure is to determine the positions of the RSU devices in global coordinate system (GCS) and save it in their internal memory blocks. To facilitate the hardware implementation, we introduced some modifications to existing (conventional) iterative algorithms used for the computation of the statistical quantities. For this purpose, we eliminated division operations, substituting them with bit shift operations. Shifting the bits may be easily realized fully asynchronously in hardware, using only a passive commutation field.

I. INTRODUCTION

It is supposed that future active safety functions, which in recent years are intensively developed in the automotive industry, will be supported by the communication between vehicles and the road infrastructure (V2I – vehicle-to-infrastructure). In systems of this type one group of the communicating devices (RSU – road side units), mounted in the road infrastructure, will provide useful data to passing vehicles.

In the literature one can find many concepts of using the V2I system as a support for advanced driver assistance system (ADAS), and for the autonomous driving of vehicles in the future. In one of the simplest forms, it may rely on providing the vehicles relevant messages about dangerous situations on the road [2], [3]. They may include, for example, the risk of collision with a pedestrian or other vehicle, a traffic jam, an accident ahead, etc. In most advanced approaches, a framework

M. Banach is with Aptiv Services Poland ul. Podgórki Tynieckie 2, 30-399, Kraków, Poland, and with Poznan University of Technology, Institute of Architecture and Spatial Planning, Nieszawska 13C, 61-021 Poznań, Poland E-mail: marzena.banach@put.poznan.pl

R. Długosz and T. Talaśka are with Aptiv Services Poland ul. Podgórki Tynieckie 2, 30-399, Kraków, Poland and with UTP University of Science and Technology Faculty of Telecommunication, Computer Science and Electrical Engineering ul. Kaliskiego 7, 85-796, Bydgoszcz, Poland, E-mail: rafal.dlugosz@gmail.com

of the RSU devices may actively participate in control of the trajectories of the autonomous vehicles. This option, however, requires a sufficiently dense network of the RSU devices playing the role of local navigation / orientation points. In this case, from a safety point of view, it is important to determine the position of these devices with respect to passing vehicles with relatively high precision. This problem is more and more frequently raised in the literature [4]–[7]. The vehicles have to distinguish whether they are recipients of the messages being transmitted over the network.

In our works, we focus on the development of the algorithms, that aim at increasing the efficiency and the quality of the positioning of the RSU devices in relation to passing vehicles. We rely on statistical methods based on repeated distance measurements to the stationary devices, carried out by passing vehicles. On the basis of a series of such measurements, it is also possible to initially calibrate the V2I system. A novel calibration algorithm of the V2I system has been recently reported by us in a patent application [1] (in progress in the European Patent Office). In contrary to state-of-the art solutions that propose the GPS as a support of such a system [4]–[6], we assume that the GPS is not necessarily required. This is due to some potential problems. One of them is often a reduced localization precision based on the GPS in dense urban area, in tunnels, etc., so that it is not always possible to rely on this system. Another question is how to effectively supply the RSU devices equipped with the GPS.

The last issue is related to the problem of energy consumption. We believe that circuits used in the RSU devices will have to be energy-efficient. This in practice means the need of a development of novel algorithms with low computational complexity. As a contribution, in one of our former works we have proposed simplified mathematical iterative methods for computing selected statistical quantities ,such as the mean and the variance [8]. Based on a wide and comprehensive simulation investigations, we showed that with a sufficiently large number of the input samples, the errors introduced by the proposed methods do not exceed a few percent compared to the methods considered as classic [9].

The aim of this work is a novel hardware implementation of the described algorithms at the transistor level.

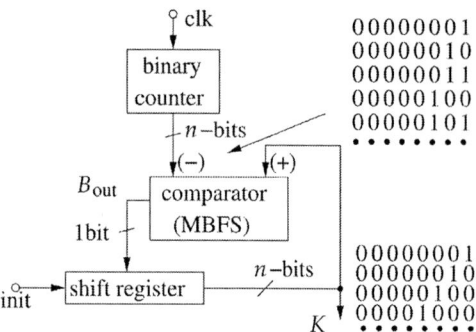

Fig. 1. A supporting circuit that controls denominator value in the mean and the variance circuits.

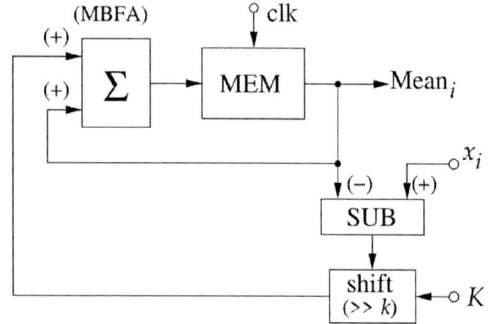

Fig. 2. The proposed, iterative circuit for the computation of the mean value.

In the comparison with conventional iterative methods, we use only simple arithmetical operations that include summing, subtracting and multiplying. The division operation has been eliminated and substituted with shifting the bits to the right by a number of positions that is automatically adapted to the number of the input samples.

II. AN OVERVIEW OF THE PROPOSED METHODS FOR COMPUTING STATISTICAL QUANTITIES

Determining the location of the RSU devices can be performed through a direct communication between the RSU devices and the vehicles, based, for example, on such technologies as the impulse radio – ultra wide band (IR-UWB) one. This technology offers a relatively high positioning accuracy [10]. For this reason, it is considered as one of the technologies that may be used in the V2I communication [11], [12]. In this technique, the position of a device may be computed on the basis of the measurements of the time that elapsed between the transmission and the reception of the signal. When the measurements are repeated, and additionally the trajectory of the moving device (associated with the car)

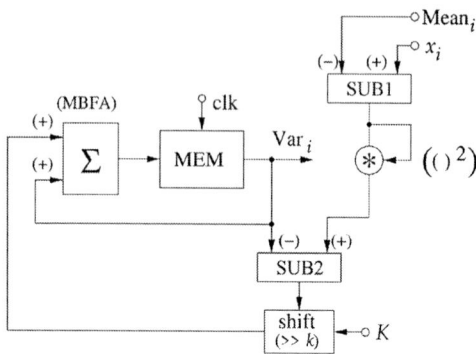

Fig. 3. The proposed, iterative circuit for the computation of the variance value.

is known, the use of the trilateration methods allow to compute the position of the RSU [11].

Theoretically, on the basis of a series of distance measurements the location of the RSU device can be unambiguously determined. In practice, various factors may affect the obtained results. The variation of external temperature and the noise may impact the response time of the devices, that will translate into a positioning error [13]. Due to these negative factors, a set of apparent positions of the RSU device is obtained. Based on them, the real location may be estimated by applying appropriate statistical computations.

In the proposed calibration algorithm [1] we assume that the statistical computations will be performed in particular RSU devices. Following, i^{th}, samples representing their apparent (x_i, y_i) positions in the global coordinate system (GCS) will be provided to these devices sequentially. In this situation iterative algorithms are the most appropriate. Due to expected hardware limitations we introduced some modifications to existing methods, as it is briefly presented below. The mean and the variance variables are initialized as follows:

$$\text{Mean}_1 = x_1 \qquad (1)$$

$$\text{Var}_1 = 0 \qquad (2)$$

where: x_1 is a first recorded sample. The same computations are performed for the y coordinate in the GCS. For each new sample, the updates of these variables are computed in accordance with the following formulas:

$$\text{Mean}_i = \text{Mean}_{i-1} + [x_i - \text{Mean}_{i-1}]/i \qquad (3)$$

$$\text{Var}_i = \text{Var}_{i-1} + [(x_i - \text{Mean}_{i-1})^2 - \text{Var}_{i-1}]/i \qquad (4)$$

These equations, consistent with [9], require a division operation by the factor equal to the number of the

978-1-7281-3420-8/19 $31.00 © 2019 IEEE

samples already processed. To eliminate the division operation in order to simplify the computation scheme, we replace the i factor with a new C variable [8]:

$$\text{Mean}_i = \text{Mean}_{i-1} + [x_i - \text{Mean}_{i-1}]/C \quad (5)$$

$$\text{Var}_i = \text{Var}_{i-1} + [(x_i - \text{Mean}_{i-1})^2 - \text{Var}_{i-1}]/C \quad (6)$$

which is initially set to 2 and then updated as follows:

$$\mathbf{if}\ (i > C)\{C = C \cdot 2;\} \quad (7)$$

As a result, the C variable is always one of the powers of 2, i.e. $C \in 2, 4, 8, 16, \dots$. This variable allows for applying a much simpler operation, in which the terms in square brackets are shifted by k positions to the right ($\gg$). This operation is the substitution of division by 2^k factor. Finally we get:

$$\text{Mean}_i = \text{Mean}_{i-1} + [x_i - \text{Mean}_{i-1}] \gg k \quad (8)$$

and

$$\text{Var}_i = \text{Var}_{i-1} + [(x_i - \text{Mean}_{i-1})^2 - \text{Var}_{i-1}] \gg k \quad (9)$$

The k variable for each iteration is computed as follows ($C = 2$, $k = 1$, for $i = 0$):

$$\mathbf{if}\ (i > C)\{C = C \ll 1; k = k + 1;\} \quad (10)$$

III. Transistor level implementation of the proposed algorithms

Formulas 8, 9 and 10 can be implemented relatively easily in the CMOS technology. Fig. 1 illustrates the circuit responsible for the adaptive change of the C and the k variables. It is based on a digital comparator, built on the basis of a multi-bit full subtractor (MBFS) and a 1-bit shift register. In Fig. 1 an additional variable $K = C/2$ is introduced, as its particular bits directly control the 'shift' block.

When, as a result of the incrementing operation the output of the binary counter (provided to the negative input of the MBFS) becomes larger than the K variable, then the B_{out} (borrow out) bit becomes '1'. This signal shifts the bits in the K variable by one position and, consequently, modifies the C variable. For example, when $K = 1$, then the expressions in square brackets in 8 and 9 are shifted by one bit, which corresponds to the division by 2. After subsequent switches of the shift register, the variable K takes values: 2 (binary: Bx10), 4 (Bx100), 8 (Bx1000), ... This means that the expressions in [] in 8 and 9 are shifted by 2, 3, 4, ... positions to the right, i.e. they are divided by 4, 8, 16, ..., respectively. The 'shift' block was implemented as a commutation field composed of switches directly controlled by particular bits of the K signal.

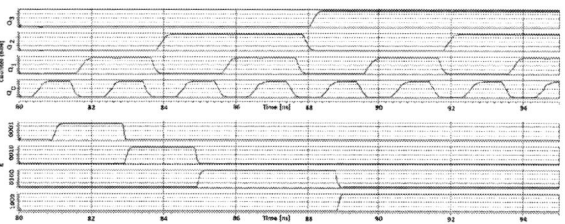

Counter (DEC)	Counter (BIN)	K (BIN)	C
0	0000	0000	1
1	0001	0001	2
2	0010	0001	2
3	0011	0010	4
4	0100	0010	4
5	0101	0100	8
6	0110	0100	8
7	0111	0100	8
8	1000	0100	8
9	1001	1000	16
10	1010	1000	16
11	1011	1000	16
12	1100	1000	16
13	1101	1000	16
14	1110	1000	16
15	1111	1000	16

Fig. 4. Selected simulation results of the circuit shown in Fig. 1

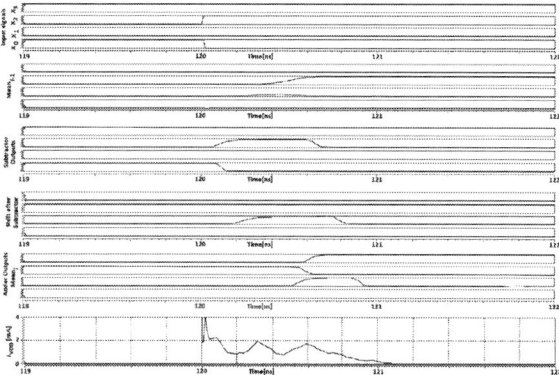

Fig. 5. Selected simulation results of the circuit shown in Fig. 2 illustrating computation time of a single cycle.

Fig. 4 presents selected transistor level simulation results of the circuit shown in Fig. 1 , as well as a table with states of particular variables described above. This circuit throughout the K variable controls the circuits shown in Figs. 2 and 3 that are responsible for computing the mean and the variance, respectively. The circuit that computes the variance uses also the mean value calculated by the circuit shown in Fig. 2.

Both these circuits work on a similar principle. Since the algorithms are iterative, therefore an accumulator (ACU) has to be applied in the proposed circuits. In 8 and 9 the ACU is represented by the summing operation (before the square brackets). The ACU consists of a summing circuit, realized as an asynchronous multi-bit full adder (MBFA) and a memory block realized in this case on the basis of the D-flip flops. The memory is the only block in these circuits that are controlled by a 1-phase clock circuit. All other components operate fully asynchronously. This is illustrated in selected transistor-level simulations shown in Fig. 5 for the mean algorithm. As can be observed, the overall computation cycle between the input of the circuit and the output of the MBFA in the ACU takes about 1 ns. In case of the variance circuit this time is longer (2-3 ns) due to some additional blocks described below.

Of these two circuits, the variance one is more complex. The current value of the mean and a new signal sample x_i are provided to the inputs of this circuit. The circuit first calculates the difference between these signals in an asynchronous subtraction system (SUB1). The subtraction result is squared using a multiplier, which is realized as an asynchronous binary tree circuit. The multiplier operates in one-quadrant mode, that simplifies its structure. It is possible as the result of the squaring operation is independent of the sign of the difference between the sample x_i and the mean. If the subtraction result is negative, which is signaled by the borrow out bit of the MBFS signal used in the SUB1 block, the absolute value of this signal is calculated. It requires negation of all bits and adding '1', as this signal is coded with two's complement code.

At the next stage of the computation chain, the SUB2 block computes a difference between the square signal and the value of the variance computed in the previous iteration, Var_{i-1}. The output from SUB2 is coded in two's complement code, which means that negative numbers have the value '1 'on the most significant bits (MSB). In this case, when shifting the bits to the right, the MSBs have to be supplemented with '1'. For a positive output of SUB2 block, the MSBs are supplemented with '0'. In practice, these bits are supplemented by the borrow out bit from the MBFS used in the SUB2 block. The final result of the described operations is added to the value stored in the ACU. In the following clock impulse the new value of the variance is stored in the memory of the ACU.

IV. Conclusions

In the paper we proposed a concept of a hardware implementation of iterative algorithms for calculating the mean and the variance statistical quantities. To facilitate this implementation, the classic algorithms have been modified so that to eliminate the division operation. This resulted in a relatively high data rates. The calculation time of a single sample of the mean and the variance does not exceed 4 ns, with an energy consumption at the level of about 4-10 pJ per single sample, depending on the resolution of the processed signals.

Acknowledgment

The work is part of the project No.: POIR.01.01.01-00-1398/15 (National Centre for Research and Development, Poland), entitled "Development of innovative active safety technologies that will be used in advanced driver assistance systems (ADAS) and in autonomous driving, intended for commercial production." The project is realized in Aptiv Services Poland company.

References

[1] Banach M., Długosz R., "Method to improve the determination of a position of a roadside unit, roadside unit and system to provide position information", Application number: 18200799.7, Date of filing: 16-Oct-2018, Applicant: Aptiv Technologies Limited, European Patent Office

[2] Vadim A. Butakov, Petros Ioannou, "Personalized Driver Assistance for Signalized Intersections Using V2I Communication", IEEE Transactions on Intelligent Transportation Systems, Vol: 17, Issue: 7, July 2016, pp.1910-1919

[3] Olaverri-Monreal C., Errea-Moreno J., Díaz-Álvarez A., "Implementation and Evaluation of a Traffic Light Assistance System Based on V2I Communication in a Simulation Framework", Journal of Advanced Transportation, Vol. 2018, Article ID 3785957, 11 pages, https://doi.org/10.1155/2018/3785957

[4] Kim Tae Won, Jung Jae Il, "GPS UWB V2x Autonomous Driving Method And System for Determining Position of Car Graft on GPS UWB And V2X," KR20170071207 (A) – 2017-06-23

[5] Jian Wang, Yang Gao, Zengke Li, Xiaolin Meng, Craig M. Hancock, "A Tightly-Coupled GPS/INS/UWB Cooperative Positioning Sensors System Supported by V2I Communication", Sensors, 16(7), 944; doi:10.3390/s16070944, 2016.

[6] Amini A., Vaghefi R.M., de la Garza M., Michael Buehrer R.M., "Improving GPS-Based Vehicle Positioning for Intelligent Transportation Systems", IEEE Intelligent Vehicles Symposium (IV), Electronic ISBN: 978-1-4799-3638-0, DOI: 10.1109/IVS.2014.6856592, Jul. 2014

[7] Fascista A., Ciccarese G., Coluccia A., Ricci G., "A Localization Algorithm Based on V2I Communications and AOA Estimation", IEEE Signal Processing Letters, Vol. 24, Issue: 1, Jan. 2017

[8] M. Banach, K. Kubiak and R. Długosz, "Calculation of descriptive statistics by devices with low computational resources for use in calibration of V2I system", 24th International Conference on Methods and Models in Automation and Robotics (MMAR), Międzyzdroje, Poland, August 2019.

[9] West D.H.D., "Updating Mean and Variance Estimates: An Improved Method", Communications of the ACM, vol.22, 9, p.532-535, 1979

[10] Martynenko D., Klymenko O., Fischer G., Kreiser D., "Implementation and Evaluation of a Two-Way-Ranging Module based on an IR-UWB Chip-Set", 12th Workshop on Positioning, Navigation and Communication (WPNC'15), Niemcy, 2015

[11] Hassan O., Adly I., Shehata K.A., "Vehicle Localization System based on IR-UWB for V2I Applications", 8th International Conference on Computer Engineering & Systems (ICCES), 26-28 Nov. 2013

[12] Llorente R., Morant M., "UWB technology for safety-oriented vehicular communications", Proc. SPIE 9807, Optical Technologies for Telecommunications 2015, 980705; doi:10.1117/12.2234430, 26 marca 2016.

[13] Banach M., Długosz R., "Real-time Locating Systems for Smart City and Intelligent Transportation Applications", 30th International Conference on Microelectronics (MIEL), Nis, Serbia, Oct. 2017, pp. 231–234.

978-1-7281-3420-8/19 $31.00 © 2019 IEEE

Absorptive Filters in the Realization of RCIED Activation Jamming

A. Lebl, M. Mileusnić, B. Pavić, and J. Radivojević

Abstract - This paper presents the realization of absorptive filters in RCIED activation jammer. The measured filter characteristics satisfactory agree with the simulated characteristics. The main goal is to achieve sufficent degree of reflected signal attenuation (absorption), and not to provide high signal attenuation in stop-band in forward direction. It is proved that for such behaviour it is enough to implement simple first-order band-pass absorptive filters. These filters are realized as lumped (*RLC*) structures for relatively lower frequencies and as microstrip filters for higher frequencies. The jammer with the embedded absorptive filters is verified for military applications.

I. INTRODUCTION

Mobile communications are moving to higher frequency bands at the present time and hostile messages are also sent over the same frequency bands together with legal transmissions. Having in mind Remote Controlled Improvised Explosive Devices (RCIEDs), the usually used activation message frequencies approach several GHz. This fact has to be taken into account in the process of RCIED activation jammer development and implementation.

Institute IRITEL has long-term experience and tradition in the development of various jamming systems for widespread applications. For the sake of illustration, three systems are referenced [1]-[3]. IRITEL also has experience in the development of mobile phone (cellular) jamming systems [4]. Following improvements in RCIED activation jamming techniques, the most modern R&D IRITEL contributions in this field are presented in [5]-[7]. Maximum jamming frequency is even 6GHz in these solutions. To fulfil all seriously requested tasks, it is necessary to use latest technologies and adequate techniques in the design. That's why the absorptive filters are applied to IRITEL development of RCIED activation jammers.

There is not a great number of jammer solutions where absorptive filters are implemented. The contribution [8] presents the implementation of absorptive filters just for a similar purpose – jammers suppression. That's the reason why solution presented in this paper is a novel implementation of an already known structure.

A. Lebl, M. Mileusnić, B. Pavić and J. Radivojević are with the Radiocommunications Department, Institute IRITEL, 11080 Belgrade, Batajnički put 23, Serbia, E-mail: lebl@iritel.com

II. THE ABSORPTIVE FILTERS IMPLEMENTATION

Absorptive filters eliminate instability, in-band ripples, and also false triggers of power-detector circuitry due to reflected harmonics. In this way absorptive filters also completely remove the risk of damage to power amplifiers due to reflected high-power outside power amplifier bandwidth. We must take into account that jamming systems are usually broadband in nature and that often contain multicarrier signals (comb) producing a lot of out-of-band intermodulation products in nonlinear power amplifiers. The internally terminated absorptive filters are used to prevent intermodulation that excites further nonlinear behaviour of the power amplifier, which results in regrow of intermodulation products.

Reflective signals generally cause instability effects within the power amplifier. The out-of-band signals are internally terminated in an absorptive filter, preventing them from generating undesirable intermodulation products and also eliminating instability of power amplifiers at out-of-band frequencies. In this way absorptive filter eliminates potential damage to power amplifiers due to reflection of high power out-of-band energies. The effect is achieved by resistors in absorptive filter structure, which dissipate reflected power [9], [10].

Absorptive filters are implemented in various technologies. Besides realizations using lumped (*RLC*) components [11], they are also applicable as waveguide filters [12], [13] and as microstrip filters [14], [15].

The principle absorptive filter schematic and signal paths are presented in Fig. 1 [16]. Impedances Z_1, Z_2 and Z_3 are specific for different filter types (low-pass, high-pass, band-pass or band-stop) [17]. These impedances consist of inductors (L) and capacitors (C). For example, the band-pass filter is formed from the serial LC connection on the place of Z_1, parallel LC connection on the place of Z_3, while each Z_2 consists of one parallel and one serial LC connection. Frequencies from pass-band are transmitted through Z_1, while the paths of the frequencies from stop-band at the filter input and output are through the impedances Z_2, resistor R, and the common impedance Z_3, respectively. The paths of attenuated pass-band and stop-band signals (leakage) are opposite, as presented in Fig. 1. The number of repeated identical LC structures in Z_2 and the number of serially cascaded identical filter structures from Fig. 1 may be variable, depending on the necessary transmission characteristics [15].

978-1-7281-3420-8/19 $31.00 © 2019 IEEE

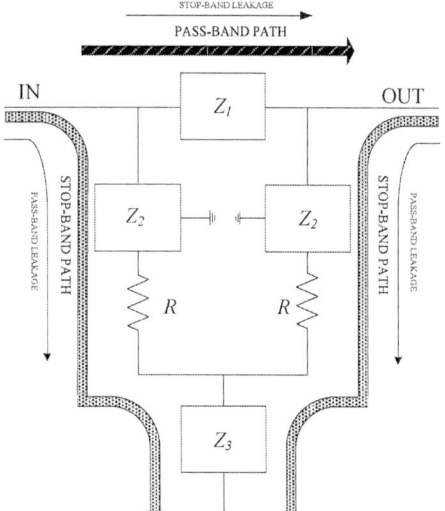

Fig. 1. Signal paths in absorptive filters realization.

III. RESULTS

Absorptive filters in the implementation of RCIED activation jammer are placed between the power amplifier and the antenna [5].

Fig. 3. Practical realization of absorptive filter in the frequency band 180-500MHz (lumped LCR elements).

Figure 2a) presents theoretical (simulated) response of absorptive filter, realized in RCIED activation jammer for frequency band 180-500MHz. Fig. 2b) presents the corresponding measured characteristic of the realized filter. Two characteristics are presented both in Fig. 2a) and Fig. 2b): signal gain as the function of frequency at the filter output (S_{21}) and the reflected signal gain at the filter input (S_{11}). After that, Fig. 3 presents practical realization of this filter. Filter is realized using lumped (RLC) elements. The filter realization is in the simplest, very cheap variant of the first-order band-pass filter [16], which is enough to achieve satisfactory results for RCIED activation jammer.

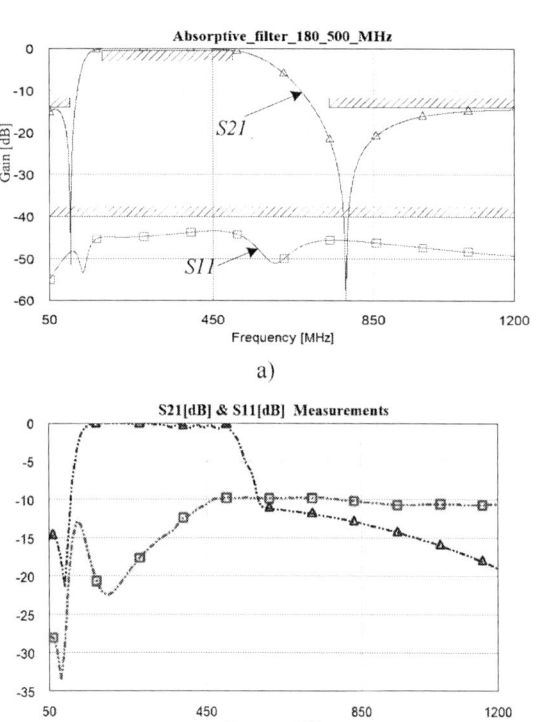

a)

b)

Fig. 2. a) Simulated and b) measured transmission characteristic of absorptive filter in the frequency band 180-500MHz (lumped LCR elements).

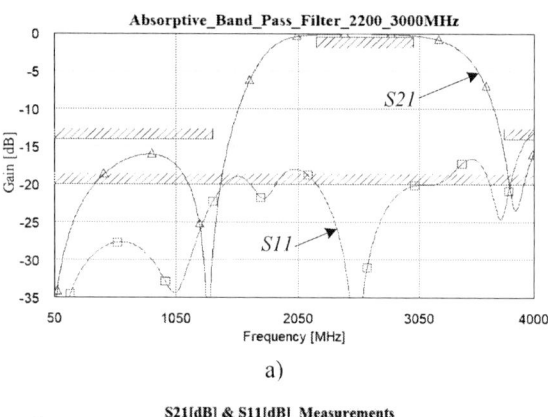

a)

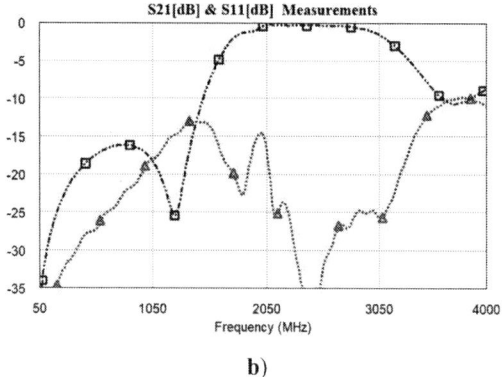

b)

Fig. 4. a) Simulated and b) measured transmission characteristic of absorptive filter in the frequency band 2200-3000MHz (microstrip realization).

Figure 4a) presents theoretical (simulated) response of absorptive filter, realized in RCIED activation jammer for frequency band 2200-3000MHz. Fig. 4b) presents the corresponding measured characteristic of the realized filter. After that, Fig. 5 presents practical realization of the filter. The filter is realized as a microstrip variant. More precisely speaking, the realization is hybrid, because capacitors are discrete elements. The realization of capacitors in microstrip technology is a problem for this specific implementation, because very near conductive lines forming a capacity have a low breakdown voltage, typically lower than the voltages present in a jammer. This filter in the band 2200-3000MHz is also realized as first-order band-pass filter.

Fig. 5. Practical realization of absorptive filter in the frequency band 2200-3000MHz (microstrip realization).

Fig. 6. RCIED activation jammer where absorptive filters are implemented.

Simulated and measured characteristics of both realized filters have a high degree of agreement. The agreement is higher when considering signal gain (S_{21}) than when considering reflected signal gain (S_{11}). In more detail, the measured S_{11} characteristic is worse than calculated characteristic for the filter in the band 180-500MHz. It reaches even -10dBm in the realized filter comparing to better than -40dBm in the calculated filter. This is the consequence of mutual coupling between inductive elements in the realized filter, which is increased when the applied frequency is increased. Due to limited space and

small physical distance, it is hard to avoid this mutual coupling, which is not considered in simulation. But, although worse, this characteristic is still satisfactory for the desired implementation. The improvement of this characteristic is possible by different printed-circuit board layout.

Figure 6 presents IRITEL's solution of RCIED activation jammer [5]-[7], where absorptive filters are implemented. The jammer has successfully passed all required tests and has been verified for its full defensive implementation. It has been also successfully presented at the Eurosatory 2018 – Defense & Security International Exhibition in Paris and at Partner 2019 – 9th International Armament and Military Equipment Fair in Belgrade. IRITEL's reliable results in the field of jammers as well as other defense products development have influenced that it is put at the list of defense leaders in the world [18].

IV. CONCLUSIONS

This paper presents the characteristics of absorptive filters, which are used in the IRITEL original solution of RCIED activation jammer. The characteristics on Fig. 2 and Fig. 4 prove satisfactory agreement between simulated and measured practical results both in lumped and microstrip realization of absorptive filters. The absorptive filter application is very important for jammer function to avoid system unstable behaviour due to generation of intermodulation products or even damage of amplifier stages as the consequence of the out-of-band high reflected power. This goal is achieved by significant attenuation i.e. absorption of the signal in the out-of-band frequency range. Absorptive filters present cheap, frequency selective alternative solution comparing to other solutions based on circulators and insulators. Circulators and insulators are convenient and significantly more expensive solution implemented at wide-band jammers. Design of absorptive filters is simpler in the sense of component values determination than it is the case with classical filters. They are more easily serially cascaded than other filter types to achieve the desired out-of-band signals suppression.

ACKNOWLEDGEMENT

The presented development is realized in the framework of the project TR32051, which is cofinanced by Ministry of Education, Science and Technological Development of the Republic of Serbia, 2011-2019.

REFERENCES

[1] "IRITEL High Frequency (HF) radio surveillance and jamming system," in the book M. Streetly, *Jane's Radar And Electronic Warfare Systems*, IHS Global Limited, 2011.
[2] "IRITEL Very/Ultra High Frequency (V/UHF) radio surveillance and jamming system," in the book M. Streetly, *Jane's radar and electronic warfare systems*, IHS Global Limited, 2011.

978-1-7281-3420-8/19 $31.00 © 2019 IEEE

[3] P. Petrović, N. Remenski, P. Jovanović, V. Tadić, B. Pavić, M. Mileusnić, B. Mišković, „WRJ 2004 wideband radio jammer against RCIEDs", tehničko rešenje – novi proizvod na projektu tehnološkog razvoja TR32051 pod nazivom „Razvoj i realizacija naredne generacije sistema, uređaja i softvera na bazi softverskog radija za radio i radarske mreže", 2011., http://www.iritel.com/images/pdf/wrj2004-e.pdf, in Serbian.

[4] N. Remenski, B. Pavić, P. Petrović, M. Mileusnić, V. Marinković-Nedelicki, "Integrisana radio-oprema za zaštitu prostora od mobilnih veza (Treća generacija radio-opreme), tehničko rešenje – novi proizvod s oznakom CJ-1P na projektu tehnološkog razvoja TR-11030 "Razvoj i realizacija nove generacije softvera, hardvera i usluga na bazi softverskog radija za namenske aplikacije", 2010., http://www.iritel.com/images/pdf/cj-1p-e.pdf, (also published in the book M. Streetly, *Jane's Radar And Electronic Warfare Systems*, IHS Global Limited, 2011.). Prva generacija radio-opreme s oznakom CJ-1 je realizovana na projektu tehnološkog razvoja TR6149B, 2006.

[5] M. Mileusnić, P. Petrović, B. Pavić, V. Marinković-Nedelicki, J. Glišović, A. Lebl, I. Marjanović, „The Radio Jammer Against Remote Controlled Improvised Explosive Devices", *25th Telecommunications Forum (TELFOR)*, November 21-22, 2017, pp. 151-154.

[6] M. Mileusnić, B. Pavić, V. Marinković-Nedelicki, P. Petrović, D. Mitić, A. Lebl, „Analysis of Jamming Successfulness against RCIED Activation", *5th International Conference IcETRAN 2018*, Palić, June 11-14, 2018., Proceedings of Papers, pp. 1206-1211.

[7] M. Mileusnić, B. Pavić, V. Marinković-Nedelicki, P. Petrović, D. Mitić, A. Lebl, „Analysis of jamming successfulness against RCIED activation with the emphasis on sweep jamming", the extended and revised version of the awarded paper from the IcETRAN 2018, *Facta Universitatis, Series Electronics and Energetics*, Vol. 32, No. 2, June 2019, pp. 211-229.

[8] W. N. Allen, D. Peroulis, „Fully Autonomous Multiple-Jammer Suppression", 2017 *IEEE MTT-S Inernational Microwave Symposium (IMS)*, Honololu, HI, USA, 4-9. June 2017.

[9] M. A. Morgan, T. A. Boyd, „Reflectionless Filter Structures", *IEEE Transactions on Microwave Theory and Techniques*, Vol. 63, No. 4, April 2015, pp. 1263-1271.

[10] S. C. Dutta Roy, „A New Lumped Element Bridged-T Absorptive Band-stop Filter", *Facta Universitatis, Series Electronics and Energetics*, Vol. 30, No. 2, June 2017, pp. 179-185.

[11] J. Lee, T. C. Lee, W. J. Chappell, „Lumped-Element Realization of Absorptive Bandstop Filter With Anomalously High Spectral Isolation", *IEEE Transactions on Microwave Theory and Techniques*, Vol 60, Issue 8, August 2012, pp. 2424-2430.

[12] T. Stander, „*High-power broadband absorptive waveguide filters*", Dissertation for the degree of Doctor of Philosophy in Engineering, University of Stellenbosch, Pretoria, South Africa, December 2009.

[13] T. Stander, P. Meyer, P. W. van der Walt, „Compact High-Power Broadband Absorptive Filters using Slotted Waveguide Harmonic Pads", *IET Microwaves Antennas & Propagation*, Vol. 8, Issue 9, June 2014., pp. 673-678.

[14] J. Breitbarth, D. Schmelzer, „Absorptive Near-Gaussian Low Pass Filter Design with Applications in the Time and Frequency Domain", 2004 *IEEE MTT-S International Microwave Symposium Digest*, 6-11. June 2004., pp. 1303-1306.

[15] D. Ulinic, „*Advanced Filter Response Based On Reflectionless Concept*", bachelor thesis, Departament de Telecommunicació i Enginyeria de Sistemes, Universitat Autònoma de Barcelona, 25. June 2018.

[16] M. A. Morgan, „Sub-network Enhanced Reflectionless Filter Topology", United States Patent No. US 9,705,467 B2, 11. July 2017.

[17] M. A. Morgan, T. A. Boyd, „Synthesis of a New Class of Reflectionless Filter Prototypes", 2010.

[18] „The Executive Guide of the Global Defense and Security Industry, United Arab Emirates 2014, Policy Making, Providing, Preventing, Protecting, Services", Frontline Publishing, pp. 85.

978-1-7281-3420-8/19 $31.00 © 2019 IEEE

Multi-Rate Signal Processing with the Use of Filter Banks Composed of Parallel FIR Filters

M. Banach and R. Długosz

Abstract—**The paper presents a concept of a hardware implementation of 2-D finite impulse response (FIR) filter banks for the application in image processing and analysis. Banks composed of low- and high-pass FIR filters are basic components of multi-stage discrete wavelet transform (DWT). The applications of such solutions that are in the scope of our interests are vision systems used in automotive active safety functions (e.g in line departure warning). Basics of the DWT are broadly described in the literature. In our work we focus on solutions supporting hardware realization of filter banks for DWT. The proposed parallel and asynchronous circuits allow to achieve the processing time for a single pixel not exceeding 2 to 4 ns, depending on the size of the mask (data for TSMC 180 nm CMOS process).**

I. INTRODUCTION

Filtering of signals is one of basic operations performed in software and electronics systems. In fact, every digital system may be considered as a filter. Filters are used to process either continuous or discrete time signals, analog or digital (quantized and sampled in time). The properties of the processed signal, as well as the target application determine the type of the used filters, their mathematical parameters, as well as the way of their implementation. The last aspect, in this work understood as a hardware structure, strongly depends on such parameters of the frequency response as the cut-off frequency, the steepness in the transient band, the attenuation in the stopband, etc.

In this work we focus on the implementation of filter banks composed of FIR filters that may be used, for example, in discrete wavelet transform (DWT). Banks of this type are commonly applied in signal processing in many engineering areas [1]. In the scope of our interests are solutions for the intelligent transportation system (ITS). They include automotive active safety (AS) functions, autonomous driving and its interaction with the intelligent road and urban infrastructure. Pattern recognition that may be supported by the DWT is used, for example, in line departure warning (LDW), camera calibration and traffic sign recognition (TSR).

M. Banach is with Aptiv Services Poland ul. Podgórki Tynieckie 2, 30-399, Kraków, Poland, and with Poznan University of Technology, Institute of Architecture and Spatial Planning, Nieszawska 13C, 61-021 Poznań, Poland E-mail: marzena.banach@put.poznan.pl

R. Długosz is with UTP University of Science and Technology Faculty of Telecommunication, Computer Science and Electrical Engineering ul. Kaliskiego 7, 85-796, Bydgoszcz, Poland, and with Aptiv Services Poland ul. Podgórki Tynieckie 2, 30-399, Kraków, Poland E-mail: rafal.dlugosz@gmail.com

In case of the application of filter banks in automotive area, it is necessary to pay attention to several important aspects. One of them is the need to process data in real time [2], so that the vehicle can quickly react to changing situation on the road. Another issue is the computational complexity, which should be kept as small as possible, taking into account the range of tasks typically performed by the vision system as well as hardware limitations that result from the final price of the system. In this context, the use of solutions based on application specific integrated circuits (ASIC) can play an important role – specialized co-processors speeding up selected signal processing tasks.

II. DISCRETE WAVELET TRANSFORM AND ITS APPLICATIONS – A BRIEF OVERVIEW

In the literature one can find examples of the application of the DWT in automotive and vision applications. One of them is a system that allows for an estimation of the occupancy of ad-hoc parking lots [3]. The system based on 2-D DWT is realized in field programmable gate array (FPGA) and exhibits low computation complexity. It demonstrates the ability of detecting objects when no background subtraction is applied. Another example is an in-vehicle monocular pre-crash vehicle detection system [4]. At one of the steps of this real-time procedure a simple Haar wavelet is used in the decomposition of the input signal for feature extraction and object classification. The reported system was successfully tested under different traffic scenarios, such as highway and urban environment and under varying weather conditions. The DWT can also be used in vehicle tracking systems in object detection on the basis of their motion [5], or more generally in optical flow applications [6]. In [5] the Gabor and the Mallat wavelets were used for improving the accuracy and speed of the vehicle detection. In [7] a robust vehicle detection system is proposed. In this case the fast wavelet transform (FWT) is used to extract image texture. Comparing the textures, different for the vehicles and the shadows allow to detect them.

The DWT is commonly realized as a cascade multi-rate operation, in which filter banks are applied alternately with downsampling operations, as illustrated in Fig. 1 for 1 dimensional (1-D) signal. In this example, the bank is composed of two complementary low-pass (LP) and high-pass (HP) FIR filters. Filters, filter

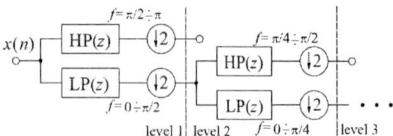

Fig. 1. Discrete wavelet transform realized as a multi-stage filtering with quadratic mirror filter (QMF) bank and down-sampling operations.

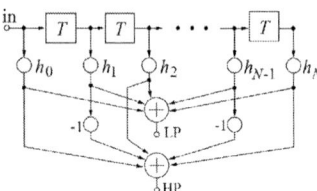

Fig. 2. Possibilities of the simplification of the QMF bank in case of symmetrical TFs.

banks and functions used in the vision systems are usually realized as functions in software systems, frequently in digital signal processors (DSP). In hardware one can find realizations in FPGAs [8], [3], [9] as well as in the full-custom style analog circuits [10].

One of important problems in hardware is the realization of the filter coefficients. Ratio between the largest and the smallest coefficient and the implementation precision directly impact the circuit complexity and the quality of the frequency response [11]. The precision here is related to a mapping of theoretical values of the coefficients into usually limited signal resolutions in a target hardware platform. In filters, that offer very large attenuation, the values at remote decimal places of the numbers representing the coefficients are important. In hardware, on the other hand, it is necessary to limit the precision, to avoid large circuit complexity. There exists a trade-off between these two factors.

In some vision applications, described above, simple Haar wavelets are used (an equivalent of the Daubechies (db1) wavelet), with 1st order masks (length $L = 2$) with equal coefficients. However, in other applications more complex transfer functions (TF) may be required. An additional problem is that many wavelets feature unsymmetrical TF, which impose an additional challenge on the hardware realization. For a symmetrical TF of the FIR filter, the HP filter may be realized on the basis of the LP filter simply by the multiplication of each odd coefficient by -1 (see Fig. 2). It is not possible in unsymmetrical TF, in which the vector of the coefficients needs to be additionally mirrored. In our work we focus on symmetrical TFs, as we found that such filters are sufficient in our application.

In the case of image analysis in the DWT, in the decomposition phase, the input signal it is divided into four subbands. The input signal is first processed line by line with $(L \times 1)$ (L)ow pass and (H)igh pass filters. The two output signals, obtained in this way, are then filtered in columns also with 1-D filters of type L and H, forming four output signals that can be designated as LL, HL, LH and HH. In the case of the symmetrical TFs of the used filters, to calculate all subbands one can use a similar solution as shown in Fig. 2.

In this work we present a concept of a hardware realization of the QMF banks composed of FIR filters with symmetrical TFs. In our work we propose a parallel and additionally asynchronous solutions, that allow for a significant simplification of the controlling clock circuit, and at the same time ensure a high data rate.

III. THE PROPOSED QMF BANK

As mentioned earlier, in hardware realizations there exists a trade-off between the circuit complexity and the functionality at the system level. For this reason, in this work we consider 2-D filters with a flat frequency response, that allow for substantial simplification of the hardware structure, while in many situations offering good properties at the signal processing level. An example here is the use of simple Haar wavelets in some applications. The Haar wavelet is the simplest solution in which filters with symmetrical TFs may be used.

We assume that all input signal samples for a given computation cycle (under the filter mask) are available at the inputs of the filter bank. In the proposed solution, the outputs of all four filters (LL, HL, LH, HH) are then computed in four clock cycles only, as each of the filters operates asynchronously. A general structure of the proposed solution is shown in Fig. 3 (a). This circuit consists of 1-D FIR filters, whose structures for selected mask sizes are shown in diagrams (b) and (c). A set of the filter masks for the (3×3) is given as follows:

$$
LL_{(3\times3)} = \begin{bmatrix} 1 & 2 & 1 \\ 2 & 4 & 2 \\ 1 & 2 & 1 \end{bmatrix} \quad LH_{(3\times3)} = \begin{bmatrix} 1 & 2 & 1 \\ -2 & -4 & -2 \\ 1 & 2 & 1 \end{bmatrix}
$$
$$
HL_{(3\times3)} = \begin{bmatrix} 1 & -2 & 1 \\ 2 & -4 & 2 \\ 1 & -2 & 1 \end{bmatrix} \quad HH_{(3\times3)} = \begin{bmatrix} 1 & -2 & 1 \\ -2 & 4 & -2 \\ 1 & -2 & 1 \end{bmatrix} \tag{1}
$$

Example filter masks for higher order filters with flat frequency response, for the (4×4), (5×5) and (7×7) cases respectively are as follows:

$$
LL_{(4\times4)} = \begin{bmatrix} 1 & 3 & 3 & 1 \\ 3 & 9 & 9 & 3 \\ 3 & 9 & 9 & 3 \\ 1 & 3 & 3 & 1 \end{bmatrix}, HH = \begin{bmatrix} 1 & -3 & 3 & -1 \\ -3 & 9 & -9 & 3 \\ 3 & -9 & 9 & -3 \\ -1 & 3 & -3 & 1 \end{bmatrix} \tag{2}
$$

$$
HH_{(5\times5)} = \begin{bmatrix} 1 & -4 & 6 & -4 & 1 \\ -4 & 16 & -24 & 16 & -4 \\ 6 & -24 & 36 & -24 & 6 \\ -4 & 16 & -24 & 16 & -4 \\ 1 & -4 & 6 & -4 & 1 \end{bmatrix} \tag{3}
$$

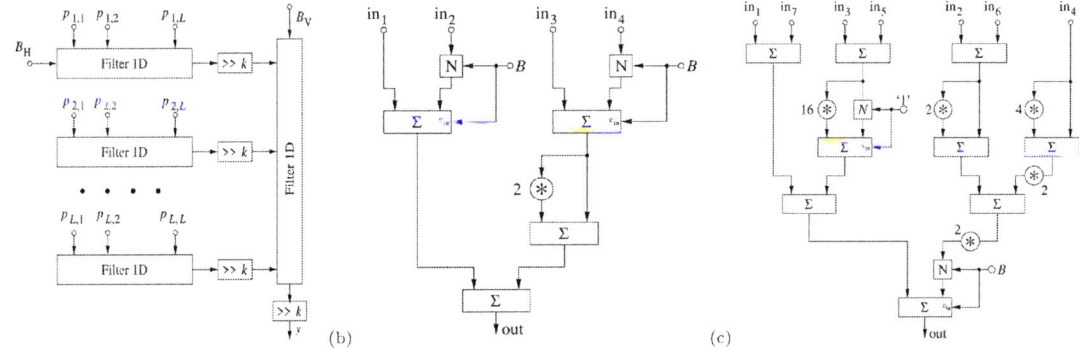

Fig. 3. The proposed solution for parallel and asynchronous filter bank: (a) a general block diagram of the 2-D filter bank, composed of a set of equal asynchronous 1-D FIR filters, (b) 1-D filter (mask: [1 (-)3 1 (-)3]) for the (4×4) case, (c) 1-D filter (mask: |1 (-)6 15 (-)20 15 (-)6 1]) for the (7×7) case.

Fig. 4. Results for proposed filter bank – LL, HL, LH and HH subbands, for L = [1 1] and H = [1 -1] 1-D filters. The results are presented for 2nd stage of the DWT (decimation by 4 in each direction). (photo: own source).

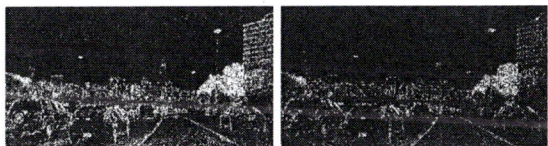

Fig. 5. Results for HH subband for (a) L = [1 2 1], H = [1 -2 1], (b) L = [1 3 3 1], H = [1 -3 -3 1].

$$LL_{(7 \times 7)} = \begin{bmatrix} 1 & 6 & 15 & 20 & 15 & 6 & 1 \\ 6 & 36 & 90 & 120 & 90 & 36 & 6 \\ 15 & 90 & 225 & 300 & 225 & 90 & 15 \\ 20 & 120 & 300 & 400 & 300 & 120 & 20 \\ 15 & 90 & 225 & 300 & 225 & 90 & 15 \\ 6 & 36 & 90 & 120 & 90 & 36 & 6 \\ 1 & 6 & 15 & 20 & 15 & 6 & 1 \end{bmatrix} \quad (4)$$

One of main aspects presented in this work is how to realize these filters in a parallel and asynchronous way. It is shown in Fig. 3 for 1-D filters. The L filters used in the rows in this circuit are equal to the one used in the column. The computation schemes for selected filters with flat frequency response are as follows:

$$\mathrm{out}_{(3 \times 1)} = (\mathrm{in}_1 + \mathrm{in}_2) + \beta \cdot (2 \cdot \mathrm{in}_3) \quad (5)$$

$$\mathrm{out}_{(4 \times 1)} = (\mathrm{in}_1 + \beta \cdot \mathrm{in}_4) + (1 + 2) \cdot (\mathrm{in}_3 + \beta \cdot \mathrm{in}_2) \quad (6)$$

$$\mathrm{out}_{(5 \times 1)} = (\mathrm{in}_1 + \mathrm{in}_5) + ((1 + 2) \cdot \mathrm{in}_3) \cdot 2 + \\ 4 \cdot (\mathrm{in}_2 + \mathrm{in}_2) \cdot \beta \quad (7)$$

$$\mathrm{out}_{(7 \times 1)} = (\mathrm{in}_1 + \mathrm{in}_7) + (\mathrm{in}_3 + \mathrm{in}_5) \cdot (16 - 1) + \\ [(\mathrm{in}_2 + \mathrm{in}_6) \cdot (2 + 1) + \\ \mathrm{in}_4 \cdot (4 + 1) \cdot 2] \cdot 2 \cdot \beta \quad (8)$$

The input and the output signals are marked as in and out, respectively. The values of the input signal are expressed in two's complementary code. The summing operation is performed by the use of a multi-bit full adder operating in an asynchronous way. The filter coefficients have been expressed by means of sums of numbers that are powers of 2. It allowed us to realize the multiplication by appropriate bit shifting and summation. For example, multiplication by a factor of 6 is realized as the sum of a given input sample and the same sample after its shifting by one bit to the left, followed by another shift by 1 position to the left $((1+2) \cdot 2)$. The multiplication by the 15 factor is realized as the difference of a given sample after the shift by 4 bits to the left (multiplication by 16) and this sample without the shift $(16 - 1)$. The presented computing diagrams have been designed in such a way, to minimize the number of the summation operations, which is important from the point of view of the circuit complexity.

An important here is to explain the meaning of the β factor and the B signals in Fig. 3. The B signals may be either '0' or '1', depending on whether a given filter coefficient is positive or negative, respectively. Bits B (index H and V for horizontal and vertical directions) are provided to some MBFAs as carry-in signals (c_{in}) in least significant 1-bit full adder. This operation, along with the negation of all bits of a given input pixel (in

978-1-7281-3420-8/19 $31.00 © 2019 IEEE

block N), are the subtraction. To simplify the functional description of the presented circuits, we used the β factor, which takes the values 1 or -1 for B equal to 0 or 1, respectively.

The negation block (N) is implemented using NOT gates and switches. The multiplication operations realized as bit shifting do not require additional elements. The signal is simply provided to the inputs of a corresponding MBFA with an appropriate offset.

Circuit complexity for a given mask depends on the number of the used MBFAs. The (3×3) mask requires 2 summing operations in each 1-D filter (10 for the overall filter). For the masks (4×4), (5×5) and (7×7) these numbers are respectively, 20 (4 in 1-D direction), 30 (5 in 1-D) and 72 (9 in 1 -D). The output of each 1-D block is divided by the sum of the absolute values of all its coefficients, which in filters with flat frequency response always is a power of 2. Thus it may be carried out by shifting all bits of the output signal to the right by a given number of positions.

A. Verification of the proposed solutions

The proposed systems have been verified by means of simulations at the signal processing level. For this purpose, the described calculation schemes have been reproduced with details in the simulator. Selected test results are shown in Figs. 4 and 5 for an example picture illustrating a road (authors' private source). The objective was to detect the boundaries of the current vehicle's line. A number of tests were carried out for the filter masks described above. The shown results are four subbands of the 2-D QMF bank at the second stage of the DWT. To better illustrate the HL, LH and HH components, the image was normalized. For the HH component, which is the most interesting here, binary thresholding has been applied. This facilitates a direct assessment and the comparison of performance of particular filters. Fig. 4 shows all subbands for the use of Haar wavelets i.e. the (2×2) mask, while Fig. 5 presents the HH component for the (3×3) and (4×4) filter masks. The observable trend is that filters with larger mask sizes (more selective) allow to achieve a higher signal-to-noise ratio – the boundaries of the line are more clearly visible on its background.

The circuit was tested in such a way that at the beginning the signals B_H and B_V were set to '0', thus the filter worked in the LL mode. After calculating the output signal, the B signals were switched successively to states 01, 10 and 11, which allowed for the computation of remaining signal components for a given set of the input samples. In case of a transistor level implementation, for each combination of the B signals, the filter further operates fully asynchronously. Thus, the data rate depends on the number of the maximum number of the summing circuits in a single path between the system inputs and its output. For a mask (7×7) for

both 1-D filtering stages, the total number of MBFAs equals 8. In the CMOS 180 nm technology a single pixel is in this case computed in less than 4 ns. For smaller masks, this time is even half as long. For newer technologies, one can expect a significant improvement in these results. As in the DWT after each stage the signal is decimated, therefore the number of the output pixels that have to be computed for a single filter is amounted to only 1/4 of pixels of the input signal.

IV. CONCLUSIONS

A concept of the 2-D filter bank for the application in image analysis in the automotive vision systems has been presented. An advantage of the proposed solutions is a parallel and asynchronous operation of the used filters that allows for achieving high data processing rates. A single filter mask can be easily switched between four filters used in the 2-D QMF bank in DWT.

REFERENCES

[1] M. Sifuzzaman1, M.R. Islam, and M.Z. Ali, "Application of Wavelet Transform and its Advantages Compared to Fourier Transform," *Journal of Physical Sciences*, Vol. 13, 2009, pp.121-134, ISSN: 0972-8791

[2] M. Komorkiewicz, T. Kryjak, K. Chuchacz-Kowalczyk, P. Skruch, M. Gorgoń, "FPGA based system for real-time structure from motion computation," *Proceedings of the DASIP 2015 Conference on Design & Architectures for Signal & Image Processing*, 23-25.09.2015, Kraków, Poland.

[3] M. Kłosowski, M. Wójcikowski, and A. Czyżewski, "Vision-based parking lot occupancy evaluation system using 2D separable discrete wavelet transform," *Bulletin of the Polish Academy of Sciences, Technical Sciences*, Vol. 63, No. 3, DOI: 10.1515/bpasts-2015-0066, 2015, pp. 569-573

[4] Zehang Sun, R. Miller, G. Bebis, D. DiMeo, "A Real-time Precrash Vehicle Detection System," *Sixth IEEE Conference: Applications of Computer Vision*, (WACV), 2002.

[5] K. Subramaniam, S.S. Dlay and F.C. Rind, "Vehicle Detection and Tracking using Wavelet Transforms," *Computers and Computational Engineering in control*, Singapore: World Scientific and Engineering Society Press, 1999, pp.335-340

[6] C.P. Bernard, "Discrete Wavelet Analysis for Fast Optic Flow Computation," *Applied and Comput. Harmonic Analysis*, Vol. 11, doi:10.1006/acha.2000.0341, 2001, pp.32-63

[7] Fuqing Zhuo, Peiqun Lin, Yumu Gu, "Vision-based Vehicle Detection in Real Traffic Environment Using Fast Wavelet Transform and Kalman Filter," *Advanced Materials Research*, pp.717-722, 2014

[8] J. Britto Pari, D. Vaithiyanathan, "An Efficient Multichannel FIR Filter Architecture for FPGA and ASIC Realizations", *International Journal of Applied Engineering Research*, Vol. 12, No. 10, pp. 2209–2220, 2017.

[9] Komorkiewicz, M., Turek, K., Skruch, P., Kryjak, T., Gorgoń, M., "FPGA-based hardware-in-the-loop environment using video injection concept for camera-based systems in automotive applications", *Design & Architectures for Signal & Image Processing* (DASIP), Book Series: J.F. Nezan (Ed.) *Conference on Design and Architectures for Signal and Image Processing*, pp.183-190, 2016, Rennes, France.

[10] A. Dąbrowski, *Multirate and Multiphase Switched-capacitor Circuits*, Chapman & Hall, ISBN-13: 978-0412724909, ISBN-10: 0412724901, London, 1997

[11] Dlugosz, R., Szulc, M., Kolasa, M., Skruch, *et al.*, "Design and optimization of hardware-efficient filters for active safety algorithms," *SAE International Journal of Passenger Cars – Electronic and Electrical Systems*, vol. 8, no. 1, 2015, doi: 10.4271/2015-01-0152.

Algorithm for Restructuring of Structurally Synthesized BDDs

L. Jürimägi and R. Ubar

Abstract – In this paper, we present a method for synthesizing Shared Structurally Synthesized BDDs (S^3BDD) for representing digital circuits with the goal of speeding up fault simulation. As the core of the method, we propose a novel algorithm to restructure the Structurally Synthesized BDDs (SSBDDs) [1], [2] into a form, which allows iterative embedding of SSBDDs, and compressing a set of SSBDDs into a single Shared SSBDD (S^3BDD) [3]. The target of such a compression of the S^3BDD model, representing a digital circuit, is to reduce the memory requirements of the model. Experiments with proposed algorithm, applied to three families of benchmark circuits, demonstrate feasibility of the proposed method, its high scalability, and considerable speedup in fault simulation compared to the initial SSBDD model.

I. INTRODUCTION

Fault simulation is one of the most critical tools in the design and test field of digital circuits, since it is needed for solving many test and reliability related tasks, such as verification, test pattern generation, fault coverage measurement, fault diagnosis, dependability evaluation etc. The efficiency of fault simulation depends directly on the number of faults to be considered, and especially, on the data structures used in simulation algorithms and tools.

Binary Decision Diagrams (BDD) are the state-of-the-art data structure in VLSI CAD [4]. BDDs were introduced first, for logic simulation in [5], and thereafter for test generation in [6], [7]. In 1986, Bryant proposed Reduced Ordered BDDs (ROBDDs) [8]. He showed simplicity of manipulation and proved the model canonicity, that made BDDs one of the most popular representations of Boolean functions [4], [9], [10]. However, the classical BDDs are not well suited as a model for use in design and test tasks, which essentially depend on structural information about the circuit.

In [1], [2], Structurally Synthesized BDDs (SSBDDs), were elaborated with the main goal to represent both, the functions and also the structural characteristics of digital circuits. Additionally, the size of the SSBDD model is linear in respect to the circuit size [11], whereas for BDDs it can be exponential [12]. In SSBDDs a subset of faults is related to each node. Hence, the number of nodes characterizes the size of the total fault set in related circuit.

L. Jürimägi and R. Ubar are with the Department of Computer Systems, Tallinn University of Technology, Akadeemia tee 15A, 12618 Tallinn, Estonia, E-mails: raiub@pld.ttu.ee, lembit.jyrimagi@gmail.com

In [13], an extension of the SSBDD model in the form of Shared SSBDDs (S^3BDDs) was introduced, which allowed significant reduction of the model size measured in number of nodes. The direct result of such compression is further reduction of the representative set of faults, and hence, increase of fault simulation speed. In [3], a formula for calculating the lower bounds was developed for the size of S^3BDDs. However, so far, no practical method has been developed, which allows achieving this bound.

In this paper, we investigate several properties of S^3BDDs, and present a novel method for equivalent transformations of SSBDDs with the goal of synthesizing S^3BDDs with the size equal to its lower bound.

Unlike a lot of optimization methods of traditional BDDs, where the target is mainly minimization of the number of nodes [14]-[18], in the case of SSBDDs, the number of nodes must remain invariant to preserve the exact mapping between SSBDDs and the related gate-level circuit. The goal of transformations of SSBDDs, representing the circuit, is to reconfigure the topologies of SSBDDs, to allow embedding a subset of related SSBDD network into a single S^3BDD. The end result of such reconfiguration of the model, is minimization of the representative fault set in the given circuit, to be used as the basis of fault simulation.

The paper is organized as follows. In Section 2 we present the concept of S^3BDDs and explain the necessity of restructuring SSBDDs. Section 3 presents the algorithm of restructuring and proposes a procedure for converting SSBDD model to a S^3BDD model with number of nodes equal to the theoretical lower bound. Section 4 demonstrates the procedure with an example. In Section 5 we present experimental results when applying the proposed algorithm to various benchmark circuits and Section 6 concludes the paper.

II. OVERVIEW OF S^3BDDs

SSBDDs are constructed by converting the gate-level netlist of a circuit into a collection of SSBDD graphs, each of which representing a Fan-out-Free Region (FFR) of the circuit [1]. The construction process represents an iterative super-positioning procedure, where a node of the current SSBDD, representing signal x in the circuit, is substituted by another SSBDD which represents a gate with output x. The construction starts from gates at outputs, (and gates with fan-outs) and moves towards inputs. This process continues for each SSBDD until fan-outs (or inputs if there are no fan-outs along the path from inputs to output).

As the result, an SSBDD model of a circuit with n FFRs represents a network of n SSBDD graphs. For S³BDDs, the super-positioning procedure continues beyond fan-outs by joint sharing the graphs.

Fig.1 presents an example where the fan-out of signal x_3 has segmented the circuit into three FFRs. The SSBDD model will consist of three SSBDD graphs with a total of 9 nodes. When constructing the S³BDD for the circuit, we continue the super-positioning procedure for both SSBDDs $G(y_1)$ and $G(y_2)$ beyond the fan-out node x_3 by substituting the nodes x_3 in these graphs with SSBDD $G(x_3)$. To avoid explosion of the number of nodes in the final model, the graph $G(x_3)$ is shared for both graphs $G(y_1)$ and $G(y_2)$.

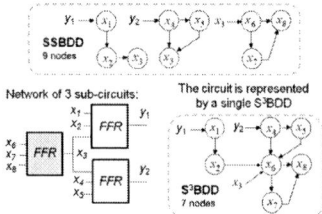

Fig. 1. Transformation of a set of SSBDDs into a single S³BDD

Note, differently from traditional BDDs, in SSBDDs, there always exists a single Hamiltonian path through all nodes, meaning that the set of nodes in the SSBDD is strongly ordered. The ranking of nodes is the prerequisite for organizing parallel simulation on SSBDDs [3].

It can be shown that, the set of nodes in the S³BDD can also be presented as the linear ordered set consisting of the Hamiltonian paths related to the original SSBDDs. An example of ordering the nodes of the S³BDD in Fig.1 is presented in Fig.2.

Fig. 2. Connected full Hamiltonian path for the S³BDD model

In Fig.1 we see that the new S³BDD model as a super-graph consists of only 7 nodes compared to the 9 nodes in the initial SSBDD model, which consists of 3 graphs. Continuing the super-positioning beyond fan-outs without sharing the sub-graph $G(x_3)$, the two graphs $G(y_1)$ and $G(y_2)$ would consist of 12 nodes.

Note that using Boolean expressions it is not possible to achieve a similar compression of the model, compared to the S³BDD approach. The Boolean "equivalent" of the S³BDD can be imagined as shown in Fig.3.

$$y_1 = x_1 \lor x_2 \land x_3$$
$$y_2 = x_4 \land x_5 \lor x_3$$
$$x_3 = (x_6 \lor x_7) \land x_8$$

$$y_1 = x_1 \lor x_2 \land ((x_6 \lor x_7) \land x_8)$$
$$y_2 = x_4 \land x_5 \lor$$

Fig. 3. Boolean expressions related to SSBDDs in Fig.1

It is easy to see that the possibility of constructing the S³BDD as shown in Fig.1, was only possible thanks to the node x_3 being the last node in graphs $G(y_1)$ and $G(y_2)$. In general, such a coincident, where a desired node happens to be the last node of an SSBDD graph, is unlikely. Therefore, the SSBDDs may need to go through suitable equivalent structural transformations, so that the desired node is moved to the position of the end node.

The circuit in Fig.1 is a very basic one with just one fan-out. In case of many embedded fan-outs and a lot of re-convergent fan-outs, a problem arises where there can be several candidates for the end node in the graph.

In [3] a method was proposed for systematic creation of the Fan-out Topology Graph (FTG) for the FFR network of the given circuit for planning the super-positioning procedure of SSBDDs.

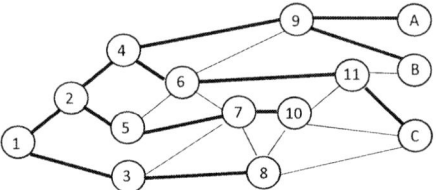

Fig. 4. Example of the Fan-out Topology Graph [3]

Consider as an example an FTG in Fig.4. Nodes of the graph represent single-output FFRs, and respectively, the SSBDDs of the given circuit. The FFRs may have additional input signals without fan-outs, and only the fan-out connections are high-lighted. The bold lines point out which node should be the last in each graph. For example, the SSBDD 11 should be linked to the last node of the SSBDD C. On the other hand, the SSBDD 9 should be linked to the last nodes of both SSBDDs A and B.

This algorithm of constructing the FTG and finding the links between SSBDDs (for colouring the related edges in FTG) is used for the experiments in section 5.

III. SSBDD RESTRUCTURING ALGORITHM

Let us use following definitions. We label the *edges* in SSBDD as follows: the right directed edges are called *1-edges* (picked when the node variable is 1), and the edges pointing down are called *0-edges* (when the node variable is 0).

A *path* $l = (m_1, m_k)$ in SSBDD is a connected set of nodes, starting at the node m_1 and ending with m_k. A path is *homogeneous* if all edges of the path have the same direction. The homogeneous paths can be either *horizontal* or *vertical*.

A homogeneous path $b(m)$ backwards from a node m is called a *borderline* controlled by m. E.g. in Fig.5, the path (x_{10}, x_6, x_5, x_2) is borderline $b(x_{10})$ controlled by x_{10}. We call the node where two (vertical and horizontal) borderlines cross as *crossroad*.

By borderlines we can mark *sub-graphs* in SSBDD like $G_y(m_R, m_E)$ where m_R is the root, and m_E is the end-node of the sub-graph. We call m^0 and m^1 the neighbors of the node m in 0-direction and 1-direction respectively.

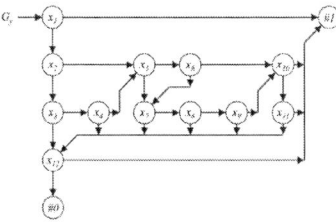

Fig. 5. An SSBDD for illustration of paths and sub-graphs

Using the properties of SSBDDs [16], we can easily prove the following Lemmas and Theorem 1.

Lemma 1. A sub-graph $G_y(m_R, m_E) \subseteq G_y$ is true, if m_R is the crossroad of two borderlines $b(m^0_E)$ and $b(m^1_E)$.

A true sub-graph in SSBDD represents a sub-function of the Boolean expression represented by the SSBDD. For example, $G^*_y(x_2, x_9) \subset G_y$ with crossroad $x_2 = b(x_{10}) \cap b(x_{12})$ is true.

Lemma 2. A horizontal homogeneous path $l = (m_1, m_k)$ with k-inputs represents AND-gate, if 0-edges of all the nodes are entering the same node. A vertical homogeneous path represents OR-gate, if 1-edges of all the nodes are entering the same node.

Lemma 3. SSBDD which contains a homogeneous path $l = (m_1, m_k)$ where the off-edges from this path are entering the same node, remains equivalent if any two nodes will exchange the places in this path.

Let us call two homogeneously connected subgraphs G_y^*, and G_y^{**}, which have a joint node m with entering edges from both graphs, as *neighboring subgraphs*.

Theorem 1. SSBDD which contains two neighboring sub-graphs $G_y(m_R, m_E)$ and $G_y(m_{R*}, m_{E*})$, remains equivalent if the sub-graphs will exchange the positions.

Based on Theorem 1, Procedure 1 was constructed, which represents the algorithm of restructuring SSBDDs.

Procedure 1.

(1) Specify the target node m.
(2) Determine a true sub-graph $G_y(m_R, m_E)$ for m, so that $m_E = m$, and m_R will be found as the crossroad of two borderlines $m_R = b(m^0_E) \cap b(m^1_E)$, according to Lemma 3.
(3) Find the neighboring sub-graph $G_y(m_{R*}, m_{E*})$, so that $m_{R*} = m_E + 1$.
(4) Exchange the position of the subgraphs $G_y(m_R, m_E)$, and $G_y(m_{R*}, m_{E*})$ to move the node m towards end of the SSBDD.
(5) Repeat the procedure till the node m has become the end-node of the SSBDD.

IV. EXAMPLE OF RESTRUCTURING ALGORITHM

In the computer memory SSBDD can be represented as a simple table where the nodes are ordered according to their Hamiltonian path. Identifying the first block with the crossroad method would require several passes of checking the various borderlines. Instead it is faster to check whether a node is located outside the crossroad boundary.

Each node is numbered and has two neighbours, right and downwards. Starting from the node m we can move towards the beginning of the graph and check the neighbours of each node. When a node has a neighbour that is positioned later in the Hamiltonian path than m^0_E or m^1_E then we have crossed the crossroad and the previous node is the root node in the block $G_y(m_R, m_E)$.

But the first node beyond the crossroad also provides an additional boundary for the second block. Let's note that boundary as m_B. For the second block, the node beyond the last node of that block is $m_{E*} < \min(m_B, m^0_E)$ for horizontal blocks or $m_{E*} < \min(m_B, m^1_E)$ for vertical blocks.

For example, in Fig5 the nodes x_1 to x_{12} have been numbered according to their position in the Hamiltonian path. The terminal nodes #1 and #0 are at position 13.

Let us pick x_9 as m. The node m^0_E would be x_{12}, and m^1_E would be x_{10}. The node across the crossroad would be x_1, because the right neighbour of x_1 is #1 with numbering 13 which is larger than 10. Node m_B is #1. Block A would be $G_y(x_2, x_9)$ and block B would be $G_y(x_{10}, x_{11})$ as $x_{11} < \min$ (#1, x_{12}).

Let's pick x_8 as m. Nodes m^0_E would still be x_{12} while m^1_E would be x_9. The first node across the boundary is x_6 as its rightward neighbour $x_{10} > x_9$. Node m_B is x_{10}. Block A would be $G_y(x_7, x_8)$ and block B would be $G_y(x_9)$ since $x_9 < \min(x_{10}, x_{12})$.

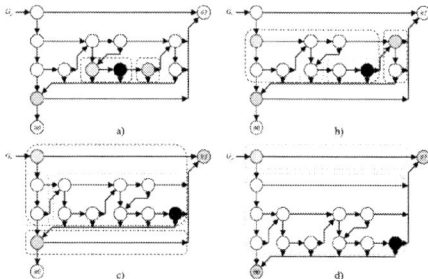

Fig. 6. Example of moving selected node to the end of SSBDD

In Fig.6, an example is depicted, how the black node m in the SSBDD in Fig.5 is moved in 4 iterations step-by-step to the end of the graph using Procedure 1. The grey nodes represent the neighbours m^0_E and m^1_E that the borderlines are calculated from.

V. EXPERIMENTAL RESULTS

Experiments were carried out with Intel Core i7-7700 3.6 GHz, 16 GB RAM, using ISCAS'85, ISCAS'89 and ITC'99 benchmark circuits. The goal of the experiments was to convert the benchmark circuits from SSBDD format to S^3BDD format using Procedure 1 and the algorithm developed in [3].

All 72 circuits were processed, 63321 graphs were selected to be reconfigured and it took 84418 transformations to achieve it. The majority of graphs required just one transformation.

978-1-7281-3420-8/19 $31.00 © 2019 IEEE

TABLE I.
EXPERIMENTAL RESULTS

Benchmarks		# wires	# gates	# nodes		# graphs			# transformations		Total
Family	Circuit			SSBDD	S³BDD	SSBDD	Transf.	Av. size	total	per graphs	time, ms
ISCAS85	c1355	1097	514	809	551	291	78	3.3	79	1.01	0.27
	c3540	2784	1446	1648	1316	378	192	6.6	256	1.33	1.16
	c6288	4864	2416	3872	2416	1488	1126	2.8	1561	1.39	3.83
	c7552	5795	2977	3552	2715	920	495	4.9	781	1.58	2.52
ISCAS89	s386	372	158	227	200	39	14	14,3	35	2,50	0,25
	s1494	1432	647	847	760	101	34	11,6	64	1,88	0,43
	s38417	34831	21736	16160	11246	6311	2042	5,1	3016	1,48	10,44
	s38584	36173	19484	19179	14941	5676	2286	6,5	4716	2,06	17,92
ITC99	b17	64711	30777	40665	32425	9657	3550	8,4	4297	1,21	23,13
	b18	222499	111241	138989	107917	34409	11276	8,5	13933	1,24	76,37
	b19	448502	224624	280352	217210	69767	23032	8,3	28321	1,23	152,56
	b20	39247	19682	23688	19053	5157	2118	6,5	3015	1,42	13,01

Results are shown in Table I. Columns 3-4 show sizes of circuits and columns 5-7 the sizes of SSBDDs. Column 8 shows the number of graphs, that needed transformations. Column 9 depicts the average numbers of nodes in transformed graphs. Columns 10 and 11 show the total and average number of transformations per graph. The last column shows the total processing time.

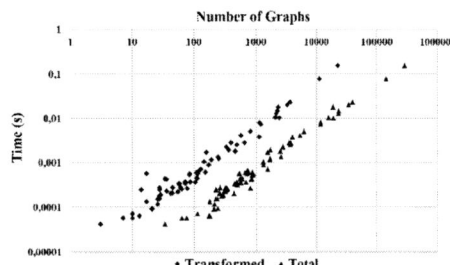

Fig. 7. Transformation times per circuit

Fig.7 shows the time cost for reconfiguring each circuit based on the number of necessary transformations and also the total number of SSBDD nodes in those circuits. As it can be seen, the time cost of transformations is linear and parallel to the circuit size.

VI. CONCLUSIONS

We have presented in this paper the theory of equivalent transformations of SSBDDs and implemented an algorithm for reordering the nodes in SSBDDs, enabling construction of S³BDDs with the size equal to theoretical lower bound. The time cost of the procedure is linear and hence, the method is well scalable.

ACKNOWLEDGEMENTS.

The work has been supported by EU HORIZON 2020 RIA project IMMORTAL, by European Structural Funds and by Estonian IT Academy program Study IT in Estonia.

REFERENCES

[1] R.Ubar, "Test Synthesis with Alternative Graphs", *IEEE Design & Test of Computers*, 1996, pp. 48-59.

[2] R.Ubar, J.Raik, H.-T.Vierhaus, "Design and Test Technology for Dependable Systems-on-Chip", *IGI Global*, 2011, p. 550.

[3] R.Ubar, et al, "Shared Structurally Synthesized BDDs for Speeding Up Fault Simulation in Digital Circuits", *NorCas*, 2015.

[4] M.Karpovsky, et al, *Spectral Logic and its App-lications for the Design of Digital Devices*, Wiley & Sons, 2008.

[5] C.Y. Lee, "Representation of Switching Circuits by Binary Decision Diagrams", *Bell Sys. Tech. J.*, 1959, v.38, No7.

[6] R.Ubar, "Test Generation for Digital Circuits Using Alternative Graphs", *Proc. Tallinn Technical University*, 1976, No409, Tallinn.

[7] S.Akers, "BDDs", *IEEE Trans. on Comp.*, 1978, Vol.27.

[8] R.Bryant, "Graph-based algorithms for Boolean function manipulation", *IEEE Trans. on Comp.*, 1986, pp. 677-691.

[9] T.Sasao, M.Fujita (eds), *Representations of Discrete Functions*, Kluwer Academic Publishers, 1996.

[10] R.Drechsler, B.Becker, *Binary Decision Diagrams*, Kluwer Academic Publishers, 1998.

[11] A.Jutman, J.Raik, R.Ubar, "SSBDDs: Advantageous Model and Efficient Algorithms for Digital Circuit Modeling, Simulation & Test", *5th WS on Boolean Problems. Freiberg*, 2002, pp.157-166.

[12] H.-T.Liaw, et al, "On the OBDD-representation of general Boolean functions", *IEEE Trans. on Comp.*,C-41, 6, 1992.

[13] R.Ubar, D.Mironov, J.Raik, A.Jutman, "Structural Fault Collapsing by Superposition of BDDs for Test Generation", *ISQED*, 2010.

[14] G.Fey, J.Shi, R.Drechsler, "BDD circuit optimization for path delay fault testability", *DSD*, 2004.

[15] J.Butler, T.Sasao, M.Matsuura, "Average Path Length of BDDs", *IEEE Trans. on Comp*, 2005, Vol.54, 9, 1041 - 1053.

[16] P.Prasad et al, "Minim. average path length in BDDs based on static variable ordering", *48th Midwest Symp. Circ. and Systems*, 2005.

[17] S.Chaudhury, A.Dutta, "Genetic algorithm based variable ordering of BDDs for multi-level logic optimization with area-power trade-offs", *2010 17th Int. Conf. on Electronics, Circuits & Systems*, 2010.

[18] M.Bansal, A.Agarwal, "Genetic algorithm for ordering and reduction of BDDs for MIMO circuits", *3rd Int. Conf. on Innovative Computing Technology - INTECH*, 2013, pp. 411-414.

Design of Non-Metastable SRAM Cells in 28 nm CMOS Technology

F. Crescioli, L. Frontini, V. Liberali, and A. Stabile

Abstract—**This paper presents the design of an SRAM cell in 28 nm, specifically designed to avoid metastability at start-up. Metastable operation is avoided by unbalancing the size of transistors. Extensive simulations have confirmed that the probability of metastable operation is greatly reduced.**

I. INTRODUCTION

The upgrade of the Large Hadron Collider (LHC) will improve the accelerator performance. Since the experiments will produce a large volume of data and only a small fraction can be stored for subsequent offline processing, a trigger system is required to select the most relevant information for long term storage. The trigger system exploits the massive parallelism of Associative Memory (AM) circuits, which store millions of pre-calculated patterns and compare them in parallel with the data coming from the detectors.

For the 2020 upgrade, the AM06 chip, a CMOS integrated circuit aiming at reducing power consumption while maintaining a high level of efficiency, has been designed and fabricated in a 65 nm CMOS technology [1]. Each AM06 chip can store 18.9 Mbit, corresponding to 2^{17} sets of track coordinates.

The next generation of associative memory circuits is being designed in 28 nm CMOS, for next ATLAS upgrade. A small 28 nm prototype, the AM07 chip, has been designed and characterized to explore the potential of technology scaling. Compared to the 65 nm node, the transistor density increases roughly by a factor of 5 [2]. However, during the characterization of AM07 prototypes, a high current intensity was observed at the start-up. The current was reduced after writing the memory, thus indicating a possible problem related to the metastability of SRAM cells.

To avoid metastability, transistor sizes in SRAM cell have been changed, and the design has been verified through extensive Monte Carlo simulations.

F. Crescioli is with the LPHNE, Paris, France, E-mail: francesco.crescioli@lpnhe.in2p3.fr

L. Frontini is with the INFN − Sezione di Milano, Milano, Italy, E-mail: luca.frontini@mi.infn.it

V. Liberali and A. Stabile are with the Department of Physics, Università degli Studi di Milano, Milano, Italy, E-mail: valentino.liberali@unimi.it, and alberto.stabile@unimi.it

II. CAM CELL IN 28 nm TECHNOLOGY

A new Content-Addressable Memory (CAM) cell was designed for the AM07 chip [3]. As the CAM is a massively parallel device, which compares the input data with the information stored in all memory cells in parallel, the dynamic power consumption due to the propagation of signals through the internal busses can be quite high.

In the AM07, the dynamic power consumption has been reduced in two ways: (i) by minimizing the interconnection parasitics of input data busses in the memory array; and (ii) by reducing the switching activity of cells, with a "kill" signal that inhibits further comparisons on one coordinate set when one of the data bit is not matching. The average energy required is less than 0.7 fJ/bit per comparison.

Figure 1 shows the schematic diagram of the new CAM cell. Each cell contains 16 transistors (7 PMOS and 9 NMOS): 6 transistors for the SRAM and 10 transistors for the combinational logic.

A. SRAM metastability

During AM07 tests, we noticed that the current consumption at power-on is higher than after writing data into the memory array. The measured current is (27 ± 6) mA at power-on, and (4.40 ± 0.01) mA after writing the 16 Ki pattern[1] memory array. Figure 2 shows the distribution of 100 current measurements at power-on of a single chip.

This behaviour is typical of SRAM memories that exhibit metastability. Indeed, if the output and the input of the two inverters of the SRAM cell are at the same voltage, the cell operates in an unstable equilibrium point, and an SRAM in that operating point consumes much more power than SRAM in a definite state (0 or 1), because all the transistors of the cell are turned on [4].

From circuit-level simulations, the current of a single SRAM cell in metastable condition is 240 nA. Since the AM07 chip contains about 2.4 million SRAM cells, from current measurements we estimate that a number

[1]1 Ki = 1204; 1 pattern = 8 words; 1 word = 18 bit.

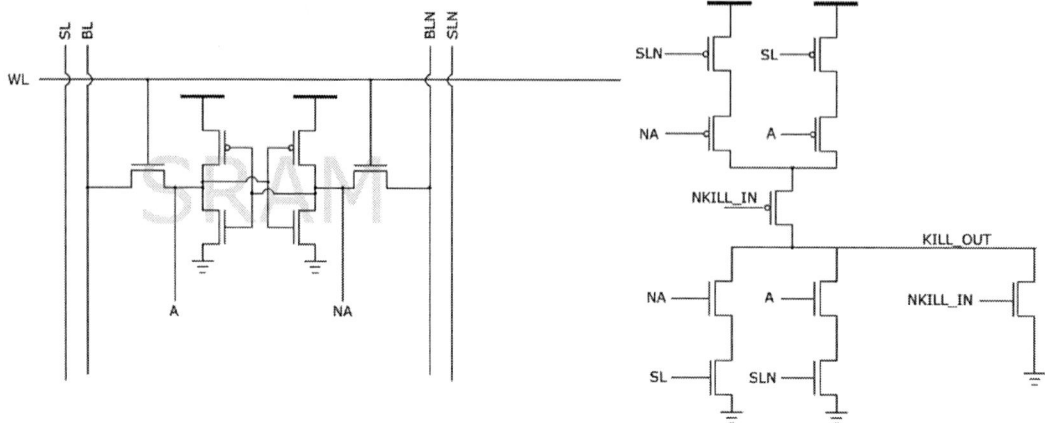

Fig. 1. Schematic diagram of the CAM cell.

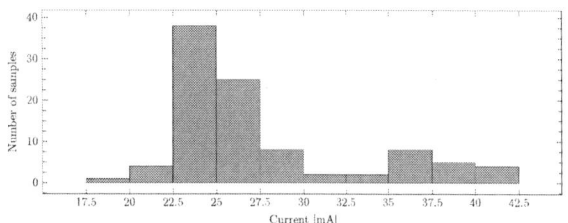

Fig. 2. Histogram of the measurements of current due to SRAM metastability at power-on. The histogram shows the distribution of 100 measurements of a single chip.

of cells ranging from 4 % to 6 % remains in metastable condition after power-on. The number of metastable cells changes every time the chip is turned on; after power-on, the current consumption remains at a constant value, until the memory is written.

The final chip forecast for the future upgrade (AM09) will contain 3×128 Ki patterns. With the same CAM cell of AM07, the power consumption of AM09 at startup due to metastability would be in the range of $(0.5$ to $0.8)$ A. Although this value of current consumption can be tolerated, a reduction of the probability to have a metastable operating point at power-on would be highly desirable.

B. Butterfly diagrams

Butterfly diagrams are commonly used to study several properties of an SRAM cell [5], [6], [7], including metastability.

Figure 3 shows an example of butterfly diagram, obtained by drawing the characteristics of the two SRAM inverters on the same plot.

An SRAM is prone to metastability when the crossing point between the two characteristics is located on

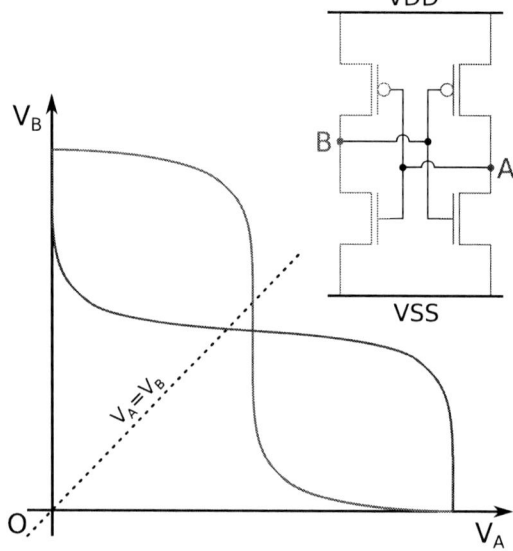

Fig. 3. Example of butterfly diagram

the bisector of the quadrant (where $V_A = V_B$). In this case, there is a non-negligible probability that the cell, during the power-on transient, remains in an 'equilibrium' state, instead of randomly selecting an operating point corresponding to a bit (0 or 1).

Figure 4 shows the Monte Carlo simulation of a completely symmetric SRAM with minimum size transistors ($W = 100$ nm, $L = 30$ nm for both PMOS and NMOS devices, in the layout; during fabrication the layout design is shrunk by a linear factor of 0.9). The parameters of the Monte Carlo simulation account for both global variations (i.e., fluctuation of the parameters in the fabrication process, which affect all the devices in the same wafer), and local variations (i.e.,

978-1-7281-3420-8/19 $31.00 © 2019 IEEE   244

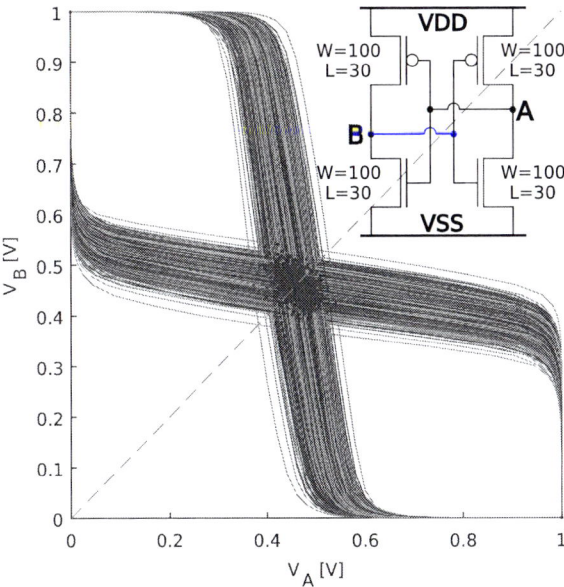

Fig. 4. Monte Carlo butterfly diagrams of the symmetric SRAM cell, with crossing points in blue (width and length of MOS transistors in nm).

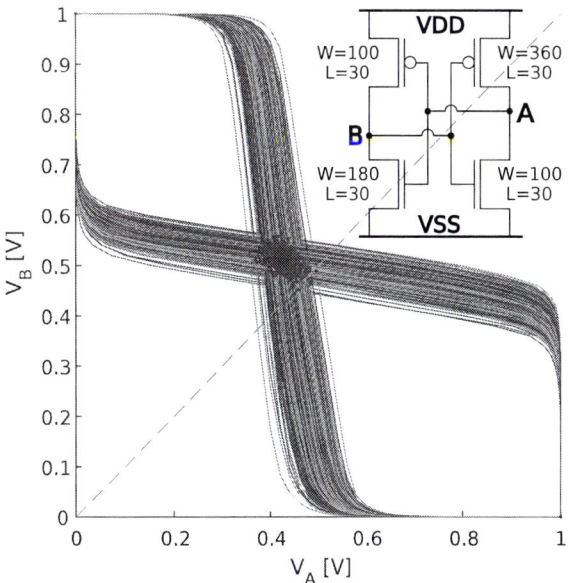

Fig. 5. Monte Carlo butterfly diagrams of the asymmetric SRAM cell, with crossing points in blue (width and length of MOS transistors in nm).

random fluctuation of parameters between different coordinates on the same wafer). The crossing points are placed across the bisector, indicating a possible metastable behavior.

III. NON-METASTABLE SRAM DESIGN

To reduce the number of SRAM cells that exhibit metastability at power-on, we decided to unbalance the SRAM cell, by enlarging the width of one PMOS transistor from 100 nm to 360 nm and the width of the opposite NMOS transistor from 100 nm to 180 nm. PMOS transistor size has been increased more than NMOS size, because there are less P-type devices and therefore there is more room for enlarging PMOS transistors in the layout.

Figure 5 shows the Monte Carlo simulation results of the asymmetric SRAM, which demonstrates that the number of points at or near the bisector is greatly reduced. Therefore, we expect that metastability will be a rare event, much less probable than in the symmetric configuration. Figure 6 compares the distribution of distances between crossing points and the bisector, for symmetric and asymmetric SRAM cells. The distribution histograms show that the asymmetric SRAM has a negligible probability to show a metastable behaviour.

Figure 7 shows the layout of a double CAM cell: (a) with minimum-size transistors, and (b) with larger transistors to reduce metastability. In Figure 7(b),

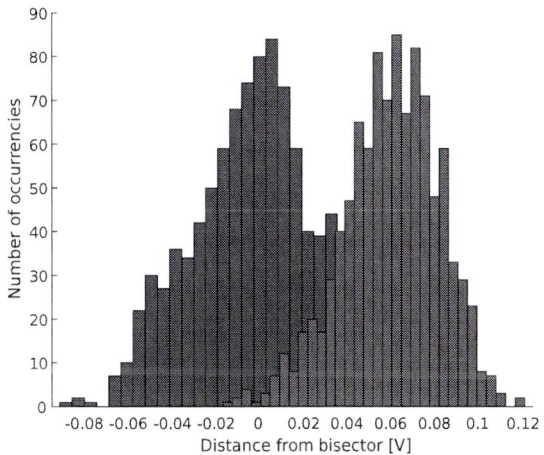

Fig. 6. Distribution of the distance of the crossing points from the bisector. Blue: symmetric SRAM cell (histogram centered at 0 V). Red: asymmetric SRAM cell (histogram centered at 0.06 V).

larger MOS transistors have been achieved by connecting elementary transistors in parallel.

The size of transistors in the writing circuit have been modified accordingly, to ensure that the memory cell is written correctly in all worst-case corners.

Each of the layout designs in Figure 7 contain two CAM cells which share the same input data signals, in order to reduce interconnection length, as explained in Sect. II.

978-1-7281-3420-8/19 $31.00 © 2019 IEEE

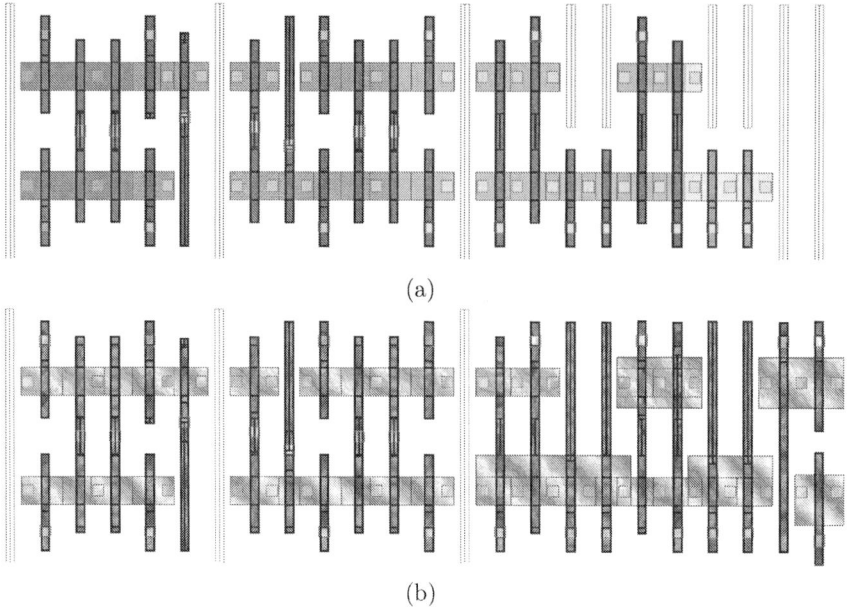

(a)

(b)

Fig. 7. Layout of the two versions of the double CAM cell: (a) symmetric version; (b) asymmetric version with larger transistors.

The silicon area of the layout in Figure 7(a) is $1.10\,\mu m^2$ per cell, while the area of the layout in Figure 7(b) is $1.16\,\mu m^2$ per cell, with an area increase of 5.4%.

IV. CONCLUSION

An SRAM cell that mitigates metastable operation at power-on has been designed. Thanks to the unbalanced size of the inverters in the cell, the probability of metastable operating point is greatly reduced, and the overall current consumption at power-on is limited. The proposed circuit is suitable for large CAM ICs dedicated to pattern recognition.

REFERENCES

[1] A. Annovi, M. Beretta, G. Calderini, F. Crescioli, L. Frontini, V. Liberali, S. Shojaii, and A. Stabile, "AM06: the Associative Memory chip for the Fast TracKer in the upgraded AT-LAS detector," *J. Instrum.*, vol. 12, pp. C04013.1–10, Apr. 2017.

[2] A. Annovi, G. Calderini, S. Capra, B. Checcucci, F. Crescioli, F. De Canio, G. Fedi, L. Frontini, M. Garci, C. Gentsos, T. Kubota, V. Liberali, F. Palla, J. Shojaii, C. L. Sotiropoulou, A. Stabile, G. Traversi, and S. Viret, "Characterization of an associative memory chip in 28 nm CMOS technology," in *IEEE Int. Symp. on Circ. and Syst. (ISCAS)*, Florence, Italy, May 2018.

[3] A. Annovi, L. Frontini, V. Liberali, and A. Stabile, "Design and characterization of new content addressable memory cells," in *IEEE Int. Symp. on Circ. and Syst. (ISCAS)*, Florence, Italy, May 2018.

[4] K. Agarwal and S. Nassif, "Statistical analysis of SRAM cell stability," in *Design Automation Conf. (DAC)*, San Francisco, CA, USA, July 2006, pp. 57–62.

[5] M. Wieckowski, D. Sylvester, D. Blaauw, V. Chandra, S. Idgunji, C. Pietrzyk, and R. Aitken, "A black box method for stability analysis of arbitrary SRAM cell structures," in *Design, Automation and Test in Europe Conference (DATE)*, March 2010, pp. 795–800.

[6] J. U. Horstmann, H. W. Eichel, and R. L. Coates, "Metastability behavior of CMOS ASIC flip-flops in theory and test," *IEEE J. Solid-State Circuits*, vol. 24, no. 1, pp. 146–157, Feb. 1989.

[7] J. Lohstroh, "Static and dynamic noise margins of logic circuits," *IEEE J. Solid-State Circuits*, vol. 14, no. 3, pp. 591–598, June 1979.

978-1-7281-3420-8/19 $31.00 © 2019 IEEE

A Highly Parametrizable Chisel HCL Generator of Single-Path Delay Feedback FFT Processors

V. M. Milovanović and M. L. Petrović

Abstract—A configurable fast Fourier transform (FFT) engines and their inverse counterparts are indispensable in modern wireless communication and radar systems. The FFT processors are usually customized per use case. Therefore, a design generator of single-path delay feedback type of an FFT processor, that permits continuous input and output data streaming has been captured inside Chisel hardware construction language. It supports a wide range of parameter settings, like input data and twiddle factor widths, FFT sizes and number of stages, three radices, different scaling and rounding methods after each butterfly or dragonfly stage, among others, thus enabling an agile design space exploration. A comparison with commercially available FFTs which were specifically tailored for the particular FPGA platforms proves that FFT generator instances can be both performance- and resource-competitive with state of the art designs.

I. INTRODUCTION

Spectral analysis finds application in both wired and wireless communications, but it is also the essential step in many other spheres like radar signal processing. The Fourier transform is at the heart of spectral analysis as it is used to decompose arbitrary function of time into its constituent frequencies. In contemporary electronic systems, that range from general-purpose computers to dedicated hardware accelerators, this frequency domain representation is actually obtained numerically exploiting the discrete Fourier transform (DFT). In practice, all DFT implementations usually employ more efficient fast Fourier transform (FFT) algorithms that greatly reduce computational complexity of the original.

While the software FFT realizations sequentially step through a single instruction at a time, the hardware implementations often exploit parallelism to achieve data throughputs that are orders of magnitude higher. Commonly, a real-time FFT data processing mandates hardware realizations simply because clock speeds could otherwise become prohibitively high. Even though a direct parallel form of the FFT is straightforward for implementation its resource utilization is immense. Pipelined in-place computation alternatives offer more efficient hardware usage at the cost of increased latency, while still offering streaming input and output data behavior. In addition, pipeline FFT processors are highly regular and can hence be easily scaled and parameterized inside hardware description languages like Verilog or VHDL.

V. Milovanović is with the Faculty of Engineering, University of Kragujevac, Sestre Janjić 6, Kragujevac, (e-mail: vlada@kg.ac.rs)
M. Petrović is with NovelIC Microsystems, Veljka Dugoševića 54/B5, Belgrade, Serbia, (e-mail: marija.petrovic@novelic.com).

In spite of its evident omnipresence, FFT engines are not only optimized for a certain application, but per use cases too. Therefore, a wide variety of similar, but sufficiently distinct FFT cores are needed in order to deliver a target optimum. A classical design methodology would result in high effort which increases both development time and the non-recurring engineering cost.

This paper presents one example of an agile [1] digital design methodology applied to an FFT processor with streaming input and output (I/O) data, leveraging open source code reuse through generator-based design and testing. The methodology supports a series of pipelined FFT engine designs which are applicable to wide variety of problems. Highly parameterized digital hardware generators and testers let users customize their implementations and quickly explore FFT design spaces thus simultaneously cutting cost and the execution time.

To capture a batch of viable designer decisions a relatively recent hardware construction language (HCL), Chisel [2] has been used to create the proposed FFT engine generator. Being embedded in a modern object-oriented and functional language such as Scala, it offers much more parametrization possibilities than a regular HDL. A collection of pipelined FFT processors based on the single-path delay feedback architecture [3] with plethora of parametrization options have been caught by the presented generator. Various generator instances have been mapped onto a modern Field-Programmable Gate Array (FPGA), tested and mutually compared.

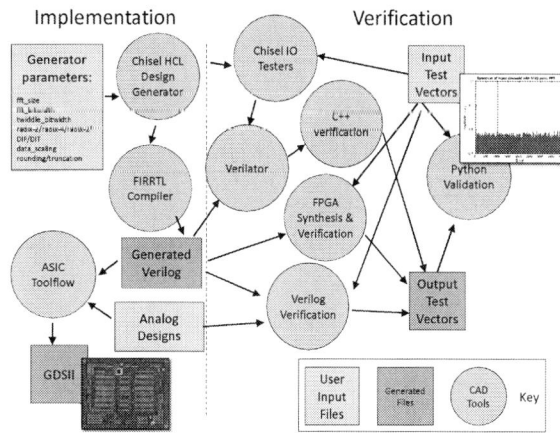

Fig. 1. Chisel HCL generator flow diagram, showing verification paths on the right and the implementation path on the left.

II. PIPELINED SDF FFT PROCESSOR TOPOLOGIES

The pipelined FFT is a conceptual compromise between area efficient memory-based time-multiplexed and direct-mapped fully parallel FFT architectures. It provides a balanced trade-off [4] between occupation of resources and the execution time. Two major types of pipelined FFT topologies are multi-path delay commutator (MDC) and Single-path Delay Feedback (SDF).

Since butterfly and dragonfly outputs share the same storage with the corresponding inputs, the SDF architecture is generally the preferred one as it requires less memory. The name is derived from the fact that computation outputs are stored in feedback shift registers.

Three most popular SDF schemes are shown in Fig. 2 in their simplified form in which additive butterflies and dragonflies have been separated from multipliers. Control signals have been omitted for clarity so that only signal-flow graph remains which clearly reveals spatially regular cascading delay feedback structure. Every arithmetic operation and all data are complex while the assumption that FFT size N is a power of four, applies.

The first two are fairly classical radix-2 and radix-4 FFT schemes and both require N-1 memory locations. While the latter uses only half of the multipliers it doubles the number of complex adders and has more complicated control. Finally, the radix-2^2 SDF [5] combines best of the two, namely it has the same amount of nontrivial multiplications at the same positions as radix-4, but the same butterfly structure of radix-2 algorithm.

III. SDF FFT CHISEL GENERATOR DESCRIPTION

The previously described SDF FFT topologies have been captured inside a Chisel 3 [6] HCL generator. The recently available open compiler Chisel 3 was chosen because it stimulates productivity through new programming language features available in Scala and the abundance of useful facilities for digital signal processing.

This enables a rapid design space exploration at the level of algorithms, architecture and implementation.

The latest Chisel generator flow for ASIC and FPGA implementation and verification is depicted in Fig. 1, it follows two paths that are basically self-explanatory.

The open-source SDF FFT generator [7] is sketched in Fig. 3 and is made available as free hardware library.

Powerful parametrization capabilities begin with the interface itself. Complex FFT input and output consist of the fixed-point real and imaginary data parts each with arbitrary total width and the number of fractional bits (or binary point location). Since Chisel's decoupled IO interface is used, additional ready and valid protocol signals are wrapped around inout data. These are used by the internal finite-state machine (FSM) to control the actual dataflow between the successive FFT stages.

A wide list of parameters further continues with the SDF FFT scheme (radix-2, radix-4 or radix-2^2) and also with the choice between decimation-in-time (DIT) or decimation-in-frequency (DIF) variants of each scheme.

A number of selected radix stages is also configurable and it directly governs the maximum number of samples that can be processed by the FFT instance, i.e., its size.

Besides, once the size is selected in compile-time, the FFT instance can process any smaller FFT size through run-time configurability via adequate control registers.

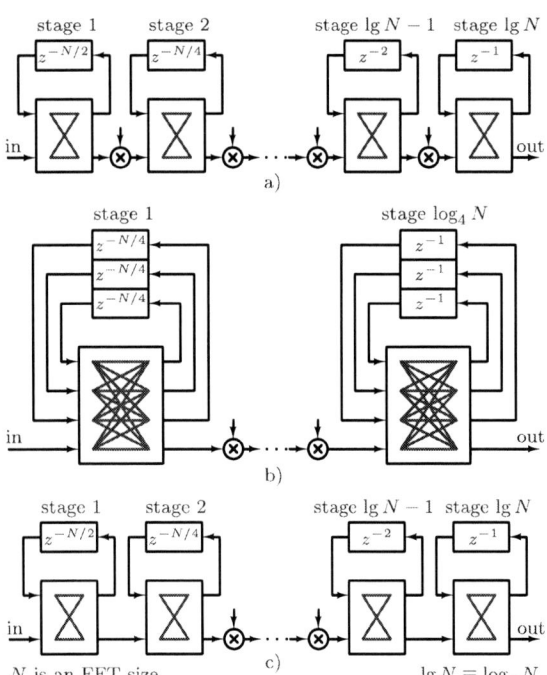

Fig. 2. Three different pipelined DIF Single-path Delay Feedback (SDF) FFT schemes: a) radix-2, b) radix-4, and c) radix-2^2.

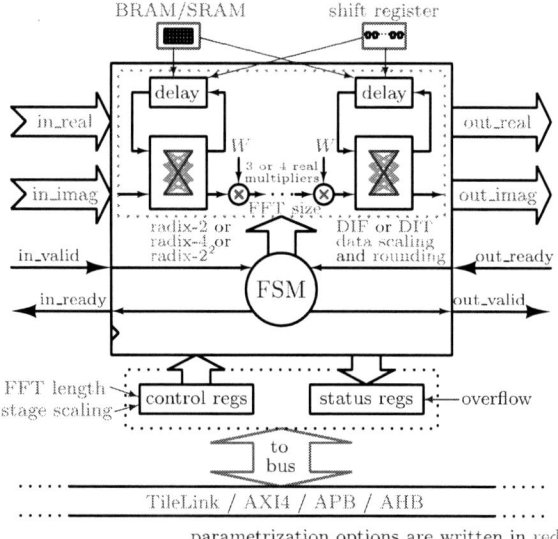

Fig. 3. Interface of the implemented SDF FFT Chisel 3 generator showing inout signals as well as control and status registers.

978-1-7281-3420-8/19 $31.00 © 2019 IEEE

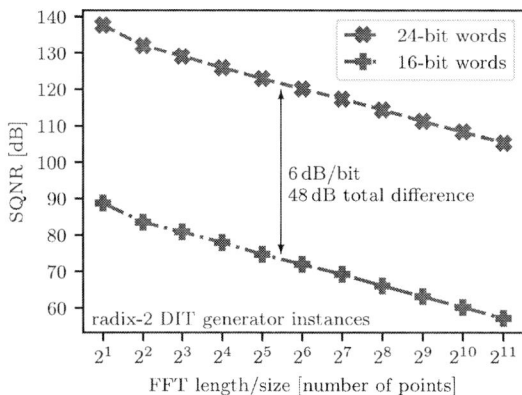

Fig. 4. Fixed-point signal-to-quantization noise ratio (vs floating point) as a function of FFT length for two different bitwidths.

Fig. 5. Fixed-point signal-to-quantization noise ratio (vs floating point) as a function of FFT size for several rounding methods.

Not only number of bits, i.e., word width, of the FFT input and output data is parametrized, but the so-called twiddle factor W bitwidth also. It will be demonstrated later that these parameters can have a profound effect on FFT's Signal-to-Quantization-Noise Ratio (SQNR).

A butterfly and dragonfly stages typically consist of add/subtract and twiddle multiplication operations. At the output of each add/subtract structure, the bitwidth would increase by a single bit. In order to handle this single-bit growth due to add/subtract operation, there is an additional data (un)scaling parameter which can either allow every growth to be carried all the way to the output or alternatively apply a user-defined scaling.

The former way maintains the highest precision, i.e., there is no loss in information, at the expense of more logic resources in the datapath spent to accommodate this bit growth, while the latter way makes a provision at the output of each butterfly/dragonfly add-subtract stage to scale the result back, by either dividing the output by two/four, thus round off the least significant bit (LSB), or by saturating the most significant bit (MSB).

A dedicated run-time configuration register is used to control this divide-by-two/four scaling operation at each stage, so that the user has full flexibility to control the signal level through the different butterfly stages. If the appropriate register bit is zero for a particular stage, then the output is saturated at the MSB, otherwise if it is one, the output is appropriately rounded at the LSB. The user can hence control the scaling at each stage.

There also exists a devoted read-only status register which indicates whether there was any clipping in any of the butterfly/dragonfly stages. This register is a sticky one that for convenience gets set whenever an overflow event occurs and remains in that state until it is cleared.

The whole SDF FFT is wrapped as generic DSP block in a diplomatic interface which is actually AXI4-Stream for inputs and outputs and memory-mapped (TileLink, AXI4, APB or AHB) for control and status registers.

Namely, a bus towards the memory-mapped configuration registers is also just a generator parameter which is accessible through `dsptools`[†], a free library of Chisel tools for Digital Signal Processing (DSP) that the proposed SDF FFT generator's context heavily relies on.

Precisely this DSP context additionally enables easily selectable rounding methods inside every FFT stage from a sequence of different options which include ceiling, floor, truncate, convergent rounding, among others.

Apart from previously explained parameters which were all feature-specific, there are only implementation-specific parameters that, even though do not have functional influence, are just as important because they allow fast, as well as area- and power- efficient instances.

The first parameter tailors complex multipliers which can either be implemented in a manner that actually exploits three real multipliers accompanied with five add/subtract structures or four real multipliers together with two add/subtract structures. While the former implementation case is less expensive in terms of resources the latter permits higher clock rates and faster designs.

The second parameter customizes the feedback data delay lines in each FFT stage which can be either implemented by shift registers or using Static RAM (SRAM) in ASICs or Block RAM (BRAM) in FPGAs. For larger FFT sizes (number of stages), this logic, if implemented through registers, quickly becomes dominant factor in resource utilization. To reduce power consumption and the total area, it is therefore advantageous to use single- or dual-port SRAM/BRAM for longer delays and shift registers only for shorter delay chains. Modern FPGA synthesis tools can even infer the optimal option for each individual stage based on its delay length/width.

The inverse FFT is also supported and it is computed through $1/N$ scaling and conjugating twiddle factors.

The whole generator core uses a single clock domain.

[†]Available: www.github.com/ucb-bar/dsptools

IV. Implementation and Experimental Testing

For comparison with floating point software FFT results, Scala Breeze numerical computing library is used.

This library is a powerful tool which can be used to create complex random inputs for testing and verification purposes of generator systems. Linked with ACED (A Hardware library for Generating DSP systems) [8] it greatly simplifies the whole process of generator development and accelerated design space exploration.

For example, important system metric such as SQNR is used to determine the system's sensitivity to different Chisel DSP generator parameters. A fixed-point SQNR (versus Scala's Breeze floating-point as the golden standard) for various FFT sizes and bitwidths are plotted in Fig. 4 which shows the expected 6 dB/bit improvement for wider inout FFT words or 48 dB total SQNR increase from 16-bit to 24-bit words. Likewise, a decrease in SQNR is observed for longer FFT lengths, i.e., more FFT stages, due to accumulation of rounding errors.

Additionally, to deeper examine the last effect, a similar plot of SQNR against FFT length, but now for several different rounding methods applied after each FFT stage, is given in Fig. 5 which demonstrates that an impact can be even larger than a single bit in word width.

This is one of the best illustrations of HCL expressiveness over classical HDLs, and design generator abilities.

Furthermore, several generator instances are functionally tested on a budget FPGA board, particularly, a Digilent's Arty S7 which is based on Xilinx Spartan-7 FPGA. Instances obtained from the proposed Chisel 3 SDF FFT generator are compared with Xilinx LogiCORE's FFT IP [9] core, specifically, to its Pipelined Streaming I/O architecture that allows continuous data processing and which supposedly uses (suspected, but nowhere publicly announced) a radix-2^2 SDF scheme.

All instances use a DIF FFT variant, all are synthesized for 100 MHz target clock frequency, use three real multipliers per complex one and feature 16-bit data and twiddle factors. Moreover, no run-time selection of FFT length is allowed, however the use of BRAM/SRAM is.

The comparison for different radix schemes and FFT lengths are summarized in Table I which gives an overall FPGA resource utilization. As anticipated, the radix-2 uses significantly more multipliers (DSP slices), but less adders (logic cells) than the equivalent radix-4 scheme.

Eventually, the radix-2^2 topology takes best of each and it is apparently the finest SDF FFT architecture. Ultimately, its resource distribution is very much alike equivalent Xilinx IP core, which proves the feasibility of generator-based FFT designs in integrated circuits, in particular, and convenience of Chisel 3 HCL in general.

V. Conclusion

A single-path delay feedback FFT processor with plentiful of parametrization options which enable quick design space exploration has been captured and implemented inside a Chisel HCL digital hardware generator.

The comparison with commercially available FFT IP core that is specifically tailored to the particular FPGA platform used for testing, proves that generators can produce performance and resource-competitive designs.

Acknowledgements

The authors would like to thank their colleagues from the University of Kragujevac and NOVELIC d.o.o. on helpful discussions and Serbian Ministry of Education, Science and Technological Development for supporting this research activities through III-41007 project grant.

References

[1] Y. Lee, A. Waterman, H. Cook, B. Zimmer, B. Keller, A. Puggelli, J. Kwak, R. Jevtić, S. Bailey, M. Blagojević, P. Chiu, R. Avizienis, B. Richards, J. Bachrach, D. Patterson, E. Alon, B. Nikolić, and K. Asanović, "An agile approach to building RISC-V microprocessors," *IEEE Micro*, vol. 36, no. 2, pp. 8–20, Mar. 2016.

[2] J. Bachrach, H. Vo, B. Richards, Y. Lee, A. Waterman, R. Avižienis, J. Wawrzynek, and K. Asanović, "Chisel: Constructing hardware in a Scala embedded language," in *49th ACM/ESDA/IEEE Design Automation Conference (DAC 2012)*, Jun. 2012, pp. 1212–1221.

[3] S. He and M. Torkelson, "Design and implementation of a 1024-point pipeline FFT processor," in *Proceedings of the IEEE 1998 Custom Integrated Circuits Conference*, May 1998, pp. 131–134.

[4] C.-H. Yang, T.-H. Yu, and D. Marković, "Power and area minimization of reconfigurable FFT processors: A 3GPP-LTE example," *IEEE Journal of Solid-State Circuits*, vol. 47, no. 3, pp. 757–768, Mar. 2012.

[5] S. He and M. Torkelson, "A new approach to pipeline FFT processor," in *Proceedings of International Conference on Parallel Processing*, Apr. 1996, pp. 766–770.

[6] E. Alon, K. Asanović, J. Bachrach, and B. Nikolić, "Open-source EDA tools and IP, A view from the trenches," in *56th ACM/ESDA/IEEE Design Automation Conference (DAC 2019)*, Jun. 2019, pp. 79.1–79.3.

[7] V. M. Milovanović and M. L. Petrović, "A single-path delay feedback (SDF) fast Fourier transform (FFT) generator," www.github.com/milovanovic/sdf-fft, accessed: 2019/08/03.

[8] A. Wang, P. Rigge, A. Izraelevitz, C. Markley, J. Bachrach, and B. Nikolić, "ACED: A hardware library for generating DSP systems," in *55th ACM/ESDA/IEEE Design Automation Conference (DAC 2018)*, Jun. 2018, pp. 61.1–61.6.

[9] Xilinx, *Fast Fourier Transform v9.1*, LogiCORE IP Product Guide — PG109, May 2019.

TABLE I

FPGA Resource Utilization for Different FFT Instances.

Generator Instance		FPGA Resources			
length	radix/IP	Slice LUTs	Slice Registers	BRAM Tiles	DSP muls
64-pt	radix-2	1153	835	0	15
256-pt	radix-2	1689	1502	2	21
1024-pt	radix-2	2551	3351	4	27
64-pt	radix-4	2411	899	0	6
256-pt	radix-4	3413	1459	3	9
1024-pt	radix-4	5150	3197	6	12
64-pt	radix-2^2	1023	689	0	6
256-pt	radix-2^2	1516	1278	2	9
1024-pt	radix-2^2	2427	3045	4	12
64-pt	Xilinx IP	1353	2271	0.5	6
256-pt	Xilinx IP	1857	3115	2	9
1024-pt	Xilinx IP	2396	3991	3.5	12

978-1-7281-3420-8/19 $31.00 © 2019 IEEE

Experimental Estimation of Input Offset Voltage Radiation Degradation Rate in Bipolar Operational Amplifiers

A. Bakerenkov, V. Pershenkov, V. Felitsyn, A. Rodin, V. Telets, V. Belyakov, A. Zhukov, and N. Gluhov

Abstract - Radiation degradation rate of input offset voltage in bipolar operational amplifiers was estimated experimentally. High degradation rate was observed in devices with high input offset voltage initial values and temperature drifts. Obtained results were discussed.

I. INTRODUCTION

In modern electronics for nuclear and space applications operational amplifiers are used [1]. Due to bipolar operational amplifiers have several advantages in comparison with CMOS devices [2] it is very important to estimate radiation hardness of bipolar operational amplifiers correctly. One of the most significant advantages of bipolar devices is low input offset voltage, especially in devices with compensated input stage [3].

A lot of research works were focused on investigation of input bias current radiation degradation of bipolar operational amplifiers [3-6]. It is the most obvious parameter susceptible to total ionizing dose (TID) effect due to it is connected with base currents of bipolar transistors of an input stage directly. There are many works, which describe physical mechanisms of current gain degradation in bipolar transistors under ionizing radiation impact due to radiation-induced increase of base currents [7, 8]. Unless input bias current is a good indicator of TID effect in bipolar operational amplifiers it has limited significance for developers of electronics for nuclear and space applications. Usually in electrical circuits operational amplifiers are used with negative feedbacks. Non inverting amplifier circuit is presented as an example in Fig.1.

Usually feedback current I_{fb} (Fig.1) in circuits with operational amplifiers is much greater then input current I_{in}, even after ionizing radiation impact. Therefore it can't affect output voltage V_{out} significantly.

A.S. Bakerenkov is with the National Research Nuclear University MEPhI (Moscow Engineering Physics Institute), Moscow, Russian Federation (corresponding author to provide phone: +7 499 324 01 84; fax: +7 (499) 324 21 11; e-mail: AS_Bakerenkov@ list.ru).

V.S. Pershenkov, V.A. Felitsyn, A.S. Rodin, V.A. Telets, V.V. Belyakov, A.I. Zhukov and N.S. Gluhov are with the National Research Nuclear University MEPhI (Moscow Engineering Physics Institute), Moscow, Russian Federation (e-mail: VSPershenkov@mephi.ru).

In a number of works [9,10] it was experimentally demonstrated, that input offset voltage of operational amplifiers is also susceptible to TID effects, as well as input bias current. For designers of electronics input offset voltage can become a very important parameter, especially in circuits with high values of voltage gain. Any changes in input offset voltage will be amplified by the voltage gain and transmitted to output terminal V_{out}. Suppose that a voltage gain value of circuit in Fig.1 is approximately 100. So the 10 mV of the offset voltage shift corresponds to 1V shift of the output voltage. Therefore radiation degradation of input offset voltage has significant effect on output voltages of electrical circuits with operational amplifiers and must be considered as a very important parameter in TID radiation tests.

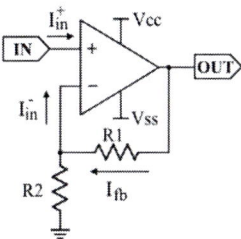

Fig. 1. Non inverting amplifier circuit.

In contract to input bias current radiation degradation of input offset voltage is a complex function of the total dose for each type of the input stage. Input offset voltage doesn't depend on total dose and base currents of bipolar transistors of the input stage directly, usually we observe significant circuit effect. There are several works, which were performed to investigate TID effect on input offset voltage in bipolar operational amplifiers [3,11,12]. In work [3] it was demonstrated that total dose dependence of input offset voltages of bipolar operational amplifiers from the same lot can be very different, even under the same bias, temperature and dose rate conditions during irradiation. It makes the issue of correct estimation of the radiation hardness of bipolar operational amplifiers more complex. It is supposed, that there are some internal factors in each device, which are responsible for input offset voltage radiation degradation rate, which can be obtained through detailed electrical characterization before irradiation.

978-1-7281-3420-8/19 $31.00 © 2019 IEEE

Possible factors are the initial value of input offset voltage and its temperature drift before irradiation. The purpose of this work is to determine and explain correlation between radiation degradation rate of input offset voltage and its initial value and temperature drift in operational amplifiers widely used for nuclear and space applications.

II. Experimental Details

Experiments were performed for two types of bipolar operational amplifiers: LM324 and LM358 manufactured by STMicroelectronics. Samples of each type were divided into two lots in three samples in each. Irradiation was provided by Co[60] at room and elevated temperatures. Detailed data about irradiation and bias conditions of each lot are presented in Table I. The schematic diagram of the bias mode of the devices under test is presented in Fig. 2. Irradiation process was interrupted 3 times due to technical feature of the experimental setup. The duration of the first interrupt was 44 minutes. The second interrupt took 20 minutes and the third - 10 minutes. The temperature and bias conditions weren't changing during any interrupt.

TABLE I
IRRADIATION AND BIAS CONDITIONS

Type	Lot number	Tempera-ture, °C	Electrical mode	Dose rate, rad(Si)/s	Total dose, krad(Si)
LM324	1	25	Active (see Fig. 2)	10,2	59,2
	2	70			30,4
LM358	1	25			57,0
	2	70			80,0

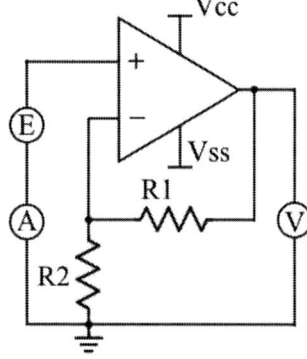

Fig. 2. The schematic diagram of the bias mode of the devices under test, where: Vcc = 5 V, Vss = -5 V, E = 0 V, A – amperemeter, V – voltmeter.

III. Experimental Results

The input offset voltage of each lot was measured under irradiation. The dependences of input offset voltage on total ionizing dose for each sample of each lot are presented in Fig. 3 – 6. Thick solid vertical lines show the interrupts, described above.

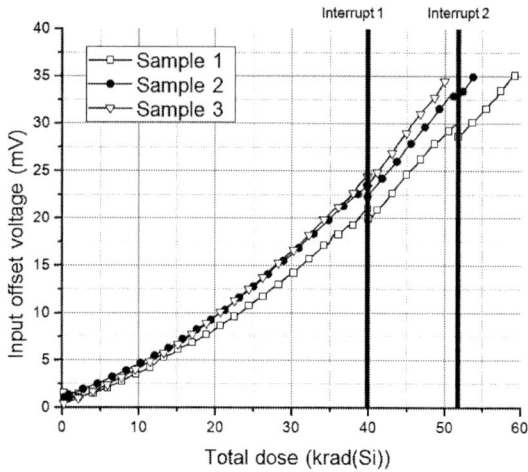

Fig. 3. The dependences of input offset voltage on total dose for the LM324 lot 1.

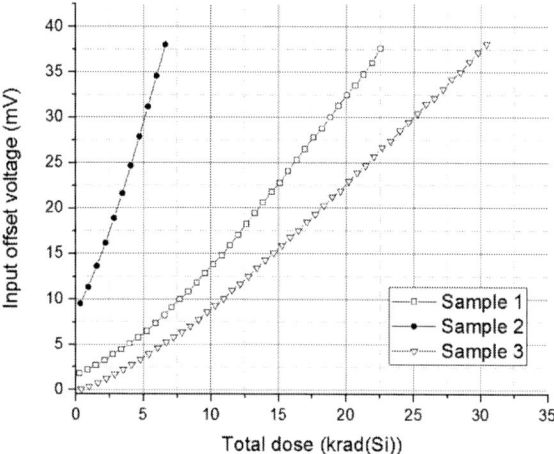

Fig. 4. The dependences of input offset voltage on total dose for the LM324 lot 2.

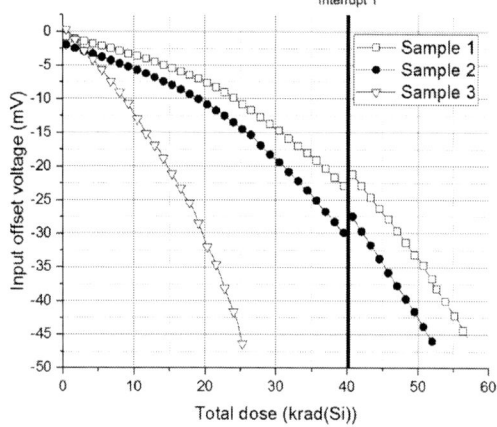

Fig. 5. The dependences of input offset voltage on total dose for the LM358 lot 1.

978-1-7281-3420-8/19 $31.00 © 2019 IEEE

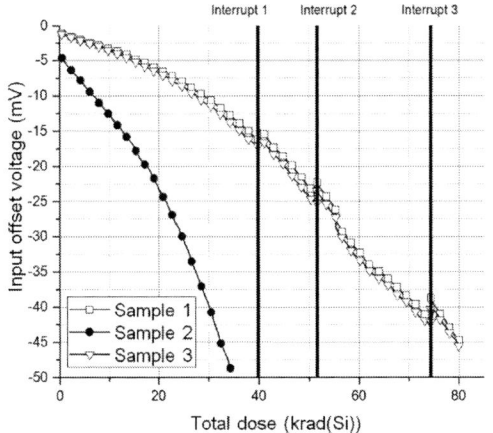

Fig. 6. The dependences of input offset voltage on total dose for the LM358 lot 2.

The dependences of input offset voltage on temperature were measured for each sample before radiation tests. According to the measurement results the values of temperature drift of the input offset voltage were calculated for each sample of each lot. The initial values of V_{io} input offset voltage and initial values of $\Delta V_{io}/\Delta T$ temperature drift of the input offset voltage are presented in Table II.

TABLE II
INITIAL VALUES OF INPUT OFFSET VOLTAGE AND TEMPERATURE DRIFT FOR EACH SAMPLE OF OPERATIONAL AMPLIFIER

Type	Lot number	Sample number	V_{io}, V	$\Delta V_{io}/\Delta T$, V/K
LM324	1	1	$-3,98 \cdot 10^{-4}$	$6,98 \cdot 10^{-6}$
		2	$1,05 \cdot 10^{-3}$	$6,84 \cdot 10^{-6}$
		3	$2,54 \cdot 10^{-4}$	$1,12 \cdot 10^{-5}$
	2	1	$1,21 \cdot 10^{-3}$	$8,03 \cdot 10^{-6}$
		2	$5,32 \cdot 10^{-3}$	$4,32 \cdot 10^{-5}$
		3	$-5,45 \cdot 10^{-4}$	$3,53 \cdot 10^{-6}$
LM358	1	1	$-1,31 \cdot 10^{-3}$	$-1,10 \cdot 10^{-6}$
		2	$-1,70 \cdot 10^{-3}$	$1,94 \cdot 10^{-6}$
		3	$8,97 \cdot 10^{-4}$	$-2,64 \cdot 10^{-5}$
	2	1	$-1,17 \cdot 10^{-3}$	$-1,81 \cdot 10^{-6}$
		2	$-3,75 \cdot 10^{-3}$	$-4,06 \cdot 10^{-6}$
		3	$-7,89 \cdot 10^{-4}$	$-5,12 \cdot 10^{-6}$

IV. DISCUSSION

As it follows from experimental results (table II) the most significant degradation rate we observe in devices with high initial values of input offset voltage or input offset voltage temperature drift. It occurs during room temperature irradiation as well as during irradiation at elevated temperature. In LM324 radiation degradation rate of input offset voltage increases during irradiation at elevated temperature in comparison with room temperature irradiation results, while in LM358 we didn't observe any significant dependence of the degradation rate on irradiation temperature.

Since the dependence of input offset voltage degradation rate on initial values is essentially the same for room and elevated temperature irradiation we can assume, that devices with high initial values of input offset voltage and input offset voltage temperature drift will demonstrate enhanced degradation rate for any irradiation conditions in comparison with devices with low initial values subjected to ionizing radiation impact at the same conditions.

We can assume, that radiation degradation rate of input offset voltage depends at any time under ionizing radiation impact on current value of input offset voltage. It is connected with dependency of the degradation rate, especially initial, on input offset voltage initial values. It can be explained by difference of emitter-base biases of input transistors of operational amplifier differential stage in linear operation mode. In this case the deference can be connected with asymmetry of a current mirror of the differential stage, as well as with difference of *I-V* characteristics of differential pair transistors.

If we observe high value of input offset voltage temperature drift it means, that there is a significant difference of *I-V* characteristics of differential pair transistors. Therefore even we have low initial value of input offset voltage we will observe high degradation rate at any irradiation temperature, as it was obtained in LM358 lot 1 sample 3. In this case low initial value of input offset voltage slightly decreases initial degradation rate.

Additionally, we suppose, that temperature during measurements can be considered as a common mode signal due to it effects on all transistors in operational amplifier circuit simultaneously. Therefore to determine devices with enhanced values of input offset voltage temperature drift before irradiation it is possible to measure Common Mode Rejection Ratio (CMRR) or Power Supply Rejection Ratio. It can simplify measure equipment and test procedures.

V. CONCLUSION

For developers of electronics for nuclear and space applications radiation degradation of input offset voltage is an important parameter of operational amplifiers. Unlike input bias current, which is a good indicator of TID effect, input offset voltage has much more significance for electrical circuits with negative feedbacks based on operational amplifiers. Input offset voltage radiation degradation can be very different in the devices from the same wafer lot even they were irradiated at the same temperature, dose rate and bias conditions. It is very important to understand the reasons of this significant difference, which is connected with internal properties of each operation amplifier sample. These initial values of input offset voltage and input offset voltage temperature drift are connected with these internal properties. In this work we experimentally obtained, that the most significant radiation degradation rate occurs in devices with high initial values of these parameters. In our experiments we performed irradiation procedures at room and elevated temperatures and obtain enhanced degradation

978-1-7281-3420-8/19 $31.00 © 2019 IEEE

in devices with high initial values of input offset voltage and input offset voltage temperature drift in both cases. Enhanced degradation in devices with high initial input offset voltages can be connected with asymmetry of a current mirror of the differential stage, as well as with difference of *I-V* characteristics of differential pair transistors. It can lead to the dependence of the degradation rate on current value of input offset voltage during irradiation due to slight radiation induced drift of electrical mode of the input stage. In devices with high initial temperature drift of input offset voltage enhanced degradation can be explained by difference of differential pair transistors *I-V* characteristics only. The difference can be connected with technology variation of emitter areas and effective base widths of the transistors. To determine devices with high degradation rate before irradiation it is possible to measure Common Mode Rejection Ratio (CMRR) or Power Supply Rejection Ratio instead of input offset voltage temperature drift because temperate variations we can consider as a common mode signal for operational amplifier.

REFERENCES

[1] V. Narayanan, S. Narayanan, J. C. Colon, Y. Christian, "Radiation Effects Characterization of TI OPA4277-SP High Precision Operational Amplifier", *IEEE Radiation Effects Data Workshop (REDW)*, 2017.

[2] P. Horowitz, W. Hill, *The Art of Electronics*, 3rd edition, Cambridge University Press, 2015.

[3] A. H. Johnston, B. G. Rax, D. Thorbourn, "Total dose effects in op-amps with compensated input stages", *9th European Conference on Radiation and Its Effects on Components and Systems (RADECS)*, 2007.

[4] Rodin A.S., Bakerenkov A.S., Pershenkov V.S., Felitsyn V.A., Miroshnichenko A.G., Glukhov N.S., "Evaluation of the post-irradiation temperature dependence of operational amplifier input bias current", *Proceedings of the European Conference on Radiation and its Effects on Components and Systems (RADECS)*, 2015.

[5] Philippe C. Adell, Ronald L. Pease, Hugh J. Barnaby, Bernard Rax, Xiao J. Chen, Steven S. McClure, "Irradiation With Molecular Hydrogen as an Accelerated Total Dose Hardness Assurance Test Method for Bipolar Linear Circuits", *IEEE Transactions on Nuclear Science*, Vol. 56, Issue 6, 2009, pp. 3326-3333.

[6] H.J. Barnaby, C.R. Cirba, R.D. Schrimpf, D.M. Fleetwood, R.L. Pease, M.R. Shaneyfelt, T. Turflinger, J.F. Krieg, M.C. Maher, "Origins of total-dose response variability in linear bipolar microcircuits", *IEEE Transactions on Nuclear Science*, Vol. 46, Issue 6, 2000, pp. 2342-2349.

[7] A.S. Bakerenkov, V.V. Belyakov, V.S. Pershenkov, A.A. Romanenko, D.V. Savchenkov, V.V. Shurenkov, "Extracting the fitting parameters for the conversion model of enhanced low dose rate sensitivity in bipolar devices", *Russian Microelectronics*, vol. 42, 2013, pp. 48-52.

[8] S.C. Witczak, R.D. Schrimpf, K.F. Galloway, D.M. Fleetwood, R.L. Pease, J.M. Puhl, D.M. Schmidt, W.E. Combs, J.S. Suehle, "Gain degradation of lateral and substrate pnp bipolar junction transistors", *IEEE Transactions on Nuclear Science*, Vol.43, Issue 6, 1996, pp. 3151-3160.

[9] Jian-Bo Liu, Yuan Liu, Jin-Li Cheng, Yun-Fei En, Ting Zhang, Yu-Juan He, Fu-Yao Dong, "Total dose irradiation effects in the µA741 operational amplifier with different biases", *International Conference on Quality, Reliability, Risk, Maintenance, and Safety Engineering (QR2MSE)*, 2013, pp. 1156 – 1159.

[10] R.L. Pease, J. Krieg, M. Gehlhausen, D. Platteter, J. Black, "Total dose induced increase in input offset voltage in JFET input operational amplifiers", *Fifth European Conference on Radiation and Its Effects on Components and Systems. RADECS 99* (Cat. No.99TH8471), 1999, pp. 569-572.

[11] Guilherme S. Cardoso, Tiago R. Balen, Marcelo S. Lubaszewski, Rafael G. Vaz, Odair L. Gonçalez, "Impact of TID-induced threshold deviations in analog building-blocks of operational amplifiers", *13th Latin American Test Workshop (LATW)*, 2012, pp. 1-6.

[12] Richard D. Harris, Bernard G. Rax, Steven S. McClure, Allan H. Johnston, "Degradation of RH1056 parameters at low total dose", *9th European Conference on Radiation and Its Effects on Components and Systems (RADECS)*, 2007, pp. 1-5.

978-1-7281-3420-8/19 $31.00 © 2019 IEEE

Linear Slot Array Centrally Fed by CPW T-junction for 5G Applications

M. Milijic and B. Jokanovic

Abstract - This paper discusses the design of an array consisting of eight identical rectangular slots for 5G mobile communication applications. The array has low-profile, high-gain, and wide bandwidth radiation properties. The design steps and electromagnetic analysis of a CPW T–junction as feeding element are presented. The radiation characteristics and other array parameters are evaluated with respect to requirements of applications in 5G frequency range 24.25-27.5 GHz.

I. INTRODUCTION

5G generation of mobile networks are intended to connect hundreds different devices creating crucial demands for services of great broadband capacities and transmission speeds [1]. Consequently, new challenges to design of millimeter band antennas have come out requiring antennas composed of dozens of radiating elements. Antenna arrays feature better radiation characteristics, combined with the reduced sizes and higher gains. CPW (coplanar waveguide) - fed antennas are considered as good candidates for applications in 5G mobile communication systems due to their wide bandwidth, low cost, light weight, small size, and ease of fabrication and integration with active components [2]-[8].

Compared with previous generations of mobile communication networks, the operation bandwidth of the proposed 5G networks will be at millimeter-wave frequencies since it is one of the ways to get rid of the overcrowded sub-6 GHz frequency range. Several promising millimeter-wave bands have been released by International Telecommunication Union (ITU) for the 5G wireless communication system that include the $24.25 - 27.5$ GHz, $37 - 40.5$ GHz, $66 - 76$ GHz bands [9]. Meanwhile, the Federal Communications Commission (FCC) has considered the spectrum of approximately 11 GHz above 24GHz for flexible, mobile and fixed wireless broadband for the next-generation 5G networks and technologies in the United States [10].

This paper presents a centrally offset fed rectangular slot array designed for wide-band applications in 5G frequency range 24.25-27.5 GHz. Aside from planar structure and inexpensive fabrication, the proposed antenna array has 1.85 GHz bandwidth, 16.5 dBi average gain, and radiation pattern corresponding series of identical slots that make this class of antennas suitable for a variety of 5G communication applications.

M. Milijic is with the Faculty of Electronic Engineering, University of Niš, 18000 Nis, Serbia, E-mail: marija.milijic@elfak.ni.ac.rs

Branka Jokanovic is with the Institute of Physics, University of Belgrade, Pregrevica 118, 11080 Pregrevica, Serbia E-mail: brankaj@ipb.ac.rs

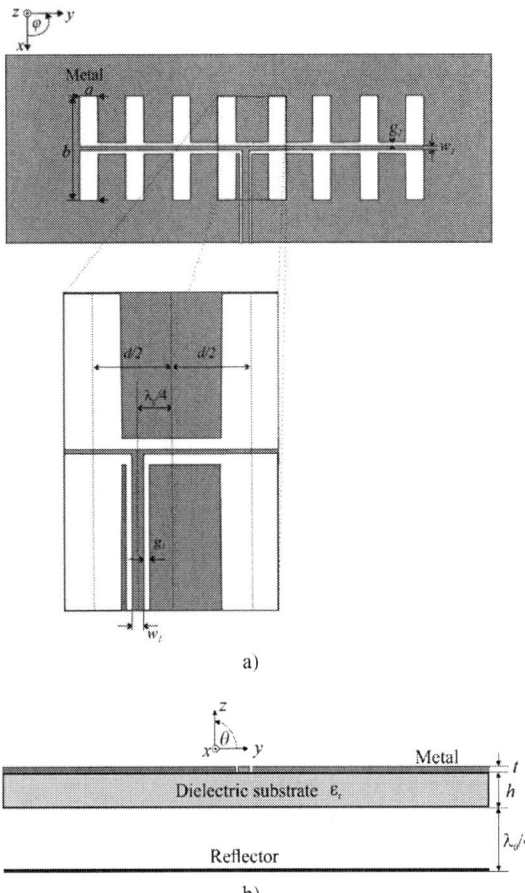

Fig. 1. Centrally offset fed linear slot array: a) Top view b) Side view.

II. THE ARRAY OF CPW-FED RECTANGULAR SLOTS

The configuration of proposed CPW-fed rectangular slot array is depicted in Fig. 1. A dielectric substrate of constant $\varepsilon_r=2.54$ and thickness $h=0.508$ mm is used. The centre frequency $f_c=25.875$ GHz is calculated as the central value of range $24.25 - 27.5$ GHz, recommended by ITU [9]. The antenna array is at distance $\lambda_0/4 =2.89$ mm from the reflector plate whose dimensions are the same as the array's dimensions (λ_0 is wavelength in vacuum at the centre frequency $f_c=25.875$ GHz). Unlike the microstrip antennas with a backside ground plane, the printed dipole and slot antennas require the reflector plane to be at a distance equal to the quarter of the free space wave-length. It should ensure that the antenna radiates only in half of the space. The array is split into two identical four-slot subarrays placed at the different distances from the CPW T-junction. In order to provide the in-phase feeding of both subarrays, the CPW T-junction is moved off the symmetry axes for $\lambda_g/4$ (see Fig. 1), where λ_g is CPW feed line wavelength at the centre frequency f_c. The difference between lengths of CPW lines between T-junction and subarrays results in difference between their S-parameters (magnitudes ($|S_{ij}|$) and for the phases ($\angle S_{ij}$) of S_{21} and S_{31} S scattering (S-) parameters for left and right subarrays) presented in Fig. 2. Analyzing shown results, it can conclude that asymmetrical CPW T junction, which is moved off the symmetry axes, provides necessary feeding for both subarrays for range between 24-27.5 GHz. In higher frequencies, the differences between magnitudes and phases of S-parameters calculated at CPW T junction are above acceptable deviations. The 60 Ω CPW line (featuring the strip w_1=0.9 mm and gap g_1=0.1 mm) is used to enable feeding for both sub - arrays dividing power into two 120 Ω CPW feed lines featuring the strip w_2=0.3 mm and gap g_2=0.375 mm for feeding every four slots sub - array.

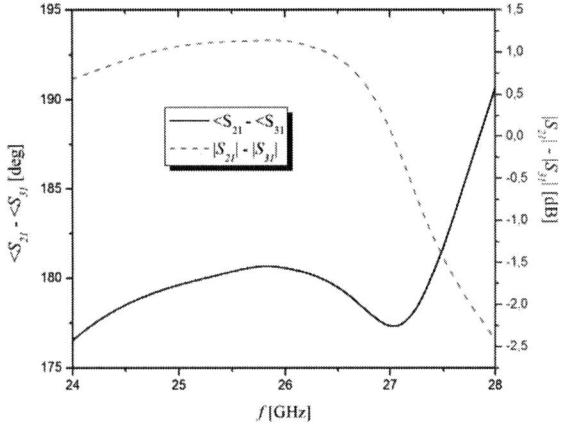

Fig. 2. The difference between insertion losses for two sub-arrays calculated at CPW T-junction. S-parameters are normalized to the impedance of CPW feeding line (120 Ω).

The dimensions of metal plate are 80 mm x 25 mm x 0.508 mm. Better antenna parameters can be achieved if a thicker metallization t (Fig. 1) is used. An additional source of loss is the dielectric loss tangent of the dielectric material. In practice, design a low-loss wideband antenna must consider all possible losses.

Each slot has rectangular shape with 2 mm width and 8.25 mm length. Also, the slots are positioned at mutual distance d=9.25 mm. The antenna array is designed, simulated and analyzed using WIPL-D software [11].

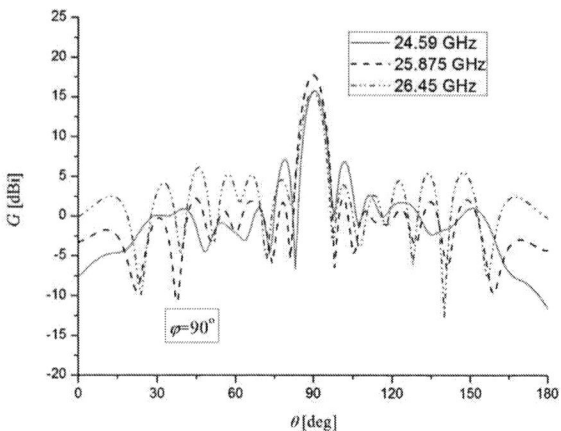

Fig. 3. Radiation pattern in the E-plane for the centrally offset fed linear slot array at 24.59 GHz, 25.875 GHz and 26.45 GHz (φ=90°).

TABLE I
CHARACTERISTICS OF THE CENTRALLY OFFSET FED LINEAR SLOT ARRAY.

Frequency [GHz] / Characteristics	24.59	25.875	26.45
Gain [dBi]	15.75	17.75	15.6
SLS [dB]	8.5	15.5	9.5
3dB-beamwidth [°]-E plane	7	8	6
3dB-beamwidth[°]-H plane	56	64	51

III. SIMULATION RESULTS

The main goal of conducted research was to obtain the best possible bandwidth of antenna array with acceptable other antenna parameters. Therefore, besides central, two more frequencies are considered: lower and upper. Their values are determined as frequencies for which the change of gain is less than 12.5 % in relation to the corresponding value at the central frequency. The radiation patterns for lower, central and upper frequency are presented in Fig. 3 resulting in the range from 24.59 GHz to 26.45 GHz. The obtained simulation results point out that arrays with central feeding feature more wideband characteristics than the series-fed arrays [12] which is one of the crucial demands for 5G communication systems. In the operating range the maximum gain fluctuation is about 2 dB in respect to the gain at the central frequency which is 17.75 dBi. Side lobe suppression varies from 9 dB at the edge frequencies to 15.5 dB at the central frequency, which is expected for an uniform antenna array. Furthermore, the comparisons between E-plane co-polar and cross-polar radiation patterns as well as H- plane radiation pattern for φ=0° for three considered frequencies are shown in Fig. 4.

The Table I presents the review of the overall characteristics of examined array at three considered frequencies.

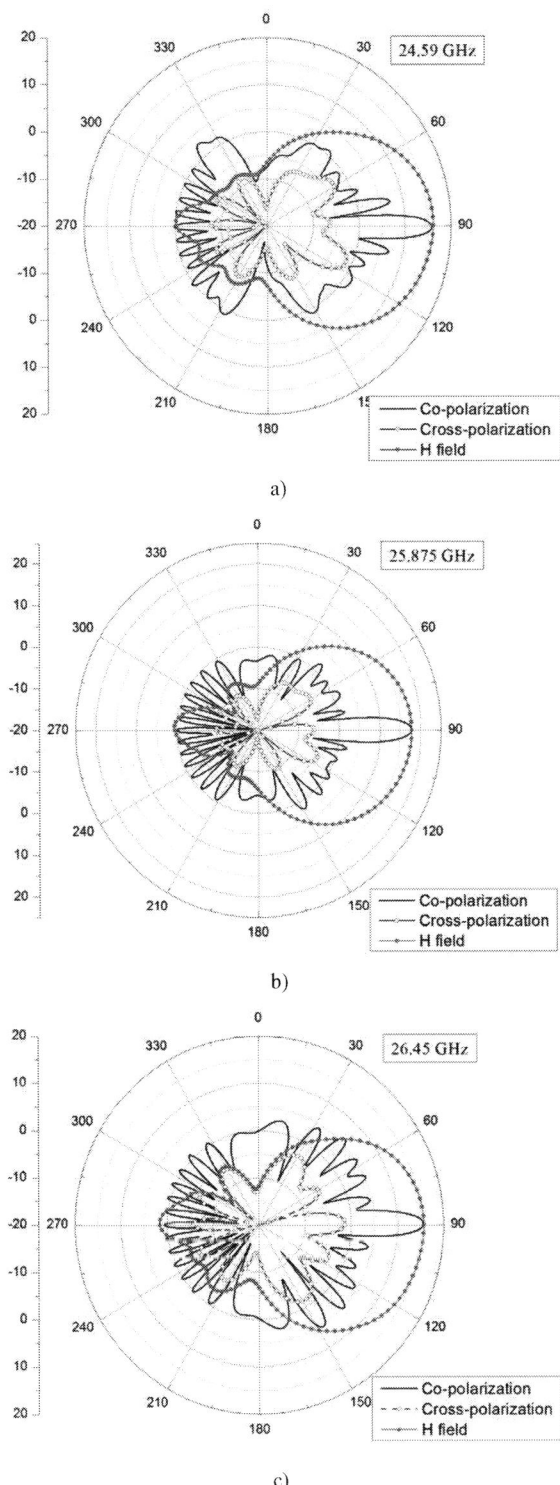

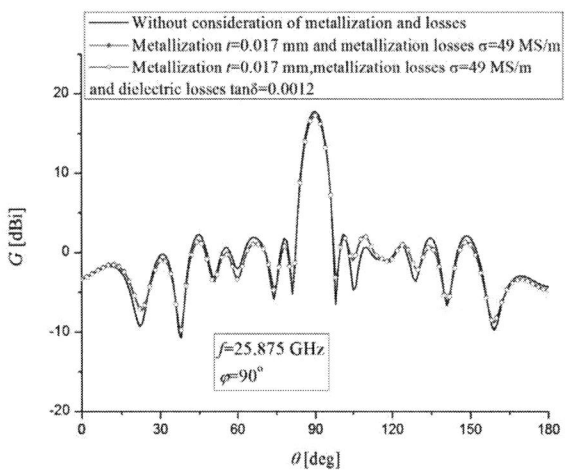

Fig. 5. Radiation pattern in the E-plane (φ=90°) for the centrally offset fed linear slot array without and with consideration of metallization and losses in metallization and dielectric.

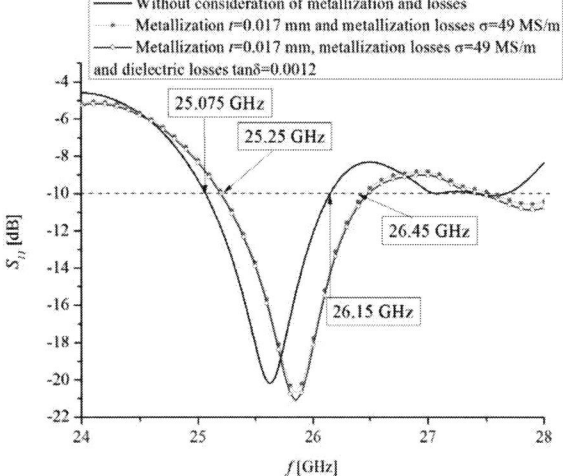

Fig. 6. S_{11} parameter of the centrally offset fed linear slot array without and with consideration of metallization and losses in metallization and dielectric.

In the future experiments, the proposed antenna array should be realized by photolithographic process, mounted above the reflector plate and measured. In order to obtain more realistic simulation results, the antenna array should be analyzed taking into account the influence of metallization thickness as well as the metal and dielectric losses on antenna parameters. The obtained results are shown in Fig. 5-6. and Table II. The values of metallization thickness and dielectric losses are referenced to Rogers RT/duroid 5870 and 5880 laminates. The change of gain and side lobe suppression considering finite metallization thickness and losses in dielectric substrate is not significant.

Fig. 4. E-plane co- and cross- polar and H-plane radiation pattern for the centrally offset fed linear slot array at: a) 24.59 GHz b) 25.875 GHz c) 26.45 GHz.

978-1-7281-3420-8/19 $31.00 © 2019 IEEE

Influence of losses in metallization is more pronounced. Furthermore, the bandwidth of considered antenna array respecting S_{11}<-10 dB does not vary, although it shifts towards higher frequencies for 0.2 GHz when metallization thickness t=17 μm is considered.

TABLE II
CHARACTERISTICS OF THE CENTRALLY OFFSET FED LINEAR SLOT ARRAY WITHOUT AND WITH CONSIDERATION OF THE METALLIZATION AND DIELECTRIC LOSSES.

Characteristics / Dielectric	Gain [dBi]	SLS [dB]	Bandwidth respecting S_{11}<-10 dB [GHz]-[GHz]
loss free	17.75	15.5	25.075-26.15
metallization t=17 μm σ=49MS/m	17.3	15.3	25.25-26.45
metallization t=17 μm, σ=49MS/m and dielectric losses $tanδ$=0.0012	17.2	15	25.2-26.45

IV. CONCLUSION

Antenna design is one of the major considerations to realize mm-wave based 5G mobile networks. Modern wireless communication system requires low profile, light weight, high gain and ease of installation, antenna systems. The design of a wideband compact antenna is a challenging task especially at the higher frequencies around 28-GHz.

In this paper, a wideband compact array of rectangular slots is proposed. The central feeding provides a wide bandwidth from 24.59 to 26.45 GHz in respect to series feeding. Furthermore, the improved simulation which takes into account metallization thickness and losses in metal and dielectric show excellent characteristics of the proposed antenna array, concerning satisfying gain and radiation characteristics.

ACKNOWLEDGEMENT

This work was supported by the Ministry of Education, Science and Technological Development of Republic Serbia under the projects No. III44009 and TR 32024.

REFERENCES

[1] V. Milosevic, B. Jokanovic, O. Boric-Lubecke, V. M. Lubecke, "Key Microwave and Millimeter Wave Technologies for 5G Radio", in *Powering the Internet of Things with 5G Networks*, V. Mohanan, R. Budiatru, I. Aldmour, Eds. IGI Global, July 2017.

[2] G. H. Elzwawi, M. Mantash and T. A. Denidni, "Improving the gain and directivity of CPW antenna by using a novel AMC surface", *2017 IEEE International Symposium on Antennas and Propagation*, San Diego, CA, 2017, pp. 2651-2652.

[3] W. Tu, "Analysis and Design of Coplanar Waveguide-Fed Capacitively Coupled Slot Antennas", *2015 International Workshop on Antenna Technology*, Seoul, Republic of Korea, 4-6 March 2015.

[4] M. Yang, X. Yin, H. Zhao, "Wideband Coplanar Waveguide-Fed Slot Antenna Array with Via-Wall Structure", *Proc. of 2016 10th European Conference on Antennas and Propagation*, Davos, Switzerland, 10-15 April 2016.

[5] P. Chaudhary, A. Kumar and R. Mittra, "Quadrilateral-Shaped Wideband Circularly Polarized CPW-Fed Monopole Antenna," *2019 URSI Asia-Pacific Radio Science Conference (AP-RASC)*, New Delhi, India, 2019, pp. 1-4.

[6] M. E. de Cos Gómez, H. Fernández Álvarez, C. García González, B. Puerto Valcarce, J. Olenick and F. Las-Heras, "Ultra-Thin Compact Flexible Antenna for IoT Applications," *2019 13th European Conference on Antennas and Propagation (EuCAP)*, Krakow, Poland, 2019, pp. 1-4.

[7] A. I. Afifi, D. M. Elsheakh, A. B. Abdel-Rahman, A. Allam and S. M. Ahmed, "Dual Broadband Coplanar Waveguide-Fed Slot Antenna for 5G Applications," *2019 13th European Conference on Antennas and Propagation (EuCAP)*, Krakow, Poland, 2019, pp. 1-3.

[8] A. Zaidi, A. Baghdad, W. A. Awan, Halima, A. Ballouk and A. Badri, "CPW Fed Wide to Dual Band Frequency Reconfigurable Antenna for 5G Applications," *2019 International Conference on Wireless Technologies, Embedded and Intelligent Systems (WITS)*, Fez, Morocco, 2019, pp. 1-3.

[9] W. Hong, K. Baek, S. Ko "Millimeter-Wave 5G Antennas for Smartphones: Overview and Experimental Demonstration", *IEEE Trans. Antennas Propag.*, December 2017, vol. 65, no. 12, pp. 6250 - 6261.

[10] M. J. Marcus, "5G and IMT for 2020 and beyond", *IEEE Wireless Commun.*, Aug. 2015, vol. 22, no. 4, pp. 2-3.

[11] WIPL-D Pro, WIPL-D Team

[12] M. Milijic, B. Jokanovic, "Radiation bandwidth of series-fed slot arrays for 5G and IoT applications", *2018 26th Telecommunications Forum (TELFOR)*, Belgrade, Serbia, November, 20-21, 2018 , pp. 462-465.

Advanced Electro-Optical Analysis of Photoplethysmogram Signal

L. Evdochim, D. Dobrescu and L. Dobrescu

Abstract − Commonly used to monitor oxygen saturation, the pulse oximeter device can be exploited further than its primary function, using an advanced analysis of the optical sensor output signal. Due to its simplicity, the cost to diagnose patient's circulatory system disease becomes low, but the effort to process the recorded signal and analyze it becomes high. The pulse oximeter output signal contains information such as the blood oxygen concentration and the cardiac rhythm period. However, because of strong connections between the mechanical, optical and electrical proprieties, a lot of valuable information can be extracted regarding arterial blood pressure and arterial stiffness. Medical data of the cardiac abnormalities or the circulatory diseases can also be extracted. In this paper an advanced analysis describe the connection between pulse oximeter signal and mechanical, optical and electrical phenomena in order to determine blood pressure value.

I. INTRODUCTION

Optical blood pressure estimation using only a signal from the finger tips is known as photoplethysmogram or PPG. Many advanced analyses have been developed [1].

II. BLOOD VESSEL MECHANICAL ANALYSIS

The following analysis is based on the supposition that a system energy can be ideally transferred, so the friction forces are removed and the blood vessel can be modeled by a segment of cylindrical shape.

In a segment of finite length L, the blood mass which flows is composed of finite number of independent particles which dissociates: one part of them follows the direction of flow and another part interacts with vessel segment's wall. In order to minimize the flow resistance the second part of them creates a local dilatation along the segment due to the elastic proprieties of blood vessels. The speed of dilatation will be referenced as $\mathbf{v_y}$ and the speed of flowing blood through artery as $\mathbf{v_x}$. A two-axis analysis (Fig. 1) is usually performed, but in this paper a radial analysis will be developed. Writing the energetic conservation equation on the input of analyzed segment:

$$E_{C1} = E_P + E_{C2} \qquad (1)$$

The input kinetic energy E_{C1} is split into the elastic potential energy E_P (stored in the elastic segment's walls)

L. Evdochim, D. Dobrescu, L. Dobrescu are with the Faculty of Electronics, Telecommunication and Information Technology, University Politehnica of Bucharest, Bucharest, Romania, E-mail: lucy_evd12@yahoo.com, lidia.dobrescu@electronica.pub.ro

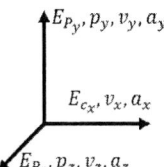

Fig. 1. Distribution of energy and physical quantities: p - pressure, v – velocity and a - acceleration

and into the kinetic energy E_{C2} which keep the flowing direction as output value. The loss of energy between the input and the output energy is stored as potential energy:

$$E_P = \frac{m_1 v^2_{1x}}{2} - \frac{m_2 v^2_{2x}}{2} = \frac{k_W y^2}{2} \qquad (2)$$

Where mass m and velocity v belong to each kinetic energy E_{C1} and E_{C2}, k_w stands for vessel's wall elastic constant and y is wall elongation from its initial state, on the radial axis. So, this amount of elastic potential energy is responsible of local dilatation y. A high value of k_w indicates that vessel is rigid and a low value denotes a soft vessel. The blood mass m of density ρ_B fills the space V_{DIL} given by elongation y (Fig. 2):

$$V_{DIL} = V_1 - V_0 = \pi L (y^2 + 2ry) \qquad (3)$$

$$S_{DIL} = S_1 - S_0 = 2\pi L y \qquad (4)$$

For the next step, the dilatation phenomena will be analyzed. The elongation on the radial axis can't be infinite because it is limited when two or more forces reach equilibrium state. So, the action force is given by blood mass which is pushing into vessel walls. From these phenomena two reaction forces appear. An obvious reaction force is the elastic force $\vec{F_E}$ which opposes against deformation. The

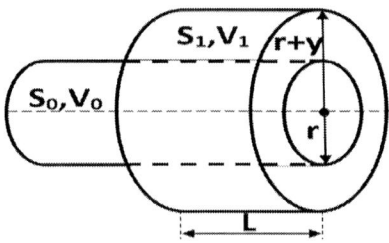

Fig. 2. Artriers surface and volume at two phase: systolic and distolic on defined length L

system reaches in equilibrium state when the two forces become equal in amount but by opposite direction:

$$\vec{F} + \vec{F}_E = 0 \qquad (5)$$

The first force is the action force:

$$ma = k_W y \qquad (6)$$

So, y term represents maximum elongation when sum of the two forces become zero (Eq. 5). If the acceleration is rewritten as distance per time square, interesting information is extracted from previous equation:

$$m \frac{y}{t^2} = k_W y \qquad (7)$$

$$k_W = \frac{m}{t^2} \qquad (8)$$

The information about the vessel's walls elastic constant k_W can be extracted from mass and time measurements. The second reaction force is described between product of pressure and surface on which pressure is pushing:

$$F = pS \qquad (9)$$

$$ma = pS \qquad (10)$$

But S surface represents dilated cover surface of wall S_1 which is given by initial radius r summed with dilatation radius y (Fig. 3). Therefore, after mathematical processing, the final pressure is given by:

$$p = \frac{m}{2\pi L t^2} \frac{y}{r+y} \qquad (11)$$

It is observed that pressure is inversely proportional to the square of time. A rapid dilatation phase (small rise time period) suggests that pressure which pushing the vessel wall is greater than a slow dilatation phase where rise time is increased. Replacing the initial radius with r_0 and the final radius of vessel wall with r_1 the pressure formula becomes more suggestive:

$$p = \frac{m}{2\pi L t^2} \left(1 - \frac{r_0}{r_1}\right) \qquad (12)$$

The greater is final radius of vessel walls, the greater is pushing pressure. The blood mass that causes this phenomenon can be extracted from PPG signal processing same as blood volume if it's know the blood density. Therefore, this is a closer approaching equation for arterial pressure extracting method.

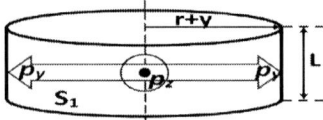

Fig. 3. Disk surface where pressure is pushing on radial plan (y and z axis)

Important final observations consist in [2]: pressure (mostly systolic type) is directly proportional to the mass of pushing blood, volume and vessel's wall elasticity constant. So, a stiffness wall implies a greater value of pressure while an elastic wall, which easily expands, implies a lower pressure.

III. ELECTRO-OPTICAL ANALYSIS OF BLOOD PULSE SIGNAL

The recorded signal comes from the LED-photodiode pair, where photodiode is connected in the photoconductive mode through a transimpedance amplifier (Fig. 4). After this configuration, the output voltage U_{OUT} will be:

$$U_{OUT} = RI_{PD} \qquad (13)$$

The photocurrent I_{PD} is given by two terms: the dark current I_S and the photocurrent given by product of radiant flux ϕ and spectral sensitivity S_{PD}. For the IR light source used in measurements, photodiode sensitivity has a value of 0.54 W/A for 900 nm wavelength:

$$I_{PD} = S_{PD}\varphi + I_S \qquad (14)$$

$$U_{OUT} = RS_{PD}\varphi + RI_S \qquad (15)$$

For the optical analysis, simplified continuity equation of light flux at boundry between air and blood vessel is:

$$\varphi_I = \varphi_A + \varphi_R \qquad (16)$$

Where ϕ_I is the total light flux transmitted by LED, ϕ_A is the absorbed flux by blood, and φ_R is the reflected light flux to the photodiode detector. Due to Beer-Lambert theorem, the thicker the luminaire layer is, the more light is absorbed. In this paper, the correlation between the absorbed light flux and the sampled blood volume V_{SMP} with density ρ_B is described by a first order equation:

$$\varphi_A = \gamma \rho_B V_{SAMPLED} \qquad (17)$$

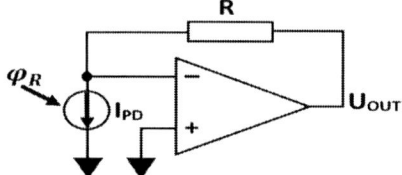

Fig. 4. Transimpedance amplifier

The γ factor represents the absorbed light flux power for a given mass of substance at a specific wavelength of light. It is measured in W/kg but for this research an appropriate unit is uW/g.

As long as recorded light flux by optical sensor is in fact the reflected one:

$$\varphi_R = \varphi_I - \varphi_A \qquad (18)$$

Replacing reflected recorded light flux with Eq. 15 where the dark current factor is lower than photocurrent and with Eq. 17:

$$\frac{U_{OUT;R}}{S_{FD}R} = \frac{U_{OUT;I}}{S_{FD}R} - \gamma m_{SMP} \qquad (19)$$

Where $U_{OUT;R}$ stands for recorded voltage for reflected light flux and $U_{OUT;I}$ stands for equivalent voltage for incident light on the photodiode active area given by LED source. The analyzed sampled blood mass given by reflected flux in relation with incident light flux:

$$m_{SMP} = \frac{U_{OUT;I} - U_{OUT;R}}{\gamma S_{FD}R} \qquad (20)$$

From Eq. 19 it is easily observed that a greater sampled blood mass will cause a lower output recorded voltage, as long as incident flux voltage factor is always constant. Therefore, at the diastolic momentum when vessel is relaxed and the sampled mass is at minimum value, the absorbed light flux in diastolic phase is less than at the systolic phase (Fig. 5):

$$\varphi_{A;SYS} > \varphi_{A;DIA} \qquad (21)$$

$$\varphi_{R;SYS} < \varphi_{R;DIA} \qquad (22)$$

$$\frac{U_{OUT;SYS}}{S_{FD}R} < \frac{U_{OUT;DIA}}{S_{FD}R} \qquad (23)$$

This is an important observation because the recorded output voltage is mirrored respecting to y axis and neglect this statement leads to wrong interpretation (Fig. 6). The peak of the recorded signal corresponds to the diastolic phase, where the arterial pressure is at minimum, and the base of the signal correspond to the systolic phase where the arterial pressure is at maximum. The sampled volume is related to recorded output voltage according to Eq. 20:

$$V_{SMP} = \pi L r^2{}_{SMP} = \frac{1}{\gamma \rho_B} \frac{U_{OUT;I} - U_{OUT;R}}{S_{FD}R} \qquad (24)$$

So, it can be extracted the sampled radius of cylindrical shape of predefined length L:

$$r_{SMP} = \sqrt{\frac{1}{\pi L \gamma \rho_B} \frac{U_{OUT;I} - U_{OUT;R}}{S_{FD}R}} \qquad (25)$$

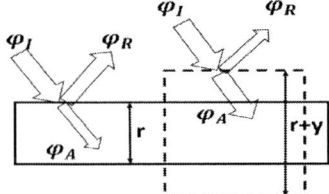

Fig. 5. Light fluxes in two moments: systolic and distolic

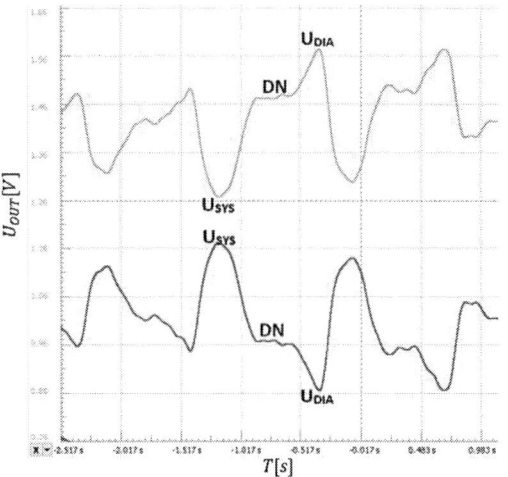

Fig. 6. Original recorded PPG signal (top) and inverted signal (bottom) in IR mode; DN – dictoric notch

Due to Eq. 12, with knowing sampled radius:

$$p = \frac{m_{DIL}}{2\pi L t^2}\left(1 - \sqrt{\frac{U_{OUT;I} - U_{OUT;R0}}{U_{OUT;I} - U_{OUT;R1}}}\right) \qquad (26)$$

Where the mass that produces local dilatation is difference between the sampled mass in systolic phase and the mass in diastolic phase.

IV. ELECTRICAL ANALYSIS OF PPG SIGNAL

The obtained PPG signal contains information about: constant of vessel's wall (Eq. 8) and arterial pressure (Eq. 12) but and other features like: augmentation index, age index density and circulation performance. So, the device which records the PPG signal needs to consider a few of variables but most important is the absorption factor which depends by skin color and by tissue layer above arteries in order to give a suitable measurements. The elastic constant can be measured from raw PPG signal as the time interval t_1 in which the maximum is reached (Fig.7):

$$k_W = \frac{U_{OUT;DIA} - U_{OUT;SYS}}{\gamma S_{FD}R t_1^2} \qquad (27)$$

In Millasseau research [3] the stiffness index is defined slightly different (Fig. 8):

$$k = \frac{U\max;_{SYS}}{t_2} \qquad (28)$$

Where t_2 is time between maximum values of the two phases. Information about an equivalent elastic constant is given by Padilla [4] but as augmentation index (AI):

$$AI = \frac{U_{\max;DIA}}{U_{\max;SYS}} \qquad (29)$$

The second derivative PPG signal is made of three pairs of maximum and minimum values: *a* and *b*, *c* and *d*, *e* and *f*. The first two pairs appear in the systolic phase and the last pair appears in the diastolic phase. Using the Eq. 26 and Eq. 27, the arterial pressure can be computed as:

$$p = \frac{a_{U_{OUT}}}{2\pi L \gamma S_{FD} R}\left(1 - \sqrt{\frac{U_{OUT;I} - U_{OUT;R0}}{U_{OUT;I} - U_{OUT;R1}}}\right) \qquad (30)$$

Where $a_{U_{OUT}}$ is peak *a* from Fig. 7. A new featured that can be extracted after the signal processing is the age index described by Sano [5]:

$$age\_index = \frac{c + d - b}{a} \qquad (31)$$

This index modifies with age, because the arterial elasticity decrease with it thus the vessel becomes rigid. Also, information about age can be extracted directly from raw PPG signal: if the dicrotic notch becomes weaker it means that patient is older. The dicrotic notch appears between systolic and the diastolic phase (Fig. 6). From the second derivate PPG can be approximated blood circulation efficiency [6], [7], [8] (Fig. 8). First signal noted with A show a good circulation, the second B shows a good circulation but deteriorated, third C shows a bad circulation and the next shows distinctively bad circulation.

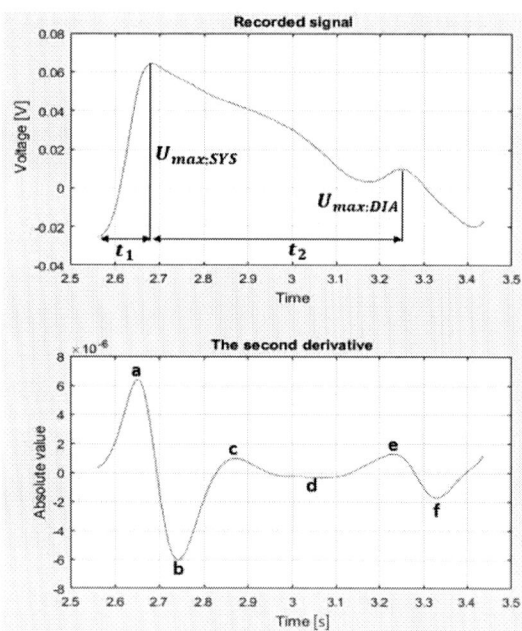

Fig. 7. Recorded signal and second derivative

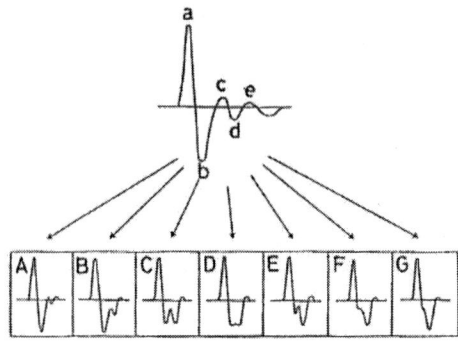

Fig. 8. Different second derivative morphology

V. CONCLUSION

Starting from the mechanical energy and forces where was joining optoelectronics with signal processing, it was proven that at least an estimated blood pressure can be extracted from the PPG signal. So, hypertension can be identified. Improving this method can lead to replacing the classical method of measuring blood pressure with the inflated cuff. Therefore, the first advantage is that optoelectronic method gives a continuous measurement, different from classical method. Also, processing the electrical signal reveals features like: vessels elastic constant, augmentation index, age index.

The last feature is that this method can identify vascular performance based on different second derivative PPG morphology.

REFERENCES

[1] X Xing M Sun, "Optical blood pressure estimation with photoplethysmography and FFT-based neural networks", *Biomedical optics express*, 2016

[2] Guyton, *Medical Physiology*, 11 edition, Elsevier, 2006

[3] Alty SR, Angarita-Jaimes N, Millasseau SC, Chowienczyk PJ., "Predicting arterial stiffness from the digital volume pulse waveform", *IEEE Trans Biomed Eng.*, 2007 Dec, 54(12), pp. 2268-2275.

[4] Padilla JM, Berjano EJ, Saiz J, Facila L, Diaz P, Merce S. "Assessment of relationships between blood pressure, pulse wave velocity and digital volume pulse", *Computers in Cardiology*, 2006. pp. 893–896.

[5] Sano Y. Kasokudo Myakuha ni kansuru Kenkyuu no Gaiyou (in Japanese) http://jsspot.org/sano/:2003.

[6] Homma S, Ito S, Koto T, Ikegami H. " Relationship between accelerated plethysmogram, blood pressure and arterior elasticity", *The Japanese Society of Physical Fitness and Sport Medicine*, 1992, 41, pp. 98–107.

[7] M.Elgendi "On the Analysis of Fingertip Photoplethysmogram Signals", *Cardiology Reviews*, 2012, pp. 14-25.

[8] S Sun, R Bezemer, X Long, J Muehlsteff, R M Aarts, "Systolic blood pressure estimation using PPG and ECG during physical exercise", *Physiol. Meas*, 37, 2016.

CIRCUIT AND SYSTEM DESIGN AND TESTING – POSTER SESSION

978-1-7281-3420-8/19 $31.00 © 2019 IEEE

VHDL-AMS Model Development for Digitally Programmable Monolithic Instrumentation Amplifiers

I. Pandiev

Abstract - A simple behavioral model has been developed for monolithic instrumentation amplifiers (in-amps) with digitally programmable gains. The model is based on the physical structure of the three - operational amplifiers in-amp (3-op amps in-amp) topology. In comparison with SPICE-based simulation standard macro-model the proposed model uses a reduced number of nodes and has shorter simulation time, without limiting the number of modeled electrical parameters. The in-amp model is implemented by using very-high hardware description language (VHDL) analog and mixed-signal language (VHDL-AMS), and accurately simulates the static and dynamic electrical parameters. The VHDL-AMS code is created on the basis of set of analytical equations and is adapted to the simulators of the SystemVision® (version 5.5) and the SystemVision® Cloud environment.

I. INTRODUCTION

The digitally programmable instrumentation amplifiers (in-amps) are important functional blocks in the complex data acquisition systems [1-4]. In order to prove the benefits of the in-amps and to ensure their quick and successive implementation in the electronic circuits, the functionality has to be fully evaluated. In the design process of new system it is necessary to evaluate various circuit variants and design solutions according to the required technical parameters and other operational conditions. The sample circuits have to be simulated to ensure the desired operation, by using simulation platforms with integrated electrical models or macro-models for discrete and integrated components. The analysis of the literature and technical documentation showed the existence of simulation models [5-6], based on the physical electronic circuit. These transistor-level models provide accurate simulation results but require longer simulation time, and in some cases additional requirements can be added to improve the convergence of the simulation process. Over the last decade, many behavioral models have been created, using analog behavior modeling (ABM) blocks [7] and hardware description languages (HDLs) such as VHDL-AMS [8], that are easier to implement and contain fewer parameters.

This paper proposes VHDL-AMS model for digitally programmable in-amps, based on the simulation model, represented in [9], but with extension to include common-mode input voltage range, non-dominant poles at differential input signals, AC common-mode rejection and maximum

I. Pandiev is with the Department of Electronics, Faculty of Electronic Engineering and Technology (FEET), Technical University of Sofia, 8 Kl. Ohridski Blvd, 1000 Sofia, Bulgaria, E-mail: ipandiev@tu-sofia.bg

gain error. Further, the original model, reported in [9], has been modified to reproduce the conversion of the differential input voltage into single-ended output voltage with respect to a reference (*ref*) terminal.

II. STRUCTURES AND ELECTRICAL PARAMETERS

Unlike the pin-programmable in-amps, in which the differential voltage gain is set by an external resistor in the input stage of the structure, for the digitally-programmable amplifiers several integrated resistors are used that are switched by analog switches through the address decoder. The control address signals (or controlled word) $D = a_0, a_1, a_2, \ldots, a_{n-1}$ are applied to the parallel input port of the address decoder. The size n of the controlled word D determines the possible number of the voltage gains equal to 2^n of the certain amplifier. Moreover, the word D is loaded into the decoder under the control of the $\overline{WR}$ write enable signal.

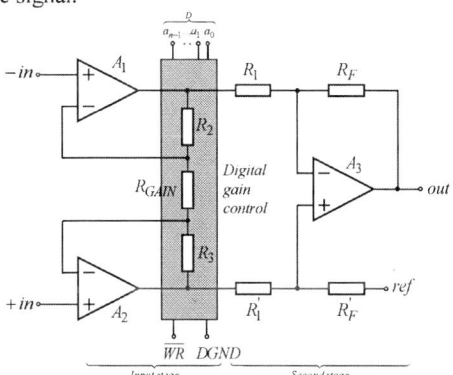

Fig. 1. Simplified block diagram of digitally programmable in-amp.

Most of the programmable monolithic in-amps use *three op amp in-amp* topology (Fig. 1). In comparison with the simple differential amplifier the electronic circuit in Fig. 1 can provide greater input impedances greater value for the common-mode rejection ratio (CMRR) and the differential voltage gain can be easily varied in wide ranges without violating the symmetry of the structure.

During the past decade the most of the commercially available programmable in-amps of various semiconductor manufactures, such as Analog Devices, Texas Instruments and Microchip Technology, are built using classic 3-op amp topology or modified structures. Based on an analysis of their structures and the operational principles in this section, all electrical characteristics and related with them static and dynamic parameters with their typical values (Table I) that are subject to reflection in the proposed simulation model for digitally programmable amplifiers are defined.

TABLE I
BASIC ELECTRICAL PARAMETERS FOR PROGRAMMABLE IN-AMPS

Name	Symbol	Range	Units
Common-mode rejection ratio	$CMRR$	from 80 to 120	dB
Pole-frequency for CM signals	f_{pcm}	from 10 to 100	Hz
Input bias current	I_{iB}	from 1 to 50	nA
Input offset current	I_{io}	from 1 to 50	nA
Input offset voltage	V_{io}	less than 1	mV
Differential input resistance	r_{id}	from 10^9 to 10^{12}	Ω
Differential input capacirance	C_{id}	from 1 to 5	pF
Common-mode input resistance	r_{icm}	from 10^9 to 10^{12}	Ω
Common-mode input capacirance	C_{icm}	from 1 to 5	pF
Output resistance	r_o	less than 1	Ω
Terminal voltage operating range	$V^{+(-)}$	± 1 to ± 18 and $+2$ to $+36$	V
Quiescent supply current	I_{SC}	from 0.1 to 5	mA
Voltage gain	A_V	1, 10, 10^2... or 1, 2, 2^2...	
Bandwidth	$B_{0.7}$	up to 10	kHz
Maximum peak output voltage swing	$V_{om}^{+(-)}$	± 1 to ± 18 and $+2$ to $+36$	V
Short-circuit current	$I_{o,max}^{+(-)}$	from 10 to 20	mA
Slew Rate	SR	from 1 to 10	V/μs
Gain error (max)	δ_A	0.03 at A_V=1 0.04 at A_V=10, 100 and 1000	%

III. MODEL DESCRIPTION

For the proposed model the approach of behavioral modeling is based on a set of analytical equations representing the input, transfer and output electrical characteristics of the monolithic in-amps. The input offset voltage of the proposed model can be represented by two components according to

− *Input stage* of a programmable in-amp:

$$V_{io} = V_{io,0} + \alpha_{io}/A_{Vi} \tag{1}$$

where $V_{io,0}$ is the input offset voltage, A_{Vi} is the i-th value of the voltage gain ($i = 1...n$ (n is the number of voltage gains)) and the constant α_{io} is used to define the effect of changing the offset voltage versus the voltage amplification. The input bias currents are represented as

$$I_B^+ = I_{iB} + 0.5 I_{io} \tag{2}$$

$$I_B^- = I_{iB} - 0.5 I_{io} \tag{3}$$

For differential and common-mode input signals the input impedances are represented as

$$i_{id} = (r_{id}^{-1} + sC_{id})v_{id} \tag{4}$$

$$i_{cm}^+ = [(2r_{cm}^+)^{-1} + sC_{cm}^+]v_{cm}^+ \tag{5}$$

$$i_{cm}^- = [(2r_{cm}^-)^{-1} + sC_{cm}^-]v_{cm}^- \tag{6}$$

where $v_{id} = v_{in}^+ - v_{in}^-$ is the differential input voltage. The input operating voltage range to the inverting or non-inverting input terminals is approximated as

$$v_{in}^+ = \begin{cases} V_{DD}^+ - V_T \ln(i_{in}^+/I_{SAT}+1) \ at \ v_{in}^+ \geq V_{DD}^+ \\ v_{cm}^+ + V_{io} \ at \ V_{SS}^- < v_{in}^+ < V_{DD}^+ \\ V_{SS}^- + V_T \ln(i_{in}^+/I_{SAT}+1) \ at \ v_{in}^+ \leq V_{SS}^- \end{cases} \tag{7}$$

$$v_{in}^- = \begin{cases} V_{DD}^+ - V_T \ln(i_{in}^-/I_{SAT}+1) \ at \ v_{in}^- \geq V_{DD}^+ \\ v_{cm}^- \ at \ V_{SS}^- < v_{in}^+ < V_{DD}^+ \\ V_{SS}^- + V_T \ln(i_{in}^-/I_{SAT}+1) \ at \ v_{in}^- \leq V_{SS}^- \end{cases} \tag{8}$$

where $V_T = kT/q$ is the thermal voltage, k is the Boltzmann's constant ($8.617 \times 10^{-5} eVK^{-1}$), T is the temperature, q is the elementary charge ($1.602 \times 10^{-19}C$) and I_{SAT} is the reverse saturation current ($1 \times 10^{-14}A$).

− *Transfer stage* of a programmable in-amp:

$$v_{amp}(s) = \begin{cases} \frac{\omega_{p1}\omega_{p2}(A_V \pm \delta_{A1})}{\omega_{p1}\omega_{p2}+(\omega_{p1}+\omega_{p2})s+s^2}v_{id} + \frac{\omega_{pcm1}}{\omega_{pcm1}+s} \times \frac{A_V \cdot v_{cm}^+}{CMRR_1} + v_{ref} \ at \ A_V=1 \\ \frac{\omega_{p1,10}\omega_{p2}(A_V \pm \delta_{A2})}{\omega_{p1,10}\omega_{p2}+(\omega_{p1,10}+\omega_{p2})s+s^2}v_{id} + \frac{\omega_{pcm2}}{\omega_{pcm2}+s} \times \frac{A_V \cdot v_{cm}^+}{CMRR_2} + v_{ref} \ at \ A_V=10 \\ \frac{\omega_{p1,100}\omega_{p2}(A_V \pm \delta_{A3})}{\omega_{p1,100}\omega_{p2}+(\omega_{p1,100}+\omega_{p2})s+s^2}v_{id} + \frac{\omega_{pcm3}}{\omega_{pcm3}+s} \times \frac{A_V \cdot v_{cm}^+}{CMRR_3} + v_{ref} \ at \ A_V=10^2 \\ \frac{\omega_{p1,1000}\omega_{p2}(A_V \pm \delta_{A4})}{\omega_{p1,1000}\omega_{p2}+(\omega_{p1,1000}+\omega_{p2})s+s^2}v_{id} + \frac{\omega_{pcm4}}{\omega_{pcm4}+s} \times \frac{A_V \cdot v_{cm}^+}{CMRR_4} + v_{ref} \ at \ A_V=10^3 \end{cases}$$

where $\omega_{p1} = 2\pi f_{p1}$, $\omega_{p1,10} = 2\pi f_{p1,10}$, $\omega_{p1,100} = 2\pi f_{p1,100}$ and $\omega_{p1,1000} = 2\pi f_{p1,1000}$ are the angular frequencies of the dominant poles at certain voltage gain A_V, $\omega_{p2} = 2\pi f_{p2}$ is the angular frequency of the secondary pole and $\omega_{pcm} = 2\pi f_{pcm}$ is the angular frequency for the common-mode input signal. For the various voltage gains the CMRR is defined as follows $CMRR_1$, $CMRR_2$, $CMRR_3$ and $CMRR_4$.

− *Output stage* of a programmable in-amp:

$$i_{ref} = v_{ref}/r_{ref} \tag{9}$$

$$v_{int} = \begin{cases} V_{DD}^+ - V_D^+ \ at \ v_{amp} > V_{DD}^+ - V_D^+ \\ v_{amp} \ (\Delta v_{amp}^+ \approx SR^+ \Delta t \ or \ \Delta v_{amp}^- \approx SR^- \Delta t) \\ V_{SS}^- + V_S^- \ at \ v_{amp} < V_{SS}^- + V_S^- \end{cases} \tag{10}$$

where SR^+ and SR^- are the positive and negative rate of change in the output voltage caused by a step-signal input signal.

$$v_{out} = v_{int} + v_{ro} \tag{11}$$

$$i_{ro} = \begin{cases} I_{o,max}^+ \ at \ i_{ro} \geq I_{o,max}^+ \\ v_{ro}/r_o \\ I_{o,max}^- \ at \ i_{ro} \leq I_{o,max}^- \end{cases} \tag{12}$$

entity PGInAmp is

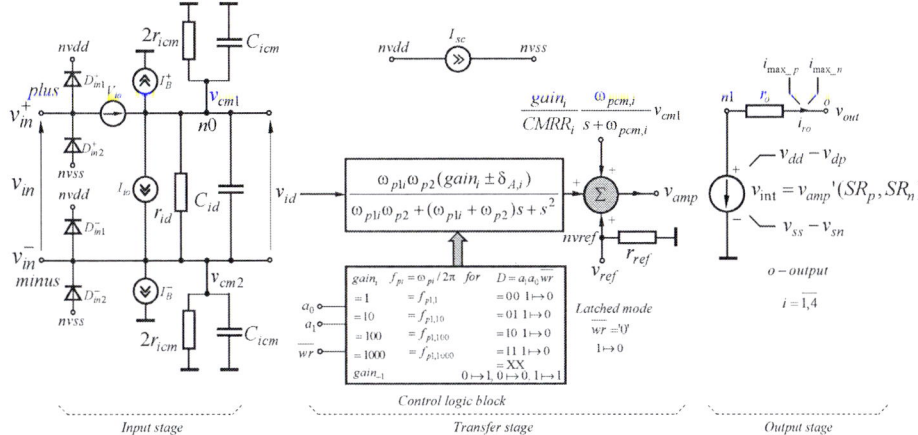

Fig. 2. Structure of a digitally programmable in-amp model.

– Description of gain modification for four values

$$A_V = \begin{cases} gain_1 & at \;\; y=\bar{a}_0\bar{a}_1 \;\; and \;\; \overline{wr} = '1' \mapsto '0' \\ gain_2 & at \;\; y=a_0\bar{a}_1 \;\; and \;\; \overline{wr} = '1' \mapsto '0' \\ gain_3 & at \;\; y=\bar{a}_0 a_1 \;\; and \;\; \overline{wr} = '1' \mapsto '0' \\ gain_4 & at \;\; y=a_0 a_1 \;\; and \;\; \overline{wr} = '1' \mapsto '0' \end{cases} \quad (13)$$

where, *gain* is the value of the parameter A_V, a_0 is the least significant bit (LSB) gain setting address pin, a_1 is the most significant bit (MSB) gain setting address pin and $\overline{wr}$ is the write enable input signal. The input address state is read only during the negative-going edge of the $\overline{wr}$ signal.

IV. Model Implementation

The proposed model is developed for monolithic in-amps with gains 1, 10, 10^2 and 10^3 through programming of a parallel port. The complete set of analytical equations is implemented according to topology in Fig. 1, by using VHDL-AMS [8] (analog and mixed signal).

A. Structuring the in-amp model

The analytical equations representing the behavior of the input, transfer and output electrical characteristics of a programmable in-amp are implemented as a schematic structure, shown in Fig. 2. As can be seen, the structure of the model is represented as a combination of structural and behavioral elements. The structural description is a net-list and it is represented by a set of appropriately connected voltage-controlled voltage sources (VCVSs), capacitors and resistors, as shown in Fig. 2. The VCVSs and passive components are standard VHDL-AMS components that are previously defined models in the VHDL simulation package IEEE.electrical_systems. In the behavioral description, the behavior of the model is specified by a set of simultaneous statements and some event-driven operators represent the gain modification process.

```
entity PGInAmp is
generic(
  --generic parameters here);
port (terminal plus, minus,output,nvdd,nvss,nvref:ELECTRICAL;
signal wr_n : in STD_LOGIC;
signal a0,a1: in STD_LOGIC);
end entity PGInAmp;
architecture behavioral of PGInAmp is
constant fp1_gain1:real:=. . .;
constant fp1_gain10:real:= . . .;
constant fp1_gain100:real:= . . .;
constant fp1_gain1000:real:= . . .;
constant fpcm --(G = 1, 10, 100, 1000)-- :real:= . . .;
constant fp2:real:= . . .;
constant wp1 :real:=math_2_pi*fp1_gain1;
constant wp1_10 :real:=math_2_pi*fp1_gain10;
constant wp1_100 :real:=math_2_pi*fp1_gain100;
constant wp1_1000 :real:=math_2_pi*fp1_gain1000;
constant wp2 :real:=math_2_pi*fp2;
constant wpcm --(G = 1, 10, 100, 1000)-- :real:=math_2_pi*fpcm;
constant tempk : temperature :=
ambienttempfromgeneric(temp_ambient, temp_units);
constant vt : real := phys_k*tempk/phys_q;
terminal n0, n1:ELECTRICAL;
quantity vin across plus to minus;
quantity vd1plus across id1plus through plus to nvdd;
quantity vd2plus across id2plus through nvss to plus;
quantity vd1minus across id1minus through minus to nvdd;
quantity vd2minus across id2minus through nvss to minus;
quantity v_io across i_io through plus to n0;
quantity vid across ii, icid, irid through n0 to minus;
quantity vcm1 across ircm1, iccm1, ibplus through n0;
quantity vcm2 across ircm2, iccm2, ibminus through minus;
quantity vdd across nvdd;
quantity vss across nvss;
quantity isc through nvdd to nvss;
quantity vref across iref through nvref;
quantity vint across in1 through n1;
quantity vro across iro through n1 to output;
quantity vout across output;
quantity vamp:VOLTAGE;
quantity iro_h:CURRENT;
signal gain:real:=1.0;
signal a0_b,a1_b:BIT;
signal wr_n_h:BIT;
signal control_in:BIT_VECTOR(0 to 1);
begin
isc==supply_current;
v_io == vio1.10e-6*(avio/gain); ii == iio/2.0;
icid == cid * vid'dot; irid == vid/rid;
ircm1 == vcm1/(2.0*ricm); iccm1 == cicm * vcm1'dot;
ircm2 == vcm2/(2.0*ricm); iccm2 == cicm * vcm2'dot;
ibplus == iib;
ibminus == iib;
id1plus == Isat*(exp(vd1plus/vt) - 1.0);
id2plus == Isat*(exp(vd2plus/vt) - 1.0);
id1minus == Isat*(exp(vd1minus/vt) - 1.0);
id2minus == Isat*(exp(vd2minus/vt) - 1.0);
if gain=1.0 use
vamp==vid'ltf((0=>wp1*wp2*(1.0+ gain_error1)),
(0=>wp1*wp2,1=>wp1+wp2,2=>1.0))+(1.0/cmrr1)*vcm1'ltf((0=>
wpcm1*1.0),(0=>wpcm1,1=>1.0))+vref;
elsif gain=10.0 use
vamp==vid'ltf((0=>wp1_10*wp2*(10.0+ gain_error2)),
(0=>wp1_10*wp2,1=>wp1_10+wp2,2=>1.0))+(10.0/cmrr2)*vcm1'ltf((0=>wp
cm10*1.0),(0=>wpcm10,1=>1.0))+vref;
elsif gain=100.0 use
```

```
vamp==vid'ltf((O=>wp1_100*wp2*(100.0+
gain_error3)),(O=>wp1_100*wp2,1=>wp1_100+wp2,2=>1.0))+(100.0/cmrr3
)*vcm1'ltf((O=>wpcm100*1.0),(O=>wpcm100,1=>1.0))+vref;
else
vamp==vid'ltf((O=>wp1_1000*wp2*(1000.0+gain_error4)),
(O=>wp1_1000*wp2,1=>wp1_1000+wp2,2=>1.0))+(1000.0/cmrr4)*
vcm1'ltf((O=>wpcm1000*1.0), (O=>wpcm1000,1=>1.0))+vref;
end use;
iro_h==vro/ro;
iref == vref/rref;
if vamp'above(vdd-vdp) use
vint==vdd-vdp;
elsif not vamp'above(vss+vsn) use
vint==vss+vsn;
else
if gain=1.0 or gain=10.0 use
vint==vamp'slew(SRp,SRn); elsif gain=100.0 use
vint==vamp'slew(SRpl1,SRnl1);
else
vint==vamp'slew(SRpl2,SRnl2);
end use; end use;
if iro_h'above(imax_p) use
iro==imax_p;
elsif not iro_h'above(imax_n) use
iro==imax_n;
else iro==iro_h; end use;
break on gain,
vamp'above(vdd-vdp), vamp'above(vss+vsn);
wr_n_b<=to_bit(wr_n);
control_in<=(to_bit(a1), to_bit(a0));
case_change:process is
procedure gain_change (signal_in:in BIT_VECTOR;
signal s:out REAL) is
begin
case signal_in(0 to 1) is
when b"00" => s <= 1.0;
when b"01" => s <= 10.0;
when b"10" => s <= 100.0;
when b"11" => s <= 1000.0;
end case;
end procedure gain_change;
begin
if (wr_n_b = '0' and wr_n_b'event) then
gain_change(control_in,gain);
else
wait until (wr_n_b = '0' and wr_n_b'event);
gain_change(control_in,gain);
end if;
wait on wr_n_b,control_in,gain;
end process case_change;
end architecture behavioral;
```

Fig. 3. VHDL-AMS model of a digitally programmable in-amp.

B. Model building

Based on the structure presented in the previous sub-section, a VHDL-AMS description (Fig. 3) for a digitally programmable in-amp is compiled. The VHDL-AMS code is adapted to the SystemVision® (version 5.5) and the SystemVision® Cloud environment, which are parts of the electronic design automation system Mentor Graphics. The model is composed by an entity and the adjacent architecture, the text written in bold indicates reserved words. The entity declares the generic model parameters and specifies six interface terminals of type electrical (non-inverting input port—plus, inverting input port—minus, port for positive supply voltage—nvdd, port for negative supply voltage—nvss, reference terminal—nref and output terminal—output). Also, the model has two inner terminals n0 and n1. There are used to specify the voltages v_io and vint. In the model description there are three terminals of type std_logic (the least significant bit (LSB) gain setting address pin—a0, most significant bit (MSB) gain setting address pin—a1 and write enable input signal—wr_n). The architecture of the model contains differential equations and some event-driven operators.

V. CONCLUSION

An analytical model for digitally programmable monolithic in-amps has been developed and implemented in the SystemVision® simulation platform, using VHDL-AMS. The model is created basis on a set of analytical equations according to the 3 - op amp in-amp topology and accurately simulates the static and dynamic electrical parameters at various voltage gains. Further, the model also takes into account the latched mode of operation thought write enable signal and the conversion of differential input voltage into single-ended output voltage with respect to a reference terminal.

The created model can be used to support the analysis and design of a wide range of electronic circuits and systems, intended for analog and mixed signal processing.

ACKNOWLEDGEMENTS

The author would like to thank the Research and Development Sector at the Technical University of Sofia for the financial support.

REFERENCES

[1] C. Kitchin and L. Counts, "A Designer's Guide to Instrumentation Amplifiers 3rd Edition," Analog Devices, Norwood, MA, USA, 2006 [Online]. Available: https://www.analog.com.

[2] Ch. Wu, G. Li, D. J. Pommerenke, V. Khilkevich, G. Hess, "Characterization of the RFI Rectification Behavior of Instrumentation Amplifiers," in *Proc. IEEE Symposium on Electromagnetic Compatibility, Signal Integrity and Power Integrity*, Long Beach, CA, USA, 2018, pp. 156-160.

[3] A. Jain, K. Kandpal, "Design of a high gain, Temperature Compensated Biomedical Instrumentation Amplifier for EEG Applications," in *Proc. 11th International Conference on Intelligent Systems and Control (ISCO)*, 2017, pp. 292-296.

[4] K. Fortunado, "Programmable Gain Instrumentation Amplifiers: Finding One that Works for You," Analog Devices, Norwood, MA, USA, Dec. 2018 [Online]. Available: https://www.analog.com/en/analog-dialogue.html.

[5] Simulation models for Programmable & variable gain amplifiers (PGA/VGA), Texas I., Dallas, TX, USA, 2018.

[6] SPICE Models, Analog Devices, Norwood, MA, USA, Dec. 2018 [Online]. Available: https://www.analog.com.

[7] Analog Behavioral models, Cadence Design Systems, CA USA, 2018. [Online]. Available: http://www.pspice.com/analog-behavioral-models.

[8] E. Christen and K. Bakalar, "VHDL-AMS—A hardware description language for analog and mixed-signal applications," *IEEE Trans. Circuits Syst.* II, vol. 46, no. 10, Oct. 1999, pp. 1263–1272.

[9] D. Martev, I. Panayotov and I. Pandiev, "Behavioural VHDL-AMS Model for Monolithic Programmable Gain Amplifiers," in *Proc. ICEST* 2010, Ohrid, Macedonia, 2010, pp. 835-838.

978-1-7281-3420-8/19 $31.00 © 2019 IEEE

Verification of VHDL-AMS Simulation Model for Digitally Programmable Monolithic Instrumentation Amplifiers

I. Pandiev

Abstract - This paper focuses on verification of VHDL-AMS model [1] for digitally programmable monolithic instrumentation amplifiers. The modeling parameters are extracted for commercially available in-amp AD8253 from Analog Devices, by analyzing semiconductor data books or through characterization measurements. The values for the static and dynamic parameters obtained by simulation of the proposed model are compared against the typical values of the corresponding parameters from the datasheets for the real device at different gains A_V and temperature 300 K. The proposed model shows good agreement between typical values and simulated results. Furthermore, the maximum value of the relative error does not exceed 8 %. The validation of the model is performed at differential and common-mode input signals by comparing the simulation results with the behavior of the SPICE compatible macro-model for in-amp AD8253.

I. INTRODUCTION

Upon completion of the model [1] development process, must be traced the static and dynamic state, their quick and effective testing, i.e. performing verification and validation [2, 3]. The purpose of the verification is to ensure the correct behavior of each element or group of elements in the model. During verification, each element or group of elements are checked whether, first, provide the behavior as predicted in the model, and second, whether their behavior is adequate to the real device, i.e. performing the evaluation of the degree of accuracy of the model. For this purpose, a comparison of the simulation results with the typical values of the instrumentation amplifier (in-amp) datasheet is performed. Here it is important to focus the attention to the fact that the datasheets do not give the exact values of the electrical parameters but define the region guaranteed by the manufacturer, in which the parameters for each sample are changed. In the process of creating the majority of simulation models are taken into account the datasheet typical values of the real devices.

The purpose of validation is to ensure the required degree of accuracy by checking whether the overall behavior of the model is adequate to the real device. The most commonly used validation methods for models are two: (1)

I. Pandiev is with the Department of Electronics, Faculty of Electronic Engineering and Technology (FEET), Technical University of Sofia, 8 Kl. Ohridski Blvd, 1000 Sofia, Bulgaria, E-mail: ipandiev@tu-sofia.bg

comparison with a real system and (2) comparison with other models that are previously validated. For the purpose of this work a comparison analysis is performed with the corresponding SPICE compatible macro-model [5], in order to highlight advantages and drawbacks of the proposed behavioral model.

II. PARAMETER EXTRACTION

The verification of the developed model, representing in [1], is performed by comparing the simulation results for the static and dynamic parameters with the typical values given in the datasheet of a programmable gain monolithic in-amp AD8253 [4]. The AD8253 use *three op amp in-amp topology* and it has a high value of the bandwidth up to 10 MHz at $A_V = 1.0$ (at supply voltages equal to ±15 V) and the THD is below -110 dB. Also, at A_V equal to 1000, the offset voltage drift and the gain drift is ≤ 1.2 μV/°C and 10 ppm/°C, respectively. In addition the common-mode (CM) rejection ratio (*CMRR*) is 100 dB (or 10^5), at gain 10^3 up to 20 kHz. The AD8253 consists of a control port that allows setting the four values of differential voltage gain via 2-bit control word (a_0 and a_1). Furthermore, the 2-bit word sent via a bus can be latched using the $\overline{WR}$ input signal or in the other words the model will be work in a latched gain mode of operation as a real device. In fact, AD8253 is a typical representative of the monolithic digitally-programmable in-amps.

The model parameters are extracted from an analysis of the curves of the electrical characteristics and using values of the parameters listed in the tables in the standard datasheets. Also, some parameters are extracted from experimental testing of sample circuits, constructed according to the test circuits and measurement condition, given in the datasheets.

All modeling parameters that give good agreement with the behavior of the real part AD8253 are presented in Table II. The last column in Table II gives the principal method of the parameter extraction, where DS is a datasheet typical value, EXP is obtained from experimental test and E is a typical estimated value from device design.

Based on the VHDL-AMS description, commented in [1], a symbol (Fig. 1) PGInAmp is created for use in the SystemVision® environment.

978-1-7281-3420-8/19 $31.00 © 2019 IEEE

TABLE I

MODELING PARAMETERS FOR PROGRAMMABLE IN-AMPS

Parameter	Name	Value	Ext. by
iib	Input bias current	5×10^{-12} A	DS
iio	Input offset current	5×10^{-12} A	DS
vio	Input offset voltage	150×10^{-6} V	DS
avio	Voltage amplification offset drift	900.0	DS
rid	Differential input resistance	$4 \times 10^{9}\,\Omega$	DS
cid	Differential input capacitance	1.25×10^{-12} F	DS
ricm	CM input resistance	$1 \times 10^{9}\,\Omega$	DS
cicm	CM input capacitance	5×10^{-12} F	DS
gain	Voltage gain $- A_V$	1, 10, 100 and 1000	Exp
gain_error1..4	Gain error at 1.0 Error at 10, 10^2, 10^3	0.03% 0.04%	DS DS
fp1_gain1	-3 dB bandwidth at $A_V =1.0$	10×10^6 Hz	Exp
fp1_gain10	-3 dB bandwidth at $A_V =10$	4×10^6 Hz	Exp
fp1_gain100	-3 dB bandwidth at $A_V =10^2$	550×10^3 Hz	Exp
fp1_gain1000	-3 dB bandwidth at $A_V =10^3$	60×10^3 Hz	Exp
fp2	Secondary pole	20×10^6 Hz	Exp
rref	Input resistance of the nvref terminal	$20 \times 10^3\,\Omega$	DS
ro	Output resistance	$0.01\,\Omega$	DS
cmrr0	CMRR at gain $A_V =1.0$	65×10^3	DS
cmrr1	CMRR at gain $A_V =10$	200×10^3	DS
cmrr2	CMRR at gain $A_V =10^2$	250×10^3	DS
cmrr3	CMRR at gain $A_V =10^3$	260×10^3	DS
fpcm_1	Pole-frequency for CM signals at $A_V =1.0$	8.5×10^3 Hz	DS
fpcm_10	Pole-frequency for CM signals at $A_V =10$	18×10^3 Hz	DS
fpcm_100	Pole-frequency for CM signals at $A_V =10^2$	30×10^3 Hz	DS
fpcm_1000	Pole-frequency for CM sig. at $A_V =10^3$	32×10^3 Hz	DS
ics	Quiescent current	4.5×10^{-3} A	DS
SRp, SRn	Slew rate at $A_V =1$ and 10	±20 V/µs	DS
SRp11, SRn11	Slew rate at $A_V =10^2$	±12 V/µs	DS
SRp12, SRn12	Slew rate at $A_V =10^3$	±2 V/µs	DS
vdp, vsn	Voltage drop at maximum/minimum output signal	1.4 V 1.3 V	DS DS
imax_p, imax_n	Maximum positive/ negative output current	37×10^{-3} A 37×10^{-3} A	DS DS
isat	Leakage current	1.0×10^{-14} A	E
vt	Thermal voltage at temperature 300K	26×10^{-3} V	E

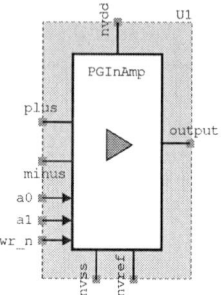

Fig. 1. SystemVision® Cloud symbol of the implemented digitally programmable in-amp model [1].

III. RESULTS AND DISCUSSION

The electrical parameters of the developed model, obtained by simulation with System Vision, are verified against the typical values of the corresponding parameters from the datasheets for AD8253 at all possible voltage gains A_V equal to 1, 10, 100 and 1000. For this purpose an electronic evaluation circuit (Fig. 2) of the model is implemented, to study its behavior in static and dynamic operational mode. The evaluation circuit consists basically of two voltage sources which generate the inverting and non-inverting input voltages and two DC voltage sources, which generate the positive and negative power_supply voltage. The control word (a₀ and a₁) and the $\overline{WR}$ input signal is simulated by three digital sequence sources, connected to a0, a1 and wr_n terminals. The input reference input voltage is obtained by one voltage source vref and the load is represented by resistor r1, connected to the output terminal according to ground.

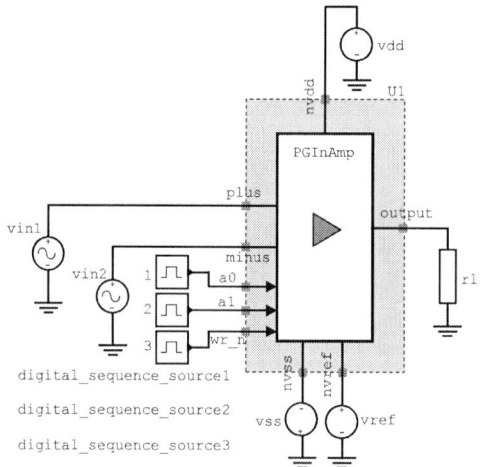

Fig. 2. The SystemVision® Cloud schematic of the evaluation circuit with the digitally programmable in-amp model, represented by block marked with U1.

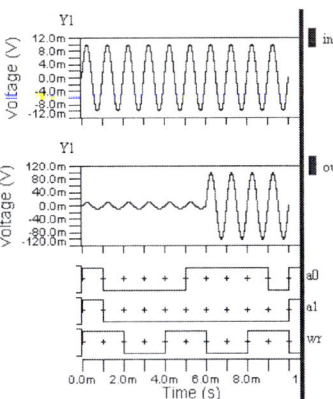

Fig. 3. The simulated dynamic performance of the implemented digitally programmable in-amp model at $A_V = 1$ and 10.

The dynamic performance is simulated by using the test circuit in Fig. 2 for the four values of the voltage gain, namely $A_V = 1$, 10, 100 and 1000. The obtained time diagrams are shown in Fig. 3 and Fig. 4. As can be seen (Fig. 3), at input signal with amplitude 10 mV the output voltage switches to the new value during only the falling edge of the $\overline{WR}$ signal. In this case, the amplitude of the output voltage became equal to 100 mV. Also, such behavior can be seen at gains equal to 100 and 1000. The time diagrams showing a coefficient switch from 100 to 1000 are represented in Fig. 4. Changes the level of a_0 or a_1 (at $t = 5$ ms), without changing of the $\overline{WR}$, does not lead to a change of the A_V.

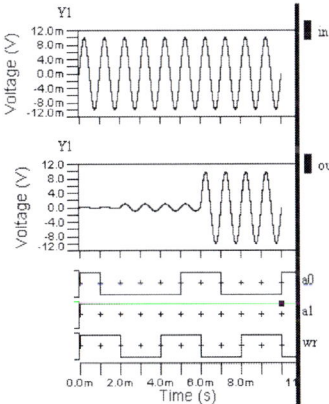

Fig. 4. The simulated dynamic performance of the implemented digitally programmable in-amp model at $A_V = 100$ and 1000.

Table II summarizes the comparison of the simulation results and datasheet typical values for IC AD8352 [4] at $V_{DD}^+ = -V_{SS}^- = 15V$, $V_{ref} = 0V$, $T_A = 25^\circ C$ and $R_L = 2k\Omega$. The analysis of the obtained results shows a relatively good agreement. The maximum value of the relative error is below 8 %. The comparatively large difference between the simulation results and the datasheet typical values for some of the parameters is due to the fact that in the modeling

process, a group of modeling parameters often determines certain electrical parameters of the real device.

TABLE II
COMPARISON BETWEEN SIMULATION RESULTS AND DS PARAMETERS

Parameter	This model for AD8253	Datasheet typical value [4]	Error
I_{iB}	5.1×10^{-12} A	5×10^{-12} A	2%
I_{io}	4.9×10^{-12} A	5×10^{-12} A	2%
V_{io}	148×10^{-6} V	150×10^{-6} V	1.3%
r_{id}	$4 \times 10^9 \Omega$	4×10^9	–
C_{id}	1.23×10^{-12} F	1.25×10^{-12} F	1.6%
r_{icm}	$1 \times 10^9 \Omega$	$1 \times 10^9 \Omega$	–
C_{icm}	4.87×10^{-12} F	5×10^{-12} F	2.6%
A_V	0.984, 9.93, 99.8 and 984	1, 10, 100 and 1000	< 2%
$B_{0.7(AV=1)}$	9.2×10^6 Hz	10×10^6 Hz	8%
$B_{0.7(AV=10)}$	3.7×10^6 Hz	4×10^6 Hz	7.5%
$B_{0.7(AV=100)}$	551×10^3 Hz	550×10^3 Hz	0.2%
$B_{0.7(AV=1000)}$	59.58×10^3 Hz	60×10^3 Hz	0.7%
$CMRR_1$	64.89×10^3	65×10^3	0.2%
$CMRR_{10}$	201.557×10^3	200×10^3	0.8%
$CMRR_{100}$	255.358×10^3	250×10^3	2.1%
$CMRR_{1000}$	262.361×10^3	260×10^3	0.9%
f_{pcm} at $G=10^3$	31.62×10^3 Hz	30×10^3 Hz	5.4%
r_o / r_{ref}	$0.01 / 20 \times 10^3 \Omega$	$0.01 / 20 \times 10^3 \Omega$	–
$V^{+(-)}$	$-14.3 / +14.3$ V	$-14 / +13.5$ V	< 6%
I_{SC}	4.5×10^{-3} A	4.5×10^{-3} A	–
$V_{om}^{+(-)}$	$-13.7 / +13.6$ V	$-13.7 / +13.6$ V	–
$I_{o,\max}^{+(-)}$	36.9×10^{-3} A	37×10^{-3} A	0.3%
SR at 1 & 10 SR at 10^2 & 10^3	$\pm 19.29 / \pm 19.62$ $\pm 11.95/\pm 1.99$ V/µs	$\pm 20 / \pm 20 /$ $\pm 12 / \pm 2$ V/µs	< 4%
δ_A	0.029 and 0.04%	0.03 and 0.04%	< 3%

For validation process the behavior of the proposed in-amp model is compared with the SPICE compatible (valid) macro-model [5] of IC AD8253. Furthermore, the overall behavior of the both models in the frequency- and time-domain is studied at various voltage gains, applying differential and common-mode signal to the input port.

The amplitude-frequency responses of the two simulation models are plotted in Fig. 5. At low frequencies (up to 10 kHz) for the A_V, the analysis of the obtained results shows a relatively good agreement. The maximum value of the relative error is below 2%. As can be seen at higher frequencies for gains 10 (20 dB), 100 (40 dB) and 1000 (60 dB), the proximity between the two characteristics is preserved, and for the frequency of the first pole (dominant pole) the relative error does not exceed 5%. For voltage gain equal to 1 (0 dB) the difference with the SPICE macro-model of AD8253 is greater than 10%, as in its characteristic a peak occurs and the frequency of the dominant pole is obtained with value higher than 10 MHz. For the proposed behavioral model, at frequency higher than 10 MHz the slope of the voltage gain is approximately equal to –40 dB per decade.

978-1-7281-3420-8/19 $31.00 © 2019 IEEE

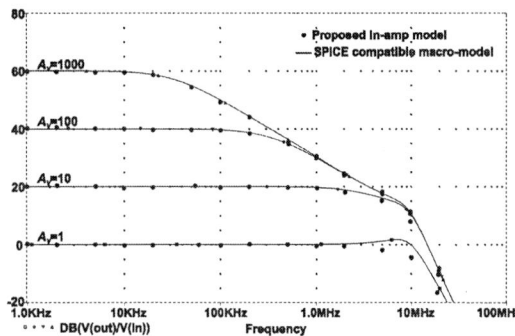

Fig. 5. Simulated amplitude-frequency responses obtained with the SPICE macro-model of AD8253 and proposed in-amp model.

The frequency dependences of the *CMRR* for the two simulation models are plotted in Fig. 6. At low frequencies (up to 1 kHz) for the various differential voltage gains, the analysis of the obtained results for the *CMRR* shows that the maximum value of the relative error is below 5%, which guarantee sufficient degree of accuracy. For frequencies higher than the corresponding pole frequency the *CMRR* decreases monotonically with slope around −20 dB per decade. This form of the frequency responses corresponds to one-pole approximation of the frequency dependence of the *CMRR* for majority of the in-amps.

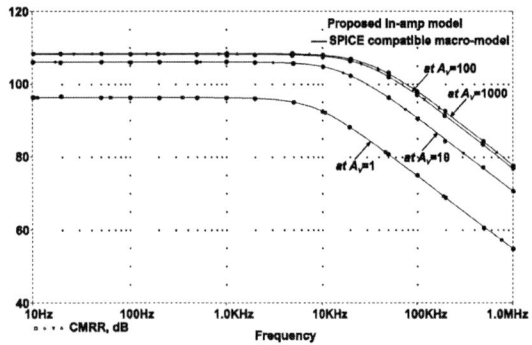

Fig. 6. Simulated *CMRR* frequency dependences with the SPICE macro-model of AD8253 and proposed in-amp model.

Table III presents a comparison of the created model with SPICE compatible macro-model, according to the slew rates, number of components and simulation parameters. As can be seen, for the created model, the minimal values of the slew rates are simulated and the value of the relative error against the datasheets [4] not exceeding 4%. The analysis of the simulation results shows that for the macro-model at gains equal to 100 and 1000, the slew rates are less than the minimal values, whereas for the gains 1 and 10 the slew rates are greater than the minimum values. The relative error for them is greater than 10%, since the equivalent circuit of the macro-model is relatively complicated, and contains a large number of passive and active component (totally 265 devices).

TABLE III
COMPARISON BETWEEN SIMULATION RESULTS FOR BOTH MODELS

Parameter	This model	SPICE compatible macro-model [5]
Large-signal pulse response (V_{out} = 10 V_{p-p} and R_L = 10 kΩ)		
Slew rate at A_V = 1	±19.29 V/μs	±27.04 V/μs
Slew rate at A_V = 10	±19.62 V/μs	±26.87 V/μs
Slew rate at A_V = 100	±12.03 V/μs	±5.37 V/μs
Slew rate at A_V = 1000	±1.99 V/μs	±1.34 V/μs
Component list		
Passive components	6	71
Diode / Transistor	–	21 / 82
Independent source	9	36
Controlled switches:	–	24
Controlled source	14	31
Simulation summary		
Simulation parameters:	eps=1.0e–7	abstol=1.0e–12
Simulation parameters:	hmin=0.9e–9	gmin=1.0e–12
Number of nodes	8	153
Performing DC analysis Number of iterations:	4	–
Simulation from 1kHz to 100MHz. CPU time:	0.46 s	1.09 s
Simulation to 10 ms. CPU time:	0.29 s	≥ 2.81 s

IV. CONCLUSION

In this paper a verification and validation process of the developed VHDL-AMS model for programmable monolithic in-amps has been presented. The proposed model shows good compliance with the chosen real device, both for differential and common-mode input signals. In comparison with the corresponding SPICE compatible macro-model the created behavioral model contains significantly smaller number of nodes and electronic components, but providing the same detail of the analysis. As a result of the simplicity, the model runs about five times as fast on transient analysis and with better DC convergence of the computation process.

ACKNOWLEDGEMENTS

The author would like to thank the Research and Development Sector at the Technical University of Sofia for the financial support.

REFERENCES

[1] I. Pandiev, VHDL-AMS Model Development for Digitally Programmable Monolithic Instrumentation Amplifiers, MIEL 2019 to be published.

[2] R. Sargent, Verification and Validation of Simulation Models, in *Proc. of the Win. Sim. Conference*, 2011, pp. 183-198.

[3] Q. Wang, *Mixed-signal SOC Verification using Analog Behavioral Models*, EDN Network, 2012, pp. 33-37.

[4] Programmable Gain Instrumentation Amplifier AD8253 – datasheet. Analog Devices, Norwood, MA, USA, 2009.

[5] AD8253 SPICE Macro Model. Analog Devices, MA, USA, Rev F., 2015.

PROC. 2019 IEEE 31st INTERNATIONAL CONFERENCE ON MICROELECTRONICS (MIEL), NIŠ, SERBIA, SEPTEMBER, 16-18, 2019

Multi-phase ring-coupled oscillator for TDC using a differential inverter with an oscillation frequency booster circuit

T. Shima, S. Kozuki, T. Otsuka, and N. Retdian

Abstract— **A multi-phase ring coupled oscillator is a time measuring component of a time-to-digital converter (TDC). The proposed multi-phase ring coupled oscillator consists of a differential inverter ring oscillator. For improving the time resolution, an enhanced sub-oscillator is added to boost the oscillation frequency. Oscillation mode analysis and the start-up technique discussed are for the multi-phase ring coupled oscillator to oscillate in a specified mode. The proposed circuit performance simulates by CADENCE hspiceD. The simulation model used is a 0.18um 1-poly 5-metal CMOS transistor model.**

I. Introduction

Improvements in semiconductor processes have enabled ultra-low voltage and high-speed operation. This improvement is suitable for digital circuit applications. However, the resulting narrow voltage operating range is undesirable for analog circuit applications. A time-to-digital converter (TDC) is mainly composed of time delay elements and flip-flops. Then the circuit components for the time-to-digital AD conversion is mainly digital circuits. Advances in CMOS process will be possible to realize sub-picoseconds TDC.

TDC shows many applications, such as time of flight measurements, IoT sensor devices, digital PLL phase detectors, PET, et.al. In the original TDC circuit [1], the delay time of the delay element determines the time resolution of TDC. The Vernier TDC [2] was developed to improve the time resolution. By using the caliper mechanism, the time resolution is improved by configuring delay lines on both the start and stop signal propagation paths.

Multi-phase ring oscillator TDC (MPTDC) has been proposed to reduce the influence of PVT fluctuation

Takeshi Shima is with the Department of Electrical, Electronics and Information Engineering, Kanagawa University, Yokohama, 221-8686 JAPAN. e-mail: shima@kanagawa-u.ac.jp.

Shun Kozuiki previously studied at the Department of Electrical, Electronics and Information Engineering, Kanagawa University, Yokohama, 221-8686 JAPAN, and now he is with Nidec Elesys Corporation, Yokohama, JAPAN, r201303906bg@jindai.jp.

Tamiki Ohtsuka previously studied at the Department of Electrical, Electronics and Information Engineering, Kanagawa University, Yokohama, 221-8686 JAPAN, and now he is now with the Department of Electronics and Electrical Engineering, Keio University, e-mail: t.ohtsuka@phot.elec.keio.ac.jp.

Nicodimus Retdian is with Shibaura Institute of Technology, Saitama City, 337-8570 JAPAN. e-mail: nico@shibaura-it.ac.jp.

using a stable external oscillator [3].

To further improve the time resolution, the time measuring components of TDC [4] utilize a multi-phase ring coupled oscillator (MPRCO). The phase difference of the output waveforms of the MPRCO is the reference delay. The time resolution of the multi-phase ring coupled oscillator is $1/(Nf_c)$, where N is the number of oscillators and f_c is the oscillation frequency. In this equation, f_c is the oscillation frequency of each oscillator. The time resolution can be improved simply by increasing N. The time resolution of the MPRCO is expected to be higher than the resolution of the so-called inverter ring oscillator.

However, the instability of the oscillation mode degrades MPRCO performance. Uncertainty in the oscillation mode means that the measurement result by the coupled ring oscillator is different every time TDC is turned on. The oscillation mode is the relationship between the temporal appearance order of each oscillation waveform and the node arrangement in space. The waveforms arranged in phase difference order are labeled W1, W2,..., respectively. The node names are marked, as shown in Fig.1. Let the waveform of n1 be W1. Mode indicates from which node the next output waveform W2 is output. For example, if W2 is n3 output waveform, Mode is 2. In MPRCO, there are several different oscillation modes observed depending on the number of stages N, power supply voltage, transistor characteristics, et.al. Therefore, TDC based on MPRCO needs to be encoded, taking into account the existing oscillation mode in [4].

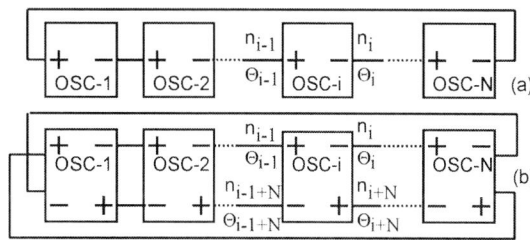

Fig. 1. (a) MPRCO using single-ended oscillators. (b) MPRCO using differential oscillators. n_i is the node name, Θ_i is the oscillation phase, $i = 1, 2, \cdots, N$, and N is the number of stages.

978-1-7281-3420-8/19 $31.00 © 2019 IEEE   273

The reference [5] proposed the differential inverter for MPRCO, where each oscillator in Fig.1(b) uses the differential inverter shown in Fig.2(a). In this circuit, two three-stage ring oscillators are appended at the differential output nodes in each differential inverter, respectively. The portion surrounded by the bold broken line is an appended ring oscillator called the sub-oscillator to boost the oscillation frequency. The part surrounded by the dashed-dotted line is called the main oscillator. An input signal, MCs control the operation of the sub-oscillator. When the control signal, MC, is 1, the sub-oscillator starts to oscillate.

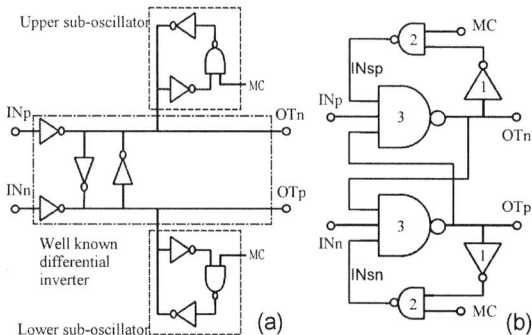

Fig. 2. Differential inverter with frequency booster for MPRCO. (a) Conventional differential inverter [5], and (b) Proposed differential inverter.

The advantage of using a differential inverter ring coupled oscillator is that the oscillator oscillates in even number N, where N is the number of differential inverters. The time resolution of the differential inverter MPRCO is $1/(2N \times f_c)$, where f_c is the oscillation frequency.

The problems with this circuit are as follows:

1. The oscillation mode depends on the logic gate size ratio of the differential inverter and sub-oscillator.
2. The output terminals of several gates are connected.
3. The sinal MC controls the operation of the sub-oscillator. However, its role is not defined.

In this paper, to solve these problems of the previous work, the advanced differential inverter MPRCO is proposed. Section 2 describes the proposed differential inverter for MPRCO.

Section 3 describes the startup procedure of the proposed circuit. Section 4 describes the circuit design method and simulation with hspiceD. Section 5 discusses PVT variations. Section 6 is the conclusion.

II. PROPOSED DIFFERENTIAL INVERTER FOR MPRCO

Solving the problems of the conventional differential inverter MPRCO, we propose the circuit in Fig.2(b), where the differential inverter consists of two 3-input

NANDs, two 2-input NANDs, and two INVERTERs. The central part of the differential inverter made of two 3-input NANDs, and the N-stage differential inverter ring oscillator, RO, is formed by this NANDs. One of the sub-oscillators configured by 3-input NAND, 2-input NAND, and an INVERTER, and plays a role of the oscillation frequency booster of RO. The N-stage differential inverters ring coupled oscillator with the sub oscillators forms MPRCO. MC uses to enable the sub-oscillators.

III. THE PROPOSED MPRCO OSCILLATION MODE

If the oscillation mode is different in every time the power is turned on, the measured result is unreliable. Oscillation mode stability is a critical issue when TDC applies MPRCO. The reference [6] formulated the locking phenomenon of the oscillator, and the reference [7] extended the formulation of the oscillator to the coupled oscillators. This section analyses the eigenvalues of the connection matrix of the MPRCO.

A. MPRCO using single-ended oscillators

The phase dynamics of the coupled oscillator in Fig.1(a) is as follows:

$$\frac{d\theta_i}{dt} = \omega_{oi} - \Delta\omega_{lock}\Sigma_{i=1}^{2N}\epsilon_{ij}\sin(\theta_i - \theta_j) \qquad (1)$$

, where ω_{lock} is the locking range, ω_{oi} is oscillation frequency, ϵ_{ij} is the connection coefficient. To derive steady-state phase in each oscillator as shown in [8], $\theta_i(t) = \phi_i + \omega_{ref}t$, $i = 1, \cdots, N$, and $\omega_{ref} = \frac{1}{N}\Sigma_1^N\omega_{oi}$. The adjacent oscillator phase relation is close to π,

$$\frac{d\phi_i}{dt} = \omega_{oi} - \omega_{ref} + \Delta\omega_{lock}\Sigma_{i=1}^{2N}\varepsilon_{ij}(\phi_i - \phi_j) \qquad (2)$$

The matrix form of the equation is

$$\frac{d[\phi]}{d\tau} = [\Delta\Omega_{tune}] + [\epsilon][\phi] \qquad (3)$$

where $\tau = \Delta_{lock}t$, $\Delta\Omega_{tune} = \frac{\omega_{oi}-\omega_{ref}}{\omega_{lock}}$. $[\epsilon]$ is the coupling magnitude matrix, and is

$$\begin{vmatrix} -1 & 0 & 0 & 0 & 0 & 0 & 0 & 0 & 0 & 1 \\ 1 & -1 & 0 & 0 & 0 & 0 & 0 & 0 & 0 & 0 \\ 0 & 1 & -1 & 0 & 0 & 0 & 0 & 0 & 0 & 0 \\ 0 & 0 & 1 & -1 & 0 & 0 & 0 & 0 & 0 & 0 \\ 0 & 0 & 0 & 1 & -1 & 0 & 0 & 0 & 0 & 0 \\ 0 & 0 & 0 & 0 & 1 & -1 & 0 & 0 & 0 & 0 \\ 0 & 0 & 0 & 0 & 0 & 1 & -1 & 0 & 0 & 0 \\ 0 & 0 & 0 & 0 & 0 & 0 & 1 & -1 & 0 & 0 \\ 0 & 0 & 0 & 0 & 0 & 0 & 0 & 1 & -1 & 0 \\ 0 & 0 & 0 & 0 & 0 & 0 & 0 & 0 & 1 & -1 \end{vmatrix}_{N \times N}$$

Eigenvalues of $[\epsilon]$ are

$$\lambda_k = -1 - e^{j\frac{(2k-1)}{N}\pi}, \qquad k = 0, 1, \cdots, N-1 \qquad (4)$$

Eigenvectors $[v_k]$ are

$$v_k = c_k[-a_k^9, a_k^8, -a_k^7, a_k^6, -a_k^5, \qquad (5)$$

$$a_k^4, -a_k^3, a_k^2, -a_k^1, 1] \qquad (6)$$

, where $a_k = e^{j\frac{(2k-1)}{N}\pi}$.

$$e^{\tau[J]} = \begin{bmatrix} e^{v_{i,0}}\lambda_0 & & \\ & \ddots & \\ & & e^{v_{i,N-1}}\lambda_{N-1} \end{bmatrix} \qquad (7)$$

where $[J] = [v]^{-1}[\epsilon][v]$, $[v] = [v_1, \cdots, v_{N-1}]$, and v_i is the eigenvectors, $[v_i] = [v_{i,0}, \cdots, v_{i,N-1}]$. The explicit form of $[\phi]$ is

$$[\phi] = e^{\tau[\epsilon]}[c] = e^{[v](\tau[J])[v]^{-1}}[c] = [v]e^{(\tau[J])}[c_1] \qquad (8)$$

Using the equation, the oscillation MODE 2 is the stable mode.

B. MPRCO using differential oscillators

The phase dynamics of the coupled oscillator in Fig.1(b) is as follows: For the phases, Θ_i, $i = 1, \cdots, N$, of N oscillators in Fig.1(b),

$$\frac{d\theta_i}{dt} = \omega_i - \frac{\omega_i}{2Q}\Sigma_{i=1}^{2N}\kappa_{ij}\sin(\Theta_j - 2\Theta_i + \Theta_{j+N}) \qquad (9)$$

The connection coefficient matrix $[\epsilon]$ is

$$\begin{vmatrix} -2 & 0 & 0 & 0 & 0 & 1 & 0 & 0 & 0 & 1 \\ 1 & -2 & 0 & 0 & 0 & 0 & 1 & 0 & 0 & 0 \\ 0 & 1 & -2 & 0 & 0 & 0 & 0 & 1 & 0 & 0 \\ 0 & 0 & 1 & -2 & 0 & 0 & 0 & 0 & 1 & 0 \\ 0 & 0 & 0 & 1 & -2 & 0 & 0 & 0 & 0 & 1 \\ 1 & 0 & 0 & 0 & 1 & -2 & 0 & 0 & 0 & 0 \\ 0 & 1 & 0 & 0 & 0 & 1 & -2 & 0 & 0 & 0 \\ 0 & 0 & 1 & 0 & 0 & 0 & 1 & -2 & 0 & 0 \\ 0 & 0 & 0 & 1 & 0 & 0 & 0 & 1 & -2 & 0 \\ 0 & 0 & 0 & 0 & 1 & 0 & 0 & 0 & 1 & -2 \end{vmatrix}_{2N \times 2N}$$

Analysis of this matrix derives oscillation mode.

IV. START-UP PROCEDURE

The start-up procedure consists of the RO oscillation phase and the MPRCO oscillation phase. In the RO oscillation phase, MC is 0, and the sub-oscillators in Fig.2(b) are not active. The inputs, INsp, and INsn of two 3-input NANDs are forced to be one, and the cross-coupled two 3-input NANDs act as a differential inverter. N-stage differential inverter RO oscillates in Mode N+1 under any condition. While this oscillation continues, the differential inputs, INp and INn repeats zero and one or one and zero, respectively, the following stage, differential inverter receives zero and one and vice versa.

In the MPRCO oscillation phase, MC is 1. Each upper or lower sub-oscillator is ready to oscillate. However, the oscillation of the sub-oscillator depends on the

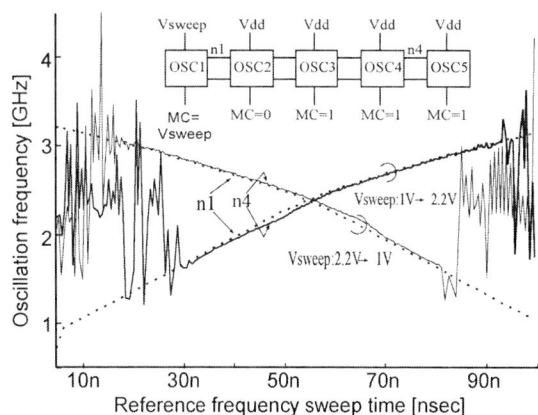

Fig. 3. Frequency pull-in phenomena of the proposed differential inverter. OSCi, i=,1,2,$\cdots$,5, are the proposed differential inverter in Fig2(b).

input INp and INn of each differential inverter. If INp in Fig.2(b) is one, the upper sub-oscillator oscillates. Conversely, INp of the differential inverter is zero, and the upper sub-oscillator does not oscillate at this moment.

Furthermore, in response to the oscillating sub-oscillator, the following stage, a non-oscillating sub-oscillator starts to oscillate. The previous stage oscillation induces the oscillation frequency and the oscillation phase of the subsequent stage sub-oscillators. Figure 3 shows the pull-in phenomenon of the proposed differential inverter. The block diagram in the figure shows the test circuit to evaluate the pull-in phenomenon. OSC1 acts as the FM signal source, and the oscillation frequency of OSC1 is swept from about 1 GHz to 3.2 GHz, or from about 3.2 GHZ to 1 GHz. OSC2 is the waveform shaping buffer, and OSC5 is the dummy load. The graph in Fig.3 explains why the oscillation of the sub-oscillator follows the oscillation of the preceding sub-oscillator [6]. The pull-in range is from about1.6 GHz to 3.0 GHz, where the free-running frequency of each oscillator is about 2.4 GHz.

As described above, half the sub-oscillators start to oscillate simultaneously at the beginning of the MPRCO oscillation phase. These sub-oscillators locate alternately and oscillate at the free-running oscillation frequency. Figure 4 shows the waveforms which simulated the starting characteristic in the MPRCO oscillation phase. In Fig.4 the waveform at the node n10 oscillates from 3 nsec. However, the waveform at the node n1 is not oscillating until about 4.9 nsec.

Figure 5 shows the steady-state oscillation waveforms of the proposed differential inverter MPRCO. The oscillation mode observed is N+1.

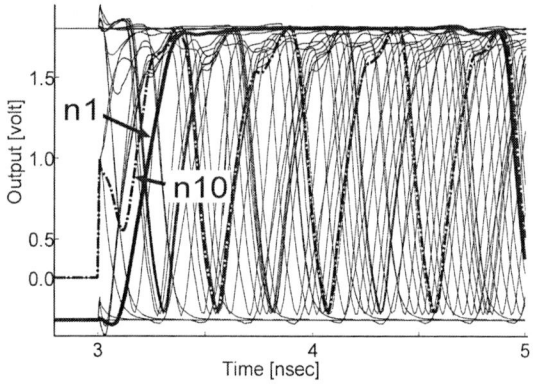

Fig. 4. Waveforms of the differential inverter MPRCO, where MC=0 in the RO oscillation phase until 3 nsec. MC=1 in the MPRCO oscillation phase started at 3 nsec.

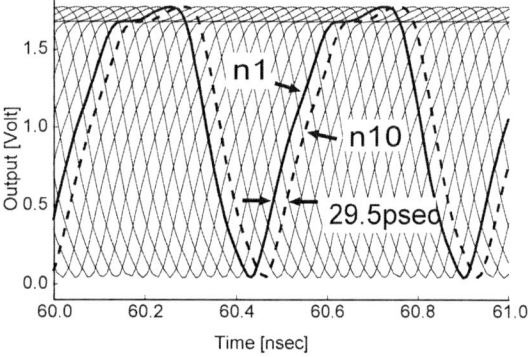

Fig. 5. Typical oscillation waveform of the proposed MPRCO, where MC=1 and the number of the stage, N, is 8.

V. SIMULATED RESULTS

8-stage proposed MPRCO simulated to evaluate the effectiveness of the proposed circuit and the start-up procedure. Table I shows the circuit parameters for simulation and the simulated oscillation frequency and the oscillation mode. The typical simulated oscillation waveform is in Fig.5. 0.18um 1-poly, 5-metal CMOS process is assumed for simulation. The power consumption of an 8-stage differential inverter MPRCO is 1.44 mW in the RO phase and is 13 mW in the MPRCO phase. The extra sub-oscillator consumes much power. However, it achieves very high oscillation frequencies.

VI. CONCLUSION

The circuit techniques to improve the TDC time resolution have been proposed to increase the oscillation frequency of the differential ring oscillator. Oscillation mode analysis and the proposed start-up procedure verify the proposed circuit oscillates at stable oscillation mode $N + 1$.

TABLE I

8-STAGE PROPOSED MPRCO SIMULATION. PARAMETER IS 3-INPUT NAND GATE SIZE. 2-INPUT NAND GATE CHANNEL WIDTHS ARE $\frac{Wp}{Wn} = \frac{2um}{2um}$ AND INVERTER GATE CHANNEL WIDTHS ARE $\frac{Wp}{Wn} = \frac{2um}{1um}$.

Test name	$\frac{Wp[um]}{Wn[um]}$	f_c [GHz]	Mode
TEST 1	6/9	2.04	9
TEST 2	4/6	2.11	9
TEST 3	2/3	2.31	9
TEST 4	1/1.5	2.01	9

ACKNOWLEDGMENT

This work was supported by JSPS KAKENHI Grant Number 16K06318. Also, it was partly supported by the VLSI Design and Education Center (VDEC), and the University of Tokyo in collaboration with Cadence Design Systems, Inc.

REFERENCES

[1] Y. Arai, and T. Baba, "A CMOS time to digital converter-VLSI for high-enaergy physics," IEEE Symposium on VLSI circuits, pp.121.-122, 1988.

[2] D. Dudek, S. Szczepanski, and J. V. Hatfield "A high-resolution CMOS time-to-digital converter utilizing a Vernier delay line," IEEE J. SSC, Vol. 35, No. 2, pp.240-247, Feb. 2007.

[3] S. Henzler, Time-to-Digital Converters. Springer 2010. M. Song, I. Jung, S. Pamarti, and C. Kim, "A 2.4 GHz 0.1-fref-bandwidth all-digital phase-locked loop with delay-cell-less TDC," IEEE Trans. on CAS-I, Vol. 60, No. 12, pp.3145-3151, Dec. 2013.

[4] S. Kozuki, N. Retdian, and T. Shima, "Experimental study of the oscillation mode of the coupled oscillator ORIGAMI for TDC", Mixed Design of Integrated Circuits and Systems 2017 MIXDES 24th International Conference, pp. 90-94, 2017.

[5] Tamiki Ohtsuka, Shun Kozuki, Takeshi Shima and Nicodimus Retdian, "Multi-phase ring-coupled oscillator for TDC using differential inverter," 3rd International Nordic-Mediterranean Workshop on Time-to-Digital Converters and Applications NoMe-TDC 2019, Vienna, May, 2019.

[6] R. Adler, "A study of locking phenomena in oscillators," Proc. of I.R.E. and Waves and Electronics, pp.351-357, June 1946.

[7] R. A. York, "Nonlinear analysis of phase relationships in quasi-optical oscillator arrays," IEEE Trans. on Microwave Theory snd Techniques, Vol.41, No. 10, pp.1799-1809, Oct. 1993.

[8] R. J. Pogorzelski, and A. Georgiadis, Coupled-oscillator based active-array antennas. JPL, Caltec., June 2011.

978-1-7281-3420-8/19 $31.00 © 2019 IEEE

A programmable current-mode digital-to-analog converter with correction of nonlinearity of input-output characteristics

J. Dalecki, R. Długosz, T. Talaśka, and G. Fischer

Abstract— The paper presents a concept of a programmable digital-to-analog converter (DAC) with the correction mechanism of a non-linearity of the input-output characteristics. The motivation behind the research presented in this work is an earlier project of a 10-bits two stage current mode DAC, whose basic block is a two-stage multi-output current mirror with binary weighted branches. The laboratory measurements of this converter were carried out, during which a certain non-linearity of the input-output characteristics was observed. Particular branches of the DAC theoretically should offer gains of the input trimming of 1, 2, 4, 8, ..., 512, respectively. In fact, in several sections this gain differs by several percent from the assumed values. The result of these observations is the proposition of a programmable correction mechanism, suitable for current mode converters.

I. INTRODUCTION

Digital-to-analog converters (DACs) play an important role in signal processing. They are a link between circuits that operate with digital and analog signals. Converters of this type are widely used in various engineering systems that include metrology [1], high data rates electronics [2], optoelectronics [3] control [4], telecommunication [5], etc. DACs are also used as components of larger circuits, e.g. as a block that provides a reference current (I_{ref}) in successive approximation register (SAR) analog-to-digital converters (ADC). They may also be used in artificial neural networks [6].

Depending on the application, different parameters play a key role. In metrology, for example, a high resolution and a high accuracy is desired. On the other hand, in wireless systems, especially in wireless sensor networks a low energy consumption is one of the most important features. However, independently on the application and thus the performance parameters, a basic feature of the DAC is the quality of its input-output characteristics. It has to be linear, which in general means, that for each following input code a difference between the resultant analog output values should be equal. In practice various problems may appear that

J. Dalecki, R. Długosz and T. Talaśka are with UTP University of Science and Technology Faculty of Telecommunication, Computer Science and Electrical Engineering ul. Kaliskiego 7, 85-796, Bydgoszcz, Poland, E-mail: rafal.dlugosz@gmail.com

G. Fischer is with IHP – Innovations for High Performance Microelectronics, Technologiepark 25, 15236 Frankfurt (Oder), Germany, Email: gfischer@ihp-microelectronics.com

lead to differential nonlinearity, integral nonlinearity, loss of the monotonicity, etc.

In this work we present selected results concerning a current-mode DAC we designed in the IHP CMOS 130 nm technology. We have made several core assumptions. The basic one was to reach as small chip area as possible. In fact we implemented a successive approximation register (SAR) analog-to-digital converter (ADC), in which the described DAC is one of the main components that provides a reference signal that in successive steps is compared with an input unknown signal (current in this case) [7].

The purpose of the mentioned assumption was to use a large number of such ADCs in a single chip. One of example applications of such ADC is a front-end ASIC used in nuclear medicine [8]. In systems of this type, even several hundred asynchronous channels operate in parallel, performing detection of X and gamma particles and registering their energy. In typical solutions of this type, analog multiplexers (MUX) are used, which redirect analog outputs from particular channels to the MUX connected then to a single high data rate ADC [9]. However, when the number of the channels is large and high data rates are needed, the ADC may become a bottleneck. For this reason, in an alternative approach, each channel may be equipped with its own ADC, which does not have to be very fast [10]. Such converters, on the other hand, should offer a low chip area, so as to enable placing a large number of them in a single chip, without a substantial growth of the overall chip area. In this approach, it is possible to use a digital MUX at the output of the chip.

Another applications, in which such converter may find the application are artificial neural networks (ANN) operating in mixed analog-digital mode. In such ANNs, the main computation tasks may be carried out based on analog blocks [11], [12], whereas the final neuron weights may be stored in a digital memory, for a longer period of time in this case. In addition, their values may be easily copied between the samples of the chip. This, however, requires the use of local low chip area DACs and ADCs. High output resolutions are not always required in this case [13].

One of our initial assumptions was to implement a 7 or 8 bit ADC. However, after preliminary study a larger

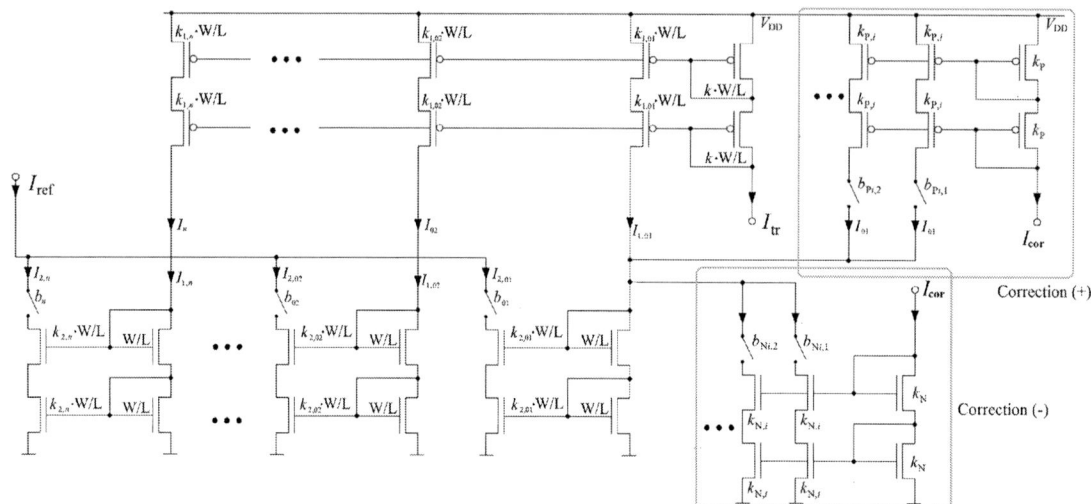

Fig. 1. Structure of the proposed 2-stage current-mode n-bits DAC with a correction mechanism.

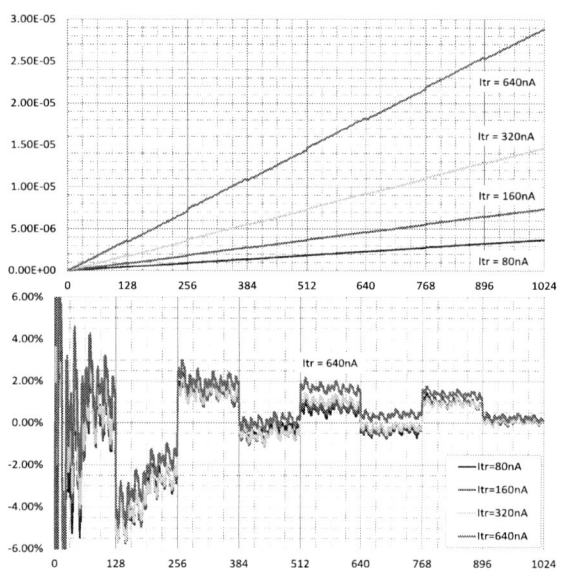

Fig. 2. Selected measurement results for $V_{DD}=1.2$ V as a function of the I_{tr} current.

resolution (10 bits) was realized, mostly for investigation purposes (the design chip is a prototype). The chip was fabricated and verified by means of laboratory tests. Selected results are presented in this paper. Since we observe some discrepancies between the theoretical and the obtained input-output characteristic, in this work we also consider some correction abilities for a future prototype of the proposed DAC.

II. THE PROPOSED CONCEPT OF A TWO-STAGE DAC

The proposed current-mode DAC has been realized as a two stage current mirror (CM) with binary weighted gains of particular branches. A general diagram of the circuit is shown in Fig. 1, here with the addition of a correction block described in next Section. The cascoded CMs have been used to increase the precision of the circuit.

A typical problem observed in CMs is a sensitivity to transistor mismatch, especially to threshold voltage mismatch. To reduce its impact on the linearity of the input-output characteristic, the sizes of particular transistors should be oversized. Unfortunately, in case of a large resolution of the DAC, it leads to very large sizes of the transistors associated with the most significant bits. In case of of the resolution of 10 bits, the width of the transistor in the branch representing the most significant bits (MSB) equals 512 unit widths (UW), compared to 1 UW in the least significant bit (LSB) branch. When the smallest transistors are oversized, the MSB transistors become excessively large. This in practice limits the resolution of the converter.

To solve the described problem, we proposed a DAC implemented as a cascade connection of two CMs. The first of them (1st stage) is a multi-output PMOS-type CM. Its particular outputs, i, are connected to its corresponding NMOS-type CMs (2nd stage). In this approach, the transistor widths in particular CMs and their particular branches do not need to be binary weighted, as it is required in conventional solutions. The gains of resultant branches are products of the gains of particular corresponding branches of intermediate CMs. This allows for keeping the spreads between transistors at relatively low level. In the implemented

978-1-7281-3420-8/19 $31.00 © 2019 IEEE

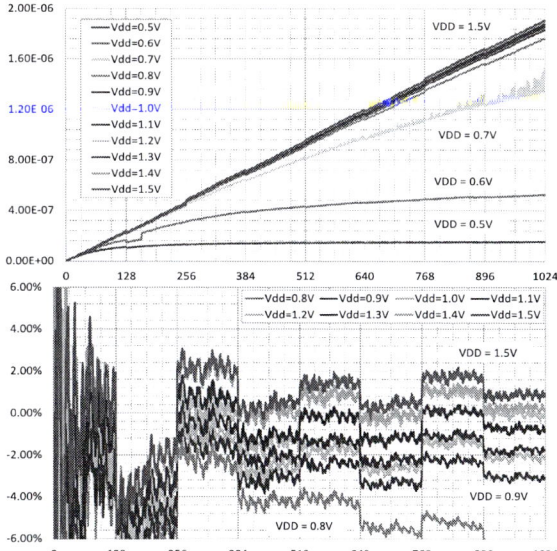

Fig. 4. Selected measurement results for I_{tr}=40 nA as a function of the V_{DD} supply voltage.

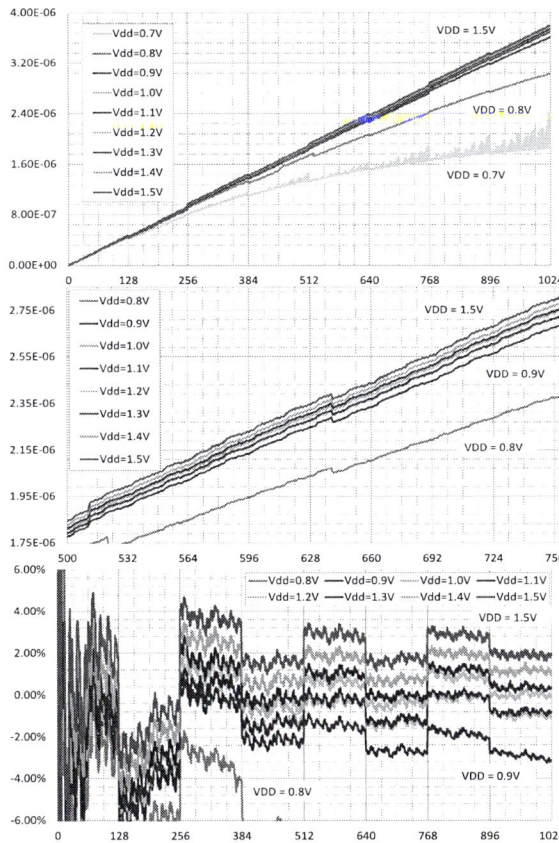

Fig. 3. Selected measurement results for I_{tr}=80 nA as a function of the V_{DD} supply voltage.

DAC, the gains of particular branches in the PMOS and the NMOS CMs are respectively, as follows:

$$k_{1,i} = \{1, 1, 2, 2, 4, 4, 8, 16, 16, 16\} \quad (1)$$

$$k_{2,i} = \{1, 2, 2, 4, 4, 8, 8, 8, 16, 32\} \quad (2)$$

The resultant gains ($k_i = k_{1,i} \cdot k_{2,i}$) and thus output currents ($I_{2,i}$) of particular branches are binary weighted.

The postlayout simulation results obtained in a detailed corner analysis were very good with gain errors not exceeding 0.1 %. However, in measurements the errors were larger, reaching even 2-3 %. We already have some suspicions, where the problem is. However, we consider also some alternative solutions.

III. Verification of the fabricated DAC and selected improvement possibilities

The measurements of the DAC were carried out on a specially prepared measuring board. The tests included several samples of the same chip prototype. In this work we present the results for one of them. To eliminate random errors, the measurements were repeated many times for each sample. The output current I_{ref} is the product of a constant (in a given test) control input current I_{tr}, a given digital code, and a constant coefficient $1/k$ (1/32 in this case). We measured the overall input-output characteristics for digital codes varying in the range from 0 to 1023 (for the resolution of 10 bits).

During the measurements we observed some nonlinearities in the input-output characteristics. It is worth mentioning that nonlinearities resulting from the technological process have fixed values for a given chip sample. In practice, this means that real values of particular coefficients $k_{1,i}$ and $k_{2,i}$ deviate by some constant values from their assumed values (Δk_i). The described problem may require modification of the size of selected transistors in the next prototype. Regardless of this, it was now possible to test the robustness of the DAC against variation of external conditions. For this reason, the circuits have been tested for different values of the supply voltage V_{DD} and different values of the I_{tr} current. Selected results are shown in Figs. 2, 4 and 3.

As can be observed in Figs. 2 – 3, regardless of the input current I_{tr} and for different values of VDD the shape of the input-output characteristics is almost the same. This is more clearly visible in the error waveforms. The differences are basically only in some DC levels. The value of the error for a given digital input code D_{inj} is computed according to following formula:

$$E_j = \frac{I_{ref j} - D_{inj} \cdot \frac{I_{max}}{2^N - 1}}{D_{inj} \cdot \frac{I_{max}}{2^N - 1}} \quad (3)$$

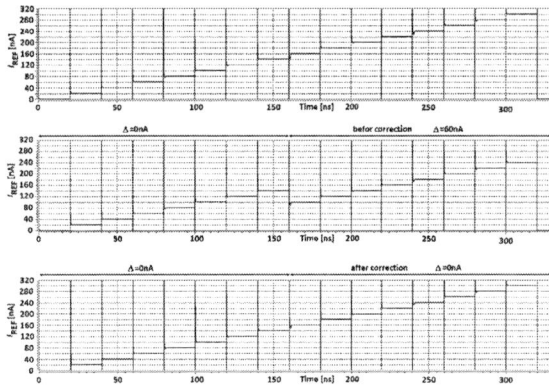

Fig. 5. Selected simulations results illustrating the correction procedure.

The $I_{\mathrm{ref}j}$ signal is a measured value of the output current of the DAC for the input code j. The value of the $I_{\max}$ current was adopted for the supply voltage of 1.2 V, which we treat here as the optimal case – it is the nominal voltage supply for this technology. As can be seen, similar error waveforms were obtained for the I_{tr} current varying by more than 10 times in particular carried out tests. This is a good illustration of a high flexibility of the circuit for such parameters as ranges of the processed input current, the power dissipation and data rate.

A. An example programmable correction mechanism for the DAC

One of possible correction possibilities is to introduce additional CMs to each branch of the DAC with small gains $K_{\mathrm{P},i}/K_{\mathrm{P}}$ and $K_{\mathrm{N},i}/K_{\mathrm{N}}$ (fractions of nominal gain values), controlled by additional bits $b_{\mathrm{P}i,l}$ and $b_{\mathrm{N}i,l}$. A preliminary concept of this solution is shown in Fig. 1. The correction block is shown only for one branch, for simplicity. The real gains of particular branches of the DAC, due to the technological process, may deviate from their nominal values in both the positive and the negative direction. To correct them, one can use two types of additional CMs, i.e. the PMOS and the NMOS ones. The first type can add some additional current to a given branch, while the second one subtracts some current from it. The additional CMs can be controlled by a separate current I_{cor} that allows for an additional calibration of the overall correction mechanism.

We have verified the described concept by means of transistor level simulations, as shown in Fig. 5 for a part of the input-output characteristic. Particular diagrams (from top to bottom) present: an ideal characteristic, a real one, as well as the one after the correction. In practice, the additional CMs can also be affected by some random errors. This causes that such a correction mechanism is not a perfect solution. However, if

the number of the branches in the correction CMs is relatively large, due to a kind of the adaptive calibration process, the parameters of the overall DAC will be improved by some value.

IV. CONCLUSIONS

The paper presents detailed measurement verification of a digital-to-analog converter designed by us in the CMOS 130 nm technology. Measurements have shown that there are some nonlinearities in the input / output characteristics. However, they remain constant for a given chip sample for different values of control signals, which is an advantage. The paper also presents an example possible nonlinearity correction system. This mechanism requires further investigations.

REFERENCES

[1] R. Kochan, A. Sachenko, V. Kochan, "Double cascade digital to analogue converter for metrology testing," *IEEE Instrumentation and Measurement Technology Conference*, Italy, 2014.

[2] Long Zhao, Ji He, Yuhua Cheng, "A 6bit 4GS/s current-steering digital-to-analog converter in 40nm CMOS with adjustable bias and DfT block," *International Conference on ASIC (ASICON)* , 2015.

[3] T. Sugihara, "Optical precoding technologies with high-speed DAC at 40G or beyond," *Asia Communications and Photonics Conference and Exhibition (ACP)*, China, 2011.

[4] H.K. Lam, "Sampled-data fuzzy-model-based control systems: stability analysis with consideration of analogue-to-digital converter and digital-to-analogue converter,", *IET Control Theory & Applications*, Vol. 4. Iss. 7, 2010.

[5] S. Khandagale, S. Sarkar, "A 6-Bit 500 MSPS segmented current steering DAC with on-chip high precision current reference," *International Conference on Computing, Communication and Automation (ICCCA)*, India, 2016.

[6] T. Talaśka, A. Rydlewski, R. Długosz, "A new realization of the conscience mechanism for self-organizing neural networks implemented in CMOS technology," *International Conference on Microelectronics (MIEL)*, Serbia, 2014.

[7] R. Długosz, and G. Fischer, "Low Chip Area, Low Power Dissipation, Programmable, Current Mode, 10-bits, SAR ADC Implemented in the CMOS 130nm Technology," *International Conference Mixed Design of Integrated Circuits and Systems (MIXDES)*, Gdynia, Poland, 2015.

[8] R. Długosz, P.A. Farine, K. Iniewski, "Power Efficient Asynchronous Multiplexer for X-Ray Sensors in Medical Imaging Analog Front-End Electronics", *Microelectronics Journal*, Vol. 42, Issue 1, 2011

[9] G. De Geronimo, P. O'Connor, J. Grosholz, "A generation of CMOS readout ASIC's for CZT detectors", *IEEE Transactions on Nuclear Science*, Vol. 47, 2000.

[10] M. Żołądź, P. Gryboś, M. Kachel, P. Kmon, R. Szczygieł, "Analogue multiplexer for neural application in 180 nm CMOS technology, *Sixteenth International Conference Mixed Design of Integrated Circuits & Systems (MIXDES)*, Poland, 2009.

[11] R. Długosz, T. Talaśka, W. Pedrycz, R. Wojtyna "Realization of the Conscience Mechanism in CMOS Implementation of Winner-Takes-All Self-Organizing Neural Networks", *IEEE Transactions on Neural Networks*, Vol. 21, Iss.6, 2010

[12] R. Długosz, T. Talaśka, W. Pedrycz, "Current-Mode Analog Adaptive Mechanism for Ultra-Low Power Neural Networks", *IEEE Transactions on Circuits and Systems–II: Express Briefs*, Vol. 58, Iss. 1, 2011.

[13] R. Długosz, M. Kolasa, M. Szulc, W. Pedrycz, P.A. Farine, "Implementation Issues of Kohonen Self-Organizing Map Realized on FPGA", *15th European Symposium on Artificial Neural Networks* (ESANN), Bruges, Belgium, 2012.

978-1-7281-3420-8/19 $31.00 © 2019 IEEE

A Parallel Adaptive LMS FIR Filter Realized in CMOS Technology

R. Długosz, T. Talaśka, T. Nikolić, and G. Nikolić

Abstract— **The paper presents an adaptive Finite Impulse Response (FIR) filter implemented at the transistor level in the CMOS technology. The filter is based on the LMS (least mean squares) adaptation algorithm. We applied several solutions that allows for simplifications of the hardware structure of the filter. One of them is to eliminate the division operation, replacing it with the bit-shifting operation. Parallel operation of the blocks representing particular filter coefficients allows to achieve operating speeds that are independent on the filter order. Main components of the filter were designed in the CMOS 130 nm technology and verified by means of laboratory tests, while selected structures of the overall filter have been examined by means of transistor-level simulations.**

I. INTRODUCTION

Adaptive filters are known for decades and thus are well described in the literature from the theoretical side [1], [2]. For this reason, especially in the field of such algorithms, as the least mean squares (LMS) one, it is difficult to expect substantial novelties. However, there is still a room for new solutions in the area of the implementation of such filters, and the fields of engineering in which such algorithms may be applied. Filters of this type are usually realized as computer programs, and more specifically as components of larger systems [3], [4]. Existing hardware implementations usually apply to FPGA [5], [6] and microcontrollers [7], [8], rarely specialized integrated circuits (ASIC) [9], [10]. In the first case, the advantage is the possibility of parallel operation of such filters, which is essential from the point of view of the speed of the system using such filters. Transistor level solutions usually concern analog realizations.

Depending on the application, adaptive FIR filters with different lengths may be used, even with several dozen or more coefficients. When realized in software, the data rate is in general inversely proportional to the length of the filter, due to sequential computations. A parallel data processing is possible both in the FPGA and the ASIC platforms. In such realizations, the data rate is almost independent on the filter length, however we pay for this advantage with the necessity of duplication of the blocks that represent particular filter coefficients. In such situations, it is necessary to look for possibilities to optimize the structure of particular filter components. It is an objective of this work.

Adaptive filters are used in a wide range of the applications. They are frequently used to identify parameters of an unknown object or in noise cancellation [11], [12]. Assuming that the object has linear properties, it may be modeled with the use of a linear filter, as it is assumed in the LMS adaptive algorithm. In case of such an object, unknown may be the order of the corresponding filter and the values of its coefficients. If the order of the model is at least equal to the order of object, the adaptation process should be convergent. As a result, particular coefficients of the model should become similar to the coefficients of the object.

II. PARALLEL HARDWARE IMPLEMENTATION OF THE LMS ALGORITHM

In this work we focus on a fully digital, parallel transistor level implementation of the LMS adaptive algorithm in the CMOS technology. In the scope of our interests here is, in particular, a block that handles computations for a single filter coefficient, h_i. A careful realization of this block is essential, as it is then used L times in the filter, where L is the filter length. In the proposed implementation, this block performs two operations alternately. One of them is calculating an update factor for a given coefficient based on the corresponding sample of the input signal X. This operation is theoretically performed in accordance with the formula below:

$$\hat{h}_i(n+1) = \hat{h}_i(n) + \mu \cdot e(n) \cdot x_i(n) \tag{1}$$

The μ parameter is a constant factor that controls the speed of the adaptation process. It plays a similar role as the learning rate η in many artificial neural networks. The adaptive filter may be viewed as a model of an unknown identified object. The $\hat{h}_i$ signals are coefficients of this model.

The second operation performed by the realized circuit is the multiplication of a coefficient $\hat{h}_i(n)$ by a corresponding sample $x_i(n)$ in moment n. The products of particular multiplication operations are then used to compute the output of the adaptive filter:

R. Długosz and T. Talaśka are with UTP University of Science and Technology, Faculty of Telecommunication, Computer Science and Electrical Engineering, ul. Kaliskiego 7, 85-796, Bydgoszcz, Poland and with Aptiv Servics Poland, ul. Podgórki Tynieckie 2, 30–399, Kraków, Poland, E-mail: rafal.dlugosz@gmail.com

T. Nikolic and G. Nikolic are with University of Niš, Faculty of Electronic Engineering, Aleksandra Medvedeva 14, Niš 18000, Serbia, Email: tatjana.nikolic@elfak.ni.ac.rs

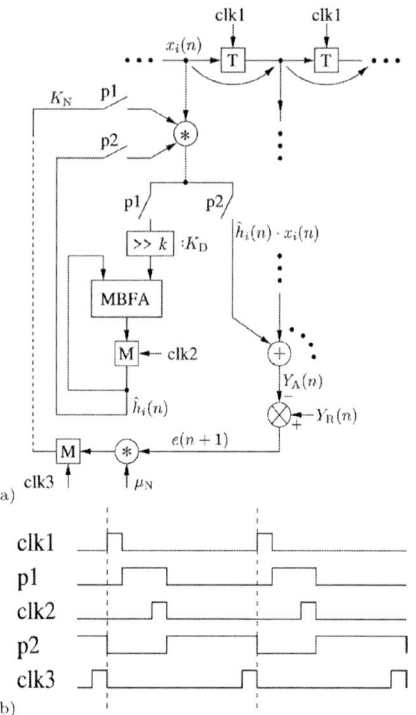

Fig. 1. The proposed solution: (a) block representing a single coefficient of the proposed adaptive filter, (b) the control clock scheme.

$$y_A(n) = \sum_{i=0}^{N} \hat{h}_i(n) \cdot x_{n-i}(n). \qquad (2)$$

In 1 and 2 we have three multiplication operations. In the CMOS implementation, depending on the signal resolutions, the circuit responsible for the multiplication is one of the most complex ones. This is why we focused on the solutions that allow for reducing the number of such operations.

The $e(n)$ factor is an error, which in this case is a difference between the output of the adaptive filter and the signal coming from an unknown process or object for a given time moment n. One can notice that both the factor μ and the signal $e(n)$ are equal, in a given moment, for all filter coefficients. For this reason, the multiplication of these quantities may be performed only once, regardless of the filter length and thus the number of the signal samples $x_i(n)$. The result of this multiplication is then provided, in parallel, to all blocks representing the filter coefficients.

The mu coefficient is a fractional number. For the purposes of the implementation in the CMOS technology, it can be expressed as:

$$\mu = \mu_N/\mu_D \qquad (3)$$

where μ_N is its numerator, whose value decreases during the adaptation process, while μ_D is a constant denominator whose value is one of the powers of the number 2. On the basis of such assumptions, a K coefficient can be introduced:

$$K = K_N/K_D \qquad (4)$$

where $K_N = e(n) \cdot \mu_N$ and $K_D = \mu_D$. The investigations performed by us show that the values of K_D may be limited to $8 - 128$ in the majority of cases, while the maximum values of μ_N is usually below 4 to 64, depending on the value of μ_D. Finally the computation of the update for a given filter coefficient may be expressed as follows:

$$\hat{h}_i(n+1) = \hat{h}_i(n) + \Delta\hat{h}_i(n) \qquad (5)$$

where:

$$\Delta\hat{h}_i(n) = K_N \cdot x_i(n)/K_D \qquad (6)$$

with two distinct operations, i.e. the multiplication and the division. Since as we assumed $K_D \in 2, 4, 8, 16, ...,$ the division operation can be performed by shifting all bits in the $K_N \cdot x_i(n)$ term by $k = \log_2(K_D)$ positions to the right.

The circuit that performs the operation given by 6, as well as a single multiplication (for i^{th} coefficient) in 2 is shown in Fig. 1 (a). Both these operations for a given coefficient $\hat{h}_i$ are performed alternately. For this reason, it is possible to make a reuse a single multiplication block per each coefficient, as shown in Fig. 1. To make it possible an appropriate clock scheme is required, as shown in Fig. 1 (b). In this figure two types of clock phases are distinguished. In phases p (process) the circuit performs some computations as described below. In phases denoted as clk the circuit updates particular memory cells with new values of particular signals computed in phases p.

In the clk1 phase we update the state of the delay line, by shifting particular samples $x_i(n)$ between delay elements (memory cells). This process is very quick and thus a relatively short period of time is reserved for this phase. Then we activate the p1 signal that allows to connect the K_N signal to the multiplier and then the product of the multiplication to the divider. All operations in 5 and 6 are in the p1 phase performed fully asynchronously. Once a new value of $\hat{h}_i$ is steady , the clk2 phase updates the memory M, and thus the updated coefficient may be used in the computation of the output signal of the filter. This operation is performed in accordance with 2 in phase p2 also fully asynchronously. In the p2 phase the circuit also computes new values of the e and then the K_N signal. Finally the K_N signal has to be also stored in a memory that is performed in the clk3 phase.

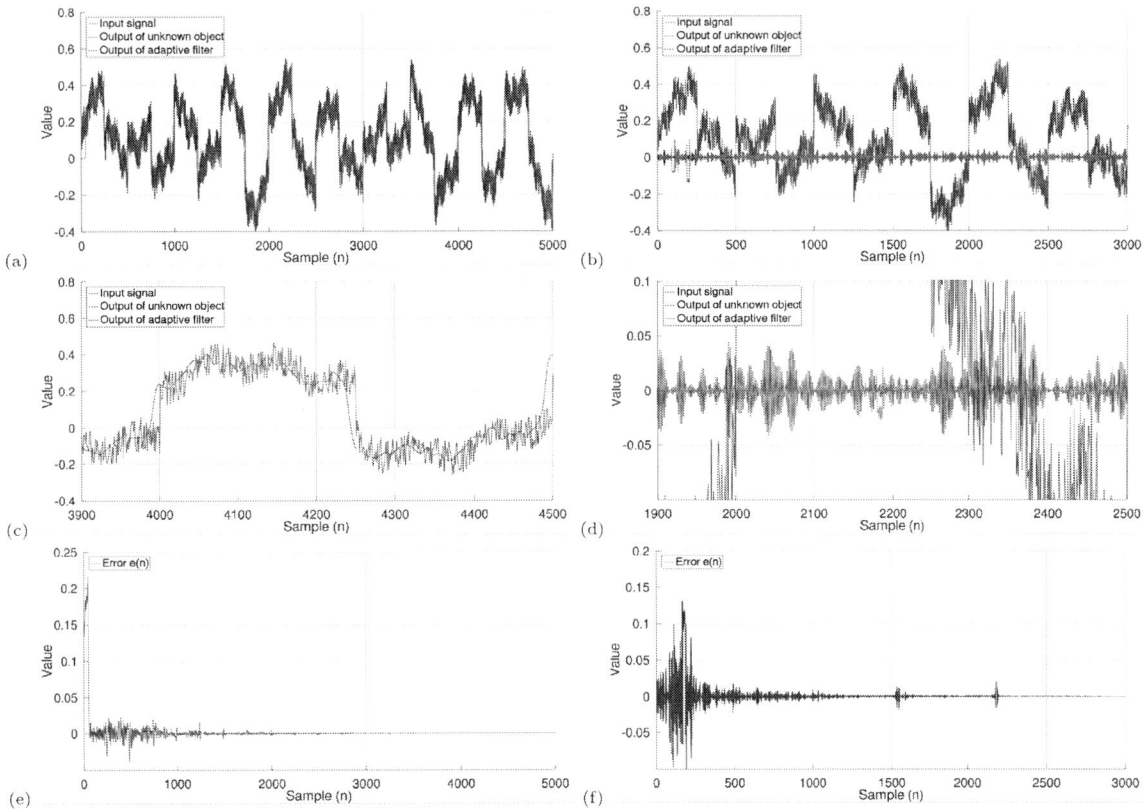

Fig. 2. Selected simulation results of the LMS adaptation process – a functional model of the proposed hardware implementation of the filter: (a, c, e) for low pass filter, (b, d, f) for high pass filter.

The sample of error signal e which will be used in the next adaptation cycle $(n + 1)$ is computed as follows:

$$e(n + 1) = Y_A(n) - Y_R(n) \qquad (7)$$

where $Y_R(n)$ is the output signal coming from an unknown object.

In a transistor level realization of the adaptive filter, the p1 and the p2 phases control blocks of switches, as particular signals are multi-bit. In Fig. 1 they are illustrated as single blocks, for a simplicity. In case of the hardware realization, it is also important to take into account the signs of particular signals. The K_N and x_i signals can be either negative or positive. For this reason they are coded in two's complement code. On the other hand, the division is performed always by the positive signal K_D. If the signs of K_N and x_i are different, then the most significant bit (MSB) of their product equals '1'. In this situation, in the division operation (shifting the bits to the right), the k oldest bits are filled with '1'. For equal signs of both multiplied signals, the k oldest bits are replaced with '0'.

The method of implementing a single filter coefficient is important here, as this block is then repeated L times

in the filter, as mentioned above. As a multiplier we used an asynchronous binary-tree circuit, described in detail in [13]. This circuit although more complex than a conventional shift-and-add circuit is much faster and strongly simplifies the structure of the clock.

Updating the filter coefficient requires an accumulation circuit, in Fig. 1 marked as ACU. It is composed of a multi-bit full adder (MBFA) operating in two's complement code and the multi-bit memory block. The output of the memory is provided to one of the multibit inputs of the MBFA. The $\Delta \hat{h}_i$ signal is provided to the second input of the MBFA. The memory block may be implemented in different ways. One of the possibilities is the use of a chain of NOT gates and switches, with data being stored in parasitic capacitances of the NOT gates. We applied such a solution in one of our prototype chips realized in the 130 nm CMOS technology. In this approach each bit requires only eight transistors – 4 in two NOT gates and 4 in two switches implemented as transmission gates, composed of PMOS and NMOS transistors. One of disadvantages of this solution is some information leakage observed in the measurements. To avoid the impact of the leakage and to keep transistor sizes small, the memory had to be refreshed with a frequency greater than 20 kHz. Another option is to use D-flip flops in the memory. It requires the use of a larger number of transistors, however, it offers the stability of the stored information. In addition, it requires only a 1-phase clock (the clk2 signal in the clock scheme described above), instead of a 2-phases clock required in the previous approach.

A. Verification of the proposed concept

The proposed concept of the hardware implementation of the adaptive LMS filter has been thoroughly verified in the software model. The computations were carried out exactly according to the sequence described above, taking into account the resolution of particular signals and coefficients. The filter was tested for different signal resolutions, different lengths L and different input signals. Exemplary simulation results are presented in Fig. 2. The results have been obtained for an input signal being a sum of several sinus waveforms with different frequencies, a pulse and a triangle waveforms as well as a white noise. We present the results for a symmetrical low-pass (LPF) and a high-pass filter with $L{=}21$. The coefficients of the model filter (LPF case) has been selected as follows:

$$\begin{aligned} H_{\mathrm{LPF}} = \{&1, 4, 10, 20, 35, 56, 80, 104, \\ &125, 140, 146, 140, 125, 104, \quad (8) \\ &80, 56, 35, 20, 10, 4, 1\}/1296. \end{aligned}$$

The results show that the adaptation process is convergent.

In the paper, we do not present transistor level verification of the proposed solution. The proposed digital filter is mostly composed of the blocks that have already been verified by means of simulation (Hspice or Cadence), as well as to a large extent in laboratory tests in a prototype chip designed CMOS 130 nm technology [14]. Hardware-implemented blocks include, for example, a multi-phase clock circuit, an asynchronous multiplier based on the binary tree, an accumulator with memory based on NOT gates, a MBFA, etc. The former simulation and the measurement results allows us to estimate the speed of the filter. In the CMOS 130 nm technology the p1 and p2 phases do not exceed 10 and 20 ns for signal resolutions of 16 bits. An additional time required for the clk1, clk2, clk3 phases does not exceed 3 ns, as these phase partially overlap with p1 and p2. Due to parallel data processing the overall computation cycle is almost independent on the filter length, so the achievable data rate may exceed 30 MSamples/s.

III. CONCLUSIONS

The paper presents a concept of a fully digital implementation of the LMS adaptive filter. One of the

assumptions was to achieve high data rate of the system that is possible through a parallel and largely asynchronous signal processing. We in particular focused on the optimization of the circuit representing a single filter coefficient, as this block is then duplicated in the filter. An appropriate signal processing sequence allowed us to use a single multiplier per coefficient and a simple clock generator. The overall filter has been verified in a software model that takes into account details of the hardware implementation. In the case of digital circuits it is acceptable at the functional level.

REFERENCES

[1] J. Lee, and Hsu-Chang Huang, "On the Step-Size Bounds of Frequency-Domain Block LMS Adaptive Filters", *IEEE Signal Processing Letters*, Vol. 20, Iss. 1, 2013.

[2] B.Kumar Das, and M. Chakraborty, "Sparse Adaptive Filtering by an Adaptive Convex Combination of the LMS and the ZA-LMS Algorithms", *IEEE Transactions on Circuits and Systems I: Regular Papers*, Vol. 61, Iss. 5, 2014.

[3] S. Omer Gilani, Y. Ilyas, and M. Jamil, "Power line noise removal from ECG signal using notch, band stop and adaptive filters", *International Conference on Electronics, Information, and Communication (ICEIC)*, 2018.

[4] G. Makwana, and L. Gupta, "De-noising of Electrocardiogram (ECG) with Adaptive Filter Using MATLAB", *International Conference on Communication Systems and Network Technologies*, 2015.

[5] A. Shiva, E. Senthilkumar, J. Manikandan, and V. K. Agrawal, "FPGA implementation of reconfigurable adaptive filters", *International Conference on Wireless Communications, Signal Processing and Networking (WiSPNET)*, 2017.

[6] C. Safarian, T. Ogunfunmi, W. J. Kozacky, and B.K Mohanty, "FPGA implementation of LMS-based FIR adaptive filter for real time digital signal processing applications", *IEEE International Conference on Digital Signal Processing (DSP)*, 2015.

[7] S. Shaikh, and S. Pujari, "Migration from microcontroller to FPGA based SoPC design: Case study: LMS adaptive filter design on Xilinx Zynq FPGA with embedded ARM controller", *International Conference on Automatic Control and Dynamic Optimization Techniques (ICACDOT)*, 2016.

[8] R.H. Standberg, P.B. Patel, and M.A. Soderstrand, "Comparison of microprocessor-based and FPGA-based adaptive sample rate notch filters", *Midwest Symposium on Circuits and Systems*, Vol. 2, 1996.

[9] S. Raghunadha Reddy, and P. JayaKrishnan, "ASIC implementation of distributed arithmetic in adaptive FIR filter", *International Conference on Circuit ,Power and Computing Technologies (ICCPCT)*, 2017.

[10] Yunzhi Dong, et all. "Adaptive digital noise-cancellation filtering using cross-correlators for continuous-time MASH ADC in 28nm CMOS", *IEEE Custom Integrated Circuits Conference (CICC)*, 2017.

[11] W. Hernandez, J. de Vicente, O. Sergiyenko, and E. Fernández, "Improving the Response of Accelerometers for Automotive Applications by Using LMS Adaptive Filters", *Sensors* 2010, 10, 313-329; doi:10.3390/s100100313

[12] S. Dixit, and D. Nagaria, "LMS Adaptive Filters for Noise Cancellation: A Review", *International Journal of Electrical and Computer Engineering* (IJECE) Vol. 7, No. 5, October 2017, pp. 2520 2529

[13] K. Kubiak, and R. Długosz, "Trade-offs and other challenges in CMOS implementation of parallel FIR filters", *International Conference Mixed Design of Integrated Circuits and Systems* (MIXDES), Gdynia Poland, Jun. 2019

[14] M. Banach, T. Talaśka, J. Dalecki, and R. Długosz, "New Technologies for Smart Cities – High Resolution Air Pollution Maps Based on Intelligent Sensors", *Concurrency and Computation: Practice and Experience*, Wiley, 2019, DOI: 10.1002/cpe.5179

Performance Evaluation of Block-Based Adaptive Algorithms

T. Nikolić, T. Talaśka, G. Nikolić, and R. Długosz

Abstract - Performance of real-time digital signal processing systems is limited by their data computing capability. Therefore, evaluation of different architectures to determine the most efficient one is an important task. An efficient architecture for the implementation of block-based least mean square (LMS) adaptive filter is presented in this paper. In order to achieve lower adaptation delay different methods for optimization are used. Proposed solution is implemented in FPGA technology. Adaptive filters with different orders and block lengths are analyzed. Simulation results indicate that, compared to sample by sample based algorithm, the adaptation delay may be reduced by up to N times, where N is the block size of input data samples.

I. INTRODUCTION

Digital FIR filters are widely employed in practical real-time digital signal processing applications [1], [2], [6]. Whereas any fixed filter is designed in advance with knowledge of the statistics of signals, the adaptive filter continuously adjusts to a changing environment through the use of adaptive algorithms that are needed in order to continuously update the filter coefficients. The adjustable parameters are dependent upon the applications. Noise cancellation is the most typical application in the previous works [3], [4]. This paper will focus on applying adaptive filtering in system identification (Fig. 1).

The selection of a technique for adaptation for a particular application depends on many factors, such as complexity, adaptation delay, ability to track rapid variations, and so on. The algorithm structure for effective implementation in hardware is major design consideration. A lot of work has been done to implement the LMS as one of the most widely used algorithms for adaptive filtering [5], [7]. The algorithm for every input sample first computes the output $y[n]$ using the current set of coefficients $\boldsymbol{h}_n[k]$, then computes the error between the desired response and the filter output, and then updates all the coefficients using the following equation:

$$\boldsymbol{h}_{n+1}[k] = \boldsymbol{h}_n[k] - \mu e[n]x[n-k]$$

Tatjana Nikolić and Goran Nikolić are with the University of Nis, Faculty of Electronic Engineering, Nis, Serbia, (e-mail: tatjana.nikolic; goran.nikolic@elfak.ni.ac.rs)
Tomasz Talaśka and Rafał Długosz are with the UTP University of Science and Technology, Faculty of Telecommunication, Computer Science and Electrical Engineering, Bydgoszcz, Poland (e-mail: rafal.dlugosz; tomasz.talaska@gmail.com)

The factor μ determines the convergence of the algorithm. Large values of μ result in fast convergence [1].

Block digital filtering involves the calculation of a block or finite set of filter outputs from a block of input values. Block LMS (BLMS) adaptive filtering is procedure in which the coefficients of the adaptive filter are changed only once for every block of input data, compared with the simple LMS that updates on a sample by sample basis. BLMS adaptive filtering provides faster processing so that supports higher sampling rate.

The main goal of this paper is performance evaluation of BLMS adaptive filter which is used in the system identification configuration. We will explore the use of embedded System-on-Chip (SoC) solutions in efficiently implementing adaptive algorithm for particular application [8], [9], [10]. Optimization techniques for hardware realization of arithmetic operations on numbers in a fixed point format will be used. Parallelism will be introduced to increase the speed of operation [11], [12]. Different architectures for the filtering will be compared.

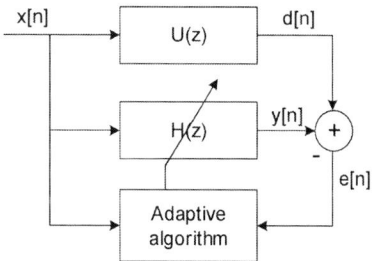

Fig. 1. System identification with adaptive filter

II. PROPOSED ARCHITECTURE

In order to increase speed of system operation and reduce adaptation delay, we modify conventional LMS adaptive filer architecture. The architecture of the BLMS adaptive filtering is depicted in Fig. 2. It consists of serial-in-parallel-out (SIPO), parallel-in-parallel-out (PIPO) and parallel-in-serial-out (PISO) buffers and the combinational computing blocks. If N is the block size and L is filter order, the SIPO buffer gathers an input block of N new data samples and feeds them in parallel to the first N locations of the PIPO buffer of size $(N+L-1)$. At the same time, the old values of these registers are bypassed by N locations to the right. The PIPO register thus holds the $(N+L-1)$ most recent samples to be processed by the combinational computing block. This combinational block performs the necessary multiplication and addition operations, to generate N output samples. PIPO buffer accepts the output samples and delivers them out in parallel to the combination block for computing the error signal. Let $\boldsymbol{h}_{n\text{-}N}$ be the array of coefficients computed for the previous block of input data samples. Firstly, the output samples $y[n]$ and the error signal $e[n]$ are computed using $\boldsymbol{h}_{n\text{-}N}$ for the current block i of data for $l = 0, ..., N\text{-}1$:

$$y[n] = \sum_{m=0}^{L-1} h_{n-N}[m]x[n-m], \text{ for } n = iN + l$$

$$e[n] = d[n] - y[n]$$

Using the error signal and the block of data samples, the array of coefficients for the current block of input data is updated as:

$$\boldsymbol{h}_n = \boldsymbol{h}_{n-N} + \mu \sum_{m=iN}^{iN+N-1} e[m]x[m]$$

These values are calculated in parallel within the combinational block for coefficient calculation. Note that the architecture of BLMS adaptive filter in the general case for k-th block of input data samples is shown in Fig 2.

Different values N and L are applied in the proposed BLMS adaptive filter and optimal filter length is determined. Our main goal is to show that a larger filter length than the one required increases the price of the adaptive filter and causes a decrease in performance that relate to operating frequency and occupied area.

The adaptation delay relates to the delay introduced by the adaptive filter structure consisting of finite impulse response filtering and the weight update process. For achieving lower adaptation delay implementation, we explore architectural design options for optimization of computational blocks such as parallel adders and multipliers. In our proposal, parallel multiplier architecture consists of three basic operations: partial product generation, partial product reduction, and computation of the final sum using a carry propagate adders (CPAs). As can be seen from Fig. 2, all the input samples are simultaneously multiplied by multiple coefficients and constant μ. Additionally, these samples are multiplied by the error signal e. Therefore, we use multiple constant multiplication technique for reducing the computational complexity of implementation of corresponding products.

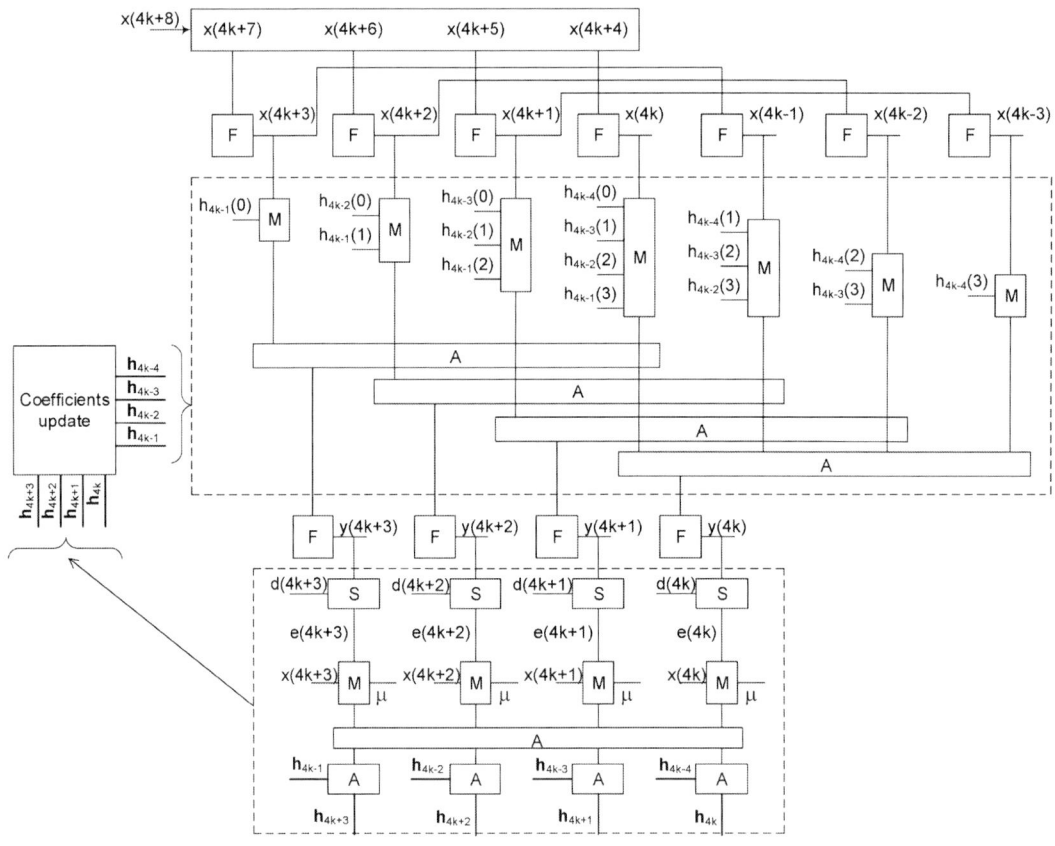

Fig. 2. Architecture of the BLMS adaptive filtering.
Notice: F – register, M – multiplier, A – adder, S – subtractor

Block processing method is used in proposed architecture to involve parallelism and to achieve a high speed of the hardware structure. A block LMS adaptive filter adjusts the weight coefficients once per block of data while LMS adaptive filter adjusts parameters once per each data sample. Due to parallelization of block processing the adaptation delay may be reduced by up to N times.

In addition, the conventional LMS algorithm can be modified to a form called the delayed LMS (DLMS) algorithm, which allows pipelined implementation of the filter. In this way, clock frequency adaptive filter is increased.

III. SIMULATION RESULTS

In order to assess the implementation results, the architectures of the conventional LMS and block LMS adaptive filter were described at register transfer level using VHDL and implemented on Virtex xc6vlx75t-3ff784 device by using Xilinx development CAD tool ISE WebPack. Design verification was performed using testbenches intended for excitation of different system architectures. For different filter orders (from 4 to 32) implementation of both mentioned architectures, LMS and BLMS, is performed. In all cases, the input sample width and coefficients width are equal 4 bits. In BLMS architecture the block size is equal to the filter order. The results generated by a CAD tool relate to number of occupied slices and minimal clock interval. Occupied silicon area expressed as the number of slices and maximal operating frequency in MHz with respect to the filter order are presented in Fig. 3 and Fig. 4, respectively.

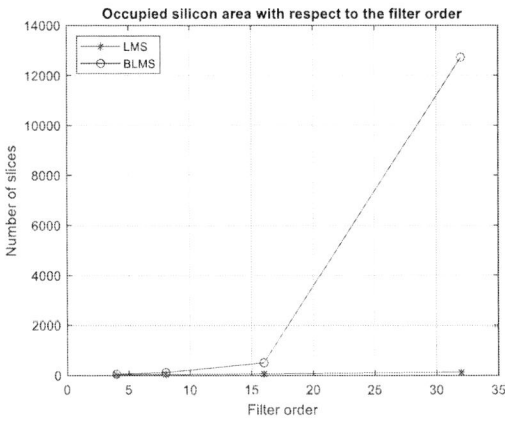

Fig. 3. Occupied silicon area with respect to the filter order

By analyzing the results presented in Fig. 3 and 4 we conclude the following:

a) With increasing filter order from 4 up to 16, occupied area of BLMS adaptive filter is increased from 2 up to 7 times compared to original LMS scheme; if the filter order is increased above 16 then area overhead of BLMS system is very significant.

b) For the filter order below 8, BLMS system operates at higher maximal frequency then LMS system; for the filter order equal to 8 maximal operating frequency of BLMS adaptive filter is smaller for 10 MHz in respect to LMS structure; with increasing filter order above 8, difference between maximal operating frequencies of BLMS and LMS adaptive filter is smaller. In addition, BLMS simultaneously processes block of input samples, so that its maximal operating frequency by sample is higher N (block size) times compared to LMS adaptive filter.

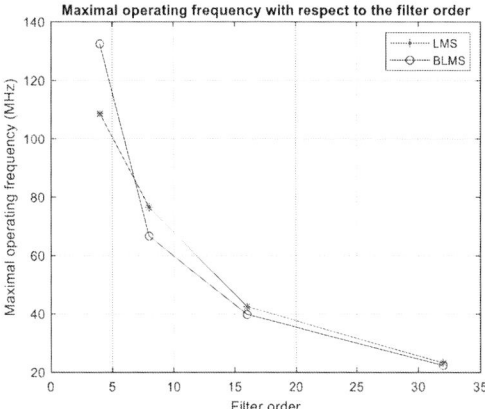

Fig. 4. Maximal operating frequency with respect to the filter order

For presentation of input data samples and coefficients numbers in a fixed point format are used. Since it is need to present them with higher number of bits. Occupied silicon area expressed as number of slices and maximal operating frequency in MHz with respect to the input sample width and coefficients width are presented in Fig. 5 and Fig. 6, respectively.

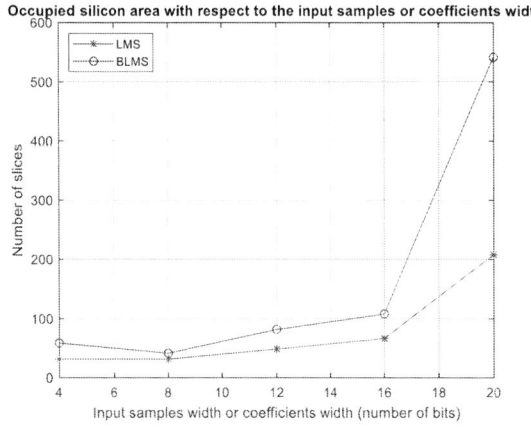

Fig. 5. Occupied silicon area with respect to the input samples width and coefficients width

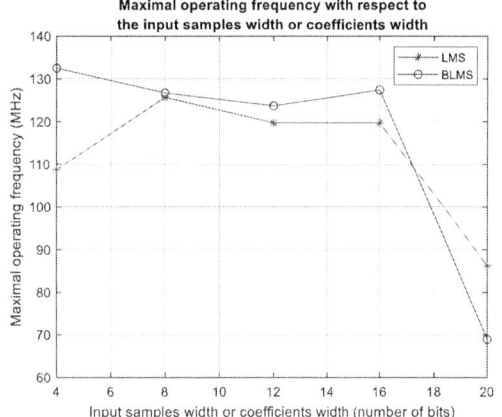

Fig. 6. Maximal operating frequency with respect to the input samples width and coefficients width

Results presented in Fig. 5 and Fig. 6 indicate the following:

a) The number of occupied slice in BLMS architecture is greater for values from 10 to 40 slices in respect to conventional LMS structure if number of input data sample bits and coefficient bits is less than 16. If the number of bits of input samples and coefficients is greater than 16, then the overhead of the BLMS filter surface in the LMS filter is much larger.

b) Maximal operating frequency of BLMS adaptive filter is higher in respect to LMS structure for all lengths of input data samples and coefficients less than 16 bits. If this number of bits is increased above 16, then the maximal operating frequency of both the filter decreases rapidly, while BLMS filter has a worse performance.

In according to previous, we conclude that BLMS adaptive filter has better performance in respect to conventional LMS adaptive filter for smaller filer order (less than 16), as well as for smaller lengths of input data samples and coefficients.

IV. CONCLUSION

In this paper we implement the adaptive LMS and block LMS filtering on FPGA chips for typical system identification applications and compare the behavior of LMS and BLMS adaptive algorithms in terms of chip area utilization and the critical path time or adaptive filter maximal operating frequency. The VHDL language is used for hardware description of proposed LMS and BLMS architectures. We consider adaptive filters with different filter order, input data sample width and coefficient width. The obtained implementation results by the CAD tool demonstrate that the BLMS algorithm is faster than LMS algorithm while it uses more chip area due to parallel structure. BLMS adaptive filter has better performance compared to conventional LMS if the filer order and lengths of input data samples and coefficients are less than 16.

ACKNOWLEDGEMENT

This work was supported by the Serbian Ministry of Education and Science, Project No TR-32009 – "Low power reconfigurable fault-tolerant platforms".

REFERENCES

[1] Shoab Ahmed Khan, *Digital Design of Signal Processing Systems, A Practical Approach*, John Wiley & Sons, Ltd., 2011.

[2] Marian Pristach, Vojtech Dvorak, Lukas Fujcik, "Enhanced Architecture of FIR Filters Using Block Memories", *IFAC Papers On Line*, 2015, Vol. 48, No. 2, pp. 306-311.

[3] Syed Ateequr Rehman , R. Ranjith Kumar, "Performance Comparison of Adaptive Filter Algorithms for ECG Signal Enhancement", *International Journal of Advanced Research in Computer and Communication Engineering*, 2012, Vol. 1, Issue 2, pp. 86-90.

[4] Ioana Homana, Irina Muresan, Marina Topa, Cristian Contan, "Fpga Implementation of LMS and NLMS Adaptive Filters for Acoustic Echo Cancellation", *Acta Technica Napocensis, Electronics and Telecommunications*, 2011, Vol. 52, No. 4, pp. 13-16.

[5] Pramod Kumar Meher, Sang Yoon Park, "Area-Delay-Power Efficient Fixed-Point LMS Adaptive Filter with Low Adaptation-Delay", *IEEE Transactions on Very Large Scale Integration (VLSI) Systems*, 2014, Vol. 22, No. 2, pp. 362-371.

[6] Mohanty B. K., Meher P. K., "A High-Performance FIR Filter Architecture for Fixed and Reconfigurable Applications", *IEEE Transactions on Very Large Scale Integration (VLSI) Systems*, 2016, 24(2), pp. 444–452.

[7] Hesam Ariyadoost, Yousef S. Kavian, Karim Ansari-As, "Performance Evaluation of LMS and DLMS Digital Adaptive FIR Filters by Realization on FPGA", *International Journal of Science & Emerging Technologies*, 2011, Vol. 1 No. 1 September, pp. 7-10.

[8] Safarian C., Ogunfunmi T., Kozacky W. J., Mohanty B. K., "FPGA implementation of LMS-based FIR adaptive filter for real time digital signal processing applications," *IEEE International Conference on Digital Signal Processing (DSP)*, 2015.

[9] Samrin Shaikh, Shashank Pujari, "Migration from microcontroller to FPGA based SoPC design: Case study: LMS adaptive filter design on Xilinx Zynq FPGA with embedded ARM controller," *International Conference on Automatic Control and Dynamic Optimization Techniques (ICACDOT)*, 2016.

[10] Shiva Ajay, Senthilkumar E., Manikandan J., Agrawal V. K., "FPGA implementation of reconfigurable adaptive filters," *International Conference on Wireless Communications, Signal Processing and Networking (WiSPNET)*, 2017.

[11] K. Kubiak, R. Długosz, "Trade-offs and other challenges in CMOS implementation of parallel FIR filters", *International Conference Mixed Design of Integrated Circuits and Systems (MIXDES)*, Gdynia Poland, Jun. 2019

[12] Manali Mukherjee, Kamarujjaman, Mausumi Maitra, "Reconfigurable architecture of adaptive median filter — An FPGA based approach for impulse noise suppression," *Proceedings of the 2015 Third International Conference on Computer, Communication, Control and Information Technology (C3IT)*, 2015.

978-1-7281-3420-8/19 $31.00 © 2019 IEEE

Comparative Analysis of Layout-Aware Fault Injection on TMR-based DMA Controllers

P. Chernyakov, A. Skorobogatov, A. Zvyagin, E. Emin, I. Danilov, A. Balbekov,
A. Shnaider Khazanova and M. Gorbunov

Abstract— **We present a comparative analysis of the layout-aware fault injection simulation results for Direct Memory Access (DMA) controllers with local, distributed, global and block Triple Modular Redundancy (TMR). The applied technique is also presented.**

I. Introduction

The improvement of the Single Event Effects (SEE) tolerance is an essential design problem [1]. Special attention is required not only for the fault-tolerance of memory arrays but also for their control circuits. Triple modular redundancy (TMR) is a very effective technique for soft error rate (SER) decrease. The experimental SER calculation by heavy ion beams is rather complicated, and the interest in fault injection simulation approaches [2] is growing.

In deep-submicron processes, a charged particle can cause upsets in several cells at once, which is impossible to simulate using only a netlist. Layout geometry must be considered at all design modeling levels: both at the SPICE [3] level and at the level of the synthesized netlist. A number of recent works [4, 5] represents different approaches of layout-based fault injection. Latter approaches [4, 5] are based on the proprietary fault injection tool provided by Cadence and incorporated in the Incisive® Enterprise Simulator, the suite of tools related to the design and verification of ASIC, systems on chip, and FPGA. Cadence fault injector does not take into account the device layout and a designer needs a special software generating configuration files for this tool.

This paper is organized as follows. Section II describes the proposed layout-aware fault injection technique. The DMA controller, which was chosen as the device under test (DUT) for validation and demonstration purposes, the test environment and a fault injection campaign introduced in Section III. The simulation results and analysis presented and discussed in Section IV.

II. Fault Injection Tool

We have developed a custom layout-aware fault injection tool for standard cell based designs. It is based on the extraction of layout information from the DEF (Design Exchange Format) file and matching it to netlist objects by hierarchical names. The tool is written in SystemVerilog and from a user point of view it is just a single SystemVerilog

class which can be used to inject faults into a DUT during a test. However, DUT's netlist and layout data are collected using C/C++ functions accessed through Direct Programming Interface, which is a part of the SystemVerilog standard. This technique is compatible with any HDL simulator, which fully supports the SystemVerilog standard.

The fault injection tool consists of several parts, the main of which are: a design information extractor, a design information database, a fault injector, and a fault injection history database. At the simulation start, one should perform the tool initialization by providing a hierarchical name of the DUT in a test environment, DEF file and LEF (Library Exchange Format) for standard cells. All necessary DUT information is extracted using this data and saved in the design information database.

During the simulation, the test environment can call injector methods to perform netlist based or layout based fault injection. Single Event Upsets (SEU) are simulated as a permanent change of a logic state to an opposite one at the sequential cell output. Single Event Transients (SET) are simulated as a temporal change of the logic state at the combinational cell output. An amount of injections through one simulation run is unlimited and defined by a test environment designer. The same is true for injection time, which is also defined by the designer. Also, an injection history is collected during the simulation, and several methods for access to it are available including reporting at the test end.

An injection target can be random or user-specified. For the netlist based injection, the target is the hierarchical name of the affected cell or cells. For the layout based injection, the target is the coordinate inside the DUT boundary. In the last case, the injector generates a square spot of a given size and defines which cells affected by this event according to the design information database using a binary search. After that, a procedure similar to the netlist based injection is performed for these cells.

The square spot is used to simplify calculations and could be treated as a pessimistic estimation. However, injector could be easily extended by a more accurate and advanced model.

It is worth to mention that the proposed technique doesn't require any DUT netlist modifications or special configuration files obtained using custom third-party tools to perform layout aware fault injection. All necessary information about DUT is collected during the simulation. Thus one could perform standard functional tests extended by fault in-

---

P. Chernyakov, A. Skorobogatov, A. Zvyagin, E. Emin, I. Danilov, A. Balbekov, A. Shnaider Khazanova and M. Gorbunov are with Scientific Research Institute of System Analysis of the Russian Academy of Sciences (SRISA), Nakhimovsky prosp. 36-1, 117218, Russian Federation, Moscow. E-mail: emin@cs.niisi.ras.ru

jection tool without significant modifications of the test environment.

III. DEVICE UNDER TEST: DMA CONTROLLER

According to the TMR methodology[6, 7] there are four types of TMR schemes:

1. Local (LTMR): Only flip-flops are triplicated. Voters are inserted and placed after the flip-flops. Clock domains are not triplicated.

2. Distributed (DTMR): The entire data-path of the design is triplicated (flip-flops and combinational logic). Clock domains are not triplicated.

3. Global (GTMR): The entire design is triplicated. Clock domains are also triplicated.

4. Block (BTMR): Three samples of the same block layout are connected via one voter.

The DMA controller was implemented as 5 different design versions: reference design without triplication (REF), DTMR, LTMR, GTMR and Block TMR. In addition, the reference design, LTMR, DTMR and GTMR versions were implemented with spatial separation (SS) of neighboring SEE-sensitive volumes. The registers were placed at the distances at least 3.2 μm from each other (without degrading the timing performance) thereby reducing the probability of multi-bit upset (MBU), i.e. the worst case scenario, when single particle strikes several registers simultaneously (see Fig. 1). Our task was also to minimize the sizes of blocks without timing performance trade-off. During physical design flow, every design has been optimized for 500MHz operating frequency with a margin of 100ps for clock jitter.

TABLE I
CHARACTERISTICS OF DMA BLOCKS

Type	Area, $\times 10^3 \mu m^2$	Density, %	Fmax, MHz	P_D, $\frac{\mu W}{MHz}$	P_L, $\frac{nW}{\mu m^2}$
REF	145.4	75.6	620	78.4	12.7
REF (SS)	145.4	75.8	597	81.7	13.0
LTMR	163.6	91.2	591	147.5	12.8
LTMR (SS)	374.1	54.0	554	191.9	14.8
DTMR	435.3	90.6	522	314.7	13.3
DTMR (SS)	489.9	75.4	534	268.8	13.1
GTMR	544.3	54.3	576	398.4	12.2
GTMR (SS)	544.3	57.5	476	370.5	12.7
BTMR	600.7	59.0	569	191.5	12.8

Table I shows the simulation results. For clarity, dynamic power at 20% switch activity and leakage power are converted to $\mu W/MHz$ and to $nW/\mu m^2$ correspondingly. In this paper, the custom designed radiation tolerant library based on TSMC (Taiwan Semiconductor Manufactur-

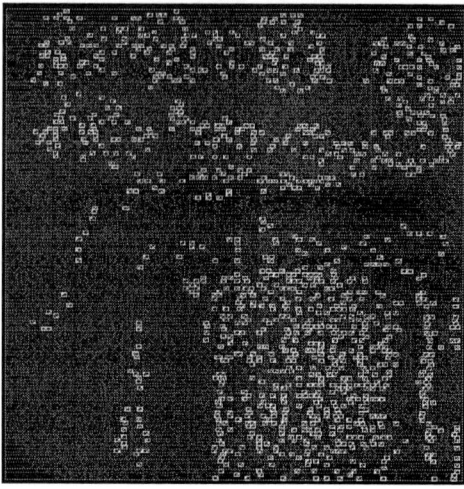

(a)

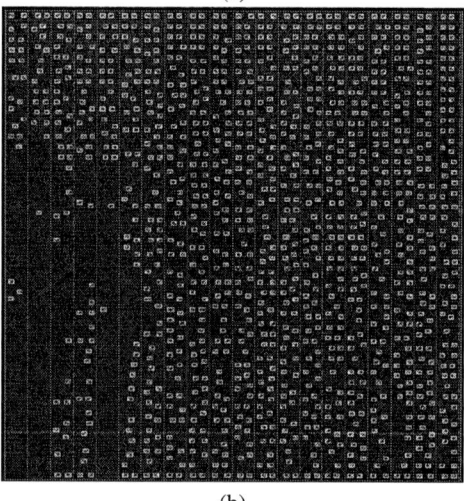

(b)

Fig. 1. Nontriplicated layout without (a) and with (b) spatial separation option with highlighted flip-flops positions.

ing Company) 65nm technology was chosen for implementation. The test system (see Fig. 2) includes:
• reference DMA controller with required test environment;
• 9 implementations of a tested DMA controller with a similar test environment;
• task generator;
• fault injection tool;
• error monitor.

First, the test task is launched on the golden model of the controller. Then, after the completion, the contents of the controller software registers, states of every device in the test environment and the test time are saved. At the next stage, the same task is launched on the controller test implementations. During the test, one fault is simulated into each controller at a random time. As soon as the test execution time expires, the results obtained from each implementation are compared with the golden ones. If the results are different,

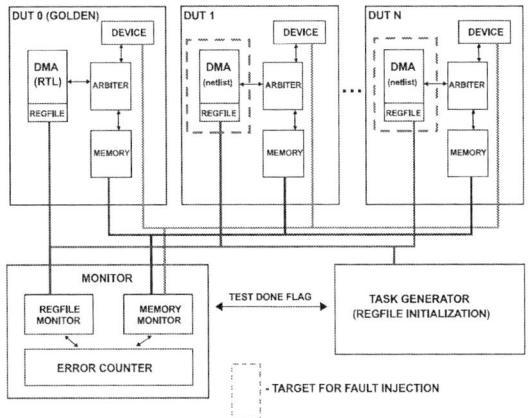

Fig. 2. Block schematic diagram of testbench system for DMA controller.

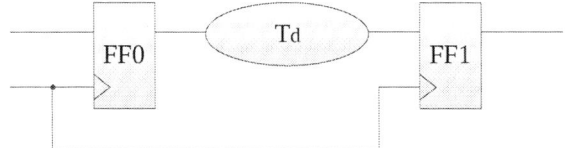

Fig. 4. FF0 to FF1 Data path.

an error is recorded. The test system tracks the following types of errors:

1. error in the transferred data (from memory to device or from device to memory);

2. error in the register file of the controller;

3. if the task execution time exceeds the reference time by more than 10 cycles.

During the work of tested controllers, the following test task is simulated on the golden DMA. This allows to run testing cyclically in order to collect statistics on faults and errors.

IV. SIMULATION RESULTS AND ANALYSIS

The above-mentioned test has been evaluated 10^5 times for each layout version. The results obtained for the events cross section for different frequency summarized in Fig. 3.

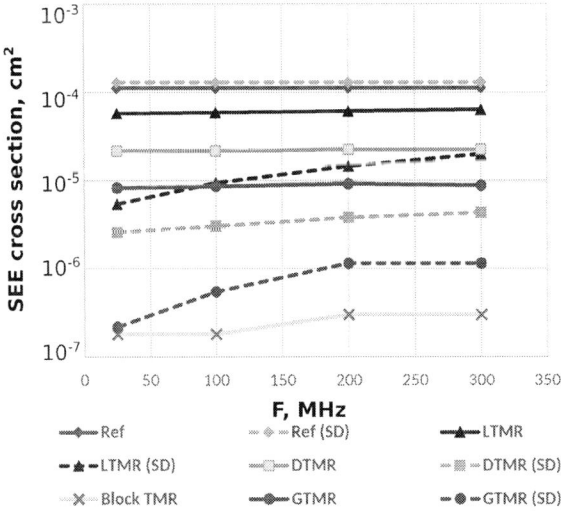

Fig. 3. SEE cross section simulation results for different TMR implementations.

We use the SEE cross section (in cm^2) as a metric for comparing mitigation strategies. The formulation of the metric is as follows: cross section σ is a number of failed tests N_{faults} to fluence Φ ratio.

The increase of the events cross section for the REF version with spatial separation compared to the same version but without spatial separation is a result of different local standard cells placement density. In the case of the version with SS the layout instances are more uniformly distributed. The difference in the distribution of flip-flops in the layout can be seen in Figures 1.

It was observed that in the REF version with spatial separation, the number of multiple faults of the flip-flops and their maximum multiplicity decreased, but the total number of tests with at least one fault has increased. This leads to general events cross section increase. Notable, that the number of tests with SET also has increased. But it has a lesser impact, because SET has a small contribution to the event cross section.

From the analysis of the event cross section frequency dependence observed in simulations, it can be noted that the number of SET induced errors increases and the number of SEU errors does not change or decreases with increasing frequency. The SEU reduction can be explained by time window shortage when fault in the flip-flop can lead to a test error. It can be explained as follows. Let us consider the case when a fault is simulated within a single period. Suppose that there are two consecutive flip-flops FF0 and FF1 (Fig. 4), the clock is ideal and reaches both FF simultaneously. A particle strikes only FF0 and inverts its logical state. FF1 captures the data from FF0 at the end of the clock cycle. It is obvious, that if incorrect data from FF0 reaches FF1 before next clock rising edge the fault will not be masked. Thus one could say that for fault propagation it is necessary that fault rise at a certain time window.

The condition of fault propagation can be formulated as inequality:

$$T_0 < T - T_d, \quad (1)$$

where T is clock period, T_0 is ion strike time inside the period provided that current clock cycle starts at zero time, T_d is combinational circuitry delay.

Therefore, the probability p of SEU fault:

$$p = (T - T_0 - T_d)/T = 1 - (T_0 + T_d)/T \quad (2)$$

Now let's consider SET which can occur somewhere from combinational circuitry on datapath from FF0 to FF1. Two conditions should be met to register fault in FF1. The first one is that the SET pulse should reach FF1 before the data capture. The second one is that the SET pulse should not disappear at the data input of FF1 before data capture event. The probability of satisfying both conditions increases with the operating frequency increase. It was also observed experimentally [8].

Thus, there are 2 trends: decreasing SEU cross section dependence and increasing SET cross section dependence on the operating frequency. There are no multiple flip-flop faults in LTMR, DTMR, GTMR with spatial separation option. A fault in a spatial separated flip-flops without simultaneous combinational logic faults does not lead to an error, since it is masked by redundant domains. Therefore, SET dependence prevails. The event cross sections for these controller implementations grow with increasing frequency. In versions without spatial separation, multiple faults are possible, which is also affects the event cross section. So in this case both trends are in effect and the frequency dependence can be more complex. Both trends in reference version with flip-flops separation are also have a significant role. Due to the separation, there are no multiple flip-flops faults. But SEU are not corrected and it make substantial contribution to the event cross section. For BTMR, the insufficient statistics obtained, which makes it irrelevant for analyzing.

It was observed, that LTMR with spatial separation has a smaller SEE cross section than GTMR without spatial separation at 25 MHz. At the same time, the area of LTMR is much smaller. Also, LTMR with spatial separation has a smaller SEE cross section than DTMR without spatial separation, at all simulated frequencies. But the difference with increasing frequency is reduced.

V. Conclusion

We present a comparative analysis of the layout-aware fault injection simulation results for Direct Memory Access (DMA) controllers with LTMR, DTMR, GTMR, BTMR and without TMR for the cases with and without spatial separation of neighboring sensitive volumes. The applied technique based on SystemVerilog is also proposed.

Most of the errors in DTMR with spatial separation caused by fault injection into the clock buffers.

The majority of GTMR errors caused by simultaneous faults in combinational cells, especially in two voters from the same domain. The increase in the number of errors with the frequency growth is driven by the probability of SET induced errors.

Almost all errors in BTMR are caused by particle hit into the clock buffers, the input port buffer or the output voter.

Block TMR has the smallest SEE cross section, but it has the largest area.

Table II shows the maximum cross sections for each DMA implementation and the frequency at which this cross section was obtained.

TABLE II

The maximum cross section for each implementation of DMA Block.

Type	Max cross section $\times 10^{-6}\ cm^2$	Frequency MHz
Ref	112.7	100
Ref (SS)	130.7	25
LTMR	63.5	300
LTMR (SS)	19.9	300
DTMR	22.3	200
DTMR (SS)	4.3	300
GTMR	9.1	200
GTMR (SS)	1.1	200
BTMR	0.3	200

Acknowledgements

The research was provided under financial support of the State Fundamental Research Program of the Russian Academy of Sciences (project No. 0065-2019-0008).

References

[1] P. E. Dodd et al., "Current and future challenges in radiation effects on cmos electronics," *IEEE Trans. on Nucl. Sci.*, vol. 57, no. 4, pp. 1747–1763, Aug. 2010.

[2] R. Velazco et al., "Predicting error rate for microprocessor-based digital architectures through c.e.u. (code emulating upsets) injection," *IEEE Trans. on Nucl. Sci.*, vol. 47, no. 6, pp. 2405–2411, Dec. 2000.

[3] A. O. Balbekov et al., "Circuit-level layout-aware modeling of single-event effects in 65-nm cmos ics," *IEEE Trans. on Nucl. Sci.*, vol. 65, no. 8, pp. 1914–1919, Aug. 2018.

[4] C. Bottoni, B. Coeffic, J.-M. Daveau, G. Gasiot, L. Naviner, and P. Roche, "A layout-aware approach to fault injection for improving failure mode prediction," in *Proc. of Workshop on SELSE*, 2015.

[5] B. Coeffic et al., "Radiation Hardening Improvement of a SerDes under Heavy Ions up to 60 mev.cm2/mg by Layout-Aware Fault Injection," in *Proc. of SELSE-12*, Mar. 2016.

[6] M. D. Berg et al. (2016). The effects of race conditions when implementing single-source redundant clock trees in triple modular redundant synchronous architectures, [Online]. Available: https://ntrs.nasa.gov/archive/nasa/casi.ntrs.nasa.gov/20160013226.pdf (visited on 07/30/2019).

[7] P. N. Osipenko et al., "Fault-tolerant soi microprocessor for space applications," *IEEE Trans. on Nucl. Sci.*, vol. 60, no. 4, pp. 2762–2767, Aug. 2013.

[8] S. Buchner et al., "Comparison of error rates in combinational and sequential logic," *IEEE Trans. on Nucl. Sci.*, vol. 44, no. 6, pp. 2209–2216, Dec. 1997.

Approximate Adder with Reduced Error

P. Balasubramanian, D. L. Maskell and K. Prasad

Abstract – A new approximate adder is proposed, which is suitable for FPGA- and ASIC-based implementations. Here, we consider an Artix-7 FPGA for the implementations using Vivado 2018.3. For 32-bit addition, the proposed approximate adder with an 8-bit least significant inaccurate sub-adder reports an improvement in the maximum frequency by 7.7% compared to the native accurate FPGA adder while consuming 22% fewer LUTs and 18.6% fewer registers. For 64-bit addition, the proposed approximate adder reports an increase in the maximum frequency by 9.1% than the accurate FPGA adder while consuming 11% fewer LUTs and 9.3% fewer registers. The power-delay product (PDP) is computed as the product of total on-chip power consumption and the minimum clock period. The proposed approximate adder achieves 14.7% and 9.3% reductions in PDP compared to the accurate FPGA adder for 32- and 64-bit additions respectively. Further, in comparison with a recent approximate adder presented in the literature, the proposed approximate adder reports a 40% reduction in the root mean square error (RMSE) while having the same design metrics.

I. INTRODUCTION

Approximate computing, also called imprecise or inaccurate computing, is considered to be a potential alternative to conventional accurate computing due to its ability to improve the speed, reduce the area, minimize the power and optimize the energy, while providing acceptably correct computation results [1]. In practical applications such as multimedia [2], low power graphics processing [3], big data analytics [4], artificial intelligence and machine learning [5], neuromorphic computing [6] etc. approximate computation results subject to a specified error bound are considered as acceptable. Approximate arithmetic circuits [7] is an important topic that has been receiving significant attention from the VLSI design community.

This paper proposes a new approximate adder that is suitable for FPGA- and ASIC-based implementations. Although several approximate adders have been presented in the literature many of them are unsuitable for a FPGA-based implementation since they incorporate custom logic which are inefficient to realize using a FPGA. This is because the fast carry chain embedded in the FPGA fabric would not be utilized in the FPGA implementation of an ASIC-oriented approximate adder, which would result in

P. Balasubramanian and D.L. Maskell are with the School of Computer Science and Engineering, Nanyang Technological University, Singapore 639798, E-mails: {balasubramanian, asdouglas}@ntu.edu.sg

K. Prasad is with the Department of Electrical and Electronic Engineering, Auckland University of Technology, Auckland 1142, New Zealand, E-mail: krishnamachar.prasad@aut.ac.nz

less speed and consumption of more FPGA resources compared to the native accurate FPGA adder, as remarked in [8]. For an example, we considered a 32-bit FPGA implementation of the ASIC-oriented error tolerant approximate adder of [9]. This approximate adder reports a maximum frequency of 454.545MHz (i.e., a clock period of 2.2ns), and requires 53 LUTs and 97 registers (flip-flops) for implementation. These design metrics were estimated post-place and route on an Artix-7 FPGA (part: xca100tcsg324-3) using Xilinx Vivado 2018.3 with the synthesis strategy set as Flow_AreaOptimized_high and with a default implementation strategy. Compared to the FPGA implementation of the approximate adder of [9], the 32-bit native accurate FPGA adder reports a 4.8% increase in the maximum frequency (476.19MHz) and requires only 32 LUTs (i.e., 40% fewer LUTs) while requiring the same number of registers. This suggests that ASIC-oriented approximate adders may not be suitable for an efficient FPGA-based implementation.

Further, approximate adders incorporating dynamic approximation where the degree of approximation may be varied and/or accurate outputs may be produced on demand may not be suitable for an efficient FPGA-based implementation either. This is because of the introduction of extra circuitry such as error detection and correction and/or carry prediction and control, and the likely need for invoking multiple clock cycles etc.

II. ACCURATE AND APPROXIMATE ADDERS

The block schematics of accurate and approximate adders are shown in Fig. 1. In Fig. 1, assuming an N-bit adder, A_{N-1} to A_0 and B_{N-1} to B_0 represent the input bits, and SUM_N to SUM_0 represents the sum output bits which includes the carry overflow. All the approximate adders in Fig. 1 incorporate an inaccurate sub-adder, which is least significant, and an accurate sub-adder which is more significant. K denotes the size of the inaccurate sub-adder and (N–K) denotes the size of the accurate sub-adder. In Fig. 1, 'Adder1' represents the accurate adder, and 'Adder2' to 'Adder6' represent the approximate adders, with 'Adder6' being the proposed approximate adder. References to the corresponding literature are provided for the approximate adders shown in Fig. 1.

As mentioned earlier, Adder1, shown in Fig. 1(a), represents the accurate adder.

Adder2 [10] is realized by combining an (N–K)-bits accurate sub-adder with a K-bits inaccurate sub-adder. The inaccurate sub-adder uses K 2-input OR functions. The OR functions perform logical disjunction of the corresponding

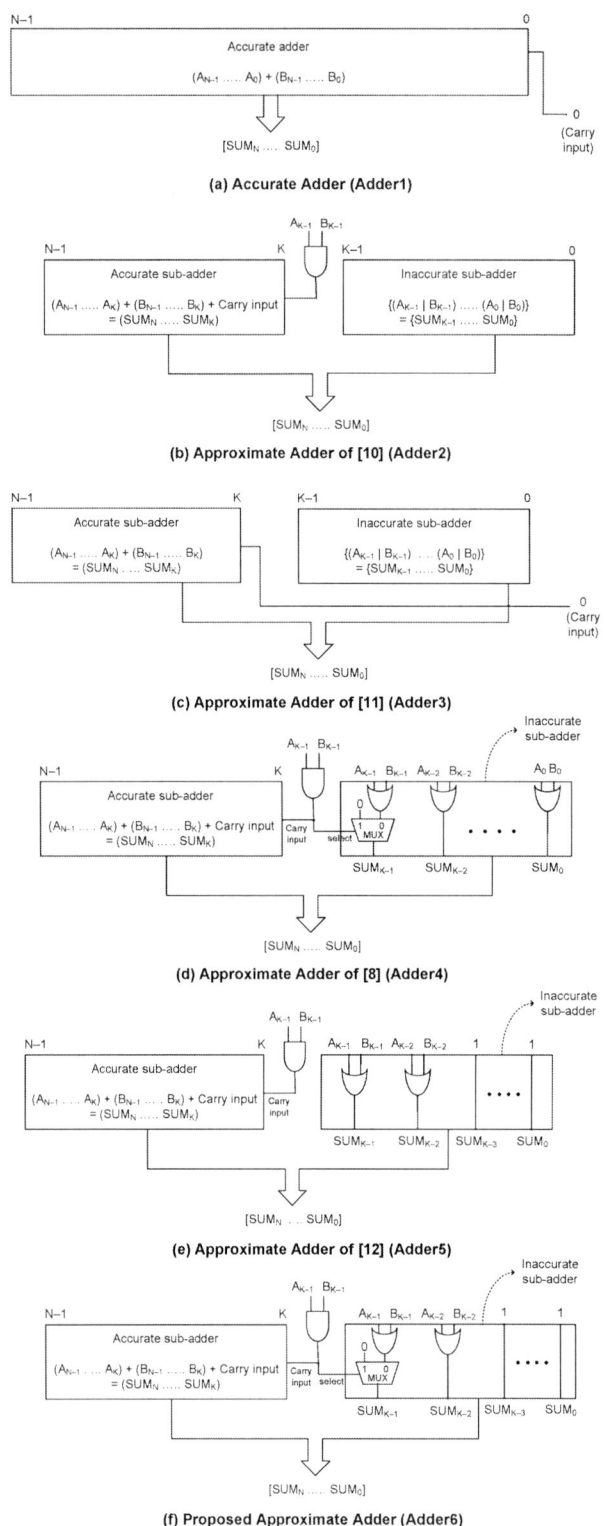

Fig. 1. (a) Accurate adder, and (b) to (f) are approximate adders. (f) is the proposed approximate adder.

978-1-7281-3420-8/19 $31.00 © 2019 IEEE

input bit-pairs to produce the respective sum output bits. Bit-wise addition is performed in parallel in the inaccurate sub-adder with no connections between the 2-input OR functions. The most significant input bit-pair of the inaccurate sub-adder is AND-ed and is given as the carry input for the accurate sub-adder.

Adder3 [11] is similar to Adder2, with the exception that there is no carry input given from the inaccurate sub-adder to the accurate sub-adder; rather, the carry input of the accurate sub-adder is set to 0.

Adder4 [8] is a modified version of Adder3, which additionally contains a 2:1 multiplexer (MUX) in the inaccurate sub-adder. The most significant input bit-pair of the inaccurate sub-adder is AND-ed and its output serves as the selection input for the MUX besides serving as the carry input for the accurate sub-adder. If the MUX selection input is 0, the most significant sum bit of the inaccurate sub-adder, i.e., SUM_{K-1} is the OR of A_{K-1} and B_{K-1}, i.e., $(A_{K-1} | B_{K-1})$. On the other hand, if the MUX selection input is 1, SUM_{K-1} would assume 0.

Adder5 [12] is a modified version of Adder2 in that only two 2-input OR functions are used to perform logical disjunction of the two most significant input bit-pairs in the inaccurate sub-adder to produce the sum bits SUM_{K-1} and SUM_{K-2}. The remaining less significant (K–2) sum bits of the inaccurate sub-adder are connected to the supply (V_{dd}) resulting in the production of a constant 1 for the (K–2) sum bits viz. SUM_{K-3} to SUM_0.

Adder6 is the proposed approximate adder, which is a modified version of the approximate adder, Adder4. The logic corresponding to the sum bits SUM_{K-1} and SUM_{K-2} of Adder6 are the same as Adder4. However, a constant 1 is set for the remaining less significant sum bits (SUM_{K-3} to SUM_0) by connecting these to V_{dd}, which would reduce the FPGA resources needed for their implementation.

III. FPGA IMPLEMENTATION RESULTS AND ERROR CHARACTERISTICS

In [13], the size of the inaccurate sub-adder considered for a practical video encoding application is limited to 8-bits. Typically, the degree of approximation is restricted to a range of 7- to 9-bits in an approximate adder. Hence, we considered K = 8 for the approximate adders shown in Figs. 1(b) to 1(f). However, the size of the inaccurate sub-adder may be varied based on need depending on a target application since the approximate adder architectures shown in Fig. 1 are generic. It may be noted that as the degree of approximation is increased, the speed of an approximate adder will increase because the critical path delay will decrease.

32- and 64-bit accurate and approximate adders were implemented on an Artix-7 FPGA (part: xc7a100tcsg324-3) using Vivado 2018.3. The synthesis strategy was set to Flow_AreaOptimized_high, and a default implementation strategy was used. All the adders were implemented with a pair of registers on the adder inputs and a register on the

adder output. The adders were successfully synthesized and placed and routed, and the design metrics estimated are given in Table I. The design metrics include minimum clock period (in ns), maximum operating frequency (in MHz), number of LUTs and flip-flops (FFs, also called registers), total on-chip power consumption and the adder power consumption in Watts. The total on-chip power and adder power consumption given in Table I are the default estimates reported by Vivado after placement and routing.

TABLE I

FPGA-BASED DESIGN METRICS OF 32- AND 64-BIT ACCURATE AND APPROXIMATE ADDERS (ESTIMATED AFTER PLACE-AND-ROUTE)

Adder Name	Clock Period	Max. Freq.	# LUTs	# FFs	Power (W)	
					Total	Adder
32-bit Adders						
Adder1	2.10	476.19	32	97	0.209	0.117
Adder2	1.94	515.46	28	97	0.200	0.109
Adder3	1.92	520.83	28	97	0.200	0.109
Adder4	1.94	515.46	28	97	0.201	0.110
Adder5	1.95	512.82	25	79	0.191	0.100
Adder6	1.95	512.82	25	79	0.192	0.101
64-bit Adders						
Adder1	2.89	346.02	64	193	0.264	0.173
Adder2	2.77	361.01	60	193	0.260	0.168
Adder3	2.77	361.01	60	193	0.258	0.167
Adder4	2.77	361.01	60	193	0.261	0.169
Adder5	2.65	377.36	57	175	0.261	0.169
Adder6	2.65	377.36	57	175	0.261	0.170

From Table I, it is seen that for both 32- and 64-bit additions, Adder5 and Adder6 consume identical and also the least FPGA resources in comparison with the accurate adder and other approximate adders. Adder1, which is the native accurate FPGA adder, requires 32 LUTs and 97 FFs for realizing 32-bit addition and 64 LUTs and 193 FFs for realizing 64-bit addition. Adder2 to Adder4 require the same number of FFs as Adder1. However, they require 4 LUTs less compared to Adder1 for realizing 32- and 64-bit additions. Adder5 and Adder6 require the least LUTs and FFs amongst all. This is mainly because the six least significant sum bits of the inaccurate sub-adder in Adder5 and Adder6 are connected to V_{dd}, and they do not consume the FPGA resources. In terms of the maximum operating frequency, Adder5 and Adder6 report increases over the accurate adder by 7.7% for 32-bit addition and 9.1% for 64-bit addition. Also, Adder5 and Adder6 report higher speed over other approximate adders for 64-bit addition.

In Table I, the total on-chip power includes the power consumed by the clock, logic, input-output (IO), signals, and the device static power component. It is noticed that the adder power consumption is approximately half of the total on-chip power consumption. We compute the power-delay product (PDP) as the product of total on-chip power consumption and the clock period, and the PDP serves as a representative metric for low power. Based on the PDP calculations, it is noted that Adder5 and Adder6 report reductions in PDP compared to the accurate FPGA adder by 14.7% for 32-bit addition and 9.3% for 64-bit addition. From the combined perspectives of all the design metrics, it is observed that Adder5 and Adder6 are preferable to the accurate and other approximate adders.

We analyzed the error range of the approximate adders by considering a K-bit inaccurate sub-adder and computed the RMSE by assuming K = 8. The error characteristics are given in Table II. We compute the root mean square error (RMSE) for the approximate adders using (1) [8], as RMSE gives a relatively higher weight to larger errors, which is more likely to impact a practical application employing approximate computer arithmetic.

$$\text{RMSE} = \sqrt{\frac{1}{2^{2N}} \sum_{j=0}^{2^{2N}-1} e_j^2} = \sqrt{\sum_{\delta} e_{\delta}^2 \cdot P_{\delta}} \quad (1)$$

Where N is the adder bit-width, e is the error (i.e., the difference between the accurate and the approximate adder outputs), P is the probability of an error value occurring, and δ is the set of all error values.

TABLE II
ERROR RANGE AND RMSE OF APPROXIMATE ADDERS

Adder Legend	Error Range	RMSE
Adder1	Not applicable	Nil
Adder2	$-(2^{K-1}-1)$ to 2^{K-1}	64
Adder3	$-(2^{K}-1)$ to 0	90.33
Adder4	$-(2^{K-1}-1)$ to 0	45.08
Adder5	$-(2^{K-1}-1)$ to $(2^{K-1}+2^{K-2}-1)$	69.13
Adder6	$-(2^{K-1}-1)$ to $(2^{K-2}-1)$	41.31

Adder1 is accurate and hence its error characteristics are nil, by default. The error range mentioned in Table II gives an indication that Adder6 has a relatively better normal error distribution with respect to zero error compared to the other approximate adders. Moreover, Adder6 has a reduced RMSE compared to the other approximate adders. It was noted from Table I that Adder5 and Adder6 feature the same design metrics for 32- and 64-bit additions, but from Table II, it is seen that Adder6 has a 40% less RMSE compared to Adder5.

IV. CONCLUSIONS

This paper presented a new approximate adder which is suitable for FPGA- and ASIC-based implementations. The new approximate adder, in comparison with the native accurate FPGA adder, reports an increase in the maximum frequency while consuming fewer LUTs and registers for 32- and 64-bit additions. Also, comparisons with other approximate adders reveal that Adder6 has optimized design metrics and a reduced error characteristic.

ACKNOWLEDGEMENT

This research is funded by the Ministry of Education (MOE), Singapore under grant MOE2018-T2-2-024.

REFERENCES

[1] K. Roy, and A. Raghunathan, "Approximate computing: an energy-efficient computing technique for error resilient applications", in *Proc. IEEE Computer Society Annual Symposium on VLSI ISVLSI 2015*, France, 2015, pp. 473-475.

[2] M.A. Breuer, "Multi-media applications and imprecise computation", in *Proc. 8th Euromicro Conference on Digital System Design DSD 2005*, Portugal, 2005, pp. 2-7.

[3] H. Zhang, M. Putic, and J. Lach, "Low power GPGPU computation with imprecise hardware", in *Proc. 51st Design Automation Conference DAC 2014*, USA, 2014, pp. 1-6.

[4] R. Nair, "Big data needs approximate computing: technical perspective", *Comm. of the ACM*, 2015, vol. 58, pp. 104.

[5] S.S. Sarwar, G. Srinivasan, B. Han, P. Wijesinghe, A. Jaiswal, P. Panda, A. Raghunathan, and K. Roy, "Energy efficient neural computing: a study of cross-layer approximations", *IEEE Journal of Emerging and Selected Topics in Circuits and Systems*, 2018, vol. 8, pp. 796-809.

[6] P. Panda, A. Sengupta, S.S. Sarwar, G. Srinivasan, S. Venkataramani, A. Raghunathan, and K. Roy, "Cross-layer approximations for neuromorphic computing: from devices to circuits and systems", in *Proc. 53rd Design Automation Conference DAC 2016*, USA, 2016, pp. 1-6.

[7] H. Jiang, C. Liu, L. Liu, F. Lombardi, and J. Han, "A review, classification, and comparative evaluation of approximate arithmetic circuits", *ACM Journal on Emerging Technologies in Computing Systems*, 2017, vol. 13, pp. 60:1-60:34.

[8] P. Balasubramanian, and D. Maskell, "Hardware efficient approximate adder design", in *Proc. IEEE Region 10 Conference TENCON 2018*, South Korea, 2018, pp. 806-810.

[9] N. Zhu, W.L. Goh, and K.S. Yeo, "An enhanced low-power high-speed adder for error-tolerant application", in *Proc. 12th International Symposium on Integrated Circuits ISIC 2009*, Singapore, 2009, pp. 69-72.

[10] H.R. Mahdiani, A. Ahmadi, S.M. Fakhraie, and C. Lucas, "Bio-inspired imprecise computational blocks for efficient VLSI implementation of soft-computing applications", *IEEE Transactions on Circuits and Systems I: Regular Papers*, 2010, vol. 57, pp. 850-862.

[11] P. Albicocco, G.C. Cardarilli, A. Nannarelli, M. Petricca, and M. Re, "Imprecise arithmetic for low power image processing", in *Proc. 46th Asilomar Conference on Signals, Systems and Computers*, USA, 2012, pp. 983-987.

[12] A. Dalloo, A. Najafi, and A. Garcia-Ortiz, "Systematic design of an approximate adder: the optimizer lower part constant-OR adder", *IEEE Transactions on VLSI Systems*, 2018, vol. 26, pp. 1595-1599.

[13] A. Raha, H. Jayakumar, and V. Raghunathan, "Input-based dynamic reconfiguration of approximate arithmetic units for video encoding", *IEEE Transactions on VLSI Systems*, 2016, vol. 24, no. 3, pp. 846-857.

978-1-7281-3420-8/19 $31.00 © 2019 IEEE

Simulation of Ternary CMOS Schemes for Many-Valued Logic Systems

A. A. Krasnyuk and A. G. Prozorova

Abstract - Of interest is the possibility of implementing many-valued logic systems using traditional CMOS technologies. We considered an example of the implementation of three-valued logic elements based on symmetric 3vL logic using the values {-,0,+}, {−1.0, + 1}, {1,0,1}, {i, 0,1} etc. From the totality of obtained results, it can be assumed that ternary CMOS logic can be fully implemented according to norms of 28-180 nm with minimal changes for design rules.

I. INTRODUCTION

In this paper, the three-valued logic analysis is not associated with the ternary number system [1]. We consider only the case of multilevel as a certain number of levels used for each category. In general, the ternary digits can be denoted by any three characters {A, B, C}, but the hierarchy of their values is indicated, for example, A <B <C. In our case, for the set of values {A,B,C} it is assumed that the values A and C are independent, and level B performs the function of the zero element.

Such logic elements, including memory elements, are an effective means of implementing Walsh-Hadamard algorithms, asynchronous and selftimed, clockless and Convention Logic (NCL) [2], multiplication / division operations for non-positional number systems, for example, RNS (Residue Number System) [3], building redundant and multichannel fault-tolerant systems for space applications.

In this case, the zero element B is understood as an element for which the equalities A + B = A and C + B = C are valid, where A and C are any significant numbers.

This allows, for example, performing the operation of subtraction to use the operation of addition with a change in the sign of the subtracted, which is implemented by XOR elements. Actually sign of the number in this system is determined by the place of the highest non-zero digit. The proposed approach can be considered as extended binary codes that simulate properties of the ternary logic [4].

II. ELEMENTS OF THREE-VALUED LOGIC

In general the 3vL CMOS model of the inverter can be represented as a functional (Fig.1a) and schematic circuit

A. G. Prozorova is with the Scientific Research Institute of System Analysis, Russian Academy of Sciences, Nakhimovsky pr. 36-1, 117218 Moscow, Russia, E-mail: prozorova@cs.niisi.ras.ru

A.A.Krasnyuk is with the National Research Nuclear University MEPhI (Moscow Engineering Physics Institute), Kashirskoe sh. 31, 115409 Moscow, Russia, E-mail: aakrasnyuk@mephi.ru

diagram (Fig.1b) [5]. The model of the corresponding symmetric trigger is shown in Fig.2b.

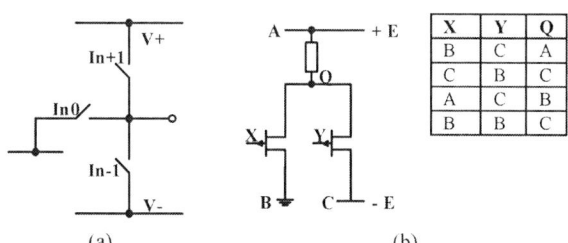

X	Y	Q
B	C	A
C	B	C
A	C	B
B	B	C

(a)   (b)

Fig. 1. 3vL inverter model: functional (a) and schematic (b).

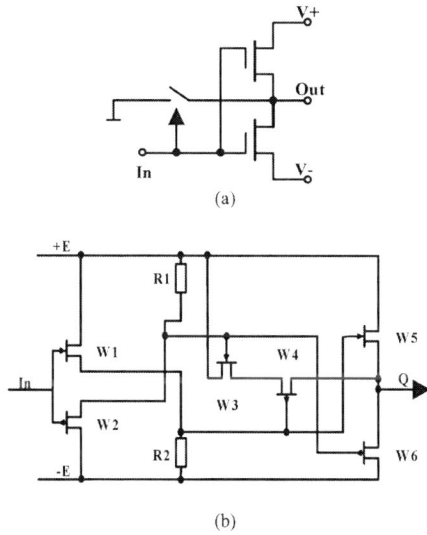

Fig. 2. CMOS inverter model (a) and trigger with one input (In) based on it (b).

III. SIMULATION OF TERNARY CMOS SCHEMES

The inverter model shown in Fig.1b was used to analyze the time and frequency characteristics. Mode "dual-rail" was used for calculations. In "dual-rail" mode signals "+E true", "-E false" and "null" are transmitted on different wires.

Simulation was performed in Cadence CAD for technological standards 28 nm and 180 nm. The dependence between output signal Q rise time (fall time) and two parameters (width of used transistors W, resistance value R) is obtained.

According to the simulation results, the dependences were obtained for three sections:

switching from + E to 0,
switching from 0 to –E,
switching from 0 to +E.

Time analysis (tran) was performed for inverter built on 28 nm technology. Output signal delay were measured (Table I). The parameters of inverter elements were obtained using parametric analysis of a wide range of all values. Wx = 300 nm, Wy = 500 nm, R = 25 kΩ.

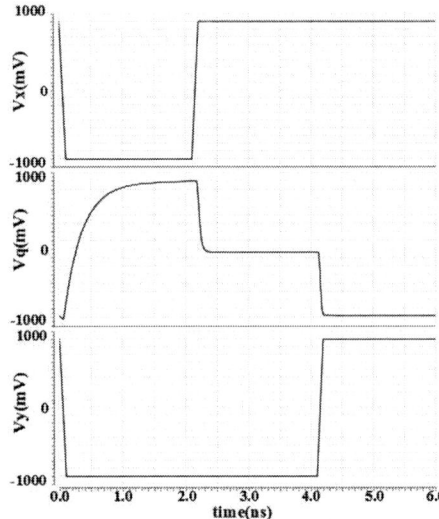

Fig. 3. Time diagrams for 28 nm-invertor.

TABLE I
OUTPUT DELAY DEPENDING ON THE INPUT

Switching	Delay	Vout(Vin)
+ E to 0	130 ps	Q (X)
0 to –E	30 ps	Q(Y)
0 to +E	60 ps	Q(Y)

Table I shows which of the input signals was used to measure output delay.

A. Switching from + E to 0

Fig. 4 shows results for the case of switching element from +E to 0 for technology 28nm. The width of transistors W has the greatest effect on fall time when the potential at the output Q decreases from + E to 0. The greater value of W, the less time it takes to achieve the required potential. In the worst case required voltage at node Q will not be reached in the allotted time (W=100 nm). The most optimal width of transistor W>300 nm.

Figure 5 shows the results obtained for the 180 nm technology. The fall time of the output signal Q when switching from + E to 0 for 180 nm technology is also more dependent on the widths of the transistors (Fig. 5b). It

can be seen from the graphs that when the resistance is less than 10 kΩ and width of transistors is less than 1,5 um, required voltage at the Q node will not be reached within the allotted time interval.

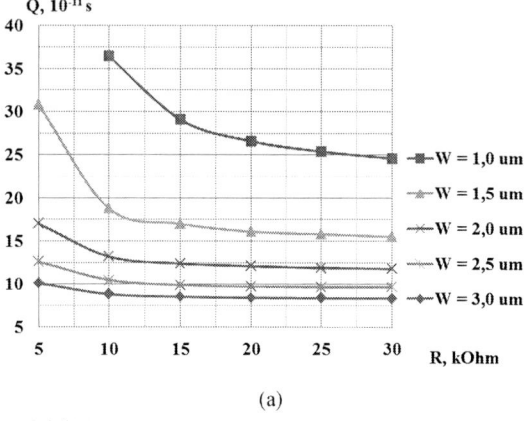

(a)

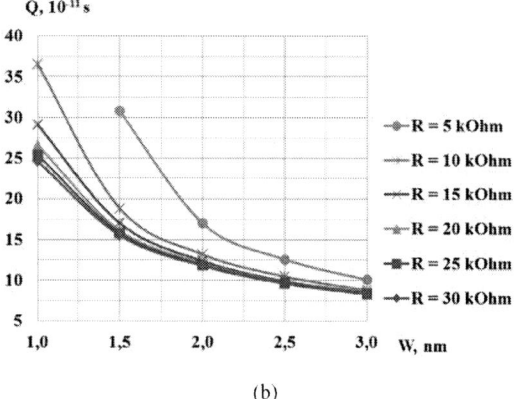

(b)

Fig. 4. The dependence between output signal Q fall time and resistance value R (a); width of used transistors W (b) for technological standard 28 nm.

B. Switching from 0 to –E

For the case of switching output signal from 0 to –E, the situation is similar for both of technologies. The widths of transistors have the greatest effect.

C. Switching from 0 to +E

The greatest effect on rise time when the potential at the output Q increases from 0 to + E has the resistance value R.

For technological standard 28 nm the greater value of R, the longer it takes to charge node required potential. In the worst case, required voltage on node Q will not be reached within the allotted time interval. At values of R> 30 kΩ, rise time becomes longer than 1 ns.

Figure 7 shows the dependences between rise time of the output signal Q and width of used transistors at the switching from 0 to + E for the 180 nm technology.

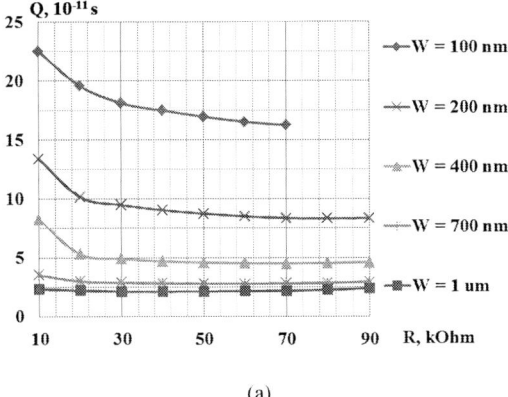

(a)

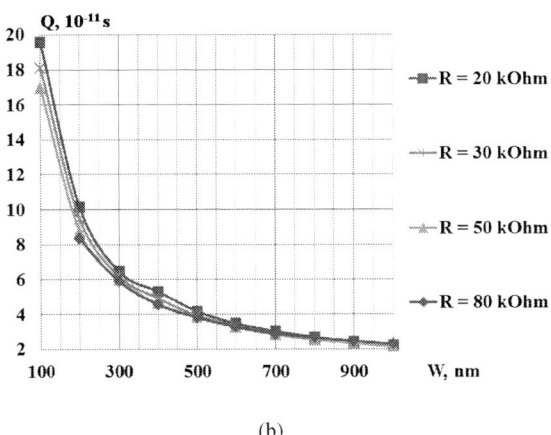

(b)

Fig. 5. The dependence between output signal Q fall time and resistance value R (a); width of used transistors W (b) for technological standard 180 nm.

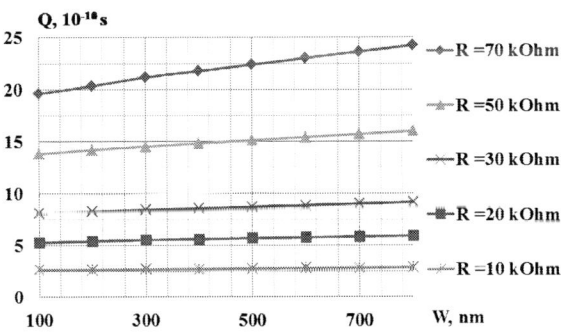

Fig. 6. The dependence between output signal Q rise time and width of used transistors W for technological standard 28 nm.

According to results of presented simulation, the optimal ratio between parameters of inverter under consideration is the choice of transistor width more than 300 nm and resistance value less than 30 kΩ for technological standard 28 nm. For design rules 180 nm the optimal width of transistor is more than 1,5 um and resistance value should be less than 30 kΩ and more than 10 kΩ.

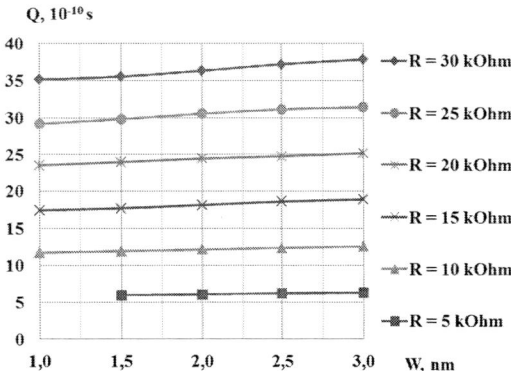

Fig. 7. The dependence between output signal Q rise time and width of used transistors W for technological standard 180 nm.

Based on the inverter presented in figure 2a, two symmetrical triggers were constructed: with one input (Fig. 2b) and with two inputs X and Y (Fig. 8a). Fig. 8b shows the DC analysis results for the symmetric trigger with two inputs.

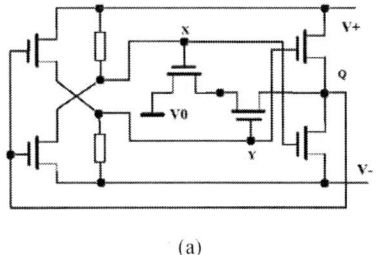

(a)

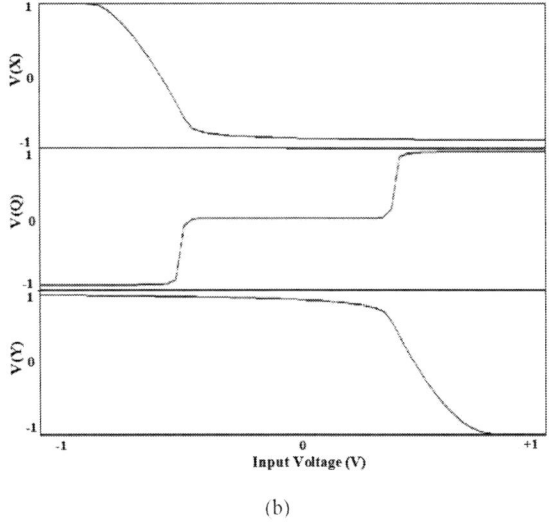

(b)

Fig. 8. Symmetric trigger with two inputs X, Y (a) and its DC analysis (b).

The simulation of the switching characteristics for CMOS-symmetric trigger with 1 input (Fig. 2b) was performed in Cadence CAD using a 65 nm planar technology. The dependence between output (Q) and input (In) voltage is shown in figure 9.

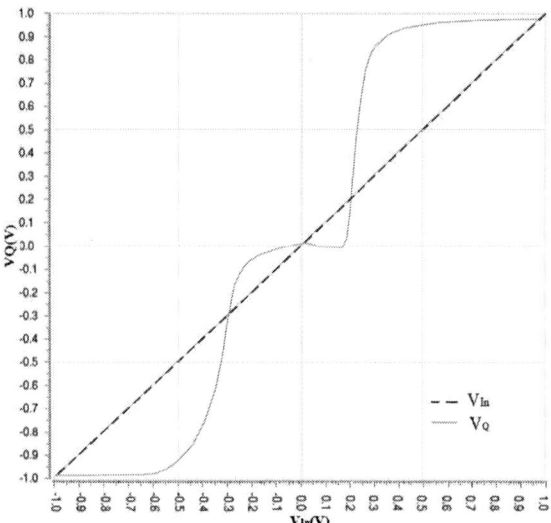

Fig. 9. Switching characteristics of symmetric 3vL CMOS trigger with 1 input.

The optimal parameters of elements used in considered trigger were obtained by parametric analysis of a wide range of all values. From set of obtained waveforms variant was chosen, where output waveform was as close as possible to the expected one. Waveform of output signal (Q) should have 3 distinct levels, logical 0 should be at 0V, as well as 1 and -1 at 1V and -1V accordingly.

TABLE II
SELECTED TRIGGER OPTIONS

Parameter	Value
W1	500 nm
W2	200 nm
W3	1,2 um
W4	650 nm
W5	300 nm
W6	200 nm
R1	8 kOhm
R2	14 .kOhm

Values of the selected parameters are presented in table II. Parameters of the elements were obtained for a specific technology. According to the simulation results, switching voltage range was determined (Table III).

TABLE III
SWITCHING VOLTAGE RANGE OF CONSIDERING TRIGGER

Input voltage In	Output voltage Q
-1V…-623mV	-1V
-220mV…170mV	0V
340mV…1V	1V

IV. CONCLUSION

From the totality of obtained results, it can be assumed that ternary CMOS logic can be fully implemented according to norms of 28-180 nm with minimal changes for design rules. Investigated switching mode of the dual-rail transistors presents practical interest to perform summation / subtraction operations both in redundant and multichannel fault-tolerant so as for non-positional residue number systems.

ACKNOWLEDGEMENT

This work is supported by Russian Foundation for Basic Research, grant № 19-07-00651\19.

REFERENCES

[1] R. Mariani, R. Roncella, R. Saletti, P. Terreni. "On the Realisation of Delay-Insensitive Asynchronous Circuits with CMOS Ternary Logic", *Third International Symposium on Advanced Research in Asynchronous Circuits and Systems (ASYNC '97)*, 1997, pp.54.

[2] Overview of Convention Logic (NCL)/ https://www.ndsu.edu/pubweb/~scotsmit/NCL_intro.pdf.

[3] A. Mohan, "P.V. Residue Number Systems", *Springer International Publishing*, 2016, pp. 351.

[4] P. Ambrož, C. Frougny, Z. Masáková and E. Pelantová "Arithmetics on number systems with irrational bases", *The Bulletin of the Belgian Mathematical Society - Simon Stevin*, 2003, Vol. 10, no. 5, pp. 641–659.

[5] A. Stakhov. "Brousentsov's Ternary Principle, Bergman's Number System and Ternary Mirrorsymmetrical Arithmetic", *The Computer Journal*, 2002, Vol. 45, no.2, pp. 221-236.

Optimization of Hsiao Decoders by Circuit-Level Minimization

K. Petrov, I. Danilov, A. Shnaider Khazanova, M. Gorbunov

Abstract—**We showed that the Hsiao decoder circuit could be minimized, resulting in the delay or area reduction without significant increase of the decoder failure ratio. We designed three versions of the decoder (full, shortened and minimized) and showed that it is possible to reduce its delay time by 13-18%, or the area by 33-57% relative to the full version. Also, we showed using fault injection simulation that the value of the failure ratio varies from -8% to +6% for shortened and minimized versions relative to the full version.**

I. INTRODUCTION

The sensitivity to ionizing radiation of the microprocessor system strongly depends on the internal system-on-chip (SoC) memory because its elements may occupy up to 90% area on a SoC [1] and data can be stored in these elements during the whole period of system operation.

Error correction coding (ECC) is widely used to increase memory resilience to failures. Hsiao codes [2] are the most commonly used to provide single error correction and double error detection (SEC-DED) in data words stored in the memory. For use the ECC, it is necessary to calculate the check bits when writing to memory as well as it is necessary to correcting and detecting errors when reading. The write path uses the coder circuit, and the read path uses the decoder. The characteristics of these two circuits influence the parameters of the entire memory because they are in the data path. Therefore, the problem of reducing delays in coders and decoders is important. Also, when using coders and decoders for small memory elements, for example, for individual processor registers, the task of reducing the area is relevant too. We consider only decoders in this work since coders are less complex elements and do not have opportunities for optimization at the circuit level.

Section II introduces the ECC memory design path and common Hsiao decoder circuit. Approaches to decoder minimizing at different design levels focusing on the circuit level described in Section III. Section IV presents three decoder versions and the results of their synthesis and simulation.

II. HSIAO DECODERS DESIGN

The designer can make several key decisions at different design levels affecting the decoder characteristics.

First, the error-correction code is chosen. The choice depends on the task. The optimal solution for memory elements with a parallel interface, small data width (up to 256 bits) and high speed is Hamming [3], orthogonal Latin

K. Petrov, I. Danilov, A. Shnaider Khazanova, M. Gorbunov are with Scientific Research Institute of System Analysis of the Russian Academy of Sciences (SRISA), 117218, Nakhimovsky prosp. 36-1, Moscow, Russian Federation. E-mail: petrovk@cs.niisi.ras.ru

square [4] or Hsiao codes [2]. Among them, the Hsiao code is preferable because it allows detecting double errors with the minimum required number of check bits.

Second, the generator and parity-check matrices for the code are created. At this stage, the designer could decide to use more check bits, for example. This approach negatively affects the compactness of the memory array because extra cells are needed to store extra check bits, but decrease the decoder propagation delays [5] or increase the number of detectable triple errors [6].

Finally, a designer can apply several circuit-level optimization options. For example, we do not consider decoders that correct the check bits in this article. Only the data bits are corrected in the case of an error. However, this means detecting correctable errors in the check bits too.

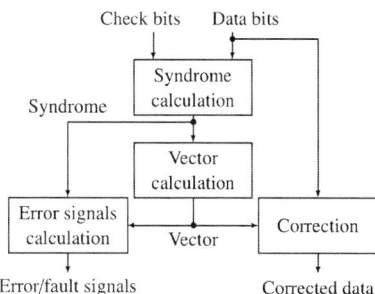

Fig. 1. Functional scheme of the Hsiao decoder.

The Hsiao decoder circuit is shown in Fig. 1 contains several composite blocks and the connections between them. The syndrome calculation block (SCB) receives data and check bits from memory to determine error syndrome. The parity-check matrix of code generally defines its structure. The SCB is based on the Hsiao coder circuit, which can be implemented in two versions. One of the versions reduces delays and area, while the other increases the failure resistance of the unit [7]. The syndrome obtained in this block is used to calculate error signals and an error vector. In this paper, we consider the versions of SCB with delay and area reduction.

The vector calculation block (VCB) determines the error vector, which is used by the correction block (CB) and the error signal calculation block (ESCB). Using the error vector, CB can correct data with a single corrupted bit. In case when two or more bits of data are corrupted, the output of CB is invalid. The ESCB uses error vector to produce the "fault" signal indicating the uncorrectable error in the input

978-1-7281-3420-8/19 $31.00 © 2019 IEEE

data and check bits. The "error" signal indicates the presence of any error in the input data and checks bits obtained by ESCB using the syndrome. The case when "error" is calculated to true, and "fault" is calculated to false treated as the presence of correctable error in the input data. However, it can be wrong when three or more bits of input data and check bits are corrupted.

The number of triple errors for information words from 16 to 64 bits detected this way is theoretically from 35 to 45% [8]. If triple error detection is not necessary, the decoder circuit can be optimized by decreasing or abandon at all its triple error detection circuitry. The approach presented in this work relies on such an optimization technique.

III. Approaches To Decoder Minimizing

We consider the following two circuit-level optimization options. The first one is not to use the vector to calculate error signals. The scheme of such a decoder differs from the one shown in the Fig. 1. The first one has no connection between the VCB and ESCB. This modification reduces the load capacity requirements of the output logical elements of VCB and reduces the number of links in the decoder and the number of logical elements in VCB and ESCB.

The second option is using only those syndrome bits that uniquely distinguish it from the others in the case of a single error in data bits. The work [9] suggested not using only one syndrome bit to calculate each vector bit. It did not allow to achieve the highest possible decoder minimization. The work [10] presents a particular case of the decoder implementation. VCB calculates each vector bit using syndrome bits that match the logical "true" in the parity-check matrix, and therefore in the VCB truth table. At the same time, nine check bits are required instead of eight standard ones to create a minimized Hsiao decoder for 64 bits information word. In this case, only three out of nine syndrome bits are needed for each vector bit. In this article, extra check bits are not applied, and five out of eight syndrome bits are needed for each vector bit.

The algorithm for removing unnecessary elements from the VCB truth table can be described in following paragraph:

1. The lower rows of the table that correspond to the check bits are deleted.

2. Rows containing the maximum number of units are selected. All zeros are removed from them.

3. In this last paragraph, "zero" bits are removed from the rows of the syndrome, which do not have the maximum number of "one" bits. The goal is to remove the same number of "zero" bits as many "one" bits were removed in the previous paragraph. This is done in a manner that numbers of "zero" bits remaining in the row of the syndrome do not fully coincide with numbers of deleted "zero" bits in none of the lines with the maximum amount of ones.

Let us give an example of using the first two paragraphs of the algorithm for the Hsiao (8, 4) decoder. Parity-check matrix for this code is

$$H = \begin{bmatrix} 0 & 1 & 1 & 1 & 1 & 0 & 0 & 0 \\ 1 & 0 & 1 & 1 & 0 & 1 & 0 & 0 \\ 1 & 1 & 0 & 1 & 0 & 0 & 1 & 0 \\ 1 & 1 & 1 & 0 & 0 & 0 & 0 & 1 \end{bmatrix}. \quad (1)$$

It corresponds to the VCB truth table presented in Table I(a). Such VCB is used in the decoder, where the vector used by ESCB. The VCD scheme for this case is shown in Fig. 2a.

VCB truth table after the first paragraph of the algorithm has been completed is presented in Table I(b). Such VCB is employed in the decoder, where the vector has not used the vector by ESCB. The VCD scheme for this case is shown in Fig. 2b.

TABLE I

VCB truth tables for Hsiao (8, 4) decoder: a) with using the vector by ESCB; b) without using the vector by ESCB; c) minimized on proposal principle

Syndrome	Vector
0111	10000000
1011	01000000
1101	00100000
1110	00010000
1000	00001000
0100	00000100
0010	00000010
0001	00000001

a)

Syndrome	Vector
0111	1000
1011	0100
1101	0010
1110	0001

b)

Syndrome	Vector
x111	1000
1x11	0100
11x1	0010
111x	0001

c)

VCB truth table after the second paragraph of the algorithm has been completed is presented in Table I(c). Syndrome bits, which are not used to calculate the vector bits corresponding to a given line of the truth table, is denoted by x. The VCD scheme for this case is shown in Fig. 2c.

Schemes in Fig. 2 are given for a better understanding of the proposed principle. A real design topology is different because CAD tools use additional buffers for design implementation. Moreover, it applies logic elements with the optimal number of inputs contained in the library, as well as optimize logical expressions duplicating each other.

The most popular values of the data width in memory are 4, 8, 16, 32 and 64 bits. The third paragraph of the algorithm is necessary only for 64-bit words. Parity-check matrix for Hsiao (72, 64) code shown in Fig. 3. The first eight rows of the VCB truth table after the third paragraph of the algorithm has been completed are presented in Table II. All three zeros are removed from the fifth row of the table (it corresponds to the fifth column of the check matrix). also three zeros are removed from the remaining rows. And they are deleted so that the remaining two zeros in each line are not in the same digits as in the fifth line.

978-1-7281-3420-8/19 $31.00 © 2019 IEEE

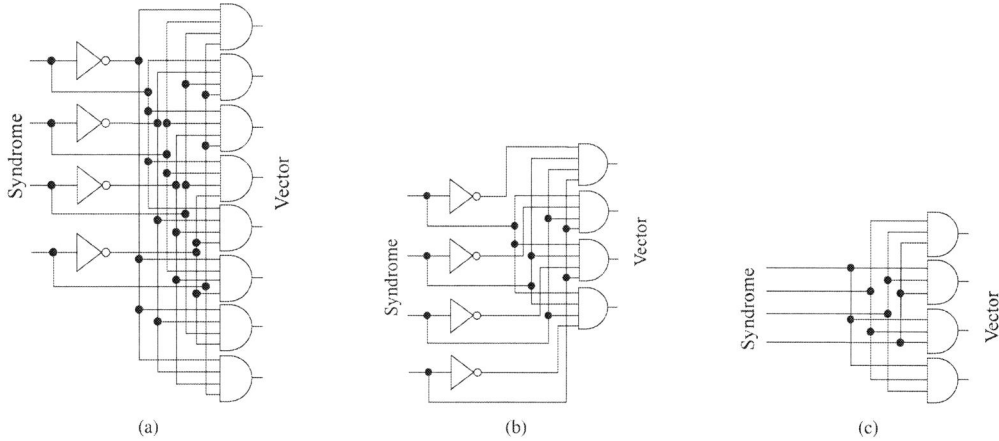

Fig. 2. VCB schemes for Hsiao (8, 4) decoder: a) with using the vector bits by ESCB; b) without using the vector bits by ESCB; c) minimized on proposal principle.

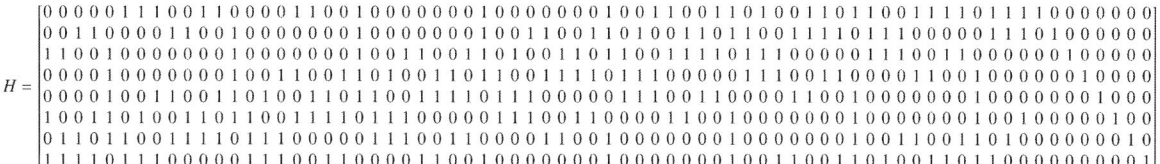

Fig. 3. Parity-check matrix for Hsiao (72, 64).

TABLE II

VCB TRUTH TABLES FOR HSIAO (72, 64) DECODER (PART - THE FIRST EIGHT LINES)

Syndrome	Vector
0x10x1x1	1000000000000000000000000000000000000000000000000000000000000000
x01x0x11	0100000000000000000000000000000000000000000000000000000000000000
x10xx011	0010000000000000000000000000000000000000000000000000000000000000
01xx01x1	0001000000000000000000000000000000000000000000000000000000000000
xx11111x	0000100000000000000000000000000000000000000000000000000000000000
1xx0x101	0000010000000000000000000000000000000000000000000000000000000000
10xx10x1	0000001000000000000000000000000000000000000000000000000000000000

IV. SYNTHESIS AND SIMULATION

Three versions of the Hsiao decoder for 8, 16, 32 and 64 data words were proposed for the simulation:

• full version, where the vector is used to calculate the error signals, which allows detecting a portion of the triple error in the data bits;

• shortened version, where the vector is not used to error signals calculation;

• minimized version, where the VCB is minimized according to the principle presented above.

All presented decoders have the same SCB and CB. Also, they do not have outputs of check bits and do not correct them.

We developed the Register-Transfer-Level (RTL) models for each of these decoders, and synthesized netlist and layout using the custom radiation tolerant standard cell library de-

veloped in SRISA and based on TSMC 65 nm CMOS technology.

A fault injection simulation was also conducted to estimate the failure ratio, which is the percentage of injected faults that caused the output values to mismatch with the ideal ones at the time of decoder work completion. We take into account the difference in area between decoders for the same information word when calculating the failure rate. Failure injection here is a change in the value at the output of a random logic element to the opposite one by 100 ps.

The synthesis and simulation results are presented in Table III. Shortened and minimized decoder versions are presented in the table twice. At first, they were synthesized with minimal delay time. At second they were additionally synthesized with a delay time of full version. This allows

TABLE III

SYNTHESIS AND SIMULATION RESULTS FOR HSIAO DECODERS

Decoder version and delay parameter	Full (72, 64), minimal delay	Shortened (72, 64), minimal delay	Minimized (72, 64), minimal delay	Shortened (72, 64), full-version delay	Minimized (72, 64), full-version delay
Delay time, ps	964	867	843	964	964
Area, μm^2	6054	4627	4656	3250	2827
Failure ratio, %	10.3	9.8	12.4	5.6	13.2
Decoder version and delay parameter	Full (39, 32), minimal delay	Shortened (39, 32), minimal delay	Minimized (39, 32), minimal delay	Shortened (39, 32), full-version delay	Minimized (39, 32), full-version delay
Delay time, ps	842	737	691	842	842
Area, μm^2	2618	2536	2314	1724	1338
Failure ratio, %	11.5	12.8	15.6	10.1	8.4
Decoder version and delay parameter	Full (22, 16), minimal delay	Shortened (22, 16), minimal delay	Minimized (22, 16), minimal delay	Shortened (22, 16), full-version delay	Minimized (22, 16), full-version delay
Delay time, ps	730	658	614	730	730
Area, μm^2	1732	1478	1107	1019	747
Failure ratio, %	14.6	12.1	13.7	10.0	8.1
Decoder version and delay parameter	Full (13, 8), minimal delay	Shortened (13, 8), minimal delay	Minimized (13, 8), minimal delay	Shortened (13, 8), full-version delay	Minimized (13, 8), full-version delay
Delay time, ps	633	570	527	633	632
Area, μm^2	657	866	691	558	440
Failure ratio, %	14.4	23.1	22.8	14.1	16.0

a separate estimation of the influence of considered circuit-level optimization options to the resulting design area.

V. CONCLUSION

Hsiao decoders can be shortened and minimized at the circuit level if it is not necessary to detect triple errors. As a result, it is possible to reduce its delay time by 13-18%, or the area by 33-57% relative to full version. Also, the simulated failure ratio varies from -8% to +6%.

The minimized decoder version allows reducing delay (by 3-8%) or space (by 13-27%) relative to the shortened version.

ACKNOWLEDGEMENT

The reported study was funded by RFBR according to the research project No. 18-37-20008.

REFERENCES

[1] J. Singh, S. Mohanty, and D. Pradhan, *Robust SRAM Designs and Analysis*, 1st ed. Springer-Verlag, 2013.

[2] M. Y. Hsiao, "A class of optimal minimum odd-weight-column SEC-DED codes," *IBM J. of Res. and Dev.*, vol. 14, no. 4, pp. 395–401, Jul. 1970.

[3] R. Hamming, "Error detecting and error correcting codes," *Bell Sys. Tech. J.*, vol. 29, pp. 147–160, 1950.

[4] M. Hsiao and other, "Orthogonal latin square codes," *IBM J. of Res. and Dev.*, vol. 14, pp. 390–394, 1970.

[5] A. Kazéminéjad, "Fast, minimal decoding complexity, systematic (13, 8) single-error-correcting codes for on-chip DRAM applications," *Electronics Letters*, vol. 37, no. 7, pp. 438–440, Mar. 2001.

[6] M. Anwar and other, "Decoder design for a new single error correcting/double error detecting code," *Proc. Wor. Ac. Sci.*, vol. 22, no. 4, pp. 247–251, 2007.

[7] J. Maestro, P. Reviriego, and other, "Fault tolerant single error correction encoders," *Journal of Electronic Testing: Theory and Applications*, vol. 22, no. 2, pp. 215–218, Jul. 2011.

[8] M. Richter, K. Oberlaender, and M. Goessel, "New linear SEC-DED codes with reduced triple bit error miscorrection probability," in *2008 14th IEEE Int. On-Line Testing Symp.*, Jul. 2008, pp. 37–42.

[9] K. Petrov and V. Stenin, "Error-correction coding in CMOS RAM resistant to the effect of single nuclear particles," *Russian Microelectronics*, vol. 44, no. 5, pp. 316–323, 2015.

[10] P. Reviriego, S. Pontarelli, and other, "Method to construct low delay single error correction codes for protecting data bits only," *IEEE Trans. on Comp.-Aid. Des. of Int. Circ. and Sys.*, vol. 32, no. 3, pp. 479–483, Mar. 2013.

978-1-7281-3420-8/19 $31.00 © 2019 IEEE

Differential Input Area Efficient Current Comparator

A. R. Serazetdinov and E. V. Atkin

Abstract - Differential input area efficient current comparator for multichannel detector (sensor) applications is presented. Comparator consists of current preamplifier, hysteresis latch, amplifier-voltage limiter and output low-voltage to CMOS translator, having built-in polarity selection switch. The latch geometry was chosen to feature non-zero hysteresis and minimum size. The key features of the proposed solution are low voltage swing before translator, low power consumption and simplicity. The comparator was developed in UMC 180 nm MMRF CMOS process. The power consumption is in range of 60 μW at 1.8 V for all PVT variations. Its layout cell was designed to be an area efficient one and occupies 1200 μm^2.

I. INTRODUCTION

Comparators are widely used builing blocks in analog signal processing [1]. They determine whether one input value is higher or lower than the other one at the specific time point or to perform the comparison in the asynchronous manner, that is to detect the time point at which the difference of the two input signal has changed the sign. The compared signals may be of any analog electrical quantity, like current, voltage, charge or time. This paper is devoted to an application specific asynchronous analog current comparator.

In this work the comparator discussed in this work has been developed for charged particles pulses detection. In such systems comparators, placed closely together, could be vulnerable for cross-channel cross-talks, interference from supply rails, and could initiate current or/and voltage pulses on the rails. In our case the speed of the comparator was not critical (propagation delay of even 100 ns was still acceptable).

Also the power consumtion is an issue and had to be restricted to 100 μW.

The comparator has to implement discrimination function and digitize signal from shaper to the counter.

The comparator input was determined by the analog front-end structure. It is presented by the differential current signal out of the shaper (filter) amplifier having dynamic signal range from 50 nA up to 2 μA. To allow class A operation for the preamplifier stage additional 3μA bias current to each differential leg was added to the output of shaper. The comparator threshold was preset by the 6-bit differential current steering DAC, having full range of sink current of about 2 μA.

Several current comparator schemes were considered as references [2-8]. Some of them were considered as the

A.R. Serazetdinov and E.V. Atkin are with the Department of Electronics, Faculty of Automation and Electronics, National Research Nuclear University MEPhI, Kashirskoe shosse 31, 115409 Moscow, Russia, E-mail: arserazetdinov@mephi.ru

basic ones: Traff's [1], Bank's[2], Wang's[3], Chasta's[4]. Nevertheless, none of references fit to specifications of interest, mostly in terms of combination speed and power consumption with reasonably small silicon area.

II. STRUCTURE, SCHEMATICS AND LAYOUT

Block-level schematics of the proposed comparator is shown in Fig.1.

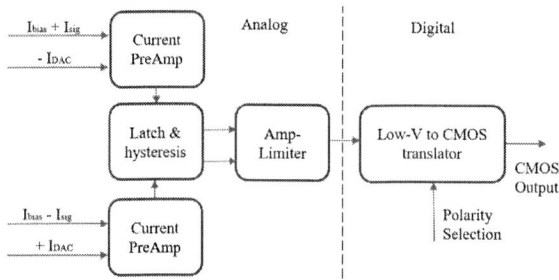

Fig. 1. Block-diagram of the current comparator.

The comparator analog part is comprised of two current mirror preamplifiers, the latch, amplifier-limiter, followed by guard-ring isolated low-voltage to CMOS translator. The latter (futher called digital part, although it's not fully digital) is additionally equipped with polarity selection capability. The biasing circuit was included to eliminate the need of an external 1200 mV voltage reference.

As the cross-leg and digital part induced interference has to be minimized, differential amplification structure was choosen. The voltage swings on intermediate amplification stage have to be restricted and current spikes need to be in the opposite directions and synchronized in time. In this sense conventional comparators, that employ single-ended architecture, are not acceptable.

The simplified transistor-level implementation of the blocks is given in Fig. 2.

The fully differential comparator structure poses the requirement that current spikes on the intermediate amplification stages are to be opposite direction and precisely synchronized in time. Therefore, working of the current preamplifiers (Fig. 1a) in the non-linear region (for example, in schemes with current subtraction for limiting of power consumption) – lead to asynchronous delays due to shut off of one of the preamplifier legs. So the preamplifier is to be biased into the class A operation point. In this case the amplification gain of the preamplifier is limited by power requirement, and was chosen to be equal to 2. The total current consumption of preamplifier stage is about 10 uA per leg.

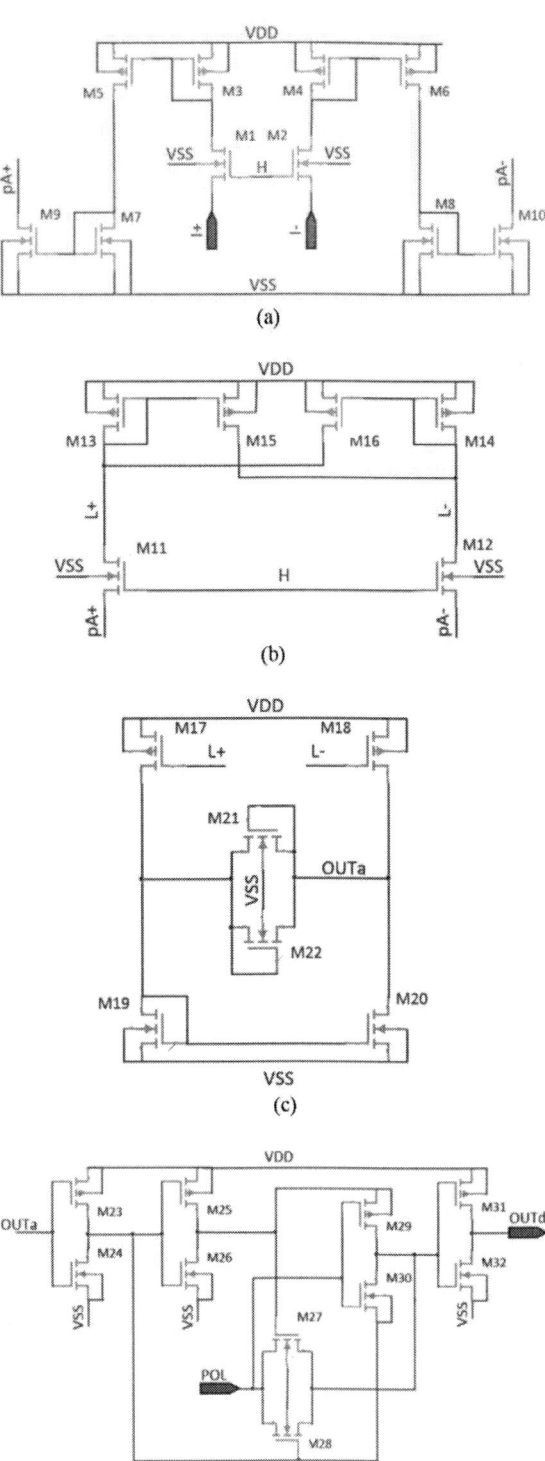

Fig. 2. Simplified transistor-level schematics blocks. a) current preamplifiers; b) latch with hysteresis; c) amplifier-limiter; d) low-voltage to CMOS translator with polarity selection pin (POL).

The hysteresis was introduced by the latch (M11-M16). The significant issue with the latch of these transistor dimensions (about 1u per 1u) is hysteresis strong dependence over latch mismatch (about 0.1 u width variation cause about 50% hysteresis change; number of fingers also influence hysteresis value), so that large sizes are preferable. This transistors should be as big as possible to provide stable hysteresis value. Although in our case it's stability is not critical as comparator would be precalibrated by DAC. The latch is primarily used in syncronous comparators as it provides positive feedback. This feedback could be easily made in such way, that it clamps the latch and the latch is reseted with timing pulse. However in this work the loop gain was choosen so as to allow asyncronous operation.

The analog part was minimized in terms of voltage swings at the nodes (see M21 and M22 in Fig. 2c). The voltage swings were limited by fast diodes at the latch output. Although voltage over the diode is dependent on current, this way is the most simple and stable with corner analysis.

The output digital part of the comparator provides CMOS levels and allows for the output polarity control, keeping power consumption at lowest level. This block is essentially the XOR gate with additional amplifiers M29/M30, which function as a buffer. Direct connection of the conventional XOR gate (i.e. at the gates of M31/M32) to the analog output lead to incorrect operation as current may sink through M36, M35 and M32 when M20 is closed.

A table with all transistors' dimensions used in analog part is shown at Table 1.

TABLE I
TRANSISTORS' DIMENSIONS, USED IN ANALOG PART.

M1, M2	5u	300n
M3, M4, M5, M6	3u	300n
M7, M8	1u	300n
M9, M10	2u	300n
M11, M12	3u	180n
M13, M14	1u	300n
M15, M16	2.14u	900n
M17, M18	1u	300n
M19, M20	300n	400n
M21, M22	1u	180n

The layout of the comparator is presented in Fig 3.

Common considerations have been used to trace analog part of the comparator, such as placing differential lines close together. Although the layout is much insensitive to exact location of each transistor, some parasitic effects take place. Most noticable is hysteresis dependence over the finger number of latched transistors. The hysteresis value change equals 25-30 nA (about 10%) for two finger case against single finger one. These transistors are denoted on Fig. 3(a) with the boundary.

978-1-7281-3420-8/19 $31.00 © 2019 IEEE   306

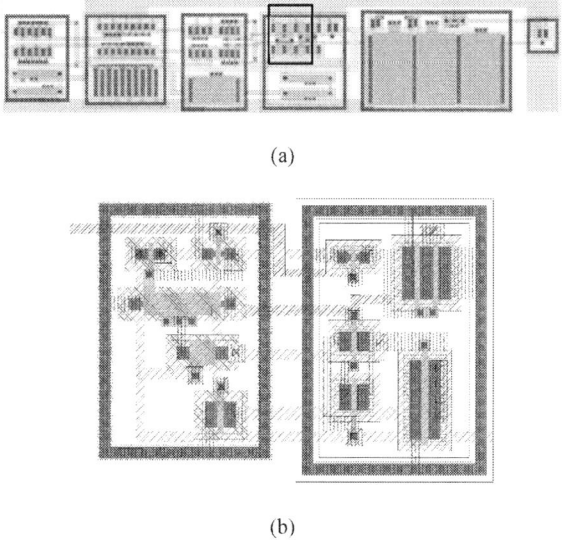

(a)

(b)

Fig. 3. Layout of the comparator. a) Analog part (70x15 um); b) Digital part (15x10 um).

III. RESULTS AND DISCUSSION

The comparator schematic and layout design and simulation were implemented using Cadence Virtuoso platform with Spectre simulator. Parasitics were extracted with Calibre PEX tool.

As a result of the simulation, following device performance under typical conditions was registered: 5.8 ns propagation delay for 100 nA current overhead over threshold voltage; analog part power consumption: 54 uW; current pulses, induced on analog power rails – less than 3 uA under all corners; layout dimensions – 30 μm per 80 μm; average hysteresis – 250 nA.

The whole comparator circuit has been studied against all PVT variations (power supply: 1.6 V – 2 V; temperature range: 0-80 °C; process variation over corners: ss, tt, ff, fnsp, snfp).

Some of the modeling results of the current comparator are given in Fig. 4.

Although the hysteresis varies significantly with Monte Carlo mismatch variations (Fig. 4a), the main requirement is to prevent it from being zero (as that would cause multiple counts on the counter). In fact hysteresis value do not act the detection resolution (nA per DAC LSB) as comparator inputs could be shifted by the constant value by DAC keeping the resolution same.

Polarity switch allows for comparator capability to work with both positive and negative signal pulses. So if DAC symmetrical structure is used, even comparator internal shift for switching threshold of 1.5 μA could be successfully dealt with.

The current consumption of the whole comparator (Fig. 4c) is mainly determined by constant part and switching currents of the digital part. The analog part consumption is

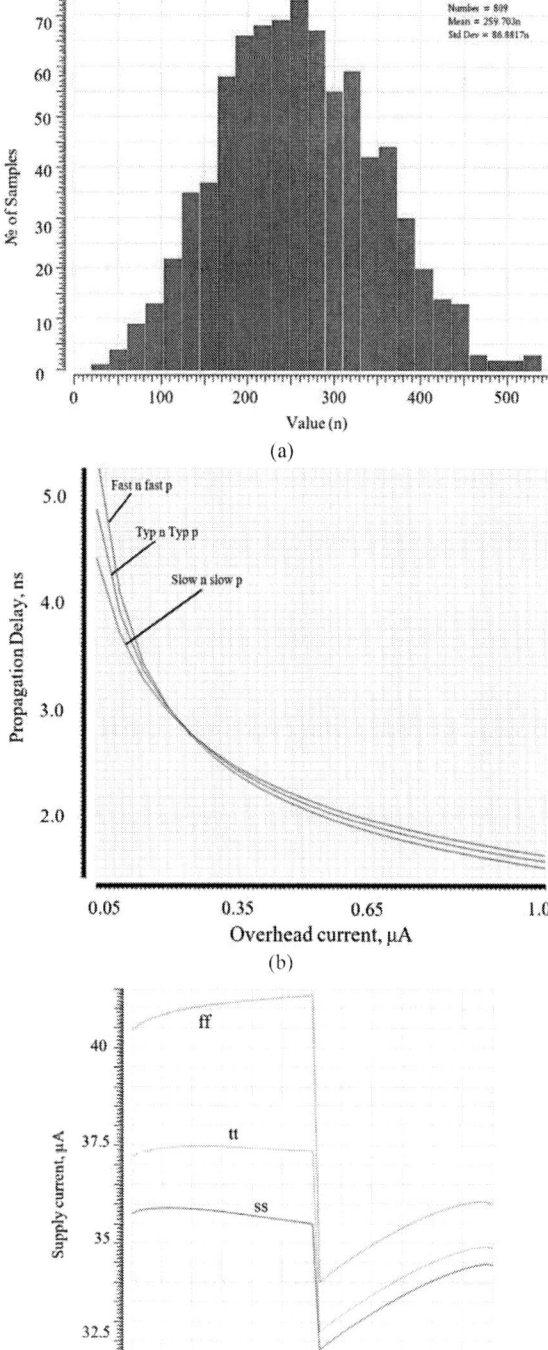

(a)

(b)

(c)

Fig. 4. Modeling results of a proposed current comparator. a) Hysteresis variation under Monte Carlo mismatch analysis; b) Propagation delay versus overhead current under corner analysis (ff, tt and ss corners, room temperature); c) Current consumption of the analogue part of the comparator (ff, tt and ss corners, room temperature).

978-1-7281-3420-8/19 $31.00 © 2019 IEEE

essentially lower, compared with similar current comparators with propagation delays of the same order [3,9,10]. It is determined mainly by contribution of linear amplifier (each leg consumes at least 6 uA static, and could be lowered by decreasing amplification gain. This would lead to increasing of the lengths of cascoded (M11/M12) and latch (M13-M16) transistors and chip size limitation should be taken into account.

IV. CONCLUSION

The paper presents the comparator for multichannel applications with schematics and layout which is implemented in 180 nm CMOS technology. The device features low voltage operation, small current spikes in the analog circuit, limited to 600 mV voltage swings under all currents on intermediate amplification stages. The design still could possibly be improved in terms of latch to mismatch stability and limitation of the voltage swings. The next step through the manufacturing process are the design of a testing board for the full chip testing and errors detection and correction. The work on the system is to be continued since the prototype chips will be delivered from the UMC's factory.

REFERENCES

[1] V. Milovanovic, H. Zimmermann, "A double-differential-input/differential-output fully complementary and self-biased asynchronous CMOS comparator", in *Proc. 26th International Conference on Microelectronics*, 2014, Serbia, pp. 355-358.

[2] H. Traff, "Novel Approach to High Speed CMOS Current Comparators", *Electronics Letters*, 1992, vol. 28, №3, pp. 310-312.

[3] D. Banks and C. Toumazou, "Low-power high-speed current comparator design", *Electronic Letters*, 2008, vol. 44, №3, pp. 171-172.

[4] Y. Wang, H. Wang and G. Wen, "Design Techniques for Ultra-Low Voltage Comparator Circuit", *Journal of Circuits, Systems and Computers*, 2015, vol. 24, №1, 1550013.

[5] N.K. Chasta, "High Speed, Low Power Current Comparator with Hysteresis", *International Journal of VLSI Design & Communication Systems (VLSICS)*, 2012, vol. 3, №1, pp. 85-96.

[6] S. Sarkar and 'S. Banerjee, "500 MHz differential latched current comparator for calibration of current steering DAC", in *Proc. IEEE Students' Technology Symposium*, India, 2014, pp. 309-312.

[7] A.T.K. Tang and C. Toumazou, "High performance CMOS current comparator", *Electronic Letters*, 1994, vol. 30, pp. 5–6.

[8] B.S. Patro, S. Biswas, In. Roy and B.Vandana, "1 GHz High Sensitivity Differential Current Comparator", *Journal of Digital Integrated Circuits in Electrical Devices*, 2017, vol. 2, pp. 7-12.

[9] M. Bchir, N. Hassen, K. Besbes, "Low voltage low power comparator circuit", in *Proc. 18th International Conference on Sciences and Techniques of Automatic Control and Computer Engineering*, Tunisia, 2017, pp. 168-172.

[10] S. Ziabakhsh, H. Alavi-Rad, M. Alavi-Rad and M. Mortazavi "The design of a low-power high-speed current comparator in 0.35-μm CMOS technology, in *Proc. 10th International Symposium on Quality Electronic Design*, USA, 2009, pp. 107-111.

Estimation of Errors of Integrated Hydrogen Sensors based on MISFET with structure Pd–Ta$_2$O$_5$ –SiO$_2$ –Si

B. Podlepetsky and A. Kovalenko

Abstract - We have estimated various types of errors of integrated hydrogen sensors based on MISFETs with structure Pd-Ta$_2$O$_5$-SiO$_2$-Si on the basis of experimental studies of the multiple sensors' hydrogen responses, taking into account operating modes. The generalized model for the calculation of these errors was proposed.

I. INTRODUCTION

The hydrogen sensors are being used in the environmental and fire-explosion safety monitoring devices and systems [1]. There are a lot of hydrogen sensors commercially available or in development: mechanical, acoustic, optical, catalytic, electrochemical, thermal conductivity, resistance and work function based [2]. The development of integrated hydrogen sensors fabricated by micro-technology is a promising area to create the small-sized gas-analysis devices and micro-systems. The technological, electro-magnetic and temperature compatibilities are important features of the integrated sensor elements. The sensitive elements based on metal-insulator-semiconductor structures (MIS-capacitors and metal-insulator-semiconductor field-effect transistors – MISFETs) possess the best compatibility with the conventional elements of integrated circuits. Therefore, such sensitive elements seem promising to develop the integrated hydrogen sensors.

The sensors based on MIS-capacitors and MISFETs have been studied by many researchers. A great contribution to the developments of gas-sensitive MIS devices has been made by the researchers at Linköping University [3]-[6], whose works describe the gas sensitivity mechanisms of MIS sensors with different materials of metals and semiconductors.

In recent years, we have been engaged in the development and research of integrated circuits' technology compatible gas-sensitive elements (in particular MISFETs) to create the small-size gas-analysis devices and systems. We have developed and investigated the number of discrete sensors (MIS-capacitors, Pd- and Pt-resistors) and two types of integrated gas sensors with structures Pd (or Pt)-

B. Podlepetsky is with the Department of Micro- and Nanoelectronics, National Research Nuclear University MEPhI (Moscow Engineering Physics Institute), 31 Kashirskoe shosse, 115409 Moscow, Russia, Email: bipod45@gmail.com

A. Kovalenko is with the Induko Ltd., 32/2 Seslavinskaia str., Moscow, 121309, Russia, Email: dir@induko.ru

SiO$_2$-Si, Pd/Ti-SiO$_2$-Si, Pd (or Pt)-Ta$_2$O$_5$-SiO$_2$-Si. Both types of integral sensors contained 4 elements. In one case, these elements were a capacitor and a MISFET with Pd-gate, thermo-sensitive diode element and the test MISFET with Al-gate (IHS-1). In another case, these elements are the gas sensitive Pd (or Pt)-resistor and MISFET with Pd (or Pt) gate, heater-resistor and temperature sensor (IHS-2). Work in this area has been going on for about 25 years. The studies of the influence of technological factors (types and thicknesses of dielectric and metal films, methods of their fabrication) on the characteristics of different types of sensors took a long time (2 – 4 years for each sensor type). Most of our articles were published in Russian journals and conference proceedings.

The experiments have demonstrated that with the same technological parameters, MISFETs have the best performance compared to MIS capacitors in IHS-1 and resistors in IHS-2. In addition, the integrated sensors of the second type, containing MISFET with Pd-Ta$_2$O$_5$-SiO$_2$-Si-structure, possess the best stability and reproducibility of characteristics. Therefore, in recent years we have investigated the characteristics of such MISFETs. The performance characteristics of these sensors and effects of chip temperature, electrical modes and irradiation were studied in our previous works [7]-[11]. However, the errors of such sensors were not specifically investigated.

The motivations of this work are to develop a methodology for estimating errors in concentration's measurements by hydrogen MISFET-based sensors, to determine the components of sensors' errors, to estimate these errors using the experimental data and to propose the models for prediction of errors in sensors and devices based on MISFETs, taking into account operating conditions. Note that part of the experimental results of this paper is published in our paper [11].

II. THE EXPERIMENTAL TECHNIQUE

The layout of integrated sensor chip is shown in Fig.1. This sensor, consisting of Pd- resistor (1), thermosensor (2), heater (3), *n*-channel Pd-gate MISFET (4) and test elements on single silicon chip (2×2 mm^2), was fabricated by means of conventional MIS-technology using laser evaporation Pd-films. Thicknesses of films Pd, Ta$_2$O$_5$ and SiO$_2$ were about 70 nm, 95 nm and 85 nm respectively. The film SiO$_2$ was prepared by oxidation of silicon in dry oxygen at temperature about 1050 ºC. The film Ta$_2$O$_5$ was fabricated

978-1-7281-3420-8/19 $31.00 © 2019 IEEE

by oxidation of tantalum film in dry oxygen at temperature 600 ºC.

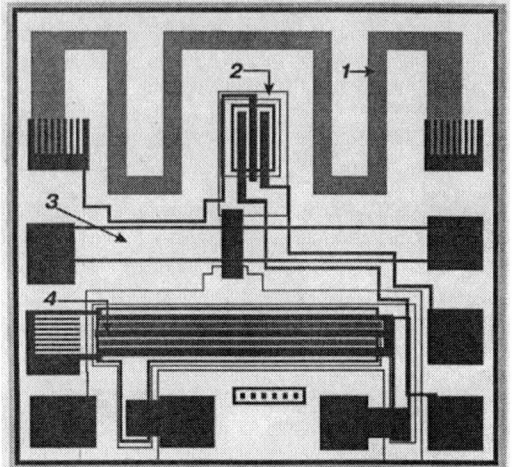

Fig. 1. The layout of integrated sensor chip.

The sensors were embedded in volt-metric circuit simplified shown in Fig. 2. The circuit provides the constant drain current I_D and source-drain voltage V_D. In this circuitry the voltage V is equal to the gate voltage V_G. The optimal for sensors chip temperature $T = 130 \pm 2$ °C [9] was supported by the temperature-stabilization circuitry with feedback loop using on-chip thermo-sensor and heater. The computerized measuring system was used for experiments.

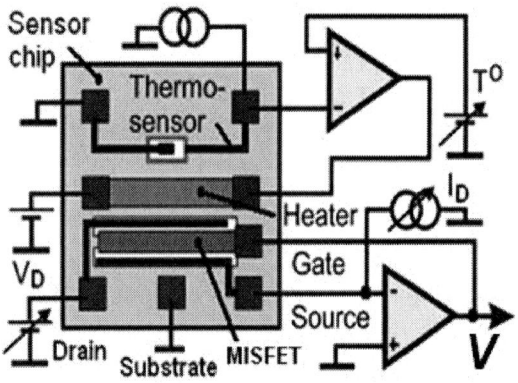

Fig. 2. Simplified view of the sensor chip and circuit.

There were repeatedly measured responses of sensors to hydrogen i-pulses with duration about 30 s for concentrations C_i (0.02; 0.05; 0.1; 0.15; 0.2) % vol. at drain currents 0.1 mA and the source-drain voltage 0.2 V. There were tested 5 sensors for 5 weeks (5 days a week in a row with 2 day breaks) by 5 of repeated hydrogen j-cycles.

III. The Experimental Results and Modeling

Experimental responses for the first cycle are presented in Fig. 3. For each j-cycle there were calculated the arithmetic means of experimental response amplitudes ΔV_{Ci}, residual values δV_{0i} (Fig. 3), variation indices $\theta_{V0}(C_i)$ and $\theta_{VC}(C_i)$ shown in Table 1.

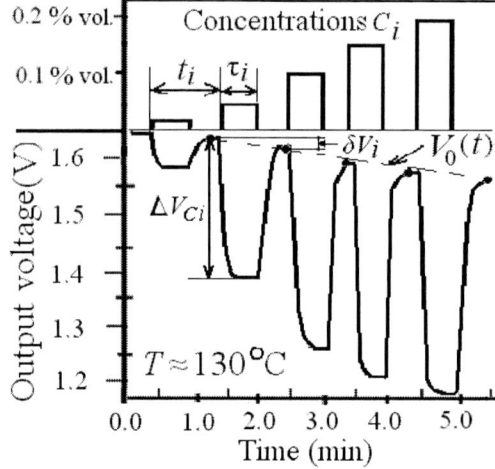

Fig. 3. Responses for the first cycle and its parameters (a fragment of Fig. 4a from our paper [11]).

TABLE I
THE AVERAGE VALUES OF COMPONENTS OF CONVERSION FUNCTION
$V(C)$ FOR FIRST/FIFTIENTH CYCLES AT DIFFERENT C

C_i, % vol.	$\Delta V_0'$, mV	$\theta_{V0}(C_i)$	$\Delta V_C'$, mV	$\theta_{VC}(C_i)$
0.02	12/7	0.16/0.28	68/56	0.04/0.05
0.05	28/20	0.07/0.1	262/215	0.01/0.014
0.1	54/41	0.04/0.05	344/282	0.009/0.01
0.15	70/46	0.03/0.04	355/290	0.008/0.01
0.2	75/48	0.04/0.06	415/340	0.01/0.012

Generalized models the conversion function $V(C, D)$ and relative error δC being equal to $\Delta C/C$ are presented as

$$V(C,D) = V_{00} + \Delta V_{0t}(t) - \Delta V_0(D) - \Delta V_C(C,D) \quad (1)$$

$$\delta C = \left| \frac{\pm \Delta V_V + \theta_V \cdot \Delta V_C + dV}{C \cdot \left(\dfrac{dV}{dC} \right)} \right| \times 100\% \quad (2)$$

Average values of output voltages $V_{ai}(C_i)$ and dispersion variation indices $\theta_{V0}(C_i)$ and $\theta_{VC}(C_i)$ were calculated as

$$V_{ai} = \frac{1}{N} \cdot \sum_{n=1}^{N} |V_{ni}| \qquad (3)$$

$$\theta_{Vi} = \frac{1}{N} \cdot \sum_{n=1}^{N} \left| \frac{V_{ni} - V_{ai}}{V_{ai}} \right| \qquad (4)$$

Indexes n and N are respectively serial numbers and quantity of sensors. The hydrogen sensitivity component is represented as

$$\Delta V_C(C, D) = \Delta V_{CM}(D) \cdot [1 - \exp(-k \times C)] \qquad (5)$$

Immediately after the turning sensors in the operating modes voltage V decrease monotonically reaching saturation value V_0 being equal to $(V_{00} - \Delta V_{0tM})$ after $3 - 4$ minutes. The average values of voltages V_{00} and ΔV_{0tM} were equal to about 1.7 V and 40 mV respectively ($T \approx 130$ °C). The initial values of V_0 of the first responses were equal to about 1.66 V. Each hydrogen exposition j- cycle started 5 minutes after the turning sensors in the operating modes. The value of ΔV_{0tM} changed in the next j -cycles within ± 5 mV. Such drift of the zero line in time did not depend on the previous hydrogen exposures.

The additive error of "zero" ΔV_{00} is determined by the sum of the two types so-called "zero-line drift"(ZLD): the initial time drift $\Delta V_{0t}(t)$ and the drift $\Delta V_0(D)$ associated with the total hydrogen dose $D = \int C(t)dt$. The models of time drift $\Delta V_{0t}(t)$ and the drift $\Delta V_0(D)$ in general and numerical forms are presented in [11].

The instrumental error ΔV_V is determined by the accuracy of the voltage measurement. The random error for this circuit dV being equal to $V \cdot (\Delta I_D / I_D)$ is associated with fluctuations in electrical parameters of the circuit (the drain current I_D in this case).

The fourth component ΔV_C of the model (1) determines the sensor integral and differential hydrogen sensitivities S_i and S_d [11]:

$$S_i = \frac{\Delta V_{Ci}}{C_i} = \Delta V_{CM}(C_i, D) \cdot [1 - \exp(-k \times C_i)] / C_i \; ;$$

$$S_d = \frac{dV}{dC} = k \cdot \Delta V_{CM} \cdot \exp(-k \times C) \qquad (6)$$

The maximum absolute error for V can be presented as

$$\Delta(V) = \left| \begin{array}{l} \Delta V_V + \theta_V(C) \cdot V_a(C) + \\ + \Delta V_p(C, \{p_k\}) + \Delta V_z(C, \{z\}) \end{array} \right| \qquad (7)$$

The components of $\Delta V_p(C, \{p_k\})$ and $\Delta V_z(C, \{z\})$ are additional errors associated with fluctuations of electrical parameters p_k of circuit and influence of external factors z respectively, which are determined experimentally [7]-[11].

IV. DISCUSSIONS

Of practical interest for sensors are the errors in determining values of C. The maximum absolute ΔC and the relative δC errors can be presented as:

$$\Delta C = |\Delta(V)/S_d| \; ; \; \delta C = |\Delta(V)/(C \cdot S_d)| \times 100\% \qquad (8)$$

The calculated plots of relative errors δC and absolute errors ΔC vs. C for different operating conditions are presented in Fig. 4. and Fig. 5.

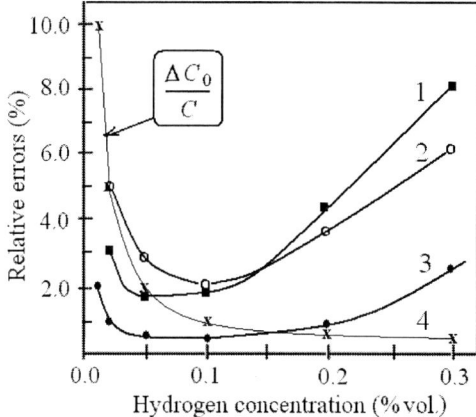

Fig. 4. Calculated dependences of relative errors δC on hydrogen concentration C: 1 – for j is equal to 1; 2 – for j is equal to 50; 3 – for ideal MISFET (the error is due only to error ΔV_V being equal to 1 mV); 4 – error determined the concentration ΔC_0 being equal to 0.001% vol.

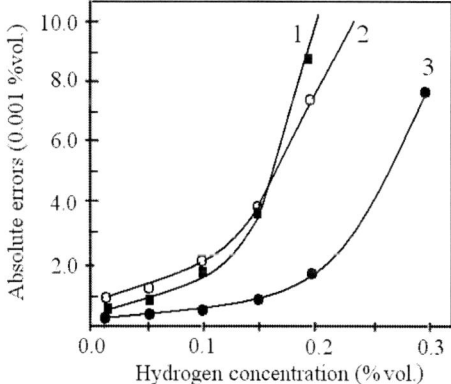

Fig. 5. The maximum absolute errors ΔC vs. C for different experimental stages: 1 – if j is equal to 1; 2 – if total j is equal to 50; 3 – if error is due only to error ΔV_V, being equal to 1 mV.

The plots of maximum response amplitudes ΔV_{CM} and ZLD parameter ΔV_0 vs. D are presented in Fig. 6. The components V_{00} and ΔV_C depend on chip temperature [9] and affecting factors (other gas and irradiation) [8]. The dependences of the operating range of the measured concentrations on the given errors are presented in Table 2.

978-1-7281-3420-8/19 $31.00 © 2019 IEEE

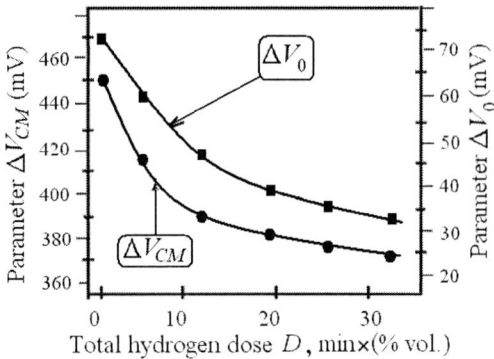

Fig. 6. Parameters ΔV_{CM} and ΔV_0 vs. D (a fragment of Fig. 6b from our paper [11]).

TABLE II

HYDROGEN CONCENTRATION RANGES C_{MIN} & C_{MAX} VS. δC

δC, % → Parameters ↓		3.0	4.0	5.0	6.0
C_{min}; C_{max}, % vol.	$j = 1$	0.02; 0.16	0.01; 0.19	0.008; 0.22	0.01; 0.24
	$j = 50$	0.05; 0.18	0.03; 0.21	0.02; 0.23	0.01; 0.3

The additive error of zero (ZLD) is determined by the initial design and technological characteristics of the sensors, and decreases during operation. Values of primary voltage V_{00} and volt-metric error ΔV_V do not depend on hydrogen concentration C.

Absolute and relative errors of concentration measurement depend on gas concentration, sensitivity to hydrogen, other gases, radiation, electrical modes, fluctuations of electrical circuits' parameters and the total dose of the hydrogen medium (see Fig. 4 – Fig. 6). In addition, depending on the concentration of errors determine the operating range of the sensor conversion for the specified errors (Table 2). For example, for a relative error of 6% the operating conversion range (C_{max} / C_{min}) is 30 in the range of 0.01 % vol. up to 0.3 % vol. The sensitivity threshold (ΔV_V /$k\Delta V_{CM}$) is about 2×10^{-4} % vol.

The hydrogen sensitivities S_i and S_{dj} reduced, if hydrogen exposition time and dose D increases. This leads to an increase in the absolute error at low and medium hydrogen concentrations.

V. CONCLUSION

There were estimated various types of errors of integrated hydrogen sensors based on MISFETs with structure Pd-Ta$_2$O$_5$-SiO$_2$-Si on the basis of previous experimental studies of the multiple sensors' hydrogen responses, taking into account operating modes [11]. A methodology for estimating errors in the measurement of hydrogen concentrations has been developed. The components of the additive zero error associated with time

drift and drift due to the degradation of palladium under the action of hydrogen have been considered. The generalized models for the calculation of absolute and relative errors were proposed. There were determined the components of sensors' errors. It is shown how the errors depend on the hydrogen concentration for different operating conditions. It is also noted that errors of concentration measurement depend on hydrogen sensitivity, other gases, radiation, electrical modes, fluctuations of electrical circuits' parameters, the total dose of the hydrogen and the instrumental error determined by the accuracy of the voltage measurement. Proposed models can be used for prediction of errors in sensors and devices based on MISFETs, taking into account operating conditions.

ACKNOWLEDGEMENT

Author acknowledges support from the NRNU MEPhI Academic Excellence Project (Contract No. 02.a03.21.0005).

REFERENCES

[1] W. Buttner, M. Post, R. Burgess, and C. Rivkin, "An overview of hydrogen safety sensors and requirements", *Int. J. Hydrogen Energy*, 2011,vol. 36, no 3, pp. 2462–2470.

[2] T. Hübert, L. Boon-Brett, G. Black, and U. Banach, "Hydrogen sensors – A review", *Sens. Actuators B*, 2011, vol. 157, pp. 329-352.

[3] I. Lundström, S. Shivaraman, C. Svensson, and L. Lundkvist, "A hydrogen-sensitive MOS field-effect transistor", *Appl. Phys. Lett.*, 1975, vol. 26, pp. 55–57.

[4] I. Lundström, M.Armgarth, A.Spetz, and F.Winquist, "Gas sensors based on catalytic metal-gate field-effect devices", *Sensors and actuators*, 1986, vol.3-4, pp. 399-421.

[5] I. Lundström, C. Svensson, A. Spetz, H. Sundgren, and F. Winquist, "From hydrogen sensors to olfactory images – twenty years with catalytic field effect devices", *Sens. Actuators B*, 1993, vol. 13–14, pp. 16-23.

[6] I. Lundström, H. Sundgren, F. Winquist, M. Eriksson, C. Krants-Rülcker, and A. Lloyd-Spets, "Twenty-five years of field effect gas sensor research in Linköping", *Sensors and Actuators B*, 2007, vol.121, pp. 247-262.

[7] B. Podlepetsky, "Integrated Hydrogen Sensors Based on MIS Transistor Sensitive Elements: Modeling of Characteristics", *Autom. Remote Control*, 2015, vol. 76, pp. 535-547.

[8] B. Podlepetsky, "Total ionizing dose effects in hydrogen sensors based on MISFET", *IEEE Transactions on Nuclear Science*, 2016, vol. 63, no. 4, pp. 2095-2105.

[9] B. Podlepetsky, M. Nikiforova, and A. Kovalenko, "Chip temperature influence on characteristics of MISFET hydrogen sensors", *Sensors and Actuators B*, 2018, vol. 254, pp. 1200-1205.

[10] B. Podlepetsky, A. Kovalenko, and M. Nikiforova, "Influence of electrical modes on radiation sensitivity of hydrogen sensors based on Pd-Ta$_2$O$_5$-SiO$_2$-Si structures", *Sensors and Actuators B*, 2018, vol. 273, pp. 999-1007.

[11] B. Podlepetsky, N. Samotaev, M. Nikiforova, and A. Kovalenko, "Performance degradations of MISFET-based hydrogen sensors with Pd-Ta$_2$O$_5$-SiO$_2$-Si structure at long-time operation", *Sensors*, 2019, 19(8), 1855, doi:10.3390/s19081855.

978-1-7281-3420-8/19 $31.00 © 2019 IEEE

Analysis and Design of Power Processing Circuits for Thin Film Piezoelectric Energy Harvesters on Flexible Polyethylene Terephthalate Substrates

I. Pandiev, M. Aleksandrova and G. Kolev

Abstract - Piezoelectric potassium niobate based harvesters with active films' thickness between 200 *nm* and 400 *nm* are fabricated by vacuum radiofrequency (RF) sputtering on plastic polyethylene terephthalate substrates. Using gold electrodes at the opposite sides of the samples, they show piezoelectric rms voltage between 300 *mV* and 600 *mV* generated from 3 *cm²* area, according to the film's thickness. The obtained current from a single harvester is up to 1 μ*A*. An AC/DC power processing circuit topology intended for this type of piezoelectric energy harvesters is proposed and investigated. A DC/DC monolithic low-power converter is connected between the output port of the bridge rectifiers with smoothing capacitor and the load (such as rechargeable battery or supercapacitor (SC)) in order to properly regulate the rectified voltage. Experimental tests, conducted by using the synthesized piezoelectric energy harvesters on flexible substrates, validate the theoretical analysis and results.

I. INTRODUCTION

Energy harvesting (EH) is advanced technology for scavenging and conversion of the waste non-electrical energy into electrical. The application of the harvesting devices is for power supply of low consuming electronics as an alternative of the conventional batteries. Among the different approaches, piezoelectric EHs (PEHs) are considered as one of the most promising after solar cells [1]. They are consisted of piezoelectric thin film (200-500 *nm*) with high dielectric permittivity ($\varepsilon_r > 17$), most often grown between two electrodes. In order to maximize the effect of mechanical loading, this multilayer structure is prepared on plastic substrate. Some of the basic characteristics of PEH are: piezoelectric coefficient (unit generated charge per unit applied tension); frequency response of the generated piezoelectric voltage and capacitance–voltage dependence. Generated power is still too small, due to the micro- and even nano-ampere range of

I. Pandiev is with the Department of Electronics, Faculty of Electronic Engineering and Technologies (FEET), Technical University of Sofia, 8 Kl. Ohridski Blvd, 1000 Sofia, Bulgaria, E-mail: ipandiev@tu-sofia.bg

M. Aleksandrova is with the Department of Microelectronics, Faculty of Electronic Engineering and Technologies (FEET), Technical University of Sofia, 8 Kl. Ohridski Blvd, 1000 Sofia, Bulgaria, E-mail: m_aleksandrova@tu-sofia.bg

G. Kolev is with the Department of Microelectronics, Faculty of Electronic Engineering and Technologies (FEET), Technical University of Sofia, 8 Kl. Ohridski Blvd, 1000 Sofia, Bulgaria, E-mail: georgi_klv@abv.bg

the current at these materials. Additionally, the produced signals suffer from distortions. Thus, suitable circuits are necessary as an intermediate unit between the PEH and the electrical load R_L, in order to enhance its performance and to process the signal in a suitable way for the power consumer.

For the last few years the analysis of the literature resources shows a wide variety of electronic circuits converting signals from PEHs [3-9]. Of particular interest are the power processing circuits in [4], [8] and [9] that operate at an input voltage with rms value less than 1 *V*, because the used thin film piezoelectric energy harvesters on flexible substrate produce an ac voltage with rms value from 300 *mV* to 600 *mV*.

In this study potassium niobate (KNbO₃) was used for piezoelectric thin films grown by vacuum RF sputtering on gold coated polyethylene terephthalate (PET) substrate. The top electrode was also gold. The optimum structural and electrical parameters of the KNbO₃ film was previously established [10]. Different deposition conditions were set, in order to find the most suitable combination of piezoelectric film growth in terms of high uniformity, controllable thickness and acceptable deposition rate. For this purpose, plasma power was varied between 84 *W* and 220 *W* and the argon sputtering pressure was varied between 10^{-1} to 10^{-4} *Torr*. Seven combinations of deposition modes were set in total, producing films with broad range of thicknesses between 25 *nm* and 600 *nm* at deposition rates between 5 and 15 *nm/min*. Although all combinations of samples were investigated, in this paper are used only those samples (Fig. 1), exhibiting the most superior results for roughness (~ 5 *nm* for a thickness of 410 *nm*) and piezoelectric voltage per mass load (~12 *mV/g.mm⁻²*).

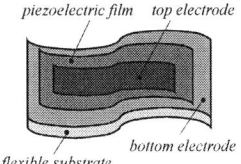 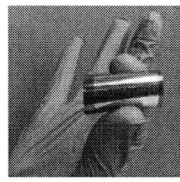

Fig. 1. Constituent layers of the piezoelectric harvesting element and photo of the prepared sample.

Prepared in this way, this energy harvesting element exhibited piezoelectric coefficient of 39 *pC/N*, which is superior, as compared to the well-known zinc oxide based and other lead-free piezoelectric transducers with similar

978-1-7281-3420-8/19 $31.00 © 2019 IEEE

design and thickness [2]. The generated current is up to 1 μA, according to the stress magnitude. As is shown in Fig. 2, produced voltage from the sample with the thickest film can vary from 178 mV at the weakest applied stress of 6 g to 424 mV for the strongest applied stress of 40 g for piezoelectric film with small thickness (210 nm).

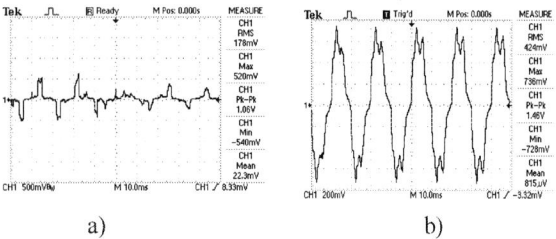

a)   b)

Fig. 2. Oscillograms of the piezoelectric voltage generated from flexible thin film (210 nm) potassium niobate based energy harvester: a) at the weakest applied stress of 6 g; b) at stress of 40 g.

III. DESIGN OF POWER PROCESSING CIRCUIT TOPOLOGY

As described above, the power processing circuit needs to transform the AC voltage from the PEHs to a DC voltage with an appropriate level and stability. Furthermore, the processing circuit has to maintain the voltage and current produced by the piezoelectric element.

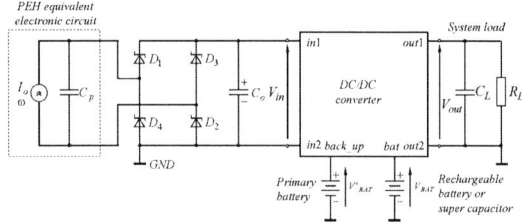

Fig. 3. An AC/DC power processing circuit diagram for thin film PEH on flexible substrate with general PEH equivalent electronic circuit.

The proposed AC/DC power processing circuit diagram used for the thin film PEH on flexible substrate applications is shown in Fig. 3. It adopts a passive diode bridge rectifier. This circuit diagram is based on the widespread AC/DC architecture for PEHs with a passive rectifier represented in [3], [4], [6] and [9].

The passive bridge rectifier, using Schottky diodes it is well known that to operate as an input stage of a power processing circuit for PEH devices. The main advantages of those rectifiers is their simple and easy implementation with discrete components on printed circuit board. Also, they can operate with relatively lower input voltages (the voltage drop over the C_p is $V_{po} > 2V_D$, where $V_D < 0.4\ V$). The disadvantage of the bridge rectifiers with output smoothing capacitor is the relatively small energy efficiency.

The second (output) stage of the proposed power processing circuit is based on the monolithic synchronous

switching DC/DC regulators. For them, a higher value of the energy efficiency can be obtained. This type of monolithic regulators combines a low powered boost regulator with a management controller adopting storage elements. They convert the output voltage with small ripples from the bridge rectifier into a voltage with constant value. The input signal for the DC/DC regulator is applying to the input port with terminals $in1$ and $in2$. Moreover, this type of regulators can store the electrical energy in the rechargeable battery or supercapacitor (SC) (or ultracapacitor), and provide electrical power to the output port with terminals $out1$ and $out2$. To the output port is connected load impedance $R_L \parallel C_L$. The energy efficiency is obtained by comparing the input power with the output power:

$$\eta = V_{in}I_{in} / V_{out}I_{out}. \tag{1}$$

The majority of the monolithic regulators control an additional power path from a primary battery cell to the electronic system. When the system operates at a condition where the stored electrical energy or harvested energy is periodically insufficient, a backup energy storage element can be connected to the BACK_UP terminal of the regulator.

A. Bridge rectifier

The easiest technique to rectify an AC voltage from PEHs is by using diode bridge rectifier with additional smoothing capacitor C_o. For this circuit the ratio of the recharge and discharge periods are implementted by charging the capacitor C_o during the positive and negative half-cycles. According to the parameters of the microgenerators, Schottky diodes, type 1N5711 or 1N5817, are chosen. For them, the maximum value of the forward voltage drop is up to 0.2 V at forward current up to 1 mA. The output capacitor C_o has to be much greater than the piezoelectric capacitor C_p. For the studied energy harvesters the capacitor C_o has to be from 1 μF to 10 μF.

The open-circuit voltage drop over the piezoelectric capacitance is given by [4]

$$V_{po} = I_o / \omega C_o, \tag{2}$$

where I_o is the average value of the current.

The non-load (open circuit) output voltage over the capacitor C_o is given by

$$V_{in} = V_{po} - 2V_D, \tag{3}$$

where the $V_D \approx 0.1,...0.2V$ is forward voltage for metal-to-silicon junction diodes. Therefore, in order to obtain an output voltage V_{in} greater than zero, it is necessary to fulfill the condition: $V_{po} > 2V_D$. The on-load output voltage depends on the internal resistance r_i of the harvester and the value of the load R_L [11]:

978-1-7281-3420-8/19 $31.00 © 2019 IEEE

$$V_{in,\infty} = V_{in}(1 - \sqrt{r_i / 2R_L}) . \tag{4}$$

The typical values for the internal resistance of the PEHs are not greater than 100 $k\Omega$. The optimal output voltage [4] for which is obtained the maximum output power is

$$V_{in} = 0.5(V_{po} - 2V_D) . \tag{5}$$

Then, for the value of the voltage defined in (5) the maximum output power, in condition that the voltage drop over the diodes is neglected, is

$$P_{\max} = I_o^2 / 2\pi\omega C_p . \tag{6}$$

As the forward voltage V_D is slightly dependent on the current I_o, the power dissipation of a single diode is given by

$$P_D = 0.5V_D I_o . \tag{7}$$

The power supplied to the load will be

$$P_L = V_{in,\infty} I_o . \tag{8}$$

B. Monolithic synchronous switching DC/DC regulator

To implement the DC/DC converter, an EVAL-ADP5090 —development board, which is built around the low-power boost regulator ADP5090 [12] with programmable voltage monitor equal to 3 V, is chosen. The IC ADP5090 uses a compact 3 mm × 3 mm LFCSP_WQ package (CP-16-33) with thermal pad on the bottom side of the package. The ADP5090 includes a "cold start-up" circuit, a synchronous boost controller with integrated MOS transistors, a charge controller with an integrated switch, and switches for the back-up power path. By using the internal cold start-up circuit, the regulator can start operating at an rms input voltage as low as 380 mV (typical value). After cold start-up, the regulator can operate with rms input voltage range from 80 mV to 3.3 V. For the cold start-up the minimum input current has to be greater than 10 μA (or the typical value of the minimum input power for cold start-up is 16 μW).

The ADP5090 is intended for charging energy storage elements such as rechargeable Li-Ion batteries, SCs, and conventional capacitors with large capacitance, for example 100 μF. For the purpose of this work we used a micro SC, type XH311HU-IV07E (from Seiko Instruments) with following parameters: (1) ∅3.8mm and height 1.1mm; (2) maximum usable voltage 3.3 V; (3) initial capacitance more than 0.022 F; (4) nominal capacity 12 μAh (measured between 3.3 V – 2.0 V) 0.035 F and (5) operating temperature range from –20 to +60 $°C$. The XH311HU-IV07E capacitors are suitable for backup power

supply element of clock and memory functions on mobile and other information devices.

A primary cell battery type CR2032 with capacity 225 mAh and nominal voltage 3 V is connected to the BACK_UP terminal of the ADP5090 and managed by an integrated control block that automatically switches the power source from the energy harvester to the battery. The thin-film battery CR2032 is chosen because it has small size, provide sufficient voltage for low-power applications and have a relatively large capacity.

IV. EXPERIMENTAL TESTING

The proposed processing circuit topology was experimentally validated, and the experimental setup is shown in Fig. 4. As can be seen to demonstrate the charging process of the chosen supercapacitor the input signal is obtained from function generator, because the possible generated current from a single PEH device is up to 1 μA. The amplitude of the input signal V_{po} is chosen equal to 1.5 V and the frequency is up to 50 Hz. For the experimental test the backup battery is not connected to the BACK_UP port. In this case the output voltage V_{OUT} is produced by the DC input voltage V_{in}, applied to the input port. The load can be connected to the output port of the circuit. The time interval for charging the capacitor to a voltage $V_{CAP} = V_{in,\infty} \approx 1.5$ V is about 5 hours.

If the backup battery is connected to the BACK_UP port, the output voltage becomes equal to the value approximately equal to $3V$. As a result, the capacity of the system output port is determined by the capacity of the battery. Moreover, the charge of the supercapacitor continues. When the voltage of the supercapacitor becomes equal to and greater than the battery voltage, the output is switched. The output terminal according to ground has a voltage dependent on the capacity of the SC.

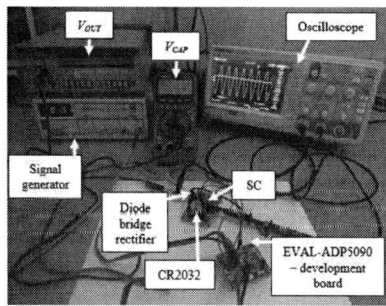

Fig. 4. Experimental setup for the power processing circuit.

During the measurements, custom PEHs were fabricated with greater film thickness of 380 nm and placed on a user-designed shaker system, which was excited at the natural frequency of the PEHs at 50 Hz. Fig. 5 shows the measured transient waveforms for different nodes in the bridge rectifier. As can be seen, value of the produced DC voltage exceeds the minimum value for the cold start-up of the switching DC/DC regulator.

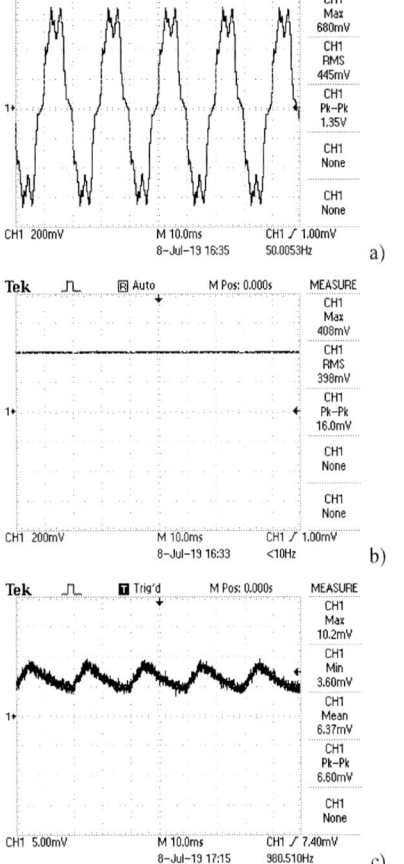

Fig. 5. Measured transient waveforms for different nodes in the rectifier: a) generated signal from single PEH at 60 g/mm^2 mass load; b) DC output voltage; c) ripples in the rectified voltage.

V. Conclusion

In this paper power processing circuit intended for thin film PEHs on flexible polyethylene terephthalate substrates has been presented. The developed electronic circuit requires minimum DC input voltage equal to 400 mV and input current greater than 10 μA. According to formula (1) the achieved energy efficiency of the circuit is greater than 80% at amplitude V_{po}=1.5V.

The future work of the authors is related to the study of voltage multiplier based circuits (voltage doubler or tripler) with parallel connection of several PEHs to the vibration system so as to increase the power of the input signal.

Acknowledgements

The authors would like to thank Bulgarian National Science Fund, grant DH 07/13 for the financial support.

References

[1] K. Uchino, and T. Ishii, "Energy Flow Analysis in Piezoelectric Energy Harvesting Systems," *Ferroelectrics*, 2010, vol. 400, pp. 305-320.

[2] M. Fraga, H. Furlan, R. Pessoa, and M. Massi, "Wide bandgap semiconductor thin films for piezoelectric and piezoresistive MEMS sensors applied at high temperatures: An overview," *Microsystem Technologies*, 2013, vol. 20, pp. 9-21.

[3] R. D'hulst, T. Sterken, R. Puers, G. Deconinck, and J. Driesen, "Power processing circuits for piezoelectric vibration-based energy harvesters," *IEEE Trans. on Ind. Electron.*, 2010, vol. 57, no. 12, pp. 4170–4177.

[4] J. Dicken, P. Mitcheson, I. Stoianov, and E. Yeatman, "Power-Extraction Circuits for Piezoelectric Energy Harvesters in Miniature and Low-Power Applications," *IEEE Trans. on Power Electron.*, 2012, vol. 27, Issue: 11, pp. 4514-4529.

[5] G. Szarka, B. Stark, S. Burrow, "Review of power conditioning for kinetic energy harvesting systems," *IEEE Trans. on Power Electron.*, 2012, vol. 27, Issue: 2, pp. 803-815.

[6] Sh. Du, Y. Jia, Ch. Zhao, G. Amaratunga, and A. Seshia, "A Passive Design Scheme to Increase the Rectified Power of Piezoelectric Energy Harvesters," *IEEE Trans. on Ind. Electron.*, 2018, vol. 65, no. 9, pp. 7095–7105.

[7] J. Liang, Y. Zhao, and K. Zhao, "Synchronized Triple Bias-Flip Interface Circuit for Piezoelectric Energy Harvesting Enhancement," *IEEE Trans. on Power Electron.*, 2019, vol. 34, no. 1, pp. 275-286.

[8] A. Shareef, W. Ling Goh, S. Narasimalu, and Y. Gao, "A Rectifier-Less AC–DC Interface Circuit for Ambient Energy Harvesting From Low-Voltage Piezoelectric Transducer Array," *IEEE Trans. on Power Electron.*, 2019, vol. 34, no. 2, pp. 1446-1457.

[9] L. Costanzo, A. Lo Schiavo, and M. Vitelli, "Power Extracted From Piezoelectric Harvesters Driven by Non-Sinusoidal Vibrations," *IEEE Trans. on Circ. and Syst.–I: Regular Papers*, 2019, vol. 66, no. 3, pp. 1291-1303.

[10] T. Tsanev, M. Aleksandrova, T. Ivanova, and G. Dobrikov, "Investigation of Lead-free Potassium Niobate Thin Films on Silicon for Piezoelectric Transducers", in *Proc. X National Conference with International Participation "Electronica 2019"*, Sofia, Bulgaria, May 16 - 17, 2019.

[11] U. Tietze, Ch. Schenk, and E. Gamm, "Power Supplies," in *Electronic Circuits. Handbook for Design and Application.* 2nd edition. New York: Springer-Verlag Berlin Heidelberg, 2008, ch. 16, pp. 885–986.

[12] "Ultralow Power Boost Regulator with MPPT and Charge Management ADP5090 – datasheet." *Analog Devices*, Norwood, MA, USA, 2009.

Towards Portable Thermal Vacuum Sensor - Consideration of Electrical Building Blocks and Compact Housing

D. V. Randjelović, P. Poljak, M. Sarajlić, M. Vorkapić, M. Frantlović,
D. Tanasković, and B. Popović

Abstract - Multipurpose ICTM thermopile-based MEMS sensors have been successfully demonstrated as gas flow sensor, vacuum sensor, thermal converter and gas type sensor. Apart from further improvement of the sensor's design and performances, one direction of research is working towards realization of portable thermal devices. The aim of this paper is to present preliminary considerations of electrical building blocks and compact housing neccessary to realize a portable thermal vacuum device.

I. INTRODUCTION

Multipurpose thermopile-based MEMS sensors developed at ICTM have been so far successfully demonstrated for several applications [1-5]: gas flow sensor, vacuum sensor, thermal converter and gas type sensor.

Apart from further improvement of the sensor's design and performances, one direction of research is working towards a design of portable thermal devices. This paper presents preliminary considerations of electrical building blocks and compact housing necessary to realize a portable thermal vacuum device. The final goal is to produce a universal compact and portable box containing electronics which will be suitable for different applications of thermal sensor, but also for other types of proprietary ICTM sensors.

The first part of the paper contains information regarding the design and performance of ICTM thermal vacuum sensors. Afterwards, the design of the portable thermal device is presented. Finally, conclusions important for further research activities are deduced.

II. THERMOPILE-BASED VACUUM SENSOR

Implementation of thermopile-based MEMS sensors for vacuum detection dates back to 1985. [6]. Output voltage of this type of sensors depends on gas pressure inside the housing which influences heat exchange between the sensor and the ambient [2]. Vacuum sensors actually measure gas pressure in the sub-atmospheric region.

D.V. Randjelović, P. Poljak, M. Sarajlić, M. Vorkapić, M. Frantlović, D. Tanasković, and B. Popović are with the Centre of Microelectronic Technologies, Institute of Chemistry, Technology and Metallurgy, University of Belgrade, Njegoševa 12, 11000 Belgrade, Serbia, E-mail: danijela@nanosys.ihtm.bg.ac.rs

Layout and the main elements of thermopile-based MEMS sensor developed at ICTM are presented in Fig. 1. The sensors were fabricated using double side polished n-Si (100) wafers with a nominal thickness of 385 μm.

Each chip contains two thermopiles, with 30 multilayer p^+Si/Al thermocouples placed on both sides of the heater. Both Al and p^+Si were used for fabrication of the heater thus forming sensors of A-type and P-type, respectively. The heater and the major part of the thermopiles are placed on thermaly isolating membrane consisting of sputtered SiO_2 (thickness d_{SiO2} = 1 μm) and residual n-Si layer (thickness d_{nSi}). Cold thermopile junctions are located on the unetched part of the chip – the rim, while hot junctions are close to the heater on the central membrane area.

Details on design and fabrication of the sensors and original method of determination of residual n-Si thickness are given elsewhere [1]. Thickness of the residual n-Si layer in fabricated chips covers the range from 25 μm down to around 3 μm. Performance of the sensors is improving with the decrease of thickness of the residual n-Si layer.

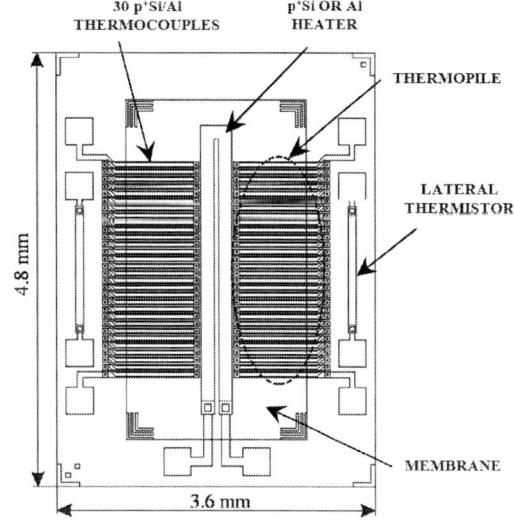

Fig. 1. Top view of the sensor with the main elements.

Fig. 2. A-type (a) and P-type (b) thermal vacuum sensor mounted on TO-8 housing.

When applied as a vacuum sensor, fabricated chips are mounted on eight-lead transistor outline packages (TO-8) as shown in Fig. 2 for both types of the sensor.

For the purpose of vacuum measurement, a special trasducer housing was designed and fabricated of stainless steel [2, 3]. It consists of two parts - the sensor mount with the multipurpose thermopile-based sensor wire bonded on TO-8 housing, and the process flange which complies with ISO (size NW 16) standard as shown in Fig. 3.

Thermal vacuum detection is a three-step process. For a given pressure and ambient temperature, the surrounding gas has a specific value of thermal conductivity. Under such conditions, a certain value of temperature difference between hot and cold thermopile junctions is established. Finally, thermoelectric conversion occurs, inducing Seebeck voltage at the thermopiles.

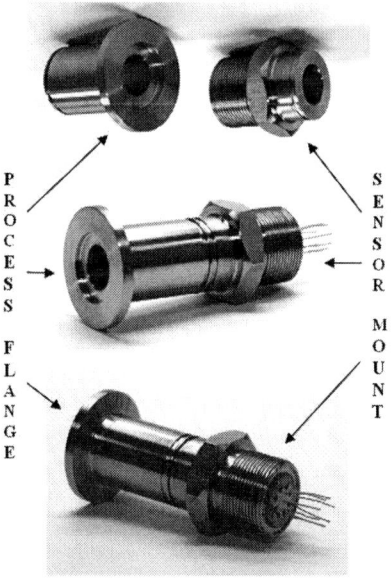

Fig. 3. Thermopile-based vaccum sensor incorporated in the transducer housing.

Sensors of A- and P-type were tested as vacuum detectors in the pressure range $(10^{-3}-10^{5})$ Pa. Experimental setup and measurement procedure are desrcibed in detail

elsewhere [2]. Dependence of output voltage on pressure measured at ambient temperature when power of 50 mW is developed at the heaters for two sensors with the best performances and thinnest residual n-Si layers are shown in Fig. 4.

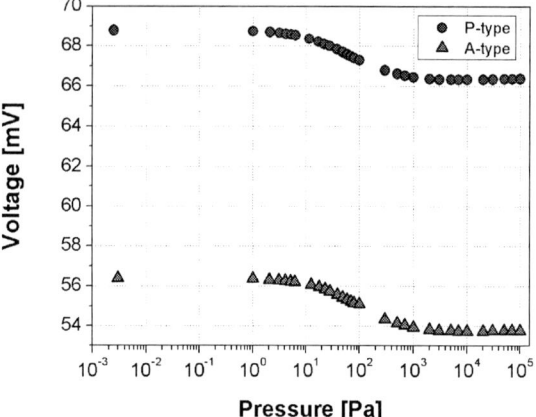

Fig. 4. Experimental results of pressure dependence of the output voltage of one thermopile measured for A-type (d_{nSi} = 3.9 μm) and P-type (d_{nSi} = 2.9 μm) sensor.

Further research activities led to realization of intelligent thermal vacuum devices of both types [3].

III. DESIGN OF THE PORTABLE THERMAL VACUUM DEVICE

A. Electrical Building Blocks

Block diagram of the portable thermal vacuum sensor is shown in Fig. 5. During the operation, the sensor's heater should be powered by a constant current supply of either 100 mA or 3 mA, depending on the chip's design. While in all previous measurements, an external current source was used, in the portable sensor current will be provided by the Sensor excitation block with programmable constant current source.

The output of the sensor is connected to the Sensor interface block whose simplified diagram is given in Fig. 6 (the thermal sensor and its constant current excitation are also shown for completeness). Since the thermal sensor contains two thermopiles, it has two voltage outputs. Depending on the application, one or both of them should be used. Two instrumentation amplifiers (IA) amplify the two corresponding voltage signals. It should be noted that variation of the signals from the sensor are of the order of 100 μV, while the offset is of the order of 50 mV in the considered application. Therefore, very low noise and low drift instrumentation amplifiers must be used, and, in order to accommodate the offset, their gain cannot be much higher than 20. Low-pass filters at the inputs and between amplifier stages (not shown in the diagram) are needed to

978-1-7281-3420-8/19 $31.00 © 2019 IEEE   318

attenuate the noise at higher frequencies. In order to maximize the dynamic range of the two amplified sensor signals, the offset should be suppressed on both channels. For that purpose, an appropriate voltage level, generated by a digital-to-analog converter (DAC), is applied to both IA reference inputs. The level is set by software, and derived from a high performance voltage reference (V_{ref}). The two amplified sensor signals with the suppressed offset are multiplexed, so any of them can be further amplified by the programmable gain amplifier (PGA), and then converted to digital data by the high-resolution (24-bit) sigma-delta analog-to-digital converter (ADC). The PGA gain equals 2^n, where $n=0,1...7$ is set by software. The ADC also uses the high performance voltage reference (V_{ref}). Some highly integrated analog-to-digital converters, such as the AD7124-8 from Analog Devices, contain MUX, PGA, V_{ref}, digital filters, and some other circuits on the same chip, thus enabling a significant reduction in size and complexity of the sensor interface block. The ADC and DAC are connected to the microcontroller via the SPI interface.

The Low noise power supply block provides the supply voltages used by the Sensor interface block with minimal noise, so that the high effective resolution of the system can be preserved. The Microcontroller block consists of a microcontroller (STM32F407, ST Microelectronics), a crystal oscillator, and all the other necessary circuitry. The microcontroller executes the firmware in order to control the hardware of the device, perform all the calculations, communicate with a remote computer system and provide the local readout. The Power isolation block and the Digital isolation block eliminate noise propagation from digital blocks to the sensitive analog circuitry in the Sensor interface block. The User interface block consists of an alphanumeric liquid crystal display and at least two buttons. It provides the measurement readout and a means for the user to choose the mode of operation of the portable vacuum sensor.

The proposed device also has a wireless communication option (the Wireless transceiver block in Fig. 5). The

wireless communication option is planned in the form of an internal add-on module.

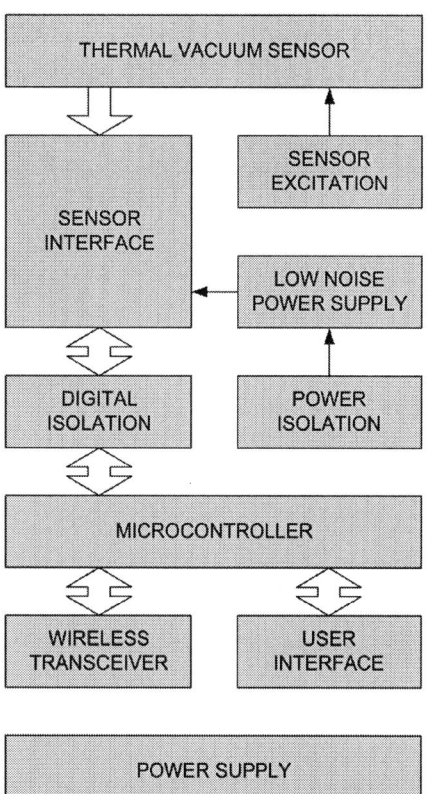

Fig. 5. Block diagram of the compact thermal vacuum measuring device with display and wireless option.

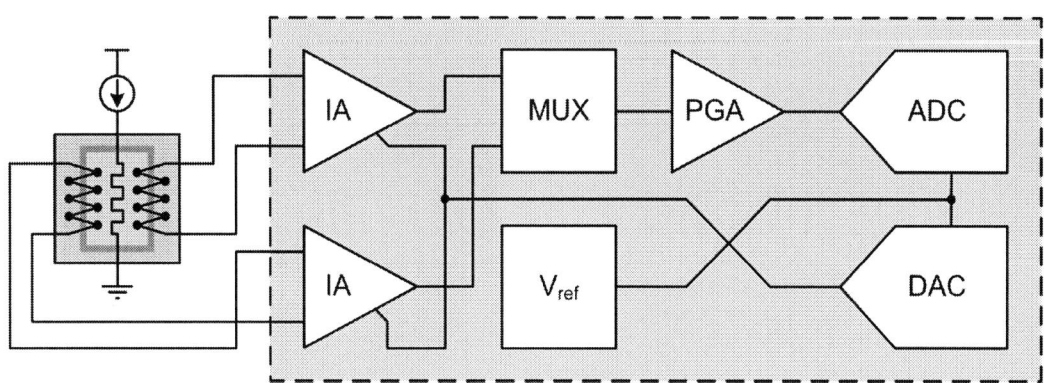

Fig. 6. Block diagram of the sensor interface block.

B. Compact Housing

Portability of the thermal vacuum measuring device will be assured by mounting the transducer housing on a compact and robust box housing which is under development.

Model of the box was created in the software package SolidWorks, and afterwards imported into the software Ultimaker Cura. Layout of the box containing electronics is illustrated in Fig. 7a.

Compact portable box was fabricated of polylactide (PLA) using 3D printer WANHAO Duplicator i3. One stage of the printing process is shown in Fig. 7b.

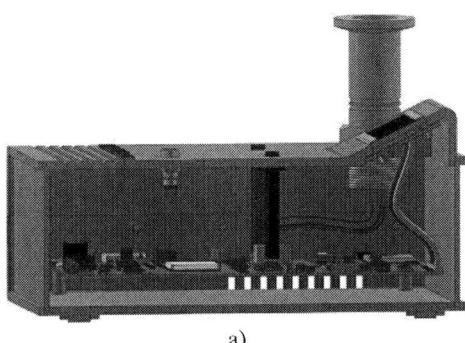

a)

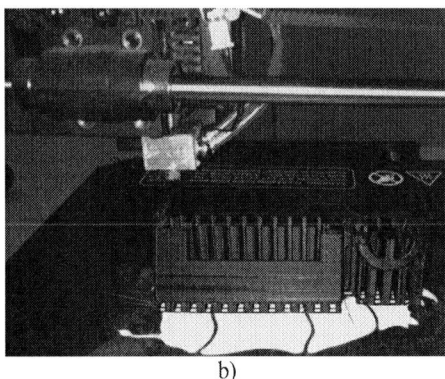

b)

Fig. 7. Portability of the thermal vacuum measuring device will be assured by a compact and robust housing containing electronics and the tranducer: a) layout of the box and (b) one stage of the printing process.

IV. CONCLUSION

The main advantages of ICTM thermal vacuum sensor are low power consumption of 50 mW and fast response. Thermal time constant is around 4 ms at atmospheric pressure. The operating range of ICTM vacuum sensor is the same as for thermal-based vacuum sensors in general and covers the range $(100 - 2 \cdot 10^3)$ Pa.

In all previous applications, an external current source was used. One of the crucial contributions of this work is making a progress towards a compact device in which the current will be provided by the Sensor excitation block.

The proposed device also has a wireless communication option. Wireless solutions are used in a variety of demanding industrial applications. Technologies such as Wireless LAN, ZigBee, LoRaWAN, Wireless HART etc. have specific characteristics and are therefore suitable for different applications and demands. The wireless communication option is planned in the form of an internal add-on module.

The final goal is to realize a universal compact and portable box containing electronics which will be suitable for different applications of thermal sensor, but also for other types of proprietary ICTM sensors like mercury vapor sensor [7]. To the best of our knowledge, a solution similar to the one proposed in this work does not exist in the available literature.

ACKNOWLEDGEMENT

This work has been partially supported by the Serbian Ministry of Education, Science and Technological Development within the framework of the Project TR32008.

REFERENCES

[1] D. Randjelović, A. Petropoulos, G. Kaltsas, M. Stojanović, Ž. Lazić, Z. Djurić, M. Matić, "Multipurpose MEMS Thermal Sensor Based on Thermopiles", *Sensors and. Actuators A - Physical*, 2008, vol. 141, pp. 404-413.

[2] D. Randjelović, V. Jovanov, Ž. Lazić, Z. Djurić, M. Matić, "Vacuum MEMS Sensor Based on Thermopiles-Simple Model and Experimental Results", in *Proc. 26th International Conference on Microelectronics MIEL 2008*, Niš, Serbia, 2008, vol. 2, pp. 367-370.

[3] D.V. Randjelović, M.P. Frantlović, B.L. Miljković, B.M. Popović, Z.S. Jakšić, "Intelligent Thermal Vacuum Sensors Based on Multipurpose Thermopile MEMS Chips", *Vacuum*, March 2014, vol. 101, pp. 118-124.

[4] D. Randjelović, Ž. Lazić, M. Popović, M. Matić, "Helium Sensing Using Multipurpose Thermopile-Based MEMS devices", in *Proc. 28th International Conference on Microelectronics MIEL 2012*, Niš, Serbia, 2012, pp. 147-150.

[5] D.V. Randjelović, B. Popović, P. Poljak, O. Jakšić, "Sensing Gas Type and Pressure with Multipurpose Device Based on Seebeck Effect", *accepted for 42nd International Semiconductor Conference CAS 2019*, Sinaia, Romania, 2019.

[6] A. W. van Herwaarden, P. M. Sarro, "Integrated vacuum sensor", *Sensors and Actuators*, Vol. 8, pp. 187-196, 1985.

[7] M. Sarajlić, P. Poljak, M. Frantlović, M. Vorkapić, D. Tanasković, "A Signal Amplifier for the Compact Mercury Vapor Sensor", in *Proc. 8th International Scientific Conference on Defensive Technologies OTEH 2018*, Belgrade, Serbia, 2018, pp. 1-4.

Electrical Characterization of Microbial Fuel Cells – Method and Preliminary Results

D. V. Randjelović, O. M. Jakšić, B. Popović, K. Joksimović,
S. Miletić, P. Poljak, and V. Beškoski

Abstract - Microbial fuel cells (MFC) present bioelectrochemical systems that allow generation of electricity during anaerobic respiration of selected bacterial species. They have very promising applications in wastewater purification systems, as biosensors or as alternative power source. This work is a result of joint multidisciplinary research and presents preliminary experimental results obtained by electrical characterization of a single-chamber MFC. The goal of research was to study activity of MFC and estimate its internal resistance.

I. INTRODUCTION

Although the concept of Microbial fuel cells (MFC) has been known for more than a century, they are still considered a new technology that has potential applications in various areas, such as ecology, medicine and agriculture [1]. At the beginning of this century, the research on methodology and technology of microbial fuel cells (MFC) was deficient in unified terminology and methods for the performance analysis of MFC related systems [2]. In time, due to the fact that it is promising as a solution to vital problems of the society (lack of energy, combat with accumulation of waste, lack of pure water, survival after natural and man-made disasters...), the research on MFC rapidly advanced [3-11], simultaneously widening the need for tools that would ease the comparison of multidisciplinary work of diverse research groups on equivalent basis. The characterization of MFC and estimation of performances of MFC related devices are still the active field of research [12-16] and it is even harder for them to be unified.

Due to all facts mentioned above, MFCs are very interesting for researchers worldwide since they can find very important applications in purifyication of wastewaters, as biosensors or as alternative power sources.

D.V. Randjelović, O.M. Jakšić, B. Popović, and P. Poljak are with the Centre of Microelectronic Technologies, Institute of Chemistry, Technology and Metallurgy, University of Belgrade, Njegoševa 12, 11000 Belgrade, Serbia, E-mail: danijela@nanosys.ihtm.bg.ac.rs

K. Joksimović is with the Innovation Center of the Faculty of Chemistry, University of Belgrade, Studentski trg 12-16, 11000 Belgrade, Serbia, E-mail: k.joksimovic31@gmail.com

S. Miletić is with the Department of Chemistry, Institute of Chemistry, Technology and Metallurgy, University of Belgrade, Njegoševa 12, 11000 Belgrade, Serbia, E-mail: srdjan@chem.bg.ac.rs

V. Beškoski is with the Faculty of Chemistry, University of Belgrade, Studentski trg 12-16, 11000 Belgrade, Serbia, E-mail: vbeskoski@chem.bg.ac.rs

Basically, MFC present bioelectrochemical systems that allow generation of electricity during anaerobic respiration of selected bacterial species. Bacterial respiration takes place in the anaerobic part of the microbial fuel cell and electrochemically coupled with electron acceptors in the cathodic region where aerobic respiration is performed. This work gives preliminary experimental results obtained by electrical characterization of a single-chamber MFC.

In the first part of the paper design and construction of a simple single-chamber MFC are presented. Afterwards, experimental setup and measurement procedure are given followed by preliminary experimental results. Finally, the obtained results are discussed and future directions of research are proposed.

II. MFC DESIGN AND CONSTRUCTION

A simple single-chamber MFC was designed and fabricated. It consists of a plastic container, two electrodes and sediment. Construction steps are shown in Fig. 1. In order to assure inert ambient for microbs, the electrodes were fabricated of inox complying with IC 316 standard. Electrodes have identical design, they consist of a rectangular part (11 cm x 7.3 cm) made of inox mesh (Fasil, Serbia), and a narrow elongated rectangular part fabricated of inox foil (Good Fellow, England) as shown in Fig. 1a). Parts fabricated of inox foil are isolated from the sediment by heat-shrink hose.

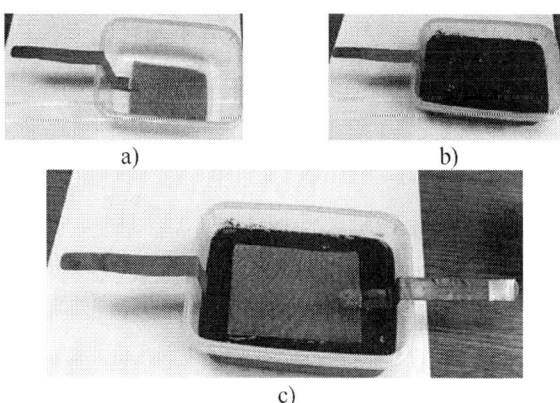

Fig. 1. Construction of a single-chamber MFC: a) positioning of anode at the bottom of the container, b) filling the container with the sediment, c) cathode floating on the surface of the sediment.

978-1-7281-3420-8/19 $31.00 © 2019 IEEE

The first step in MFC construction is positioning of anode at the bottom of the plastic container. Afterwards, sediment is poured inside container and finally, cathode is placed on top of the sediment so that its upper surface is exposed to air.

It should be noted that prior to formation of the MFC it was necessary to perform physico-chemical and microbiological characterization of the sediment sampled from the confluence of the Sava River into the Danube.

III. METHOD FOR DETERMINATION OF INTERNAL RESISTANCE OF MFC

As already mentioned, the goal of this research was to study activity of MFC and to determine its internal resistance. Among different methods for internal resistance determination, method of maximum power was chosen. This method is based on the assumption that MFC can be approximated as a non-ideal voltage source thus allowing a simple voltage-divider circuit to be implemented for determination of internal resistance of MFC [17].

Simplified schematics of this method is shown in Fig. 2a). MFC is represented as a serial connection of ideal voltage source generating electromotive force, E_{MFC}, and has internal resistance R_i. Voltage is successively measured at an array of external resistors $R_1 - R_n$.

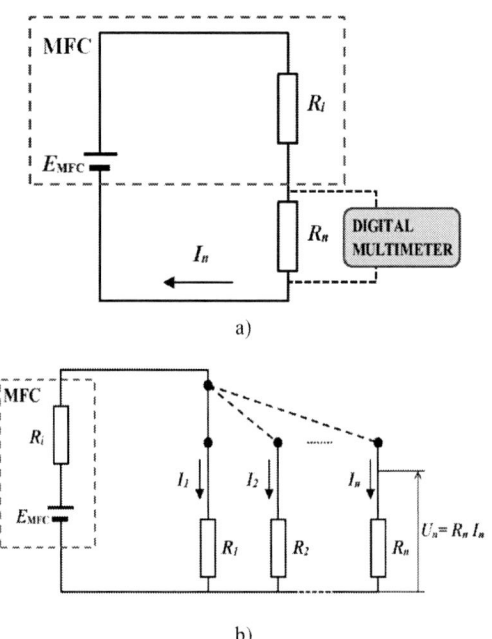

Fig. 2. Shematics of the method for the determination of internal resistance of MFC using peak power detection: a) simplified circuit showing MFC as a voltage generator and array of external resistors, b) circuit with the configuration of a voltage divider, adapted for the measurement on a specific resistor.

When a specific resistor, R_n, is connected to MFC, circuit shown in Fig. 2a) is reduced to the one shown in Fig. 2b). It is well known that such configuration presents a so called voltage divider. Using basic theory of electrical circuits for the resistor R_n, through which current I_n is flowing, power developed as a result of MFC activity is calculated as

$$P_n = U_n I_n = E_{MFC}^2 * R_n / (R_i + R_n)^2. \quad (1)$$

Condition for maximum transfer of power in the case of voltage divider is that external resistance should be equal to the internal resistance of the voltage source. Measurement aims to determine the value of the external resistance that satisfies that condition. Results show the match of that external resistance and the internal resistance of the MFC [18] as we shall see in the following considerations.

$$R_i = R_n. \quad (2)$$

Therefore, maximum power at one of the external resistors will be developed when its resistance equals the internal resistance of MFC. Under such conditions maximum power is given by:

$$P_n^{max} = E_{MFC}^2 / 4 R_n. \quad (3)$$

IV. EXPERIMENTAL SETUP AND MEASUREMENT PROCEDURE

Experimental setup for electrical characterization of MFC is shown in Fig 3. The main elements of the setup are: MFC with electrodes, breadboard SD35N (Velleman, Belgium) with 2420 holes and digital multimeter Peak Tech 2025. The electrodes of MFC are connected to the binding posts of the breadboard. In order to cover wide enough resistance range from 10 Ω do 10 MΩ sufficient number of resistors was chosen. One lead of each resistor is short-circuited and connected to the anode of MFC, while the other lead of resistors is successively connected to the cathode. During each measurement step, one of the resistors is connected to digital multimeter for the purposes of measuring resistance or voltage.

Measurements are performed successively at an array of resistors. First, it is necessary to check the actual resistance, which is in general different from the nominal value. After that, the resistor is connected to MFC and voltage is measured. It was proved empirically that it takes 3 to 10 minutes for the voltage at the resistor to stabilize. Therefore, the voltage was read at each minute till it reaches the so called quasi-stationary state when voltage change is less than 0.5 mV/min.

It was observed that for lower values of resistances it was sufficient to wait for 3 minutes, while measurements at higher resistances required more time, so the value read after 10 min was taken into account.

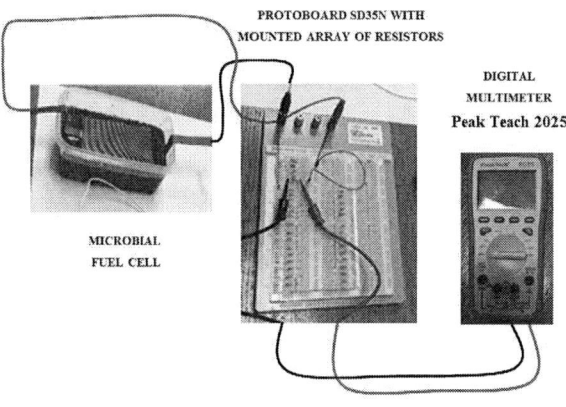

Fig. 3. Experimental setup for electrical characterization of MFC.

V. PRELIMINARY RESULTS

Measurements were performed during 11 days starting from the day when MFC was constructed. Every day, voltage was measured at each external resistor and current flowing through it was calculated. Afterwards, power was determined.

Fig. 4. shows that the voltage is increasing with the rising value of the resistance. The highest voltage values were read at 10 MΩ resistor. Maximum voltage was around 180 mV at day 1 while it reached 454 mV on day 5, which was the peak value during the experiment.

Using the measured voltage and resistances current flowing through each resistor was calculated. Results given in Fig. 5. show that, in general, current through resistors is falling with the rise of resistance. This is valid if we consider day 1 as unstable period during which MFC activity starts establishing. Every day, current value is highest at the resistor of nominal resistance 10 Ω. Maximum current of 30.6 μA is reached on day 11.

The dependences, given below, show voltage and current on different resistances, with time as parameter. The results obtained in that way served for the determination of the power developed at those resistors used in successive measurements. As illustrated in Fig. 6., the maximum power of 422 nW was developed on the fifth day, and it was obtained for the 33 kΩ resistor. Peak in power graph is clearly observed each day. It can also be seen that the lowest power was obtained during the day 1, while there is abrupt increase in MFC activity during day 2. Power values for day 11 are lower than those obtained on day 5.

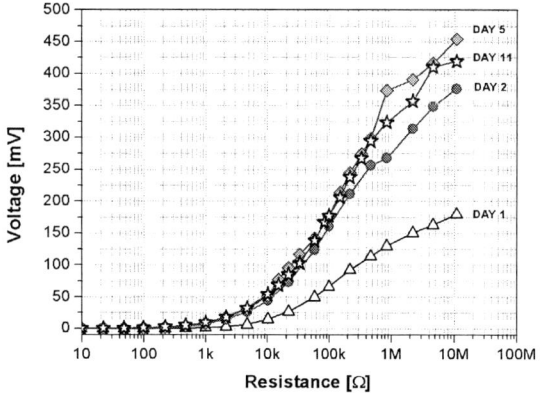

Fig. 4. Voltage measured at each resistor for the chosen days.

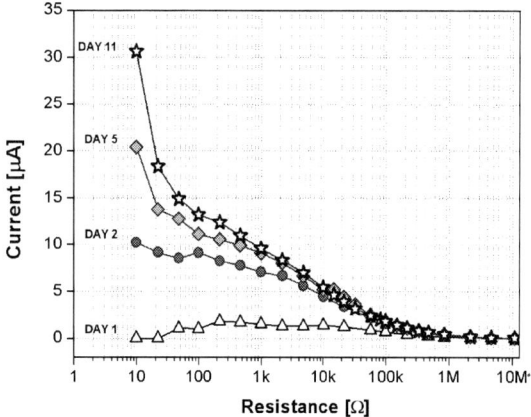

Fig. 5. Calculated values of current through the resistors for the chosen days.

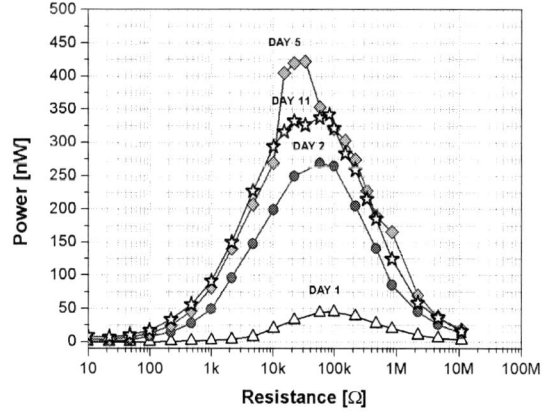

Fig. 6. Power developed at each resistor for 4 selected days.

VI. CONCLUSION

A simple single-chamber MFC was designed and realized using a plastic container filled with sediment and two electrodes fabricated of inox. The sediment was sampled from the confluence of the Sava River into the Danube. MFC activity and internal resistance were studied for 11 days. For this purpose, experimental setup was formed and measurement procedure was established.

MFC was approximated as a non-ideal voltage source thus allowing a simple voltage divider circuit to be implemented for determination of internal resistance. As predicted by voltage divider theory, peak of power vs. resistance was observed for each day. The internal resistance of MFC was changing from 100 kΩ on day 1 to 33 kΩ on day 5. Apart from that, we concluded that power developed on resistors was rising till day 5 and dropped on day 11. The highest value of power density was 5,25 nW/cm2, which is two orders of magnitude lower compared with literature due to the fact that the presented results were obtained during preliminary studies of MFC. The upper limit of MFC's activity observed on day 5 could be a consequence of degradation of microbial activity due to loss of substrate in the sediment over time.

Preliminary tests described in this work served for optimization of the resistance range and values of resistors in the set which will be suitable for application in the future experiments.

Further research activities will include study of the influence of the cell design (thickness of sediment - distance between the electrodes) and geometry and construction of electrodes (carbon clothes). Moreover, optimization of measurements day to day schedule is needed in order to cover all changes in activity. On the other hand, it is expected that substrate composition will have major influence on the performance of MFC therefore it will be one of the main research directions.

ACKNOWLEDGEMENT

This work has been partially supported by the Serbian Ministry of Education, Science and Technological Development within the framework of the Projects TR32008 and III43004.

REFERENCES

[1] B.E. Logan, *Microbial Fuel Cells*, John Wiley & Sons, Inc, Hoboken, New Jersey, 2008.

[2] B. Logan, B. Hamelers, R. Rozendal, U Schröder, J. Keller, S. Freguia, P. Aelterman, W. Verstraete and K. Rabeay, "Microbial Fuel Cells: Methodology and Technology. Critical Review", *Environmental Science & Technology*, 2006, vol 40, no. 17, pp 5181-5192.

[3] C. Santoro, C. Arbizzani, B. Erable, and I. Ieropoulos, "Microbial fuel cells: From fundamentals to applications. A review," *J. Power Sources*, 2007, vol. 356, pp. 225–244.

[4] M. Rahimnejad, A. Adhami, S. Darvari, A. Zirepour, and S. E. Oh, "Microbial fuel cell as new technol ogy for bioelectricity generation: A review," *Alexandria Eng. J.*, 2015, vol. 54, no. 3, pp. 745–756.

[5] C. Xia, D. Zhang, W. Pedrycz, Y. Zhu, and Y. Guo, "Models for Microbial Fuel Cells: A critical review," *J. Power Sources*, 2018, vol. 373, no. November 2017, pp. 119–131.

[6] I. Ieropoulos et al., "The practical implementation of microbial fuel cell technology," *Microb. Electrochem. Fuel Cells*, 2006, pp. 357–380.

[7] B. S. Yildiz, "Water and wastewater treatment: biological processes," *Metrop. Sustain.*, 2012., pp. 406–428.

[8] F. Ivars-Barceló, A. Zuliani, M. Fallah, M. Mashkour, M. Rahimnejad, and R. Luque, "Novel Applications of Microbial Fuel Cells in Sensors and Biosensors," *Appl. Sci.*, 2018, vol. 8, no. 7, p. 1184.

[9] Z. Ge, L. Wu, F. Zhang, and Z. He, "Energy extraction from a large-scale microbial fuel cell system treating municipal wastewater," *J. Power Sources*, 2015, vol. 297, pp. 260–264.

[10] F. Calignano, T. Tommasi, D. Manfredi, and A. Chiolerio, "Additive Manufacturing of a Microbial Fuel Cell - A detailed study," *Sci. Rep.*, 2015, vol. 5, pp. 1–10..

[11] G. Drendel, E. R. Mathews, L. Semenec, and A. E. Franks, "Microbial Fuel Cells, Related Technologies, and Their Applications," *Appl. Sci.*, 2018, vol. 8, no. 12, p. 2384.

[12] S. R. Higgins, C. Lau, P. Atanassov, S. D. Minteer, and M. J. Cooney, "Standardized characterization of a flow through microbial fuel cell," *Electroanalysis*, 2011 vol. 23, no. 9, pp. 2174–2181.

[13] V. Svoboda et al., "Standardized characterization of electrocatalytic electrodes," *Electroanalysis*, 2008, vol. 20, no. 10, pp. 1099–1109.

[14] A. P. Borole, "Microbial Electrochemical Cells and Biorefinery Energy Efficiency," *Biotechnol. Biofuel Prod. Optim.*, pp. 449–472, Jan. 2016.

[15] K. Scott, "Electrochemical principles and characterization of bioelectrochemical systems," *Microb. Electrochem. Fuel Cells*, pp. 29–66, Jan. 2016.

[16] J. Li, "An Experimental Study of Microbial Fuel Cells for Electricity Generating: Performance Characterization and Capacity Improvement", *Journal of Sustainable Bioenergy Systems*, 2013, 3, pp. 171-178.

[17] J.M. Kamau, D.N. Mbui, J.M. Mwaniki, F.B. Mwaura, G.N. Kamau, "Microbial Fuel Cells: Influence of External Resistors on Power, Current and Power Density", *Journal of Thermodynamics and Catalysis*, 2017, vol. 8, no. 1, pp. 1000182.

[18] D. J. Roulston, N. D. Arora, and S. G. Chamberlain, "Modeling and Measurement of Minority-Carrier Lifetime versus Doping in Diffused Layers of n$^+$p Silicon Diodes", *IEEE Transactions on Electron Devices*, 1982, vol. 29, no. 2, pp. 284-291.

Simulation Results of 2.45 GHz Coaxial Antenna with a Ring Slot for Microwave Ablation of a Cancer

K. Cocic, A. Davidovic and D. L. Sekulic

Abstract - Nowadays cancer is one of the leading causes of morbidity and mortality worldwide. Microwave ablation, as minimally invasive procedure, plays an increasingly important role in the treatment of cancer. In this thermal therapy, microwave energy delivery to a cancer tissue is enabled by electric fields radiated by interstitial antenna operating at microwave frequency. In this study, microwave ablation is numerically evaluated in the context of a liver cancer treatment using 2.45 GHz coaxial antenna with a ring slot. In order to design this antenna prototype and analyze its performance, simulations based on finite element method were done with the COMSOL Multiphysics software. *In silico* ablation experiments with the proposed antenna were performed using 10 W, 30 W and 50 W input microwave powers and different ablation times. Numerical results showed that the proposed ablation antenna has the ability to achieve a thermal lesion with a preferably spherical shape and clinically relevant volume with a maximum diameter of 3.5 cm using 30 W input microwave power during 10 minutes.

I. INTRODUCTION

In cancer treatment, surgical resection and chemotherapy are still physicians' first choices. However, for some patients removal of cancers with open surgery is not possible or involves too high a risk due to the poor condition of the patient [1]. Hence, new minimally invasive therapies are required. Recent studies show that thermal ablation techniques such as radiofrequency ablation (RFA), microwave ablation (MWA) and laser-induced interstitial thermotherapy (LITT) can be used in interventional oncology for the cancer treatment of various organs such as liver, kidney, lungs, and bones [2]-[4]. Common to all of these thermal therapies is the destruction of cancer cells based on the denaturation of proteins and the subsequent loss of function leading to cell death. It is well-known that proteins begin to denature already at temperatures about 40°C, but irreversible coagulation necrosis of cells occurs at cytotoxic temperatures between 60°C and 100°C [2], [5].

As relatively new technique in constant development, MWA offers some advantages in comparison to other thermal ablation technologies. Namely, this procedure achieves more uniform and larger volumes of cellular necrosis, faster direct heating rates, higher temperatures delivered to the target lesion, and shorter ablation times [2],

Katarina Cocic, A. Davidovic, and D. L. Sekulic are with the Department of Power, Electronic and Telecommunication Engineering, Faculty of Technical Sciences, University of Novi Sad, Trg Dositeja Obradovića 6, 21000 Novi Sad, Serbia, E–mails: katarinacocic95@gmail.com, adavidovic6@gmail.com, and dalsek@uns.ac.rs

[6]. As significantly less invasive therapy compared to surgical resection, MWA leads to lower risks for bleedings and infections, as well as a much shorter recovery time for patients [7]. Although MWA has promising application prospects, temperature distributions are still difficult to predict and monitor precisely [5], which increases the risk of potential complication.

Currently available MWA systems for clinical use mostly operate at 915 MHz and 2.45 GHz with input microwave powers usually between 10 W to 100 W [7], [8], while systems operating at higher frequencies are still under investigation. In general, an antenna is the most important component of these systems that governs microwave energy distribution during cancer tissue ablation. These interstitial antennas are typically fabricated from 50 Ω coaxial lines of diameters between 1.5 mm and 2.5 mm, since smaller antenna diameters are preferred for percutaneous applications [9]. In most MWA procedures, the antenna is inserted into the center of the targeted tissue of cancer, and the ablation zone grows radially outward [10]. According to the dielectric heating mechanism [4], microwave energy radiated from the antenna excites polar molecules, primarily water, to oscillate in the lossy medium of cancer tissue. This effect produces friction, and thus the applied electromagnetic energy is converted into heat, which induces cell death [11]. In addition, MWA procedure is typically performed with X-ray computed tomography (CT) or ultrasound to help localize the cancer tissue, provide antenna placement information, and facilitate post-ablation confirmation of the treated zone [12].

This paper presents the simulation results of 2.45 GHz coaxial antenna with a ring slot for potential application in MWA of liver cancer less than 3 cm. This antenna aims to create near-spherical ablation zone and proper impedance matching to the target cancer tissue without damaging the surrounding healthy liver tissues. By using COMSOL Multiphysics software, performance evaluation of the proposed antenna for MWA was carried out by using three different input microwave powers (10 W, 30 W and 50 W) during usual 10 minutes of ablation time.

II. METHODOLOGY

A. Coaxial Antenna Design

Fig. 1 shows the schematic illustration of the proposed 2.45 GHz coaxial antenna with a ring slot in longitudinal section. This ablation antenna model is based on a coaxial

978-1-7281-3420-8/19 $31.00 © 2019 IEEE

structure with low-loss polytetrafluoroethylene as a dielectric between two copper conductors. Its dimensions were determined in accordance with matching the antenna impedance to the 50 Ω feed lines for good power transfer, as well as the condition that the reflection coefficient on the operating frequency of 2.45 GHz is minimal ($S_{11} < 10$ dB). Additionally, it is necessary that the antenna dimensions be as small as possible, but physically feasible, since the MWA is used for treatments of cancer whose diameter is up to 4 cm [13]. Accordingly, the diameter of the inner conductor is 0.29 mm, while the inner and outer diameters of the outer conductor have been estimated to be 0.94 mm and 1.19 mm, respectively. Further, the inner and outer conductors were short-circuited at the tip of the coaxial antenna. The position of 1 mm wide ring-shaped slot on the outer conductor from the short-circuited tip was determined on the basis of wavelength in liver tissue. At the antenna operating frequency of 2.45 GHz, the effective wavelength λ in liver tissue [10] was calculated to be about 19 mm. Accordingly, the slot was placed in a distance of 0.25λ from the tip to achieve a stronger electrical field at the output [2]. Finally, the antenna is encased in a PTFE catheter of 1.79 mm in diameter to prevent adhesion of the antenna to desiccated ablated tissue [9], [11], as well as for hygienic and guidance purpose.

B. Computational Model

The finite element method implemented in COMSOL Multiphysics 5.3a software was used to simulate the proposed antenna performance for MWA of a liver carer. The electric field distribution in ablated cancer tissue was determined by numerically solving the Helmholtz wave equation in a lossy medium [14]:

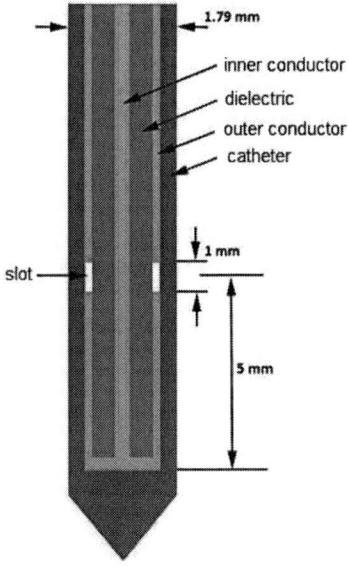

Fig. 1. Design of proposed 2.45 GHz coaxial antenna with a ring slot for MWA of a liver cancer.

$$\nabla^2 \mathbf{E} - k_0^2 \left(\varepsilon_r - j \frac{\sigma}{\omega \varepsilon_0} \right) \mathbf{E} = 0. \tag{1}$$

Here, $\mathbf{E}$ [V·m^{-1}] is the electric field, k_0 [m^{-1}] is the free-space wavenumber, ε_r is the relative permittivity, σ [S·m^{-1}] represents the effective electrical conductivity, ω [rad·s^{-1}] is the angular frequency, and ε_0 [F·m^{-1}] is the permittivity of free-space. The dielectric properties of all materials at 2.45 GHz used in simulations are given in Table I.

Heat transfer during MWA procedure of a liver cancer tissue was accurately described by the well-known Pennes' bio-heat equation [15]:

$$\rho C \frac{\partial T}{\partial t} = \nabla \cdot (k \cdot \nabla T) + Q, \tag{2}$$

where ρ is the density of liver tissue (1079 kg·m^{-3}), C represents the heat capacity of liver tissue at constant pressure (3540 J·kg^{-1}·K^{-1}), while k is the thermal conductivity of liver tissue (0.52 W·m^{-1}·K^{-1}) [10], [12]. The initial temperature T of liver tissue was set to the physiological value of 37°C. Further, parameter Q [W·m^{-3}] in Eq. 2 represents the absorbed electromagnetic energy that was computed from the electric field distribution in liver cancer tissue using the following relation [14]:

$$Q = \frac{1}{2} \sigma |\mathbf{E}|^2. \tag{3}$$

Microwave heating produced by the ablation antenna causes the tissue temperature to rise, but it cannot directly determine the final tissue temperature distribution. The temperature increment in tissue is caused by power and time [15].

III. Simulation Results and Analysis

The efficiency of proposed antenna for MWA of liver cancer was studied for various values of input microwave power (10 W, 30 W and 50 W) and the usual ablation time of 10 minutes. The simulation results for 2D distribution of the electric field radiated by designed coaxial antenna in the liver cancer tissue are given in Fig. 2. These results were obtained for 70 mm long designed ablation antenna placed in the center of 60 mm diameter cylinder with properties of liver tissue. In order to achieve best accuracy, size of mesh elements was chosen to be less than 1/10 of the wavelength in liver tissue. As can be seen from Fig. 2, the electric field distribution is uniform around MWA

TABLE I
DIELECTRIC PROPERTIES AT 2.45 GHz USED IN SIMULATION [10]

Material	Relative permittivity ε_r	Conductivity σ [S/m]
Liver	43	1.69
PTFE	2.03	0

antenna and approximately spherical shape. In addition, the field weakens with the distance from 2.45 GHz antenna, which is in line with the generally known fact that the amplitude of microwaves decreases by the exponential law in the lossy dielectric media such as tissues.

The ablation zone induced by microwave radiation power of 30 W was numerically estimated as a function of time and temperature by using an Arrhenius model [10]. The 2D fraction distribution of necrotic cancer tissue for different ablation times is presented in Fig. 3. Clinically, it is necessary that the applied MWA antenna produces near-spherical thermal lesions of up to 4 cm in diameter, where the temperature needs to be elevated to a value higher than 60°C to cause instant cell death. For such a 4 cm lesion, the diameter of the treated cancer would be around 3 cm with a

5-10 mm safe margin [13]. On the basis of our numerical results, it is noticed that designed antenna with 30 W input power causes a complete necrosis of liver cancer cells in a radius of about 20 mm from the antenna slot for about 15 minutes of ablation time. Using the proposed antenna, shorter ablation time or lower input power is necessary for the treatment of liver cancer whose dimension is smaller.

Additionally, fraction of necrotic cancer tissue at different distances from designed coaxial antenna for 30 W input microwave power and 10 minutes of ablation time is given in Fig. 4. For these parameters of MWA procedure, the obtained results show that the proposed antenna has the ability to achieve a thermal lesion with a preferably near-spherical shape and clinically relevant volume with a maximum diameter of 3.5 cm.

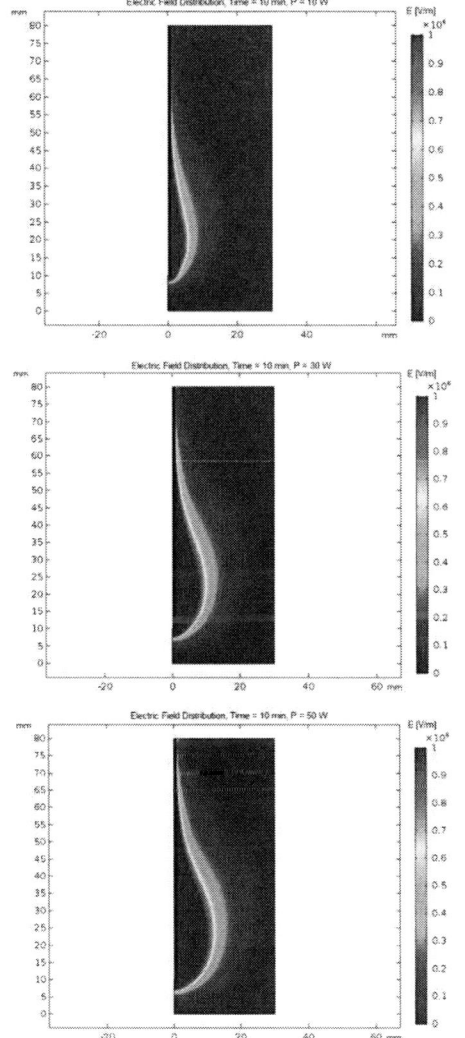

Fig. 2. 2D electric field distribution in liver cancer tissue for 10 W, 30 W and 50 W input microwave powers and usual ablation time of 10 minutes.

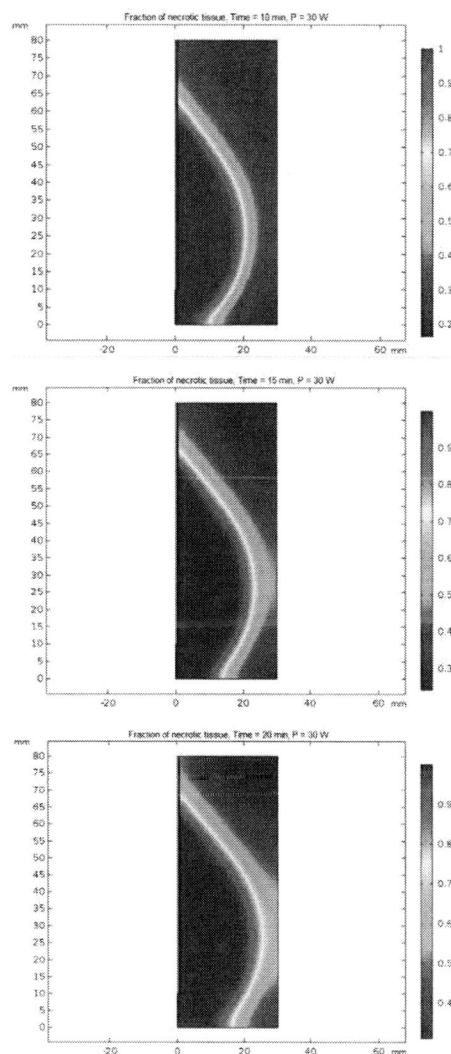

Fig. 3. 2D fraction distribution of necrotic cancer tissue for 30 W input power radiated by proposed MWA antenna at different values of ablation time (10, 15 and 20 minutes).

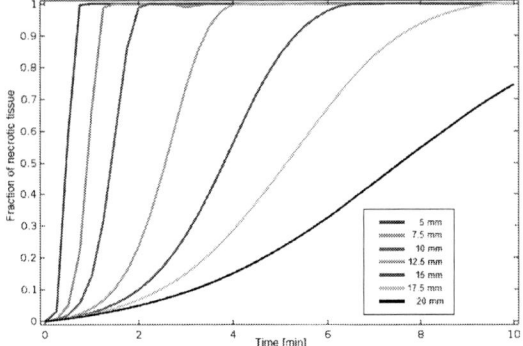

Fig. 4. Fraction of necrotic cancer tissue at different distances from the designed MWA antenna for 30 W input microwave power and 10 minutes of ablation time.

IV. CONCLUSION

In summary, the proposed coaxial 2.45 GHz coaxial antenna with a ring slot offers the high heating efficiency of liver cancer tissue and near-spherical ablation zone indicating that it has potential to become a clinically useful for MWA. Namely, simulation results, based on finite element method, showed that this antenna with only 30 W input microwave power causes a clinical relevant ablation zone of up to 3.5 cm in diameter for usual 10 minutes of ablation time. However, detailed experimental validations of the proposed antenna on the human subjects are required in the future.

ACKNOWLEDGEMENT

This research was financially supported by the Provincial Secretariat for Higher Education and Scientific Research of AP Vojvodina (Project No. 142-451-2162/ 2019-01/02).

REFERENCES

[1] G. B. Gentili, and C. Ignesti, "Dual applicator thermal ablation at 2.45 GHz: A numerical comparison and experiments on synchronous versus asynchronous and switched-mode feeding", *International Journal of Hyperthermia*, 2015, vol. 31, pp. 528-537.

[2] F. Hubner, R. Schreiner, C. Reimann, B. Bazrafshan, B. Kaltenbach, M. Schußler, R. Jakoby, and T. J. Vogl, "Ex vivo validation of microwave thermal ablation simulation using different flow coefficients in the porcine liver", *Medical Engineering and Physics*, 2019, vol. 66, pp. 56-64.

[3] A. Rosen, M. A. Stuchley, and A. V. Vorst, "Applications of RF/microwaves in medicine", *IEEE Transactions on Microwave Theory and Techniques*, 2002, vol. 50, pp. 963-974.

[4] F. Izzo, V. Granata, R. Grassi, R. Fusco, R. Palaia, P. Delrio, G. Carrafiello, D. Azoulay, A. Petrillo, and S. A. Curley, "Radiofrequency ablation and microwave ablation in liver tumors: An update", *The Oncologist*, 2019, vol. 24, pp. 1-16.

[5] H. Gao, X. Wang, S. Wu, Z. Zhou, and Y. Bai, "2450-MHz microwave ablation temperature simulation using temperature dependence feedback of characteristic parameters", *International Journal of RF and Microwave Computer-aided Engineering*, 2019, vol. 29, pp. 1-10.

[6] Y. Jiang, J. Zhao, W. Li, Y. Yang, J. Liu, and Z. Qian, "A coaxial slot antenna with frequency of 433 MHz for microwave ablation therapies: Design, simulation, and experimental research", *Medical and Biological Engineering and Computing*, 2017, vol. 55, pp. 2027-2036.

[7] C. H. nee Reimann, B. Bazrafshan, M. Schußler, S. Schmidt, C. Schuster , F. Hubner, T. J. Vogl, and R. Jakoby, "A dual-mode coaxial slot applicator for microwave ablation treatment", *IEEE Transactions on Microwave Theory and Techniques*, 2019, vol. 67, pp. 1255-1264.

[8] K. Pena, M. Ishahak, S. Arechavala, R. J. Leveillee, and N. Salas, "Comparison of temperature change and resulting ablation size induced by a 902-928MHz and a 2450MHz microwave ablation system in in-vivo porcine kidneys", *International Journal of Hyperthermia*, 2019, vol. 36, pp. 313-321.

[9] C. L. Brace, "Microwave tissue ablation: Biophysics, technology and applications", *Critical Reviews in Biomedical Engineering*, 2010, vol. 38, pp. 65-78.

[10] B. T. McWilliams, E. E. Schnell, S. Curto, T. M. Fahrbach, and P. Prakash, "A directional interstitial antenna for microwave tissue ablation: Theoretical and experimental investigation", *IEEE Transactions on Biomedical Engineering*, 2015, vol. 62, pp. 2144-2150.

[11] R. Ortega- Palacios, C. J. Trujillo-Romero, M. F. J. C. Rubio, A.Vera, L. Leija, J. L. Reyes, M. C. Ramirez-Estudillo, F. Morales-Alvarez, and M. A. Vega-Lopez, "Feasibility of using a novel 2.45GHz double short distance slot coaxial antenna for minimally invasive cancer breast microwave ablation therapy: Computational model, phantom, and *in vivo* swine experimentation", *Journal of Healthcare Engineering*, 2018, Article ID 5806753, pp. 1-10.

[12] G. Deshazer, P. Prakash, D. Merck, and D. Haemmerich, "Experimental measurement of microwave ablation heating pattern and comparison to computer simulations", *International Journal of Hyperthermia*, 2017, vol. 33, pp. 74-82.

[13] C. P. Hancock, N. Dharmasiri, M. White, and A. M. Goodman, "The design and development of an integrated multi–functional microwave antenna structure for biological applications", *IEEE Transactions on Microwave Theory and Techniques*, 2013, vol. 61, pp. 2230-2241.

[14] H. Fallahi, D. Clausing, A. Shahzad, M. O'Halloran, M. C. Dennedy, and P. Prakash, "Microwave antennas for thermal ablation of benign adrenal adenomas", *Biomedical Physics and Engineering Express*, 2019, vol. 5, pp. 1-13.

[15] M. Ge, H. Jiang, X. Huang, Y. Zhou, D. Zhi, G. Zhao, Y. Chen, L. Wang, and B. Qiu, "A multi-slot coaxial microwave antenna for liver tumor ablation", *Physics in Medicine and Biology*, 2018, vol. 63, pp. 1-13.

Dual Mode Ion Mobility Spectrometer High Voltage Formation Circuit

E. Gromov, M. Matusko, Y. Shaltaeva, V. Pershenkov, V. Belyakov, A. Golovin,
E. Malkin, I. Ivanov, and V. Vasilyev

Abstract - It was designed dual mode ion mobility spectrometer high voltage formation circuit. The circuit design includes simulation of the dynamic formation of a sinusoidal signal. The simulation of the dynamic formation of a sinusoidal signal is investigated. It is characterized by a high voltage multiplier and several variants of a voltage smoothing filter.

I. INTRODUCTION

The method of ion mobility spectrometry, is widely used for substances identification in socially important fields as prevention terrorism and traffick of drug [1] - [3], preventing the spread of hazardous and toxic chemicals [4], [5], medical diagnostics [6] - [9], quality control of food products [10] and industrial materials [11] for detection and identification of various classes of substances that form positively and negatively charged ions. The analysis of the ion mobilities in constant electric field is the basis of the ion mobility spectrometry method (IMS). The simultaneous detection of substances of different classes requires the means of joint detection of ions of both positive and negative polarity. It has been investigated and developed the system for generating a stable high voltage (up to 3 kV) in the drift region of the two polar ion mobility spectrometer with the possibility of a fast (no more than 10 ms), permanent polarity switching.

The time-of-flight ion mobility spectrometer consists of an ionization chamber and a drift chamber. The molecules of sampled substance are ionized by a corona discharge in ionization chamber. Then ions under the action of a constant electric field are separated by ion mobility in a drift chamber. The ion mobility is specific and depends on the mass, form of the ion. The structural diagram of the ion mobility spectrometer [12] and the principle of its operation are described in detail in the literature [13-17].

The ionization of molecules reaction requires a constant high voltage for the formation of an ionizing corona discharge on the discharge gap of the ionization source.

Evgeni A. Gromov, Maxim A. Matusko, Yuliya R. Shaltaeva, Vecheslav S. Pershenkov, Vladimir V. Belyakov, Anatoly V. Golovin, Evgeni K. Malkin, Igor A. Ivanov and Valery K. Vasilyev are with the Institute of Nanotechnology in Electronics, Spintronics and Photonics, National Research Nuclear University MEPhI (Moscow Engineering Physics Institute), 115409, Russian Federation, Moscow, Kashirskoe shosse, 31, E-mail: gromov-ea@yandex.ru

II. SOURCE OF HIGH VOLTAGE WITH FAST POLARITY SWITCHING

The development of a high voltage generation unit involves the following tasks: switching the voltage of 3 kV for a time not exceeding 10 ms, maintaining a stable high voltage level with fluctuations of no more than 0.1% after switching. High accuracy of voltage reproduction is required to create uniform conditions for the ionization of substances and to ensure sensitivity at the ppb-ppt nanogram level. A high voltage unit consists of several nodes: a sinusoidal modulator based on a high voltage transformer, a voltage multiplier with switchable polarity, a smoothing filter. Several variants of circuit solutions are modeled and tested.

The simplest model of the filter is shown in Fig. 1.a. Sources V1-V3 emulate the operation of a high voltage generator. The idea of filtering is as follows: the initial voltage HV is set larger than the required value (2.5 kV), and a high-voltage divider (shown as a resistor R2) is sequentially placed into a current source. Since the divider is a resistive circuit, the fixed current flowing through it will create the necessary constant voltage HV1 (2.5 kV), and the voltage difference together with the pulsations will be absorbed on the elements of the current source. The current source is included in the diode bridge circuit to provide bipolar mode of operation.

The first 6 ms control current (source I1) is 100 µA, which corresponds to a fully open filter, therefore HV and HV1 are equal. Then the control current is reduced to 20 µA, which corresponds to the normal mode of operation. The voltage HV that changes over time is greater than HV1, their difference falls on the current source, which ensures the stability of the voltage level HV1 Measurements of voltages HV and HV_1 are shown in fig. 2.

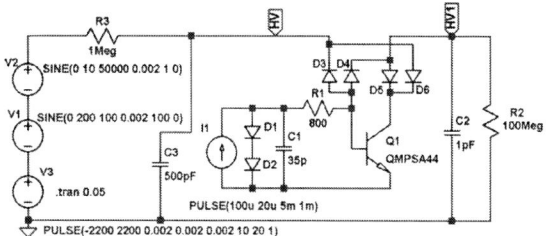

Fig. 1 Electrical circuit of a bipolar smoothing filter based on a current source, modeled in LT spice.

978-1-7281-3420-8/19 $31.00 © 2019 IEEE

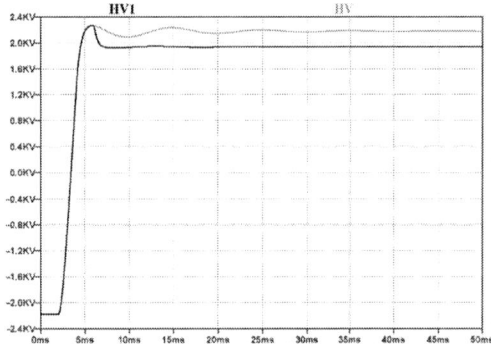

Fig. 2. Voltage plots HV and HV$_1$.

Using positive simulation results, a high-voltage circuit design version was developed. In fig. 2.a shows a high voltage unit assembly with a filtering current source based on a Darlington transistor. The FIL_CTL signal sets the reference signal for the current source based on the U4 operational amplifier, since the current supplied to the inputs of the optocouplers U20-U22 4N25 flows through the resistor R45 and creates a voltage equal to the reference one.

The current transfer ratio of the optocouplers and the gain of the transistors Q5, Q6 is known, which allows you to set the required current. On the output connector X18, a voltage of HV1 (2.5 kV) is created. Element D42 (bidirectional protective diode) serves surge voltage protection.

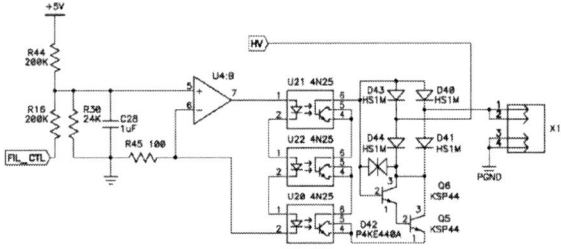

Fig. 3. Electrical circuit of a high voltage filter based on a Darlington transistor, modeled in the LT spice program

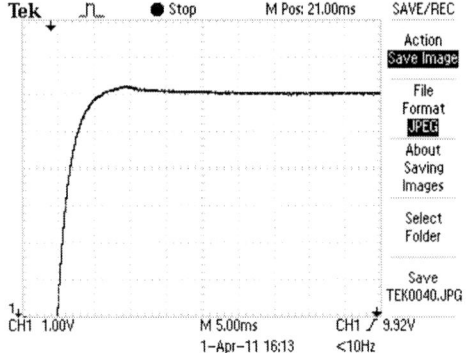

Fig. 4. High voltage establishment oscillogram.

Fig. 5. High voltage generation unit.

The process of switching voltage is shown in the waveform obtained during operation of the unit. From the analysis of Figure 3 and . Figure 4 it is clear that this circuit variant has a disadvantage, since there is a voltage surge after switching, as the transistors go out of saturation mode.

A more promising high-voltage filter circuit design solution is the one shown in fig. 6.

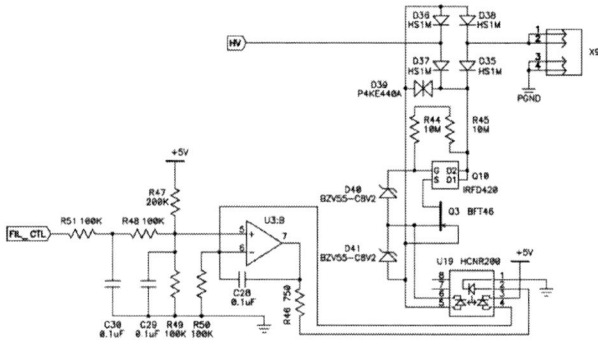

Fig. 6. Electrical circuit of a high-voltage smoothing filter based on a field-effect transistor with a control p-n junction, modeled in the LT spice program.

The current source based on the operational amplifier U3 operates in a manner similar to the previous scheme. The differences between the schemes are in the use of the analog high-linearity optocoupler HCNR200, consisting of a high-quality AlGaAs-based LED emitter and two closely-spaced photodiodes. optocoupler HCNR200 is used in various applications requiring a high level of stability and linearity. The output of the second photodiode can be used to monitor and stabilize the optical power of the LED. As a result of the optical power stabilization the nonlinearity and variation of the LED parameters are eliminated. The current at the output of the photodiode is linearly dependent on the radiation power of the LED. Closely-located photodiodes inside the case ensure high linearity and stable performance of the HCNR200 optocoupler.

978-1-7281-3420-8/19 $31.00 © 2019 IEEE   330

Resistors R44 and R45 provide a bias voltage at the Zener diode of D40 equal to 8.2 V and sufficient for the Q10 transistor to begin to open. In this case, the current flowing through the field-effect transistor (FET) Q3 creates a voltage drop, which is subtracted from the gate voltage of the transistor Q10, which prevents its further opening. The process continues until the flowing current becomes equal to the predetermined value.

The oscillogram of high voltage switching received during operation of this unit, obtained using a Tektronix TDS1001B digital oscilloscope and a 1: 200 divider, is shown in fig. 7.

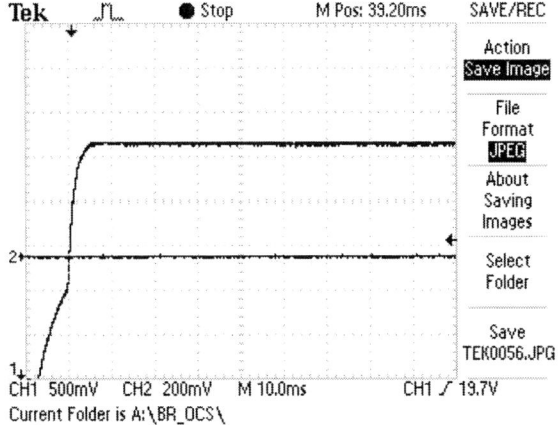

Fig. 7. Oscillograph pattern of the voltage at the output of the high-voltage unit.

Fig. 8. High voltage generation unit with filter based on field transistor with control p-n-junction.

It can be seen from the waveform analysis that the high voltage smoothing filter successfully performs its task, and the high voltage block meets the requirements.

III. CONCLUSION

As a result of the analysis of three options for the implementation of a high-voltage source with a fast polarity switching, a promising scheme was proposed for the ion mobility spectrometer. The scheme work operation includes the generation of control signals by an external controller with key elements based on MOS transistors. It was also investigated the operation of a high voltage multiplier with switchable polarity, based on the Cockroft Generator – Walton scheme. A design, excluding voltage surges after switching the polarity of the device, was developed on the study of various designs of a smoothing voltage filter.

The development method used in the design allows the composition of high-precision analytical instruments that allow one to measure pico-ampere current changes. In this work the approbation of the high voltage forming unit showed that the time for establishing a high voltage on the conducting electrodes of the drift tube, ion gate and collector grid does not exceed 10 ms for positive and negative polarities, which is sufficient for alternate, continuous detection of substances in both spectrometric modes. In the future, it is planned further improvement of the high voltage source.

ACKNOWLEDGEMENT

This work was supported by the Competitiveness Program of NRNU MEPhI.

REFERENCES

[1] A.A. Sysoev, S.S. Poteshin, D.M. Chernyshev, A.V. Karpov, Y.B. Tuzkov, V.V. Kyzmin, A. A. Sysoev, "Analysis of new synthetic drugs by ion mobility time-of-flight mass spectrometry", *European Journal of Mass Spectrometry*, 2014, vol. 20 (2), pp. 185-192.

[2] A.B. Kanu, H.H. Hill, "Identity confirmation of drugs and explosives in ion mobility spectrometry using a secondary drift gas", *Talanta*, 2007, vol. 73, pp. 692-699.

[3] E.Geraghty, C.Wu, W.McGann, "Effective screening for "club drugs" with dual mode ion trap mobility spectrometry", *Int. J. Ion Mobility Spectrom.*, 2002, vol. 5, pp. 41–44.

[4] I.A. Buryakov, "Express analysis of explosives, chemical warfare agents and drugs with multicapillary column gas chromatography and ion mobility increment spectrometry", *J. Chromatogr. B: Anal. Technol. Biomed. Life Sci.*, 2004, vol. 800, pp. 75-82.

[5] S. Sielemann, F. Li, H.Schmidt, J.I. Baumbach, "Ion mobility spectrometer with UV-ionization source for determination of chemical warfare agents", *Int. J. Ion Mobil. Spectrom.*, 2001, vol. 4, pp. 81–84.

[6] J.B. Fenn, "Electrospray wings for molecular elephants (Nobel lecture)", *Angew. Chem.*, 2003, vol. 42, pp. 3871–3894, 2003

[7] J.I. Baumbach, "Ion mobility spectrometry coupled with multi-capillary columns for metabolic profiling of human breath", *J. Breath Res*, 2009, 3, 034001.

[8] A. A. Bykova, A. L. Syrkin, F.Yu. Kopylov, P.Sh. Chomahidze, V.S. Pershenkov, V.V. Beljakov, N.N. Samotaev, "Breath acetone in diagnostic of heart failure", *European Journal of Heart*, 2014, vol. 16, Supplement: 2, pp. 177-177.

[9] C. Uetrecht, R.J. Rose, E. van Duijn, K. Lorenzen, A.J.R. Heck, "Ion mobility mass spectrometry of proteins and protein assemblies", *Chem. Soc. Rev.*, 2010, vol. 39, pp. 1633–1655.

[10] W. Vautz, J.I. Baumbach, J.Jung, "Beer fermentation control using ion mobility spectrometry - results of a pilot study", *J. Institute Brewing*, 2006, vol. 112(2), pp. 157-164.

[11] W. Vautz, W. Mauntz, S. Engell, J.I.Baumbach, "Monitoring of emulsion polymerisa- tion processes using ion mobility spectrometry—a pilot study", *Macromol. React. Eng.*, 2009, vol. 3(2–3), pp. 85–90.

[12] G.A. Eiceman, Z. Karpas, H.H. Hill, *Ion Mobility Spectrometry*, Taylor & Francis Group, 2013.

[13] A. A. Sysoev, D. M. Chernyshev, S. S. Poteshin, A. V. Karpov, O. I. Fomin, and A. A. Sysoev, "Development of an Atmospheric Pressure Ion Mobility Spectrometer–Mass Spectrometer with an Orthogonal Acceleration Electrostatic Sector TOF Mass Analyzer", *Analytical Chemistry*, 2013, vol. 85 (19), pp. 9003-9012.

[14] S. Zimmermann, N. Abel, W. Baether, S.Barth, "An ion-focusing aspiration condenser as an ion mobility spectrometer", *Sens. Actuators B Chem.*, 2007, vol. 125(2), pp. 428-434.

[15] R.A. Miller, G.A. Eiceman, E.G.Nazarov, "A micro-machined high-field asymmetric waveform-ion mobility spectrometer (FA-IMS)", *Sensor Actuators B Chem.*, 2000, vol. 67, pp. 300-306.

[16] H. Borsdorf, T. Mayer, M. Zarejousheghani, G.A. Eiceman, "Recent developments in ion mobility spectrometry", *Appl. Spectrosc. Rev.*, 2011, vol. 46, pp. 472–521.

[17] Shvartsburg,A.A., Creese,A.J., Smith,R.D., Cooper,H.J., "Separation of peptideisomers with variant modified sites by high-resolution differential ion mobility spectrometry", *Anal. Chem.*, 2010, vol. 82(19), pp. 8327–8334.

Fast Switching of the Polarity of Dual Mode Ion Mobility Spectrometer

V. Pershenkov, V. Belyakov, Y. Shaltaeva, E. Malkin, A. Golovin, I. Ivanov,
V. Vasilyev, M. Matusko, and E. Gromov

Abstract - The ion mobility spectrometer with a fast switching of the polarity of the drift field is designed for alternate, incessant detection of positive and negative ions from non-radioactive active ionization source. It were also solved the problems of dielectric absorption on the collector and voltage stabilization on the protective ion grid.

I. INTRODUCTION

Modern standards and competitional conditions in the analytical equipment require compliance with increasingly high requirements for sensitivity, selectivity, reliability and compactability. You can find recent advances in gas analysis in papers on mass spectrometry [1, 2], chromatography [3, 4] and in articles on gas sensors [5, 7]. The ion mobility spectrometry (IMS) [8, 9] is the technology used along with the mentioned ones to detect and identify various classes of substances. This method based on the analysis of gas mixtures by comparison of mobility of ions in a constant electric field in a gaseous medium. The simultaneous detection of substances of different classes requires the means of joint detection of ions of both positive and negative polarity. It is possible to identify the composition from tabular values recorded in the database of substances of the ion mobility spectrometer. The magnitudes of the ion mobility, reduced to standard values of temperature and atmospheric pressure.

The spectrometric detector consists of an ionization chamber, spatially separated by an electric gate from a drift chamber with a uniform, longitudinal electric field and an ion detection unit. Ion mobility spectrometers operating in the unipolar mode require a calibration each time when the polarity is switched. The switching cause additional time losses, complicates operation and significantly limits the capabilities of the detector. Whereas a bipolar instrument is sufficient to calibrate for a mixture of substances, or one substance, forming both positive and negative ions.

Evgeni A. Gromov, Maxim A. Matusko, Yuliya R. Shaltaeva, Vecheslav S. Pershenkov, Vladimir V. Belyakov, Anatoly V. Golovin, Evgeni K. Malkin, Igor A. Ivanov and Valery K. Vasilyev are with the Institute of Nanotechnology in Electronics, Spintronics and Photonics, National Research Nuclear University MEPhI (Moscow Engineering Physics Institute), 115409, Russian Federation, Moscow, Kashirskoe shosse, 31, E-mail: gromov-ea@yandex.ru.

II. THE IMS STRUCTURE

The ion mobility spectrometer consists of a large number of control units. Fig. 1 shows the general structure of the spectrometric detector and the interaction between the main units.

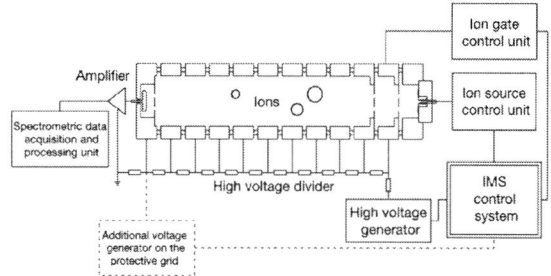

Fig. 1. The IMS structural diagram.

The ion source ionize gas molecules by a corona discharge. The ion shutters inject ions to the drift chamber where ions drift to the collector under the action of a constant electric field. The ion mobility depends on mass and shape of ions.

The key element of the ion mobility spectrometer is the ion current detection unit. The structure of the detection unit is shown in Fig. 2.

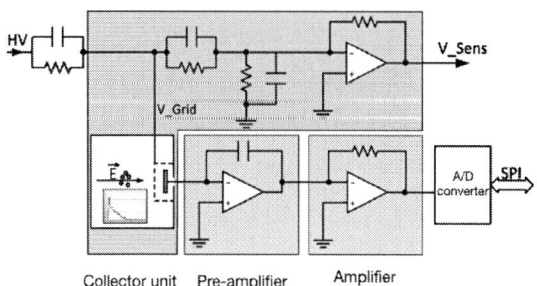

Fig. 2. General structure of the detection unit.

Ions under the influence of the drift chamber field reach the collector unit and are absorbed by the Faraday plate. A flowing charge generates an electrical current that flows into the first stage of the amplifier (preamplifier). The ion current entering the detection unit is of the order of

(10 ... 100) PA value, and it is preferable to use an input integration link in the preamplifier stage to register it. The received signal is additionally amplified by the second stage, which is an operational amplifier in the differentiation mode. The signal is then sent to the analog-to-digital converter. The digitized signal is transmitted to the data processing unit via the SPI interface.

The gain of the transimpedance amplifier is expressed as the ratio of the output voltage to the input current and therefore has a resistance dimension. It is necessary to develop an amplifier (ion current converter) with a transfer coefficient of about 5 G" and a small input impedance. The necessary requirement for the amplifier under development is linearity with an amplitude distortion characteristic of no more than 10% for a pulse duration of 0.5 ms of the input type signal of the gaussian shape. The phase characteristic of the developed ion current converter is the time shift of the output signal not more than 200 μs for the input pulse of more than 0.5 ms, with the minimum delay time dispersion for the pulses of the gaussian form with a width at half height in the range (1 ... 5) ms. The reverse amplitude ejection should not exceed 10% at a pulse with a steep edge of about 10 μs.

Current converter in combination with analog-to-digital converter is an ion current digitizing system. The requirement for the digit capacity of this system is 65535 quanta, i.e. 16 bits. The necessary feature of the developed amplifier is bipolarity, i.e. the possibility to work out the pulses of ion current of both positive and negative polarity. The noise level requirement for this system is (max) 50 quanta.

The collector unit includes a sensor of high voltage of the drift chamber (Fig. 2. top part). The current flowing in the high voltage divider flows from the last link to the sensor resistor. To ensure that the electronic control and measurement circuitry does not affect the sensor resistor voltage, a voltage repeater with high input and low output resistance is used, since the current flowing in the high voltage divider is about 15 μA.

III. MODELING AND CONSTRUCTION OF THE ION CURRENT COLLECTOR

In a spectrometric detector, ions that have crossed the drift region fall on a collector [10–12] and produce current pulses with an amplitude in range of several picoampers. The interference current on the collector from a bunch of ions flying toward it begin to influence at such small currents during the processing of the spectrometric signal. The protective collector grid is used in the conventional scheme of the ion mobility spectrometer to ensure the correctness of measurements and the elimination of interference. The grid is installed at a distance of several millimeters from the collector and separates the collector from the drift region. The ion current collector is connected to the input of the amplifier. Since there is a capacitive coupling between the protective grid and the amplifier

input, even a small voltage jitter on the protective grid can cause interference to the output signal of the ion current amplifier.

It is necessary to minimize the impact of the collector unit on the amplifier to ensure the operation of the data collection and processing unit. It is required that the signal level at the amplifier output does not change more than 100 divisions (ADC quanta), which is 1.5 PA at 15 fA/quantum amplifier sensitivity.

The parasitic current of electrical interference on the collector can be calculated using the formula.

$$I = \frac{\partial Q}{\partial t} = \frac{\partial CU}{\partial t} = U\frac{\partial C}{\partial t} + C\frac{\partial U}{\partial t} \qquad (1)$$

The first term characterizes mechanical vibrations, the second term characterizes voltage pulsations. Both sources of interference must be taken into account when designing the detector and collector units. Formula 1 can be used to make an equation.

$$I = C\frac{\partial U}{\partial t} = 0.5nF\frac{\partial U}{\partial t} = 1.5nA \qquad (2)$$

The solution of equation (2) allows us to obtain the permissible voltage change rate on the collector mesh.

$$\frac{\partial U}{\partial t} = 3\ V/s \qquad (3)$$

The obtained solution of equation (3) 3 V/s at a spectrum duration of about 50 ms is equivalent to 150 mV, which is less than 0.1% of the collector mesh voltage. In order to ensure the required level of voltage stability in the ion mobility spectrometer operating in the same polarity, large electrical capacitances are installed in the voltage forming unit on the protective mesh.

In the ion mobility spectrometer operating in the single polarity, the voltage shaping unit on the protective grid represents the last link of the high-voltage divider of the drift region (Fig. 3a). The characteristic resistance and capacitance of the high-voltage divider link are, respectively, about 10 MΩ and 1 μF. Thus, the RC time constant of the high-voltage divider circuit is 10 s, which limit the speed of establishing a stable voltage level on the protective grid when switching the high-voltage polarity (for operation in another mode). The realization of the fast-sequential detection mode of negative and positive ions is complicated by long time of stabilization of the voltage level. To solve the problem, it is used an additional high voltage generator on the protective grid. This construction allows to eliminate the RC circuit with a large time constant (Fig. 3b).

The Resistors R1, R2.1, R2.2 forms the reference voltage for the comparator. The values of resistors R3, R4 relate to each other with the same coefficient as R2.1 and R2.2. The selection of resistor values allows to create feedback for the generator to maintain the exact (working) value of high voltage. The generator can recharge the capacitance with sufficient speed, since the generator has a low output impedance and a sufficient output current. The high voltage signal applied to the protective grid of the collector from the output of the additional high voltage generator is shown in fig. 4. The waveform is obtained with a Tektronix TDS1001B digital oscilloscope and a 1:20 divider.

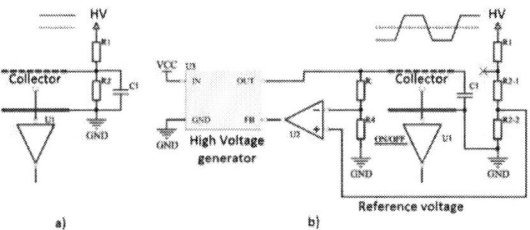

Fig. 3. The electrical circuit diagram of the formation of voltage on the protective grid of the collector (a) in a unipolar device, (b) in a bipolar device (with an additional generator).

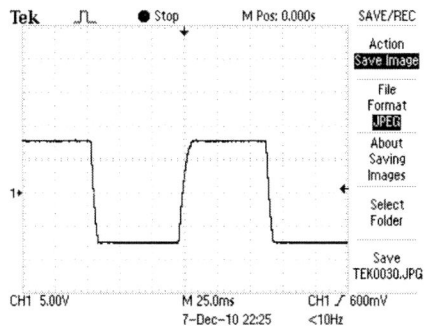

Fig. 4. a) The oscillogram of voltage on the protective grid of the collector (divider 1:20).

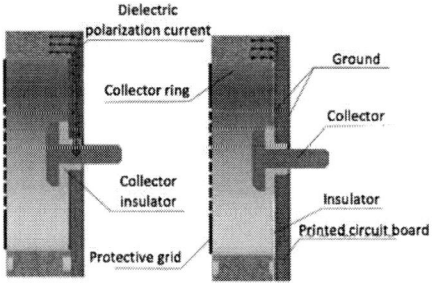

Fig. 5. left) the view of the collector node in the cross section right) the view of the collector unit in the section with grounded metal polygons that prevent the dielectric from overcharging.

Another variant of the device for generating a switching voltage on the protective grid of the ion current collector is to replace the voltage follower with a separate controlled bipolar voltage source with fast switching the polarity of the output voltage in the range from -150 V to

+150 V. The source can be implemented based on two independent controlled voltage sources with different voltage polarities U1 and U2 combined a switching system. The electromagnetic relays, optocouplers and other suitable elements can be used as switching elements K1 and K2.

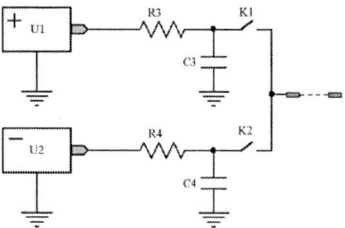

Fig. 6. Electrical diagram of the voltage forming device on the protective grid of the ion current collector with separate positive and negative voltage sources.

The source U1 and the switching element K1 are used for forming a positive voltage on the grid. In addition the source of U2 and the switching element K2 for forming of negative voltage. The low-pass filters with large time constants are used to minimize voltage ripples at the outputs of sources U1 and U2. The resistor R3 and capacitor C3 filter is used for a positive voltage source U1 and the resistor R4 and capacitor C4 for a negative source U2. The voltage at the output of the U2 source is regulated in the range from −50 V to −150 V. The process of polarity switching is concluded in apply of the trip signal of the previously included key K1 or K2, waiting for the completion of the trip, determined by the dynamic parameters of the key, and then giving the key circuit signal for the required polarity. The use of separate sources U1 and U2 makes it possible to control the voltage level on the protective grid independently for positive and negative polarities for flexible adjustment of the electric field in the collector area and to optimize the collection of ions of different polarities. The advantage of this method is a high switching speed, determined by the switching speed of the keys K1 and K2, and the absence of voltage drift on the protective grid after switching the polarity. An independent system of forming a voltage on a protective grid reduces the requirements for frequency compensation and the time required to establish potentials on a high-voltage divider.

The principal problem of the efficiency of the detector, the collector of ion current is due to the effect of dielectric absorption (dielectric absorption). In the construction of an ion mobility spectrometer operating in the single polarity (Fig. 5, left), a dielectric is located between the drift chamber and collector board, in which, after a change in voltage, residual polarization currents that can flow for some time and as consequence partly fall to collector electrode (tens of milliseconds even for a high-quality dielectric). A similar effect can be observed on almost all types of dielectrics. The intensity of the manifestation of this effect depends mainly on the properties of the dielectric. Following non-polar dielectrics have the lowest dielectric absorption: polytetrafluoroethylene (fluoroplast), polystyrene, polypropylene, etc.

The quantitative value of absorption is usually characterized by the absorption coefficient, which is determined under standard conditions. A dielectric could be modelled as a set of consecutive RC-chains with different time constant connected in parallel way. The simplest model of this effect is shown in Fig. 7 with parameters taken from tables with characteristic coefficients for the dielectric.

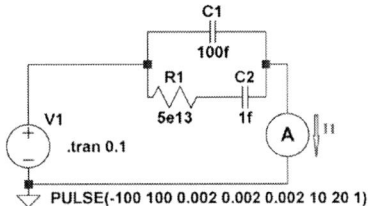

Fig. 7. The electrical diagram of the dielectric absorption model.

The simulation results with appointed current I1 flowing into the collector are shown in Fig. 8. The figure analysis allows to conclude that after the switching process, for a long time a current flows with a value greater than the permissible level of pA units. Based on the analysis of the graphs of the model, it can be derived that the collector node needs to be modernized in order to ensure the bipolar mode.

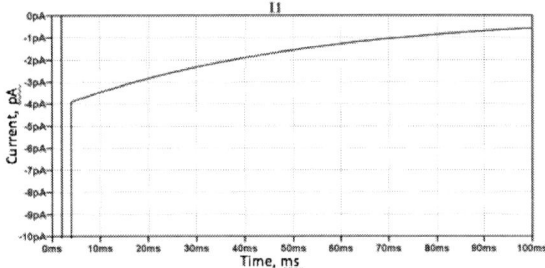

Fig. 8. The graph of current I1 flowing through an ammeter.

The use of grounded metal polygons in design (Fig. 5. right) prevents the recharge of the dielectric when the polarity changes and eliminates parasitic currents on the collector electrode.

IV. Conclusion

Accelerated stabilization of the voltage on the protective grid and solving the problem of dielectric absorption at the collector to ensure the simultaneous detection of positive and negative ions of substances of different classes allow to improve the portable ion mobility spectrometers and contribute valuable input to the general methods of constructing similar devices. It is planned to continue the research work to improve the measuring and operational characteristics of the ion mobility spectrometer serial devices.

Acknowledgement

This work was supported by the Competitiveness Program of NRNU MEPhI.

References

[1] Shaltaeva, Y.R., Podlepetsky, B.I., Pershenkov, V.S, "Detection of gas traces using semiconductor sensors, ion mobility spectrometry, and mass spectrometry" *European Journal of Mass Spectrometry*, 2017, vol. 23 (4), pp. 217-224.

[2] A. Sysoev, S.S. Poteshin, D.M. Chernyshev, A.V. Karpov, Y.B. Tuzkov, V.V. Kyzmin, Alexander A. Sysoev, "Analysis of new synthetic drugs by ion mobility time-of-flight mass spectrometry", *European Journal of Mass Spectrometry*, 2014, vol. 20 (2), pp. 185-192.

[3] H. Alinoori, S. Masoum, "Multicapillary Gas Chromatography - Temperature Modulated Metal Oxide Semiconductor Sensors Array Detector for Monitoring of Volatile Organic Compounds in Closed Atmosphere Using Gaussian Apodization Factor Analysis", *Analytical Chemistry*, 2018, vol. 90 (11), pp. 6635-6642.

[4] H. Yuan, X. Du, H. Tai, M. Xu, "Temperature-programmed multicapillary gas chromatograph microcolumn for the analysis of odorous sulfur pollutants", *Journal of Separation Science*, 2018, vol. 41 (4), pp. 893-898.

[5] Podlepetsky, Y. Sukhoroslova, "Influence of Electrical Modes on Sensitivity of MISFET Ionizing Radiation Dose Sensors", *Procedia Engineering*, 2016, vol. 168, pp. 741-744.

[6] Podlepetsky, N. Samotaev, "Hazardous gases sensing: Influence of ionizing radiation on hydrogen sensors", *Lecture Notes of the Institute for Computer Sciences, Social-Informatics and Telecommunications Engineering, LNICST*, 2016, vol. 170, pp. 217-222.

[7] G.A. Eiceman, Z. Karpas, H.H. Hill, *Ion Mobility Spectrometry*, Taylor & Francis Group, 2013.

[8] H. Borsdorf, T. Mayer, M. Zarejousheghani, G.A. Eiceman, "Recent developments in ion mobility spectrometry", *Appl. Spectrosc. Rev.*, 2011, vol. 46, pp. 472–521.

[9] Bykova, A. L. Syrkin, F.Yu. Kopylov, P.Sh. Chomahidze, V.S. Pershenkov, V.V. Beljakov, N.N. Samotaev, "Breath acetone in diagnostic of heart failure", *European Journal of Heart*, 2014, vol. 16, Supplement: 2, pp. 177-177.

[10] J. Puton, B. Siodłowski, "Generation of current pulses in collector electrode of IMS detectors", *International Journal of Mass Spectrometry*, 2010, vol. 298 (1-3), pp. 55-63.

[11] Q. Zhou, L. Peng, D. Jiang, X. Wang, H. Wang, H. Li, "Detection of nitro-based and peroxide-based explosives by fast polarity-switchable ion mobility spectrometer with ion focusing in vicinity of Faraday detector", *Scientific Reports*, 2015, vol. 5, no. 10659.

[12] F. Tang, X.-H. Wang, L. Zhang, " Array micro Faraday cup ion current detector for FAIMS", *Guangxue Jingmi Gongcheng/Optics and Precision Engineering*, 2010, vol. 18 (12), pp. 2597-2602.

Microcontroller's Sensitivity to Voltage Pulse Series in Comparison with a Single Voltage Pulse

A. N. Shemonaev, K. A. Epifantsev, P. K. Skorobogatov, and A. Y. Nikiforov

Abstract - The paper presents the results of ARM 32-bit Cortex-M0 and M4 CMOS microcontroller's sensitivity to a series of voltage pulses in comparison to a single pulse – all with damage subthreshold energy. The effect of pulses amount on the device voltage overstress threshold value was found and analyzed.

I. INTRODUCTION

The effect of a single electromagnetic pulse on an electronic parts operation is well known [1-5]. The sources of electromagnetic pulses may be lightning discharges, the influence of powerful radar, and electrical equipment. In most cases, electromagnetic pulses induce high voltage pulses in the internal circuit interconnections, which may cause electronic components damage. Most common voltage overstress effects that may cause IC's catastrophic failures are the following:

- p-n junction thermal breakdown;
- metallization burnout;
- gate oxide breakdown.

Thermal damage [4,6] is the result of the excessive heat generated during the electrical overstress (EOS) event, which is a result of resistive heating in the connections within a device. High currents experienced during an EOS event can generate by localized overheating even in normally low resistance paths (fig.1).

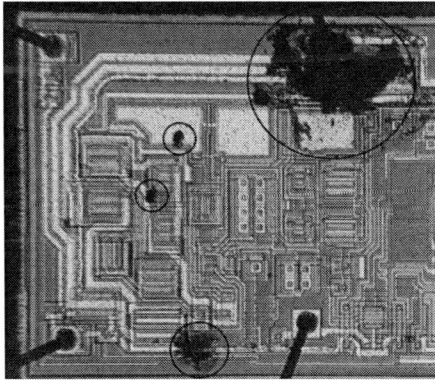

Fig. 1. IC damage due to EOS overheating.

Overheating causes destructive damage to materials within a device's construction [2]. Some sources of

A.N. Shemonaev, K.A. Epifantsev P.K. Skorobogatov and A.Y.Nikiforov are from the National Research Nuclear University (MEPhI), Moscow, Russia, E-mail: anshem@spels.ru

electromagnetic radiation may cause a voltage pulses series in electronic components and it is useful to compare effects from single voltage pulse (induced by electromagnetic pulse) and voltage pulses series.

II. EXPERIMENTAL SET-UP

A single voltage pulse EMI-0502 generator (NRNU MEPhI, SPELS) [7] was used to EOS experiments. A special multiple voltage pulse generator GSI-01 was designed for electronic components testing under voltage pulses series (NRNU MEPhI, SPELS). The technical data parameters of these generators are depicted in Tables 1 and 2.

TABLE I.
EMI-0502 PARAMETERS

Parameter	Value
Form	Two-exponential
Amplitude of Impulse, V	$4 < Um < 5000$
Pulse rise time, ns	<5
Pulse width, µs:	0.1, 1.0, and 10
Polarity	Positive and negative
Output Impedance, Ω	50

TABLE II.
GSI-01 PARAMETERS

Parameter	Value
Pulse Form	Two-exponential
Pulse Amplitude, V	$4 < Um < 5000$
Pulse rise time (0.1-0.9), ns	<5
Pulse width, µs:	1.0
Output Impedance, Ω	50
Polarity	Positive
Frequency, Hz	1-100

EMI-0502 generates one voltage pulse with an amplitude of 4V - 5kV at the output of the tested ICs with a pause interval of 10 seconds, while GSI-01 generates pulses with the same amplitude at a frequency of up to 100 Hz (up to 100 pulses per second). The structural scheme of both generators is presented in fig.2. The main difference between EMI-0502 and GSI-01 installation is the capacitor charge current value. EMI-0502 high voltage source provides the capacitor charge current value 30mA, while GSI-01 high voltage source provides the capacitor charge current value 14 Amps.

Resistors R1 and R2 are used to form a two-exponential pulse, as well as to monitor the pulse amplitude by a 1/50 voltage divider (point B, fig.2). Resistor R3 is used to provide generator's 50 Ω output impedance. Resistor R4 is to measure a current pulse in DUT using an oscilloscope (fig.3).

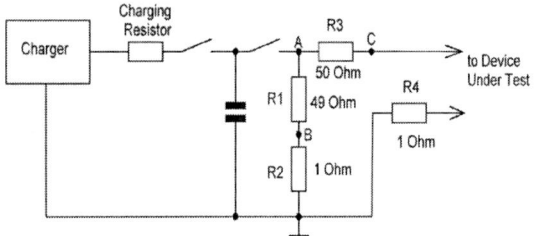

Fig. 2. EMI-0502 and GSI-01 structural scheme.

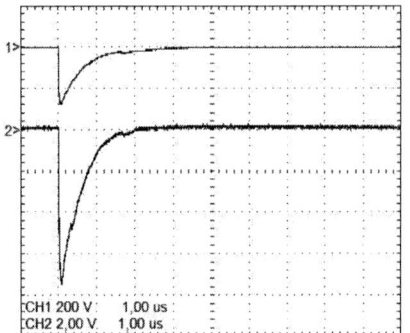

Fig. 3. Ch1 - the test voltage pulse (point B in Fig.2), Ch2 – DUT's current pulse (measured by resistor R4).

III. Test Procedure

Time intervals between single voltage pulses in a sequence were either 5 minutes or 1 second - during this time, DUT parameters were checked and monitored.

Samples were affected by a two-exponential single pulse (1) with a pulse rise time 5 ns (τ_1). and series pulses of positive polarity with a pulse duration at a half-height of 1 µs (τ_2).

$$U(t) = U_m \cdot \left[-\exp(-\frac{t}{\tau_1}) + \exp(-\frac{t}{\tau_2}) \right]$$

(1)

All samples were manufactured and packaged in one batch. Before carrying out the experiment, functional and parametric monitoring (FPM) tests have been performed. Then samples were exposed to voltage pulses with an FPM after each pulse. If FPM demonstrated DUT failure, it was excluded from the further experiment. Then the pulse amplitude was increased by 20% at each step. In case of failure the last one pulse amplitude was recorded and the sample was no longer used for research.

The next sample was exposed to the pulse amplitude, which caused a failure of the previous sample. If there was no failure of a sample, then the amplitude of the pulse was increased by 20% at each step and testing continued until a new failure occurred.

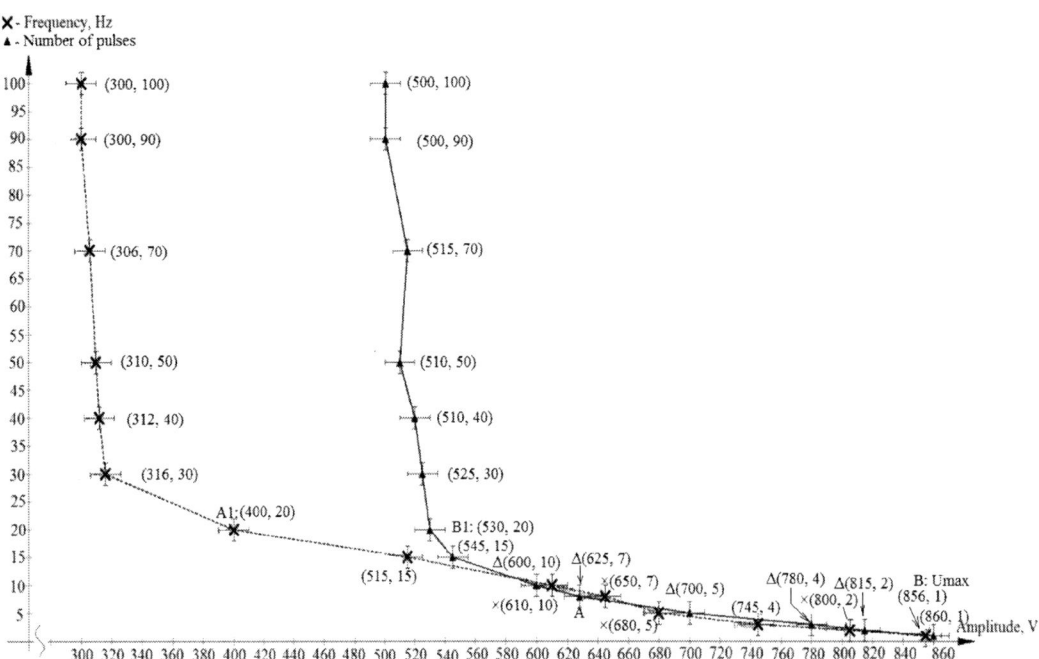

Fig. 4. The pulses number dependencies on the single voltage pulse the threshold amplitudes when exposed to pulses with a 1Hz to 100 Hz frequency and the series from 1 to 100. (Value A: 625V; B: 856V; A1: 400V; B1: 530V)

This procedure was repeated on 'fresh' samples until the failure of the sample did not occur when exposed to the first voltage pulse (the pulse amplitude value was at that threshold - U_{max}) (point B at Fig. 4).

Five samples were used to determine the threshold amplitude more precisely.

Subsequently, samples were exposed to pulses with an amplitude less than 10 % from value at point B (fig 4) and the number of affecting pulses necessary for a parametric and/or functional failure was determined.

IV. TEST RESULTS

As a result of this procedure, the data presented in fig.4 was obtained, which shows the number of pulses to obtain the failure at a particular pulse amplitude (a failure voltage pulse amplitude value).

Then the same experiment was carried out at the GSI-01 to confirm the results obtained at the EMI-0502, which subsequently allowed a reduction in the duration of time for the test for determining the number of pulses. Control of DUT parameters were carried out after each series of pulses at a frequency of 10 Hz. It was found that a failure voltage pulse amplitude value at 10 Hz frequency has a little difference as at ten single pulses with a time interval of five minutes.

The failure voltage pulse amplitude values significant difference was observed when twenty single pulses were applied from EMI-0502 installation and one series of pulses with a 20 Hz frequency from GSI-01 installation. With the number of pulses increase, the electric hardness of DUT dropped by 10%, while DUT electric hardness decreased by 15-20% with increasing frequency for every 10 Hz. In the example at Fig. 4: point A1: 20 pulses per second from GSI-01 installation; point B1: 20 single pulses submitted with an interval of two minutes from EMI-0502. One can see that with increasing voltage pulses frequency (from 10 Hz to 20 Hz), a failure voltage pulse amplitude value is reduced by 40-45%.

This damage character was also found in some high-power field effect and IGBT transistors tests for the drain-source circuit, as well as some DC-DC converters for the output circuit and digital isolator interfaces [9].

The voltage pulse amplitude values 30-32% difference at the points A and B (fig.4) indicates that there are additive effects [10,11] in ICs and semiconductors impacted by voltage pulses series with 10 Hz frequency.

A samples impacted by voltage pulse with an amplitude less than 10 % from value at point B (fig 4) may be damaged by voltage pulse with an amplitude less than 70 % from value at point B.

Figure 4 also shows that for a single pulse, IC failure occurs at 860 V pulse amplitude value. For 10 pulses with a pause between pulses of both 5 minutes and 6 seconds, a failure voltage pulse amplitude value is about 600 V. With smaller pause values a failure voltage pulse amplitude

value becomes even lower. So for a pause of 600 ms the failure voltage pulse amplitude value is already 300 V.

A simplified general view of damage pattern is presented in Figure 5.

This graph shows how electronic components vulnerability differ under EMI-0502 and GSI-01 pulses. This simplified graph is divided into 4 parts:

• The first area shows that no matter how many pulses are applied, the sample retains its electric hardness and is fully operable.

• The second and third areas show the area of additive effect - the accumulation of damage is a result of exposure to voltage pulses. The difference between Region II and III shows that DUT failure occurs only at 20-100 Hz frequency. To obtain the same DUT failure at 1 Hz frequency, it is required to increase pulse amplitude (region III).

• There are an irreversible failure to function and irreversible failures parameters (catastrophic failure) at fourth area.

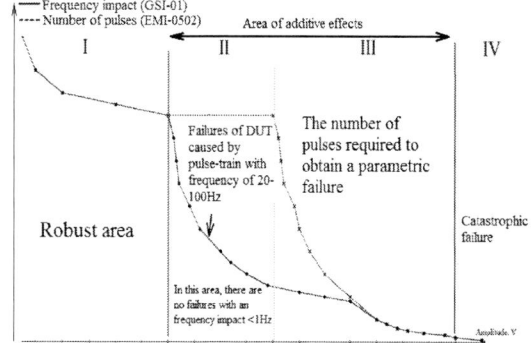

Fig. 5. A simplified DUT damage pattern

According to Figure 5 data, it can be assumed that the region of additive effects, which is below 10 pulses, depends on frequency: the higher frequency of voltage pulses - the wider the range of additive effects. However, confirmation of this dependence requires additional research at frequency above 100 Hz

V. DISCUSSION

Hidden defects dew to the impact of voltage pulse series can be characterized in three categories:

• Parameter degradation is comparatively insignificant and the device fully complies the technical specification. Thus, product reliability is not reduced.

• A damaged element of the device complies the specifications but parameter values slightly out of norms, but the product is operational. In this case, the device reliability may be reduced;

• The device is operational, but does not meet all the requirements for it. The reliability of the product is significantly reduced and the declared lifetime is not guaranteed.

Additive effects can be associated with the thermal effects that occur when IC exposed to a series of voltage pulses. The less payse between pulses - the less time it takes to dissipate the heat. During local heating, damage occurs due to the effect of exceeding the energy in a certain element (pn-transition, metallization area, etc.) and the destruction of the element leads to a change in the parameters of the device and its performance.

VI. CONCLUSION

The results of ARM 32-bit Cortex-M0 and M4 CMOS microcontroller's sensitivity to a series of voltage pulses in comparison to a single pulse show that:

1. The dependence is found of voltage pulse amplitude failure threshold on voltage pulses series frequency.

2. When U_{max} pulse amplitude is applied to DUT at a frequency from 1 Hz to 100 Hz, as compared to a same single pulse series from 1 to 100 pieces with large pauses between pulses, there is a difference voltage pulse amplitude failure threshold values when the number of pulses is more than 10. Example, voltage pulse amplitude failure threshold at 100 single pulses amount with time pause not less two minutes is 500V and at 100 pulses with frequency 100 Hz is 300V.

3. When single pulses series and high frequency pulses series is applied to DUT, there are four damage areas. The first area is robust area and fourth area is catastrophic failure area. There are additive effects in second (pulses series with frequency effects) and third area (single pulses series effects). It is assumed that pulses series frequency increase reduce a second area lower boundary, but it is necessary to perform some additional research for confirmation

4. The are no effect of the time interval between voltage pulses on the voltage pulse amplitude failure threshold when the time interval is more than 6 seconds.

REFERENCES

[1] P.K.Skorobogatov, "Test Method for IC Electrical Overstress Hardness Estimation", in *Proc. 4th European Conf. on Radiations and Its Effects on Components and Systems (RADECS 97)*, Sept. 15-19, 1997, Palm Beach, Cannes, France, pp. 174-177.

[2] P.K.Skorobogatov, O.A.Gerasimchuk, K.A.Epifantsev, and V.A.Telets, "Specifics of electromagnetic radiation effects on integrated circuits", *Russian Microelectronics*, vol. 46, no. 3, 2017, pp. 166-170.

[3] P.K.Skorobogatov, K.A.Epifantsev, and O.A.Gerasimchuk, "Simulating the exposure of ICs to voltage surges caused by nuclear explosions," *Russian Microelectronics*, vol. 38, no. 4, 2009, pp. 260-272.

[4] O.A.Gerasimchuk, K.A.Epifantsev, T.V.Pavlova, and P.K.Skorobogatov, "Electrothermal behavior of the elements of SOS CMOS ICs," *Russian Microelectronics*, vol. 40, no. 3, 2011, pp. 215-224.

[5] A.N.Shemonaev, K.A.Epifantsev, and P.K.Skorobogatov, "Experiments on electrical overstress influence on digital ICs depending on the input/output port configuration," *in Proc. 30th Int. Conf. on Microelectronics, MIEL 2017*; Nis, Serbia, October 2017, pp. 321-323.

[6] P.K.Skorobogatov, K.A.Epifantsev, and O.A.Gerasimchuk, "An analysis of the temperature effect on the impulse electric strength of CMOS chips," *Russian Microelectronics*, vol. 44, no. 1, 2015, pp. 40-43.

[7] O.A.Gerasimchuk, K.A.Epifantsev, and P.K.Skorobogatov, "The influence of ambient temperature on ICS electrical overstress pulse hardness", *RAD Conference Proceedings*, vol. 2015-June, 2015, pp.459-462.

[8] A.N.Shemonaev, K.A.Epifantsev, and P.K.Skorobogatov, "The specialized pulse voltage generator EMI-0502," in *Proc. 30th Int. Conf. on Microelectronics, MIEL 2017*; Nis, Serbia, October 2017, pp. 325-327.

[9] P.K.Skorobogatov, K.A.Epifantsev, A.N.Shemonaev, V.A. Telets "Method and mean of IC's testing under multiple electrical overstress," *Seventh International Conference on Radiation in Various Fields of Research (RAD-2019)*, June 2019.

[10] P.K. Skorobogatov, and K.A.Epifantsev, "Latent effects in digital ICs under electrical overstress pulses," in *Proc. 29th Int. Conf. on Microelectronics, MIEL 2014*, Belgrade, Serbia, May 2014, pp. 315-317.

[11] N.S. Diatlov, P.K.Skorobogatov, and K.A.Epifantsev, "Additive effects under the series of EOS in space application VLSI circuits," *MATEC Web of Conferences*, vol. 102, 2017.

Correlation between Temperature and Dose Rate Dependences of Input Bias Current Degradation in Bipolar Operational Amplifiers

A. Bakerenkov, V. Pershenkov, V. Felitsyn, A. Rodin, V. Telets, V. Belyakov, A. Zhukov, and N. Gluhov

Abstract - It was demonstrated experimentally that in ELDRS-susceptible operational amplifiers elevated temperature irradiation increases degradation rate of input bias current, while in ELDRS-free devices degradation rates at room and elevated temperatures are approximately equal.

I. INTRODUCTION

It is well known that degradation of electrical parameters of bipolar and CMOS integrated circuits (IC) under ionizing radiation impact depends on irradiation dose rate, as well as on total dose level [1-5]. In early research works it was experimentally demonstrated, that decreasing of irradiation dose rate leads to increasing of radiation degradation of electrical parameters of bipolar devices at fixed total dose level [1]. Today we know this effect as enhanced low dose rate sensitivity (ELDRS) of bipolar devices. The opposite effect in CMOS devices was described in work [1]. Based on results of these researches in was concluded, that bipolar devices are essentially ELDRS-susceptible and CMOS devices are ELDRS-free.

In recent works it was demonstrated that bipolar devices can be ELDRS-free, as well as ELDRS-susceptible [2,6]. Therefore the issue of distinguishing of ELDRS-free and ELDRS-susceptible devices become very important. Accelerated tests at elevated temperature are proposed by different test methods for estimation of radiation hardness of integrated circuits at low dose rate. There are a number of research works addressing issue of elevated temperature irradiation [8-10]. Several recent works were focused on research of low temperature irradiation of different types of integrated circuits [2,10,11]. Nevertheless correlation of dose rate dependence of radiation degradation and temperature dependence of radiation degradation rate wasn't investigated for devices from the same manufacture lot in wide temperature range, including low temperatures. The

A. Bakerenkov is with the National Research Nuclear University MEPhI (Moscow Engineering Physics Institute), Moscow, Russian Federation (corresponding author to provide phone: +7 499 324 01 84; fax: +7 (499) 324 21 11; e-mail: AS_Bakerenkov@ list.ru).

V. Pershenkov, V. Felitsyn, A. Rodin, V. Telets, V. Belyakov, A. Zhukov and N. Gluhov are with the National Research Nuclear University MEPhI (Moscow Engineering Physics Institute), Moscow, Russian Federation (e-mail: VSPershenkov@mephi.ru).

propose of this work is to determine a correlation between radiation responses of electrical parameters of several types of bipolar operational amplifiers at low, elevated and room temperatures and at low dose rate. The results of this research can be useful for understanding of radiation degradation mechanisms of bipolar devices at different temperatures and dose rates.

Detailed understanding of these mechanisms can improve modern techniques for selection of ELDRS-free and ELDRS-susceptible devices.

II. EXPERIMENTAL DETAILS

The radiation hardness of two types of bipolar operational amplifiers (LM324 and LM358) was studied under ionizing radiation impact with different dose rate in a wide operational temperature range in this work. Five lots of three samples were prepared for the research. The lots were irradiated by Co^{60} ionizing radiation. Four lots were irradiated at different temperatures (one lot was irradiated at a low temperature, two lots were irradiated at room temperature and one lot was irradiated at elevated temperature) and at high dose rate. Detailed data about temperature and bias conditions of each lot are presented in Table I. Electrical parameters of all test samples were measured before and after irradiation at room temperature ($25^{\circ}C$). The schematic diagram of the operational amplifier during the irradiation process is shown in Fig. 1.

From experimental results was obtained that the radiation-sensitive parameter of the test samples is an input bias current I_{ib}. This parameter was controlled in all samples during radiation exposure at the temperature at which the irradiation was performed.

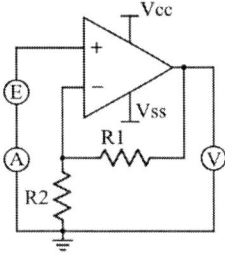

Fig. 1. The schematic diagram of the bias mode of the devices under test, where: Vcc = 5 V, Vss = -5 V, E = 0 V, A is amperemeter, V is voltmeter.

TABLE I
IRRADIATION AND BIAS CONDITIONS

Type	Lot number	Temperature, °C	Electrical mode	Dose rate, rad(Si)/s	Total dose, krad(Si)
LM324	1	25	Active (see Fig. 1)	10,2	59,2
	2	70		10,2	30,4
	3	25		8,1	9,7
	4	0		7,6	45,4
	5	25		$8 \cdot 10^{-1}$	2,63
LM358	1	25		10,2	57,0
	2	70		10,2	80,0
	3	25		8,1	9,7
	4	0		7,6	45,4
	5	25		$8 \cdot 10^{-1}$	2,63

III. EXPERIMENTAL RESULTS

The first and the second lots of both types of operational amplifiers (LM324 and LM358) were irradiated by ionizing radiation at room and elevated temperature respectively. The dependences of input bias current of the first lot of LM324 operational amplifier on total dose, which were irradiated at room temperature are presented in Fig. 2. Also in Fig. 2 the dependences of input bias current of the second lot of LM324 operational amplifiers which were irradiated at elevated temperature (70 °C) on total dose and values of input bias current of the second lot samples, measured at room temperature are presented. Similar results for LM358 operational amplifier are shown in Fig. 3.

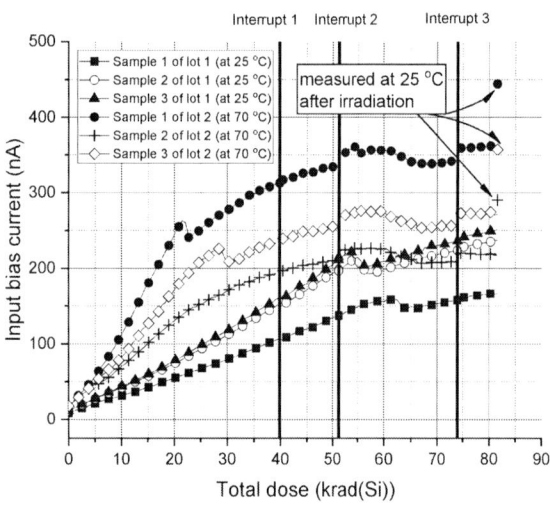

Fig. 2. The dependences of input bias current on total dose at room and elevated temperatures for LM324.

The third and the fourth lots of two type operational amplifiers (LM324 and LM358) were irradiated at room and low temperatures respectively. The dependences of input bias current of the third lot of LM324 on total dose at room temperature and dependences of input bias current of

the fourth lot at low temperature (0 °C) on total dose are presented in Fig. 4. Also in Fig. 4 the values of input bias currents of the fourth lot of LM324 which were measured after irradiation at room temperature are presented. Similar results for LM358 operational amplifier are shown in Fig. 5.

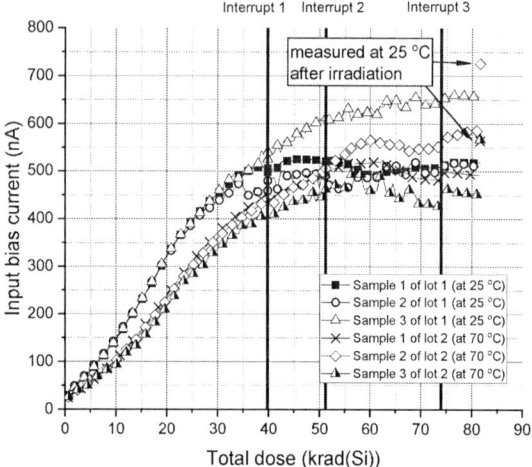

Fig. 3. The dependences of input bias current on total dose at room and elevated temperatures for LM358.

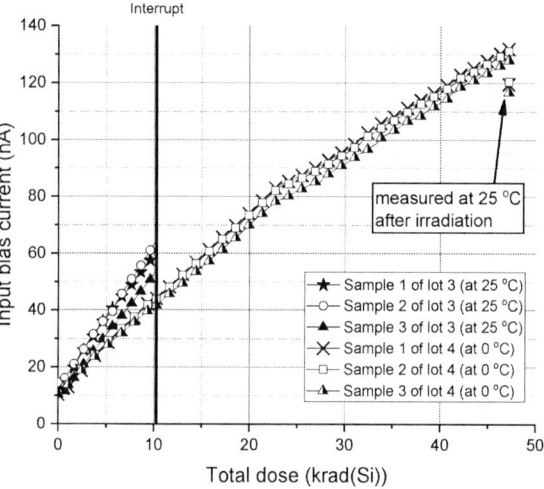

Fig. 4. The dependences of input bias current on total dose at room and low temperatures for LM324.

The fifth lot of each type was irradiated at room temperature. The dependences of input bias current of LM324 and LM358 on total dose and results obtained at exposure to ionizing high dose rate radiation on LM324 and LM358 operational amplifiers at room temperature are presented in Fig.6 and Fig.7. The values of input current of control samples of two types of operational amplifiers which are measured simultaneously with the input current values of the fifth lot which were low dose rate irradiated were presented in Fig. 6 and Fig.7.

978-1-7281-3420-8/19 $31.00 © 2019 IEEE   342

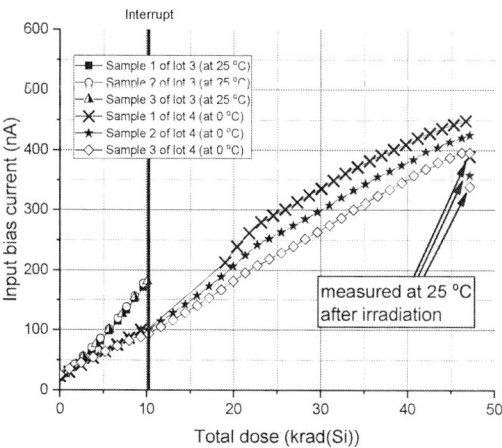

Fig. 5. The dependences of input bias current on total dose at room and low temperatures for LM358.

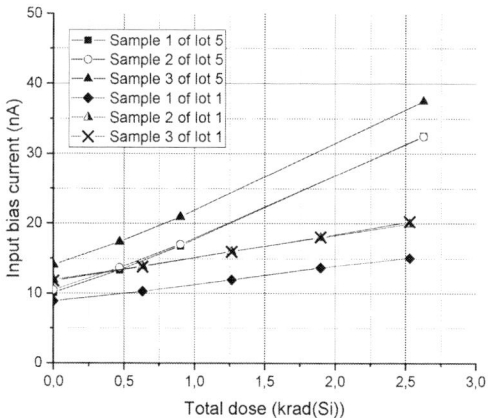

Fig. 6. The dependences of input bias current on total dose at low and high dose rates at room temperature for LM324 lot 5 (at 0.8 mrad(Si)/s) and lot 1 (at 10.2 rad(Si)/s)

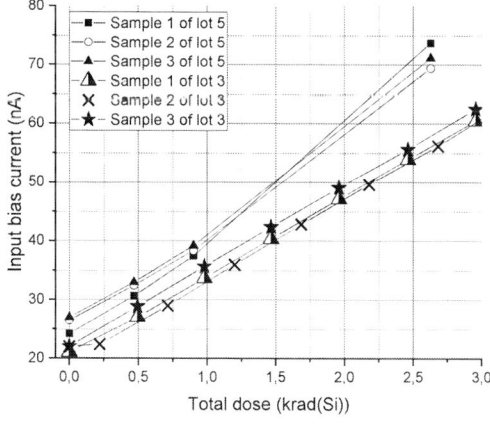

Fig. 7. The dependences of input bias current on total dose at low and high dose rates at room temperature for LM358 lot 5 (at 0.8 mrad(Si)/s) and lot 3 (at 8.1 rad(Si)/s).

IV. DISCUSSION

As it follows from presented experimental results ELDRS effect is observed only in LM324. LM358 devices are ELDRS-free. All devices demonstrate decreasing of input bias current degradation rate at low temperature irradiation. In both types of operational amplifiers the degradation rate is approximately two times lower at low temperature irradiation, than during irradiation at room temperature. Elevated temperature irradiation increases degradation rate of LM324 devices and doesn't produce significant impact on degradation rate of LM358 in comparison with results of room temperature irradiation. Form presented results we can conclude that increasing of the degradation rate at elevated temperature in comparison with degradation rate at room temperature irradiation can be considered as the feature of ELRDS-susceptibly. Experimental results, presented in [12], enable to obtain the same conclusion.

There are several research works subjected to describe physical mechanisms of temperature dependences of the radiation degradation of bipolar devices [8,9]. In [9] the dependence of the radiation degradation on temperature during irradiation is explained using temperature dependence of radiation-induced charge yield in the oxide. In another works [12] it is supposed, that the dependence of the degradation on temperature is connected with temperature dependence of the conversion rate of oxide trapped charge to interface traps.

In ELDRS-free LM358 we observe decreasing of the degradation rate at low temperature irradiation, while irradiation at elevated temperature doesn't produce any significant effect in comparison with room temperature irradiation. Let´s suppose that it is connected with temperature dependence of the radiation-induced charge yield only. In this case we must observe increasing of the degradation rate at elevated temperature irradiation as well as decreasing of the degradation rate at low temperature irradiation in comparison with room temperature irradiation results. In our experiments we observed decreasing of the degradation rate at low temperature irradiation only. Increasing of the degradation rate in ELDRS-free LM358 devices at elevated temperature irradiation we didn't observe. In this case the decreasing of the degradation rate at low temperature irradiation can be explained by decreasing of the conversion rate of oxide trapped charge to interface traps. We didn't observe increasing of the degradation rate in LM358 at low dose rate at room temperature. It means that a room temperature is sufficient for full conversion of all oxide trapped charge during high dose rate irradiation process. Moreover the same full conversion will be at low dose rate when irradiation time increases. In [2] it was demonstrated, that devices, which are ELDRS-free at room temperature, can demonstrate significant ELDRS effect at low temperature irradiation. Using the experimental data we can suppose the same effect in LM358. To demonstrate this effect in is necessary

978-1-7281-3420-8/19 $31.00 © 2019 IEEE   343

to perform low temperature irradiation of additional sample lot of LM358 devices at low dose rate.

In ELDRS-susceptible LM324 we observe decreasing of the degradation rate at low temperature irradiation and increasing of the degradation rate at elevated temperature irradiation in comparison with room temperature radiation response. In can be explained by temperature dependence of oxide charge yield, as well as by the dependence of the oxide charge conversion rate on temperature. The increasing of the degradation rate at low dose rate irradiation can be connected with low positive charge conversion rate, as well as with dose rate dependence of the radiation induced oxide charge yield.

V. CONCLUSION

Two types of bipolar operational amplifiers were irradiated at high dose rate and different temperatures and at low dose rate to research a correlation between temperature and dose rate dependences of input bias current degradation. It was obtained that in ELDRS-susceptible devices elevated temperature irradiation increases degradation rate of input bias current, while in ELDRS-free devices degradation rates at room and elevated temperatures are approximately equal. The degradation rate decreases at low temperature irradiation in both types of operational amplifiers in two times in comparison with degradation rate at room temperature. In can be connected with dependence of oxide charge yield on irradiation temperature, as well as with temperature dependence of the conversion rate of oxide trapped charge to interface traps. From obtained experimental results we can conclude, that ELDRS susceptibility of bipolar devices can depend on irradiation temperature, as it was shown in [2]. The results of performed experiments can be used to develop a universal model for description of total ionizing dose effects in bipolar devices in wide range of temperatures and dose rates.

REFERENCES

[1] R. L. Pease, R. D. Schrimpf, D. M. Fleetwood, "ELDRS in Bipolar Linear Circuits: A Review", *IEEE Transactions on Nuclear Science*, vol. 56, Issue: 4, 2009, pp. 1894 – 1908.

[2] A. S. Bakerenkov, V. S. Pershenkov, V. A. Felitsyn, A. S. Rodin, V. A. Telets, V. V. Belyakov, V. V. Shurenkov, "ELDRS Susceptibility of Bipolar Transistors and Integrated Circuits During Low-Temperature Irradiation", *IEEE Transactions on Nuclear Science*, vol. 64, Issue: 8, 2017, pp. 2227 – 2234.

[3] Z. Yuzhan, L. Wu, R. Diyuan, G. Qi, "A new accelerated method for evaluating the ELDRS of bipolar operational

amplifiers: Temperature switching approach", *14th European Conference on Radiation and Its Effects on Components and Systems (RADECS)*, 2013, pp. 1 – 4.

[4] D. M. Fleetwood, "Total Ionizing Dose Effects in MOS and Low-Dose-Rate-Sensitive Linear-Bipolar Devices", *IEEE Transactions on Nuclear Science*, vol. 60, Issue: 3, 2013, pp. 1706 – 1730.

[5] I. S. Esqueda, H. J. Barnaby, P. C. Adell, B. G. Rax, H. P. Hjalmarson, M. L. McLain, R. L. Pease, "Modeling Low Dose Rate Effects in Shallow Trench Isolation Oxides", *IEEE Transactions on Nuclear Science*, vol. 58, Issue: 6, 2011., pp. 2945 – 2952.

[6] J. Boch, A. Michez, M. Rousselet, S. Dhombres, A. D. Touboul, J.-R. Vaillé, L. Dusseau, E. Lorfèvre, N. Chatry, N. Sukhaseum, F. Saigné, "Dose Rate Switching Technique on ELDRS-Free Bipolar Devices", *IEEE Transactions on Nuclear Science*, vol. 63, Issue: 4, 2016, pp. 2065 – 2071.

[7] J. Boch, F. Saigne, R.D. Schrimpf, D.M. Fleetwood, R. Cizmarik, D. Zander, "Elevated temperature irradiation at high dose rate of commercial linear bipolar ICs", *IEEE Transactions on Nuclear Science*, vol. 51, Issue: 5, 2004, pp. 2903 – 2907.

[8] D. R. Hughart, R. D. Schrimpf, D. M. Fleetwood, B. R. Tuttle, S. T. Pantelides, "Mechanisms of Interface Trap Buildup and Annealing During Elevated Temperature Irradiation", *IEEE Transactions on Nuclear Science*, vol. 58, Issue: 6, 2011, pp. 2930 – 2936.

[9] S.C. Witczak, R.D. Schrimpf, D.M. Fleetwood, K.F. Galloway, R.C. Lacoe, D.C. Mayer, J.M. Puhl, R.L. Pease, J.S. Suehle, "Hardness assurance testing of bipolar junction transistors at elevated irradiation temperatures", *IEEE Transactions on Nuclear Science*, vol. 44, Issue: 6, pp. 1989 – 2000.

[10] A. H. Johnston, R. T. Swimm, D. O. Thorbourn, "Total Dose Effects on Bipolar Integrated Circuits at Low Temperature", *IEEE Transactions on Nuclear Science*, vol. 59, Issue: 6, 2012, pp. 2995 – 3003.

[11] P. C. Adell, I. S. Esqueda, H. J. Barnaby, B. Rax, A. H. Johnston, "Impact of Low Temperatures (<125 K) on the Total Ionizing Dose Response and ELDRS in Gated Lateral PNP BJTs", *IEEE Transactions on Nuclear Science*, vol. 59, Issue: 6, 2012, pp. 3081 – 3086.

[12] V. S. Anashin, V. S. Pershenkov, A. S. Bakerenkov, P. A. Chubunov, A. V. Solomatin, A. S. Rodin, V. A. Felitsyn, "Experimental Technique for Determination of ELDRS-Free Devices", *15th European Conference on Radiation and Its Effects on Components and Systems (RADECS)*, 2015, pp. 1 – 4.

[13] J. Boch, F. Saigne, R.D. Schrimpf, J.-R. Vaille, L. Dusseau, E. Lorfevre, "Physical Model for the Low-Dose-Rate Effect in Bipolar Devices", *IEEE Transactions on Nuclear Science*, vol. 53, Issue: 6, 2006, pp. 3655 – 3660.

[14] V. S. Pershenkov, D. V. Savchenkov, A. S. Bakerenkov, V. N. Ulimov, A. Y. Nikiforov, A. I Chumakov, A. A. Romanenko, "The conversion model of low dose rate effect in bipolar transistors", *European Conference on Radiation and Its Effects on Components and Systems*, 2009, pp. 290 – 297.

Software and Hardware System for Charge Coupled Devices with Interline Transfer of Charge Parameters Monitoring During Radiation Tests

V. P. Lukashin, M. E. Cherniak, A. O. Akhmetov, A. Y. Nikiforov, and A. V. Ulanova

Abstract - the paper presents a method of device monitoring during radiation testing for charge coupled devices with interline transfer (hereinafter — CCD). The results of heavy ions, dose rate and total ionizing dose tests are presented together with the description of the developed software and hardware set-up based on the National Instruments platform and on the designed specialized equipment adapted for radiation tests.

I. INTRODUCTION

CCDs are widely used in onboard equipment of space vehicles [1, 2] (star sensors, orbital telescopes, imaging equipment of Earth remote sensing satellites) and scientific equipment (RFNC-VNIIEF in Russia, NIF in USA and LMJ in French) intended for research in high energy density physics. This includes the application for laser thermonuclear fusion, where CCDs are affected by pulsed ionizing radiation. The necessity of heavy ion tests is perpetrated by the use of CCDs in open space in different orbits. Therefore, it is necessary to carry out tests to assess the radiation hardness of CCDs to total ionizing dose, transients and heavy charged particles.

II. DEVICE UNDER TEST AND ITS PARAMETERS

CCD for research has a format of 1024×2048 pixels, the size of a photosensitive pixel is 11×11 micrometers, the visible range is 450 - 1000 nanometers. The CCD includes a Peltier thermoelectric module for chip cooling. The structure includes epi-Si substrate with a gate dielectric layer on the surface. CCDs radiation test results [3] introduced a limited set of radiation-dependent parameters that characterizes normal operation of a device under test (DUT) and the possibility of its application in space equipment:

- dynamic range, dB (decrease of saturation voltage, increase of dark current due to the radiation induced defects formation and charge accumulation);
- dark signal nonuniformity and sensitivity nonuniformity, % (these parameters are strongly dependent on the number of hot pixels and RTS - pixels - with a random telegraph signal);

Vladislav P. Lukashin, Maxim E. Cherniak, Alexey O. Akhmetov, Alexander Y. Nikiforov, Anastasia V. Ulanova are with the National Research Nuclear University (NRNU) "MEPHI", and JSC "SPELS", Kashirskoe shosse 31, Moscow, Russia. E-mail: vpluk@spels.ru

- supply currents of constant voltage channels, A (degradation of CMOS structures due to total ionizing dose (TID) effects);
- number of defects (pixels with a constant brightness value) or structural damage, pcs (this parameter has a strong effect on the dark signal, nonuniformity, and caused by heavy ion irradiation; an interesting study is the identification of the dependence of the number of defects on fluence of particles with a certain linear energy transfer - LET).

III. HARDWARE AND SOFTWARE SYSTEMS

There is a problem of commercially available front-end control modules radiation degradation due to their location near a test sample during radiation tests. So, we used National Instruments equipment [4, 5], which allowed to create back-end electronics for control signals generation and data processing (located 2-3 meters away from the object and irradiation zone).

The developed specialized equipment consists of:

- low- and high-level drivers for control signals;
- analog-to-digital converter (AD9826) with correlated double sampler for measuring the voltage difference between the signal and reference levels of CCD output signal (Fig. 1);

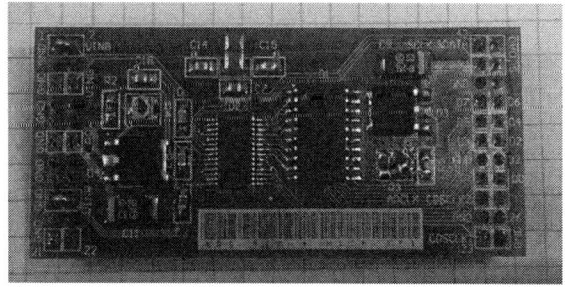

Fig. 1. ADC module (AD9826).

- high-speed transistor switches (IRF7309) for corresponding control signal outputs switching to high and low levels (overlapping «phases» that allow move light-generated photoelectrons in the right direction) (Fig. 2);

Fig. 2. Vertical charge transfer module (IRF7309).

- low noise voltage regulators for constant potential circuits (drain of output transistor, peripheral diffusion; drain of output register);
- a device for pulse LED illumination with selectable pulse width up to 1 ms for test samples that do not have an electronic shutter.

If it is necessary due to the layout of the test facility, we can use a cable adapter (about 2-3 meters long) to increase the distance between the CCD and test equipment with an individual shielding of wires to protect against interference and influence on analog output signal.

The hardware and software system based on National Instruments platform consists of:
- NI PXI-1033 chassis (5 slots);
- NI PXI-4110 power supplies with current measurement capability;
- a configurable board based on Xilinx FPGA with integrated NI PXI-7953R 128 MB DRAM;
- NI 6581 I/O module for NI PXI-7953R;
- a program in LabVIEW which enables a fast calculation of parameters listed above based on received frames during the process of radiation studies.

The structural diagram of the developed set-up is shown in Fig. 3, its appearance - in Fig. 4.

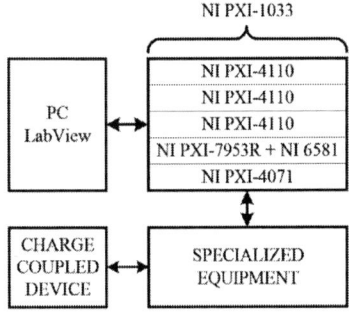

Fig. 3. The structural diagram of the developed test set-up.

The software was developed in the Lab View and allowed to perform the following:
- reception of a picture from a CCD (1 fps for 512×1024 format);
- adjustment of the accumulation time and the duration of the pulse illumination;

- frame analysis – mean value of the whole frame and in a limited area;
- calculation of dynamic range, dark signal nonuniformity and sensitivity nonuniformity, number of defects;
- parrying single event latch-ups, logging currents, saving received frames;
- initialization and tuning of the ADC.
- The software desktop appearance is shown in Fig.5.

Fig. 4. The developed test set-up photo.

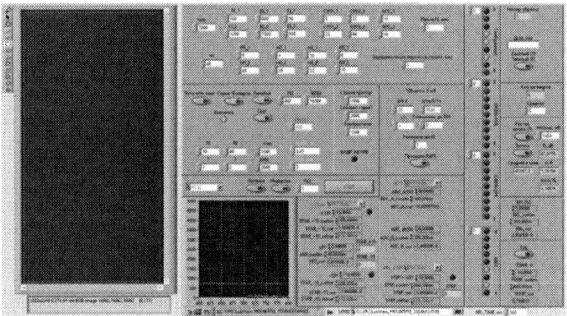

Fig. 5. The software appearance.

IV. TRANSIENT RADIATION EFFECTS

The vulnerable parameter – dark current signal (mV/s). The dark signal evaluation method is based on CCD average output voltage measurement at different charge accumulation times:

$$U_T = \frac{|U_{Tj} - U_{Ti}|}{|T_j - T_i|}, \qquad (1)$$

U_{Tj} – average output voltage at accumulation time T_j, U_{Ti} – average output voltage at accumulation time T_i.

Dark signal monitoring showed that this parameter does not go beyond the specified upper limit (180 mV/s) for dose rates up to 1,0E+07 rad/s.

The graph of dark signal (mV/s) versus dose rate for two samples is shown in Fig. 6. Radiation research revealed a dose rate threshold ($7,7 \cdot 10^6$ rad/s) at which the dark signal exceeds the specified limit. The device was in accumulation mode during irradiation.

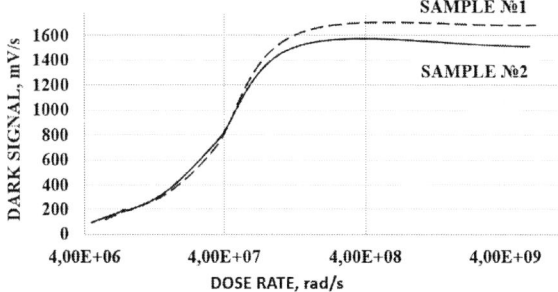

Fig. 6. Dark signal vs. dose rate dependence.

Figures 7 and 8 show frames for 165 (below the threshold) and 187 mV/s (above threshold).

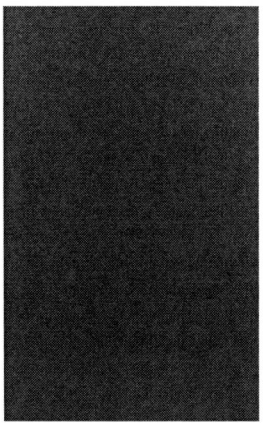

Fig. 7. Dark signal equal 165 mV/s.

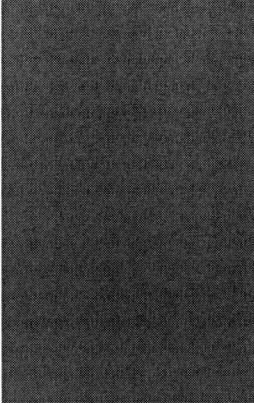

Fig. 8. Dark signal equal 187 mV/s.

V. TOTAL IONIZING DOSE (TID) EFFECTS

An integrated Peltier thermoelectric module (TEM) allows us to set the operating temperature of the photosensitive region of the crystal (down to – 60 °C) during irradiation. This leads to a decrease in the intrinsic noise of the device and the relative intensity of the defective pixels.

Tests were conducted at two currents TEM values: I_{TEM} = 300mA and 800mA. The average square noise voltage (ASNV) behavior will vary at different currents TEM (therefore, at different temperatures of the sensitive area) and the dynamic range will vary in different ways.

The ASNV calculation formula (for two reading cycles):

$$\sigma U_{\text{ш}} = \sqrt{\frac{1}{2(N-1)}\sum_{i=1}^{N}|U_i' + U_i'' - \frac{1}{N}\sum_{i=1}^{N}(U_i' - U_i'')|^2}, (2)$$

U_i' - output voltage for i photosensitive element (first reading cycle); U_i''- output voltage for i photosensitive element (second reading cycle); N - number of photosensitive elements.

Experimental data showed that at current I_{TEM} = 300mA for every ~ $2 \cdot 10^4$ rad an increase of I_{TEM} by ~ 10mA (Fig. 9) is required to keep ASNV in a condition, in which the dynamic range does not exceed the specified value (60 dB).

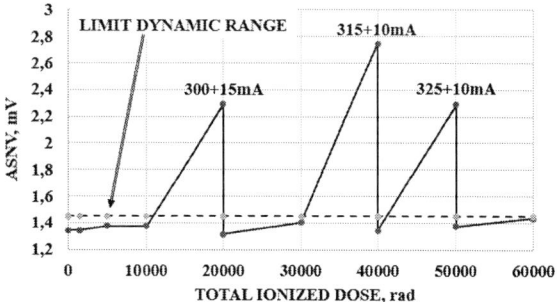

Fig. 9. ASNV vs. TID dependence, I_{TEM} = 300 mA.

ASNV increase was not observed until 60 krad TID at current I_{TEM} = 800mA (fig. 10).

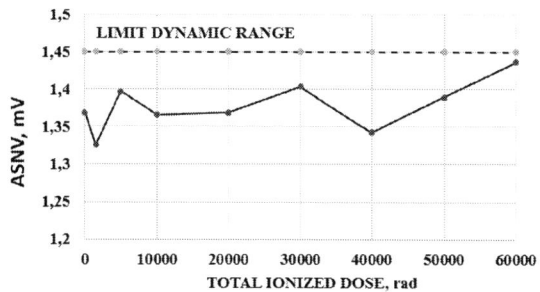

Fig. 10. ASNV vs. TID dependence, I_{TEM} = 800 mA.

VI. SINGLE EVENT EFFECTS

During heavy ion irradiation, currents are monitored for single event latch-up (SEL) parrying, and frames are also recorded on a computer for analyzing the number of particle tracks (impulse response - IR) with various LET. If the particles flux accedes a certain time and the number of tracks in a frame is known, then it is possible to calculate the probability of a pulsed response appearance (see figures 11 and 12).

IR monitoring results under Xenon and Neon ions (65 MeV·cm²/mg and 4 MeV·cm²/mg respectively) at 25°C are shown in figures 11 and 12. In picture 11, we can see more fuzzy tracks of particles due to the fact that Xenon ions produce more charge along the track in the crystal.

The defect generation rate dependence on particle fluence with different LET was obtained. Defects control was carried out by obtaining several dark frames before and after exposure with different fluences.

Figures 13 - 16 show the dependences of the defects number on the particles fluence for different LET values at different accumulation times (5, 200, 500 и 1000 ms).

The measurements were carried out directly after certain fluence was reached, during a time of not more than 1 minute, therefore, the annealing process can be neglected.

It was found that at an accumulation time of 5 ms, no defects were detected upon irradiation with Xenon ions (LET ≈ 65 MeV·cm²/mg). But with an accumulation time of 1000 ms, defects were detected upon irradiation with neon ions (LET ≈ 4 MeV·cm²/mg).

Fig. 11. IR,
LET ≈ 65 MeV·cm²/mg.

Fig. 12. IR,
LET ≈ 4 MeV·cm²/mg.

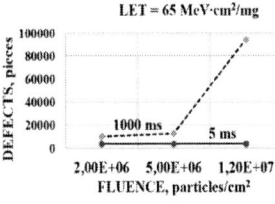

Fig.13 – Dependence of defects on the fluence,
LET ≈ 65 MeV·cm²/mg.

Fig. 14 – Dependence of defects on the fluence,
LET ≈ 44 MeV·cm²/mg.

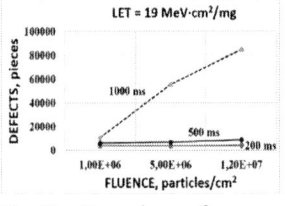 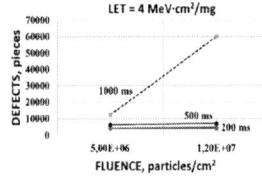

Fig. 15 – Dependence of defects on the fluence,
LET ≈ 19 MeV·cm²/mg.

Fig. 16 – Dependence of defects on the fluence,
LET ≈ 4 MeV·cm²/mg.

VII. CONCLUSION

Radiation tests of CCD with an integrated Peltier thermoelectric module were carried out. The CCD parameters radiation sensitivity was confirmed such as dynamic range, dark signal (mV/s), dark voltage, defects. SELs as well as gate dielectric breakdowns (SEGR – single event gate rupture) were not detected at the maximum allowed TEM current (minimum temperature in the sensitive area). The dark and light signals nonuniformity varied within acceptable limits (only when exposed to heavy ions), which is due to the formation of defects (spikes).

The use of TEM in the equipment can significantly increase both TID and heavy ion hardness levels, since it allows us to reduce defect formation rate and the intrinsic noise of the device (or noise caused by the TID).

In the future, we are going to investigate a CCD radiation hardness for various values of control voltages (phase range, nominal displacements) within specified limits.

REFERENCES

[1] Holst GC, Lomheim TS (2007) CMOS/CCD Sensors and Camera Systems, Bellingham, Wash.; The International Society for Optical Engineering; JCD Publishing

[2] Marshall C. J., Marshall P.W., "CCD Radiation Effects and Test Issues for Satellite Designers", *Review Draft* 1.0, 2003, 43.

[3] M. E. Cherniak, A. V. Ulanova, and A. Y. Nikiforov, "Analysis of the effect of the interline CCD-sensor dark signal increasing during gamma-irradiation", *Journal of Physics: Conference Series*, vol. 737, no. 1, 2016.

[4] M.E. Cherniak, A.A. Pechenkin, R.K.Mozhaev, A.V. Ulanova, and A.Y. Nikiforov, "Automated measurement system for optoelectronic devices based on National Instruments PXI-platform," in *Proc. 2017 International Siberian Conference on Control and Communications, SIBCON 2017*, S. Seifullin Kazakh Agrotechnical University, Astana; Kazakhstan; June 29 -30, 2017.

[5] E.V. Petrova, N.A. Komarova, M.E.Cherniak, A.V. Ulanova, and A.Y. Nikiforov "Hardware/software solution for optocouplers with output MOSFET transistors based on National Instruments PXI-platform", in *Proc. 2016 International Siberian Conference on Control and Communications, SIBCON 2016*, National Research University "Higher School of Economics" Moscow; Russian Federation; May 12 -14, 2016.

Comparative Assessment of Digital and UHF Optoelectronic Transceivers Radiation Hardness

R. K. Mozhaev, M. E. Cherniak, A. A. Pechenkin, A. V. Ulanova, and A. Y. Nikiforov

Abstract – A method for radiation hardness evaluation of digital and microwave transmitting-receiving optoelectronic modules is presented. The technical aspects of parameters monitoring during exposure are described. The most vulnerable components of optoelectronic modules are identified.

I. Introduction

As complexity of space equipment increases, the amount of information transmitted between modules is growing. A perspective way to increase bandwidth of information channels is the use of fiber-optic communication lines (FOC).

Receiver and transmitter modules are FOC main elements, which are used in various applications, including high-altitude aviation and spacecraft equipment [1]. Besides, there is a growing demand for radiation- tolerant optical transmission and reception systems in high-energy physics experiments, including LHCs. To further enhance the physics potential, the LHC will be upgraded to even higher luminosity. The particle detectors will be upgraded to fully exploit the new physics potential. The optical data transmission system will thus need to be upgraded to handle the higher data transmission speed. [2]. In this paper, two types of optical transceivers are distinguished. «Analog» ultra-high frequency (UHF) optoelectronic modules (which convert microwave signals into optical) are used to transmit broadband (1 GHz -12 GHz) information signals, including utilization in advanced and modernized radar systems. «Digital» optoelectronic modules (convert low-voltage differential signals - LVDS) are used in many technological branches, such as onboard aviation and satellite digital information exchange networks. Vertical-cavity surface-emitting laser (VCSEL) commonly used in such systems is a promising type of laser emitters for transmitting modules due to the symmetric radiation pattern and high consistency with the fiber-optic communication line.

Special applications impose certain requirements on the device to withstand the harsh radiation environment (ionization and/or structural damage caused by space particles).

Roman K. Mozhaev, Maksim E. Cherniak, Alexander A. Pechenkin, Anastasia V. Ulanova and Alexander Y. Nikiforov are with the National Research Nuclear University (NRNU) "MEPHI" and JSC «SPELS», Kashirskoe shosse 31, Moscow, Russia, E-mail: rkmozh@spels.ru.

Therefore, to estimate the suitability of the device it is necessary to perform radiation tests.

II. Devices under Test

This paper presents a radiation behavior comparison of two types of optoelectronic modules. The first device under test (DUT) was a transceiver optical module (TROM) with a wavelength of 850 nm, designed to operate in systems with an exchange data rate of at least 1 Gbit/s. The transmitting part of the module consists of a VCSEL (GaAs / AlGaAs), a laser driver IC with a memory element and a segment of optical fiber. The receiver contains a p-i-n GaAs-based photodiode, amplifiers, LVDS transformer and, a fiber segment. The receiver and transmitter of TROM are enclosed in a single case. A typical block diagram of an optical CMOS transceiver with 250 Mbit/s data rate [3] is shown in Figure 1.

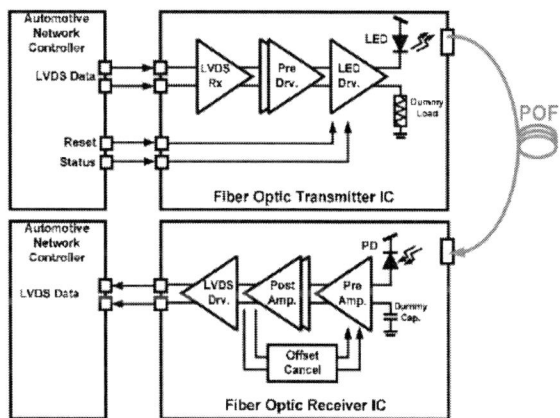

Fig. 1. General block diagram of digital (CMOS) optoelectronic transceiver

The second DUT is a pair of transmitting (TOM) and receiving (ROM) microwave optoelectronic modules. The transmitter includes a standard distributed-feedback edge-emitting laser diode with 1550 nm wavelength, a thermionic cooler (the Peltier element) and a feedback thermistor. The receiver consists of a p-i-n photodiode manufactured using the same process as the emitter (AlGaAs/InP). Thermal stabilization is not provided for the receiving module.

978-1-7281-3420-8/19 $31.00 © 2019 IEEE

III. DUT PARAMETERS AND CHARACTERISTICS

An important parameter of the digital transceiving system is the bit error rate (BER). Distortion of information can occur due to the impact of charged particles (electrons, protons, heavy ions) when operating in hardware units in space applications and gradually increase with the level of the total ionizing dose (TID).

The main parameter of the TOM-ROM pair, in addition to the emitter power and receiver sensitivity, is the gain-frequency characteristic with a specific operating point. A typical amplitude-frequency response of the TOM-ROM pair with an operating point of 5 GHz.

IV. EXPERIMENTAL RESEARCH

Radiation hardness research was carried out for the following effects: single event effects (SEE), transient radiation effects (TRE), total ionizing dose (TID) effects and structural damage.

A. Single event effects

Irradiation was carried out on cyclotron facility U-400M (JINR, Dubna, Moscow Region) with high-energy xenon, krypton and argon ions.

Ion irradiation of TROM was carried out on a cyclotron using krypton and argon ions. During irradiation, the samples operated at the receiving and transmitting modes at a speed of 1 Gbit/s/=. The temperature was +85°C – that is the most critical temperature for single-event effects (SEE). Before and after irradiation, the intensity of the vertical laser was monitored with an optical power meter.

Registration of SETs was implemented using a pseudo-random code generator able to check received information and compare it with transmitted one. In case of error detection, the system generates a strobe with a duration of 8 ns, which is interpreted as a discrepancy between the sent and received information. As the bit error rate, the inverse of the smallest time value without errors was considered.

During the tests, no failures were detected. A decrease in the intensity of the laser radiation and the photosensitivity of the receiving module was not observed. The results of TROM ion irradiation are shown in table 1.

TABLE I
RESULTS OF UPSETS MONITORING OF TROM UNDER THE FLUENCE OF KR AND AR IONS

Ion LET, MeV cm²/mg	Path, um	Sample number	Fluence, см⁻²	BER, bit/s
19.8 (Kr)	326	TROM 1	$3.5 \cdot 10^6$	68.3
		TROM 2	$2.7 \cdot 10^6$	455.3
7.4 (Ar)	270	TROM 2	$2.7 \cdot 10^6$	27.1
		TROM 3	$2.8 \cdot 10^6$	663.5

Ion irradiation of TOM-ROM pair was carried out on the same facility using xenon (68 MeV cm²/mg) and argon ions (28 MeV cm²/mg). Transmitter and receiver modules were irradiated in turn. TOM and ROM were optically locked on each other, the converted optical signal was read out from the output of the ROM, which was operating at 5 V supply voltage while TOM was operating at dynamically adjusted 40 mA pump.

During irradiation of TOM and ROM, neither SET nor failures were observed.

B. Transient radiation effects

During the impact of a gamma-pulse on TOM and ROM modules, no significant responses of TOM output optical level depending on dose rate were observed. During ROM irradiation an abrupt response was noticed. Such behavior was not unexpected and is typical for photodetectors. The response to a gamma pulse at the output of the photodetector module during the transmission of a variable signal is shown in Figure 2.

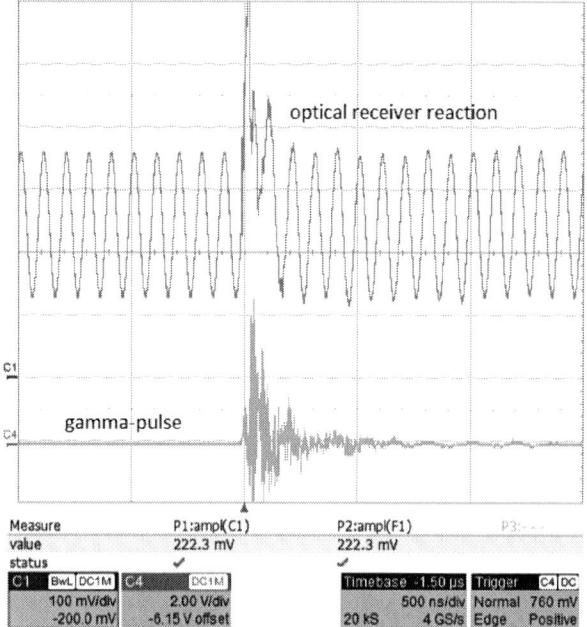

Fig. 2. Response oscillogram of receiver modules periodical signal (10 MHz) during and after gamma-pulse.

Since the TOM-ROM pair had no active elements other than laser-emitters and photodiodes, it was sufficient to test both modules as a pair.

The TROM includes both input and output preamps, as well as a memory unit with settings of the laser diode mode (driver). To separate the possible effects on microelectronic components and effects of gamma pulse on optoelectronics, tests were performed on individual components of the receiving module. As a result of the impact of pulsed gamma radiation on laser diode driver, there was a step-shaped increase in current consumption.

The TROM photodiode demonstrated a similar response with the photodiodes of the receiving module of ROM - a sharp increase in the photocurrent during the pulse. The slight difference was in duration and amplitude.

The results of the laser-emitting element (VCSEL of TROM) tests under gamma-pulse were more interesting. A sharp short-term decrease in the level of output of optical power was observed, in contrast to the case when TOM laser diode was exposed to gamma-pulse. The response, which is not typical to classical edge-emitting quantum-well lasers, probably is due to the specificity of the VCSEL device: one of the most critical values (determined by the manufacturing process) is the resonator reflection coefficient R.

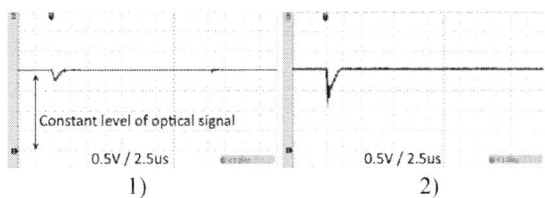

Fig. 3. The response of a constant level of output optical power of VCSEL under a gamma pulse with different dose-rates (dose rate of pulse 2 = 2× pulse 1.)

Standard edge-emitting laser diodes on quantum wells do not have a pronounced sensitivity to the reflection coefficient of the resonator (R> 0.75), while for VSCEL, given the small length of the resonator and active body, it is necessary that R> 0.995 [1, 5]. The general diagram of edge-emitting and a vertical-emitting laser is shown in Figure 4.

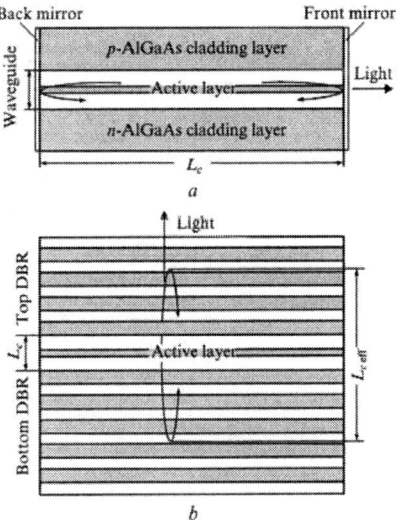

Fig. 4. Schematic structure of semiconductor edge-emitting laser (a) and surface-emitting laser (VCSEL)

One of the the main loss mechanisms for AlGaAs / GaAs based mirrors is caused by absorption on free carriers and interband transitions in the p-layers of GaAs [1, 4]. When exposed to a gamma pulse, energy is absorbed by electrons, which in the case of VCSELs significantly reduces the reflection coefficient (formula 1), which is critical for lasing.

$$R_{max} = 1 - \lambda_c \alpha \frac{n_{high}^2 + n_{low}^2}{n_{GaAs}\left(n_{high}^2 - n_{low}^2\right)}, \qquad (1)$$

where α is the absorption coefficient on free carriers.

Since VCSEL is a relatively new and promising type of emitters in the composition of optoelectronic modules, one should pay special attention to the difference in the radiation behavior of this type of lasers from other common semiconductor lasers.

C. Total ionizing dose effects

The TOM-ROM were irradiated using an electron accelerator (2.4 MeV) in bremsstrahlung mode and isotope Co^{60} facility. Samples were divided into groups and irradiated at normal, elevated (+55°C) and lowered temperatures (-55°C)., the reference source of the optical signal was outside irradiation zone during irradiation of the receiving module, the same with the irradiated transmitting module. Irradiation was carried out to the level of 2 Mrad, and no significant changes in parameters and characteristics of the transmitting and receiving modules were observed.

At the time of the exposure, the photocurrent of the built-in photodiode and the receiving module was monitored. In between the radiation stages, the output of optical power level was monitored with photosensor PD300IR (Ophir) and the microwave-spectrum analyzer N9918A (Keysight).

During TROM radiation tests of, the following parameters were monitored: the average optical power, laser drive current, receiver current and the number of errors per gamma-exposure session (at a data transmitting rate f = 1,0 Gbit/s). Preliminary, as in the case of gamma-pulse, the separate components of the module have been tested. The research has shown a large margin of radiation hardness of optical components (laser diode and photodiode) and, practically, zero sensitivity to TID effects up to levels 10^7 rad, as is the case with the TOM and ROM transmitter and receiver.

Irradiation of the whole TROM module was carried out using an electron accelerator operating in bremsstrahlung mode. The operation was monitored using a pseudo-random code generator/analyzer. The result of monitoring the loss of information during transmission is presented in Figure 5. The tests were carried out in two temperature modes: +25°C and +85°C. The average power level of the laser diode emission during signal transmission was 0.75mW. Throughout the test exposure, the laser radiation power of the laser diode varied in an insignificant range (± 0.01mW). The test result was the devices malfunction of at dose levels 75krad and 45 krad at +25°C and +85°C respectively. Due to the great margin of optoelectronic component radiation hardness, it is possible to unequivocally link the failure with the failure of one of the microelectronic components of the module [4], in particular, the laser diode driver's chip.

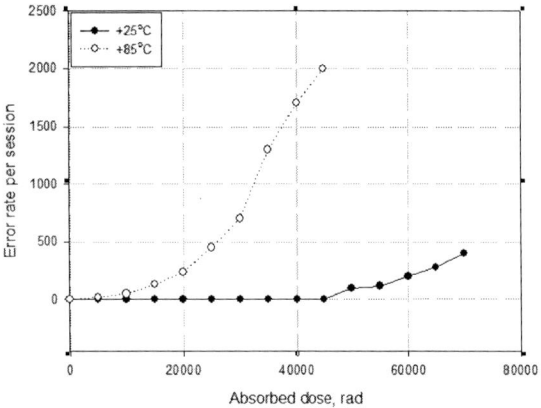

Fig. 5. Dependence of error rate per session during data transmission on absorbed dose

D. Structural damage effects

The structural damage research was carried out on a pulsed reactor of fast neutrons with an energy of 14 MeV. Devices were in unbiased mode during irradiation. Parametric monitoring after exposure ($F = 10^{13}$n/cm^2) showed that both TOM-POM and TROM have a significant hardness to this effect (no parameters degradation and functioning failure was observed).

In the relatively simple GaAs-based lasers, the increase in threshold current started to become significant at fluence levels of 10^{13} to 10^{14} n/cm^{-2}, and fluences up to about 10^{14} n/cm^{-2} did not cause catastrophic failure. Partial annealing of neutron damage can be achieved by operating in lasing mode after irradiation. Thermal annealing occurs at moderately elevated temperatures [1]. The carried-out tests confirmed it. Data on emitters and receivers' characteristics after neutron exposure is presented in the preliminary paper [5].

The amplitude-frequency characteristic of the TOM-ROM pair, measured after exposure (Figure 6) differed slightly from that have been measured before exposure and the value of the operating point at 5 GHz, even taking into account fluctuations fits into the normalized range, indicating sufficient resistance of the TOM-ROM pair to this effect.

V. CONCLUSION

The research and comparative radiation hardness analysis of optoelectronic transceivers was carried out. It was found out that the transceivers optical components radiation hardness is significantly higher as compared to microelectronic components hardness, and the rad-hardness of «analog» transceivers are more than an order of magnitude higher than «digital» ones to most ionizing effects.

In future research, we are going to analyze the behavior of a TROM with rad-hard electronic components and 1310 nm optical components. The comparison will demonstrate the electronic component selection importance for a transceiver design and will also answer the question of whether the optical part process affects the radiation hardness in such devices or not.

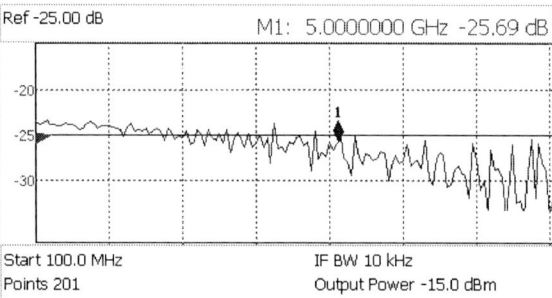

Fig. 6. The measured frequency response of a TOM-ROM pair with a power value at the operating point of 5 GHz.

REFERENCES

[1] P. W. Marshall, C. J. Dale, and E. A. Burke, "Space radiation effects on optoelectronic materials and components for a 1300 nm fiber-optic data bus", *IEEE Trans. Nucl. Sci.*, 1992, vol. 39, pp. 1982–1989.

[2] K. Gan, P. Buchholz, S. Heidbrink, H. P. Kagan, R. Kass, J. Moore, S. L. Smith, M. Vogt, M. F. Ziolkowski "10 Gb/s radiation-hard VCSEL array driver", *IEEE Nuclear Science Symposium and Medical Imaging Conference (NSS/MIC)*, 2014.

[3] Myung-Geun J. et al. "A 250Mb/s CMOS optoelectronic transmitter and receiver IC for next-generation in-vehicle networks", *International Conference on Connected Vehicles and Expo (ICCVE)*, 2013, pp. 842-846.

[4] A. Y. Borisov et al. "Analog ASIC TID behavior in a temperature range", in *Proc. 30th Int. Conf. on Microelectronics*, MIEL 2017, Nis, Serbia, October 2017, pp. 109-112.

[5] R. K. Mozhaev and M. E. Cherniak. "Research of Quantum Well Laser Diode's and Heterostructural P-I-N Photodiode's of Fiber-Optic Modules Radiation Hardness to Gamma-ray and Neutron Irradiation", *KnE Energy*, 2018, pp. 393–399.

Analysis of a Bridgeless Single Stage PFC based on LLC Resonant Converter for Regulating Output Voltage

S. Esmailirad, R. Beiranvand, S. Salehirad, and S. Esmailirad

Abstract - This paper has analysis a single stage bridgeless power factor correction converter based on a LLC resonant converter. The analysis is taken under DCM condition under the resonant frequency in BOOST region. Operation under DCM condition leads into natural power factor correction, in addition the operation under the resonant frequency leads into zero voltage switching (ZVS) for the MOSFETs and zero current switching (ZCS) for the diodes. The frequency modulation technique is synchronized with pulse width modulation technique in which they are both used in the converter analysis. In addition, the both mentioned techniques have achieved more open wide control degree in system designing in contrary with conventional methods. The converter has been analyzed in three subintervals of the duty-cycle because of DCM mode. This converter was implemented in laboratory by 100 DC volt output voltage and AC input voltage by 200 watt output load.

I. INTRODUCTION

Power factor correction (PFC) is one of the important aspects in rectifying in which prevent harmonic injection to network and therefore, many structures have been applied to optimize the PFC converters [4], [5]. This paper presents a single level bridgeless PFC based on a LLC resonant converter as illustrated in Fig. 1 [15], where the converter input voltage is the single phase ac voltage with effective extent of 220 V and 50 Hz frequency, and the output voltage is DC voltage with extent of 100 V. In another word, this converter rectifies the ac voltage and does the power factor correction in a same time. In this article the boost converter has been chosen for the PFC utility and hence the boosting operation is in action. The LLC resonant converter carry out the voltage regulation and it is placed after the boost converter [6]. The both boost converter and the PFC share one switching network contains of two MOSFETs. The LLC resonant converter has some specifics in contrary with the other resonant converters such as high efficiency because of the soft switching in a wide range of load, and input/output voltage variations [7], output voltage regulation due to little variance in switching frequency when wide range of input

Sadegh Esmailirad, Reza Beiranvand and Shervin. Salehirad are with the Department of Power Electronic, Faculty of Electrical and computer Engineering, Tarbiat Modares University, Jalal al Ahmad st, Tehran, Iran, E-mail: sadegh_es2010@yahoo.com, beiranvand@modares.ac.ir, shervin.salehirad92@gmail.com

Sajad Esmailirad is with the Department of Electrical Engineering, Shahrekord University, Shahrekord, Iran, E-mail: sajad_es2013@yahoo.com

voltage and load current varies [8], both increasing and decreasing conversion ratio [9], using the magnetic and leakage reactance as the resonant component and no need of adding any other resonant component as result [10], using the resonant capacitor to eliminate the DC input current component and preventing the transformer saturation [11].

In [2], the LLC has been analyzed in different switching frequency operation by considering the state characteristic. However, in the method which has been used have applied some limitations for the resonant current and pulse width has to be symmetrical for that matter. In [1], a soft start-up strategy has been applied by using a variable duty-cycle per optimized current limits curve. In [1], the start-up state is controllable but there is no control for the steady-state condition. In [3], unsymmetrical pulse width modulation with moderate switching frequency with dual operation has been proposed for the LLC full-bridge resonant converter. In [3] the resonant current has been decreased, the efficiency has been increased, and the converter transformer dimensions has been increased either. However, four switches have been applied, the converter operates in constant frequency, and it suffers from no synchronization between the frequency and the pulse width. The main purpose of using PFC is to synchronize the input voltage and current phase as the power factor tends to one. In [13] the operation of the PFC has been analyzed under the DCM mode in which has used the bridgeless PFC to enhance the efficiency by reducing conduction losses. In [12] and [14], different structures and characteristics of the PFCs have been undertaken.

In this paper different operation state of a single level bridgeless FCC has been studied in DCM mode under its resonant frequency. By considering the current and voltage waveforms of the converter in three duty-cycle intervals, all the possible states are studied in Sec. II. In Sec. III the voltage gain has been derived per switching frequency, pulse width, and the quality factor. The simulation results have been illustrated and fully discussed in Sec. IV, and in Sec. V the experimental result has been dedicated. In Sec VI this literature is concluded.

II. ANALYSIS OF THE CONVERTER DIFFERENT OPERATIONS

By extracting the data from the simulation results of the converter, the converter waveforms suggest to study the converter states in three duty-cycle intervals.

First state (D<0.45): the converter waveforms due to different switching frequencies and loads are illustrated in Fig. 2. With pulse widths lower than 0.45, only one of the output diodes (D_2) conducts where the other one (D_1) is not conducting.

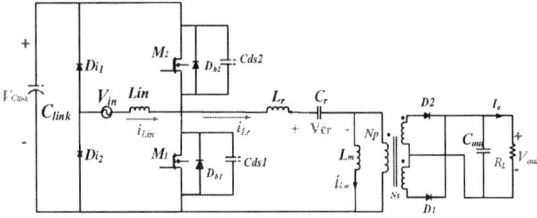

Fig. 1. This is an example of the figure caption which must come after the figure.

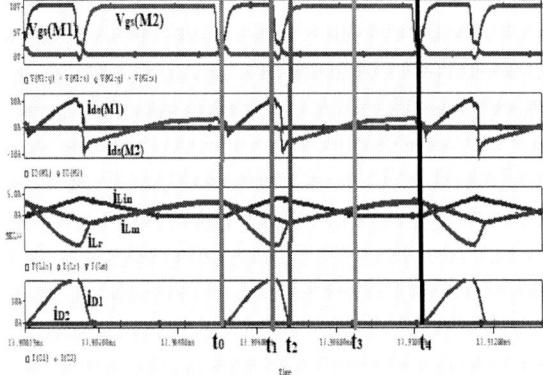

Fig. 2. Waveforms of voltages and currents in the converter in the first state.

Second state (0.45<D<0.55): the converter typical waveforms are illustrated in Fig. 3 and it is perfectly clear that the waveforms of Fig. 3 are not identical with the ones in the first state. Both diodes (D_1 and D_2) are conducting in this state.

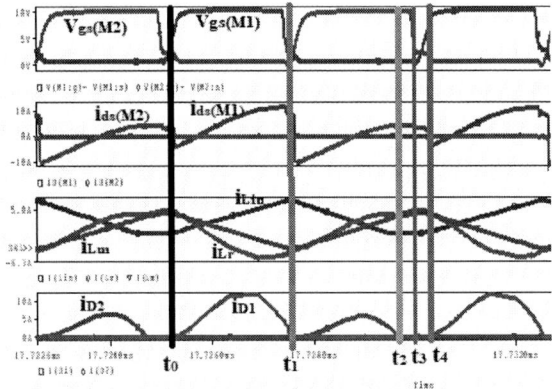

Fig. 3. Waveforms of voltages and currents in the converter in the state II.

Third state (D>0.55): the extracted subintervals from this state are actually similar to state I, only in here D_1 is conducting instead of D_2. Typical waveforms due to this state are illustrated in Fig. 4

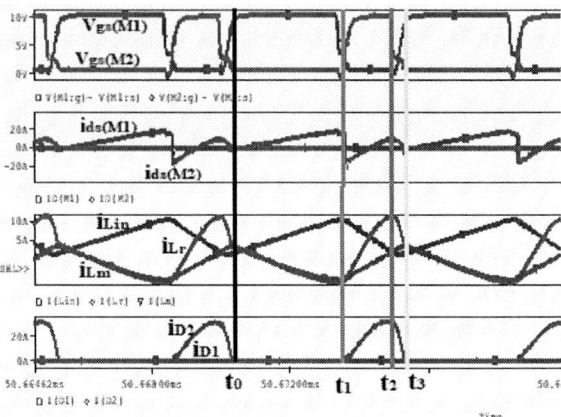

Fig. 4. Related waveforms for the state of (D>0.55).

TABLE I
RELATED PARAMETERS FOR USING IN SIMULATION AND
IMPLEMENTATION

V_{in}	220 V
V_o	100 V
f_s	200 kHz
f_{smax}	400 kHz
f_r	350 kHz
P_{out}	200 W
L_r	33 μH
C_r	15 nF
L_m	114 μF
C_{link}	52 μH
L_{in}	330 μH

The voltage gain relation has been extracted from the converter state equations which is a long process and is away from this paper aspect. Table 1 shows the parameters used for the LLC converter in which by replacing these parameters into the voltage gain relation the voltage gain behavior in the scale of parameters variation is derived out in the three mentioned duty-cycle intervals. The result is illustrated in Fig. 5 for the duty-cycle range from 0.1 up to 0.9.

Fig. 5 clearly describes the voltage gain behavior which is decreasing as the duty-cycle increases in any of the three duty-cycle intervals. This issue has been resulted in the same way by simulations in next section.

III. THE SIMULATION RESULTS

The LLC converter has been simulated by the table 1 parameters. The output voltage is demonstrated in Fig. 6,

978-1-7281-3420-8/19 $31.00 © 2019 IEEE

the typical waveforms of gate-source, drain-source and drain current is illustrated in Fig.7.

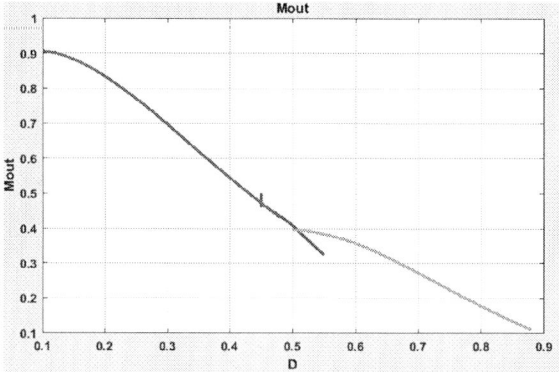

Fig. 5. The behavior of voltage gain based on duty cycle variation.

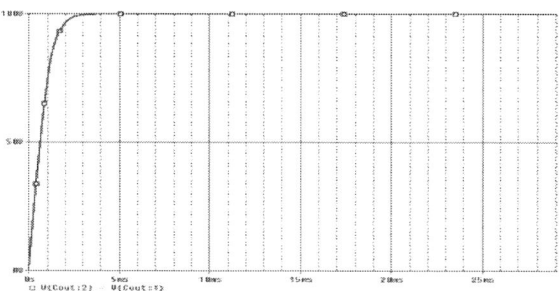

Fig. 6. Output voltage based on time variation.

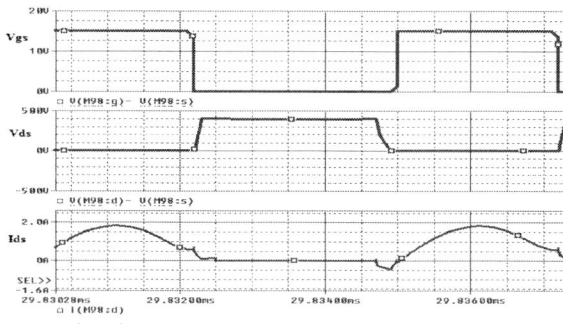

Fig. 7. Waveforms of gate-source, drain-source and drain current.

By considering the voltage and current waveforms of the MOSFETs, it is perfectly clear that ZVS operation is involved to all the MOSFETs and the ZCS operation is applied to all the diodes.

Fig. 8 shows the output voltage in different duty-cycles. By taking a closer look to Fig. 15 it is right to say that the output voltage is reduced as the duty-cycle increases. The same conclusion was derived out in the last section by using the converter state equations which shows that the theory analysis of this converter is coincides with the simulation results.

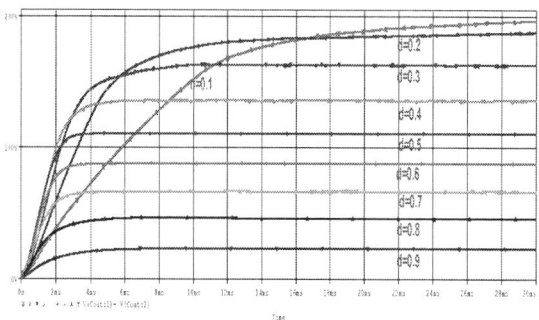

Fig. 8. The output voltage in different duty-cycles.

IV. THE EXPERIMENTAL RESULTS

Fig. 9 provides the experimental converter in which is developed by table 1 parameters. In Fig. 10 M_2 drain-source voltage is shown among with the resonant inductor current and the output diodes voltage under the symmetrical condition.

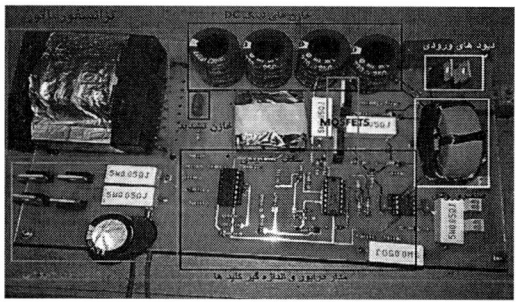

Fig. 9. The experimental converter in which is developed by table 1.

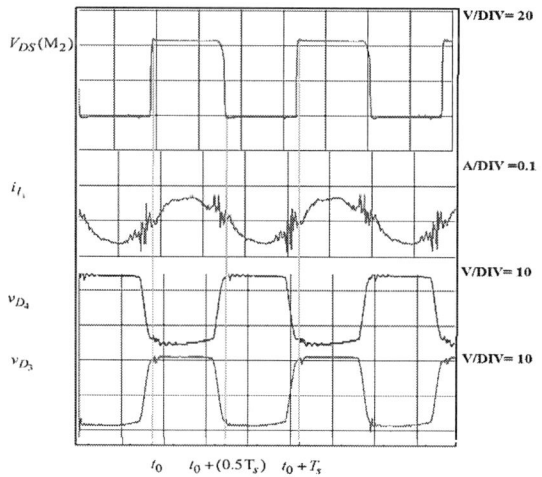

Fig. 10. M$_2$ drain-source voltage with the resonant inductor current and the output diodes voltage under the symmetrical condition.

V. CONCLUSION

By using the PFC converter in the front end of the main converter, the input power factor current has been improved. Power factor correction is natural characteristic in boost converter when it operates in DCM mode. This characteristic has been chosen for the converter control demand instead of using complicated control method in CCM operation mode. As it is derived out from the experimental result, the input current has been shaped in sinusoidal waveform without applying any feedback route.

Using both frequency modulation and pulse width modulation method by considering the state equations, the resonant LLC voltage gain has achieved per frequency, pulse width, and the quality factor when these parameters are variable in wide range. In addition, the ZVS approach has been achieved due to resonant converter soft switching in which reduces the conduction loss and as a result the converter efficiency is enhanced.

Using both frequency modulation and pulse width modulation method benefits the converter in two fold options, as output voltage and DC link voltage can be regulated among it. This capability is missing by only using frequency control. That is why the other researchers have limited the input voltage to 110 V where in this article the DC link voltage is reached up to 480 V in worst conditions when 220 V input voltage is applied.

REFERENCES

[1] D. Yang, C. Chen, S. Duan, J. Cai, L. Xiao, "A Variable Duty Cycle Soft Startup Strategy for LLC Series Resonant Converter Based on Optimal Current-Limiting Curve," *IEEE Transactions on Power Electronics*, 2016, vol. 31, pp. 7996 - 8006.

[2] L. Qiqi, et al, "Constant Resonant Current Limiting Strategy for LLC Converter Without Current Sensing," *IEEE Transactions on Power Electronics*, 2016, pp. 6756 - 6764.

[3] S. Zong, et al, "Asymmetrical Duty Cycle-Controlled LLC Resonant Converter With Equivalent Switching Frequency Doubler," *IEEE Transactions on Power Electronics*, 2016, pp. 4963-4973.

[4] V. Skanda, "Power Factor Correction in Power Conversion Applications Using the dsPIC® DSC," *Microchip Technology Inc*, 2007.

[5] J. Turchi, D. Dalal, P. Wang, and L. Jenck, "Power Factor Correction Handbook, choosing the right power factor controller solution," *ON Semiconductor*, 2014.

[6] F. Musavi and D. S. Gautam, "Control Strategies for Wide Output Voltage Range LLC Resonant DC–DC Converters in Battery Chargers," *IEEE Transactions on Vehicular Technology*, 2014, vol. 63, No. 3.

[7] Z. Fang, S. Duan, C. Chen, X. Chen, and J. Zhang, "Optimal design method for LLC resonant converter with wide range output voltage," in *Applied Power Electronics Conference and Exposition (APEC)*, 2013, pp. 2106-2111.

[8] J. Deng, S. Li, S. Hu, C. C. Mi, and R. Ma, "Design Methodology of LLC Resonant Converters for Electric Vehicle Battery Chargers," *IEEE Transactions on Vehicular Technology*, 2014.

[9] W. Inam, "High efficiency resonant dc/dc converter for solar power applications," *Massachusetts Institute of Technology*, 2013.

[10] C.-H. Chang, H.-Y. Chen, C.-T. Cho, and J.-Y. Chiu, "A novel single-stage LLC resonant AC-DC converter with power factor correction feature," in *Industrial Electronics and Applications (ICIEA)*, 2011, pp.2191-2196.

[11] B. Yang, F. C. Lee, A. J. Zhang, and Guisong Huang, "LLC resonant converter for front end DC/DC conversion," in *Proc. IEEE Applied Power Electronics Conference and Exposition*, 2002, pp. 1108-1112.

[12] Huber, L., Y. Jang, and M.M. Jovanovic, "Performance evaluation of bridgeless PFC boost rectifier," *Applied Power Electronics Conference, APEC Twenty Second Annual IEEE*, Aug. 2007.

[13] X. Zhang and J.W. Spencer, "Analysis of boost PFC converters operating in the discontinuous conduction mode", *Power Electronics, IEEE Transactions on*, 2011, pp. 3621-3628.

[14] M. Gopinath and S. Ramareddy, "A brief analysis on bridgeless boost PFC converter," *IEEE Transactions on Industrial Electronics*, 2011, pp.1789-1798.

[15] C-A. Cheng, C-H. Chang, T-Y. Chung and F-L. Yang, "Design and Implementation of a Single-Stage Driver for Supplying an LED Street-Lighting Module with Power-Factor-Corrections," *IEEE Transactions on Industrial Electronics*, 2017, vol. 64, no. 7, pp. 5766-5776.

978-1-7281-3420-8/19 $31.00 © 2019 IEEE

AUTHOR INDEX

Akhmetov A. O. 107, 345
Aleksandrova M. 125, 141, 313
Amburkin K. M. 99
Atkin E. V. 305
Attarimashalkoubeh B. 79
Azoulay M. 55
Bahchedzhiev H. 129
Bakerenkov A. 185, 251, 341
Balasubramanian P. 293
Balbekov A. 289
Banach M. 227, 335
Beiranvand R. 353
Belhimer L. 213
Belyakov V. 185, 251, 329, 333, 341
Benfdila A. 213
Bernstein J. 55
Beškoski V. 321
Bobrovsky D. V. 107
Bogdanov Yu. I. 193
Bogdanova N. A. 193
Boychenko D. V. 71, 103, 107
Brindić B. 169
Čajko K. O. 173
Chen Z. 3
Cherniak M. E. 103, 345, 349
Chernyakov P. 289
Chirkov N. A. 197
Chitanov V. 129
Cholakova T. 129
Chukov G. V. 99
Cocic K. 325
Crescioli F. 243
Dalecki J. 277
Danilov I. 289, 301
Danković D. 177
Davidovic A. 325
Davidović V. 59, 67, 181
Davydov G. G. 71
Denishev K. 141
Dimitrijev S. 31
Dimitrijević B. 45
Djorić-Veljković S. 59
Djouder M. 213
Djurić Z. 161
Długosz R. 227, 335, 277, 281, 285
Dobrescu D. 259

Dobrescu L. 259
Dong S. 3
Đorđević M. 117
Elesin V. V. 99
Emin E. 289
Epifantsev K. A. 337
Escobedo-A J. 83, 121
Esmailirad S. 353
Etrekova M. 153
Evdochim L. 259
Fastovets D. V. 193
Felitsyn V. 185, 251, 341
Filipovic L. 9
Fischer G. 277
Frantlović M. 161, 317
Frontini L. 243
Gieva E. 205, 209
Gluhov N. 251, 341
Golan G. 55
Golovin A. 329, 333
Gomez-B Y. 121
Gorbunov M. 289, 301
Gorshkova A. 157
Grimalsky V. 83, 121
Gromov E. 329, 333
Gromova P. S. 71
Gupta R. 75
Gusev A. 219
Han J. 31
Ilić S. 67
Ivanov I. 329, 333
Ivanov Tz. 59
Ivanova A. 157
Ivetić T. B. 173
Jakšić O. 87, 91, 165
Jakšić O. M. 321
Jakšić Z. 87, 91, 165
Jevtić A. 67
Jokanovic B. 255
Jokić I. 161, 165
Joksimović K. 321
Jovanovic Dolecek G. 223
Jović V. 133
Jürimägi L. 239
Kakanakov R. 129
Katunin Yu. V. 189, 201

Kessi M.	213
Kolaklieva L.	129
Kolev G.	141, 313
Kolosova A. S.	71
Koshevaya S.	83, 121
Kovalenko A.	309
Kozuki S.	273
Krasnyuk A. A.	297
Krstajić P. M.	161
Krstić M.	45
Kržanović N.	181
Kuznetsov A. G.	99
Lakhelef A.	213
Lamovec J.	133
Lazarevic Z. Z.	95
Lebl A.	231
Leblebici Y.	79
Liberali V.	243
Litvinov A.	153
Luchinin V. V.	71
Lukashin V. P.	345
Lukić–Petrović S. R.	173
Malkin E.	329, 333
Maskell D. L.	293
Matusko M.	329, 333
Miletić S.	321
Mileusnić M.	231
Milijic M.	255
Milovanović V. M.	247
Mitić V.	117
Mitrović N.	177
Mladenov M.	205
Mladenović I.	87, 133
Moskovskaya Y. M.	193
Mozhaev R. K.	103, 197, 349
Nadjdjerdj L.	181
Nicholls J.	31
Nikiforov A. Y.	107, 193, 337, 345, 349
Nikolić D.	181
Nikolić G.	281, 285
Nikolić T.	281, 285
Nikolov G.	205, 209
Nikolov R.	169
Nikolova B.	205, 209
Novikov S. V.	103
Novkovski N.	63
Oblov K.	153, 157
Obradov M.	87, 91, 133, 165
Osipov D.	219
Otsuka T.	273
Pandiev I.	141, 265, 269, 313

Pantić D.	169
Paskaleva A.	59
Paul St.	219
Paunović V.	117
Pavić B.	231
Pechenkin A. A.	197, 349
Pejović M.	113
Pershenkov V.	185, 251, 329, 333, 341
Petrov K.	301
Petrov K. A.	189
Petrović D. M.	173
Petrović M. L.	247
Podlepetsky B.	137, 157, 309
Poljak P.	317, 321
Popović B.	317, 321
Prasad K.	293
Prijić A.	37
Prijić Z.	37, 117, 177
Prozorova A. G.	201, 297
Radivojević J.	231
Radojević V.	133
Radulović K.	133, 161
Randjelović D.	165
Randjelović D. V.	317, 321
Retdian N.	273
Rodin A.	185, 251, 341
Romcevic N. Z.	95
Ruskova I.	205, 209
Salehirad S.	353
Samotaev N.	137, 153, 157
Sanchez-S J.	83
Sarajlić M.	317
Sekulic D. L.	95, 173, 325
Selberherr S.	9
Serazetdinov A. R.	305
Shaltaeva Y.	329, 333
Shemonaev A. N.	337
Shemyakov A. E.	107
Shima T.	273
Shnaider Khazanova A.	289, 301
Shumikhin V.	219
Skeparovski A.	63
Skorobogatov A.	289
Skorobogatov P. K.	337
Smolin A. A.	107
Sogojan A. V.	193
Sorokoumov G. S.	107
Sotskov D. I.	99
Spassov D.	59
Stabile A.	243
Stamenković Z.	45

Stanchev T.	59	Usachev N. A.	99
Stanimirović I.	145, 149	Vaid R.	75
Stanimirović Z.	145, 149	Vasiljević Radović D.	91, 133, 165
Stanković A.	169	Vasilyev V.	329, 333
Stanković S.	59, 67, 181	Veljković S.	113
Stenin V. Ya.	189, 201	Videkov V.	125
Stojadinović N.	59, 177	Vorkapić M.	317
Stojčev M.	45	Vračar Lj.	37
Talaśka T.	227, 277, 281, 285	Vucheva Y.	141
Tanasković D.	87, 91, 165, 317	Vyuginov V. N.	71
Tanner P.	31	Wachutka G.	19
Tararaksin A. S.	71	Wong H.	3
Telets V.	185, 251, 341	Zhukov A.	251, 341
Titova M. I.	99	Zidkov N. M.	99
Tittelbach-Helmrich K.	45	Živanović E.	113
Tsanev T.	125	Živković M.	113
Ubar R.	239	Zlatković I.	169
Ukolov D. S.	197	Zvyagin A	289
Ulanova A. V.	345, 349		